Modern Engineering Mathematics

Modern Engineering Mathematics

Abul Hasan Siddiqi
Mohamed Al-Lawati
Messaoud Boulbrachene

CRC Press
Taylor & Francis Group
Boca Raton London New York

CRC Press is an imprint of the
Taylor & Francis Group, an **informa** business
A CHAPMAN & HALL BOOK

CRC Press
Taylor & Francis Group
6000 Broken Sound Parkway NW, Suite 300
Boca Raton, FL 33487-2742

Printed on acid-free paper by CPI Group (UK) Ltd, Croydon, CR0 4YY
Version Date: 20171116

International Standard Book Number-13: 978-1-4987-1205-7 (Hardback)

Visit the Taylor & Francis Web site at
http://www.taylorandfrancis.com

and the CRC Press Web site at
http://www.crcpress.com

Prof. A. H. Siddiqi would like to dedicate the book to his wife, Prof. Azra H.Siddiqi

Prof. M. Al-Lawati dedicates to his spouse.

Prof. M. Messaoud dedicates to his family.

Contents

Preface

This book entitled *Modern Engineering Mathematics* is a compendium of fundamental mathematical concepts, methods, models and their wide range of applications in diverse fields of engineering. It comprises essentially a contemporary coverage of those areas of mathematics which provide foundation to electronic, electrical, communication, petroleum, chemical, civil, mechanical, biomedical, software and financial engineering. The rapid growth of technology and engineering science based on mathematical concepts is driving force for enlarging the domain of modern engineering mathematics.

The book gives a fairly extensive treatment to some of the recent developments in mathematics which have found very significant applications to engineering problems. Rich illustrative engineering examples and the integration of MATLAB and MATHEMATICA further support readers of the book. Essential topics of mathematics required by all types of engineers are covered clearly and concisely and fundamental ideas are highlighted and developed using illustrative examples. Extensive exercises and self-assessment questions create confidence among students to apply mathematics to engineering problems.

Contemporary topics used by engineers to solve their problems are included along with well thought applications appropriate for students of undergraduate engineering. Unlike most engineering mathematics books, the presentation, solved examples and unsolved examples are tailored (designed) particularly for engineering students. A special feature of this book is the assumption of bare minimum knowledge of pre-engineering mathematics along with evidence of theoretic results not required by engineers. Theoretic results are discussed in Appendices. The following special features are the hallmarks of the book which make it distinct from the existing books on the theme.

- Some new areas relevant to exploration and production of hydrocarbons (petroleum engineering), biomedical engineering (biomedical images, particularly, CT scans, MRI, EEG), financial engineering (Black-Sholes model), biometric for identification of persons, brain-machine interface, software engineering and information technology.

- Foundations and updated special mathematical techniques most appropriate to students of electrical, electronic, communication, computer, software, petroleum, chemical and financial engineering.

- Lucid presentation of topics like wavelet transforms, inverse problems,

Radon transform, fractals, compression sensing and besides well non-classical topic and their applications is an attractive feature. Another interesting feature is a wealth of examples to provide clues for tackling engineering problems.

- Visualization of ODEs (ordinary differential equations) and PDEs (partial differential equations) by MATLAB and MATHEMATICA are presented. We also cover topics like matrices, ODEs, PDEs, and complex analyses along with their applications. Suggestions for further reading are given at the ends of each chapter. Such lists are useful for those who intend to pursue specific topics at more advanced levels.

The book comprises eleven chapters and five appendices. Each chapter is divided into sections and subsections. Chapter 1 is devoted to matrices and their applications to engineering problems. Applications of matrix analysis to traffic flow, electrical networks, chemical equations, cryptography, transform coding, web searches, and ranking are notable themes of this chapter. MATHEMATICA and MATLAB are introduced. Chapter 2 begins with motivation for studying differential equations by future engineers. Applications of MATLAB and MATHEMATICA are also presented in this chapter along with modeling of physical phenomena such as population dynamics, radioactive decay, Newton's law of cooling, spread of disease, series circuits, falling bodies, draining a tank, carbon dating and spring systems by differential equations. Classical methods of solutions of these models, namely underdetermined coefficients, variation of parameters and Cauchy-Euler methods are presented. The Laplace transform and its applications to solve linear differential equations as well as series solutions of two well known differential equations: Legendre and Bessel are discussed.

Chapter 3 deals with vector calculus providing basic properties including three fundamental theorems: Green theorem, divergence theorem of Gauss, Stokes theorem and application of these results to engineering problems.

Concepts of orthonormal systems along with several concrete examples, Fourier expansion, Fourier transform, Fourier integral and fast Fourier transform are explained in Chapter 4. Applications to signal analysis are introduced, providing foundations of information technology. Shanon sampling theorem is explained. Heat, wave and Laplace equations are used in different branches of engineering. Solutions of these equations with appropriate initial and boundary conditions applying Fourier series, Fourier transform, and Laplace transform are discussed in Chapter 5. Visualization of solutions by MATLAB is presented and for visualization by MATHEMATICA we refer to reference [1] of Chapter 5.

Chapter 6 is devoted to algorithmic optimization. Besides well-known classical topics, new concepts like variational inequalities, Krylov method, and multigrid methods are introduced. Numerical methods for solving engineering problems such as finite difference method, finite element method are discussed in Chapter 7. Several case studies from diverse fields are presented.

Basic results of complex analysis essential for understanding and solving engineering problems are discussed in Chapter 8. Major challenging problems of engineering are in the form of inverse problems, for example, location of a tumor by CT scan or X-ray, determination of past events from observation, finding the nature of an accessible region from measurements on the boundary and the reconstruction of past events from observation of the present state. This problem is explained in Chapter 9 with the help of examples of engineering problems. This is an emerging area and gaining popularity in various branches of engineering all over the world. Chapter 10 is devoted to wavelets a topic introduced in the early 1980 s and quite popular among engineers. It is a refinement of Fourier series. Introduction of this field is presented along with various applications to engineering problems.

There are several topics introduced in the last few decades which have been used by engineers and scientists to solve challenging problems of the real world. Some of these topics such as fractals, interaction of fractals and wavelets, fractal image compression, study of physical phenomena through time series and wavelet fractal methods, graph theory, applications of Lozi maps, neural networks, large networks, fuzzy logic, ANFIS, splines, stochastic calculus, compression sensing are introduced in Chapter 11.

Appendix A is devoted to basic concepts of calculus. Appendix B deals with pre-requisite topics. Proofs of selected theorems are given in Appendix C. Appendix D presents basic concepts of medical imaging. Solutions of selected exercises will be given as Appendix E. Selected references are given at the end of each chapter.

MATLAB® is a registered trademark of The Mathworks, Inc. For product information please contact:

The Mathworks, Inc.
3 Apple Hill Drive
Natick, MA, 01760-2098 USA
Tel: 508-647-7000
Fax: 508-647-7001
E-mail:info@mathworks.com
Web:www.mathworks.com

A.H.Siddiqi
M.Al- Lawati
M. Boulbrachene

The Book has been
Completed
With
Active support
Of
Prof. P. Manchanda,
Pooja
Mamta Rani
Prof. R. Bhardwaj
Dr. Noor Zahra
Ashima Bangia
Sana Arif

Acknowledgments

The idea of writing this book was proposed during the 2014 Spring Semester while A. H. Siddiqi served as a consultant in the Department of Mathematics and Statistics at Sultan Qaboos University, Muscat, Oman. The proposal was finalized in a series of meetings in the office of Prof. M. Al-Lawati who headed the department. Taylor & Francis Group then accepted the proposal and agreed to publish the book. We appreciate the efforts of Aastha Sharma, the publisher's acquisition editor who arranged the contract and provided support throughout the writing process.

We take this opportunity to thank the Government of Oman's Sultan Qaboos University and Sharda University, Greater Noida, India, specially its Chancellor, Mr. P. K. Gupta for providing immense support for this book writing project. We also want to thank Prof. P. Manchanda who heads the Department of Mathematics at Guru Nanak Dev University, Amritsar, India for her valuable help in enabling us to complete this book.

A.H.Siddiqi

Chapter 1

Matrices for Engineers

1.1 Vectors in R^2 and R^3

In engineering, physics and other applied fields a quantity having magnitude and direction is called a **vector** or a vector quantity. A physical quantity having only magnitude is called a **scalar** and is represented by a single number using appropriate units.

Force, weight, velocity and acceleration are examples of vector quantities while mass, speed, and temperature are examples of scalar quantities.

Vectors in R^2
A vector whose initial point or end is A and whose terminal point or tip is B is $\overrightarrow{AB}$. The magnitude of a vector is written as $\|\overrightarrow{AB}\|$. Two vectors that have the same magnitude and the same direction are said to be equal. Thus in Figure 1.1 we have $\overrightarrow{AB} = \overrightarrow{CD}$.

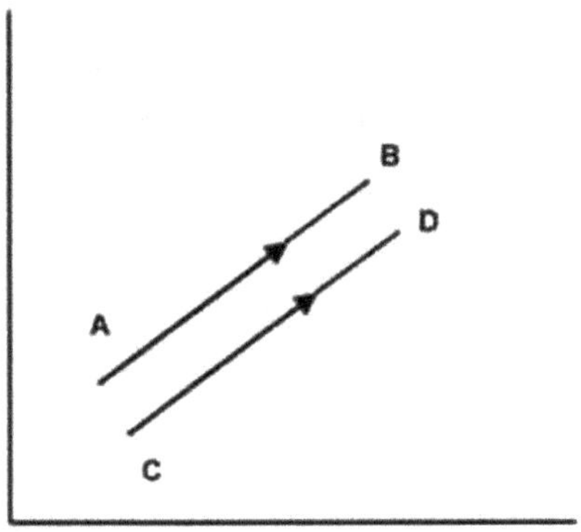

FIGURE 1.1: Equal Vectors

The negative of a vector $\overrightarrow{AB}$, written as $-\overrightarrow{AB}$ is a vector that has the same magnitude but is opposite in direction. If $\alpha \neq 0$ is a scalar, $\alpha\overrightarrow{AB}$ is called the scalar multiple of vector $\overrightarrow{AB}$ and it is a vector that is $|\alpha|$ times as long as $\overrightarrow{AB}$. If $k > 0$ then $k\overrightarrow{AB}$ has the same direction as $\overrightarrow{AB}$. For $k < 0$, then direction of $k\overrightarrow{AB}$ is opposite to $\overrightarrow{AB}$, see Figure 1.2.

Addition and Subtraction
Two vectors are said to have common initial points such as in Figure 1.3. Vector $\overrightarrow{AD}$ is the sum of $\overrightarrow{AB}$ and $\overrightarrow{AC}$. Thus diagonal $\overrightarrow{AD}$ of the parallelogram having sides $\overrightarrow{AB}$ and $\overrightarrow{AC}$ represents the sum of vectors $\overrightarrow{AB}$ and $\overrightarrow{AC}$. Difference of vectors $\overrightarrow{AB}$ and $\overrightarrow{AC}$ is defined as

$$\overrightarrow{AB} - \overrightarrow{AC} = \overrightarrow{AB} + (-\overrightarrow{AC}).$$

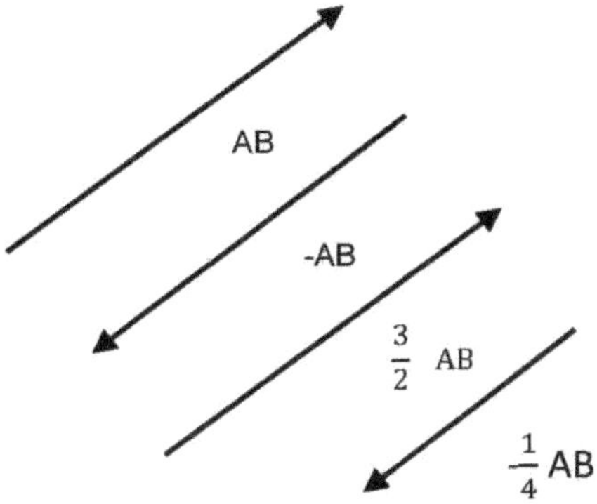

FIGURE 1.2: Parallel Vectors

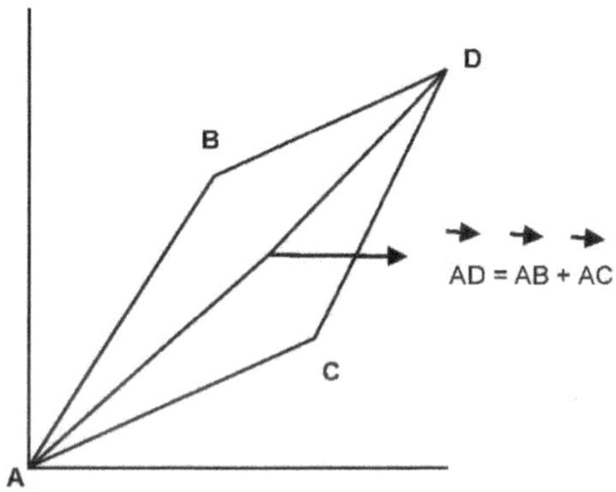

FIGURE 1.3: Addition of Vectors

If $k = 0$, $0\overrightarrow{AB} = \overrightarrow{0}$, then $\overrightarrow{0}$ is called a zero vector that can be assigned any direction. A vector $\overrightarrow{AB}$ in R^2 is often written as $\overrightarrow{AB} = \mathbf{a} = (a_1, a_2)$, where a_1 and a_2 are real numbers that is, a vector in R^2 is represented by ordered pairs of real numbers.

Definition 1. Let $\mathbf{a} = (a_1, a_2)$ and $\mathbf{b} = (b_1, b_2)$ be two vectors in R^2, then

(i) Addition: $\mathbf{a} + \mathbf{b} = (a_1 + b_1, a_2 + b_2)$

(ii) Scalar multiplication: $\alpha\mathbf{a} = (\alpha a_1, \alpha a_2)$

(iii) Equality: $\mathbf{a} = \mathbf{b}$ if and only if $a_1 = b_1$, $a_2 = b_2$.

Remark 1. (i) $-\mathbf{a} = (-a_1, -a_2)$

(ii) $\mathbf{a} - \mathbf{b} = (a_1 - b_1, a_2 - b_2)$

(iii) $\mathbf{a} + \mathbf{b} = \mathbf{b} + \mathbf{a}$ for all vectors $\mathbf{a}$ and $\mathbf{b}$

(iv) $\mathbf{a} + (\mathbf{b} + \mathbf{c}) =(\mathbf{a} + \mathbf{b} + \mathbf{c})$ for all vectors $\mathbf{a}$, $\mathbf{b}$ and $\mathbf{c}$

(v) $\mathbf{a} + 0 = \mathbf{a}$

(vi) $\mathbf{a} + (-\mathbf{a}) = 0$

(vii) $k(\mathbf{a} + \mathbf{b}) = (k\mathbf{a} + k\mathbf{b})$

(viii) $(k_1 + k_2)\mathbf{a} = k_1\mathbf{a} + k_2\mathbf{a}$

(ix) $k_1(k_2\mathbf{a}) = (k_1k_2)\mathbf{a}$, k_1 and k_2 are scalars.

(x) $1\mathbf{a} = \mathbf{a}$

(xi) $0\mathbf{a} = 0$.

Magnitude of $\mathbf{a}$ is denoted by $\|\mathbf{a}\|$ and is defined as

$$\|\mathbf{a}\|^2 = (a_1^2 + a_2^2)^{\frac{1}{2}}.$$

A vector $\mathbf{u} = (u_1, u_2)$ is called unit vector if $\|\mathbf{u}\| = 1$.
It may be observed that $\mathbf{u} = \dfrac{\mathbf{a}}{\|\mathbf{a}\|}$ is a unit vector: $\|\mathbf{u}\| = \| \dfrac{a}{\| a \|} \| = 1$.
Let $i = (1, 0)$, $j = (0, 1)$, then $\mathbf{a} = (a_1, a_2)$ can be written as

$$\begin{aligned} \mathbf{a} &= a_1 i + a_2 j \\ R.H.S &= a_1(1,0) + a_2(0,1) \\ &= (a_1, 0) + (0, a_2) \\ &= (a_1, a_2) \\ &= \mathbf{a} \\ &= L.H.S. \end{aligned}$$

Example 1. (a) Find the magnitude of vectors $2i + 4j$, and $(2, 0)$.

(b) Given $\mathbf{a} = (4, -1)$, form a unit vector.

(c) Find $5\mathbf{a}$, $\mathbf{a} + \mathbf{b}$, $\mathbf{a} - \mathbf{b}$, $\|\mathbf{a} + \mathbf{b}\|$ and $\|\mathbf{a} - \mathbf{b}\|$ if $\mathbf{a} = 4i + 8j$, $\mathbf{b} = -2 + 8j$.

(d) Find a vector in the opposite direction of $\mathbf{a} = (4, 10)$ but $\dfrac{3}{4}$ as long.

Solution: (a) $\|2i + 4j\| = \|(2, 4)\| = (4 + 16)^{\frac{1}{2}} = (20)^{\frac{1}{2}} = 2\sqrt{5}$

$$\|(2, 0)\| = \sqrt{4} = 2.$$

Vectors in R^3

A vector in R^3 is represented by an ordered triple of numbers a_1, a_2, a_3. As in R^2, the magnitude is defined as $\|\mathbf{a}\| = (|a_1|^2 + |a_2|^2 + |a_3|^2)^{\frac{1}{2}}$.
Sum of two vectors **a** and **b** in R^3 is defined as

$$\mathbf{a} + \mathbf{b} = (a_1 + b_1, a_2 + b_2, a_3 + b_3)$$

where $\mathbf{a} = (a_1, a_2, a_3)$ and $\mathbf{b} = (b_1, b_2, b_3)$.
The distance between two vectors **a** and **b** in R^3 is defined as

$$d(\mathbf{a}, \mathbf{b}) = \|\mathbf{a} - \mathbf{b}\| = \sqrt{(a_1 - b_1)^2 + (a_2 - b_2)^2 + (a_3 - b_3)^2}.$$

Remark 2. The concept of sum of vectors can be extended for vectors in R^3, for example for $\mathbf{a} = (a_1, a_2, a_3)$, $\mathbf{b} = (b_1, b_2, b_3)$, $\mathbf{a} + \mathbf{b} = \mathbf{b} + \mathbf{a}$, and $1\mathbf{a} = \mathbf{a}$, $0\mathbf{a} = 0 = (0, 0, 0)$.
Let $i = (1, 0, 0)$, $j = (0, 1, 0)$, $k = (0, 0, 1) \in R^3$, then $\mathbf{a} = (a_1, a_2, a_3) = a_1 i + a_2 j + a_3 k$

$$\begin{aligned} L.H.S. &= a_1(1,0,0) + a_2(0,1,0) + a_3(0,0,1) \\ &= (a_1, 0, 0) + (0, a_2, 0) + (0, 0, a_3) \\ &= (a_1, a_2, a_3) \\ &= \mathbf{a} \\ &= R.H.S. \end{aligned}$$

Example 2. (a) Find the magnitude of $\mathbf{a} = \left(-\frac{4}{7}, \frac{6}{7}, \frac{12}{7}\right)$.

(b) Express the vector $\mathbf{a} = (14, -15, 26)$ in terms of unit vectors i, j, k.

(c) Find $\mathbf{a} + (\mathbf{b} + \mathbf{c})$, $\mathbf{b} + 2(\mathbf{a} - 3\mathbf{c})$, $\|\mathbf{a} + \mathbf{b}\|$, $\|\mathbf{a} - \mathbf{c}\|$ and $\|\mathbf{b}\|\|\mathbf{c}\|$ where $\mathbf{a} = (1, -3, 2)$, $\mathbf{b} = (-1, 1, 1)$ and $\mathbf{c} = (2, 6, 9)$

1.1.1 Dot and Cross Products

Let $\mathbf{a} = (a_1, a_2, a_3)$ and $\mathbf{b} = (b_1, b_2, b_3)$ be two vectors in R^3. Then the **dot product** of **a** and **b** denoted by **a.b** is defined as

$$\mathbf{a}.\mathbf{b} = a_1 b_1 + a_2 b_2 + a_3 b_3. \tag{1.1}$$

The **cross product** of **a** and **b** denoted by $\mathbf{a} \times \mathbf{b}$ is defined as

$$\begin{aligned} \mathbf{a} \times \mathbf{b} &= \begin{vmatrix} i & j & k \\ a_1 & a_2 & a_3 \\ b_1 & b_2 & b_3 \end{vmatrix} \qquad (1.2) \\ &= (a_2 b_3 - a_3 b_2) i - (a_1 b_3 - a_3 b_1) j + (a_1 b_2 - a_2 b_1) k. \end{aligned}$$

It is clear that magnitude of **a.b** is denoted by

$$\|\mathbf{a}.\mathbf{b}\| = |\mathbf{a}.\mathbf{b}| = ((a_1b_1)^2 + (a_2b_2)^2 + (a_3b_3)^2)^{\frac{1}{2}}$$

and

$$\|\mathbf{a}\times\mathbf{b}\| = ((a_2b_3 - a_3b_2)^2 + (a_1b_3 - a_3b_1)^2 + (a_1b_2 - a_2b_1)^2)^{\frac{1}{2}}.$$

The distance between **a** and **b** denoted by $\|\mathbf{a}-\mathbf{b}\|$ is given by

$$\begin{aligned}\|\mathbf{a}-\mathbf{b}\| &= \|(a_1-b_1)i + (a_2-b_2)j + (a_3-b_3)k\| \\ &= ((a_1-b_1)^2 + (a_2-b_2)^2 + (a_3-b_3)^2)^{\frac{1}{2}} \\ \|\mathbf{a}\times\mathbf{b}\|^2 &= (\mathbf{a}\times\mathbf{b}).(\mathbf{a}\times\mathbf{b}) \\ &= \mathbf{a}.[\mathbf{b}\times(\mathbf{a}\times\mathbf{b})]\end{aligned}$$

by the relation $(\mathbf{a}\times\mathbf{b}).\mathbf{c} = \mathbf{a}.(\mathbf{b}\times\mathbf{c})$ or

$$\begin{aligned}\|\mathbf{a}\times\mathbf{b}\|^2 &= \mathbf{a}.[(\mathbf{b}.\mathbf{b})\mathbf{a} - (\mathbf{b}.\mathbf{a})\mathbf{b}] \\ &= (\mathbf{a}.\mathbf{a})(\mathbf{b}.\mathbf{b}) - (\mathbf{b}.\mathbf{a})(\mathbf{b}.\mathbf{a}) \\ &= \|\mathbf{a}\|^2\|\mathbf{b}\|^2 - (\mathbf{a}.\mathbf{b})^2.\end{aligned}$$

We know that $\mathbf{a}.\mathbf{b} = \|\mathbf{a}\|\|\mathbf{b}\|\cos\theta$ where θ is the angle between vectors **a** and **b** and so $\|\mathbf{a}\times\mathbf{b}\|^2 = \|\mathbf{a}\|^2\|\mathbf{b}\|^2 - \|\mathbf{a}\|^2\|\mathbf{b}\|^2\cos^2\theta$ or

$$\begin{aligned}\|\mathbf{a}\times\mathbf{b}\|^2 &= \|\mathbf{a}\|^2\|\mathbf{b}\|^2(1-\cos^2\theta) \\ &= \|\mathbf{a}\|^2\|\mathbf{b}\|^2\sin^2\theta\end{aligned}$$

or

$$\|\mathbf{a}\times\mathbf{b}\| = \|\mathbf{a}\|\|\mathbf{b}\|\sin\theta, 0 \le \theta \le \pi. \tag{1.3}$$

Let $i = (1,0,0)$, $j = (0,1,0)$, $k = (0,0,1)$, then $\mathbf{a} = a_1i + a_2j + a_3k$ and $\mathbf{b} = b_1i + b_2j + b_3k$.
It may be noticed that $\mathbf{a}\times\mathbf{b}$ is a vector while **a.b** is a scalar quantity.

Theorem 1. *For vectors **a**, **b**, **c** and α a scalar the following properties hold*

(a) $\boldsymbol{a}\times\boldsymbol{b} = \boldsymbol{b}\times\boldsymbol{a}$

(b) $\boldsymbol{a}\times(\boldsymbol{b}\times\boldsymbol{c}) = \boldsymbol{a}\times\boldsymbol{b} + \boldsymbol{a}\times\boldsymbol{c}$

(c) $(\boldsymbol{a}+\boldsymbol{b})\times\boldsymbol{c} = \boldsymbol{a}\times\boldsymbol{c} + \boldsymbol{b}\times\boldsymbol{c}$

(d) $\boldsymbol{a}\times(\alpha\boldsymbol{b}) = (\alpha\boldsymbol{a})\times\boldsymbol{b} = \alpha(\boldsymbol{a}\times\boldsymbol{b})$, *$\alpha$ a scalar*

(e) $\boldsymbol{a}\times\boldsymbol{b} = 0$ *if either* $\boldsymbol{a} = 0$ *or* $\boldsymbol{b} = 0$

(f) $\boldsymbol{a}\times\boldsymbol{a} = 0$

(g) $\boldsymbol{a}.(\boldsymbol{a} \times \boldsymbol{b}) = 0$

(h) $\boldsymbol{b}.(\boldsymbol{a} \times \boldsymbol{b}) = 0$.

The dot product is also known as inner product or **scalar product**. It can be also defined as $\mathbf{a}.\mathbf{b} = \|\mathbf{a}\|\ \|\mathbf{b}\| \cos\theta$ where θ is the angle between the vectors.

Example 3. (a) Let $i = (1, 0, 0)$, $j = (0, 1, 0)$ and $k = (0, 0, 1)$ then $i.i = 1, j.j = 1, k.k = 1$, $\|i\| = \|j\| = \|k\| = 1$ and in each case $\cos\theta = 1$.

(b) $\|\mathbf{a}\| = \sqrt{\mathbf{a}.\mathbf{a}} = (a_1^2 + a_2^2 + a_3^2)^{\frac{1}{2}}$.

(c) $i.j = j.i$, $j.k = k.j = 0$, $k.i = i.k = 0$.

Solution: (a) $i.i = (1, 0, 0).(1, 0, 0) = 1+0+0 = 1$. Similarly $j.j = 1, k.k = 1$.
$\|i\| = \sqrt{1^2 + 0^2 + 0^2} = 1$.
(b) $\|\mathbf{a}\| = (a_1^2 + a_2^2 + a_3^2)^{\frac{1}{2}}$.
$\sqrt{\mathbf{a}.\mathbf{a}} = \sqrt{(a_1, a_2, a_3)(a_1, a_2, a_3)} = \sqrt{a_1^2 + a_2^2 + a_3^2}$
Hence $\|\mathbf{a}\| = \sqrt{\mathbf{a}.\mathbf{a}}$.
(c) $i.j = (1, 0, 0).(0, 1, 0) = 0 + 0 + 0 = 0$. $j.i = (0, 1, 0).1, 0, 0) = 0$.
Similarly $j.k = k.j = 0$, $k.i = i.k = 0$.

Definition 2. Vectors **a** and **b** are said to be **orthogonal** (perpendicular) if $\mathbf{a}.\mathbf{b} = 0$ (dot product of **a** and **b** is zero).

Definition 3. (a) Angle between two vectors $\mathbf{a} = a_1 i + a_2 j + a_3 k$ and $\mathbf{b} = b_1 i + b_2 j + b_3 k$ is defined as

$$\theta = \cos^{-1} \frac{a_1 b_1 + a_2 b_2 + a_3 b_3}{\|\mathbf{a}\|\|\mathbf{b}\|}.$$

(b) For non-zero vector $\mathbf{a} = a_1 i + a_2 j + a_3 k$ in R^3, the angle α, β, γ between **a** and the unit vectors i, j and k, respectively are called direction angles of **a** and defined as

$$\cos\alpha = \frac{\mathbf{a}.i}{\|\mathbf{a}\|\|i\|} = \frac{a_1}{\|a\|}, \quad \cos\beta = \frac{a_2}{\|a\|}, \quad \cos\gamma = \frac{a_3}{\|a\|}$$

and $\cos\alpha, \cos\beta$ and $\cos\gamma$ are called **direction cosines** of **a**. The direction cosines of a non-zero vector **a** are similarly the components of the unit vector $\left(\frac{1}{\|\mathbf{a}\|}\right)\mathbf{a}$

$$\frac{1}{\|\mathbf{a}\|}\mathbf{a} = \frac{a_1}{\|\mathbf{a}\|}i + \frac{a_2}{\|\mathbf{a}\|}j + \frac{a_3}{\|\mathbf{a}\|}k = \cos\alpha i + \cos\beta j + \cos\gamma k.$$

Definition 4. The component of **a** on an arbitrary vector **b** denoted by $comp_{\mathbf{b}}\mathbf{a}$ is defined as

$$comp_{\mathbf{b}}\mathbf{a} = \mathbf{a}.\left(\frac{1}{\|\mathbf{b}\|}\mathbf{b}\right) = \frac{\mathbf{a}.\mathbf{b}}{\|\mathbf{b}\|}.$$

Example 4. (a) Find the angle between $\mathbf{a} = i + j + k$ and $\mathbf{b} = -i + j + k$.

(b) Find the direction cosines of the vectors $\mathbf{a} = i + 2j + 3k$.

(c) Find $comp_{\mathbf{b}}\mathbf{a}$, where $\mathbf{a} = 4i + 6j - 4k$ and $\mathbf{b} = i + j + 4k$.

Solution: (a) $\cos\theta = \dfrac{-1+1+1}{\sqrt{3}\sqrt{1+1+1}} = \dfrac{1}{3}$

$$\theta = \cos^{-1}\frac{1}{3}.$$

(b)
$$\begin{aligned}
\cos\alpha &= \frac{a_1}{||a||} = \frac{1}{\sqrt{1+4+9}} = \frac{1}{\sqrt{14}} \\
\cos\beta &= \frac{a_2}{||a||} = \frac{2}{\sqrt{14}} \\
\cos\gamma &= \frac{a_3}{||a||} = \frac{3}{\sqrt{14}}
\end{aligned}$$

(c) $comp_{\mathbf{b}}\mathbf{a} = \dfrac{\mathbf{a.b}}{||\mathbf{b}||} = \dfrac{4+6-16}{\sqrt{1+1+16}} = \dfrac{-6}{\sqrt{18}}.$

It may be observed that the physical interpretation of the dot product is the work done when a constant force of magnitude F moves an object a distance d in the same direction of the force, that is, $W = F.d$.

Physical meaning of cross product

Let vector **a** denote side AC and **b** denote side AB respectively of a parallel-

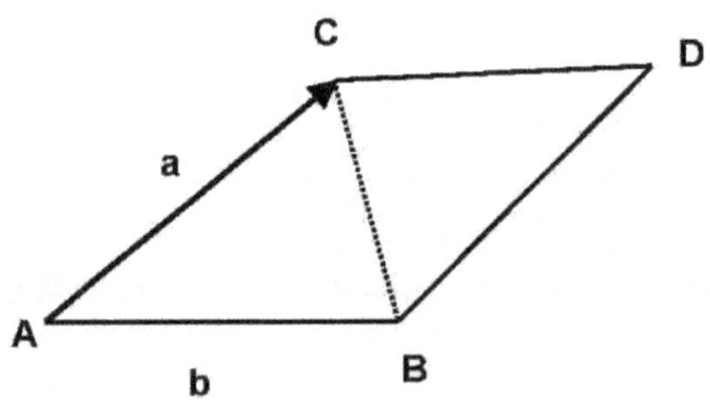

FIGURE 1.4: Geometrical Meaning of Cross Product

ogram as shown in Figure 1.4. Then the area of the parallelogram $= ||\mathbf{a} \times \mathbf{b}||$ = magnitude of the vector **a** cross **b**.

Area of the triangle ABC $= \dfrac{1}{2}||\mathbf{a} \times \mathbf{b}||$.

Definition 5. (a) Two vectors **a** and **b** are parallel if $\mathbf{a} \times \mathbf{b} = 0$.
(b) Three vectors **a**, **b** and **c** are **coplanar** (lie in the same plane) if

$$\mathbf{a}.(\mathbf{b} \times \mathbf{c}) = 0.$$

Example 5. (a) Show that $i \times j = k, j \times k = i, k \times i = j, j \times i = -k, k \times j = -i, i \times k = -j$.
(b)

$$\mathbf{a}.(\mathbf{b} \times \mathbf{c}) = \begin{vmatrix} a_1 & a_2 & a_3 \\ b_1 & b_2 & b_3 \\ c_1 & c_2 & c_3 \end{vmatrix}$$

where $\mathbf{a} = (a_1, a_2, a_3),\ \mathbf{b} = (b_1, b_2, b_3),\ \mathbf{c} = (c_1, c_2, c_3)$.
Show also that $\mathbf{a}.(\mathbf{b} \times \mathbf{c}) = (\mathbf{a} \times \mathbf{b}).\mathbf{c}$.

Example 6. (i) Find the area of a triangle whose adjacent sides are given by vectors $\mathbf{a} = (1, 2, 3), \mathbf{b} = (1, -3, 5)$. (ii) For $\mathbf{a} = (5, -2, 1), \mathbf{b} = (2, 0, -7)$ show that $\mathbf{a} \times \mathbf{b} = 0, \mathbf{a}.(\mathbf{a} \times \mathbf{b}) = 0$.

1.1.2 Linear Dependence and Independence

A subset S of R^3 is called a subspace if $x, y \in S$ implies that $x + y \in S$ and $\alpha x \in S$ for any real number α.

Definition 6. A set of vectors $\{\mathbf{a}, \mathbf{b}, \mathbf{c}\}$ is said to be **linearly independent** or vectors $\mathbf{a}, \mathbf{b}$ and **c** are called **linearly independent** if the only constants satisfying the equation

$$\alpha a + \beta b + \gamma c = 0$$

are $\alpha = 0, \beta = 0, \gamma = 0$. If $\mathbf{a}, \mathbf{b}$ and **c** are not linearly independent then the set of these vectors is linearly dependent and these vectors are also linearly dependent.

Example 7. (a) Show that $i = (1,0,0)$, $j = (0,1,0)$ and $k = (0,0,1)$ are linearly independent.
(b) Show that the vectors $(1,1)$ and $(1,2)$ are linearly independent.
(c) Examine whether vectors $\mathbf{a} = (1,1,1)$ and $\mathbf{b} = (2,-1,4)$ and $(5,2,7)$ are linearly independent or linearly dependent.

Solution: (a) Let $\alpha(1,0,0)+\beta(0,1,0)+\gamma(0,0,1) = 0$ then $(\alpha, \beta, \gamma) = 0$ which implies that $\alpha = \beta = \gamma = 0$. Hence these vectors are linearly independent.
(b) Let $\alpha(1,1) + \beta(1,2) = 0$ then $(\alpha + \beta, \alpha + 2\beta) = 0$ or

$$\alpha + \beta = 0, \quad \alpha + 2\beta = 0$$

implying $\beta = 0$ and consequently $\alpha = 0$. Thus these vectors are linearly

independent.
(c) Let $\alpha a + \beta b + \gamma c = 0$. Then

$$\alpha(1,1,1) + \beta(2,-1,4) + \gamma(5,2,7) = 0$$

or

$$(\alpha + 2\beta + 5\gamma, \alpha - \beta + 2\gamma, \alpha + 4\beta + 7\gamma) = 0$$

$$\alpha + 2\beta + 5\gamma = 0 \Rightarrow \alpha = -2\beta - 5\gamma$$

$$\alpha - \beta + 2\gamma = 0 \Rightarrow -2\beta - 5\gamma - \beta + 2\gamma = 0$$

or

$$-3\beta - 3\gamma = 0 \text{ or } \beta = -\gamma$$

$$\alpha + 4\beta + 7\gamma = 0$$

$$\alpha = 3\beta.$$

Hence these three vectors are linearly dependent.

Definition 7. (Dimensions) A set of elements in R^3 is called **basis** if these elements are linearly independent and every element of R^3 is a linear combination of basis elements. The number of elements in a basis is called the **dimension**. Dimension of R^3 is three as i, j, k are linearly independent and every element of R^3 is a linear combination of i, j, k (for each $x \in R^3, x = \alpha i + \beta j + \gamma k$).

Definition 8. (Orthogonal set of vectors) A set of vectors in R^3, $\{a, b, c\}$ is called **orthogonal** if $a.b = 0$, $b.c = 0$, $a.c = 0$.

1.1.3 Gram-Schmidt Orthogonalization Process

Notion of R^3 can be extended for any natural number n. Joïgen Pedersen Gram (1850-1916) of Denmark, an important functionary of an insurance company, and Erhard Schmidt (1876-1959) of Germany, a distinguished faculty member of well known universities, developed an algorithm for constructing orthonormal bases for R^n in particular and any vector space in general.
Let $S = \{u_1, u_2, \ldots, u_n\}, m \leq n$ be a basis for R^n. Then $\{v_1, v_2, \ldots v_m\}$, where

$$\begin{aligned}
v_1 &= u_1 \\
v_2 &= u_2 - \left(\frac{\langle u_2, v_1\rangle}{\langle v_1, v_1\rangle}\right) v_1 \\
v_3 &= u_3 - \left(\frac{\langle u_3, v_2\rangle}{\langle v_2, v_2\rangle}\right) v_2 \\
\vdots &= \quad \vdots \\
v_m &= u_m \left(\frac{\langle u_m, v_1\rangle}{\langle v_1, v_1\rangle}\right) v_1 - \left(\frac{\langle u_m, v_2\rangle}{\langle v_2, v_2\rangle}\right) v_2 - \cdots - \left(\frac{\langle u_m, v_{m-1}\rangle}{\langle v_{m-1}, v_{m-1}\rangle}\right) v_{m-1},
\end{aligned}$$

$$B^n = \{\frac{1}{\|v_1\|}v_1, \frac{1}{\|v_2\|}v_2, \ldots, \frac{1}{\|v_m\|}v_m\}$$

is an orthogonal basis in R^m.

Remark 3. Gram-Schmidt orthogonalization can be extended for any vector space.

Example 8. Let $S = \{u_1, u_2\}$, where $u_1 = (3,1)$, $u_2 = (1,1)$ is a basis for R^2. Obtain an orthonormal basis, say $\{w_1, w_2\}$ of R^2.

Solution: Let $v_1 = u_1 = (3,1)$. Then

$$\begin{aligned} v_2 &= u_2 - \left(\frac{\langle u_2, v_1\rangle}{\langle v_1, v_1\rangle}\right) v_1 \\ &= (1,1) - \frac{4}{10}(3,1) \\ &\qquad\qquad (\text{as } \langle u_2, v_1\rangle = 4 \text{ and } \langle u_2, v_1\rangle = 10) \\ &= \left(-\frac{1}{5}, \frac{3}{5}\right). \end{aligned}$$

$\{v_1, v_2\} = \{(3,1), (-\frac{1}{5}, \frac{3}{5})\}$ is an orthogonal basis and $\{w_1, w_2\}$, where

$$w_1 = \frac{1}{\|v_1\|}v_1, \quad w_1 = \left(\frac{3}{\sqrt{10}}, \frac{1}{\sqrt{10}}\right)$$

$$w_2 = \frac{1}{\|v_2\|}v_2, \quad w_2 = \left(\frac{-1}{\sqrt{10}}, \frac{3}{\sqrt{10}}\right)$$

is an orthonormal basis.

1.2 Basic Concepts of Matrices

In this section we present an introduction of matrices and their relationship with system of linear equations frequently occurring in diverse fields of study like physics, biology, chemistry, social science, medical science and different branches of engineering. Matrix theory or study of matrices provides compact way for representing systems of linear equations and yields efficient procedure for finding solutions. In Section 1.2.1 we describe properties of matrices which we need in subsequent sections.

Definition 9. An $m \times n$ matrix $A = (a_{ij}), i = 1, 2, \ldots, n$ and $j = 1, 2, \ldots, n$ is a rectangular array of mn real or complex numbers arranged in m horizontal rows and n vertical columns:

$$A = (a_{i,j}) = \begin{pmatrix} a_{11} & a_{12} & \cdots & a_{1n} \\ a_{21} & a_{22} & \cdots & a_{2n} \\ \vdots & \cdots & \cdots & a_{ij} \\ a_{m1} & a_{m2} & \cdots & a_{mn} \end{pmatrix}. \tag{1.4}$$

The ith row of A is $(a_{i1}a_{i2}\ldots a_{in})$, $1 \le i \le m$ while jth column of A is

$$\begin{pmatrix} a_{1j} \\ a_{2j} \\ \vdots \\ a_{mj} \end{pmatrix}.$$

Note that A is m by n (written as $m \times n$). If $m = n$, we say that A is a **square matrix** of order n, and that the numbers $a_{11}, a_{22}, a_{33}, \ldots, a_{nn}$ form the main diagonal of A. We call the number a_{ij}, which is in ith row and jth column of A, as the i,jth element of A, or (i,j) entry of A, and we often write 1.4 as $A = (a_{ij})$.

Example 9. Let (i) $A = \begin{pmatrix} 1 & 2 \\ -1 & 0 \end{pmatrix}$ (ii) $B = \begin{pmatrix} i & 4i+1 \\ 2+3i & -4 \end{pmatrix}$ (iii) $C = \begin{pmatrix} 1 \\ 2 \\ 3 \end{pmatrix}$ (iv) $D = \begin{pmatrix} 1 & -1 & 0 \\ 2 & 0 & 1 \\ 4 & -1 & 2 \end{pmatrix}$ (v) $E = \begin{pmatrix} 50 \end{pmatrix}$ (vi) $A = \begin{pmatrix} -2 & 3 & 5 \end{pmatrix}$.

(a) Write a_{21} and a_{22} of (i).

(b) Write b_{12} and b_{22} of (ii).

(c) Write c_{11} and c_{31} of (iii).

(d) Write d_{11} and d_{32} of (iv).

(e) Write e_{11} of (v).

(f) Write f_{12} and f_{13} of (vi).

(g) Write the main diagonal of D.

Solution: (a) $a_{21} = -1, a_{22} = 0$.
(b) $b_{12} = 4i + 1, b_{22} = -4$.
(c) $c_{11} = 1, c_{31} = 1$.
(d) $d_{22} = 0, d_{32} = -1$.
(e) $e_{11} = 5$.
(f) $f_{12} = 3, f_{13} = 5$.
(g) The main diagonal of D is 1,0,2.
It may be pointed out that in this chapter we concentrate on matrices with

real entries but matrices with complex numbers and functions as entries have been studied.

An $n \times 1$ matrix is also called an $n-$ vector, namely $\begin{pmatrix} a_{11} \\ a_{21} \\ a_{31} \\ \vdots \\ a_{n1} \end{pmatrix}$ is called an $n-$ vector or column vector.

$u = \begin{pmatrix} 1 \\ 3 \\ -1 \\ 0 \end{pmatrix}$ is a 4− vector and $v = \begin{pmatrix} 2 \\ -2 \\ 3 \end{pmatrix}$ is a 3− vector. The $n-$ vector all of whose entries are zero is denoted by $\mathbf{0}$.

If A is an $n \times n$ matrix, then rows of A are $1 \times n$ matrices and the columns of A are $n \times 1$ matrices. The set of all $n-$ vectors with real entries is denoted R_1^n $1 \times n$ matrix and also called a row vector.

Example 10. (Matrix of distances between cities given in kilometers)

	London	Madrid	New York	Tokyo	Delhi
London	0	785	3469	5959	6704
Madrid	785	0	3593	6706	7255
New York	3469	3593	0	6757	11766
Tokyo	5959	6706	6757	0	5840
Delhi	6704	7255	11766	5840	0

Example 11. (Production matrix table)
Assume that a company has five plants, each of which produces four products. Let us assume that a_{ij} denotes the number of units of product i produced by plant j in one week, based on 4×5 matrix given below.

	Plant 1	Plant2	Plant 3	Plant 4	Plant 5
Product 1	660	460	480	480	0
Product 2	440	550	520	520	100
Product 3	380	370	310	480	480
Product 4	385	375	312	310	385

Example 12. (Windchill matrix table)
A combination of air temperature and wind speed makes a body feel colder than the actual temperature. For example when temperature is 12^oF and wind is 18 km per hour, a heat loss equal to that when temperature is -20^oF with no wind occurs.

Definition 10. Two $m \times n$ matrices $A = (a_{ij})$ and $B = (b_{ij})$ are said to be equal if they agree entry by entry, that is $a_{ij} = b_{ij}$ for $i = 1, 2, \ldots, m$ and $j = 1, 2, \ldots, n$.

TABLE 1.1

kmph	17	12	7	2	−3	−8
5	12	7	0	−5	−10	−15
10	−3	−9	−15	−22	−27	−34
18	−11	−20	−25	−31	−38	−45
20	−17	−24	−31	−39	−46	−53

Example 13. $A = \begin{pmatrix} 1 & 2 & -1 \\ 3 & 3 & 2 \\ 0 & -4 & 5 \end{pmatrix} B = \begin{pmatrix} 1 & 2 & -1 \\ 3 & 3 & 2 \\ 0 & -4 & 5 \end{pmatrix}$.

A and B are equal.

Definition 11. Let $A = (a_{ij})$ and $B = (b_{ij})$ be two matrices, then the **sum** $A+B$ is an $m \times n$ matrix $C = (c_{ij})$ denoted by $c_{ij} = a_{ij} + b_{ij}$, $i = 1, 2, 3, \ldots, m$ and $j = 1, 2, 3, \ldots, n$.

Note: We merely add corresponding entries.

Example 14. (a) Let $A = \begin{pmatrix} 4 & 2 & 1 \\ -3 & 5 & 6 \end{pmatrix}$ and $B = \begin{pmatrix} 2 & 1 & -3 \\ 2 & 1 & 3 \end{pmatrix}$ then $A + B = \begin{pmatrix} 6 & 3 & -2 \\ -1 & 6 & 9 \end{pmatrix}$.

(b) Suppose a production company manufactures items A, B and C. Each item is partially made in factory F_1 in India and then finished in factory F_2 in Germany. The total cost of each item consists of the manufacturing cost and the shipping cost. Then the costs at each factory (in dollars) can be described by 3×2 matrices F_1 and F_2:

$$F_1 = \begin{pmatrix} 32 & 80 \\ 100 & 160 \\ 140 & 40 \end{pmatrix} \begin{matrix} \text{Item 1} \\ \text{Item 2} \\ \text{Item 3} \end{matrix}$$

(columns: Manufacturing cost, Shipping cost)

$$F_2 = \begin{pmatrix} 80 & 120 \\ 100 & 100 \\ 260 & 40 \end{pmatrix} \begin{matrix} \text{Item 1} \\ \text{Item 2} \\ \text{Item 3} \end{matrix}$$

(columns: Manufacturing cost, Shipping cost)

The matrix $F_1 + F_2 = \begin{pmatrix} 112 & 200 \\ 200 & 260 \\ 400 & 80 \end{pmatrix}$ provides the total manufacturing and shipping costs.

Remark 4. (a) Sum A and B, that is, $A+B$ is defined if they have the same number of rows and the same number of columns, in other words if they are of the same size.
If $A = \begin{pmatrix} 2 & 3 & -1 \\ 1 & 3 & 4 \end{pmatrix}$ and $B = \begin{pmatrix} 1 & 1 \\ -1 & 2 \end{pmatrix}$ then $A+B$ cannot be defined.
(b) Let $\mathbf{a} = (a_1, a_2, a_3, \ldots, a_n)$ then $\mathbf{a} + 0 = \mathbf{a}$.

Definition 12. (Scalar multiplication) Let $A = (a_{ij})$ be an $m \times n$ matrix and α be a real number, then the scalar multiplication of A by α denoted by αA is the $m \times n$ matrix $C = (c_{ij})$, where $c_{ij} = \alpha a_{ij}, i = 1, 2, 3, \ldots, m$ and $j = 1, 2, 3, \ldots, n$, that is C is obtained by multiplying each entry of A by α:

$$C = (\alpha a_{ij}) = \begin{pmatrix} \alpha a_{11} & \alpha a_{12} & \cdots & \alpha a_{1n} \\ \alpha a_{21} & \alpha a_{22} & \cdots & \alpha a_{2n} \\ \vdots & \vdots & \vdots & \vdots \\ \alpha a_{m1} & \alpha a_{m2} & \cdots & \alpha a_{mn} \end{pmatrix}.$$

Example 15. Let $A = \begin{pmatrix} 2 & 3 & -1 \\ 1 & 1 & 1 \\ -1 & 6 & 5 \end{pmatrix}$ and $4A = \begin{pmatrix} 8 & 12 & -4 \\ 4 & 4 & 4 \\ -4 & 24 & 20 \end{pmatrix}$.

Example 16. Let $A = \begin{pmatrix} 2 & 0 & 3 & 4 \\ 1 & 3 & 4 & 2 \\ 2 & -2 & 1 & 1 \end{pmatrix}$ and $B = \begin{pmatrix} -1 & 2 & 3 & 1 \\ 4 & 6 & 8 & -1 \\ 1 & 1 & 2 & 1 \end{pmatrix}$.

Find (i) $A + \alpha B$ (ii) $A - \alpha B$ (iii) $A - B$.

Solution: (i)
$$\alpha B = \begin{pmatrix} -\alpha & 2\alpha & 3\alpha & \alpha \\ 4\alpha & 6\alpha & 8\alpha & -\alpha \\ \alpha & \alpha & 2\alpha & \alpha \end{pmatrix}.$$

$$A + \alpha B = \begin{pmatrix} 2-\alpha & 2\alpha & 3+3\alpha & 4+\alpha \\ 1+4\alpha & 4+6\alpha & 4+8\alpha & 2-\alpha \\ 2+\alpha & -2+\alpha & 1+2\alpha & 1+\alpha \end{pmatrix}$$

for $\alpha = 3$,

$$= \begin{pmatrix} -1 & 6 & 12 & 7 \\ 13 & 22 & 28 & -1 \\ 5 & -1 & 7 & 4 \end{pmatrix}.$$

$$A - \alpha B = \begin{pmatrix} 2+\alpha & -2\alpha & 3-3\alpha & 4-\alpha \\ 1-4\alpha & 4-6\alpha & 4-8\alpha & 2+\alpha \\ 2-\alpha & -2-\alpha & 1-2\alpha & 1-\alpha \end{pmatrix}$$

$$= \begin{pmatrix} 5 & -6 & -6 & 1 \\ -10 & -14 & -20 & 5 \\ -1 & -5 & -5 & -2 \end{pmatrix}.$$

$$A - B = \begin{pmatrix} 3 & -2 & 0 & 3 \\ -3 & -2 & -4 & 1 \\ 1 & -3 & -1 & 0 \end{pmatrix}.$$

$A - B$ is called the difference between A and B.

Example 17. (Inventory control) Suppose that a general merchant store is dealing with 50 different items. The inventory at the beginning of the month can be described by the inventory vector $u = (a_{11}, a_{12}, a_{13}, \ldots, a_{150})$ in R^{50}. The number of items sold at the end of the month is given by the vector $v = (b_{11}, b_{12}, b_{13}, \ldots, b_{150})$. The number of items unsold at the end of month is represented by the vector $u - v = (a_{11} - b_{11}, a_{12} - b_{12}, a_{13} - b_{13}, \ldots, a_{150} - b_{150})$. If the store receives a fresh shipment of goods, represented by the vector $w = (c_{11}, c_{12}, c_{13}, \ldots, c_{150})$, then the new inventory of the store would be $u - v + w = (a_{11} - b_{11} + c_{11}, a_{12} - b_{12} + c_{12}, \ldots, a_{150} - b_{1150} + c_{150})$.

Summation Notation: By $\Sigma_{i=1}^{n} a_i$ we mean $a_1 + a_2 + a_3 + \cdots + a_n$. The letter i is the index of summation. It is a dummy variable that can be replaced by another letter. Thus, we can write $\Sigma_{i=1}^{n} a_i$ as $\Sigma_{j=1}^{n} a_j$ or $\Sigma_{k=1}^{n} a_k$.

$$\Sigma_{i=1}^{n} a_i = a_1 + a_2 + a_3 + \cdots + a_n$$

$$\Sigma_{j=1}^{n} a_i = a_1 + a_2 + a_3 + \cdots + a_n.$$

This summation notation satisfies the following properties:

(i) $\Sigma_{i=1}^{n} (\alpha_i + \beta_i) a_i = \Sigma_{i=1}^{n} \alpha_i a_i + \Sigma_{i=1}^{n} \beta_i b_i$

(ii) $\Sigma_{i=1}^{n} \gamma(\alpha_i a_i) = \gamma \Sigma_{i=1}^{n} (\alpha_i a_i)$

(iii) $\Sigma_{i=1}^{n} \Sigma_{i=1}^{m} a_{ij} = \Sigma_{i=1}^{m} \Sigma_{j=1}^{n} a_{ij}$.

Remark 5. (i) The right hand side matrix is obtained by adding all the entries in each row then adding all the resulting numbers.

(ii) The left hand side is obtained by adding all the entries in each column and then adding all the resulting numbers.

(iii) Let $A_1, A_2, A_3, \ldots, A_k$ be $m \times n$ matrices and $c_1, c_2, \ldots, c_k$ are real numbers then $\Sigma_{i=1}^{k} c_i A_i = c_1 A_1 + c_2 A_2 + c_3 A_3 + \cdots + c_k A_k$ is called a linear combination of $m \times n$ matrices while $c_1, c_2, c_3, \ldots, c_k$ are called coefficients.

Example 18. Let $p = \begin{pmatrix} 20 \\ 100 \\ 200 \end{pmatrix}$ be a 3– vector that represents the current prices of three items at a general merchant store. Suppose that the store announces a sale so that the price of each item is reduced by 20%.

(a) Determine a 3– vector that gives the price changes for three items.

(b) Determine a 3− vector that gives new price changes of the items.

Solution: (a) Since the prices of three item are reduced by 20%, the 3− vector

$$-20\%p = \begin{pmatrix} -54 \\ -20 \\ -40 \end{pmatrix} = -\begin{pmatrix} 4 \\ 20 \\ 40 \end{pmatrix}.$$

(b) The new prices of the items are given by the expression

$$\begin{aligned} p - 20\% &= \begin{pmatrix} 20 \\ 100 \\ 200 \end{pmatrix} - 20\%p = \begin{pmatrix} 20 \\ 100 \\ 200 \end{pmatrix} - \begin{pmatrix} 4 \\ 20 \\ 40 \end{pmatrix} \\ &= \begin{pmatrix} 16 \\ 18 \\ 160 \end{pmatrix} = 0.80 \begin{pmatrix} 20 \\ 100 \\ 200 \end{pmatrix} = -0.80\, p. \end{aligned} \tag{1.5}$$

Definition 13. Let $A = (a_{ij})$ be an $m \times n$ matrix; then the **transpose** A, denoted by A^T, is the $n \times m$ matrix defined by $a^T_{ij} = a_{ji}$.

It is clear that transpose matrix A^T is obtained from A by interchanging the rows and columns of A.

Example 19. (i) Let $A = \begin{pmatrix} 4 & 3 & 2 \\ -5 & 6 & 8 \\ 2 & 1 & -1 \end{pmatrix}$, then $A^T = \begin{pmatrix} 4 & -5 & 2 \\ 3 & 6 & 1 \\ 2 & 8 & -1 \end{pmatrix}$.

(ii) Let $B = \begin{pmatrix} 4 & 3 & 1 \\ 0 & 6 & -3 \end{pmatrix}$, then $B^T = \begin{pmatrix} 4 & 0 \\ 3 & 6 \\ 1 & -3 \end{pmatrix}$.

(iii) Let $D = \begin{pmatrix} 2 \\ -3 \\ -1 \end{pmatrix}$, then $D^T = \begin{pmatrix} 2 & -3 & -1 \end{pmatrix}$.

(iv) Let $E = \begin{pmatrix} 2 & 5 & 7 \end{pmatrix}$, then $E^T = \begin{pmatrix} 2 \\ 5 \\ 7 \end{pmatrix}$.

The concept of R^n can be extended for any finite number n. This means the set of all set of $n-$ vectors is denoted by R^n. It is also called the column matrices.

If $a = \begin{pmatrix} a_1 \\ a_2 \\ \vdots \\ a_n \end{pmatrix}$ and $b = \begin{pmatrix} b_1 \\ b_2 \\ \vdots \\ b_n \end{pmatrix}$, then the dot product or inner product of a and b denoted by $a.b$ is defined as

$$\begin{aligned} a.b &= a_1b_1 + a_2b_2 + \cdots + a_nb_n \\ &= \sum_{i=1}^{n} a_ib_i. \end{aligned}$$

Example 20. (i) Let $a = \begin{pmatrix} 1 \\ 3 \\ 1 \\ -1 \end{pmatrix}$ and $b = \begin{pmatrix} 2 \\ -3 \\ 2 \\ -1 \end{pmatrix}$ then $a.b = (2 - 9 + 2 + 1) = -4$.

(ii) Let $a = \begin{pmatrix} x \\ 2 \\ 3 \end{pmatrix}$ and $b = \begin{pmatrix} 8 \\ 2 \\ 4 \end{pmatrix}$. If $a.b = -8$, find x.

Solution:

$$\begin{aligned} a.b &= 8x + 4 + 12 \\ 8x + 16 &= -8 \\ 8x &= -24 \\ x &= -3. \end{aligned}$$

Example 21. Suppose that an instructor uses four grades to determine a student's course average: quizzes, two hourly exams, and a final exam. These are weighted as 10%, 20%, 20%, 50% respectively. If a student achieved grades of 75, 85, 65 and 95 respectively, find the course average.

Solution: If $u = \begin{pmatrix} 0.10 \\ 0.20 \\ 0.20 \\ 0.50 \end{pmatrix}$ and $v = \begin{pmatrix} 75 \\ 85 \\ 65 \\ 85 \end{pmatrix}$ then the course average is given by the dot product $u.v$, that is

$$\begin{aligned} u.v &= (0.10)(75) + (0.20)(85) + (0.20)(65) + (0.50)(85) \\ &= 0.75 + 17.0 + 13 + 42.50 \\ &= 73.25. \end{aligned}$$

Multiplication of two matrices

Definition 14. If $A = (a_{ij})$ is an $m \times p$ matrix and $B = (b_{ij})$ is a $p \times n$ matrix, then the product of A ad B, denoted by AB, is the $m \times n$ matrix $C = c_{ij}$, defined by

$$\begin{aligned} c_{ij} &= a_{i1}b_{ij} + a_{i2}b_{2j} + a_{i3}b_{3j} + \cdots + a_{ip}b_{pj} \\ &= \sum_{k=1}^{p} a_{ik}b_{kj}, \quad 1 \leq i \leq m, \ 1 \leq j \leq m. \end{aligned} \tag{1.6}$$

Remark 6. The product of two matrices A and B is defined only when the number of rows of B is exactly the same as the number of columns of A.

Example 22. (i) Let $A = \begin{pmatrix} 1 & 1 & -1 \\ 3 & 2 & 4 \end{pmatrix}$ and $B = \begin{pmatrix} -2 & 4 \\ 4 & -3 \\ 1 & 2 \end{pmatrix}$. Find

AB.

(ii) Let $A = \begin{pmatrix} 1 & x & 3 \\ 2 & -1 & 1 \end{pmatrix}$ and $B = \begin{pmatrix} 4 \\ 8 \\ y \end{pmatrix}$. If $AB = \begin{pmatrix} 24 \\ 12 \end{pmatrix}$, find x and y.

(iii) Let $A = \begin{pmatrix} 1 & 2 \\ -1 & 3 \end{pmatrix}$ and $B = \begin{pmatrix} 2 & 1 \\ 0 & 0 \end{pmatrix}$. Show that $AB \neq BA$.

Properties of Matrix Addition, Multiplication and Transpose

Properties of Matrix Addition

Let A, B, C be $m \times n$ matrices, then

(a) $A + B = B + A$.

(b) $A + (B + C) = (A + B) + C$.

(c) There is a unique $m \times n$ matrix $O = \begin{pmatrix} 0 & 0 & \cdots & 0 \\ 0 & 0 & \cdots & 0 \\ \vdots & \vdots & \vdots & \vdots \\ 0 & 0 & \cdots & 0 \end{pmatrix}$ such that $A + O = A$. O is called the zero matrix.

(d) For each $m \times n$ matrix A, there is a unique $m \times n$ matrix D such that

$$A + D = O.$$

we can write $D = -A$ so as

$$A + (-A) = O.$$

$-A$ is called a negative of A, that is each entry of A is multiplied by -1.

Properties of Matrix Multiplication

If A, B, C are matrices of the appropriate sizes (say $m \times n$), then

(a) $A(BC) = (AB)C$.

(b) $(A + B)C = AC + BC$.

(c) $C(A + B) = CA + CB$.

Properties of Scalar Multiplication

If α and β re real numbers and A and B are matrices of size $m \times n$, then

(a) $\alpha(\beta A) = (\alpha\beta)A$.

(b) $(\alpha + \beta)A = \alpha A + \beta A$.

(c) $\alpha(A + B) = \alpha A + \beta A$.

(d) $A(\alpha B) = \alpha(AB)$.

Properties of Transpose

If A and B be $m \times n$ matrices and α be a scalar (real number), then

(a) $(A^T)^T = A$.

(b) $(A+B)^T = A^T + B^T$.

(c) $(AB)^T = B^T A^T$.

(d) $(\alpha A)^T = \alpha A^T$.

Remark 7. (i) As mentioned earlier AB need not be equal to BA.

(ii) AB may be the zero matrix with $A \neq 0$ and $B \neq 0$.

(iii) AB may be equal to AC with $B \neq C$.

1.2.1 Linear Equations and Matrices

In this section we show that solving the linear system of m equations in n unknowns is equivalent to solving a matrix equation of the form $Ax = b$, where $A = (a_{ij})$ is an $m \times n$ matrix, $x = \begin{pmatrix} x_1 \\ x_2 \\ \vdots \\ x_n \end{pmatrix}$ and $b = \begin{pmatrix} b_1 \\ b_2 \\ \vdots \\ b_n \end{pmatrix}$.

Let us consider the linear system of m equations in n unknowns

$$\begin{array}{rcl} a_{11}x_1 + a_{12}x_2 + \cdots + a_{1n}x_n &=& b_1 \\ a_{21}x_1 + a_{22}x_2 + \cdots + a_{2n}x_n &=& b_2 \\ \vdots \quad\quad \vdots \quad\quad\quad \vdots \quad\quad \vdots && \vdots \\ a_{m1}x_1 + a_{m2}x_2 + \cdots + a_{mn}x_n &=& b_m. \end{array}$$

Let

$$A = \begin{pmatrix} a_{11} & a_{12} & \cdots & a_{1n} \\ a_{21} & a_{22} & \cdots & a_{2n} \\ \vdots & \vdots & \cdots & \vdots \\ a_{m1} & a_{m2} & \cdots & a_{mn} \end{pmatrix}, \; x = \begin{pmatrix} x_1 \\ x_2 \\ \vdots \\ x_n \end{pmatrix}, \; b = \begin{pmatrix} b_1 \\ b_2 \\ \vdots \\ b_m \end{pmatrix}.$$

Then

$$Ax = \begin{pmatrix} a_{11} & a_{12} & \cdots & a_{1n} \\ a_{21} & a_{22} & \cdots & a_{2n} \\ \vdots & \vdots & \cdots & \vdots \\ a_{m1} & a_{m2} & \cdots & a_{mn} \end{pmatrix} \begin{pmatrix} x_1 \\ x_2 \\ \vdots \\ x_n \end{pmatrix}$$

$$= \begin{pmatrix} a_{11}x_1 + a_{12}x_2 + \cdots + a_{1n}x_n \\ a_{21}x_1 + a_{22}x_2 + \cdots + a_{2n}x_n \\ \vdots \quad \vdots \quad \vdots \quad \vdots \\ a_{m1}x_1 + a_{m2}x_2 + \cdots + a_{mn}x_n \end{pmatrix} \tag{1.7}$$

$$= \begin{pmatrix} b_1 \\ b_2 \\ \vdots \\ b_m \end{pmatrix}.$$

Thus, we have

$$Ax = b.$$

The matrix A is called the coefficient matrix of the linear system (1.6), and the matrix $\begin{pmatrix} a_{11} & a_{12} & \cdots & a_{1n} & | & b_1 \\ a_{21} & a_{22} & \cdots & a_{2n} & | & b_2 \\ \vdots & \vdots & \vdots & a_{2n} & | & \vdots \\ a_{m1} & a_{m2} & \cdots & a_{mn} & | & b_m \end{pmatrix}$ obtained by adjoining column b to A, is called the **augmented matrix** of the linear system (1.6). The augmented matrix of the linear system is denoted by $(A|b)$. If $b_1 = b_2 = \cdots_m = 0$ in (1.7) then the linear system is called a **homogeneous system**. A homogeneous system is written as $Ax = 0$, where A is a coefficient matrix.

Example 23. (a) Write the augmented matrix of the linear system:

$$\begin{aligned} -2x + z &= 7 \\ 2x + 3y - 4z &= 9 \\ 3x + 2y + 2z &= 5. \end{aligned}$$

(b) $Ax = b$ is called consistent if b can be expressed as a linear combination of columns of matrix A. Give an example.

Solution: (a) $A = \begin{pmatrix} -2 & 0 & 1 \\ 2 & 3 & -4 \\ 3 & 2 & 2 \end{pmatrix}$, $x = \begin{pmatrix} x \\ y \\ z \end{pmatrix}$, $b = \begin{pmatrix} 7 \\ 9 \\ 5 \end{pmatrix}$.

Augmented matrix is $\begin{pmatrix} -2 & 0 & 1 & \vdots & 7 \\ 2 & 3 & -4 & \vdots & 9 \\ \cdots & \cdots & \cdots & \vdots & \cdots \\ 3 & 2 & 2 & \vdots & 5 \end{pmatrix}$.

(b) Let $Ax = b$ where $A = \begin{pmatrix} 3 & 1 & 2 \\ 4 & -5 & 6 \\ 0 & 7 & -3 \\ -1 & 2 & 0 \end{pmatrix}$

$$x = \begin{pmatrix} x_1 \\ x_2 \\ x_3 \end{pmatrix} \text{ and } b = \begin{pmatrix} 4 \\ 1 \\ 0 \\ 2 \end{pmatrix}.$$

Then $x_1 \begin{pmatrix} 3 \\ 4 \\ 0 \\ -1 \end{pmatrix} + x_2 \begin{pmatrix} 1 \\ -5 \\ 7 \\ 2 \end{pmatrix} + x_3 \begin{pmatrix} 2 \\ 6 \\ -3 \\ 0 \end{pmatrix} = \begin{pmatrix} 4 \\ 1 \\ 0 \\ 2 \end{pmatrix}$.

Hence $Ax = b$ is consistent.

1.2.2 Rank of Matrix

Definition 15. The **rank** of $m \times n$ matrix A denoted by rank A is the maximum number of linearly independent row vectors in A.

Example 24. Find the rank of $A = \begin{pmatrix} 1 & 1 & -1 & 3 \\ 2 & -2 & 6 & 8 \\ 3 & 5 & -7 & 8 \end{pmatrix}$.

Solution: Let $a = (1\ \ 1\ -1\ \ 3)$, $b = (2\ -2\ \ 6\ \ 8)$ and $c = (3\ \ 5\ -7\ \ 8)$. It is clear that $4a - \frac{1}{2}b + c = 0$ which implies that a, b, c are linearly dependent; a and b are nearly independent as they are not constant multiples of each other. Hence rank $A = 2$.

1.2.3 Special Classes of Matrices

Here we present special forms of matrices known as diagonal matrix, zero matrix, identity matrix, scalar matrix, self adjoint matrix,triangular matrix, symmetric matrix, skew matrix, nonsingular (invertible) matrix, sub-matrix, partitioned matrix, orthogonal and diagonalizable matrix.

$A = \begin{pmatrix} 0 & 0 & 0 & \cdots & 0 \\ 0 & 0 & 0 & \cdots & 0 \\ \vdots & \vdots & \vdots & \vdots & \vdots \\ 0 & 0 & 0 & \cdots & 0 \end{pmatrix}$ having all 0 entries is called a **zero matrix.**

$A = (a_{ij})$ is called a **diagonal matrix** if $a_{ij} = 0$ for $i \neq j$.

For example $A = \begin{pmatrix} 3 & 0 & 0 \\ 0 & 2 & 0 \\ 0 & 0 & 4 \end{pmatrix}$ is a diagonal matrix.

A **scalar matrix** is a diagonal matrix whose diagonal elements are equal. The scalar matrix $I_n = (d_{ij})$, where $d_{ij} = 1$ for $i = j$ and $d_{ij} = 0$ for $i \neq j$ is called the $n \times n$ **identity matrix**.

$A = \begin{pmatrix} 2 & 0 & 0 \\ 0 & 3 & 0 \\ 0 & 0 & 4 \end{pmatrix}$ is a diagonal matrix.

$B = \begin{pmatrix} 3 & 0 & 0 \\ 0 & 3 & 0 \\ 0 & 0 & 3 \end{pmatrix}$ is a scalar matrix.

$I_3 = \begin{pmatrix} 1 & 0 & 0 \\ 0 & 1 & 0 \\ 0 & 0 & 1 \end{pmatrix}$ is the identity matrix.

Let A be an $m \times n$ matrix; then $AI_n = A$ and $I_nA = A$ where I_n is an $n \times n$ identity matrix.
An $n \times n$ matrix $A = (a_{ij})$, $i = 1, 2, 3, \ldots, n$, $j = 1, 2, 3, \ldots, n$ is called **upper triangular** if $a_{ij} = 0$ for $i > j$. It is called **lower triangular** if $a_{ij} = 0$ for $i < j$. It may noted that a diagonal matrix is both upper triangular and lower triangular.

Example 25. The matrix $A = \begin{pmatrix} 1 & 5 & 5 \\ 0 & 5 & 6 \\ 0 & 0 & 2 \end{pmatrix}$ is upper triangular matrix and $B = \begin{pmatrix} 5 & 0 & 0 \\ 5 & 6 & 0 \\ 5 & 6 & 3 \end{pmatrix}$ is lower triangular.

A matrix $A = (a_{ij})$ with real entries is called **symmetric** if $A^T = A$. It is called **skew symmetric** if $A^T = -A$.

For $A = \begin{pmatrix} 1 & 1 & 2 \\ 1 & -3 & 3 \\ 2 & 3 & 1 \end{pmatrix}$, $A^T = \begin{pmatrix} 1 & 1 & 2 \\ 1 & -3 & 3 \\ 2 & 3 & 1 \end{pmatrix}$ and $A = A^T$, A is symmetric.
The matrix $B = \begin{pmatrix} 0 & 1 \\ -1 & 0 \end{pmatrix}$ is skew symmetric as

$$B^T = \begin{pmatrix} 0 & -1 \\ 1 & 0 \end{pmatrix} = -\begin{pmatrix} 0 & 1 \\ -1 & 0 \end{pmatrix} = B.$$

If we begin by an $m \times n$ matrix $A = (a_{ij})$ and then cross out some, but not all, of its rows or columns, we obtain a sub-matrix.

Let $A = \begin{pmatrix} 2 & 4 & 6 & 8 \\ -4 & 8 & -6 & 10 \\ 6 & 0 & 10 & -6 \end{pmatrix}$. $B = \begin{pmatrix} 2 & 4 & 6 & 8 \\ 6 & 0 & 10 & -6 \end{pmatrix}$ is a sub-matrix of A as well as $C = \begin{pmatrix} 2 & 4 & 6 \\ 6 & 0 & 10 \end{pmatrix}$ is a sub-matrix of A.

Non-singular Matrices

Definition 16. An $n \times n$ matrix A is called **non-singular** or **invertible**, if there is an $n \times n$ matrix B such that $AB = BA = I_n$. B is called an inverse of A if it has this property. If there is no such B then A is called **singular** or **non-invertible**.

Remark 8. If A and B are $n \times n$ matrices such that $AB = I_n$, then $BA = I_n$ and hence $B = A^{-1}$. For verification, see Kolman and Hill [5].

Example 26. Let $A = \begin{pmatrix} 2 & 3 \\ 2 & 2 \end{pmatrix}$ and $B = \begin{pmatrix} -1 & \frac{3}{2} \\ 1 & -1 \end{pmatrix}$. It can be checked that $AB = BA = I_2$ which implies that $B = A^{-1}$.

Remark 9. The inverse of a matrix, if it exists, is unique.

Verification: Let B and C be inverse of A. Then

$$AB = BA = I_n, \ AC = CA = I_n.$$

$$A = BI_n = B(AC) = (BA)C = I_nC = C.$$

This verifies that the inverse of a matrix is unique if it exists. Clearly $AA^{-1} = A^{-1}A = I_n$.

Example 27. Let $A^{-1} = \begin{pmatrix} a & b \\ c & d \end{pmatrix}$.

$$\begin{aligned} AA^{-1} &= \begin{pmatrix} 2 & -1 \\ -1 & 1 \end{pmatrix}\begin{pmatrix} a & b \\ c & d \end{pmatrix} \\ &= \begin{pmatrix} 2a - c & 2b + d \\ a + 2c & b + 2d \end{pmatrix} \\ &= \begin{pmatrix} 1 & 0 \\ 0 & 1 \end{pmatrix} \end{aligned}$$

$$\begin{aligned} 2a - c &= 1 \\ a + 2c &= 0 \\ b + 2d &= 0 \\ b + 2d &= 1. \end{aligned}$$

From these four equations we get

$$a = \frac{2}{5}, \quad b = \frac{-1}{3}, \quad c = \frac{-1}{5}, \quad d = \frac{2}{3}.$$

Orthogonal Matrix: A real matrix is called orthogonal if $A^TA = AA^T = I_n$ or $A^T = A^{-1}$. Examples of orthogonal matrices follow.
Diagonal Matrix: An $n \times n$ matrix A is diagonalizable if there is an invertible matrix P and $n \times n$ diagonal matrix D such that $P^{-1}AP = D$. We say that P diagonalizes A and we say that P is a diagonalizing matrix for A. $\begin{pmatrix} 0 & 1 & 0 \\ 1 & 0 & 0 \\ 0 & 0 & 1 \end{pmatrix}$, $\begin{pmatrix} 0 & 0 & 1 \\ -\frac{12}{13} & \frac{5}{13} & 0 \\ \frac{5}{13} & \frac{12}{13} & 0 \end{pmatrix}$ and $\begin{pmatrix} 0 & \frac{1}{\sqrt{5}} & \frac{2}{\sqrt{5}} \\ 1 & 0 & 0 \\ 0 & \frac{2}{\sqrt{5}} & -\frac{1}{\sqrt{5}} \end{pmatrix}$ are orthogonal matrix.

1.2.4 Echelon Form of Matrix

Definition 17. An $m \times n$ matrix A is said to be in reduced row **echelon form** if it satisfies the following properties:

(a) All zero rows appear at the bottom of the matrix.

(b) The first non-zero entry from the left of a non-zero row is a 1. This entry is leading one of its row.

(c) For each non-zero, the leading one appears to the right and below any leading ones in preceding rows.

(d) If a column contains a leading one, then all other entries in the column are zero.

Remark 10. A matrix in reduced row echelon from appears as a staircase (echelon) pattern of leading once descending from the upper left corner of the matrix. An $m \times n$ matrix satisfying properties (a), (b), and (c) is said to be in row echelon form. There may be no zero rows.
A similar definition may be given for a reduced column echelon form and column echelon form.

Examples of reduced row echelon form are:

$$A = \begin{pmatrix} 1 & 2 & 0 & 0 & 1 \\ 0 & 0 & 1 & 2 & 3 \\ 0 & 0 & 0 & 0 & 0 \end{pmatrix} \text{ and } B = \begin{pmatrix} 1 & 0 & 0 & 0 \\ 0 & 1 & 0 & 0 \\ 0 & 0 & 1 & 0 \\ 0 & 0 & 0 & 1 \end{pmatrix}.$$

Definition 18. An elementary row (column) operation on matrix A is any one of the following:

(a) Type I: Interchange any two rows (columns).

(b) Type II: Multiply a row (column) by a non-zero number.

(c) Type III: Add a multiple of one row (column) to another.

An $m \times n$ matrix B is said to be row (column) equivalent to an $m \times n$ matrix A if B can be produced by applying a finite sequence of row operation.

Theorem 2. *Every non-zero $m \times n$ matrix $A = (a_{ij})$ is row (column) equivalent to a matrix in row (column) echelon form.*

First column having a non-zero entry is called the **pivot column**.

Theorem 3. *Every non-zero $m \times n$ matrix $A = (a_{ij})$ is row equivalent to a unique matrix in reduced row (column) echelon form.*

Theorem 4. *Let $Ax = b$ and $Cx = d$ be two linear systems, each of m equations in n unknowns. If augmented matrices $(A|b)$ and $(C|d)$ are row equivalent, then the linear systems are equivalent, that is, they have exactly the same solution.*

For proof of these theorems we refer to [5].

1.3 Applications of Matrices to Real World Problems

1.3.1 Modeling of Temperature Distribution by Matrices

A simple model for estimating the temperature distribution on a square plate can be written in the form of a linear system of equations which can be expressed in the form of a matrix equation. Thus finding distribution of temperature on a square plate is equivalent to solving a matrix equation.

Example 28. Determine the temperature at the interior points $T_i, i = 1, 2, 3, 4$ for the plate shown in Figure 1.5.
First we construct the linear system for estimating the temperatures. The points at which we want to estimate temperature are indicated by dots in Figure 1.5. Using the averaging rule we get

$$T_1 = \frac{60 + 100 + T_2 + T_3}{4} \quad \text{or} \quad 4T_1 - T_2 - T_3 = 160.$$

$$T_2 = \frac{T_1 + 100 + 40 + T_4}{4} \quad \text{or} \quad -T_1 + 4T_2 - T_4 = 140.$$

$$T_3 = \frac{60 + T_1 + T_4 + 0}{4} \quad \text{or} \quad -T_1 + 4T_3 - T_4 = 60.$$

$$T_4 = \frac{T_3 + T_2 + 40 + 0}{4} \quad \text{or} \quad -T_2 - T_3 + 4T_4 = 40.$$

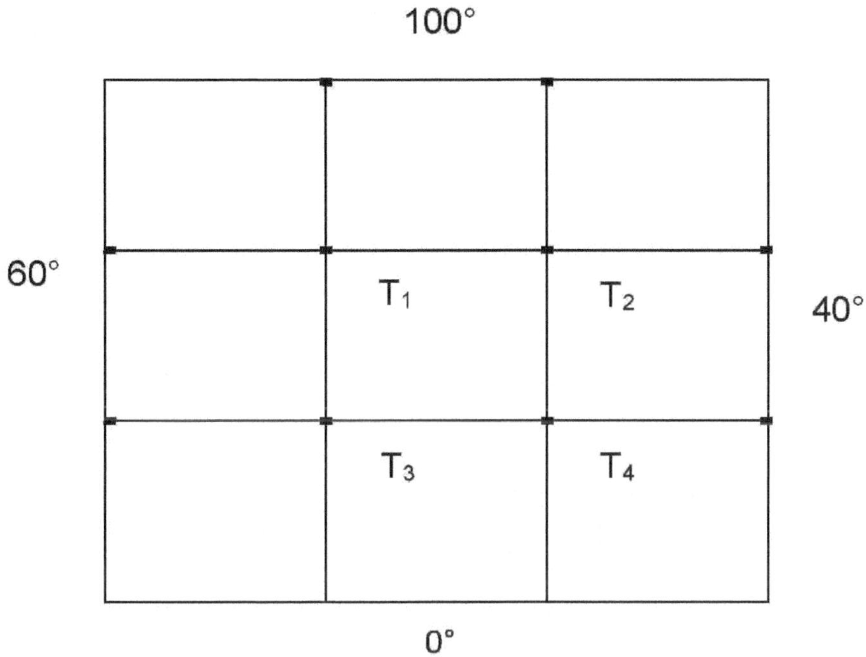

FIGURE 1.5: Temperature Distribution

This leads to the augmented matrix for this linear system:

$$A = \begin{pmatrix} 4 & -1 & -1 & \vdots & 160 \\ -1 & 4 & -1 & \vdots & 160 \\ -1 & 4 & -1 & \vdots & 60 \\ -1 & -1 & 4 & \vdots & 40 \end{pmatrix}.$$

By solving the model $Ax = y$ where

$$A = \begin{pmatrix} 4 & -1 & -1 \\ -1 & 4 & -1 \\ -1 & 4 & -1 \\ -1 & -1 & 4 \end{pmatrix} \text{ and } y = \begin{pmatrix} 160 \\ 160 \\ 60 \\ 40 \end{pmatrix},$$

we get $x = (T_1, T_2, T_3, T_4)$ where $T_1 = 65^o$, $T_2 = 60^o$, $T_3 = 40^o$, $T_4 = 35^o$.

1.3.2 Modeling of Traffic Flow by Matrices

At rush hours, traffic congestion is encountered at the street intersections shown in Figure 1.6.

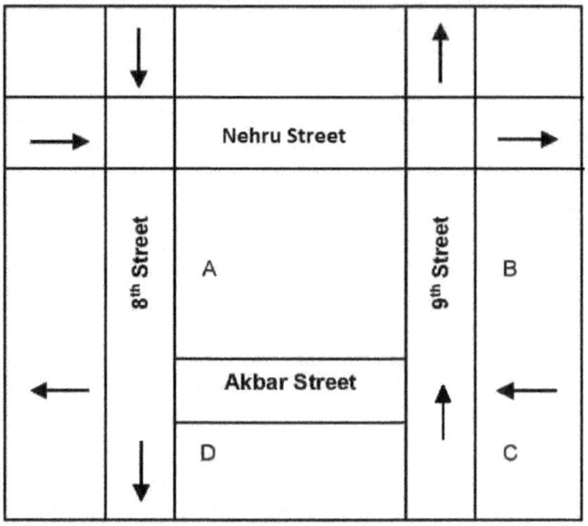

FIGURE 1.6: Traffic Flow

1. Corner A
   700 cars an hour come down Nehru street to intersection A.
   1300 cars an hour come down $8th$ to intersection A.

2. Corner B
   200 cars an hour leave intersection B on Nehru Street.
   900 cars an hour leave intersection B on $9th$ Street.

3. Corner C
   400 cars an hour enter on Akbar street to intersection C.
   400 cars an hour come down $10th$ street to intersection C.

4. Corner D
   200 cars an hour leave intersection D on Akbar Street.
   400 cars an hour leave intersection D on $8th$ Street to intersection A.

Let x_1 denote the number of cars leaving corner A on Nehru Street towards corner B.
Let x_2 denote the number of cars arriving to corner B on $9th$ Street from corner C.
Let x_3 denote the number of cars leaving corner C on Akbar Street towards

corner D.
Let x_4 denote the number of cars arriving to corner D on $8th$ Street from corner A.
Assumptions. To solve this problem we assume the following:

1. To speed the traffic flow every car that arrives to a given corner must also leave; hence at any corner, the number of cars arriving is equal to number of cars leaving.

2. All streets are one-way.

3. All variables, x_1, x_2, x_3 and x_4, are positive integers since they represent numbers of cars.

Equation: Using assumption 1 for each corner, we obtain the following equations:

At corner A	$x_1 + x_4 = 700 + 300$
At corner B	$x_1 + x_2 = 900 + 200$
At corner C	$x_2 + x_3 = 400 + 300$
At corner D	$x_3 + x_4 = 400 + 200$

These four equations form a system of linear equations that can be solved using the Gauss-Jordan method (row reduction of the augmented matrix).

$$\begin{aligned} x_1 + x_4 &= 1000 \\ x_1 + x_2 &= 1100 \\ x_2 + x_3 &= 700 \\ x_3 + x_4 &= 600 \end{aligned}$$

$$A = \begin{pmatrix} 1 & 0 & 0 & 1 \\ 1 & 1 & 0 & 0 \\ 0 & 1 & 1 & 0 \\ 0 & 0 & 1 & 1 \end{pmatrix}$$

$$y = \begin{pmatrix} 1000 \\ 1100 \\ 700 \\ 600 \end{pmatrix}.$$

By solving the matrix equation, we get x_1, x_2, x_3, x_4.

1.3.3 Matrices for Chemical Balance Equations

Mathematical equations may describe chemical reactions. In chemical reactions two sides are separated by arrow instead of equal to symbols. Let sodium hydroxide ($NaOH$) react with sulfuric acid (H_2SO_4) to form sodium sulfate and water H_2O. The chemical reaction is denoted as

$$NaOH + H_2SO_4 \to Na_2SO_4 + H_2O.$$

To balance this equation we insert unknowns x, y, z and w to get an equation

$$xNaOH + yH_2SO_4 \to zNa_2SO_4 + wH_2O.$$

By comparing the number of sodium (Na), oxygen (O), hydrogen (H) and sulfur (S) atoms on left and right sides, we get four linear equations:

$$\begin{aligned} Na &: x = 2z \\ O &: x + 4y = 4z + w \\ H &: x + 2yz = 2w \\ S &: y = z \end{aligned}$$

It may be noted that we make use of the subscript in writing this system of equations because they indicate the number of atoms of a particular element. Thus we have a homogeneous linear system of four unknowns as

$$Ax = y$$

where

$$A = \begin{pmatrix} 1 & 0 & -2 & 0 \\ 1 & 4 & -4 & -1 \\ 1 & 2 & 0 & -2 \\ 0 & 1 & -1 & 0 \end{pmatrix}$$

or this matrix equation can be written as augmented matrix

$$\begin{pmatrix} 1 & 0 & -2 & 0 & \vdots & 0 \\ 1 & 4 & -4 & -1 & \vdots & 0 \\ 1 & 2 & 0 & -2 & \vdots & 0 \\ 0 & 1 & -1 & 0 & \vdots & 0 \end{pmatrix}.$$

The reduced row echelon form of this augmented matrix is

$$\begin{pmatrix} 1 & 0 & 0 & -1 & \vdots & 0 \\ 0 & 1 & 0 & \frac{-1}{2} & \vdots & 0 \\ 0 & 0 & 1 & \frac{-1}{2} & \vdots & 0 \\ 0 & 0 & 0 & 0 & \vdots & 0 \end{pmatrix}.$$

This gives us $x = w$, $y = \frac{1}{2}w$, and $z = \frac{1}{2}w$. We choose $w = 2$ so that all unknowns x, y, z are positive ($x = 2$, $y = 1$ and $z = 1$). Thus the balanced equation is

$$2NaOH + H_2SO_4 \rightarrow Na_2SO_4 + 2H_2O.$$

1.3.4 Modeling by Matrix Equation in Business

Example 29. A construction company uses three types of steel A_1, A_2 and A_3 for constructing three types of houses H_1, H_2 and H_3. The steel required for each type of house is given below:

TABLE 1.2

Steel	H_1	H_2	H_3
A_1	2	3	4
A_2	1	1	2
A_3	3	2	1

Find the number of houses of each type of steel which can be produced using 29, 13 and 16 tons of steel.

Solution: Let x, y and z denote the number of houses that can be constructed of each type. Then we have three equations:

$$\begin{aligned} 2x + 3y + 4z &= 29. \\ x + y + 2z &= 13. \\ 3x + 2y + z &= 16. \end{aligned}$$

This system of three equations can be written as a matrix equation $Ax = y$ where $A = \begin{pmatrix} 2 & 3 & 4 \\ 1 & 1 & 2 \\ 3 & 2 & 1 \end{pmatrix}$ and $y = \begin{pmatrix} 29 \\ 13 \\ 16 \end{pmatrix}$ or $\begin{pmatrix} 2 & 3 & 4 \\ 1 & 1 & 2 \\ 3 & 2 & 1 \end{pmatrix}\begin{pmatrix} x \\ y \\ z \end{pmatrix} = \begin{pmatrix} 29 \\ 13 \\ 16 \end{pmatrix}$. This equation can be solved by the Gauss Jordan elimination method. By applying operation $R_1 \leftrightarrow R_2$ (Row R_1 is interchanged by R_2) the above system takes the form:

$$\begin{pmatrix} 1 & 1 & 2 \\ 2 & 3 & 4 \\ 3 & 2 & 1 \end{pmatrix}\begin{pmatrix} x \\ y \\ z \end{pmatrix} = \begin{pmatrix} 13 \\ 29 \\ 16 \end{pmatrix}.$$

Now applying $R_2 \to R_{\frac{3}{2}} - 2R_1$ and $R_3 - 3R_1$, the above system is equivalent to

$$\begin{pmatrix} 1 & 1 & 2 \\ 0 & 1 & 0 \\ 0 & -1 & -5 \end{pmatrix} \begin{pmatrix} x \\ y \\ z \end{pmatrix} = \begin{pmatrix} 13 \\ 3 \\ -23 \end{pmatrix}.$$

Applying $R_3 \to R_3 + R_2$ makes the above system equivalent to

$$\begin{pmatrix} 1 & 1 & 2 \\ 0 & 1 & 0 \\ 0 & -1 & -5 \end{pmatrix} \begin{pmatrix} x \\ y \\ z \end{pmatrix} = \begin{pmatrix} 13 \\ 3 \\ -20 \end{pmatrix}$$

or

$$\begin{aligned} x + y + 2z &= 13 \\ y &= 3 \\ -5z &= -20. \end{aligned}$$

Therefore, $z = 4$, $y = 3$ and putting these values in the first equation we get $x = 2$.

Example 30. A manufacturing company produces 3 items every day. The total production on a particular day is 45 tons. It is observed that the production of the third product exceeds the production of the first by 8 tons while the total production of the first and third products is thrice that of the second product. Determine the level of products of each item.

Solution: Let the production levels of the three items be x, y and z respectively. This situation is modeled by the equations

$$\begin{aligned} x + y + z &= 45 \\ z &= x + 8 \quad \text{or} \quad -x + 0y + z = 8 \\ x + z &= 2y \quad \text{or} \quad x - 2y + z = 0. \end{aligned}$$

The matrix equation

$$Au = v$$

where $A = \begin{pmatrix} 1 & 1 & 1 \\ -1 & 0 & 1 \\ 1 & -2 & 1 \end{pmatrix}$, $v = \begin{pmatrix} 45 \\ 8 \\ 0 \end{pmatrix}$.

By solving this matrix equation we get

$$x = 11, \; y = 15, \; z = 19.$$

Production level of first item =11
Production level of second item =15
Production level of third item =19.

1.3.5 Role of Matrices in Electrical Networks

The electrical network shown in Figure 1.7 contains batteries (voltage sources) and resistors (light bulbs or heaters) joined together by conductors (wires).

A voltage source provides an electromotive force E measured in volts, which moves electrons through the network. The rate at which electrons flow along a conductor is called **current** I and is measured in amperes (amps). The resistor acts to retard the flow of electrons, using up the current and lowering the voltage. According to Ohm's law, the potential difference E volts across a resistor is given by $E = IR$, where I is the current measured in amperes and R is the resistance measured in ohms (Ω).

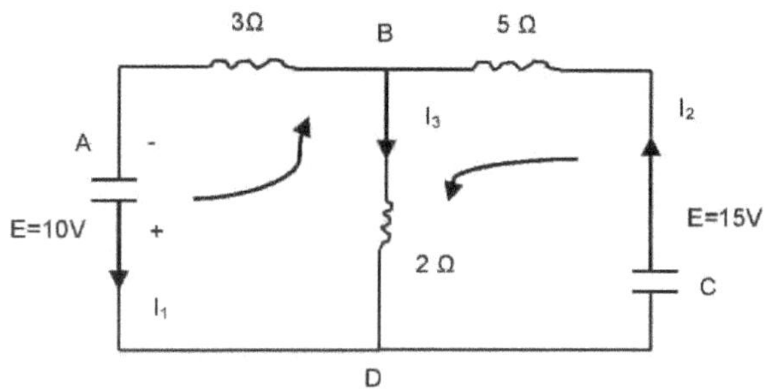

FIGURE 1.7: Electrical Network

Network analysis is based on Kirchhoff's two laws that apply to the closed voltage loops and nodes in the the network. A node is a junction of three or more conductors. The network in Figure 1.7 has two nodes B and D and three closed voltage loops defined as

$$\begin{aligned} A \to D \to B \to A &\qquad \text{(Loop } ADBA\text{)} \\ C \to B \to D \to C &\qquad \text{(Loop } CBDC\text{)} \\ A \to D \to C \to B \to A &\qquad \text{(Loop } ADCBA\text{).} \end{aligned}$$

If $R_{11}, R_{12}, \ldots, R_{mm}$ and $V_1, V_2, \ldots, V_m$ are constants the electric network can be written as a matrix equation by Kirchhoff's law for each loop.

$$Ax = y$$

where

$$A = \begin{pmatrix} R_{11} & R_{12} & \cdots & R_{1m} \\ R_{21} & R_{22} & \cdots & R_{2m} \\ \cdots & \cdots & \cdots & \cdots \\ R_{m1} & R_{m2} & \cdots & R_{mm} \end{pmatrix}$$

$$y = \begin{pmatrix} v_1 \\ v_2 \\ v_3 \end{pmatrix}, \quad x^T = \begin{pmatrix} I_1 \\ I_2 \\ \vdots \\ I_m \end{pmatrix}.$$

Example 31. Set up and solve the system of equations (matrix equations) of the networks given in Figure 1.8 and Figure 1.9.

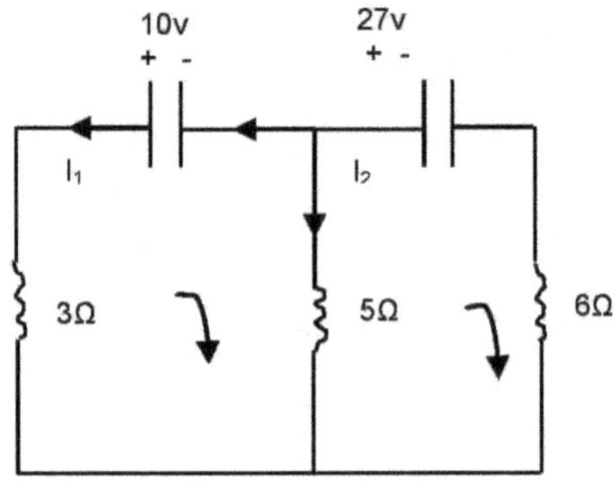

FIGURE 1.8: Electrical Network

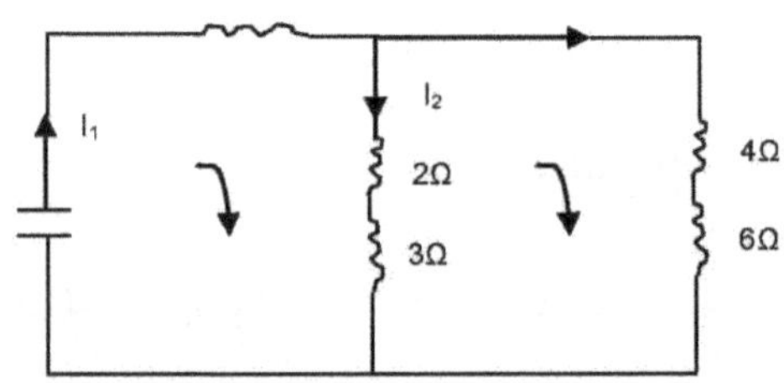

FIGURE 1.9: Electrical Network

The system of equations is

$$\begin{aligned} -i_1 + i_2 - i_3 &= 0 \\ 10 - 3_i + 5i_3 &= 0 \\ 27 - 6i_2 - 5i_3 &= 0 \end{aligned}$$

or

$$\begin{aligned} -i_1 + i_2 - i_3 &= 0 \\ 3_i - 5i_3 &= 10 \\ 6i_2 + 5i_3 &= 27. \end{aligned}$$

Gaussian elimination gives

$$\left(\begin{array}{ccc|c} -1 & 1 & -1 & 0 \\ 3 & 0 & -5 & 10 \\ 0 & 6 & 5 & 27 \end{array}\right) \xrightarrow[\text{row operations}]{} \left(\begin{array}{ccc|c} 1 & -1 & 1 & 0 \\ 0 & 1 & \frac{-8}{3} & \frac{10}{3} \\ 0 & 0 & 51 & \frac{1}{3} \end{array}\right).$$

The solution is $i_1 = \frac{35}{9}, i_2 = \frac{38}{9}, i_3 = \frac{1}{3}$.
The system of equations is

$$\begin{aligned} i_1 - i_2 - i_3 &= 0 \\ 5i_2 - i_1 - 5i_2 &= 0 \\ -10i_3 + 5i_2 &= 0 \end{aligned}$$

or

$$\begin{aligned} i_1 - i_2 - i_3 &= 0 \\ 5i_2 i_1 + 5i_2 &= 5i_2 \\ 5i_2 - 10i_3 &= 0. \end{aligned}$$

Gauss elimination gives

$$\left(\begin{array}{ccc|c} 1 & -1 & -1 & 0 \\ 1 & 5 & 0 & 52 \\ 0 & 5 & -10 & 0 \end{array}\right) \xrightarrow[\text{row operations}]{} \left(\begin{array}{ccc|c} 1 & -1 & -1 & 0 \\ 0 & 1 & \frac{1}{6} & \frac{26}{3} \\ 0 & 0 & 1 & 4 \end{array}\right).$$

The solution is $i_1 = 12, i_2 = 8, i_3 = 4$.

1.4 Determinants

1.4.1 Introduction

Let A be an $n \times n$ matrix, Associated with A is a number called the determinant of A denoted by $detA$. Symbolically, we distinguish a matrix A

from its determinant by replacing the parentheses by vertical bars:

$$A = \begin{pmatrix} a_{11} & a_{12} & \cdots & a_{1n} \\ a_{21} & a_{22} & \cdots & a_{2n} \\ \vdots & \vdots & \vdots & \vdots \\ a_{n1} & a_{n2} & \cdots & a_{nn} \end{pmatrix}$$

and

$$detA = \begin{vmatrix} a_{11} & a_{12} & \cdots & a_{1n} \\ a_{21} & a_{22} & \cdots & a_{2n} \\ \vdots & \vdots & \vdots & \vdots \\ a_{n1} & a_{n2} & \cdots & a_{nn} \end{vmatrix}.$$

A determinant of an $n \times n$ matrix is said to be a **determinant of order** n. We begin by defining the determinant of 1×1, 2×2 and 3×3 matrices.

1.4.2 Determinants of Matrices

Definition 19. Determinant of a 1×1 matrix $A = (a)$ denoted by $detA$ is defined by

$$detA = |a|.$$

If $A = (-4)$, then $detA = |-4| = 4$.

Definition 20. Determinant of 2×2 matrix $A = \begin{pmatrix} a_{11} & a_{12} \\ a_{21} & a_{22} \end{pmatrix}$ is defined by

$$\det A = \begin{vmatrix} a_{11} & a_{12} \\ a_{21} & a_{22} \end{vmatrix} = a_{11}a_{22} - a_{12}a_{21}.$$

Let $A = \begin{pmatrix} 5 & -1 \\ 5 & 4 \end{pmatrix}$, then

$$detA = \begin{vmatrix} 5 & -1 \\ 5 & 4 \end{vmatrix} = 20 - (-5) = 25.$$

Definition 21. Let $A = \begin{pmatrix} a_{11} & a_{12} & a_{13} \\ a_{21} & a_{22} & a_{23} \\ a_{31} & a_{32} & a_{33} \end{pmatrix}$. Then

$$\begin{aligned} detA &= \begin{vmatrix} a_{11} & a_{12} & a_{13} \\ a_{21} & a_{22} & a_{23} \\ a_{31} & a_{32} & a_{33} \end{vmatrix} \\ &= a_{11}(a_{22}a_{33} - a_{23}a_{32}) + a_{12}(-a_{21}a_{33} + a_{23}a_{31}) + a_{13}(a_{21}a_{32} - a_{22}a_{31}) \\ &= a_{11}\begin{vmatrix} a_{22} & a_{23} \\ a_{32} & a_{33} \end{vmatrix} + -a_{12}\begin{vmatrix} a_{22} & a_{23} \\ a_{31} & a_{33} \end{vmatrix} + a_{13}\begin{vmatrix} a_{21} & a_{22} \\ a_{31} & a_{32} \end{vmatrix}. \end{aligned}$$

Thus $detA = a_{11}c_{11} + a_{12}c_{12} + a_{13}c_{13}$, where

$$c_{11} = \begin{vmatrix} a_{22} & a_{23} \\ a_{32} & a_{33} \end{vmatrix}, \quad c_{12} = -\begin{vmatrix} a_{21} & a_{23} \\ a_{31} & a_{33} \end{vmatrix}, c_{11} = \begin{vmatrix} a_{21} & a_{22} \\ a_{31} & a_{32} \end{vmatrix}.$$

The cofactor of a_{ij} is generally the determinant

$$c_{ij} = (-1)^{i+j} M_{ij} \tag{1.8}$$

where M_{ij} is the determinant of the sub-matrix obtained by deleting the ith row and jth column of A.

Example 32. Find $detA$ and $detB$ if

$$A = \begin{pmatrix} -5 & 0 & 0 \\ 0 & 7 & 0 \\ 0 & 0 & 3 \end{pmatrix} \quad \text{and} \quad B = \begin{pmatrix} 0 & 7 & 0 \\ 4 & 0 & 0 \\ 0 & 0 & -2 \end{pmatrix}.$$

Solution:

$$\begin{aligned} detA &= \begin{vmatrix} -5 & 0 & 0 \\ 0 & 7 & 0 \\ 0 & 0 & 3 \end{vmatrix} \\ &= -5(21) - 0(-15 - 0) \\ &= -105 \\ detB &= 0(0 - 0) + 7(-8) \\ &= -56. \end{aligned}$$

1.4.3 Properties of Determinants

We state properties of determinants in the forms of the following theorems

Theorem 5. *$detA = detA^T$, where A^T denotes adjoint of $n \times n$ matrix $A = (a_{n \times n})$.*

Theorem 6. *If two rows or columns of an $n \times n$ matrix $A = (a_{n \times n})$ are the same, then $detA = 0$.*

Theorem 7. *$detA = 0$ if all entries of the matrix $A = (a_{n \times n})$ are zero.*

Theorem 8. *Let B be a matrix obtained by interchanging any two columns (rows) of an $n \times n$ matrix $A = (a_{n \times n})$, then $detB = -detA$.*

Theorem 9. *$detB = kdetA$, where B is a matrix obtained by multiplying by k (non-zero real number) all entries of a column or row.*

Theorem 10. *Let A and B be $n \times n$ matrices, then*

$$det(AB) = det(A)det(B).$$

Theorem 11. *Let A be an $n \times n$ triangular matrix (upper or lower). Then*

$$detA = a_{11}a_{22}\ldots a_{nn}$$

where a_{11}, $a_{22}\ldots$ and a_{nn} are the entries on the main diagonal of A.

Proofs of Theorems, 6, 7, 9 and 11 can be found in Appendix C.

Remark 11. Physical meaning of determinant in R^2 and R^3
In R^2 determinant $\begin{vmatrix} a_1 & a_2 \\ b_1 & b_2 \end{vmatrix}$ is an indicator of whether the vectors (a_1, a_2) and (b_1, b_2) are co-linear. If the determinant is zero, then vectors (a_1, a_2) and (b_1, b_2) are co-linear.

$$(a_1, a_2, a_3), (b_1, b_2, b_3), (c_1, c_2, c_3) \in R^3$$

and $(a \times b).c =$ Area of the parallelepiped in R^3.
The determinant of zero means the volume is zero and the three vectors are coplanar.

Example 33. Let $A = \begin{pmatrix} 1 & 1 & 1 \\ 2 & -1 & 1 \\ 3 & 1 & -1 \end{pmatrix}$. Then $detA = detA^T$.

Solution: $A^T = \begin{pmatrix} 1 & 2 & 3 \\ 1 & -1 & 1 \\ 1 & 1 & -1 \end{pmatrix}$

$$\begin{aligned} detA &= \begin{pmatrix} 1 & 1 & 1 \\ 2 & -1 & 1 \\ 3 & 1 & -1 \end{pmatrix} \\ &= 1(1-1) - 1(-2-3) + 1(2+3) \\ &= 0+5+5 \\ &= 10 \\ detA^T &= 0 - 2(-1-1) + 1(2+4) \\ &= 4+6 \\ &= 10. \end{aligned}$$

Hence $detA = detA^T$.

Example 34. Let $A = \begin{pmatrix} 2 & -1 & 1 \\ 3 & 1 & -1 \\ 0 & 2 & 2 \end{pmatrix}$ and $B = \begin{pmatrix} 2 & 1 & 5 \\ 4 & 3 & 8 \\ 0 & -1 & 0 \end{pmatrix}$.
Show that $detAB = detA\ detB$.

Solution: $AB = \begin{pmatrix} 0 & -2 & 2 \\ 10 & 7 & 23 \\ 8 & 4 & 16 \end{pmatrix}$

$$\begin{aligned} detAB &= \begin{pmatrix} 0 & -2 & 2 \\ 10 & 7 & 23 \\ 8 & 4 & 16 \end{pmatrix} \\ &= 0 + 2(160 - 184) + 2(40 - 56) \\ &= -80 \\ detA\ detB &= -80. \end{aligned}$$

Example 35. Let $A = \begin{pmatrix} 5 & 1 & 1 \\ 5 & 1 & 1 \\ 3 & 2 & 7 \end{pmatrix}$, then $detA = 0$.

Solution:
$$\begin{aligned} detA &= 5(7-2) - 1(35-3) + 1(10-3) \\ &= 25 - 32 + 7 \\ &= 0. \end{aligned}$$

Example 36. Let $A = \begin{pmatrix} 1 & 2 & 5 \\ 0 & 0 & 0 \\ 3 & 4 & 6 \end{pmatrix}$ then $detA = 0$.

Solution:
$$\begin{aligned} detA &= 1(0-0) - 2(0-0) + 5(0-0) \\ &= 0. \end{aligned}$$

Example 37. Let $A = \begin{pmatrix} 2 & 3 & 1 \\ 0 & 1 & 0 \\ 1 & 2 & 3 \end{pmatrix}$ and $B = \begin{pmatrix} 0 & 1 & 0 \\ 2 & 3 & 1 \\ 1 & 2 & 3 \end{pmatrix}$ then $-detA = detB$.

Solution:
$$\begin{aligned} detA &= 2(3-0) - 3(0-0) + 1(0-1) \\ &= 6 - 1 \\ &= 5 \\ detB &= 0(9-2) - 1(6-1) + 0(4-3) \\ &= -5. \end{aligned}$$

Thus $-detA = detB$.

Example 38. Let $A = \begin{pmatrix} 4 & 0 & 0 \\ 3 & 2 & 0 \\ 5 & 1 & -1 \end{pmatrix}$ then $detA = -8$.

Solution:
$$\begin{aligned} detA &= 4(-2-0) - 0(-3-0) + 0(3-10) \\ &= -8. \end{aligned}$$

Example 39. Show that $detA = 0$, where $A = \begin{pmatrix} 3 & 2 & 1 \\ 2 & 6 & 3 \\ 5 & -8 & -4 \end{pmatrix}$, using properties of determinants.

Solution: $detA = kdetB$, where

$$B = \begin{pmatrix} 3 & 1 & 1 \\ 2 & 3 & 3 \\ 5 & -4 & -4 \end{pmatrix}$$

$$A = \begin{pmatrix} 3 & 2.1 & 1 \\ 2 & 2.3 & 3 \\ 5 & 2.(-4) & 4 \end{pmatrix}.$$

This holds by Theorem 9 and $detB = 0$ by Theorem 6.

Example 40. Let $A = \begin{pmatrix} 3 & -4 \\ 1 & 2 \end{pmatrix}$ and $B = \begin{pmatrix} 7 & 4 \\ -1 & -5 \end{pmatrix}$. Show that $det(A + B) \neq detA + detB$.

Solution:

$$\begin{aligned} \det(A+B) &= \begin{vmatrix} 10 & 0 \\ 0 & -3 \end{vmatrix} \\ &= -30 \\ detA &= 6+4 \\ &= 10 \\ detB &= -35+4 \\ &= -31 \\ detA + detB &= -21. \end{aligned}$$

Hence $det(A + B) \neq detA + detB$.

Example 41. Let A be an $n \times n$ matrix such that $A^2 = I$, where $A^2 = A.A$ and I is the identity matrix. Show that $detA = \pm 1$.

Solution: By Theorem 10:

$$det(A.A) = detA^2 = detA.detA.$$

Since $A^2 = I$, $(detA)^2 = detI = 1$ or $detA = \pm 1$.

1.4.4 Cramer's Rule: Application of Determinants to Solve Matrix Equations

Swiss mathematician Gabriel Cramer (1704-1752) discovered a method to solve a linear system of n equations in n unknowns using determinants. This

method is known as Cramer's rule.
For a system of n linear equations in n unknowns,

$$Ax = y$$

where

$$\begin{aligned} A &= \begin{pmatrix} a_{11} & a_{12} & \cdots & a_{1n} \\ a_{21} & a_{22} & \cdots & a_{2n} \\ \vdots & \vdots & \cdots & \vdots \\ \vdots & \vdots & \cdots & \vdots \\ a_{n1} & a_{n2} & \cdots & a_{nn} \end{pmatrix} \\ x^T &= (x_1, x_2, \ldots, x_n), \\ y &= \begin{pmatrix} b_1 \\ b_2 \\ \vdots \\ b_n \end{pmatrix} \end{aligned}$$

or

$$\begin{aligned} a_{11}x_1 + a_{12}x_2 + \cdots + a_{1n}x_n &= b_1 \\ a_{21}x_1 + a_{22}x_2 + \cdots + a_{2n}x_n &= b_2 \\ \cdots\cdots\cdots\cdots\cdots\cdots\cdots\cdots \\ a_{n1}x_1 + a_{n2}x_2 + \cdots + a_{nn}x_n &= b_n. \end{aligned}$$

If $detA \neq 0$, then the solution of the matrix equation is given by

$$x_1 = \frac{detA_1}{detA}, \; x_2 = \frac{detA_2}{detA}, \ldots, \; x_n = \frac{detA_n}{detA}$$

where $A_k, k = 1, 2, 3, \ldots, n$ is defined as

$$A_k = \begin{pmatrix} a_{11} & a_{12} & \cdots & b_1 & a_{1(k+1)} & a_{1n} \\ a_{21} & a_{22} & \cdots & b_2 & a_{2(k+1)} & a_{2n} \\ \cdots & \cdots & \cdots & \cdots & \cdots & \cdots \\ a_{n1} & a_{n2} & \cdots & b_n & a_{n(k+1)} & a_{nn} \end{pmatrix}.$$

In fact, A_k is A except where the kth column of A is replaced by the entries of column matrix

$$y = \begin{pmatrix} b_1 \\ b_2 \\ \vdots \\ b_n \end{pmatrix}.$$

Example 42. Solve the following system of equations applying Cramer's rule

(i)
$$\begin{aligned} a_{11}x_1 + a_{12}x_2 &= b_1 \\ a_{21}x_1 + a_{22}x_2 &= b_2. \end{aligned}$$

(ii)
$$\begin{aligned} x_1 + x_2 &= 4 \\ 2x_1 - x_2 &= 2. \end{aligned}$$

(iii)
$$\begin{aligned} x_1 - x_2 + 6x_3 &= -2 \\ -x_1 + 2x_2 + 4x_3 &= 9 \\ 2x_1 + 3x_2 - x_3 &= \frac{1}{2}. \end{aligned}$$

Solution: (i)

$$\begin{aligned} a_{11}x_1 + a_{12}x_2 &= b_1 \\ a_{21}x_1 + a_{22}x_2 &= b_2 \end{aligned}$$
$$\Leftrightarrow$$
$$Ax = y$$

where $A = \begin{pmatrix} a_{11} & a_{12} \\ a_{21} & a_{22} \end{pmatrix}$, $x = \begin{pmatrix} x_1 \\ x_2 \end{pmatrix}$, $y = \begin{pmatrix} b_1 \\ b_2 \end{pmatrix}$

$$detA = \begin{vmatrix} a_{11} & a_{12} \\ a_{21} & a_{22} \end{vmatrix} = a_{11}a_{22} - a_{12}a_{21}.$$

$$A_1 = \begin{pmatrix} b_1 & a_{12} \\ b_2 & a_{22} \end{pmatrix}, \quad detA_1 = \begin{vmatrix} b_1 & a_{12} \\ b_2 & a_{22} \end{vmatrix} = b_1a_{22} - a_{12}b_2$$
$$A_2 = \begin{pmatrix} a_{11} & b_1 \\ a_{21} & b_2 \end{pmatrix}, \quad detA_2 = \begin{vmatrix} a_{11} & b_1 \\ a_{21} & b_2 \end{vmatrix} = a_{11}b_2 - b_1a_{21}$$

$$\begin{aligned} x_1 &= \frac{detA_1}{detA} = \frac{b_1a_{22} - a_{12}b_2}{a_{11}a_{22} - a_{12}a_{21}} \\ x_2 &= \frac{a_{11}b_2 - b_1a_{21}}{a_{11}a_{22} - a_{12}a_{21}}. \end{aligned}$$

(ii)
$$a_{11} = 1, a_{12} = 1, b_1 = 4, a_{21} = 2, a_{22} = -1, b_2 = 2$$

$$x_1 = \frac{4(-1) - 1 \times 2}{-1 - 1 \times 2} = \frac{-6}{-3} = 2$$

and

$$x_2 = \frac{-6}{-3} = 2.$$

(iii) $$A = \begin{pmatrix} 1 & -1 & 6 \\ -1 & 2 & 4 \\ 2 & 3 & -1 \end{pmatrix} \text{ and } b = \begin{pmatrix} -2 \\ 9 \\ \frac{3}{2} \end{pmatrix}$$

$$detA = -63,\ detA_1 = \begin{pmatrix} -2 & -1 & 6 \\ 9 & 2 & 4 \\ \frac{3}{2} & 3 & -1 \end{pmatrix} = 173, detA_2 = -136,$$

$$detA_3 = \frac{-61}{2}, x_1 = \frac{detA_1}{detA} = \frac{-173}{63}, x_2 = \frac{136}{63}, x_3 = \frac{61}{126}.$$

Example 43. The magnitudes T_1 and T_2 of the tensions are depicted by the support wires shown in Figure 1.10, satisfying the equations:

$$\begin{aligned} \cos(25^o)T_1 - \cos(15^o)T_2 &= 0 \\ \sin(25^o)T_1 + \sin(15^o)T_2 &= 300. \end{aligned}$$

Use Cramer's rule to find T_1 and T_2.

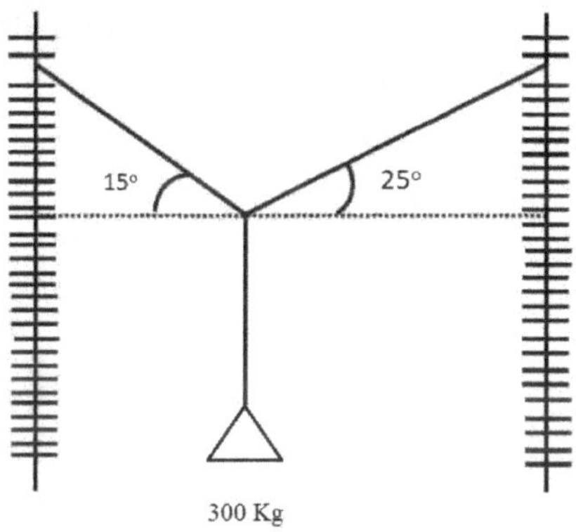

FIGURE 1.10: Tension Distribution

Solution: $A = \begin{pmatrix} \cos(25^o) & -\cos(15^o) \\ \sin(25^o) & \cos(15^o) \end{pmatrix},\quad y = b = \begin{pmatrix} 0 \\ 300 \end{pmatrix}.$

$$\begin{aligned} detA &= \begin{vmatrix} \cos(25^o) & -\cos(15^o) \\ \sin(25^o) & \sin(15^o) \end{vmatrix} \approx 0.6428 \\ detA_1 &= \begin{vmatrix} 0 & -\cos(15^o) \\ 35^o & \sin(15^o) \end{vmatrix} \approx 289.8 \\ detA_2 &= \begin{vmatrix} \cos(25^o) & 0 \\ \sin(25^o) & 300 \end{vmatrix} \approx 271.9 \end{aligned}$$

$$T_1 = \frac{detA_1}{detA} \approx \frac{289.8}{0.6428} \approx 450.8$$
$$T_2 = \frac{detA_2}{detA} \approx \frac{271.9}{o.6428} \approx 423.$$

1.5 Inverse of Matrix and Its Computation

In the real number system a real number b is called the **inverse** of a if $ab = 1$. Similarly a matrix B is called the inverse of a matrix A and is denoted by $B = A^{-1}$ if $AB = BA = I$, where I denotes the identity matrix. A is called a **non-singular** or **invertible matrix** if A^{-1} exists.

We know that every non-zero real number is invertible but in the case of matrices it is untrue. There are matrices which are not invertible. For invertible matrices the matrix equation $Ax = y$ can be easily solved. The main goal of this section is to give a characterization of an invertible matrix in term of its determinant and formula for the inverse in terms of its determinant and adjoint.

Properties of Inverse

Theorem 12. *Let A and B be invertible (non-singular) matrices*

(i) $(A^{-1})^{-1} = A.$

(ii) $(AB)^{-1} = B^{-1}A^{-1}.$

(iii) $(A^T)^{-1} = (A^{-1})^T.$

Characterization of Invertible Matrices:

Theorem 13. *An $n \times n$ matrix is invertible if and only if $detA \neq 0$.*

We require the following formula for computing inverses in the proof of Theorem 13.

Lemma 1. Let A be an $n \times n$ matrix, **adjoint** of A, denoted by **adj**A and defined as

$$\begin{pmatrix} c_{11} & c_{12} & \cdots & c_{1n} \\ c_{21} & c_{22} & \cdots & c_{2n} \\ \vdots & \vdots & \vdots & \vdots \\ c_{n1} & c_{n2} & \cdots & c_{nn} \end{pmatrix}^T = \begin{pmatrix} c_{11} & c_{21} & \cdots & c_{n1} \\ c_{12} & c_{22} & \cdots & c_{n2} \\ \vdots & \vdots & \vdots & \vdots \\ c_{1n} & c_{2n} & \cdots & c_{nn} \end{pmatrix}$$

and $detA \neq 0$. Then

$$A^{-1} = \left(\frac{1}{detA}\right) adj A.$$

Remark 12. For a 3×3 invertible matrix

$$A = \begin{pmatrix} a_{11} & a_{12} & a_{13} \\ a_{21} & a_{22} & a_{23} \\ a_{31} & a_{32} & a_{33} \end{pmatrix}$$

$$c_{11} = \begin{vmatrix} a_{22} & a_{23} \\ a_{32} & a_{33} \end{vmatrix}, c_{12} = -\begin{vmatrix} a_{21} & a_{23} \\ a_{31} & a_{33} \end{vmatrix}, c_{11} = \begin{vmatrix} a_{21} & a_{22} \\ a_{31} & a_{32} \end{vmatrix}$$

$$A^{-1} = \frac{1}{detA} \begin{pmatrix} c_{11} & c_{21} & c_{31} \\ c_{12} & c_{22} & c_{32} \\ c_{13} & c_{23} & c_{33} \end{pmatrix}.$$

Proof of theorem 13. Let $detA \neq 0$. Then A is invertible by Lemma 1 as $A^{-1} = \dfrac{1}{detA} AdjA$ can be computed. Let us assume that the converse holds, that is, A is invertible and A^{-1} exists. Then $AA^{-1} = A^{-1}A = I$. By Theorem 10:

$$det(I) = det(AA^{-1} = A^{-1}A = I) = detA.det(A^{-1}).$$

or

$$detA.detA^{-1} = 1 \neq 0.$$

This means that $detA \neq 0$.

Example 44. Show that the matrix $B = \begin{pmatrix} 2 & -1 & 2 \\ 1 & -1 & 2 \\ -3 & 2 & -3 \end{pmatrix}$ is the inverse of the matrix $A = \begin{pmatrix} 1 & -1 & 0 \\ 3 & 0 & 2 \\ 1 & 1 & 1 \end{pmatrix}$.

Solution: We are required to show that $AB = I$. We have

$$\begin{aligned} AB &= \begin{pmatrix} 1 & -1 & 0 \\ 3 & 0 & 2 \\ 1 & 1 & 1 \end{pmatrix} \begin{pmatrix} 2 & -1 & 2 \\ 1 & -1 & 2 \\ -3 & 2 & -3 \end{pmatrix} \\ &= \begin{pmatrix} 2-1 & -1+1 & -2+2 \\ 6-6 & -3+4 & 6-6 \\ 2+1-3 & -1-1+2 & 2+2-3 \end{pmatrix} \\ &= \begin{pmatrix} 1 & 0 & 0 \\ 0 & 1 & 0 \\ 0 & 0 & 1 \end{pmatrix} \\ AB &= I. \end{aligned}$$

So B is the inverse of A.

Example 45. Examine whether the following matrices are invertible (non-singular) or non-invertible (singular). If invertible, compute the inverses.

(i) $\begin{pmatrix} 6 & 0 \\ -3 & 2 \end{pmatrix}$.

(ii) $\begin{pmatrix} -2\pi & -\pi \\ -\pi & \pi \end{pmatrix}$.

(iii) $\begin{pmatrix} 0 & 2 & 0 \\ 0 & 0 & 1 \\ 8 & 0 & 0 \end{pmatrix}$.

(iv) $\begin{pmatrix} 5 & 3 \\ 5 & 3 \end{pmatrix}$.

(v) $\begin{pmatrix} 1 & 2 & 3 \\ 0 & 1 & 4 \\ 0 & 0 & 8 \end{pmatrix}$.

Solution: (i) Since $det\begin{pmatrix} 6 & 0 \\ -3 & 2 \end{pmatrix} = 12 \neq 0.$, this matrix is invertible by Theorem 13. By Lemma 1

$$\begin{aligned} A^{-1} &= \frac{1}{12}\begin{pmatrix} 6 & -3 \\ 0 & 2 \end{pmatrix} \\ &= \begin{pmatrix} \frac{1}{2} & \frac{-1}{4} \\ 0 & \frac{1}{6} \end{pmatrix}. \end{aligned}$$

(ii) $\begin{vmatrix} -2\pi & -\pi \\ -\pi & \pi \end{vmatrix} = -3\pi^2 \neq 0$. By Lemma 1 it is invertible and its inverse is $\begin{pmatrix} \frac{-1}{3\pi} & \frac{-1}{3\pi} \\ \frac{-1}{3\pi} & \frac{2}{3\pi} \end{pmatrix}$.

(iii) Let $A = \begin{vmatrix} 0 & 2 & 0 \\ 0 & 0 & 1 \\ 8 & 0 & 0 \end{vmatrix}$.

$$a_{11} = 0, a_{12} = 2, a_{13} = 0, a_{21} = 0, a_{22} = 0, a_{23} = 1, a_{31} = 8, a_{32} = 0, a_{33} = 0.$$

$$c_{11} = \begin{vmatrix} 0 & 1 \\ 0 & 0 \end{vmatrix} = 0, c_{12} = \begin{vmatrix} 0 & 1 \\ 8 & 0 \end{vmatrix} = -8, c_{13} = \begin{vmatrix} 0 & 0 \\ 8 & 0 \end{vmatrix}.$$

$$detA = 0(0-0) - 2(-8) + 0 = 16.$$

$$A^{-1} = \frac{1}{16}\begin{pmatrix} 0 & 0 & 2 \\ 8 & 0 & 0 \\ 0 & 16 & 0 \end{pmatrix} = \begin{pmatrix} 0 & 0 & \frac{1}{8} \\ \frac{1}{2} & 0 & 0 \\ 0 & 1 & 0 \end{pmatrix}.$$

(iv) Since $detA$ where $A = \begin{pmatrix} 5 & 3 \\ 5 & 3 \end{pmatrix}$, is zero, A is singular by Theorem 13.

(v) It is invertible as $det\begin{pmatrix} 1 & 2 & 3 \\ 0 & 1 & 4 \\ 0 & 0 & 8 \end{pmatrix} \neq 0$ and $A^{-1} = \begin{pmatrix} 1 & -2 & \frac{5}{8} \\ 0 & 1 & -\frac{1}{2} \\ 0 & 0 & \frac{1}{8} \end{pmatrix}$.

1.6 Eigenvalue Problems for Square Matrices

In this section we discuss eigenvalues and eigenvectors of square matrices. In fact similar studies could be carried out for linear transformations in general and matrix transformations in particular. Let A be an $m \times n$ matrix and u be a non-zero $n-$ vector; then the matrix product Au is an $n-$ vector. A mapping T on R^n into R^n is denoted by $T : R^n \to R^n$. A matrix transformation is a function $T : R^n \to R^n$ defined by $Tu = Au$. The vector Tu in R^n is called the image of u and the set of all images of vectors in R^n is called the **range** of T. The subset of R^n on which T is defined is called the **domain** of T. We limit discussion to matrices and vectors with real entries but similar discussion is possible for complex entries. The main goal is to find those vectors u for which Au and u have the same direction, that is, those vectors for which $Au = \lambda u$ where A is an $n \times n$ matrix, u is called an **eigenvector** and λ is called an **eigenvalue**. Eigenvalues and eigenvectors are also known as characteristic values or proper values and characteristic vectors or proper vectors.

The concepts of eigenvalues and eigenvectors have been utilized in different branches of engineering such as control theory, vibration analysis (modeling a mechanical system or the energy states of an atom), electric circuits, advanced dynamics and quantum mechanics.

1.6.1 Eigenvalues and Eigenvectors

Definition 22. Let A be an $n \times n$ matrix. A number λ is called an eigenvalue of A if there is a non-zero vector u such that

$$Au = \lambda u. \tag{1.9}$$

The solution vector u of the linear system (1.9) is an **eigenvector** corresponding to the eigenvalue λ.

Remark 13. (i) Non-zero scalar multiples of eigenvectors are also eigenvectors.

(ii) The geometrical meaning of eigenvectors is that the image of eigenvector u under A is parallel to u.

(iii) According to definition, a zero vector cannot be an eigenvector, but an eigenvalue can be 0.

(iv) Let 0 be an eigenvalue of $n \times n$ matrix A. Then $Ax = 0x = 0$ which implies that A is not invertible. A is invertible if and only if 0 is not an eigenvalue of A.

Procedure for finding eigenvalues and eigenvector:

$$\begin{aligned} Au = \lambda u \text{ or } Au - \lambda u &= 0. \\ (Au - \lambda I u) &= 0. \\ (A - \lambda I)u &= 0. \end{aligned}$$

λ is an eigenvalue of A if and only if

$$det(A - \lambda I) = 0. \tag{1.10}$$

The equation (1.10) is called the **characteristic equation** of A. The eigenvalues of A are roots of the characteristic equation. To find an eigenvector corresponding to an eigenvalue λ solve

$$(A - \lambda I) = 0. \tag{1.11}$$

Example 46. Find eigenvalues and associated eigenvectors of the following matrices:

(i) $A = \begin{pmatrix} 1 & 2 \\ 2 & 4 \end{pmatrix}$ (ii) $B = \begin{pmatrix} 3 & 4 \\ -1 & 7 \end{pmatrix}$ (iii) $C = \begin{pmatrix} 1 & -2 \\ 2 & 0 \end{pmatrix}$

(iv) $D = \begin{pmatrix} -3 & 1 & 1 \\ 0 & 0 & 0 \\ 0 & 0 & 0 \end{pmatrix}$ (v) $E = \begin{pmatrix} 0 & 0 & -1 \\ 1 & 0 & 0 \\ 1 & 1 & -1 \end{pmatrix}$.

Solution: (i) $det(A - \lambda I) = \begin{vmatrix} 1-\lambda & 3 \\ 2 & 4-\lambda \end{vmatrix} = (1-\lambda)(4-\lambda) - 4 = \lambda^2 - 5\lambda = 0.$
This yields eigenvalues $\lambda_1 = 0$ and $\lambda_2 = 5$. The eigenvector corresponding to $\lambda_1 = 0$ yields

$$(A - 0I)u = \begin{pmatrix} 1 & 2 \\ 2 & 4 \end{pmatrix} \begin{pmatrix} u_1 \\ u_2 \end{pmatrix} = \begin{pmatrix} 0 \\ 0 \end{pmatrix}.$$

This implies that $\begin{pmatrix} u_1 \\ u_2 \end{pmatrix} = \begin{pmatrix} 2 \\ -1 \end{pmatrix}$ is an eigenvector corresponding to $\lambda_1 = 0$.

(ii) The characteristic equation of $A = \begin{pmatrix} 3 & 4 \\ -1 & 7 \end{pmatrix}$ is

$$\begin{aligned} det(A - \lambda I) &= \begin{vmatrix} 3-\lambda & 4 \\ -1 & 7-\lambda \end{vmatrix} \\ &= (3-\lambda)(7-\lambda)+4 \\ &= \lambda^2 - 10\lambda + 25 \\ &= (\lambda-5)^2 \\ &= 0. \end{aligned}$$

We see that $\lambda_1 = 5 = \lambda_2$ is an eigenvalue of multiplicity 2.

$$\begin{aligned} (A-5I)u &= \begin{pmatrix} 3-5 & 4 \\ -1 & 7-5 \end{pmatrix}\begin{pmatrix} u_1 \\ u_2 \end{pmatrix} = 0 \\ \text{or} \quad -2u_1 + 4u_2 &= 0 \\ -u_1 + 2u_2 &= 0. \end{aligned}$$

It is clear that $u_1 = 2u_2$. Thus if we choose $u_2 = 1$ we find the single eigenvector

$$u_1 = \begin{pmatrix} 2 \\ 1 \end{pmatrix}.$$

(iii) The characteristic equation is

$$det(A - \lambda I) = \begin{pmatrix} \lambda-1 & 2 \\ -2 & \lambda \end{pmatrix} = \lambda^2 - \lambda + 4 = 0.$$

Eigenvalues are roots of the equation:

$$\lambda_1 = \frac{1+\sqrt{15}i}{2} \quad \text{and} \quad \lambda_2 = \frac{1-\sqrt{15}i}{2}.$$

Eigenvector corresponding to λ_1 is obtained by solving $(A - \lambda I)U = 0$:

$$\begin{aligned} \left[\begin{pmatrix} 1 & -2 \\ 2 & 0 \end{pmatrix} - \begin{pmatrix} \lambda_1 & 1 & 0 \\ & 0 & 1 \end{pmatrix}\right] u &= 0 \\ \text{or} \quad \frac{-1+\sqrt{15}i}{2}u_1 + 2u_2 &= 0 \\ -2u_1 + \frac{1+\sqrt{15}i}{2}u_2 &= 0. \end{aligned}$$

We obtain the general solution of this system to be $\alpha \begin{pmatrix} 1 \\ \frac{1-\sqrt{15}i}{4} \end{pmatrix}$, and this is an eigenvector associated with eigenvalue $\frac{1+\sqrt{15}i}{2}$ for any non-zero scalar α. Similarly we find that $\beta\begin{pmatrix} 1 \\ \frac{1+\sqrt{15}i}{4} \end{pmatrix}$ is an eigenvector corresponding to eigenvalue $\frac{1-\sqrt{15}i}{2}$ for any $\beta \neq 0$.

(iv) $D = \begin{pmatrix} -3 & 1 & 1 \\ 0 & 0 & 0 \\ 0 & 1 & 0 \end{pmatrix}$

$$\begin{aligned} det(D-\lambda I) &= \left| \begin{pmatrix} -3 & 1 & 1 \\ 0 & 0 & 0 \\ 0 & 1 & 0 \end{pmatrix} - \lambda \begin{pmatrix} 1 & 0 & 0 \\ 0 & 1 & 0 \\ 0 & 0 & 1 \end{pmatrix} \right| \\ &= \begin{vmatrix} -3-\lambda & 1 & 1 \\ 0 & -\lambda & 0 \\ 0 & 1-\lambda & -\lambda \end{vmatrix} \\ &= (-3-\lambda)\lambda^2 - 1(0) + 1(0) \\ &= -(3+\lambda)\lambda^2 \\ &= 0. \end{aligned}$$

$u_1 = \begin{pmatrix} 1 \\ 0 \\ 3 \end{pmatrix}$ is only one eigenvector for eigenvalue $\lambda_2 = \lambda_1 = 0$ and $u_2 = \begin{pmatrix} 1 \\ 0 \\ 0 \end{pmatrix}$ is only one eigenvector for eigenvalue $\lambda_3 = -3$.

(v) $E = \begin{pmatrix} 0 & 0 & -1 \\ 1 & 0 & 0 \\ 1 & 1 & -1 \end{pmatrix}$

$$\begin{aligned} det(E-\lambda I) &= \begin{vmatrix} -\lambda & 0 & -1 \\ 0 & -\lambda & 0 \\ 1 & 1 & -1-\lambda \end{vmatrix} \\ &= -(\lambda+1)(\lambda^2+1) \\ &= 0. \end{aligned}$$

For $\lambda_1 = -1$ we have

$$\left(\begin{array}{ccc|c} 1 & 0 & -1 & 0 \\ 1 & 1 & 0 & 0 \\ 1 & 1 & 0 & 0 \end{array}\right) \Rightarrow \left(\begin{array}{ccc|c} 1 & 0 & -1 & 0 \\ 0 & 1 & 1 & 0 \\ 0 & 0 & 0 & 0 \end{array}\right)$$

so that $u_1 = u_2$ and $u_2 = -u_3$. If $u_3 = 1$ then $u = \begin{pmatrix} 1 \\ -1 \\ 1 \end{pmatrix}$.

For $\lambda_2 = i$ we have

$$\left(\begin{array}{ccc|c} -1 & 0 & -1 & 0 \\ 1 & -i & 0 & 0 \\ 1 & 1 & -1-i & 0 \end{array}\right) \Rightarrow \left(\begin{array}{ccc|c} 1 & 0 & -i & 0 \\ 0 & 1 & -1 & 0 \\ 0 & 0 & 0 & 0 \end{array}\right)$$

so that $u_1 = iu_3$ and $u_2 = u_3$.

If $u_3 = 1$ then $v = \begin{pmatrix} 1 \\ 1 \\ 1 \end{pmatrix}$ and $w = \overline{v} = \begin{pmatrix} -1 \\ 1 \\ 1 \end{pmatrix}$.

Definition 23. Let A be $n \times n$ matrix and A^T denote its adjoint matrix. A is called **symmetric** if $A = A^T$. A is called **orthogonal** matrix if $A^T A = I$.

Remark 14. (i) It is clear that an $n \times n$ invertible matrix A is orthogonal if $A^{-1} = A^T$.

(ii) Let $A = \begin{pmatrix} a_{11} & a_{12} & \cdots & a_{1n} \\ a_{21} & a_{22} & \vdots & a_{21} \\ \vdots & \vdots & \vdots & \vdots \\ a_{n1} & a_{n2} & \cdots & a_{nn} \end{pmatrix}$.

A is orthogonal matrix if and only if its columns

$$\begin{pmatrix} a_{11} \\ a_{21} \\ \vdots \\ a_{n1} \end{pmatrix} = X_1, \begin{pmatrix} a_{12} \\ a_{22} \\ \vdots \\ a_{n2} \end{pmatrix} = X_2, \ldots, \begin{pmatrix} a_{1n} \\ a_{2n} \\ \vdots \\ a_{nn} \end{pmatrix} = X_n$$

form an orthogonal set.

(iii) Let A be a symmetric matrix with real entries. Then eigenvalues of A are real.

(iv) The eigenvectors corresponding to distinct eigenvalues of an $n \times n$ symmetric matrix are orthogonal.

Example 47. (i) Show that the identity matrix is orthogonal.

(ii) Let $A = \begin{pmatrix} 0 & -1 & 0 \\ -1 & -1 & 1 \\ 0 & 1 & 0 \end{pmatrix}$. Show that it is symmetric and its eigenvectors associated with three distinct eigenvalues are orthogonal and construct an orthogonal matrix related to A.

Solution: (i) $I = \begin{pmatrix} 1 & 0 & 0 \\ 0 & 1 & 0 \\ 0 & 0 & 1 \end{pmatrix}$

$$I^T = \begin{pmatrix} 1 & 0 & 0 \\ 0 & 1 & 0 \\ 0 & 0 & 1 \end{pmatrix}$$

and $I^T I = I$. Hence I is orthogonal.
(ii) It can be checked that $\lambda_1 = 0, \lambda_2 = 1$, and $\lambda_3 = -2$ and corresponding eigenvectors are

$$u_1 = \begin{pmatrix} 1 \\ 0 \\ 1 \end{pmatrix}, u_2 = \begin{pmatrix} -1 \\ 1 \\ 1 \end{pmatrix}, u_3 = \begin{pmatrix} 1 \\ 2 \\ -1 \end{pmatrix}.$$

Since all the eigenvalues are distinct, we have

$$u_1^T u_2 = \begin{pmatrix} 1 & 0 & 1 \end{pmatrix} \begin{pmatrix} -1 \\ 1 \\ 1 \end{pmatrix} = 1.(-1) + 0.1 + 1.1 = 0$$

$$u_1^T u_3 = \begin{pmatrix} 1 & 0 & 1 \end{pmatrix} \begin{pmatrix} 1 \\ 2 \\ -1 \end{pmatrix} = 1.1 + 0.2 + 1.(-1) = 0$$

$$u_2^T u_3 = \begin{pmatrix} -1 & 1 & 1 \end{pmatrix} \begin{pmatrix} 1 \\ 2 \\ -1 \end{pmatrix} = (-1).1 + 1.2 + 1.(-1) = 0$$

$$u_1 = \begin{pmatrix} 1 \\ 0 \\ 1 \end{pmatrix}, u_2 = \begin{pmatrix} -1 \\ 1 \\ 1 \end{pmatrix}, u_3 = \begin{pmatrix} 1 \\ 2 \\ -1 \end{pmatrix}.$$

They are eigenvectors of the given matrix which are orthogonal. Now norms of u_1, u_2 and u_3 are respectively

$$||u_1|| = \sqrt{u_1^T u_1} = \sqrt{2}, ||u_2|| = \sqrt{u_2^T u_2} = \sqrt{3}, ||u_3|| = \sqrt{u_3^T u_3} = \sqrt{6}.$$

Thus, an orthogonal set of vectors is

$$\begin{pmatrix} \frac{1}{\sqrt{2}} \\ 0 \\ 1 \end{pmatrix}, \begin{pmatrix} -\frac{1}{\sqrt{3}} \\ \frac{1}{\sqrt{3}} \\ \frac{1}{\sqrt{3}} \end{pmatrix}, \begin{pmatrix} \frac{1}{\sqrt{6}} \\ \frac{2}{\sqrt{6}} \\ -\frac{1}{\sqrt{6}} \end{pmatrix}.$$

Hence $\begin{pmatrix} \frac{1}{\sqrt{2}} & -\frac{1}{\sqrt{3}} & \frac{1}{\sqrt{6}} \\ 0 & \frac{1}{\sqrt{3}} & \frac{2}{\sqrt{6}} \\ 1 & \frac{1}{\sqrt{3}} & -\frac{1}{\sqrt{6}} \end{pmatrix}$ is an orthogonal matrix. We can check that $P^T = P^{-1}$.

Example 48. Determine whether the given matrix is orthogonal:

(i) $\begin{pmatrix} 0 & 1 & 0 \\ 1 & 0 & 0 \\ 0 & 0 & 1 \end{pmatrix}$ (ii) $\begin{pmatrix} 0 & 0 & 0 \\ 0 & 0 & 0 \\ 0 & 0 & 0 \end{pmatrix}$.

Example 49. Construct an orthogonal matrix from eigenvectors of the given symmetric matrix: (i) $\begin{pmatrix} 7 & 0 \\ 0 & 4 \end{pmatrix}$ (ii) $\begin{pmatrix} 0 & 1 & 1 \\ 1 & 1 & 1 \\ 1 & 1 & 0 \end{pmatrix}$.

Properties of orthogonal and symmetric matrices:

(i) An $n \times n$ matrix A is orthogonal if and only if A^T is orthogonal.

(ii) The determinant of an $n \times n$ orthogonal matrix A is $+1$ or -1, that is, $detA = \pm 1$.

(iii) Let A be real symmetric matrix. Then eigenvectors associated with distinct eigenvalues are orthogonal.

(iv) Let A be real symmetric matrix. Then there is a real orthogonal matrix that diagonalizes A.

(v) An $n \times n$ matrix A is orthogonal if and only if the row vectors and column vectors form an orthonormal set of vectors in R^n.

1.6.2 Diagonalization and Similar Matrices

Definition 24. A square matrix have all off diagonal elements equal to zero is called a **diagonal matrix**.

$$\begin{pmatrix} 1 & 0 \\ 0 & 1 \end{pmatrix}, \begin{pmatrix} 1 & 0 & 0 \\ 0 & 1 & 0 \\ 0 & 0 & 1 \end{pmatrix}, \begin{pmatrix} 1 & 0 & 0 & 0 \\ 0 & 1 & 0 & 0 \\ 0 & 0 & 1 & 0 \\ 0 & 0 & 0 & 1 \end{pmatrix}$$

are examples of diagonal matrices.

Properties of diagonal matrix:

Let $A = \begin{pmatrix} \alpha_1 & 0 & 0 & \cdots & 0 \\ 0 & \alpha_2 & 0 & \cdots & 0 \\ \cdots & \cdots & \cdots & \cdots & \cdots \\ 0 & 0 & 0 & 0 & \alpha_n \end{pmatrix}$ and $B = \begin{pmatrix} \beta_1 & 0 & 0 & \cdots & 0 \\ 0 & \beta_2 & 0 & \cdots & 0 \\ \cdots & \cdots & \cdots & \cdots & \cdots \\ 0 & 0 & 0 & 0 & \beta_n \end{pmatrix}$.

Then

(i) $AB = BA = \begin{pmatrix} \alpha_1\beta_1 & 0 & 0 & \cdots & 0 \\ 0 & \alpha_2\beta_2 & 0 & \cdots & 0 \\ \cdots & \cdots & \cdots & \cdots & \cdots \\ 0 & 0 & 0 & 0 & \alpha_n\beta_n \end{pmatrix}$.

(ii) $|A| = \alpha_1\alpha_2 \ldots \alpha_n$.

(iii) A is invertible if and only if each main diagonal element is non-zero.

(iv) If each $\alpha_j \neq 0$, then

$$A^{-1} = \begin{pmatrix} \frac{1}{\alpha_1} & 0 & 0 & \cdots & 0 \\ 0 & \frac{1}{\alpha_2} & 0 & \cdots & 0 \\ \cdots & \cdots & \cdots & \cdots & \cdots \\ 0 & 0 & 0 & 0 & \frac{1}{\alpha_n} \end{pmatrix}.$$

(v) The eigenvalues of A are its main diagonal elements $\alpha_1, \alpha_2, \ldots, \alpha_n$.

(vi) An eigenvector associated with α_j is

$$\begin{pmatrix} 0 \\ 0 \\ 0 \\ 1 \\ 0 \\ \vdots \\ 0 \end{pmatrix}$$

with 1 in row j and all other elements are zero.

Definition 25. The $n \times n$ matrices A and B are said to be similar if there is an invertible $n \times n$ matrix P such that $A = PBP^{-1}$.

Properties:

(i) $n \times n$ similar matrices A and B have the same characteristics as polynomials and hence the same eigenvalues (with the same multiplicities).

(ii) Two row equivalent matrices A and B are not necessarily similar. Note for example matrices $A = \begin{pmatrix} 2 & 0 \\ 0 & 1 \end{pmatrix}$ and $B = \begin{pmatrix} 1 & 0 \\ 0 & 1 \end{pmatrix}$ (A and B are row equivalent but they are not similar: $A = PBP^{-1}$ leads to a contradiction. Since B is the identity matrix, $A = PIP^{-1} = PP^{-1} = I$ is a contradiction as A is not an identity matrix).

Definition 26. A square matrix A is said to be diagonalizable if it is similar to a diagonal matrix. In other words, a diagonal matrix A has the property of an invertible matrix P and a diagonal matrix D such that $A = PDP^{-1}$.

Remark 15. (i) If A is diagoalizable, then

$$A^3 = PD^3P^{-1}$$

and in general

$$A^k = PD^kP^{-1}, k = 1, 2, 3, 4, \cdots.$$

Verification:

$$\begin{aligned} A^3 &= (PDP^{-1})^3 = (PDP^{-1})(PDP^{-1})(PDP^{-1}) \\ &= PDP^{-1}PDP^{-1}PDP^{-1} = PD(PP^{-1})D(PD^{-1})DP^{-1} \\ &= PDDDP^{-1} = PD^3P^{-1}. \end{aligned}$$

(ii) If $D = \begin{pmatrix} 7 & 0 & 0 \\ 0 & -2 & 0 \\ 0 & 0 & 3 \end{pmatrix}$ then $D^3 = \begin{pmatrix} 7^3 & 0 & 0 \\ 0 & (-2)^3 & 0 \\ 0 & 0 & 3^3 \end{pmatrix}$.

(iii) An $n \times n$ matrix is diagonalizable if and only if it has n linearly independent eigenvectors.

Theorem 14. *(diagonalization theorem)*

(i) An $n \times n$ matrix A is diagonalizable if and only if A has n linearly independent eigenvectors.

(ii) If $v_1, v_2, \ldots, v_n$ are linearly independent eigenvectors of A and $\lambda_1, \lambda_2, \ldots, \lambda_n$ are their corresponding eigenvalues, then $A = PDP^{-1}$, where

$$P = (v_1, v_2, \ldots, v_n)$$

and

$$D = \begin{pmatrix} \lambda_1 & 0 & \cdots & 0 \\ 0 & \lambda_2 & \cdots & 0 \\ \vdots & \vdots & \vdots & \vdots \\ 0 & 0 & \cdots & \lambda_n \end{pmatrix}.$$

(iii) If $A = PDP^{-1}$ and D is a diagonal matrix, then the columns of P must be linearly independent eigenvectors of A and the diagonal entries of D must be their corresponding eigenvalues.

Remark 16. Theorem 14 tells us that if we can find n linearly independent eigenvectors for $n \times n$ matrix A, then it is diagonalizable. Furthermore, we can use those eigenvectors and their corresponding eigenvalues to find the invertible matrix P and diagonal matrix D required to show that A is diagonalizable.

1.6.3 Orthogonalization and Diagonalization of 2×2 Matrices

We discuss here the orthogonalization of 2×2 matrices $A = \begin{pmatrix} a & b \\ c & d \end{pmatrix}$ and their diagonalization. In order that A is orthogonal, the set of rows (a, b)

and (c, d) must be orthogonal and must have length 1, that is

$$ac + bd = 0. \tag{1.12}$$

$$a^2 + b^2 = 1. \tag{1.13}$$

$$c^2 + d^2 = 1. \tag{1.14}$$

Case 1: Multiply Equation (1.12) by d to get

$$acd + bd^2 = 0.$$

Substitute $ad = 1 + bc$ ($|A| = ad - bc = \pm 1$ or $ad = 1 + bc$) into this equation to get

$$c(1 + bc) + bd^2 = 0$$

or

$$c + b(c^2 + d^2) = 0.$$

But $c^2 + d^2 = 1$ so $c + b = 0$ or $c = -b$. Inserting this value in (1.12) yields $ab - bd = 0$ or $b(a - d) = 0$. This means either $b = 0$ or $a = d$.
Case 1(a): $b = 0$. Then $c = -b = 0$ so

$$A = \begin{pmatrix} a & 0 \\ 0 & d \end{pmatrix}.$$

But each row vector has length 1, so $a^2 = 1 = d^2$. Further, $detA = ad = 1$ in the present case, so $a = d = 1$ or $a = d = -1$. In these cases $A = I_2$, or $A = -I_2$.
Case 1(b): $b \neq 0$. Then $a = d$, so

$$A = \begin{pmatrix} a & b \\ -b & a \end{pmatrix}.$$

Since $a^2 + b^2 = 1$, there is some θ in $[0, 2\pi]$ such that $a = \cos\theta, b = \sin\theta$. Then

$$A = \begin{pmatrix} \cos\theta & \sin\theta \\ sin\theta & \cos\theta \end{pmatrix}.$$

This includes the two results of case 1(a) by choosing $\theta = 0$ or $\theta = \pi$.
Case 2:

$$ad - bc = -1.$$

By an analysis similar to that above, we find that for some θ

$$A = \begin{pmatrix} \cos\theta & \sin\theta \\ \sin\theta & -\cos\theta \end{pmatrix}.$$

These two cases gives all the 2×2 orthogonal matrices. For $\theta = \dfrac{\pi}{4}$ we get the orthogonal matrices $\begin{pmatrix} \frac{1}{\sqrt{2}} & \frac{1}{\sqrt{2}} \\ -\frac{1}{\sqrt{2}} & \frac{1}{\sqrt{2}} \end{pmatrix}$ and $\begin{pmatrix} \frac{1}{\sqrt{2}} & \frac{1}{\sqrt{2}} \\ \frac{1}{\sqrt{2}} & -\frac{1}{\sqrt{2}} \end{pmatrix}$ and for $\theta = \dfrac{\pi}{6}$

we get $\begin{pmatrix} \frac{\sqrt{3}}{\sqrt{2}} & \frac{1}{2} \\ -\frac{1}{2} & \frac{\sqrt{3}}{2} \end{pmatrix}$ and $\begin{pmatrix} \frac{\sqrt{3}}{\sqrt{2}} & \frac{1}{2} \\ \frac{1}{2} & -\frac{\sqrt{3}}{2} \end{pmatrix}$ orthogonal matrices.
We can recognize the orthogonal matrices

$$\begin{pmatrix} \cos\theta & \sin\theta \\ -\sin\theta & \cos\theta \end{pmatrix}$$

as rotations in the plane. If the positive x, y system is rotated counterclockwise θ radians to find new x', y' system, the coordinates in the two systems are rotated by

$$\begin{pmatrix} x' \\ y' \end{pmatrix} = \begin{pmatrix} \cos\theta & \sin\theta \\ -\sin\theta & \cos\theta \end{pmatrix} \begin{pmatrix} x \\ y \end{pmatrix}.$$

Diagonalization of a 2×2 real symmetric matrix

Consider the most general real symmetric 2×2 matrix

$$A = \begin{pmatrix} a & c \\ c & b \end{pmatrix}$$

where a, b and c are arbitrary real numbers. We compute the eigenvalues and eigenvectors of A, and then find the real orthogonal matrix that diagonalizes A.
The eigenvalues are the roots of the characteristic equation:

$$\begin{vmatrix} a-\lambda & C \\ c & b-\lambda \end{vmatrix} = (a-\lambda)(b-\lambda) - c^2 = \lambda^2 - \lambda(a+b) + (ab - c^2) = 0.$$

The two roots, λ_1 and λ_2, can be determined from the quadratic formula. Noting that $(a+b)^2 - 4(ab-c^2) = (a-b)^2 + 4c^2$, the two roots can be written as:

$$\lambda_1 = \frac{1}{2}\left[a + b + \sqrt{(a-b)^2 + 4c^2}\right]$$
$$\text{and} \quad \lambda_2 = \frac{1}{2}\left[a + b - \sqrt{(a-b)^2 + 4c^2}\right] \tag{1.15}$$

where by convention $\lambda_1 \geq \lambda_2$.
Since $(a-b)^2 + 4c^2 \geq 0$ (as the sum of two squares must be non-negative), Equation (1.15) implies that λ_1 and λ_2 are real. We next work out the two eigenvectors and demonstarate that they are orthogonal. It is convenient to define

$$D \equiv \sqrt{(a-b)^2 + 4c^2}. \tag{1.16}$$

We first solve the eigenvalue equation,

$$\begin{pmatrix} a & c \\ c & b \end{pmatrix} \begin{pmatrix} x \\ y \end{pmatrix} = \frac{1}{2}(a+b+D) \begin{pmatrix} x \\ y \end{pmatrix}.$$

This yields two equations:

$$
\begin{aligned}
ax + cy &= \frac{1}{2}(a + b + D)x, \\
cx + by &= \frac{1}{2}(a + b + D)y,
\end{aligned}
$$

which can be rewritten as:

$$\frac{1}{2}(a - b - D)x + cy = 0 \tag{1.17}$$

$$cx + \frac{1}{2}(b - a - D)y = 0. \tag{1.18}$$

One can show that Equation (1.18) is a multiple of Equation (1.17) since the rank of the matrix $A - \lambda_1 I$ is one]. Simply multiply Equation (1.18) by $(a - b - D)/(2c)$ to obtain

$$
\begin{aligned}
&\frac{1}{2}(a - b - D)x + \frac{(a - b - D)(b - a - D)y}{4c} \\
&= \frac{1}{2}(a - b - D)x + \frac{[D^2 - (a - b)^2]y}{4c} = 0.
\end{aligned}
$$

Using Equation (1.16), $D^2 - (a - b)^2 = 4c^2$, and the above equation reduces to

$$\frac{1}{2}(a - b - D)x + cy = 0,$$

which is equivalent to Equation (1.17). Solving for y yields

$$y = \frac{(b - a + D)x}{2c},$$

which means that the eigenvector corresponding to eigenvalue λ_1 is given by

$$\begin{pmatrix} x \\ y \end{pmatrix}_1 = \frac{x}{2c}\begin{pmatrix} 2c \\ b - a + D \end{pmatrix}.$$

Since λ_2 differs from λ_1 by changing the sign of D, it follows without further computation that the eigenvector corresponding to eigenvalue λ_2 is given by

$$\begin{pmatrix} x \\ y \end{pmatrix}_2 = \frac{x}{2c}\begin{pmatrix} 2c \\ b - a - D \end{pmatrix}.$$

To show that the two eigenvectors are orthogonal, we evaluate the dot product of $(x\ y)_1$ and $(x\ y)_2$, which is equal to $x_1x_2 + y_1y_2$. Inserting the corresponding vector components, we end up with:

$$
\begin{aligned}
\frac{x^2}{4c^2}\left[4c^2 + (b - a + D)(b - a - D)\right] &= \frac{x^2}{4c^2}\left[4c^2 + (a - b)^2 - D^2)\right] \\
&= \frac{x^2}{c^2}\left[4c^2 - 4c^2\right] \\
&= 0,
\end{aligned}
$$

after making use of $D^2 - (a-b)^2 = 4c^2$ [cf. Equation (1.16)].
We now propose to find the real orthogonal matrix that diagonalizes A. The most general 2×2 real orthogonal matrix S with determinant equal to 1 must have the following form:

$$S = \begin{pmatrix} \cos\theta & -\sin\theta \\ \sin\theta & \cos\theta \end{pmatrix}.$$

Using this result, we determine θ in terms a, b and c such that

$$S^{-1}AS = \begin{pmatrix} \lambda_1 & 0 \\ 0 & \lambda_2 \end{pmatrix}$$

where λ_1 and λ_2 are the eigenvalues of A obtained in Equation (1.15). The most straight-forward approach is to compute $S^{-1}AS$ explicitly. Since the off-diagonal terms must vanish, one obtains a constraint on the angle θ.

$$\begin{aligned} S^{-1}AS &= \begin{pmatrix} \cos\theta & \sin\theta \\ -\sin\theta & \cos\theta \end{pmatrix}\begin{pmatrix} a & c \\ c & b \end{pmatrix}\begin{pmatrix} \cos\theta & -\sin\theta \\ \sin\theta & \cos\theta \end{pmatrix} \qquad (1.19) \\ &= \begin{pmatrix} \cos\theta & \sin\theta \\ -\sin\theta & \cos\theta \end{pmatrix}\begin{pmatrix} a\cos\theta + c\sin\theta & -a\sin\theta + c\cos\theta \\ c\cos\theta + b\sin\theta & -c\sin\theta + b\cos\theta \end{pmatrix} \end{aligned}$$

$$= \begin{pmatrix} a\cos^2\theta + 2c\cos\theta\sin\theta + b\sin^2\theta & (b-a)\cos\theta\sin\theta + c(\cos^2\theta - \sin^2\theta) \\ (b-a)\cos\theta\sin\theta + c(\cos^2\theta - \sin^2\theta) & a\sin^2\theta - 2c\cos\theta + b\cos^2\theta \end{pmatrix}$$

$$= \begin{pmatrix} \lambda_1 & 0 \\ 0 & \lambda_2 \end{pmatrix}. \qquad (1.20)$$

The vanishing of the off-diagonal elements of $S^{-1}AS$ implies that:

$$(b-a)\cos\theta\sin\theta + c(\cos^2\theta - \sin^2\theta) = 0.$$

Using $\sin 2\theta = 2\sin\theta\cos\theta$ and $\cos 2\theta = \cos^2\theta - \sin^2\theta$, we can rewrite the above equation as

$$\frac{1}{2}(b-a)\sin 2\theta + c\cos 2\theta = 0.$$

It follows that:

$$\tan 2\theta = \frac{2c}{a-b} \qquad (1.21)$$

after writing $\tan 2\theta = \sin 2\theta / \cos 2\theta$.
Let us now consider the range of the angle θ. You might think that $0 \le \theta < 2\pi$. However, since

$$\cos(\theta+\pi) = -\cos\theta, \quad \text{and} \quad \sin(\theta+\pi) = -\sin\theta,$$

it follows that shifting $\theta \to \theta + \pi$ simply multiplies S by an overall factor of -1. Thus, $S^{-1}AS$ is unchanged. Hence, without loss of generality, we may assume

that $0 \leq \theta < \pi$. Unfortunately, Equation (1.21) does not distinguish between the two intervals $0 \leq \theta < \pi/2$ or $\pi/2 \leq \theta < \pi$, since $\tan 2\theta = \tan(2\theta + \pi)$ is unchanged if $\theta \to \theta + \pi/2$.
However, we have not yet used all the available information. In particular, the diagonal elements of Equation (1.20) also provide some information on the possible values of θ. Summing the diagonal terms of the matrices in Equation (1.20) yields:

$$\begin{aligned}\lambda_1 + \lambda_2 &= (a\cos^2\theta + 2c\cos\theta\sin\theta + b\sin^2\theta) \\ &\quad +(a\sin^2\theta - 2c\cos\theta\sin\theta + b\cos^2\theta) \\ &= (a+b)(\cos^2\theta + \sin^2\theta) \\ &= a+b,\end{aligned}$$

which is independent of θ. This is surprising since we know that

$$TrA = \lambda_1 + \lambda_2 = a + b.$$

However, $\lambda_1 - \lambda_2$ does not depend on θ:

$$\begin{aligned}\lambda_1 - \lambda_2 &= (a\cos^2\theta + 2c\cos\theta\sin\theta + b\sin^2\theta) \\ &\quad -(a\sin^2\theta - 2c\cos\theta\sin\theta + b\cos^2\theta) \\ &= (a-b)(\cos^2\theta - \sin^2\theta) + 4c\sin\theta\cos\theta \\ &= (a-b)\cos 2\theta + 2c\sin 2\theta\end{aligned} \tag{1.22}$$

From Equations (1.15) and (1.21), we obtain

$$\lambda_1 - \lambda_2 = \sqrt{(a-b)^2 + 4c^2} = (a-b)\cos 2\theta + 2c\sin 2\theta \tag{1.23}$$

Using Equation (1.21) to write:

$$a - b = \frac{2c}{\tan 2\theta} = \frac{2c\cos 2\theta}{\sin 2\theta},$$

and inserting this on the left hand side of Equation (1.23), the latter reduces to:

$$\begin{aligned}(a-b)\cos 2\theta + 2c\sin 2\theta &= 2c\frac{\cos^2\theta}{\sin 2\theta} + 2c\sin 2\theta \\ &= \frac{2c}{\sin 2\theta}(\cos^2 2\theta + \sin^2\theta) = \frac{2c}{\sin 2\theta}.\end{aligned}$$

Substituting this result back into Equation (1.23) and solving for $\sin 2\theta$, we find:

$$\sin 2\theta = \frac{2c}{\sqrt{(a-b)^2 + 4c^2}}. \tag{1.24}$$

We can also obtain $\cos 2\theta$ using eqs. (1.21) and(1.24):

$$\cos 2\theta = \frac{a-b}{\sqrt{(a-b)^2+4c^2}}. \tag{1.25}$$

Equation (1.24) tells us in which quadrant θ lives. If $0 < \theta < \pi/2$, then $\sin 2\theta > 0$, which implies that $c > 0$. If $\frac{1}{2}\pi < \theta < \pi$, then $\sin 2\theta < 0$, which implies that $c < 0$. Thus , the sign of c determines the quadrant of θ. Equation (1.25) provides additional information. For $c > 0$, the sign of $a-b$ determines whether $0 < \theta < \frac{1}{4}\pi$ or $\frac{1}{4}\pi < \theta < \frac{1}{2}\pi$. The former corresponds to $a-b > 0$ while the latter corresponds to $a-b < 0$. Likewise, if $c < 0$, the sign of $a-b$ determines whether $\frac{1}{2}\pi < \theta < \frac{3}{4}\pi$ or $\frac{3}{4}\pi < \theta < \pi$. The former corresponds to $a-b < 0$ while the latter corresponds to $a-b > 0$. The borderline cases are likewise determined:

$$\begin{array}{lll} c=0 & \text{and} & a > b \Rightarrow \theta = 0, \\ c=0 & \text{and} & a < b \Rightarrow \theta = \frac{1}{2}\pi, \\ a=b & \text{and} & c > 0 \Rightarrow \theta = \frac{1}{4}\pi, \\ a=b & \text{and} & c > 0 \Rightarrow \theta = \frac{3}{4}\pi. \end{array}$$

If $c = 0$ and $a = b$, then $A = \mathbf{I}$ and it follows that $S^{-1}AS = S^{-1}S = \mathbf{I}$, which is satisfied for any invertible matrix S. Consequently, in this limit θ is undefined.

Example 50. Consider the rotation $R_\theta : R^2 \to R^2$ by an angle $\theta \in [0, 2\pi]$ given by the matrix

$$R_\theta = \begin{pmatrix} \cos\theta & -\sin\theta \\ \sin\theta & \cos\theta \end{pmatrix}.$$

We get eigenvalues by solving polynomial equation

$$p(\lambda) = (\cos\theta - \lambda)^2 + \sin^2\theta = \lambda^2 - 2\lambda\cos\theta + 1 = 0$$

where we have used $\cos^2\theta + \sin^2\theta = 1$.

$$\begin{aligned} \lambda &= \cos\theta \pm \sqrt{\cos^2\theta - 1} = \cos\theta \pm \sqrt{-\sin^2\theta} \\ &= \cos\theta \pm i\sin\theta = e^{\pm i\theta}. \end{aligned}$$

Thus we see that R_θ has eigenvalues when $\theta = 0$ or $\theta = \pi$. However if we interpret the vector $\begin{pmatrix} x_1 \\ x_2 \end{pmatrix} \in R^2$ as a complex number $z = x_1 + ix_2$, z is an eigenvector of $R_\theta : C \to \theta$ maps $z \to \lambda z = e^{\pm i\theta} z$, C is the set of complex numbers. Multiplication by $e^{\pm i\theta}$ corresponds to rotation by the angle $\pm\theta$, see Chapter 8.

1.7 Miscellaneous Applications

In this section we discuss the relevance of matrices in digital image processing image compression, cryptography, transform coding, information theory and literature.

1.7.1 Digital Image Processing

Photographs taken by mobile cameras and illustrations on internet pages are typical examples of digital images. Such type of images can be represented by matrices. For example, the small image of "Felix the cat" cartoon character can be represented by a 35×35 matrix whose elements are the numbers 0 and 1 that specify the color of each pixel: a 0 indicates black and a 1 indicates white. Digital images using only two colors are called binary images or Boolean images. A pixel is the smallest graphical element of a matrical image which can take only one color at a time.
Let an image $f(x, y)$ be represented by an $n \times n$ matrix

$$\begin{pmatrix} f(1,1) & f(1,2) & \cdots & f(1,n) \\ f(2,1) & f(2,2) & \cdots & f(2,n) \\ \vdots & \vdots & \vdots & \vdots \\ f(n,1) & f(n,2) & \cdots & f(n,n) \end{pmatrix} \tag{1.26}$$

where $f(n, n) = a_{i,j}, i = 1, 2, \ldots, n, j = 1, 2, \ldots, n$. Each entry of this matrix is called a pixel. The quality of an image will vary along with variations of i and j. Gray scale and color images are also represented by matrices. For convenience, most of the current digital files use integer numbers between 0 (to indicate black, the color of minimal intensity) and 255 (to indicate white maximum intensity) giving a total of $256 = 2^8$ different levels of gray (This quantity of levels of gray is appropriate for Web page images but certain types such as medical images need more levels of gray).

Color images are represented by three matrices. Each matrix specifies the amount of red, green and blue that make up the image. This color system is known as RGB. There are many other color systems that may be used, depending on applications. The elements of those matrices are integers between 0 and 255 and they determine the intensity of the pixel with respect to the color of the matrix. Thus in the RGB system, it is possible to represent $256^3 = 2^{24}$ different colors. Once a digital image is represented by matrices, the pertinent question arise how operations on their elements affect the corresponding image.

(i) Let $A = (a_{ij})$ represent the image in Figure 1.11 and the matrix adjoint of A, that is $A^T = a_{ji}$, represents the image in Figure 1.12.

FIGURE 1.11: Image Representation by Matrix

FIGURE 1.12: Image Representation by Adjoint Matrix

(ii) Let a color image be represented by matrices $A = (a_{ij}), B = (b_{ij})$ and $AC = (c_{ij})$ Then we get a gray scale version of the image (non-integer values are rounded to the nearest integer) by taking the standard arithmetic mean of the matrices A, B and

$$C\left(d\frac{A+B+C}{3} = \left(\frac{a_{ij}+b_{ij}+c_{ij}}{3}\right)_{ij}\right).$$

(iii) The operation of multiplication by a scalar and sum of matrices, allows us to create an image transition effect commonly used, for instance, in power point presentation and slide shows. For example $M(t) = (1 - t)A + tB$ where matrices A nd B represent gray scale images of the same size. It is clear that $M(0) = A$ and $M(1) = Z$ and, for each t between 0 and 1, the elements of the matrix $M(t)$ are between the elements of the matrices A and B. Therefore when t varies from 0 to 1, the matrix varies from A to Z.

Important ingredients of digital image processing include image modeling, image enhancement, image smoothing, image restoration image compression, denoising and transmission of information under a secure environment.

1.7.2 Matrices in Digital Image Compression

Figure 1.13 and 1.14 show the sequence involved in digital image compression. A compression technique which provides error-free compressed images

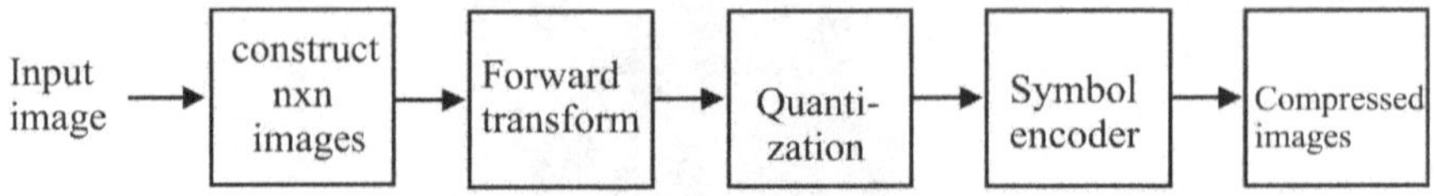

FIGURE 1.13: Transform Coding System for Input Image

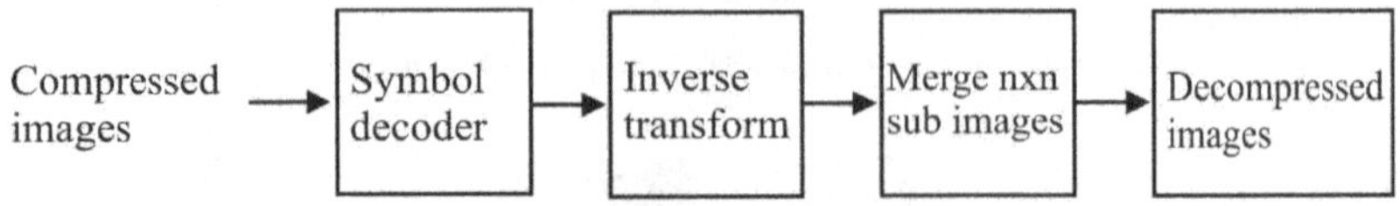

FIGURE 1.14: Transform Coding System for Compressed Image

is called error free compression. Let $E = (n, n)$ denote the **error** between the original digitized image A and decompressed image B. Then

$$E(n,n) = \sum_{i=1}^{n}\sum_{j=1}^{n}[a_{ij} - b_{ij}].$$

$$\text{The } \textbf{mean absolute} = \frac{1}{n^2}\sum_{i=1}^{n}\sum_{j=1}^{n}|a_{ij} - b_{ij}| = em_s.$$

$$\text{The } \textbf{mean square} = \frac{1}{n^2}\sum_{i=1}^{n}\sum_{j=1}^{n}[a_{ij} - b_{ij}].$$

The **root mean square** $= \sqrt{\text{Mean} \quad \text{square}}.$

The **mean square signal to noise ratio** of the output image denoted by SNR_{ms} is defined as

$$SNR_{ms} = \frac{\sum_{i=1}^{n}\sum_{j=1}^{n}(b_{ij})^2}{\sum_{i=1}^{n}\sum_{j=1}^{n}[a_{ij} - b_{ij}]^2}.$$

$\sqrt{SNR_{rms}}$ is called the root-mean square signal to noise ratio.
The peak signal-to-noise ration $PSNR$ is defined as

$$PSNR = 10\log_{10}\left(\frac{255 \times 255}{em_s}\right).$$

The digital image processing main task is to find methods which give minimum value of mean absolute error em_s and SNM_{rms}.
For more details, see Neunzert and Siddiqi [9], Gonzalez and Woods [3], Gomes and Velho [2] and Lay [7].

1.7.3 Cryptography with Matrices

Cryptography is a combination of the Greek words *cry* meaning hidden or secret and *grapho*, which means writing. Thus cryptography is the study of secret writings or codes. We discuss here a system of encoding and decoding messages that requires both sender and the receiver to know specified rules of correspondence between a set of symbols like letters of the alphabet and punctuation marks from which messages are composed and a set of integers forming a specified invertible matrix A.

Problem:

(i) Encode a message into a string of numbers via matrix multiplication.

(ii) Decode the string of numbers into a written message.

Consider the following message :

$$HI\ THERE\ BUDDY$$

We shall require the number 0 to represent a space and use the numbers 1 through 26 to represent 26 letters. We permute the digits in some way instead of making direct substitution. See Table 1.3. We choose an invertible $n \times n$

TABLE 1.3

A	B	C	D	E	F	G	H	I	J	K	L	M
2	1	4	3	6	5	5	7	10	9	12	11	14
N	O	P	Q	R	S	T	U	V	W	X	Y	Z
13	16	15	18	17	20	19	22	21	24	23	26	25

matrix which will be used to encode the message. Let us use

$$A = \begin{pmatrix} 3 & 2 & 3 \\ -6 & -5 & -4 \\ 9 & 8 & 7 \end{pmatrix}.$$

We first convert the message to numeric script using $H = 7, I = 10$, blank $= 0$, $T = 19, H = 7, E = 6, R = 17, E = 6$, Blank $= 0$, $B = 1, U = 22, D = 3, Y = 26$. This numeric form of message is

$$7100197617601223326.$$

Then we convert this script into an $m \times 3$ matrix (3 columns because our A is 3×3). The number of rows m depends on the length of the message. We fill the last row with more 0's as necessary:

$$B = \begin{pmatrix} 7 & 10 & 0 \\ 19 & 7 & 6 \\ 17 & 6 & 0 \\ 1 & 22 & 3 \\ 3 & 26 & 0 \end{pmatrix}.$$

Here B is 5×3. Now

$$\begin{aligned} C &= BA \\ &= \begin{pmatrix} 7 & 10 & 0 \\ 19 & 7 & 6 \\ 17 & 6 & 0 \\ 1 & 22 & 3 \\ 3 & 26 & 0 \end{pmatrix} \begin{pmatrix} 3 & 2 & 3 \\ -6 & -5 & -4 \\ 9 & 8 & 7 \end{pmatrix} \\ &= \begin{pmatrix} -39 & -36 & -19 \\ 69 & 51 & 71 \\ 15 & 4 & 27 \\ -102 & -84 & -64 \\ -147 & -124 & -95 \end{pmatrix}. \end{aligned}$$

We convert matrix C into a string that becomes the coded message, see Table 1.4.

It may be observed that H has been sent to two different values. The first H

TABLE 1.4

−39	−36	−19	69	51	71	15	4	27	−102	−84	−64	−147	−124
H	*I*		*T*	*H*	*E*	*E*	*E*		*B*	*U*	*D*	*D*	*Y*

has been encoded as −39, and the second H has been encoded as 51. Likewise the three blanks assigned different values as are the two E s and two D s. The person at the other end receives only the coded message:

−39 −36 −19 69 51 71 15 4 27 −102 −84 −64 −147 −124 −95.

The receiver knows the alpha-numeric code and the coding matrix A. Thus, the coded script is converted back into an $m \times 3$ matrix C and then multiplied by A^{-1} to undo the original matrix multiplication. Since $C = BA$, we get $B = CA^{-1}$, and then we convert the string back to the alphabet.

Example 51. Encode and decode the following message using the given matrix:

(i) Message: SEND HELP, $A = \begin{pmatrix} 1 & 2 \\ 1 & 1 \end{pmatrix}$.

(ii) Message: MADAME X HAS THE PLANS

$$A = \begin{pmatrix} 1 & 2 & 3 \\ 1 & 1 & 2 \\ 0 & 1 & 2 \end{pmatrix}.$$

A natural correspondence between the first 27 non-negative integers and the letters of the alphabet and blank spaces (to separate words) is given by Table 1.5.

TABLE 1.5

0	1	2	3	4	5	6	7	8	9	10	11	12	13
space	a	b	c	d	e	f	g	h	i	j	k	l	m
0	14	15	16	17	18	19	20	21	22	23	24	25	26
space	n	o	p	q	r	s	t	u	v	w	x	y	z

Solution: (i) $B = \begin{pmatrix} 19 & 5 & 14 & 4 & 0 \\ 8 & 5 & 12 & 16 & 0 \end{pmatrix}$

$$\begin{aligned} C &= AB \\ &= \begin{pmatrix} 1 & 2 \\ 1 & 1 \end{pmatrix}\begin{pmatrix} 19 & 5 & 14 & 4 & 0 \\ 8 & 5 & 12 & 16 & 0 \end{pmatrix} \\ &= \begin{pmatrix} 35 & 15 & 38 & 36 & 0 \\ 27 & 10 & 26 & 20 & 0 \end{pmatrix}. \end{aligned}$$

C corresponds to the encoded message.

$$\begin{aligned} B &= A^{-1}C \\ &= \begin{pmatrix} -1 & 2 \\ 1 & -1 \end{pmatrix}\begin{pmatrix} 35 & 15 & 38 & 36 & 0 \\ 27 & 10 & 26 & 20 & 0 \end{pmatrix} \\ &= \begin{pmatrix} 19 & 5 & 14 & 4 & 0 \\ 8 & 5 & 12 & 16 & 0 \end{pmatrix} \end{aligned}$$

represents the decoded message.

(ii) $B = \begin{pmatrix} 7 & 15 & 0 & 14 & 15 & 18 & 20 \\ 8 & 0 & 15 & 14 & 0 & 13 & 1 \\ 9 & 14 & 0 & 19 & 20 & 0 & 0 \end{pmatrix}.$

$$\begin{aligned} C &= AB \\ &= \begin{pmatrix} 1 & 2 & 3 \\ 1 & 1 & 2 \\ 0 & 1 & 2 \end{pmatrix}\begin{pmatrix} 7 & 15 & 0 & 14 & 15 & 18 & 20 \\ 8 & 0 & 15 & 14 & 0 & 13 & 1 \\ 9 & 14 & 0 & 19 & 20 & 0 & 0 \end{pmatrix} \\ &= \begin{pmatrix} 50 & 57 & 30 & 99 & 75 & 44 & 22 \\ 33 & 43 & 15 & 66 & 55 & 31 & 21 \\ 26 & 28 & 15 & 52 & 4 & 13 & 1 \end{pmatrix}. \end{aligned}$$

C represents the encoded message:

$$\begin{aligned} B &= A^{-1}C \\ &= \begin{pmatrix} 0 & 1 & -1 \\ 2 & -2 & -1 \\ -1 & 1 & 1 \end{pmatrix}\begin{pmatrix} 50 & 57 & 30 & 99 & 75 & 44 & 22 \\ 33 & 43 & 15 & 66 & 55 & 31 & 21 \\ 26 & 28 & 15 & 52 & 4 & 13 & 1 \end{pmatrix} \end{aligned}$$

$$= \begin{pmatrix} 7 & 15 & 0 & 14 & 15 & 18 & 20 \\ 8 & 0 & 15 & 14 & 0 & 13 & 1 \\ 9 & 14 & 0 & 19 & 20 & 0 & 0 \end{pmatrix}$$
$$= \text{decoded message.}$$

1.7.4 Transform Coding

Let a transformation A be defined with the help of an $n \times n$ matrix as follows:

$$y = Ax \tag{1.27}$$

where

$$x = \begin{pmatrix} x_1 \\ x_2 \\ \vdots \\ x_n \end{pmatrix}, y = \begin{pmatrix} y_1 \\ y_2 \\ \vdots \\ y_n \end{pmatrix}$$

$$A = \begin{pmatrix} a_{11} & a_{12} & \cdots & a_{1n} \\ a_{21} & a_{22} & \cdots & a_{2n} \\ \vdots & \vdots & \vdots & \vdots \\ a_{n1} & a_{n2} & \cdots & a_{nn} \end{pmatrix}.$$

In general, A is not invertible. In this case, the vector of pixels x is transformed into a vector of coefficients y. For some set of vectors, transformations of fewer bits are required to encode the n coefficients of y than the n pixels of x. In particular, if the elements $x_1, x_2, \ldots, x_n$ are highly correlated and the transformation matrix A is chosen such that the coefficients $y_1, y_2, \ldots, y_n$ are less correlated than the y_i's and can be individually coded for efficiently, explained below, if we choose A as

$$A = \begin{pmatrix} 1 & 0 & 0 & 0 & 0 & 0 \\ 1 & -1 & 0 & 0 & 0 & 0 \\ 0 & 1 & -1 & 0 & 0 & 0 \\ 0 & 0 & 1 & -1 & 0 & 0 \\ 0 & 0 & 0 & 1 & -1 & 0 \\ 0 & 0 & 0 & 0 & 1 & -1 \end{pmatrix} \tag{1.28}$$

in equation (1.27). The first element of y is $y_1 = x_1$ and all subsequent coefficients are given by $y_i = x_{i-1} - x_i$. If the grey levels of the adjacent pixels are similar, then the difference $y_i = x_{i-1} - x_i$ will, on the average, be smaller than the grey level that should require fewer bits to code them. This mapping

is invertible.
If A is a unitary matrix, then A^{-1} exists and $A^{-1} = A^t$, whereas if A^t is the transpose of A, then (1.27) can be written as

$$x = A^t y. \tag{1.29}$$

It is clear from (1.27) that each coefficient y_k is a linear combination of n pixels; that is,

$$y_k = \Sigma_{i=1}^n a_{ki} x_i, \tag{1.30}$$

for $k = 1, 2, 3, \ldots, n$.
Similarly, by Equation (1.29), each pixel x_i is a linear combination of all the pixels

$$x_i = \Sigma_{k=1}^n b_{ik} y_k, \tag{1.31}$$

for $i = 1, 2, 3, \ldots, n$.
Equations (1.30) and (1.31) are similar to the expressions defining the forward and the inverse transformation kernels, respectively, whereas a_{ki} is the forward formation kernel and b_{ik} is the inverse transformation kernel.
For the two-dimensional case, (1.30) and (1.31) take the forms

$$y_{kl} = \Sigma_{i=1}^n \Sigma_{j=1}^n x_{ij} a_{ijkl} \tag{1.32}$$

and

$$x_{ij} = \Sigma_{k=1}^n \Sigma_{l=1}^n y_{kl} b_{ijkl}. \tag{1.33}$$

Here, a_{ijkl} and b_{ijkl} are forward and inverse transformation kernels, respectively.
In equations (1.30) and (1.31) the Fourier, Walsh and Hadmard transforms are commonly used for encoding in this setting and they produce fairly good results. For example, the Fourier kernel is given by

$$a_{ijkl} = \frac{1}{N} e^{-j2\pi(ik+jl)/N}.$$

Further interpretation of (1.33) is possible. Let us write (1.33) in the form

$$X = \Sigma_{k=1}^n \Sigma_{l=1}^n y_{kl} B_{kl}. \tag{1.34}$$

Interpret this as a series expansion of the $n \times n$ subimage X into an $n^2 n \times n$ basis gives

$$B_{kl} = \begin{pmatrix} b_{kl11} & b_{kl12} & \cdots & b_{kl1n} \\ b_{kl21} & b_{kl22} & \cdots & b_{kl2n} \\ \cdots & \cdots & \cdots & \cdots \\ \cdots & \cdots & \cdots & \cdots \\ b_{kln1} & b_{kln2} & \cdots & b_{klmn} \end{pmatrix} \tag{1.35}$$

where the y_{kl} for $k, l = 1, 2, \ldots, n$ represents the coefficient (weights) of the expansion. Therefore (1.35) gives the image X as a weighted sum of the basis B_{kl}. The coefficients of the expansion are given by (1.32) which may be written in the form

$$y = A_{kl}X, \tag{1.36}$$

where A_{kl} is formed in the same manner as B_{kl} except that the forward kernel is used. Let us consider the number of all possible values of the coefficients in the Equation (1.27) where

$$y_i = \Sigma_{j=1}^{n} a_{ij} x_j.$$

Each element x_j can have any of 2^m different values, each $a_{ij}x_j$ term can also have any of 2^m different values and the sum of n such terms could have any of 2^{mn} different values. Therefore, a natural binary representation would give mn bit words to assign a unique word to each of the possible 2^{mn} values of y_i. Since only m bit words would be required to code any x_j, and our objective is to use fewer bits to code y_i, we must round off y_i to a fewer number of allowed levels.
A **quantizer** is a device whose output can have only a limited number of possible values. Each input is forced to one of the permissible output values. For more details, we refer to Gonzales and Woods [3].

1.7.5 Markov Matrix and Markov Process

Definition 27. A real $n \times n$ matrix $A = (a_{ij}), i = 1, 2, \ldots, n$ $j = 1, 2, \ldots, n$ is called a **Markov matrix** or **row stochastic matrix** if

(i) $a_{ij} \geq 0$ for $0 \leq i, j \leq n$

(ii) $\Sigma_{j=1}^{n} a_{ij} = 1$ for $1 \leq i \leq n$.

A Markov matrix is also known a **transition matrix** or **probability matrix**. In other words an $n \times n$ matrix is Markov if all entries are non-negative and the sum of each column vector is equal to 1.
This matrix is named in honour of Andrei Andreyevich Markov (1856-1922) who was born in Ryazan, Russia, and died in St. Petersburg, Russia. He also discovered a process called Markov chain or Markov process of which the Markov matrix is an important ingredient. The five greatest applications of Markov chains are presented by Phillopp Von Hilgers and Amy N. Langvill [10] and listed in chronological order

(i) AA. Markov's application to Eugene Onegin

(ii) C.E. Shannon's application to information theory

(iii) Baum's application to hidden Markov models

(iv) Scherr's application to computer performance evaluation

(v) Brin and Page's application to Web searches

In order of importance the list is:

1. Scherr's application to computer performance evaluation
2. Brin and Page's application to Web searches
3. Baum's application to hidden Markov models
4. Shannon's application to information theory
5. Markov's application to Eugene Onegin

Definition 28. Let there be a physical and mathematical system that has n possible states. At any one time, the system is in one and only one of its n states. Also assume that at a given observation period, say the kth period, the probability of the system being in a particular state depends only on its status at $k-1$th period. Such a system is called **Markov Chain** or **Markov process**.

In this process the probability of the system being in a particular state at a given observation period depends only on its state during the immediately preceding observation period.
Suppose that the system has n possible sates. For each $i = 1, 2, \ldots, n$ and $j = 1, 2, \ldots, n$, let p_{ij} be the probability that if the system is in state j at a certain observation period, it will be in state i at the next observation period. p_{ij} is called transition probability. Since p_{ij} is a probability, we must have

$$0 \leq p_{ij} \leq 1, \quad 1 \leq i, j \leq n.$$

Also, if the system is in state j at a certain observation period, it must be in one of the n states (it may remain in state j) at the next observation period.Thus, we have

$$p_{1j} + p_{2j} + p_{3j} + \cdots + p_{nj} = 1.$$

It is clear that p_{ij} is a Markov matrix.

Definition 29. A Markov matrix is called **regular** if, for some integer r, all entries of T^r are positive. A Markov chain or **process** is called **regular** if its Markov matrix is regular.
Important properties of Markov matrices and chains are summarized below:

(i) If T is a regular Markov matrix, then as n approaches infinity, $T^n \to S$, where S is a matrix of the form $(v_1, v_2, \ldots, v_n)$ with v being a constant vector.

(ii) If T is a regular Markov matrix of a Markov chain or process, and if X is a state vector, then as n approaches infinity, $T^n X \to p$, where p is a fixed probability vector (the sum of its entries 1), all of whose entries are positive.

(iii) Let T be a regular Markov matrix of a Markov chain or process and let $\lim_{n\to\infty} T^n = S$. then $\lim_{n\to\infty} T^n X = SX = p$, p is called the **steady state vector** of the system.

(iv) $S = TS$. We know that $T^{n+1} = TT^n$, $\lim_{n\to\infty} T^{n+1} = S$ and $\lim_{n\to\infty} T^n = S$. This implies that $S = TS$.

(v) We observe that $Tp = p$ for any column p of T. Therefore, the steady state vector of a regular Markov chain with Markov matrix T is the unique probability vector p satisfying $Tp = p$.

(vi) A steady-state vector of a regular Markov chain is an eigenvector for the Markov matrix corresponding to the eigenvalue 1.

Example 52. An unofficial study of the weather in a city of Germany in early spring yields the following observations;

(i) It is almost impossible to have two nice days in a row.

(ii) If we have a nice day, we are just as likely to have snow or rain the next day.

(iii) If we have snow or rain, we have an even chance to have the same weather the next day?

(iv) If there is a change from snow or rain only half of the time, is this a change to a nice day?

(a) Write the Markov matrix to model this system.

(b) If today is nice, what is the probability of nice weather after one week?

(c) Find the long term behavior of the weather.

Solution: Since the weather tomorrow depends only on today, prediction is a Markov chain (process). The Markov or transition matrix of this system is

$$\begin{array}{c} \\ N \\ R \\ S \end{array} \begin{array}{c} \begin{array}{ccc} N & R & S \end{array} \\ \begin{pmatrix} 0 & 0.25 & 0.25 \\ 0.5 & 0.5 & 0.25 \\ 0.5 & 0.25 & 0.5 \end{pmatrix} \end{array}$$

where the letters N, R, and S represent nice, rain, snow respectively. If today is nice, the initial state vector is

$$X_0 = \begin{pmatrix} 1 \\ 0 \\ 0 \end{pmatrix}.$$

After 7 days (one week), the state vector would be

$$\begin{aligned} X_7 &= T^7 X_0 \\ &= \begin{pmatrix} 0.1999511 & 0.2000122 & 0.2000122 \\ 0.4000244 & 0.4000244 & 0.3999633 \\ 0.4000244 & 0.3999633 & 0.4000244 \end{pmatrix} \begin{pmatrix} 1 \\ 0 \\ 0 \end{pmatrix} \\ &= \begin{pmatrix} 0.1999511 \\ 0.4000244 \\ 0.4000244 \end{pmatrix} \end{aligned}$$

so, there is about 20% chance of nice weather in one week. We observe that our Markov matrix is regular so existence of the steady state vector is guaranteed. To find the steady state vector we solve the homogeneous system $(T-I)X = 0$ which has the following coefficient matrix:

$$\begin{pmatrix} -1 & 0.25 & 0.25 \\ 0.5 & -0.5 & 0.5 \\ 0.5 & 0.5 & 0.5 \end{pmatrix}.$$

Its reduced echelon form is

$$\begin{pmatrix} 1 & 0 & -0.5 \\ 0 & 1 & -1 \\ 0 & 0 & 0 \end{pmatrix}.$$

The general solution of this system is

$$\begin{pmatrix} 0.5t \\ t \\ t \end{pmatrix}, t \in R.$$

Since $0.5\ t + t + t = 1$, we ge $t = 0.4$. Thus the steady state vector is

$$p = \begin{pmatrix} 0.2 \\ 0.4 \\ 0.4 \end{pmatrix}.$$

In the long term, there is 20% chance of having a nice day, 40% chance of a rainy day and 40% chances of a snowy day.

Example 53. Suppose weather in a city in India is either dry or rainy. Meteorological data shows that the probability of a rainy day following a dry day is $\frac{1}{3}$ and probability of a rainy day following a rainy day is $\frac{1}{2}$.

Solution: Let state D be a dry day and state R be rainy day. Then the Markov matrix of this Markov process is

$$(p_{ij}) = \begin{array}{c} \begin{matrix} D & R \end{matrix} \\ \begin{pmatrix} \frac{2}{3} & \frac{1}{2} \\ \frac{1}{3} & \frac{1}{2} \end{pmatrix} \end{array} \begin{matrix} D \\ R \end{matrix} .$$

Since all entries in p_{ij} are positive, the system is a regular Markov chain. Let us begin our observations (day 0). The weather is dry so the initial state vector is

$$X_0 = \begin{pmatrix} 1 \\ 0 \end{pmatrix},$$

a probability vector. The the state vector on day 1 (the day after we begin our observations) is

$$X_1 = TX_0 = \begin{pmatrix} 0.67 & 0.5 \\ 0.33 & 0.5 \end{pmatrix} \begin{pmatrix} 1 \\ 0 \end{pmatrix} = \begin{pmatrix} 0.67 \\ 0.33 \end{pmatrix}$$

where $\frac{2}{3}$ is written as 0.67 and $\frac{1}{3}$ as 0.33 respectively for the sake of convenience. The probability of no rain on day 1 is 0.67, and the probability of rain on that day is 0.33. Similarly

$$X_2 = TX_1 = \begin{pmatrix} 0.67 & 0.5 \\ 0.33 & 0.5 \end{pmatrix} \begin{pmatrix} 0.67 \\ 0.33 \end{pmatrix} = \begin{pmatrix} 0.614 \\ 0.386 \end{pmatrix},$$

$$X_3 = TX_2 = \begin{pmatrix} 0.67 & 0.5 \\ 0.33 & 0.5 \end{pmatrix} \begin{pmatrix} 0.614 \\ 0.386 \end{pmatrix} = \begin{pmatrix} 0.604 \\ 0.396 \end{pmatrix},$$

$$X_4 = TX_3 = \begin{pmatrix} 0.67 & 0.5 \\ 0.33 & 0.5 \end{pmatrix} \begin{pmatrix} 0.604 \\ 0.396 \end{pmatrix} = \begin{pmatrix} 0.603 \\ 0.397 \end{pmatrix}.$$

From the fourth day onward the state vector remains the same: $\begin{pmatrix} 0.603 \\ 0.397 \end{pmatrix}$. This is the steady state vector which means that from fourth day onward, it is dry about 60% of the time, and it rains about 40% of the time.

Additional Properties of Markov Matrices

(a) The product of two Markov matrices is a Markov matrix

(b) Every eigenvalue λ of a Markov matrix satisfies $|\lambda| \leq 1$, that is, absolute value of an eigenvalue of a Markov matrix is less than 1.

(c) Absolute value of every eigenvalue of a positive Markov matrix (having only positive elements) is 1.

(d) A Markov matrix may have zero eigenvalues and determinant : $A = \begin{pmatrix} 0 & 0 \\ 1 & 1 \end{pmatrix}$.

(e) A Markov matrix can have several eigenvalues 1: $A = \begin{pmatrix} 1 & 0 \\ 0 & 1 \end{pmatrix}$.

1.8 Introduction to MATLAB® and MATHEMATICA

In this section we introduce two computer softwares, namely MATLAB and MATHEMATICA. MATLAB stands for MATRIX LABoratory developed by Mathworks Corporation, Natick, MA (http://www.mathwork.com). Its invention is credited to Clever Moler in early 1970. MATLAB is a high performance language for technical computing. It integrates computation, visualization and programming in an easy-to-use environment where problems and solutions are expressed in familiar mathematical notions.

All themes discussed in previous sections can carried out easily by MATLAB. It is the favorite software of mathematicians and engineers while MATHEMATICA is more attractive for physicists. MATHEMATICA is a powerful programming language invented by Wolfram Research, Inc with its formal announcement on 23 June, 1988 (www.stephenwolfram.com/about.). It is a contemporary software of MATLAB and performs better in many situations. Stephen Wolfram is the creator of Mathematica.

1.8.1 MATLAB for Matrices

MATLAB has a wide range of capabilities. We introduce here few of its features. We refer to Chapter 9 and Chapter 10 of Kolman and Hill [5] for detailed discussion on MATLAB for linear algebra.

Matrix Input: To enter a matrix into MATLAB, first type the entries of the matrix within square brackets, with space between entries and commands. Use semicolons to indicate end of rows. The example below

$$\begin{pmatrix} 9 & -8 & 7 \\ -6 & 5 & -4 \\ 11 & -12 & 0 \end{pmatrix}$$

is entered by typing $[9\ \ -8\ 7; -6\ 5\ 0.4; 11\ \ -12\ 0]$ and the accompanying display is

$$\text{ans} = \begin{matrix} 9 & -8 & 7 \\ -6 & 5 & -4 \\ 11 & -12 & 0 \end{matrix}$$

No brackets are displayed and MATLAB has assigned this matrix the name *ans*. Every matrix in MATLAB must have a name. If name is not assigned, then MATLAB assigns *ans*, which is known as a default variable. To assign a matrix name, we use the assignment operator, for example

$$A = (4\ 5\ 8; 0\ -1\ 6)$$

is displayed as

$$\mathrm{B} = \begin{matrix} 4 & 5 & 8 \\ 0 & -1 & 6 \end{matrix}$$

Important Elements of MATLAB

MATLAB has many advantages as compared with conventional computer languages or programmes for solving technical problems which can be summarized as follows:

1. **Easy to Use:** MATLAB is an interpreted language that is very easy to use. It has built-in functions that are optimized for specific problems and thus makes it easy to evaluate certain mathematical expressions using the values in their functions. Complex mathematical problems can be solved easily with a few lines of MATLAB code while other languages may require several hundred lines of code.

2. **Platform Independence:** The MATLAB software is supported by various operating systems such as Win XP/Vista, UNIX and MAC OS X. The program written in one platform will run on other platforms.

3. **Powerful Graphics:** This is the one of the important features of MATLAB for technical data analysis and interpretation. MATLAB supports a wide variety of data so that it can be interpreted well. It supports colored 2D and 3D plots and animation videos that make it unique.

4. **Graphical User Interface:** MATLAB includes tools that allow a programmer to develop a graphical user interface (GUI) for this program. A programmer can design sophisticated data analysis programs that can be operated by inexperienced users.

5. **MATLAB Help:** The help system is powerful and user friendly. It provides effective help that addresses most topics and commands. Users can obtain help offline and online through the Internet. MATLAB provides hundreds of example demonstrations for solving various problems users may encounter when writing programs.

6. **MATLAB Compiler:** Flexibility and platform independence are achieved by compiling MATLAB programs into a device-independent

p code and p instructions for use at run time. However, these steps cause slow execution of programs and a MATLAB compiler is available to compile program executable (exe) files that run faster.

How to Run MATLAB?

Let us start with the MATLAB environment. We can open the window by double clicking the MATLAB icon on the desktop. The main window (1.8.1) that opens will be divided into three parts: current directory, command history, and command window.

The workspace shows the variables in use and their maximum and minimum values. The functions of the current directory and command history are self-explanatory. The command window is the function that allows a user to write MATLAB commands.

MATLAB File Formats:

MATLAB can read and write various types of files, However, there are five main types of files for storing data and/or programs that one will use frequently. They are:

(a) **M Files:** These are standard ASCII text files with $.m$ extensions written in MATLAB editor. Our main program files are written in M formats. We will later show an example of an M file.

(b) **MAT Files:** These are binary files with $.mat$ extensions created when we variables of MATLAB workspace.

(c) **Fig Files:** These are binary files with $.fig$ extensions used to store graphics (i.e., curves obtained by plotting data).

(d) **P Files:** These are M files with $.p$ extensions. They are used mainly for distribution of MATLAB programs with hidden MATLAB codes.

(e) **MEX Files:**These are callable programs with $.mex$ extensions and are used to interface MATLAB with C and FORTRAN.

Starting MATLAB:

Variables, Operators and Matrices

Let us start with simple commands related to matrices. To enter a matrix, simply enter the following when the command prompt appears in the command window: $>> a = [1\ 2\ 3;\ 5\ 7\ 4;\ 9\ 8\ 6]$, then press Enter. The following will appear in the command window: $>> a =$

$$\begin{matrix} 1 & 2 & 3 \\ 5 & 7 & 4 \\ 9 & 8 & 6 \end{matrix}.$$

It shows that, our 3×3 matrix is stored in variable a. We can see that a semicolon is used to separate the rows. Now let us do some operations on this matrix. Simply type $a-$ on the command prompt ($>> -a$) and press enter. We get

$$ans = \begin{pmatrix} 1 & 5 & 9 \\ 2 & 7 & 8 \\ 3 & 4 & 6 \end{pmatrix}.$$

Obviously, it is a transpose of matrix a. Here ans is the default variable provided by MATLAB if we don't specify our own variable. Similarly, we can perform other operations. Students can try following commands on command prompt

1. >> **Inv(a)** [Will give the inverse of matrix a]
2. >> **det(a)** [Determinant of matrix a]
3. >> **eig(a)** [Eigenvalue and eigenvector of a]
4. >> **diag(a)** [Shows elements of main diagonal]
5. >> **rank(a)** [Finds rank of matrix a]

Now let us do some operations on two matrices such as summation, subtraction or multiplication. Matrix a is already stored in memory, so we enter another matrix b: $>> b = [2\ 4\ 5; 1\ 6\ 8; 3\ 4\ 6]$
Notice that we have put one semicolon at the end of matrix b. The function of the semicolon is to suppress output in the workspace window, which becomes useful when we generate very large matrices.
Students are advised to try following operations on matrices a and b.

1. $>> c = a + b$ [Summation of matrices a and b; result is stored in c]
2. $>> c = a - b$ [Subtraction of matrix b from a]
3. $>> c = a * b$ [Multiplication of matrices a and b]
4. $>> c = a. * b$ [Element-by-element multiplication of matrices a and b]
5. $>> c = a./b$ [Element-by-element division of matrices a and b]

Another operator called left division (\) is used in solving sets of linear algebraic equations (see MATLAB help for detail).
MATLAB offers relational and logical operators also. Please use MATLAB help for this.

Writing Script Files ($M-$ files):

In the previous section, we have seen some basic operations performed by writing commands on command prompts in command windows. It was easy to perform single and easy operations on a variable. When we have to perform several complicated and interdependent operations, then it is not possible to use a command prompt. MATLAB provides given an editor in which we can write a program called Script which can be compiled and run and results are displayed in the command window or figure window, if graphs are requested. Here we are giving one very simple example of writing an M file to have a set of linear algebraic equations. The equations are given below:

$$\begin{aligned} x + 2y - z &= 10. \\ 4x + 6y + z &= 20. \\ x - 8y + 3z &= 8. \end{aligned}$$

These three equations can be written in matrix form as $[A].[x] = [B]$

$$\begin{pmatrix} 1 & 2 & -1 \\ 4 & 6 & 1 \\ 1 & -8 & 3 \end{pmatrix} \begin{pmatrix} x \\ y \\ z \end{pmatrix} = \begin{pmatrix} 10 \\ 20 \\ 8 \end{pmatrix}.$$

The solution can be obtained by $[x] = [A]^{-1}[B]$. (In MATLAB syntax it is $x = \text{inv}(A) * B$.)
The solution can be found by writing an M file in the editor and executing it. To open editor window, go to the File menu of MATLAB window and select New > M File and write following code.
% Solution of linear algebraic equations.

$$\begin{aligned} A &= [12 - 1; 461; 1 - 83]; \\ B &= [10208]; \\ xyz &= \text{inv}(A) * B; \\ xyz. & \end{aligned}$$

After executing it, we get following result

$$xyz = \begin{pmatrix} 8.6667 \\ -1.6667 \\ -4.6667 \end{pmatrix}.$$

The first line which starts with % is a comment line and it is not executed.

Introduction to MATHEMATICA

Advantages and Disadvantages

- Powerful
- Easy to learn and use
- Powerful symbolic capabilities
- Coherent and unified
- Symbolic and uniform representation

- Numerical routines for complex problems
- Lots of add-on packages and books
- Scripted dynamic language
- easy to misuse

Arithmetic Operations

Basic Addition, Multiplication etc.

Commands for expressions which should be evaluated are entered into the input cells and displayed in bold.

Example 54. (Figures 1.15 though 1.17) Evaluate $-5+3.(-2)+\dfrac{2^{-3}+3.(5-7)}{-\dfrac{2}{3}+\dfrac{3}{4}.(-9+3)}$.

```
-5 + 3 * (-2) + (2^-3 + 3 * (5 - 7)) / (-2 / 3 + (3 / 4) * (-9 + 3))
```

$-\dfrac{1223}{124}$

FIGURE 1.15

If we want a numerical value of the last expression with, say 20 digits, then in the cell we can type

```
N[%, 20]
-9.8629032258064516129
```

FIGURE 1.16

```
%2
-9.8629032258064516129
```

FIGURE 1.17

Algebraic Calculations

Mathematica can perform algebraic calculations too using several functions.

```
Expand[(x - 1) (2 x - 3) (x + 9)]
27 - 42 x + 13 x^2 + 2 x^3
```

FIGURE 1.18

Example 55. (Figure 1.18) Expand $(x-1)(2x-3)(x+9)$.

Example 56. (Figure 1.19) Factorize $[x^3+2x^2-5x-6]$.

```
Factor[x^3 + 2 x^2 - 5 x - 6]
(-2 + x) (1 + x) (3 + x)
```

FIGURE 1.19

Example 57. (Figure 1.20 though Figure 1.24) To simplify $[(x^2-3x)(6x-7)-(3x^2-2x-1)(2x-3)]$.

```
Simplify[(x^2 - 3 x) (6 x - 7) - (3 x^2 - 2 x - 1) (2 x - 3)]
-3 + 17 x - 12 x^2
```

FIGURE 1.20

```
Together[2 x - 3 + (3 x - 5) / (x^2 + x + 1)]
```

$$\frac{-8+2x-x^2+2x^3}{1+x+x^2}$$

FIGURE 1.21

```
Apart[%8]
```

$$-3+2x+\frac{-5+3x}{1+x+x^2}$$

FIGURE 1.22

Commands on the same line within a cell must be separated by semicolons. The output of any command that ends with a semicolon is not displayed.

```
x = 5 / 4 - 3 / 7; y = 9 / 5 + 4 / 9 - 9 x; z = 2 x² - 3 y²
```

FIGURE 1.23

If we have to evaluate an expression with different values each time, the substitution command symbolized with a slanted bar and a period is useful.
Using Functions

```
z = 2 x^2 - 3 x y + 2 y^2 /. {x → 1, y → 1}
1
```

FIGURE 1.24

Built-in Functions

All built-in Mathematical functions and constants have full names that begin with a capital letter. The arguments of a function are enclosed by brackets.

- For example the familiar functions from calculus $\sin x$, $\cos x$, e^x, $\ln x$ in Mathematica are sin[x], cos[x],*Exp*[x], *Log*[x].
- The constant π in mathematica is *Pi*, while the Euler number e in Mathematica is *E*.

Using Functions
User Defined Functions
In Mathematica the function $f(x) = x^2 + 2x$ is defined by:
We can evaluate $f(a+b) + f(a-b)$.

```
f[x_] := x^2 + 2 x;
```

FIGURE 1.25

Similarly, for defining functions of several variables: $g(x,t) = \sin(\pi x)\cos(2\pi t)$

```
f[a + b] + f[a - b]
```

FIGURE 1.26

User defined functions
Define a piecewise function

```
g[x_, t_] := Sin[Pi x] Cos[2 Pi t];
```

FIGURE 1.27

$$h(x) = \begin{cases} x & \text{if } -\pi \leq x < 0 \\ x^2 & 0 \leq x < \pi \\ 0 & \text{elsewhere} \end{cases}$$

Graphics

```
h[x_] := Piecewise[{{x, -Pi ≤ x < 0}, {x^2, 0 ≤ x < Pi}}, {0}];
h[2]

4
```

FIGURE 1.28

Basic Plotting

Mathematica is exceptionally good at creating two– and three–dimensional graphs.

- The general syntax is Plot[$f, \{x, x_{min}, x_{max}\}$].
- The plot of the function $h(x)$ above on the interval $[-\pi, \pi]$ is obtained by typing the following (Figure 1.28 through Figure 1.30:

```
Plot[h[x], {x, -Pi, Pi}]
```

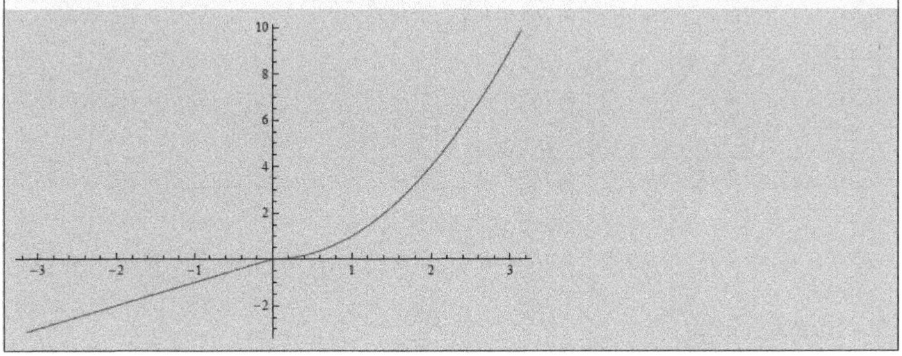

FIGURE 1.29

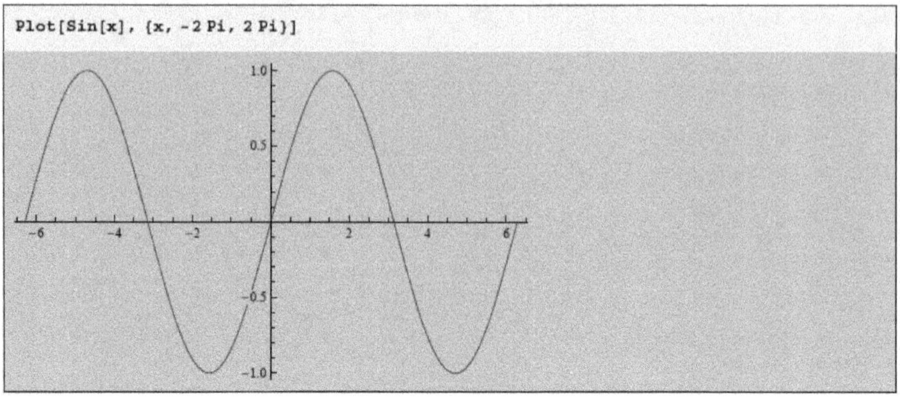

FIGURE 1.30

Multiple Plotting

MATHEMATICA can handle multiple plots.

- Syntax: Plot[$\{f1\ \ f2\}, \{x, x_{min}, x_{max}\}$].

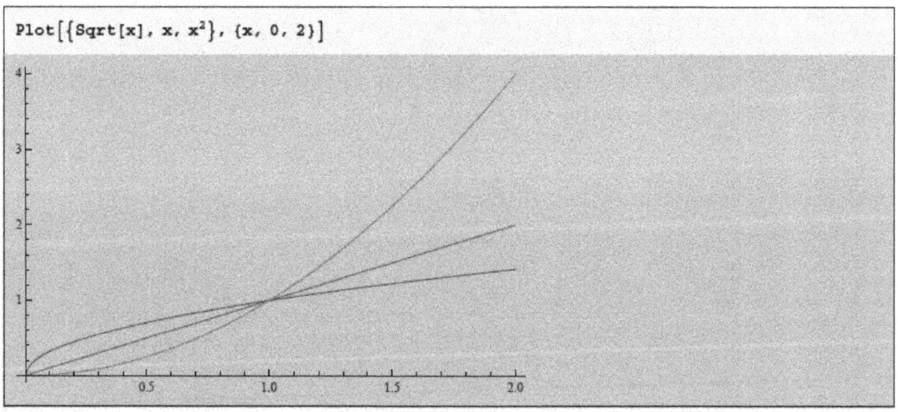

FIGURE 1.31

Plotting Graphs

Plot a dotted curve on the plane joining the points (Figure 1.31 and 1.32):
Plot both graphs together(Figure 1.33).
Plot $p1$ next to $p2$ (Figure 1.34).
Plot a Network/Graph Plot (Figure 1.35).

```
list = Table[{x, Sin[x]}, {x, -2 Pi, 2 Pi, 0.1}];
p1 = ListPlot[list]
```

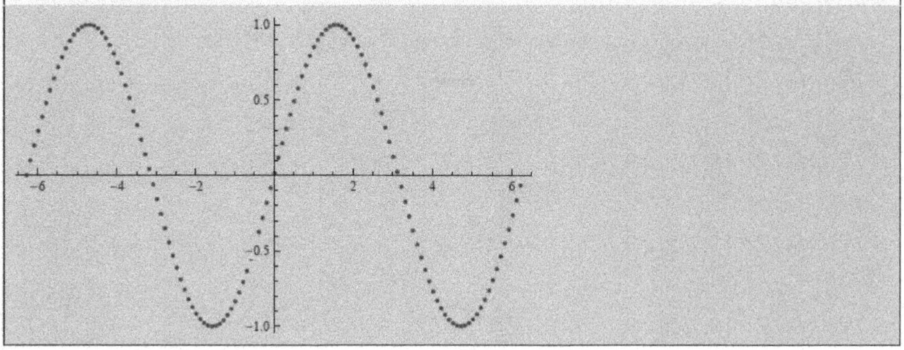

FIGURE 1.32

```
p2 = ListLinePlot[list]
```

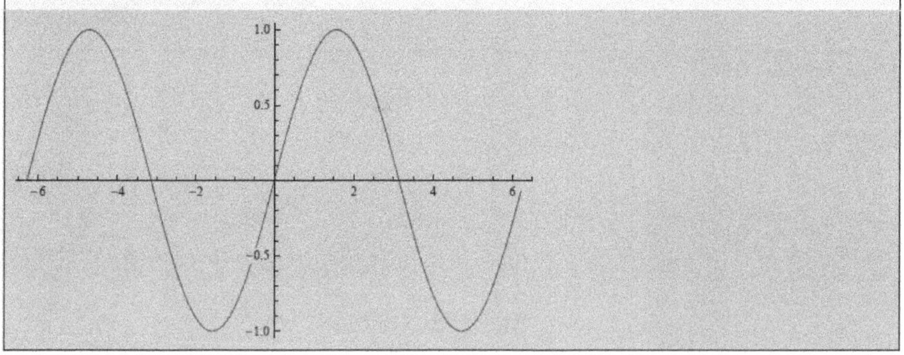

FIGURE 1.33

Plot function of one variable (Figure 1.36).
Plotting exponential graphs on log scaled vertical axis (Figure 1.37 and 1.38).
Plotting on log-scaled vertical and horizontal axis (Figure 1.39). Plot Parametric Equations (Figure 1.40). Plot function of two or more variables (Figure 1.41).
Plotting $\sin(x)\sin^2(y)$ (Figure 1.42).

```
Show[p1, p2]
```

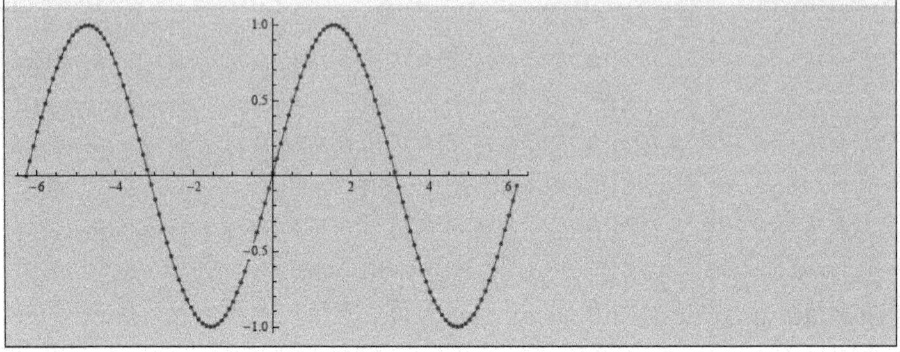

FIGURE 1.34

```
Show[GraphicsGrid[{{p1, p2}}]]
```

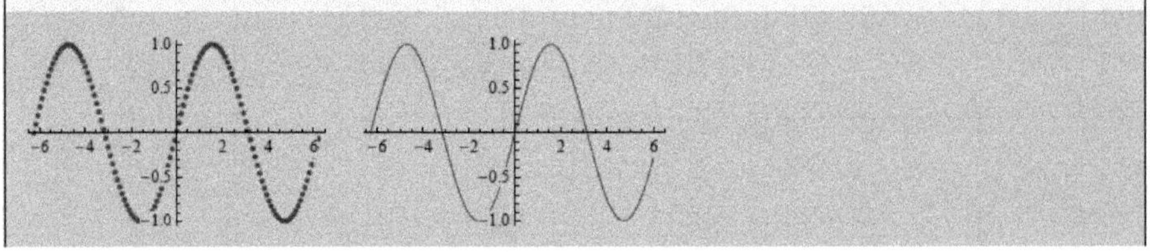

FIGURE 1.35

```
GraphPlot[{1 → 2, 1 → 4, 2 → 4, 3 → 4, 3 → 2, 3 → 5, 5 → 1}, VertexLabeling → True]
```

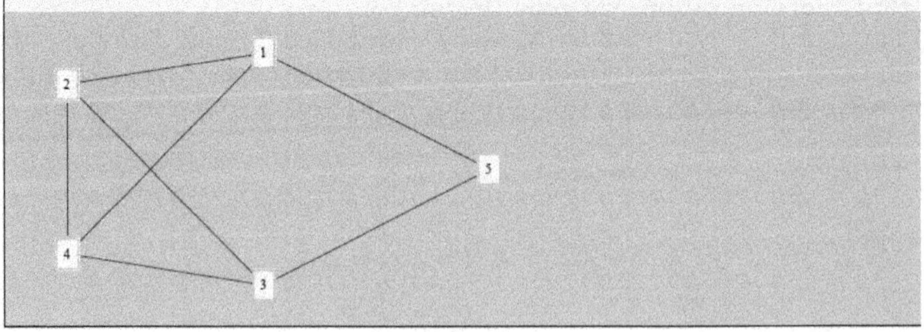

FIGURE 1.36

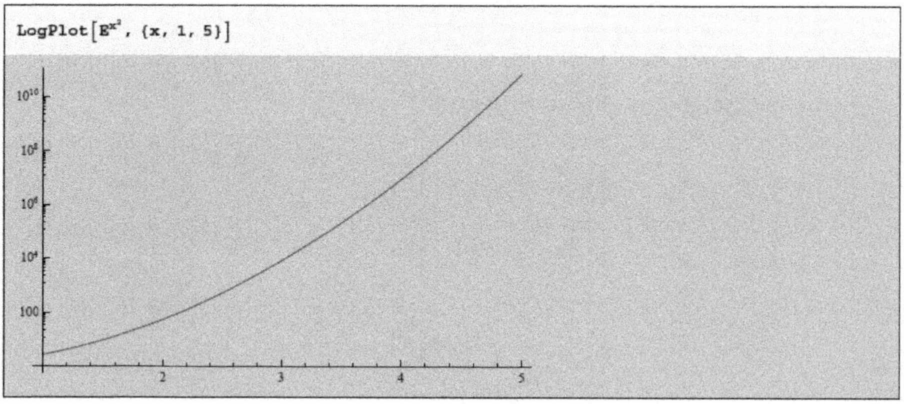

FIGURE 1.37

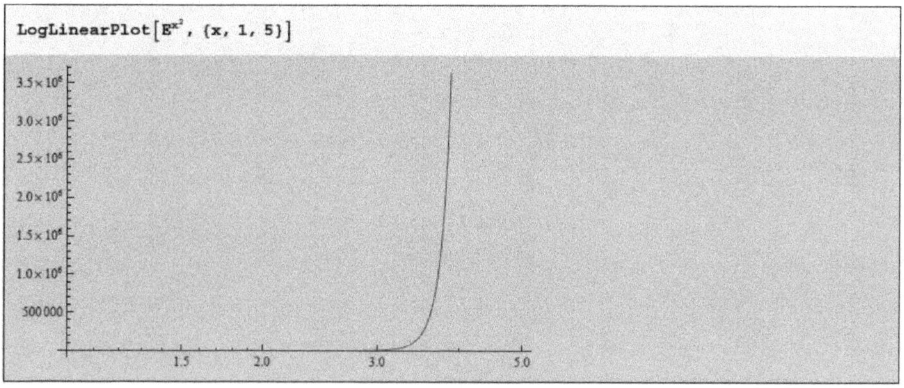

FIGURE 1.38

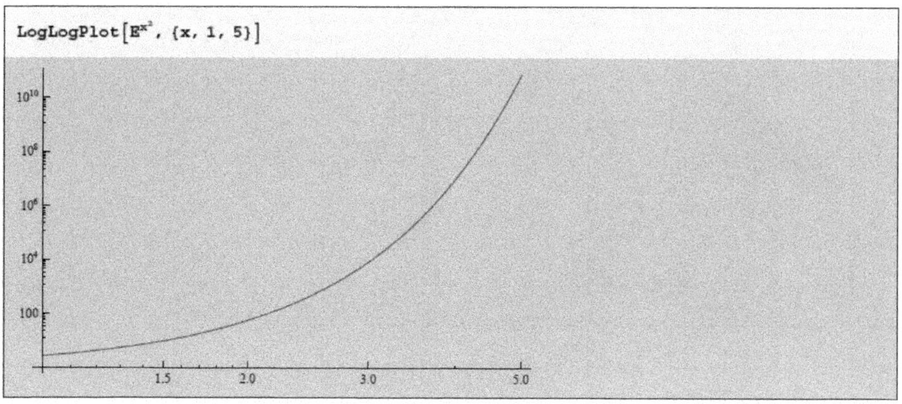

FIGURE 1.39

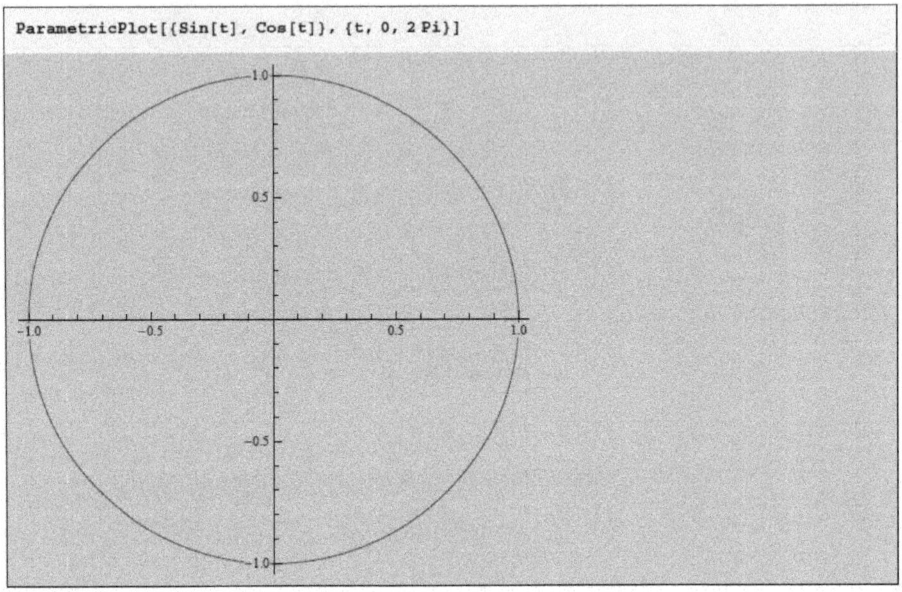

FIGURE 1.40

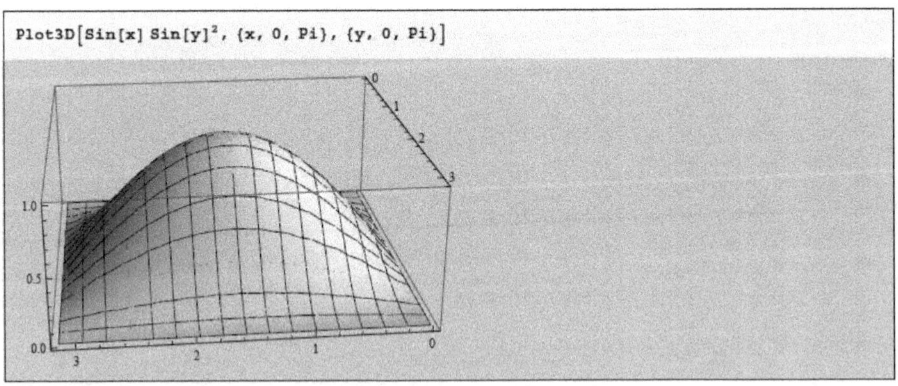

FIGURE 1.41

Plot Data

Plotting data by making table and customizing plot (Figure 1.43).
Fun with Plots

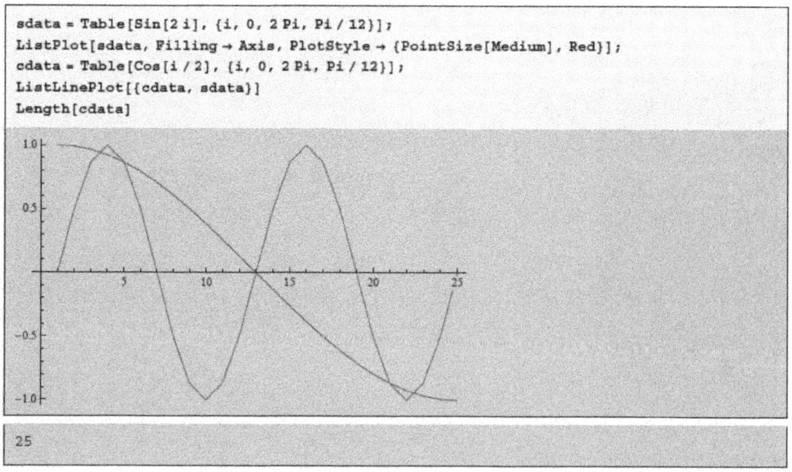

FIGURE 1.42

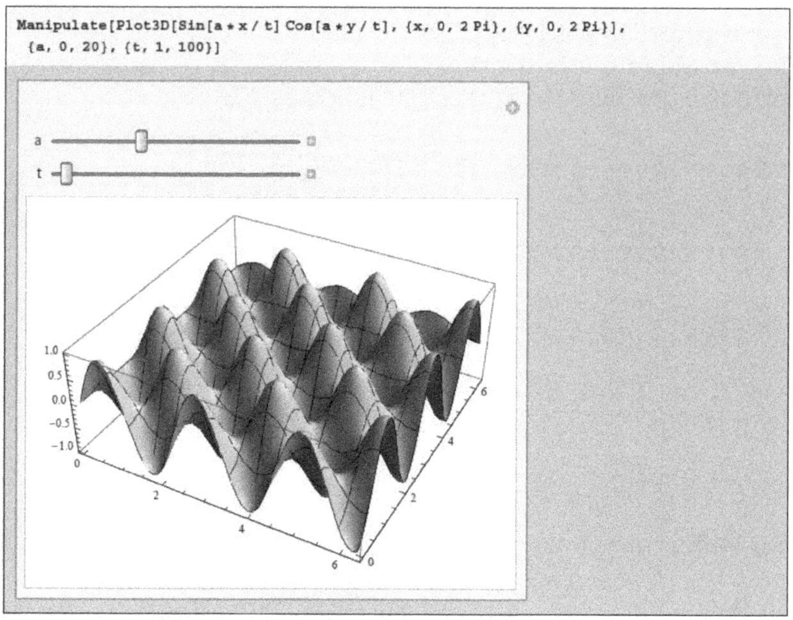

FIGURE 1.43

```
testdata = Table[{x, Sqrt[x]}, {x, 0, 4, 0.2}];
Manipulate[ListPlot[testdata, PlotMarkers → {Automatic, Size}],
 {Size, {Tiny, Small, Medium, Large}}, SaveDefinitions → True]
```

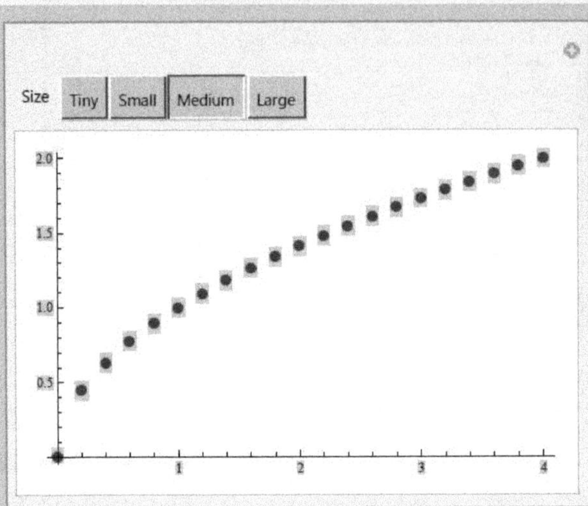

FIGURE 1.44

Differentiation Operations
Single Variable Differentiation
n^{th} derivative of a function $f(x)$: $D[f(x), \{x, n\}]$ (Figure 1.44).

Example 58. Find $\dfrac{d}{dx}[\sin(1 + 4x^2)]$ (Figure 1.45).

```
D[Sin[1 + 4 x^2], x]
```

```
8 x Cos[1 + 4 x^2]
```

FIGURE 1.45

Example 59. Find $\dfrac{d^2}{dx^2}[\sin(1 + 4x^2)]$ (Figure 1.46).

Partial Differentiation

Example 60. Find $\dfrac{d^3}{dxdy^2}[x^3y^2 + x\sin[y^2]]$ (Figure 1.47).

Integration Operations
Single Variable Integration

```
D[Sin[1 + 4 x^2], {x, 2}]

8 Cos[1 + 4 x^2] - 64 x^2 Sin[1 + 4 x^2]
```

FIGURE 1.46

```
D[x^3 y^2 + x Sin[y^2], {x, 1}, {y, 2}]

6 x^2 + 2 Cos[y^2] - 4 y^2 Sin[y^2]
```

FIGURE 1.47

Indefinite integral: Integrate$[f[x], x]$ (Figure 1.48).
Definite integral: Integrate $[f[x], \{x, a, b\}]$ (Figure 1.49).

Example 61. Evaluate $\int \sin(1 + 4x)dx$.

```
Integrate[Sin[1 + 4 x], x]

  1
- -- Cos[1 + 4 x]
  4
```

FIGURE 1.48

Example 62. Evaluate $\int_{x=0}^{\pi} \sin(1 + 4x)dx$.

```
Integrate[Sin[1 + 4 x], {x, 0, Pi}]

0
```

FIGURE 1.49

Multivariable Integration
Integrate $[f[x, y], \{x, a, b\}, \{y, c, d\}]$

Example 63. Evaluate $\int\int(x^2 \sin y + xy)dxdy$ (Figure 1.50 and 1.51)

Example 64. Evaluate $\int_{y=0}^{\pi} \int_{x=1}^{2}(x^2 \sin y + xy)dxdy$.

Solving Single and Quadratic Equations

Example 65. Solve $2x^2 - 3x + 1 = 0$ for x (Figure 1.52).

```
Integrate[x^2 Sin[y] + x * y, x, y]
```

$$\frac{1}{12} x^2 \left(3 y^2 - 4 x \operatorname{Cos}[y]\right)$$

FIGURE 1.50

```
Integrate[x^2 Sin[y] + x * y, {x, 1, 2}, {y, 0, Pi}]
```

$$\frac{14}{3} + \frac{3 \pi^2}{4}$$

FIGURE 1.51

```
Solve[2 x^2 - 3 x + 1 == 0, x]
```

$$\left\{\left\{x \to \frac{1}{2}\right\}, \{x \to 1\}\right\}$$

FIGURE 1.52

```
Solve[x^3 + x + 1 == 0, x][[1]]
```

$$\left\{x \to -\left(\frac{2}{3\left(-9+\sqrt{93}\right)}\right)^{1/3} + \frac{\left(\frac{1}{2}\left(-9+\sqrt{93}\right)\right)^{1/3}}{3^{2/3}}\right\}$$

FIGURE 1.53

cubic

Example 66. cubic (Solve $x^3 + x + 1 = 0$ for x (Figure 1.53).

Example 67. Solve a linear system for x and y.

$$\begin{aligned} ax - 2y &= -1 \\ bx - y &= 1 \end{aligned}$$

```
Solve[a x - 2 y == -1 && b x - y == 1, {x, y}]
```

$$\left\{\left\{x \to -\frac{3}{a - 2b}, y \to -\frac{a+b}{a - 2b}\right\}\right\}$$

FIGURE 1.54

Solving Single Equations Numerically
NSolve[expr,vars] – solving numerically (Figure 1.54)
NSolve[expr,vars,Reals] – attempts to find numerical approximations over the real number domain (Figure 1.55).

Example 68. Solve $x^5 - 2x + 1 = 0$ for x over the real number domain. Find Root $f(x), \{x, x_0\}]-\succ$ numerical root of $f(x)$ starting from x_0 (Figure 1.55 and 1.56).

```
NSolve[x^5 - 2 x + 1 == 0, x, Reals]
{{x → -1.29065}, {x → 0.51879}, {x → 1.}}
```

FIGURE 1.55

Example 69. Find all roots of $f(x) = x - \sin x$ near $x = \pi$.

```
FindRoot[x - Sin[x], {x, Pi}]
{x → 2.80119 × 10^-8}
```

FIGURE 1.56

Example 70. Find solution of $e^x = -x$ near $x = 0$.

```
FindRoot[Exp[x] == -x, {x, 0}]
{x → -0.567143}
```

FIGURE 1.57

Differential Equations
DSolve[eqn,y,{x1,x2..}] solves equations for the function $y = y(x1, x2..)$ (See Figures 1.58 through 1.60)

Example 71. Find general solution of $y"(x)+y'(x) = e^x$ subject to boundary conditions $y"(x) + y'(x) = e^x$ at $y(0) = 0, y'(0) = 1$.

Example 72. Solve: $2Z_{xx}(x,t) - 2Z_{tt}(x,t) = 1$.

Generating Matrices
Defining a Matrix (Figure 1.61).
Each individual list is the row of a matrix (Figure 1.62).

```
DSolve[D[y[x], {x, 2}] + y[x] == Exp[x], y[x], x]
```

$$\left\{\left\{y[x] \to C[1]\,\mathrm{Cos}[x] + C[2]\,\mathrm{Sin}[x] + \frac{1}{2} e^{x}\left(\mathrm{Cos}[x]^2 + \mathrm{Sin}[x]^2\right)\right\}\right\}$$

FIGURE 1.58

```
DSolve[{D[y[x], {x, 1}] + y[x] == Exp[x], y[0] == 0, y'[0] == 1}, y, x]
```

$$\left\{\left\{y \to \mathrm{Function}\left[\{x\}, \frac{1}{2} e^{-x}\left(-1 + e^{2x}\right)\right]\right\}\right\}$$

FIGURE 1.59

```
DSolve[3 D[Z[x, t], {x, 2}] - 2 D[Z[x, t], {t, 2}] == 1, Z, {x, t}]
```

$$\left\{\left\{Z \to \mathrm{Function}\left[\{x, t\}, \frac{x^2}{6} + C[1]\left[t - \sqrt{\frac{2}{3}}\,x\right] + C[2]\left[t + \sqrt{\frac{2}{3}}\,x\right]\right]\right\}\right\}$$

FIGURE 1.60

```
A = {{1, 0, 3, 4}, {3, 2, 0, 2}, {1, 1, 1, 1}}
{{1, 0, 3, 4}, {3, 2, 0, 2}, {1, 1, 1, 1}}
```

FIGURE 1.61

```
A = {{1, 0, 3, 4}, {3, 2, 0, 2}, {1, 1, 1, 1}}
{{1, 0, 3, 4}, {3, 2, 0, 2}, {1, 1, 1, 1}}
```

FIGURE 1.62

```
B = MatrixForm[Table[1 / (i + j), {i, 2}, {j, 3}]]
```

$$\begin{pmatrix} \frac{1}{2} & \frac{1}{3} & \frac{1}{4} \\ \frac{1}{3} & \frac{1}{4} & \frac{1}{5} \end{pmatrix}$$

FIGURE 1.63

Converting Data into Matrix (Figure 1.63 and 1.64).
Perform required addition and multiplication operations.

```
X = {{1, 3, 4}, {3, 0, 2}, {1, 1, 1}}; Y = {{2, 1, 4}, {1, 1, 3}, {1, 2, 1}};
X + Y
{{3, 4, 8}, {4, 1, 5}, {2, 3, 2}}
```

FIGURE 1.64

```
X.Y
{{9, 12, 17}, {8, 7, 14}, {4, 4, 8}}
```

FIGURE 1.65

Matrix Customization
Determinant of matrix (Figure 1.65)
Inverse of a Matrix (Figure 1.66)

```
A = {{1, 0, 3}, {3, 2, 2}, {1, 1, 2}}
{{1, 0, 3}, {3, 2, 2}, {1, 1, 2}}
```

```
Det[A]
5
```

FIGURE 1.66

Transpose of a Matrix (Figure 1.67)

```
Inverse[A]
```

$$\left\{\left\{\frac{2}{5}, \frac{3}{5}, -\frac{6}{5}\right\}, \left\{-\frac{4}{5}, -\frac{1}{5}, \frac{7}{5}\right\}, \left\{\frac{1}{5}, -\frac{1}{5}, \frac{2}{5}\right\}\right\}$$

FIGURE 1.67

Solving Linear Systems using Inverse Matrices

Example 73. Solve $3x_1 - 2x_2 + x_3 = -1$ $2x_1 + x_2 - x_3 = 0$ $3x_1 - 2x_2 - 3x_3 = 5$ (Figure 1.68).

```
Transpose[A]

{{1, 3, 1}, {0, 2, 1}, {3, 2, 2}}
```

FIGURE 1.68

```
A = {{3, -2, 1}, {2, 1, -1}, {3, -2, -3}}; b = {{-1}, {0}, {5}};
MatrixForm[b]
```

$$\begin{pmatrix} -1 \\ 0 \\ 5 \end{pmatrix}$$

FIGURE 1.69

Alternative Solution: Computing Fourier Series with MATHEMATICA

Refer to MATHEMATICA book titled. Let f be a function which on the

```
B = Inverse[A]
```

$$\left\{\left\{\frac{5}{28}, \frac{2}{7}, -\frac{1}{28}\right\}, \left\{-\frac{3}{28}, \frac{3}{7}, -\frac{5}{28}\right\}, \left\{\frac{1}{4}, 0, -\frac{1}{4}\right\}\right\}$$

```
Solution = B.b
```

$$\left\{\left\{-\frac{5}{14}\right\}, \left\{-\frac{11}{14}\right\}, \left\{-\frac{3}{2}\right\}\right\}$$

FIGURE 1.70

```
LinearSolve[A, b]
```

$$\left\{\left\{-\frac{5}{14}\right\}, \left\{-\frac{11}{14}\right\}, \left\{-\frac{3}{2}\right\}\right\}$$

FIGURE 1.71

interval $[-\pi, \pi]$ is given by $f(x) = |x|$.
Using MATHEMATICA solve the following problems

(a) Plot the function f on the interval $[-3\pi, 3\pi]$.

(b) Find the Fourier coefficients of f.

(c) Plot several Fourier partial sums $S_N f$.

Solution: (a) Define the function.
Replicate function over several periods:

```
ab[x_] := Piecewise[{{Abs[x], Abs[x] ≤ Pi}}]
Plot[ab[x], {x, -3 Pi, 3 Pi}, PlotRange → All]
```

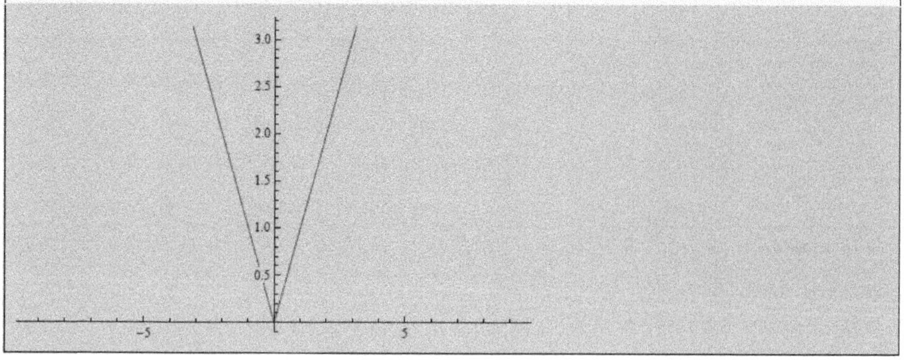

FIGURE 1.72

```
periodicextension = Sum[ab[x + 2 k Pi], {k, -4, 4}];
Plot[periodicextension, {x, -3 Pi, 3 Pi}, PlotRange → All]
```

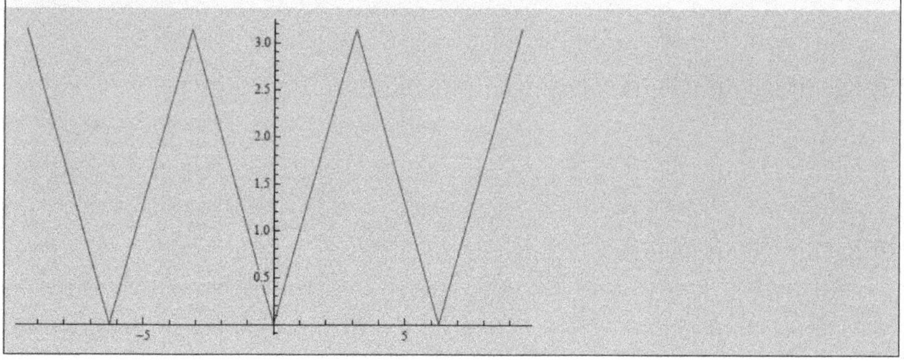

FIGURE 1.73

Solutions (b) and (c)
Fourier basis and inner product

Fourier Coefficients
Calculating nth partial sums
Plot of first and third partial sums of Fourier series of f, along with plot of function f.
Working with spreadsheets

```
s[n_, x_] := Sin[n x];
c[n_, x_] := Cos[n x];
IP[f_, g_] := 1 / Pi Integrate[f * g, {x, -Pi, Pi}];
```

FIGURE 1.74

```
a0[func_] := IP[func[x], c[0, x]];
aFC[func_, n_] := IP[func[x], c[n, x]];
bFC[func_, n_] := IP[func[x], s[n, x]];
```

FIGURE 1.75

```
fourierSeries[func_, N_, x_] :=
  a0[func] / 2 + Sum[aFC[func, n] c[n, x] + bFC[func, n] s[n, x], {n, 1, N}];
fs1[x_] = fourierSeries[ab, 1, x];
fs3[x_] = fourierSeries[ab, 3, x];
```

FIGURE 1.76

```
Plot[{periodicextension, fs1[x], fs3[x]}, {x, -2 Pi, 2 Pi}, PlotRange → All]
```

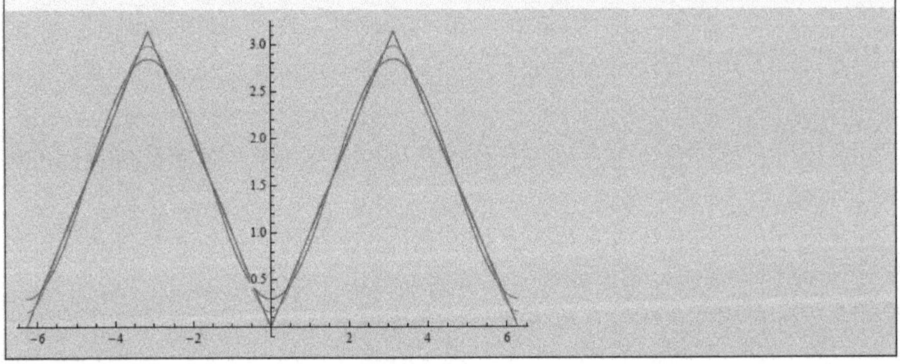

FIGURE 1.77

You can import spreadsheets created in a variety of formats to take advantage of MATHEMATICA's rich data manipulation and visualization capabilities. A spreadsheet is included in the Mathematica documentation folder.

Import only the first 3 rows of the spreadsheet: Import only row 3 of the spreadsheet:

```
s = Import["ExampleData/population.xls"]

{{{1.31397×10^9, China}, {1.09535×10^9, India},
  {2.98444×10^8, United States}, {2.45453×10^8, Indonesia},
  {1.88078×10^8, Brazil}, {1.65804×10^8, Pakistan}, {1.47365×10^8, Bangladesh},
  {1.42894×10^8, Russia}, {1.3186×10^8, Nigeria}, {1.27464×10^8, Japan}}}
```

FIGURE 1.78

```
s[[1]] // Grid

1.31397×10^9    China
1.09535×10^9    India
2.98444×10^8    United States
2.45453×10^8    Indonesia
1.88078×10^8    Brazil
1.65804×10^8    Pakistan
1.47365×10^8    Bangladesh
1.42894×10^8    Russia
1.3186×10^8     Nigeria
1.27464×10^8    Japan
```

FIGURE 1.79

```
Import["ExampleData/population.xls", {"Data", 1, {1, 2, 3}}]

{{1.31397×10^9, China}, {1.09535×10^9, India}, {2.98444×10^8, United States}}
```

FIGURE 1.80

```
Import["ExampleData/population.xls", {"Data", 1, 3}]

{2.98444×10^8, United States}
```

FIGURE 1.81

Data Manipulation

Plot data with explicit date values:

```
data =
  {{{2006, 10, 1}, 10}, {{2006, 10, 15}, 12}, {{2006, 10, 30}, 15}, {{2006, 11, 20}, 20}};
DateListPlot[
 data]
```

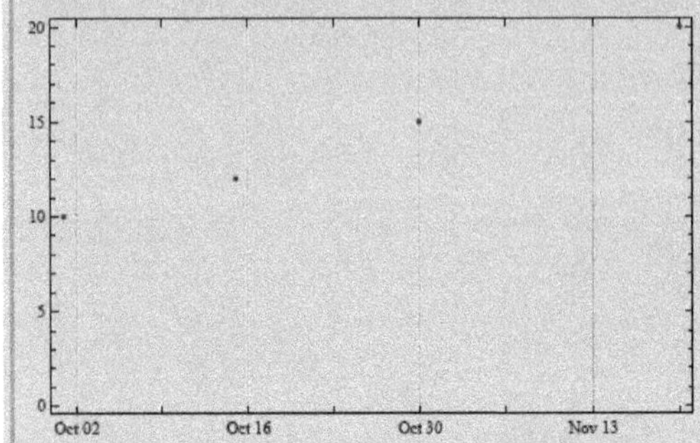

FIGURE 1.82

Plot monthly values starting in August 2000:

```
DateListPlot[{1, 1, 2, 3, 5, 8, 11}, {2000, 8}]
```

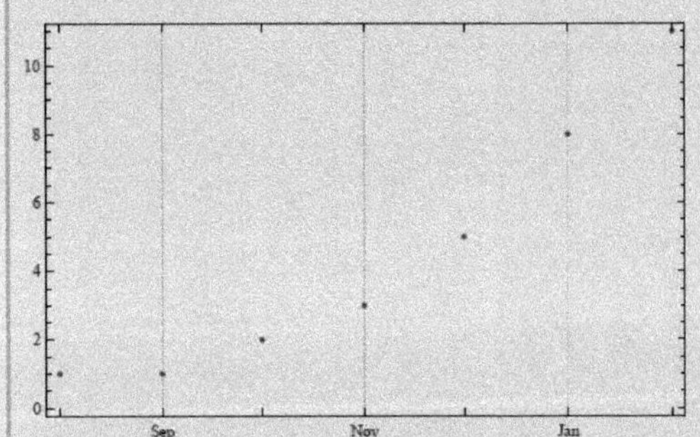

FIGURE 1.83

Plot multiple datasets: Retrieve and plot a historical stock price:

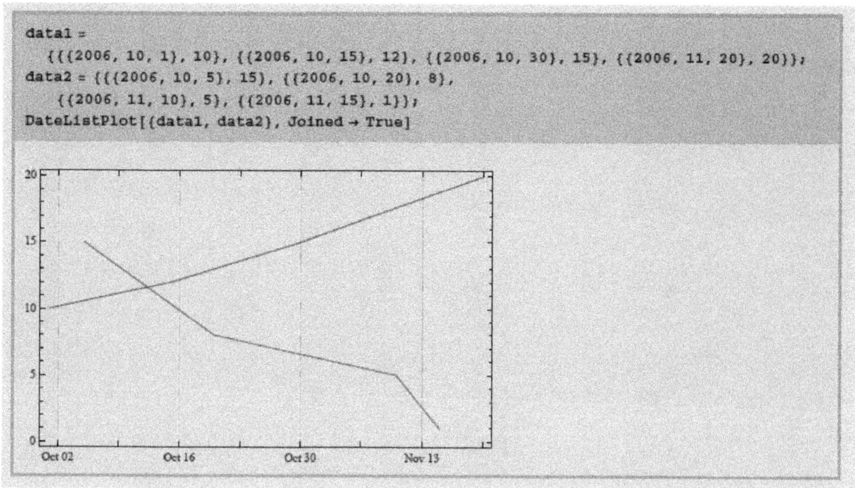

FIGURE 1.84

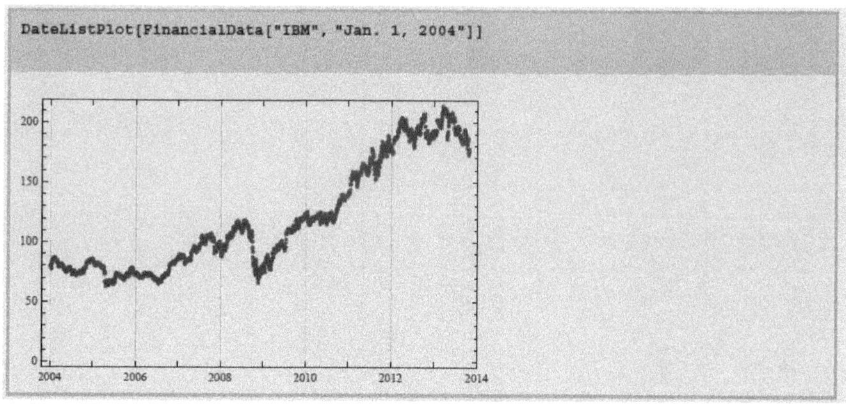

FIGURE 1.85

Scope

Dates given as DateString specifications:

Dates given as elided DateList specifications:

```
data = {{"June 2006", 10}, {"August 2006", 12},
   {"November 2006", 15}, {"January 2007", 20}};
DateListPlot[
 data]
```

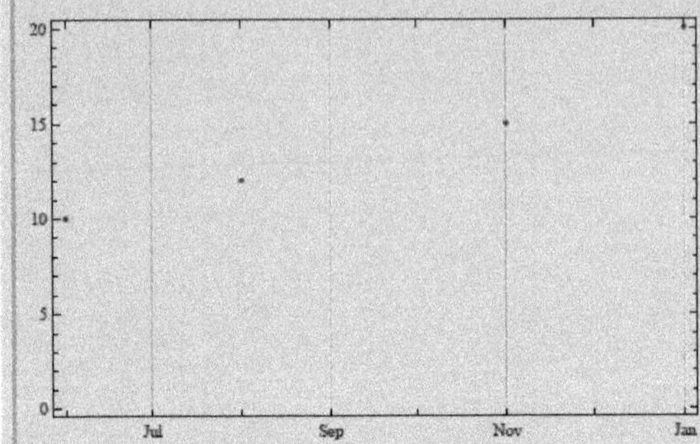

FIGURE 1.86

```
data = {{{2006, 6}, 10}, {{2006, 8}, 12}, {{2006, 11}, 15}, {{2007, 1}, 20}};
DateListPlot[data]
```

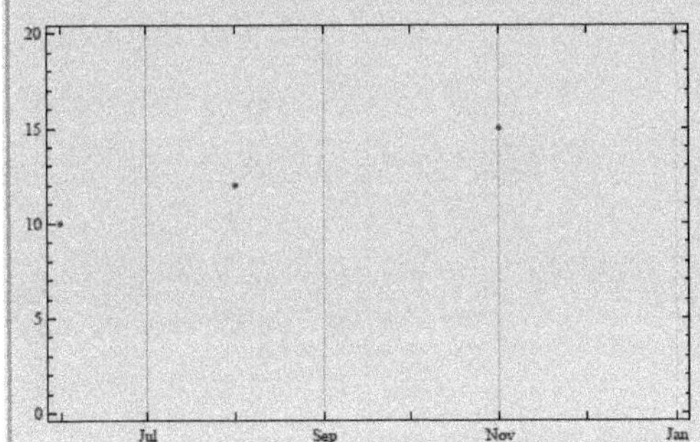

FIGURE 1.87

Plot a series of data using an initial starting date or time.

Applications

Get stock price data.

```
DateListPlot[Sin[Range[100] / (2 Pi)], "August 15, 2006"]
```

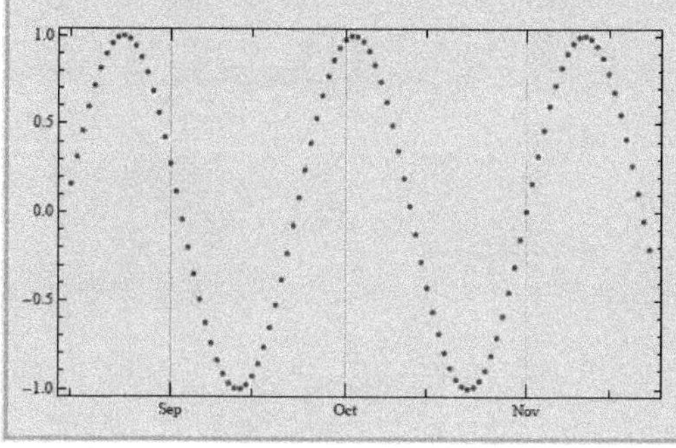

FIGURE 1.88

```
DateListPlot[Sin[Range[100] / (2 Pi)], {2006, 8, 15, 12, 15, 0}]
```

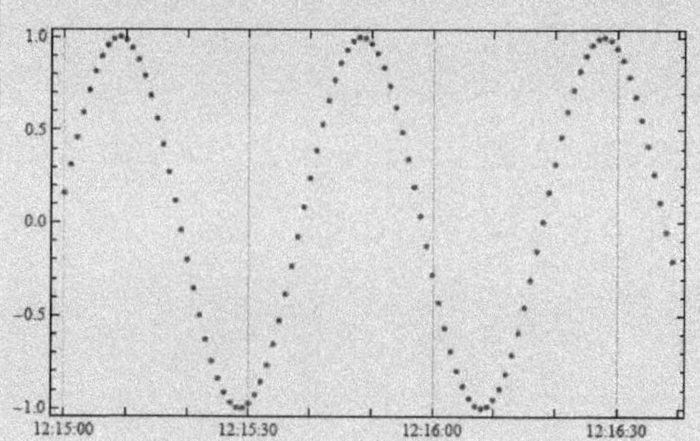

FIGURE 1.89

```
data = FinancialData["GE", {2000, 1, 1}];
```

FIGURE 1.90

A sample data point:
Plot data gathered at regular intervals and stored without explicit dates:

Annual oil consumption since 1980:

```
data[[1]]

{{2000, 1, 3}, 33.14}
```

FIGURE 1.91

```
DateListPlot[data, Joined → True, Filling → Bottom]
```

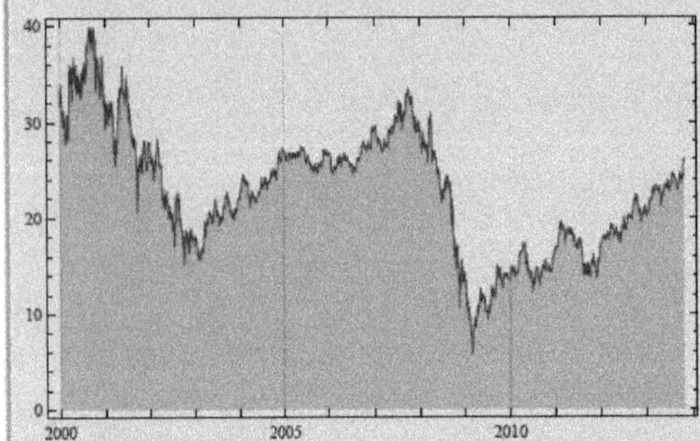

FIGURE 1.92

```
data = {56.1, 60.7, 51.6, 52., 57.5, 56.7, 67.4, 69.9, 72.9, 69.7, 70.3, 72.1};
DateListPlot[data, {{2006, 6, 1, 8}, Automatic, "Hour"}]
```

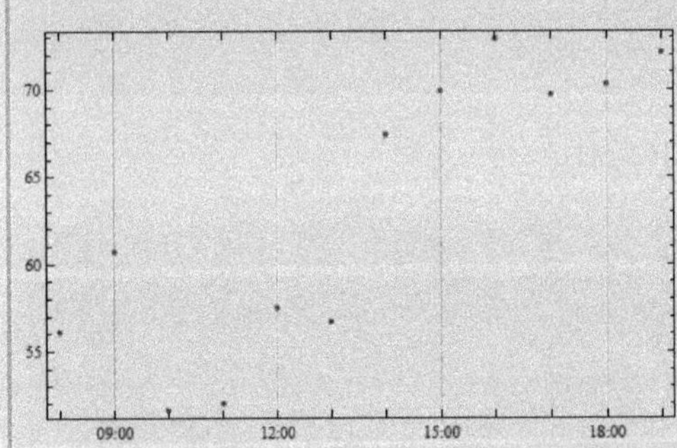

FIGURE 1.93

1.9 Exercises

1.1. Let $\mathbf{a} = (1, -3, 2)$, $\mathbf{b} = (-1, 1, 1)$ and $\mathbf{c} = (2, 6, 9)$.
Find (i) $\mathbf{a} + (\mathbf{b} + \mathbf{c})$ (ii) $4\mathbf{a} - (\mathbf{b} + \mathbf{c})$ (iii) $\mathbf{c} + 2(\mathbf{a} - 3\mathbf{b})$

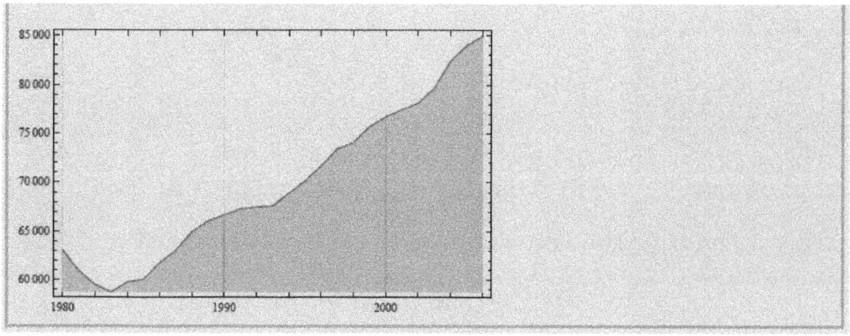

FIGURE 1.94

(iv) $4(\mathbf{a}+2\mathbf{c})-6\mathbf{b}$ (v) $||\mathbf{a}-\mathbf{c}||$ (vi) $||\mathbf{a}||||2\mathbf{c}||$
(vii) $||\frac{\mathbf{a}}{||\mathbf{a}||}||+||\frac{\mathbf{b}}{||\mathbf{b}||}||$ (viii) $||\mathbf{c}||\mathbf{a}+||\mathbf{b}||\mathbf{a}$.

1.2. (i) Find a unit vector in the opposite direction of $\mathbf{a}=(5,-5,20)$.
(ii) Find a unit vector in the same direction as $\mathbf{a}=i-3j+2k$.
(iii) Find a vector $\mathbf{b}$ that is 5 times as long as $\mathbf{a}=i-j+k$ in the same direction as a.

1.3. Let $\mathbf{a}=(2,-3,4)$, $\mathbf{b}=(-1,2,5)$ and $\mathbf{c}=(3,6,-1)$.

(i) Find $\mathbf{b}.\mathbf{a}=\langle\mathbf{b},\mathbf{a}\rangle$
(ii) $\mathbf{a}.(6\mathbf{b})=\langle a,6b\rangle$
(iii) $\mathbf{b}.\mathbf{b}$ and $\mathbf{c}.\mathbf{c}$
(iv) $\mathbf{a}.(\mathbf{a}+\mathbf{b}+\mathbf{c})$
(v) $\left(\frac{\mathbf{a}.\mathbf{b}}{\mathbf{b}.\mathbf{b}}\right)\mathbf{b}$.

1.4. Find the orthogonal vectors from the following vectors:
(i) $(2,0,1)$ (ii) $2i-j-k$ (iii) $(-4,3,8)$
(iv) $i-4j+6k$ (v) $(1,-1,1)$ (vi) $(4,-3,8)$
Indicate the pair of vectors which are non-orthogonal.

1.5. Find a scalar α so that given vectors are orthogonal:
(i) $\mathbf{a}=2i-\alpha j+3k$, $\mathbf{b}=3i+2j+4k$.
(ii) $\mathbf{a}=\alpha i+\frac{1}{2}j+\alpha k$, $\mathbf{b}=-3i+4j+\alpha k$.

1.6. Show that the vector $\mathbf{c}=\mathbf{b}-\frac{\mathbf{a}.\mathbf{b}}{||\mathbf{a}||^2}\mathbf{a}$ is orthogonal to the vector $\mathbf{a}$.

1.7. Find the angle between following vectors

(i) $\mathbf{a}=3i-k$, $\mathbf{b}=3i+3k$.

(ii) $\mathbf{a} = 2i + 4j + 0k$, $\mathbf{b} = -i - j + 4k$.

(iii) $\mathbf{a} = \left(\frac{1}{2}, \frac{1}{2}, \frac{3}{2}\right)$, $\mathbf{b} = (2, -4, 6)$.

1.8. Let $\mathbf{a} = (1, -1, 3)$ and $\mathbf{b} = (2, 6, 3)$. Find
(i) $comp_{\mathbf{b}}\mathbf{a}$ (ii) $comp_{\mathbf{a}}\mathbf{b}$ (iii) $comp_{\mathbf{a}}\mathbf{b} - \mathbf{a}$.

1.9. Let $\mathbf{a}$ and $\mathbf{b}$ be two vectors and α and β be scalars, then show that

$$||\alpha\mathbf{a} + \beta\mathbf{b}||^2 = \alpha||\mathbf{a}||^2 + \beta||\mathbf{b}||^2 + 2\alpha\beta\mathbf{a}.\mathbf{b}.$$

1.10. Find the angle between two vectors $\mathbf{a} = (-1, 3, 1)$ and $\mathbf{b} = (0, 2, -4)$.

1.11. Let $\mathbf{a} = (5, 2, -1)$, $\mathbf{b} = (2, 0, -7)$. Compute $\mathbf{a}.(\mathbf{a} \times \mathbf{b})$ and $\mathbf{b}.(\mathbf{a} \times \mathbf{b})$.

1.12. Examine whether the following expression are scalar or vector
(i) $i \times (-3k)$ (ii) $i \times (j \times k)$ (iii) $(2i - j + 5k) \times i$
(iv) $k.(j \times k)$ (v) $||4j - 5(i \times j)||$ (vi) $(i \times k) \times (j \times i)$.

1.13. Determine whether the following vectors are linearly independent or linearly dependent:

(i) (4,-8), (-6,12) in R^2.

(ii) (1,1), (0,1), (2,5) in R^2.

(iii) 1, $(x + 1)$, $(x + 1)^2$ in P_2.

1.14. Construct an orthonormal basis for R^2 from the given basis:

(i) $S = \{(-3, 2), (-1, -1)\}$.

(ii) $S = \{(1, 1), (1, 0)\}$.

(iii) $S = \{(5, 7), (1, -2)\}$.

1.15. Using Gram-Schmidt orthogonalization, construct an orthonormal basis for R^3 from the following bases:

(i) $S = \{(1, 1, 0), (1, 2, 2), (2, 2, 1)\}$.

(ii) $S = \{(1, 1, 1), (9, -1, 1), (-1, 4, -2)\}$.

1.16. State the size of the three matrices below:

(i) $\begin{pmatrix} 1 & 2 & 3 & 18 \\ 5 & 6 & 0 & 2 \end{pmatrix}$ (ii) $\begin{pmatrix} 1 & 2 \\ 8 & 4 \\ 5 & 12 \end{pmatrix}$ (iii) $(6, 8, -30)$ (iv) $\begin{pmatrix} 2 \\ 5 \\ -6 \\ 0 \\ 7 \\ -11 \\ 2 \\ 12 \end{pmatrix}$.

1.17. Determine whether the following matrices are equal:
$$\begin{pmatrix} 1 & 2 & 3 \\ 4 & 5 & 6 \end{pmatrix}, \begin{pmatrix} 1 & 4 \\ 2 & 5 \\ 3 & 6 \end{pmatrix}, \begin{pmatrix} 1 & 2 \\ 0 & 1 \end{pmatrix}, \begin{pmatrix} 1 & 0 \\ 2 & 1 \end{pmatrix}, \begin{pmatrix} 1 & 0 \\ 0 & 1 \end{pmatrix}.$$

1.18. Find values of x and y for which the following matrices are equal:
$$\begin{pmatrix} 1 & x \\ y & -3 \end{pmatrix} \text{ and } \begin{pmatrix} 1 & y-2 \\ 3x-2 & -3 \end{pmatrix}.$$

1.19. Find entries $c_{2,3}$ and $c_{1,2}$ for the matrix $C = 2A - 3B$, where

(i) $A = \begin{pmatrix} 2 & 3 & -1 \\ -1 & 6 & 0 \end{pmatrix}, B = \begin{pmatrix} 4 & -2 & 6 \\ 1 & 3 & -3 \end{pmatrix}$

(ii) $A = \begin{pmatrix} 1 & -1 & 1 \\ 2 & 2 & 1 \\ 0 & -4 & 1 \end{pmatrix}, B = \begin{pmatrix} 2 & 0 & 4 \\ 0 & 4 & 0 \\ 3 & 0 & 7 \end{pmatrix}.$

1.20. If $A = \begin{pmatrix} 2 & -3 \\ -5 & 4 \end{pmatrix}$ and $B = \begin{pmatrix} -1 & 6 \\ 3 & 2 \end{pmatrix}$, then find

(i) AB (ii) $A^2 = AA$ (iii) $B^2 = BB$ (iv) BA.

1.21. Suppose that $A = \begin{pmatrix} 2 & 4 \\ -3 & 2 \end{pmatrix}$ and $B = \begin{pmatrix} 4 & 10 \\ 2 & 5 \end{pmatrix}$. Check the following properties for those two matrices:

(i) $(A^T)^T = A$ (ii) $(A+B)^T = A^T + B^T$

(iii) $(AB)^T = B^T A^T$ (iv) $(6A)^T = 6A^T$

1.22. Find two matrices A and B such that $AB = 0$ but $A \neq 0$ and $B \neq 0$.

1.23. Write the systems of equations for:
$$\begin{aligned} 2x_1 + 6x_2 + x_3 &= 7 \\ x_1 + 2x_2 - x_3 &= -1 \\ 5x_1 + 7x_2 - 4x_3 &= 9 \end{aligned}$$

as a matrix equation $AX = B$, where X and B are column vectors.

1.24. Solve the following systems of equations

(i) $10x_1 + 15x_2 = 1.$
$3x_1 + 2x_2 = -1.$

(ii) $x_1 - x_2 - x_3 = 8.$
$x_1 - x_2 + x_3 = 3.$
$-x_1 + x_2 + x_3 = 4.$

1.25. Balance the following chemical equations

(i) $Na + H_2O \rightarrow NaOH + H_2$

(ii) $C_5H_8 + O_2 \rightarrow CO_2 + H_2O$

1.26. Find the rank of the following matrices:

(i) $\begin{pmatrix} 2 & -2 \\ 0 & 0 \end{pmatrix}$ (ii) $\begin{pmatrix} 1 & 1 & 1 \\ 1 & 0 & 4 \\ 1 & 4 & 1 \end{pmatrix}$ (iii) $\begin{pmatrix} 3 & -1 & 2 & 0 \\ 6 & 2 & 4 & 5 \end{pmatrix}$.

1.27. An elementary matrix E is obtained by performing a single row operation on the identity matrix I. Verify that the following matrices are elementary.

(i) $\begin{pmatrix} 0 & 1 & 0 \\ 1 & 0 & 0 \\ 0 & 0 & 1 \end{pmatrix}$ (ii) $\begin{pmatrix} 1 & 0 & 0 \\ 0 & 1 & 0 \\ 0 & 0 & 1 \end{pmatrix}$ (iii) $\begin{pmatrix} 1 & 0 & 0 & 1 \\ 0 & 1 & 0 & 0 \\ 0 & 0 & 1 & 0 \\ 0 & 0 & 0 & 1 \end{pmatrix}$.

1.28. Set up and solve the system of equation for the currents in the branches of the network shown below.

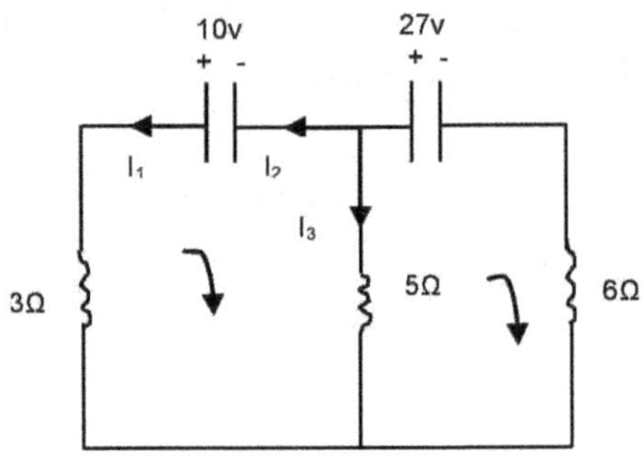

FIGURE 1.95: Electric Network

1.29. Examine whether the given set of vectors is linearly dependent or linearly independent.

(i) $\mathbf{u}_1 = (1, -1, 3, -1),\ \mathbf{u}_2 = (1, -1, 4, 2),\ \mathbf{u}_3 = (1, -1, 5, 7)$.

(ii) $\mathbf{u}_1 = (1, 2, 3),\ \mathbf{u}_2 = (1, 0, 1),\ \mathbf{u}_3 = (1, -1, 5)$.

(iii) $\mathbf{u}_1 = (2, 1, 1, 5), \mathbf{u}_2 = (2, 2, 1, 1), \mathbf{u}_3 = (3, -1, 6, 1), \mathbf{u}_4 = (1, 1, 1{-}1)$.

1.30. Let $A = \begin{pmatrix} 2 & 3 & 4 \\ 1 & -1 & 2 \\ -1 & 3 & 5 \end{pmatrix}$. Find the indicated minor, determinant or cofactor: $M_{12}, 2M_{32}, c_{13}, c_{22}, c_{23}$.

1.31. Let $A = \begin{pmatrix} 0 & 2 & 0 \\ 3 & 0 & 1 \\ 0 & 5 & 8 \end{pmatrix}$, $B = \begin{pmatrix} 1 & -1 & -1 \\ 2 & 2 & -2 \\ 1 & 1 & 9 \end{pmatrix}$. Find $detA$ and $detB$.

1.32. Let $A = \begin{pmatrix} a_{11} & a_{12} \\ a_{21} & a_{22} \end{pmatrix}$ and $B = \begin{pmatrix} a_{21} & a_{22} \\ a_{11} & a_{12} \end{pmatrix}$ Show that $detA = -detB$.

1.33. Let $A = \begin{pmatrix} a_{11} & a_{12} \\ a_{21} & a_{22} \end{pmatrix}$. Show that $detA = detA^T$.

1.34. Let $A = \begin{pmatrix} 2 & -1 & 1 \\ 3 & 1 & -1 \\ 0 & 2 & 2 \end{pmatrix}$ and $B = \begin{pmatrix} 2 & 1 & 5 \\ 4 & 3 & 8 \\ 0 & -1 & 0 \end{pmatrix}$ Verify that $detAB = detAdetB$.

1.35. Suppose that A is $n \times n$ matrix such that $A^2 = I$ where $A^2 = AA$. Show that $detA = \pm 1$.

1.36. Check whether the following matrices are invertible, if invertible find the inverse.

(i) $\begin{pmatrix} 5 & -1 \\ 4 & 1 \end{pmatrix}$ (ii) $\begin{pmatrix} 6 & -2 \\ 0 & 4 \end{pmatrix}$

(iii) $\begin{pmatrix} 1 & 2 & 3 \\ 0 & -4 & 2 \\ -1 & 5 & 1 \end{pmatrix}$ (iv) $\begin{pmatrix} 1 & 0 & 0 & 0 \\ 0 & 0 & 1 & 0 \\ 0 & 0 & 0 & 1 \\ 0 & 1 & 0 & 0 \end{pmatrix}$.

1.37. If $A^{-1} = \begin{pmatrix} 4 & 3 \\ 3 & 2 \end{pmatrix}$, then find A.

1.38. Find a value of α such that the matrix $A = \begin{pmatrix} 4 & -3 \\ \alpha & -4 \end{pmatrix}$ is its own inverse.

1.39. Find the inverse of $A = \begin{pmatrix} \sin\theta & \cos\theta \\ -\cos\theta & \sin\theta \end{pmatrix}$.

1.40. Solve the following systems applying Cramer's rule or indicate that the rule does not apply

(i)
$$\begin{aligned} 15x_1 - 4x_2 &= 5 \\ 8x_1 + x_2 &= -4 \end{aligned}$$

(ii)
$$\begin{aligned} 8x_1 - 4x_2 + 3x_3 &= 0 \\ x_1 + 5x_2 - x_3 &= -5 \\ -2x_1 + 6x_2 + x_3 &= -4 \end{aligned}$$

(iii) $x_1 + x_2 - 3x_3 = 0$

$$x_2 - 4x_3 = 0$$

$$x_1 - x_2 - x_3 = 5$$

(iv) $x_1 - 2x_2 - 3x_3 = 3$

$$x_1 + x_2 - x_3 = 5$$

$$3x_1 + 2x_2 = -4$$

1.41. (a) Find the eigenvalues of the matrix given below.
(b) Corresponding to each eigenvalue, find an eigenvector

(i) $\begin{pmatrix} -2 & 0 \\ 1 & 4 \end{pmatrix}$ (ii) $\begin{pmatrix} 1 & -6 \\ 2 & 2 \end{pmatrix}$ (iii) $\begin{pmatrix} 0 & 1 \\ 0 & 0 \end{pmatrix}$

(iv) $\begin{pmatrix} 2 & 0 & 0 \\ 1 & 0 & 2 \\ 0 & 0 & 3 \end{pmatrix}$ (v) $\begin{pmatrix} 1 & -2 & 0 \\ 0 & 0 & 0 \\ -5 & 0 & 7 \end{pmatrix}$.

1.42. Find the general terms of all 2×2 matrices with real elements having eigenvalues 4 and -2.

1.43. For each matrix below, produce a matrix that diagonalizes it or show that this matrix is not diagonalizable.

(i) $\begin{pmatrix} 5 & 3 \\ 1 & 3 \end{pmatrix}$ (ii) $\begin{pmatrix} 1 & 0 \\ -4 & 1 \end{pmatrix}$ (iii) $\begin{pmatrix} 5 & 0 & 0 \\ 1 & 0 & 3 \\ 0 & 0 & -2 \end{pmatrix}$.

1.44. Find the eigenvalues of the matrices below and for each value, a corresponding eigenvector. Verify that eigenvectors associated with distinct eigenvalues are orthogonal. Find an orthogonal matrix that diagonalizes each matrix.

(i) $\begin{pmatrix} 4 & -2 \\ -2 & 1 \end{pmatrix}$ (ii) $\begin{pmatrix} -13 & 1 \\ 1 & 4 \end{pmatrix}$ (iii) $\begin{pmatrix} 5 & 0 & 2 \\ 0 & 0 & 0 \\ 2 & 0 & 0 \end{pmatrix}$ (iv) $\begin{pmatrix} 1 & 3 & 0 \\ 3 & 0 & 1 \\ 0 & 1 & 1 \end{pmatrix}$.

1.44. Discuss the use of matrices in rating world cities based on a number of variables such as cost of living, quality of education, availability and security of jobs, extent of cultural and recreation opportunities, transportation, health facilities, air quality, safety and crime, and climate.

1.45. Show that Markov matrix

$$A = \begin{pmatrix} 0 & 1 & 0 \\ 0 & 0 & 1 \\ 1 & 0 & 0 \end{pmatrix}$$

has complex eigenvalues and is orthogonal.

1.46. Prove that eigenvalue λ of a Markov matrix satisfies $|\lambda| \leq 1$.

1.47. Let A be Markov matrix with all positive elements, then show that λ is the only eigenvalue of modulus 1 and $(A - I_n) = 1$.

1.48. Describe application of Markov chain in A. S. Pushkin's "Yevgeniy Onegin" poem.

1.49. Discuss relevance of Markov matrix and Markov chain to Shannon's communication theory.

1.50. Write an essay on application of Markov chain and Markov matrices to compute performance evaluation, page rank and Web search.

1.10 Suggestion for Further Reading

Computational science, also called scientific computing or scientific computation, is a growing multidisciplinary field that uses advanced computing capabilities to explain complex problems. The scientific computing programme aims to produce highly skilled computational scientists and engineers capable of applying numerical methods and critical evaluation of their results to their field. Scientific computing is introduced clearly by Heath [5].

Image processing is an important ingredient of modern engineering and technology. Interested readers may pursue references [2] and [3] for acquiring deeper knowledge. References [1, 4, 8, 11] are also useful.

Bibliography

[1] G. Burril, J. Burril, J. Landwehr, J. Witmer, *Advanced Modeling and Matrices*, Dale Seymour Publications, 1988.

[2] J. Gomes, L. Velho, *Image Processing for Computer Graphics and Vision*, Springer, 2008.

[3] R. C. Gonzalaez, R.E. Woods, *Digital Image Processing*, Third Edition, Prentice Hall, 2007.

[4] K. Hardy, *Liner Algebra for Engineers and Scientists Using MATLAB*, Pearson, Prentice Hall, 2005.

[5] M. T. Heath, *Scientific Computing: An Introductory Survey*, Mcgraw Hill, Second Edition, 2002.

[6] B. Kolman and D. R. Hill, *Elementary Linear Algebra with Applications*, Ninth Edition, Pearson International, 2008.

[7] D. Lay, *Linear Algebra, and Its Applications*, Fourth Edition, Addison Wesley, 2011.

[8] S. J. Leon, *Linear Algebra with Applications*, Seventh Edition, Pearson Prentice Hall, 2006.

[9] H. Neunzert and A. H. Siddiqi, *Topics in Industrial Mathematics: Case Studies and Related Mathematical Methods*, Kluwer Academic Publishers, 2000.

[10] P. V. Hilgers and A. N. Langville, *The Five Greatest Applications of Markov Chains*, Langvillea. people.cofe.edu/Mcapps7pdf.

[11] D. G. Zill and M. R. Cullen, *Advanced Engineering Mathematics*, Jones and Barlett Publishers, 2012.

Chapter 2

Differential Equations

2.1 Introduction

Mathematics has been considered the language of nature since the days of Galileo and in modern times it is considered the mother of all technologies. Differential equations were invented in the course of investigations of the laws that govern the physical world. Sir Isaac Newton (1642-1727) first solved them in the 17th century and called them fluxional equations. Gottfried Leibnitz, a contemporary of Newton credited with the invention of calculus, renamed fluxional equations as differential equations, a name that remains in use today.

Modern methods for solving differential equations and their practical applications constitute major portions of mathematics curricula for engineering and science students. Leibnitz invented some of the methods but the general theory for differential equations was developed by an electrical engineer, Augustin-Louis Cauchy (1789–1857).

In the 20th century, several Nobel Prizes were awarded for research that involved differential equations. For example, William Libby received the prize for chemistry in 1960 for the invention of carbon dating. The 1952 physics prize was awarded to Felix Block and Edward Purcell. They calculated the rates of change over time of magnetic moments of nuclei at each point in a sample. The 1979 prize for medicine and physiology went to Allan Cormack and Godfrey Hounsfield for their work on computerized tomography (CT). Robert Merton and Myron Scholes received the 1997 economics prize for their work with Fisher Black on the Black-Scholes option pricing model.

Differential equations play a central role in describing phenomena that involve changes. The main goal of this chapter is to introduce basic methods for solving ordinary differential equations and developing modeling techniques for application to real world problems. Methods for dealing with separation of variables, undetermined coefficients, parameter variations, Cauchy-Euler and Laplace transform methods, and series solutions are some of the classical methods covered in this chapter. MATLAB and MATHEMATICA solutions of differential equations are also discussed.

These methods are more efficient than analytical approximations based on power series. Therefore Section 2.3 (MATLAB and MATHEMATICA for Differential Equations) may be preferred over Section 2.8 (Series Solution of Differential Equations) in some situations. The applications of modeling techniques in various fields such as environmental problems, nuclear physics, electricity and electronics, mechanics, and biomedical engineering are presented in Section 2.6.

2.1.1 Definitions and Terminology

Definition 30. Differential Equation
An equation containing the derivatives of one or more dependent variables is said to be *differential.* Many principles underlying behaviors in the natural world involve the rates at which events (chemical reactions and disease progression for example) happen. Mathematical formulations of laws containing derivatives are differential equations (DEs). The study of DEs is essential to the proper understanding of fluid motion, current flows through electric circuits, heat dissipation, detection of seismic waves, population dynamics, spreading of diseases and even spreading of rumors. DEs that represent physical processes are often called mathematical models and they are discussed throughout this chapter.

Definition 31. Ordinary Differential Equation
A differential equation is said to be an ordinary differential equation (ODE) if it contains only ordinary derivatives of one or more dependent variables with respect to a single independent variable.

Definition 32. Partial Differential Equation
An equation involving the partial derivatives of one or more dependent vari-

ables of two or more independent variables is called a **partial differential equation** (PDE).

Example 74.

$$\frac{dy}{dx} + 20y = e^x, \quad \frac{d^2y}{dx^2} + 10y = \cos x$$

$$3\frac{d^2y}{dx^2} - 9\frac{dy}{dx} + 6y = 0, \quad \frac{dx}{dt} + \frac{dy}{dt} = 2x + y$$

are examples of differential equations.

Example 75.

$$\frac{\partial^2 u}{\partial t^2} = \frac{\partial^2 u}{\partial x^2}, \quad \frac{\partial u}{\partial x} = \frac{\partial^2 u}{\partial t^2}, \quad \frac{\partial^2 u}{\partial x^2} = -\frac{\partial^2 u}{\partial y^2}$$

$$\frac{\partial^2 u}{\partial x^2} = \frac{\partial^2 u}{\partial t^2} - 2\frac{\partial u}{\partial t}, \qquad \frac{\partial u}{\partial y} = -\frac{\partial u}{\partial x}, \quad \text{and} \quad \frac{\partial^2 u}{\partial x^2} = \frac{\partial^2 u}{\partial t^2} - 2\frac{\partial u}{\partial t}$$

are examples of partial differential equations.

Definition 33. Order of a Differential Equation
The **order** of differential equation (ODE or PDE) is the order of the highest derivative in the equation.

Example 76. (i) Order of the differential equation

$$\frac{d^2y}{dx^2} + 5\left(\frac{dy}{dx}\right)^4 - 4y = e^x \text{ is } 2.$$

(ii) Order of the differential equation

$$\frac{d^3y}{dx^3} + \frac{d^2y}{dx^2} - 4y = e^{-x} \text{ is } 3.$$

Definition 34. The **degree** of a differential equation is the degree of the highest order derivative in the equation.

Example 77. (i) The degree of ODE

$$y\frac{dy}{dx} - \sqrt{x}\left(\frac{dy}{dx}\right)^3 - 5 = 0 \text{ is } 3.$$

(ii) The degree of ODE

$$\left(\frac{d^2y}{dx^2}\right)^3 - \left(\frac{dy}{dx}\right)^2 + 6y + 10 = 0 \text{ is } 3.$$

Remark 17. (i) Very often notations $y', y'', y''', \ldots, y^n$ are respectively used for

$$\frac{dy}{dx}, \frac{d^2y}{dx^2}, \frac{d^3y}{dx^3}, \ldots, \frac{d^ny}{dx^n}.$$

(ii) In symbols we can express an nth order ordinary differential equation in one dependent variable by general form:

$$F(x, y, y', y'', \ldots, y^{(n)}) = 0, \quad \text{where } y^{(n)} = \frac{d^ny}{dx^n} \tag{2.1}$$

where F is a real valued function of $n+2$ variables.

Definition 35. Linear and Non-linear Differential Equations
An nth order ordinary differential equation is said to be **linear** in y if it can be written in the form

$$a_n(x)y^{(n)} + a_{n-1}(x)y^{(n-1)} + \cdots + a_1(x)y + a_0(x)y = f(x)$$

where $a_0, a_1, a_2 \ldots a_n$ and f are functions of x on some interval, and $a_n(x) \neq 0$ on that interval. The functions $a_k(x),\ k = 0, 1, 2, \ldots, n$ are called coefficient functions. A differential equation that is not linear is called **non-linear**.

Example 78. (a) $y'' - 4y' + 3y = x^4$ and $xy'' + ye^x + 6 = 0$ are linear differential equations.

(b) $(y-x)dx + 4xdy = 0$, $x^3\frac{d^3y}{dx^3} - 4x\frac{dy}{dx} + 12y = e^x$, $y'' - 4y' + y = 0$ are linear ordinary differential equations.

(c) ODE $(1+y)y' + 2y = e^x$, $\frac{d^2y}{dx^2} + \cos y = 0$, $\frac{d^3y}{dx^3} + y^2 = 0$ are non-linear.

(d) A general form of the first order ordinary differential equation is

$$a_1(x)\frac{dy}{dx} + a_0(x)y = g(x).$$

(e) A general form of second order ordinary linear differential equation is

$$a_2(x)\frac{d^2y}{dx^2} + a_1(x)\frac{dy}{dx} + a_0(x)y = g(x).$$

Remark 18. An ordinary differential equation is linear if the following conditions are satisfied.

(i) The unknown function and its derivative occur in the first degree only.

(ii) There are no products involving either the unknown function and its derivatives or two or more derivatives.

(iii) There are no transcendental functions involving the unknown function or any of its derivatives.

Definition 36. Solutions: (i) A **solution** or **general solution** of an nth order differential equation of the form (2.1) on an interval $I = [a, b] = \{x \in R/a \leq x \leq b\}$ is any function possessing all the necessary derivatives, which when substituted for $y, y', y'', \ldots, y^{(n)}$, reduces the differential equation to an identity. In other words an unknown function is a solution of a differential equation if it satisfies the equation.

(ii) A solution of a differential equation of order n will have n independent arbitrary constants. A solution obtained by assigning particular numerical values to some or all constants is called a **particular solution.**

(iii) A solution of a differential equation that is not obtainable from a general solution by assigning particular numerical values is called a **singular solution.**

(iv) A real function $y = \phi(x)$ is called an explicit solution of the differential equation $F(x, y, y', \ldots, y^{(n)}) = 0$ on $[a, b]$ if

$$F(x, \phi(x), \phi'(x), \ldots, \phi^{(n)}(x)) = 0 \text{ on } [a, b].$$

(v) A relation $g(x, y) = 0$ is called an **implicit solution** of the differential equation $F(x, y, y', \ldots, y^{(n)}) = 0$ on [a,b] if $g(x, y) = 0$ defines at least one real function f on [a,b] such that $y = f(x)$ is an explicit solution on this interval.

We now illustrate these concepts through the following examples:

Example 79. (i) $y = c_1 e^{x+c_2}$ is a solution of the equation $y'' - y = 0$. This ODE is of order 2 and so its solution involves two arbitrary constants c_1 and c_2. It is clear that

$$y' = c_1 e^{x+c_2}, \quad y'' = c_1 e^{x+c_2} \quad \text{and} \quad \text{so} \quad c_1 e^{x+c_2} - c_1 e^{x+c_2} = 0.$$

Hence $y = c_1 e^{x+c_2}$ is a general solution or simply a solution.
(ii) $y = ce^{2x}$ is a solution of ODE $y' - 2y = 0$, because $y' = 2ce^{2x}$ and $y = ce^{2x}$ satisfy the ODE. Since given ODE is of order 1, the solution contains only one constant.
(iii) $y = cx + \frac{1}{2}c^2$ is a solution of the equation

$$\frac{1}{2}(y')^2 + xy' - y = 0.$$

To verify the validity, we note that $y = c$, and therefore

$$\frac{1}{2}(c)^2 + cx - (cx + \frac{1}{2}c^2) = 0.$$

(iv) $y = c_1 e^{2x} + c_2 e^{-x}$ is a general solution of the differential equation $y'' -$

$y' - 2y = 0$ of order 2. To check the validity, we compute y' and y'' and put values in these equations:

$$y' = 2c_1e^{2x} - c_2e^{-x}, \quad y'' = 4c_1e^{2x} + c_2e^{-x}.$$

The left hand side of the given ODE $y'' - y' - 2y = 0$ is $= (4c_1e^{2x} + c_2e^{-x}) - (2c_1e^{2x} - c_2e^{-x}) = -2(c_1e^{2x} + c_2e^{-x}) = 0$.

Example 80. (i) Choosing $c = 1$ we get a particular solution of the differential equation considered in Example 79(iii).
(ii) For $c_1 = 1$ we get a particular solution of the differential equation in Example 79(i) that is, $y = e^{x+c_2}$ is a particular solution of $y'' - y = 0$.

Example 81. (i) $y = -\frac{1}{2}x^2$ is a singular solution of the given differential equation in Example 79(iii). $y = -\frac{1}{2}x^2$ is not obtainable from the general solution $y = cx+\frac{1}{2}c^2$. However it is a solution of the given differential equation, can be checked. For $y = -x$, by putting values of y and y' into the right hand side of the equation we get

$$\frac{1}{2}(-x)^2 + x(-x) - \left(-\frac{1}{2}x^2\right) = x^2 - x^2 = 0.$$

(ii) $y = 0$ is a singular solution of $y' = xy^{1/2}$.

Verification: The general solution of this equation is $y = \frac{x^2}{4} + c$. For $c = 0$, we do not get the solution $y = 0$. Therefore, the solution $y = 0$ of the equation is not obtainable from the general solution. Hence $y = 0$ is a singular solution.

Example 82. (i) $y = \sin 4x$ is an explicit solution of $y'' + 16y = 0$ for all real x.
Verification: $y' = -4\cos 4x$, $y'' = -16\sin 4x$. Putting the value of y and y'' in terms of x into the right hand side of equation we get $-16\sin 4x + 16\sin 4x = 0$.
Hence the equation is satisfied for $y = \sin 4x$ and $y = \sin 4x$ is an explicit solution of the given equation.

(ii) $y = c_1e^x + c_2e^{-x}$ is an explicit solution of the equation $y'' - y = 0$.
Verification: $y' = c_1e^x - c_2e^{-x}, y'' = c_1e^x + c_2e^{-x}$. Using values of y and y'' in the right hand side of the given equation to get $(c_1e^x + c_2e^{-x}) - (c_1e^x + c_2e^{-x}) = 0$.

Example 83. (i) The relation $x^2 + y^2 = 4$ is an implicit solution of the differential equation

$$\frac{dy}{dx} = -\frac{x}{y} \quad \text{on the interval } (-2, 2).$$

Verification: By implicit differentiation of the relation we get

$$-2x = 2y\frac{dy}{dx} \quad \text{or} \quad \frac{dy}{dx} = -\frac{x}{y}.$$

Further $y = \sqrt{4-x^2}$ and $y = -\sqrt{4-x^2}$ satisfy the relation ($y^2 = 4 - x^2$ and $y = \pm\sqrt{4-x^2}$) and are solutions of the differential equation

$$\frac{dy}{dx} = -\frac{x}{y}.$$

It is clear that

$$y_1 = \frac{1}{\sqrt{4-x^2}}(-2x)\frac{1}{2} = -\frac{x}{y_1}$$

and

$$y_2' = -\frac{1}{\sqrt{4-x^2}}(-2x)\frac{1}{2} = -\frac{x}{y_2}.$$

(ii) The relation $y^2 + x - 4 = 0$ is an implicit solution of $2yy' + 1 = 0$ on the interval $(-\infty, 4)$.
Verification: Differentiating $y^2 + x - 4 = 0$ with respect to x, we obtain $2y\dfrac{dy}{dx} + 1 = 0$ or $2yy' + 1 = 0$, which is the given differential equation. Hence $y^2 + x + 4 = 0$ is an implicit solution if it defines a real function on $(-\infty, 4)$. Solving the equation $y^2 + x - 4 = 0$ for y, we get $y = \pm\sqrt{4-x}$. Both $y_1 = \sqrt{4-x}$ and $y_2 = -\sqrt{4-x}$ and their derivatives are functions defined for all x in the interval $(-\infty, 4)$. We conclude that $y^2 + x - 4 = 0$ is an implicit solution on this interval.

Remark 19. Note that a relation $g(x, y) = 0$ can reduce a differential to an identity without constituting an implicit solution of the differential equation. For example $x^2 + y^2 + 1 = 0$ satisfies $yy' + x = 0$, but it is not an implicit solution as it does not define a real-valued function. This is clear from the solution of the equation $x^2 + y^2 + 1 = 0$ or $y = \pm\sqrt{1-x^2}$, imaginary number.

The relation $x^2 + y^2 = 1$ is called a formal solution of $yy' + x = 0$. That is, it appears to be a solution. Very often we look for a formal solution rather than an implicit solution.

Differential Equation of Family of Curves

Let us consider an equation containing n arbitrary constants. Then by differentiating it successively n times we get n more equations containing n arbitrary constants and derivatives. Now by eliminating n arbitrary constants from the above $(n+1)$ equations and obtaining an equation which involves derivatives up to the nth order, we get a differential equation of order n. The concept of obtaining differential equations from a family of curves is illustrated in the following examples.

Example 84. Find the differential equation of the family curves $y = ce^{2x}$.

Solution: Consider

$$y = ce^{2x}. \tag{2.2}$$

Differentiating Equation (2.2) we get

$$y' = 2ce^{2x} = 2y$$

or

$$y' - 2y = 0. \tag{2.3}$$

Thus, arbitrary constant c is eliminated and Equation (2.3) is the required differential equation of the family of curves given by Equation (2.2).

Example 85. Find the differential equation of the family of curves

$$y = c_1 \cos x + c_2 \sin x. \tag{2.4}$$

Solution: Differentiating (2.4) twice we get

$$y' = -c_1 \sin x + c_2 \cos x \tag{2.5}$$

$$y'' = -c_1 \cos x - c_2 \sin x \tag{2.6}$$

c_1 and c_2 can be eliminated from (2.4) and (2.6) and we obtain the differential equation

$$y'' + y = 0. \tag{2.7}$$

(2.7) is the differential equation of the family of curves given by (2.4).

Initial Value and Boundary Value Problems

A general solution of an nth order ordinary differential equation contains n arbitrary constants. To obtain a particular solution, we are required to specify n conditions on solution function and its derivatives and thereby expect to find values of n arbitrary constants. There are two well known methods for specifying auxiliary conditions. One is called initial conditions and other is boundary conditions

It may be observed that an ordinary differential equation does not have a solution or unique solution. However, by imposing initial and boundary conditions, uniqueness can be ensured for certain classes of differential equations.

Definition 37. Initial Value Problem. If the auxiliary condition for a given differential equation relates to a single x value, the conditions are called **initial conditions**, the differential equation with its initial conditions is called an **initial value problem.**

Definition 38. Boundary Value Problem. If the auxiliary conditions for a given differential equation related to two or more x values, the conditions are called **boundary conditions** or **boundary values.** The problem of finding solution of a differential equation with its boundary conditions is called a **boundary value problem.**

Example 86. (i) $y' + y = 3,\ y(0) = 2$ is a first-order initial value problem. The order of the initial value problem is nothing but the order of the given equation and thus $y(0) = 2$ is an initial value condition.
(ii) $y'' + 2y = 0,\ y(1) = 2,\ y'(1) = -3$ is a second order initial value problem. Initial conditions are $y(1) = 2$ and $y'(1) = -3$. Values of function $y(x)$ and its derivative are specified for value $x = 1$.
(iii) $y'' - y' + y = x^3,\ y(0) = 4,\ y'(1) = -2$ is a second order boundary value problem. Boundary conditions are specified at two points, namely $x = 0$ and $x = 1$. One may specify boundary conditions for different values of x, say $x = 2$ and $x = 5$. In this case the boundary value problem is

$$y'' - y' + y = x^3, \quad y(2) = 4, \quad y'(5) = -2.$$

The following questions are relevant as boundary value and initial value problems represent important phenomena in nature:

Problem 1. When does a solution exist? That is, does an initial value problem or a boundary value problem necessarily have a solution?
Problem 2. Is the known solution unique? That is, is there only one solution of an initial value problem or a boundary value problem?

The following theorem states that under the specified conditions, a first order initial value problem has a unique solution.

Theorem 15. Let f and $f_y = (\frac{\partial f}{\partial y})$ be continuous functions of x and y in some rectangle R of the xy-plane and let (x_0, y_0) be a point in that rectangle. Then on some interval centred at x_0 there is a unique solution $y = f(x)$ of the initial value problem:

$$\frac{dy}{dx} = f(x, y), \quad y(x_0) = y_0.$$

Example 87. (i) $y = 4e^x$ is a solution of the initial value problem:

$$y' = y, \quad y(0) = 4.$$

This means that the solution of the differential equation $y' = y$ passes through point $(0, 4)$.

Verification: Let $y = ce^x$, where c is an arbitrary constant. Then $y' = ce^x = y$ and $y' = ce^x = y$. Thus $y = ce^x$ is a general solution of the given equation $y' = y$. By applying the initial condition we get $4 = y(0) = ce^0$ or $c = 4$. Therefore $y = 4e^x$ is a solution of the given initial value problem.

(ii) Find a solution of the initial value problem $y' = y,\ y(1) = 3$. That is,find a solution of differential equation $y' = y$ which passes through the point $(1, 3)$.
Solution: As seen in part (i) $y = ce^x$ is a solution of the given equation and by imposing the given initial condition we get

$$3 = y(1) = ce^1 \text{ or } c = 3e^{-1}.$$

Therefore $y = 3e^{-1}e^x = 3e^{x-1}$ is a solution of the initial value problem.

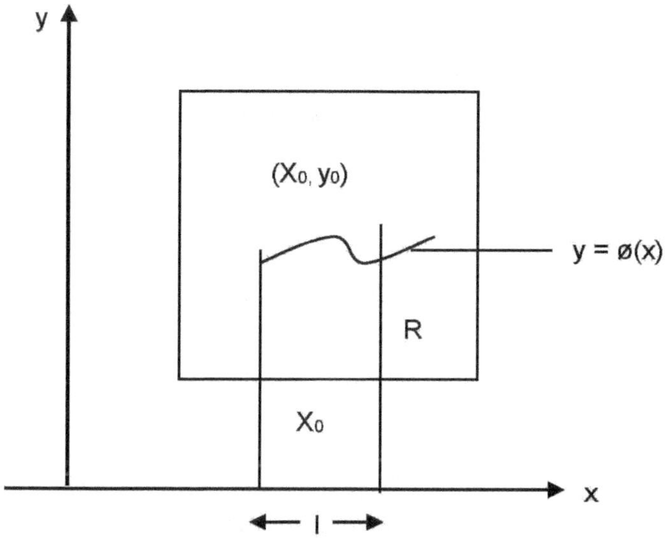

FIGURE 2.1: Geometrical Illustration of Theorem 15

Example 88. $\dfrac{dy}{dx} = xy^{1/2}$, $y(0) = 0$ has at least two solutions, namely $y = 0$ and $y = x^4/16$.

Example 89. (i) Does a solution of the boundary value problem $y'' + y = 0,\ \ y(0) = 0,\ \ y(\pi) = 2$ exist?
(ii) Show that the boundary value problem

$$y'' + y = 0, \quad y(0) = 0, \quad y(\pi) = 0$$

has infinitely many solutions.

Solution: (i) $y = c_1 \cos x + c_2 \sin x$; we get

$$\begin{aligned} 0 &= c_1 \cos 0 + c_2 \sin 0 \\ \text{and} \quad 2 &= c_1 \cos \pi + c_2 \sin \pi. \end{aligned}$$

The first equation yields $c_1 = 0$ and the second yields $c_1 = -2$ which is absurd, hence no solution exists.
(ii) The boundary values yield

$$0 = c_1 \cos 0 + c_2 \sin 0$$

and

$$0 = c_1 \cos \pi + c_2 \sin \pi.$$

Both of these equations lead to the fact that $c_1 = 0$. The constant c_2 is not assigned a value and therefore takes arbitrary values. Thus there are infinitely many solutions represented by $y = c_2 \sin x$.

Example 90. Examine the existence and uniqueness of a solution of the following initial value problems:
(i) $y' = y/x, \quad y(2) = 1$.
(ii) $y' = y/x, \quad y(0) = 3$.
(iii) $y' = -x, \quad y(0) = 2$.

Solution: We examine whether conditions of Theorem 15 are satisfied. To check (i), we observe that

$$f(x, y) = \frac{y}{x} \quad \text{and} \quad \frac{\partial f}{\partial y}(x, y) = \frac{1}{x}.$$

Both functions are continuous except at $x = 0$. Hence f and $\frac{\partial f}{\partial y}$ satisfy the condition of the theorem in any rectangle R that does not contain any part of the y-axis ($x = 0$). Since the point $(2, 1)$ is not on y-axis, there is a unique solution. One can check that $y = \frac{1}{2}x$ is the only solution. (ii) In this problem neither f nor $\frac{\partial f}{\partial y}$ is continuous at $x = 0$. Hence Theorem 15 is not applicable.

Definition 39. A first order differential equation of the form $\frac{dy}{dx} = g(x)h(y)$, where $g(x)$ and $h(y)$ are functions of x and y only, respectively is called **separable** or said to have **separable variables**.

2.2 Introduction to Mathematical Modelling

A mathematical description of a law, a process, a system or a phenomenon is called a **mathematical model**. Mathematical models could be demonstrated in terms of functions, matrices and differential equations with or without initial value and boundary conditions. In general a mathematical model may not have a solution and even it has one, it may not be unique. Appropriate initial value and boundary value conditions ensure existence and uniqueness of the solution of a model. Construction of a mathematical model begins with identification of variables that are responsible for change in the situation under consideration. For the sake of simplicity one may consider only few of such variables. Once the process or the system is understood all remaining variables or few more could be incorporated. Since the assumptions about a process or system often involves a rate of change of one or more of the variables, the mathematical representation of all these assumptions may be one or more equations containing derivatives that is, mathematical model may be a differential equation or system of differential equations. Once a model is obtained (Figure 2.2) we look for existence and uniqueness of its solution. If we can solve it, we claim that the model is reasonable if its solution is consistent with experimental data or known facts about the process or system. If the prediction by the solution is of poor quality, we may increase number of variables or make alternative assumptions in the system. The steps of the modelling are shown in the following diagram:

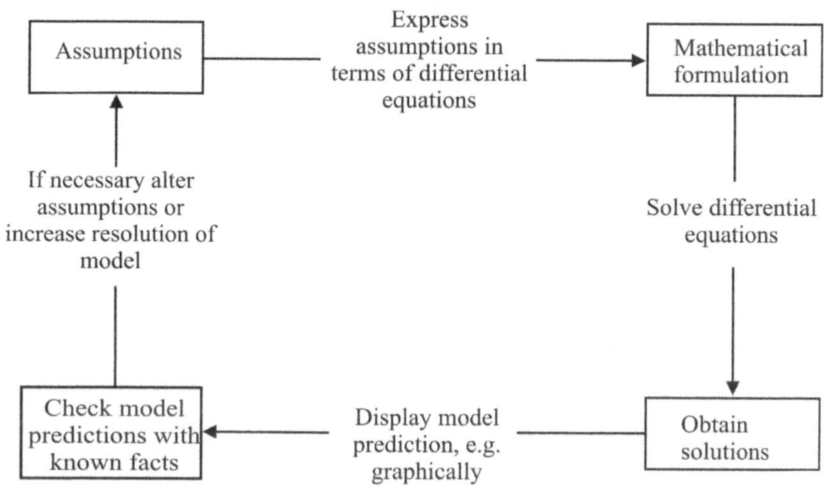

FIGURE 2.2: Steps in Modelling

It may be noted that a mathematical model of a physical system frequently involves the variable time t. A solution of such a model gives state of the system,that is, for appropriate value of t, the values of the dependent variable or variables describe the system in the past, present and future.

2.2.1 Population Dynamics (Exponential and Logistic Model)

In 1798,the British economist Thomas Malthus made the first attempt to devise a mathematical model of human population growth. He made the assumption that the rate at which a population of a country grows at a certain time is proportional to the total population of the country at that time. In mathematical language, if $P(t)$ denotes the total population at time t, then this assumption can be expressed as

$$\frac{dP}{dt} \propto P \quad \text{or} \quad \frac{dP}{dt} = \lambda P, \tag{2.8}$$

where λ is constant of proportionality. This simple model does not take into account many factors such a immigration and emigration that can influence populations to grow or decline. However, it predicts correctly the population in many cases, for example the population of the United States of America during the period 1790 to 1860. Around 1840 the Belgium mathematician cum biologist P. F. Verhulst discovered a mathematical model to predict the human population of a country or city. This model is a generalization of Malthusian model and is known as the logistic equation and its solution is called logistic function. The graph of a logistic function is called a logistic curve. Let $P(t)$ denote the population at time t and let $a > 0,\ b > 0$ be constants whose values depend on the population. The logistic model of the population growth is given as

$$\frac{dP(t)}{dt} = aP(t) - bP^2(t). \tag{2.9}$$

Verhulst had argued that the rate of change of the population $P(t)$ with respect to t should be influenced by growth factors such as population itself and also factors retarding the population, namely limitations of food and space. First term $aP(t)$ represents growth factor while $-bp^2(t)$ represents retardation factors. In applications, very often a is much larger than b. It is clear that (2.9) reduces to (2.8) if b is neglected.

2.2.2 Radioactive Decay

The nucleus of an atom consists of combinations of protons and neutrons. The atoms decay or transmute into atoms of another substance. Such nuclei are called radioactive. For instance, the radioactive radium, Ra$-$226, transmutes into a radioactive gas, Rn$-$222 over a certain period of time. In modeling the phenomenon of radioactive decay the material is pressurized so that the rate of change at which nuclei of a substance decays is proportional to the amount $A(t)$ of the substance remaining at time t, that is,

$$\frac{dA}{dt} \propto A(t) \quad \text{or} \quad \frac{dA}{dt} = \lambda A, \tag{2.10}$$

(2.10) is similar to (2.8) which tells us a single differential equation can serve as a mathematical model for different physical phenomena. In physics the half life is simply the time it takes for one-half of the atoms in an initial amount to disintegrate or transmute into atoms of another element. For example the half life of highly radioactive radium, Ra$-$226, is about 1700 years. The half life of uranium isotope U$-$238 is approximately 4,500,000,000 years. $\lambda > 0$ is called growth constant and $\lambda < 0$ is called decay constant.

2.2.3 Carbon Dating

Solution of Equation (2.10) is based on important technique developed by the chemist Willard Libby in 1950 to determine the ages of certain artifacts. Libby was awarded the 1960 Nobel Prize in chemistry for this work. The process of estimating the age of an artifact or fossil is called carbon dating. The theory of carbon dating is essentially based on the knowledge that the half life of radioactive C$-$14 is approximately 5600 years. Libby's method has been used to date furniture in Egyptian tombs, to decide Van Meegeren art forgeries and to date different civilizations through archaeological excavation.

2.2.4 Newton's Law of Cooling

Newton's law of cooling states that the rate of change at which a body cools is proportional to the difference between the temperature of the body and the temperature of the surrounding medium. Let $T(t)$ represent the temperature of a body at time t, T_m be the temperature of the surrounding medium, and $\frac{dT}{dt}$ the rate at which the temperature of the body changes. The mathematical formulation of Newton's law of cooling is

$$\frac{dT}{dt} \propto T - T_m \quad \text{or} \quad \frac{dT}{dt} = \lambda(T - T_m) \tag{2.11}$$

where λ is a constant of proportionality.

2.2.5 Spread of Diseases and Rumors

While modeling spread of a contagious disease like a flu virus, it is quite reasonable to assume that rate of the spread of disease is proportional to the number of persons $P(t)$ who have contracted the disease and the number of persons $N(t)$ who have not been exposed. If the rate of spread is $\frac{dP(t)}{dt}$ then

$$\frac{dP(t)}{dt} = \lambda P(t)N(t) \tag{2.12}$$

where λ is the constant of proportionality. Assume that one infected person enters into a fixed population of M persons, then P, N and M are related to $P + N = M + 1$ or $N = M - P + 1$. Putting the value of N in terms of M and P in (2.12), we obtain

$$\frac{dP(t)}{dt} = \lambda P(t)(M - P + 1). \tag{2.13}$$

A solution of the following initial value problem provides the number of infected persons at any time t:

$$\frac{dP(t)}{dt} = \alpha P(t)(M + 1 - P(t)) \tag{2.14}$$

$$P(0) = 1. \tag{2.15}$$

Model (2.14) can be used for spread of a rumor; see Brenan [5, p.717].

2.2.6 Series Circuit

Let us consider the single-loop series circuit having an inductor as (L), resistor (R treated as constant) and capacitor (C) shown in Figure 2.3.

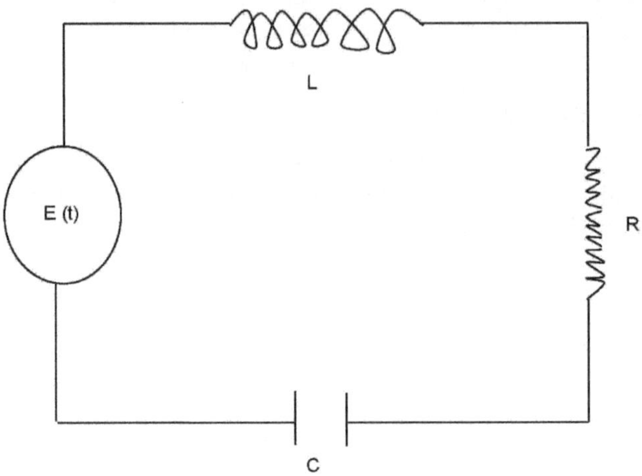

FIGURE 2.3: Series Circuit

Let $E(t)$ denote impressed voltage on a closed loop, $\theta(t)$ denote the change on the given capacitance and $I(t)$ denote the amount of current after the switch is closed. By Kirchhoff's second law, the impressed voltage $E(t)$, on a closed loop must equal the sum of the voltage drops in the loop. Current $I(t)$ is related $\theta(t)$ on the capacitor by

$$I(t) = \frac{d\theta(t)}{dt}.$$

By adding the three complicated drops $L\frac{dI}{dt} = L\frac{d^2\theta(t)}{dt}$ (inductor), $IR = R\frac{d\theta(t)}{dt}$ (resistor) and $\frac{1}{C}\theta(t)$ (capacitor) and equating the sum to the impressed voltage $E(t)$, we get

$$L\frac{d^2\theta}{dt^2} + R\frac{d\theta}{dt} + \frac{1}{C}\theta = E(t). \tag{2.16}$$

We shall discuss the solution of the model given by (2.16) in Section 2.6.5. If $E(t) = 0$, the electric vibrations of the circuit are said to be **free vibrations.** A circuit is called **over damped** if $R^2 - 4LIC > 0$, **critically damped** if

$R^2 - 4LIC = 0$, and **under damped** if $R^2 - 4LIC < 0$. For a series circuit containing only a resistor and inductor, Kirchhoff's second law states that the sum of voltage drop across the inductor $L\frac{dI}{dt}$ and the voltage drop across the resistor (IR) are the same as the impressed voltage $(E(t))$ on the circuit:

$$L\frac{dI}{dT} + RI = E(t).$$

The above equation is the model for current flow where L and R are constants known as the inductor and resistance, respectively. The current $I(t)$ is also called the response of the system.

2.2.7 Draining Tank

In Figure 2.4 a cylindrical container having a constant cross section area A is shown. The orifice at the base of the container has a constant cross sectional area B. If the container is filled with water to a height h, water will flow out through the orifice.

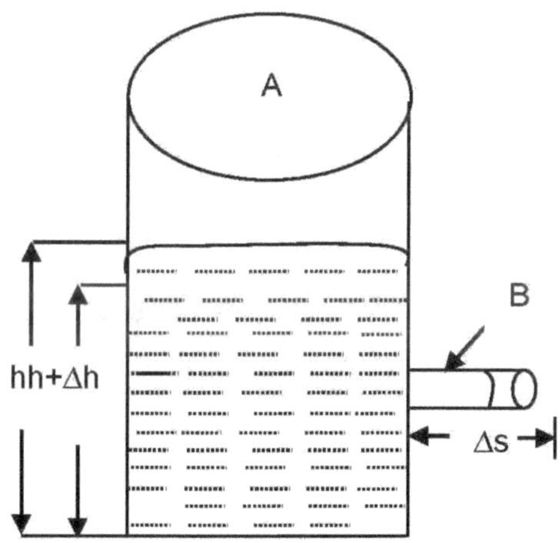

FIGURE 2.4: Draining a Tank

Let h be the height at time t and $h + \Delta h$ be the height at $t + \Delta t$. The volume of the water loss when the level drops by Δh equals the volume of water that goes out from the orifice. The volume of the water going out in time Δt is given by $-A\Delta h$ (negative sign indicates volume loss). The volume of water flowing through the orifice in time Δt is the volume contained in a cylinder of cross section B and length Δs. Thus we have

$$-A\Delta h = B\Delta s.$$

Dividing by Δt and taking limit as $\Delta t \to 0$, we get

$$-A\frac{dh}{dt} = B\frac{ds}{dt} \tag{2.17}$$

where $\frac{ds}{dt}$ is the velocity of h water through the orifice. According to Torricelli's law of hydrodynamics, the velocity of water issuing from the orifice:

$$\frac{ds}{dt} = \sqrt{2gh}. \tag{2.18}$$

By (2.17) and (2.18) we get

$$\frac{dh}{dt} = -\frac{B}{A}\sqrt{2gh}. \tag{2.19}$$

Differential Equation (2.18) represents the rate at which water level is dropping.

2.2.8 Spring and Mass System

Let an external force $f(t)$ act on a vibrating mass on a spring. This could represent a driving force causing an oscillatory vertical motion of the support of the spring, see Figure 2.5.

If $f(t)$ is considered a part of the formulation of Newton's law, the differential equation

$$m\frac{d^2x}{dt^2} = -kx - \beta\frac{dx}{dt} + f(t)$$

where β is a positive damping constant and negative sign indicates that it acts in a direction opposite to the motion.

$$\text{or} \qquad \frac{d^2x}{dt^2} + 2\lambda\frac{dx}{dt} + \omega^2 x = F(t) \tag{2.20}$$

where $F(t) = \frac{f(t)}{m}$, $2\lambda = \frac{B}{m}$ and $\omega^2 = \frac{K}{m}$, represents forced motion. It may be

observed that (2.20) is a general form of the equation

$$\frac{d^2x}{dt^2} + \omega^2 x = 0 \tag{2.21}$$

describing simple harmonic motion or free undamped motion. $x(t) = c_1 \cos\omega t + c_2 \sin\omega t$ is the general solution of (2.20). The period $T = \frac{2\pi}{\omega}$ and the frequency $f = \frac{\omega}{2\pi}$.

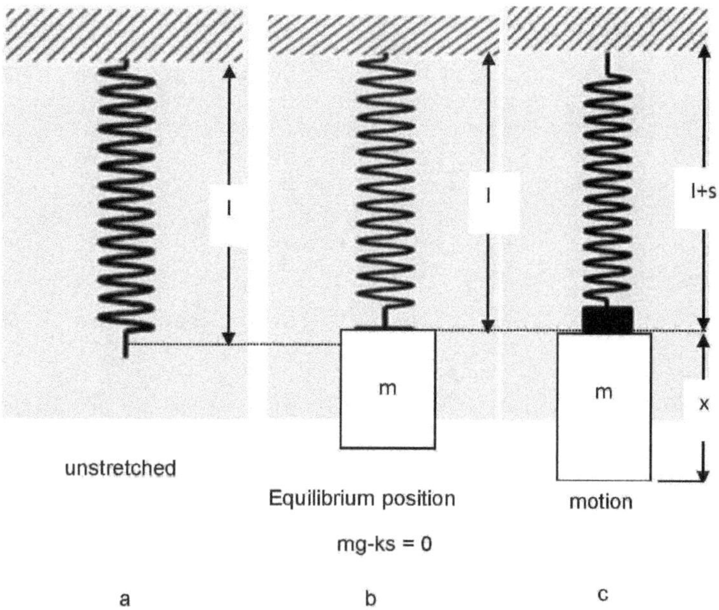

FIGURE 2.5: Spring and Mass System

2.2.9 Mixture of Salt and Payment of Loan

Let us assume that a mixing tank as shown in Figure 2.6 contains m gallons of water in which salt has been dissolved. Suppose that another brine solution is pumped into the tank at the rate of n gallons per minute and after proper mixing, the mixture is pumped out at the same rate. The concentration of salt in the inflow is p kg per gallon.

Let $A(t)$ be the amount of salt measured in kg in the tank at any time. It is assumed that salt is neither created nor destroyed in the tank. Therefore, variations in the amount of salt are due solely to the flows in and out of the tank. Therefore, the rate of change of salt in the tank, $\frac{dA(t)}{dt}$, is equal to the rate at which salt is flowing in minus the rate at which it is flowing out, more precisely

$$\begin{aligned}\frac{dA(t)}{dt} &= \text{(Rate of substance entering in)-(Rate of flowing out)} \\ &= R_{in} - R_{out}. \end{aligned} \tag{2.22}$$

The rate at which the salt is entering into the tank in kg per minute $= np$. The rate at which the salt is flowing out from the tank in kg per minute $=$

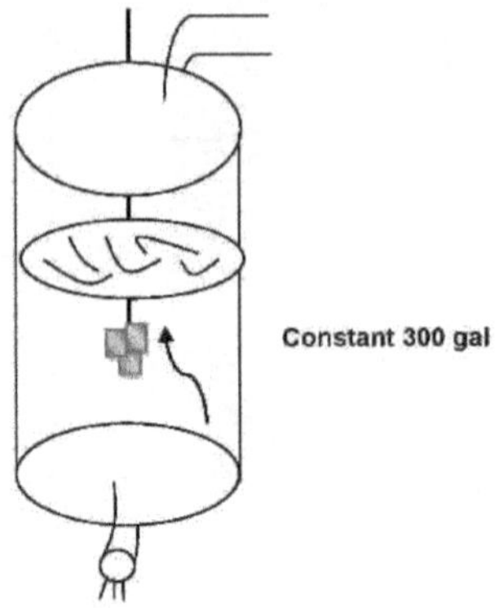

FIGURE 2.6: Mixture of Salt

$\frac{n}{m}$A(t). Thus we obtain

$$\frac{dA(t)}{dt} = np - \frac{n}{m}A(t). \tag{2.23}$$

Differential Equation (2.23) models concentration of a chemical mixture and its solution provides the amount of salt at any time t. Concrete examples and their solution will be presented in Section 2.6.9.

Remark 20. The fundamental balance Equation (2.22) or differential Equation (2.23) can be applied in different situations. This model can be used to determine pollutant in a lake or a drug in an organ of the body, monthly payment of a loan, see Section 2.6.9.

2.2.10 Predator-Prey Model

Suppose two different species of animals interact within the same environment, and assume further that the first species eats only vegetation and the second species eats the first. First species is called a predator and the second one is prey. For example, assume predators are foxes and the prey are rabbits. Let $x(t)$ and $y(t)$ denote, respectively, the fox and rabbit population at time t. If there were no rabbits, it would be expected that fox numbers would decline due to lack of an adequate food supply according to

$$\frac{dx}{dt} = -\alpha x, \;\; \alpha > 0. \tag{2.24}$$

Assume rabbits are present in the environment and encounters between these two species are proportional to the product $x(t)y(t)$. When rabbits are present there is a supply of food and so foxes are added to the system at a rate βxy, $\beta > 0$. Adding this rate to (2.24), we get a model for the fox population:

$$\frac{dx}{dt} = -\alpha x + \beta xy. \tag{2.25}$$

On the other hand, if there are no foxes, the rabbit population will grow at a rate proportional to the number of rabbits present at time t:

$$\frac{dy}{dt} = \mu y, \quad \mu > 0. \tag{2.26}$$

If foxes are present, a model for the rabbit population is (2.26) decreased by λxy, $\lambda > 0$, the rate at which the rabbits are eaten during their encounters with the foxes:

$$\frac{dy}{dt} = \mu y - \lambda xy. \tag{2.27}$$

Equations (2.25) and (2.27) constitute a system of non-linear differential equations

$$\begin{aligned} \frac{dx}{dt} &= -\alpha x + \beta xy \\ \frac{dy}{dx} &= \mu y - \lambda xy \end{aligned} \tag{2.28}$$

where α, β, μ and λ are positive constants and (2.28) is called the **Lotka-Volterra predator-prey model.**

Remark 21. Except for two constant solutions, $x(t) = 0$, $y(t) = 0$ and $x(t) = \frac{\mu}{\lambda}$, $y(t) = \frac{\alpha}{\beta}$, the non-linear system (2.28) cannot be solved in terms of elementary functions.

Interested readers may find comprehensive discussion on systems of equations (2.28) in Brenan and Boyce [5] and Edwards and Penney [7].

2.2.11 Model of Groundwater Contaminant Source

Chlorinated solvents are common causes of environmental contamination at industrial plants. Chlorinated organics, collectively referred to as dense nonaqueous phase liquids (DNAPLS), are denser than water. DNAPLS tend to accumulate as a separate phase below the water table and cause long term ground water contamination.
Let As = cross sectional area of the source region, vd = Darcy groundwater velocity, mt = total DNAPL mass in source region, $c_s(t)$ = concentration (flow averaged) of dissolved contaminant leaving the source region, m_0 = initial DNAPL mass in source region and c_0 = source zone concentration corresponding to an initial source zone mass of m_0.

The equation describing the rate of DNAPL mass discharge from the source region

$$\frac{dm}{dt} = -Asv_d c_s(t) \tag{2.29}$$

while an algebraic relationship between $c_s(t)$ and $m(t)$ is assumed in the form of a power law,

$$\frac{c_s(t)}{c_0} = \left(\frac{m(t)}{m_0}\right)^{\gamma} \tag{2.30}$$

where γ is empirically determined. (2.29) and (2.30) yield a differential equation

$$\frac{dm}{dt} = -\alpha m^{\gamma} \tag{2.31}$$

that models the dissolution of DNAPL into the ground water flowing through the source region. For more details see Falta, Rao and Basu in [5].

2.2.12 Heart Pacemaker

Consider a heart pacemaker as shown in Figure 2.7 consisting of a battery, capacitor, and the heart as resistor. When the switch is at P, the capacitor charges; when S is at Q, the capacitor discharges, sending an electrical stimulus to the heart. During this period, the voltage E applied to the heart is given by the differential equation

$$\frac{dE}{dt} = \frac{1}{RC}E, \; t_1 < t < t_2, \tag{2.32}$$

where R and C are constants.
Solution of this equation gives voltage applied to heart.

$$E(t) = E_0 e^{\dfrac{(t_1 - t)}{RC}},$$

where $E_0 = E(0)$.

2.2.13 X-Ray and Beer's Law

Beer's law describes the behavior of X-rays, that is, the rate of change of intensity per millimeter of a non-refractive, monochromatic, zero-width X-ray beam passing through a medium is jointly proportional to the intensity of the beam and to the attenuation coefficient of the medium. This can be expressed by the differential equation

$$\frac{dI}{dx} = -A(x)I(x) \tag{2.33}$$

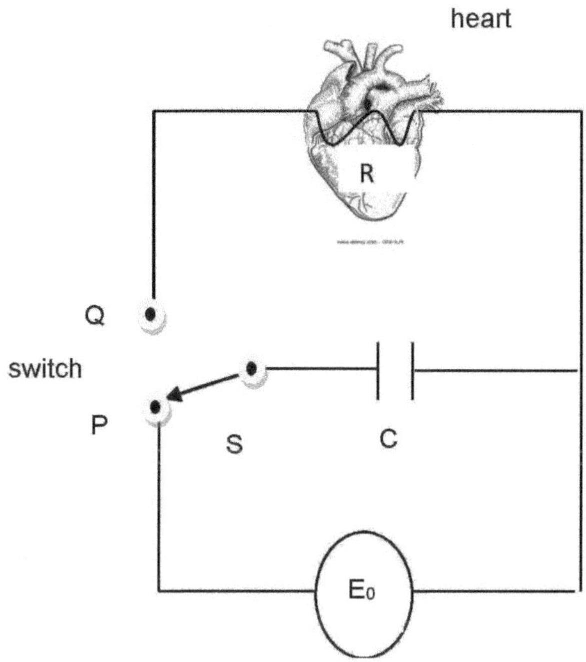

FIGURE 2.7: Model of Pacemaker

where the intensity of the X-ray beam $I(x) = E.N(x)$, E is the energy level of each photon, $N(x)$ denotes the number of photons per second passing through a point x and $A(x)$ is attenuation coefficient of the medium through which X-ray is passing.

We assume that an X-ray beam is composed of photons and that each photon has the same energy level E. Furthermore every substance through which an X-ray passes has the property that each millimeter of the substance absorbs a certain portion of the photons through it. This proportion which is specific to each substance, is called the attenuation coefficient of that material.

2.2.14 Model of Spreading Information

Let $M(t)$ be the population in which we want to propagate information. Let $N(t)$ be the number of persons who know particular information at time t. We assume that those who know the information spread it randomly in the population and those who are provided the information become propagators of the information. Furthermore it is assumed that each person having the information gives it to k individuals per unit of time. It is just possible that some of these k persons may already have the information. Over a unit of time, each of $N(t)$ persons will provide information to k persons. Therefore, the total number of persons who are given the information over the unit of time is $N(t)k$. We are interested only in new persons to whom information is provided. The proportion of the total population M that is unaware of the information over the unit of time is $\frac{M-N}{M}$. Hence, the total number of new persons who are given the information is

$$N(t)k\left(\frac{M-N}{M}\right) = \frac{k}{M(t)}N(t)(M(t)-N(t)).$$

Therefore

$$\frac{dN}{dt} = \frac{k}{M}N(M-N)$$

is a model for propagation of information among a population of size $M(t)$.

2.2.15 Model for Circulation of Money

Suppose that a country has U.S. \$ 200 billion of paper money in circulation. Each week \$ 50 million is brought into the banks for deposit, and the same amount is paid out. The government decides to issue new paper money whenever old money comes to the banks. Old money is destroyed and replaced by new money. Let $A(t)$ be the amount of old money(in millions of U.S. dollars) in circulation at time t (in weeks). Then this process is modeled by the differential equation

$$\frac{dA}{dt} = -0.025A.$$

Solution of this equation provides us the amount of the old money backed to the banks in a certain time. With the help of this model we can roughly determine the week in which 95% old paper money will be deposited to the banks.

2.3 MATLAB and MATHEMATICA for Differential Equations

2.3.1 Solving Ordinary Differential Equations (ODEs) by MATLAB

The differential equation is a very important part of calculus as it has a role in applied areas of mathematics; one of them is modeling of physical systems. In MATLAB, we can solve differential equations symbolically or numerically. First we start with symbolic solution of ODEs.

Symbolic Solution:

The function dsolve computes symbolic solutions to ordinary differential equations (Figure 2.8 and 2.9). The equations are specified by symbolic expressions containing the letter D to denote differentiation. The symbols $D2, D3, \ldots, DN$, correspond to the second, third,..., Nth derivatives respectively. Thus, $D2y$ is equivalent to d^2y/dt^2. We can specify the initial conditions also. For example, let us solve $\dfrac{dx}{dt} = x + t$ with initial condition $x(0) = 0$. We can then write following code

```
>> x= dsolve('Dx=x+t', 'x(0)=0')
x =
0-1-t+exp(t)
```

We can also plot this solution with the ezplot function.
We can solve higher order differential equations also. The differential equation to be solved is $\dfrac{d^2y}{dt^2} = \cos 2t - y$, with initial conditions $y(0) = 1$ and $\dfrac{dy(0)}{dt} = 0$. To solve it write the following code

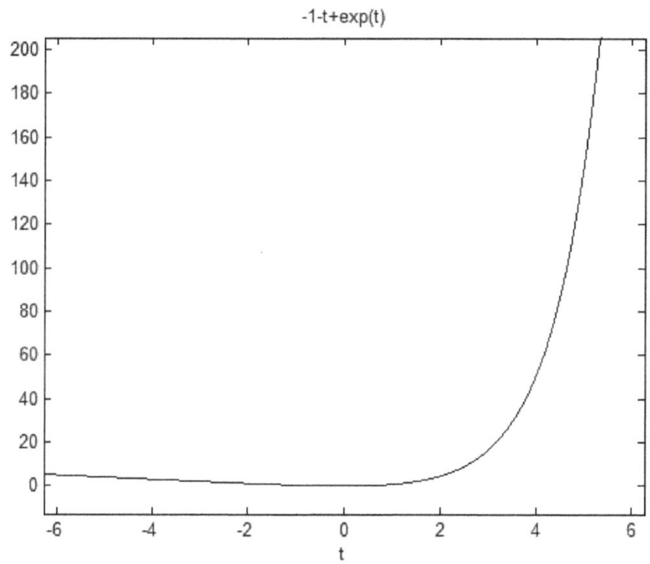

FIGURE 2.8: Solution of ODE

```
>> y = dsolve('D2y=\cos(2*t)-y', 'y(0)=1','Dy(0)=0')
y =
4/3*\cos(t)-1/3*\cos(2*t)
```

TABLE 2.1

Solver	Solves These Kinds of Problems	Method
ode45	Nonstiff differential equations	Runge-Kutta
ode23	Nonstiff differential equations	Runge-Kutta
ode113	Nonstiff differential equations	Adams
ode15s	Stiff differential equations and DAEs	NDFs (BDFs)
ode23s	Stiff differential equations	Rosenbrock
ode23t	Moderately stiff differential equations and DAEs	Trapezoidal rule
ode23tb	Stiff differential equations	TR-BDF2
ode15i	Fully implicit differential equations	BDFs

Numerical Solution:

MATLAB has several built-in functions to solve various types of ODEs numerically. The detail is given below.

We will use one simple example of solve ode45(). The general syntax is given as

```
>>[time, solution] = ode 45('myfunction', tspan, xo),
```

where myfunction is the ODE. In the syntax, tspan is the time interval for which we want a solution and xo is the initial condition. Let us use same ODE, which we used above for a symbolic solution. This ODE can be solved in two ways by writing a function file describing an ODE or by using inline function. The code is given below:

```
>> myode=inline('x+t');
>> [t,x]=ode45(myode, [0 2], 0);
>> plot (t,x)
```

We can code in alternate form by writing a function file. The code is given below.

```
% function file saved as myode.m
function dxdt= myode (t,x)
dx dt=x+t;
```

On command prompt, we will write the following:

```
>> [t,x]=ode45('myode', [0 2], 0);
>> plot (t,x)
```

We can solve higher order ODEs also. We must represent them as a set of first order ODEs. For example, let us solve $\frac{d^2y}{dt^2} - (1-y^2)\frac{dy}{dt} + y = 0$ with initial condition $y(0) = 2$ and $\frac{dy(0)}{dt} = 0$. Before writing, we have to express second order ode into set of first order ode. The ode can also be written as $\ddot{y} = (1-y^2)\dot{y} - y \ldots (1)$. Let $y = x1$ and $x2 = \dot{x}1 \ldots (2)$. Then equation in y can be written as $\dot{x}2 = (1 - x1^2)x2 - x1 \ldots (3)$.

The equations (2) and (3) can be written as

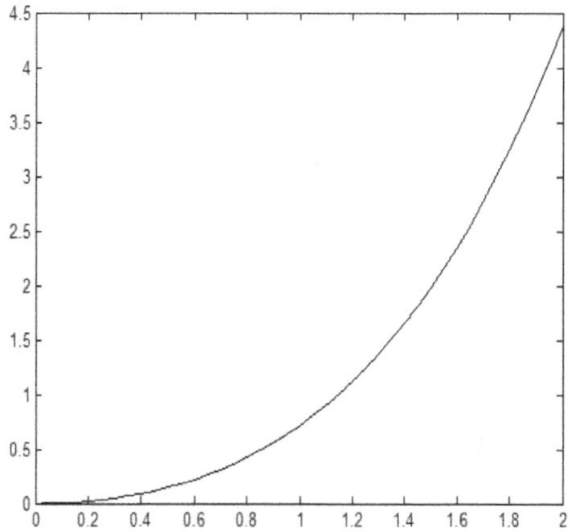

FIGURE 2.9: Numerical Solution of ODE

$$\begin{bmatrix} \dot{x}1 \\ \dot{x}2 \end{bmatrix} = \begin{bmatrix} x2 \\ (1 - x1^2)x2 - x1 \end{bmatrix} \ldots (4),$$

and (4) is of the form $\dot{\mathrm{x}} = f(x)$.

Now we write a function file which will describe Equation (4). The code is given below:

```
function xdot = myode(t,x)
xdot = [x(2); (1-x(1)^2)*x(2)-x(1)];
```

Then write the following commands on the command prompt:

```
>> [t,x]=ode45('myode',[0 20], [2; 0]);
>> plot(t,x)
```

2.3.2 Solving Differential Equation by MATHEMATICA

Solving Differential Equations Symbolically

The MATHEMATICA command DSolve[eqn,$y; x$] solves the differential equation for the function $y(x)$. The command DSolve[{eqn$_1$, eqn$_2$,...}, $\{y_1, y_2, \ldots, x\}$] solves a system of differential equations. The command DSolve[eqn, y; $\{x_1, x_2, \ldots\}$] solves a partial differential equation for the function $y = y(x_1, x_2, \ldots)$.

Example 91. Find the general solution of $y''(x) + y(x) = e^x$.

```
In[ ]:=DSolve[y''[x] + y[x] == Exp[x]; y[x]; x]
Out[ ] = {{y[x]- > c_1 cos x + c_2 sin x
        + \frac{1}{2}e^xcos^2 x+sin^2 x)}}
```

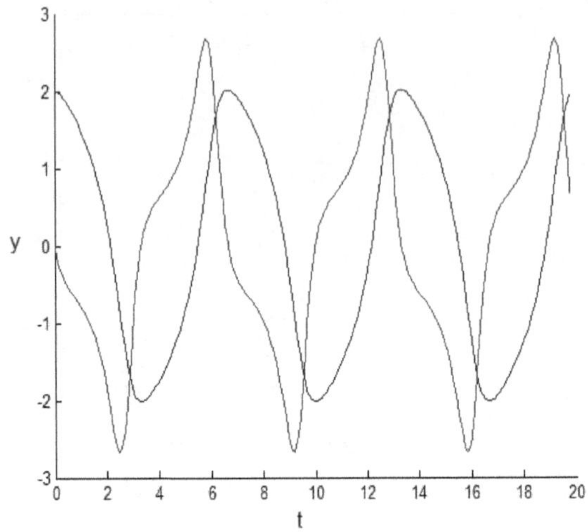

FIGURE 2.10: Symbolic Solution of ODE by MATHEMATICA

Example 92. Find the general solution of $y''(x) + y(x) = e^x$ subject to the boundary conditions $y(0) = 0, y(0) = 1$.

```
In[ ]:=DSolve[{y[x] + y[x] == Exp[x], y[0] == 1, y[0] == 1},
      y, x]
Out[ ] = {{y - > Function[{x}; \frac{1}{2}(cos x + e^x cos^2 x
        + sin x + e^x sin^2 x)]}}
```

Solving Differential Equations Numerically

The MATHEMATICA command NDSolve[$eqns, y, \{x, a, b\}$] finds a numerical solution to the ordinary differential equations *eqns* for the function $y(x)$ in the interval $[a, b]$. The command NDSolve[$eqns, z, \{x, a, b\}; \{t, c, d\}$] finds a numerical solution of the partial differential equation *eqns*.

Example 93. Solve numerically the initial value problem (ordinary differential equation

$$y'(t) = y\cos(t^2 + y^2), \quad y(0) = 1.$$

```
In[ ]:=nsol+NDSolve[{y'[t]==y[t]cos[t^2+y[t]^2],
      y(0)==1},y,{t,0,20}]
Out[ ]={{y -> InterpolatingFunction [{{0,20}},<>]}}
```

We plot this equation with

```
In[ ]:= Plot [Evaluate[y[t]/.nsol],{t,0,20},
```

```
Ticks -> {{0,5,10,15,20},{0,1}}, PlotRange -> All,
      AxesLabel -> {x,t}]
```

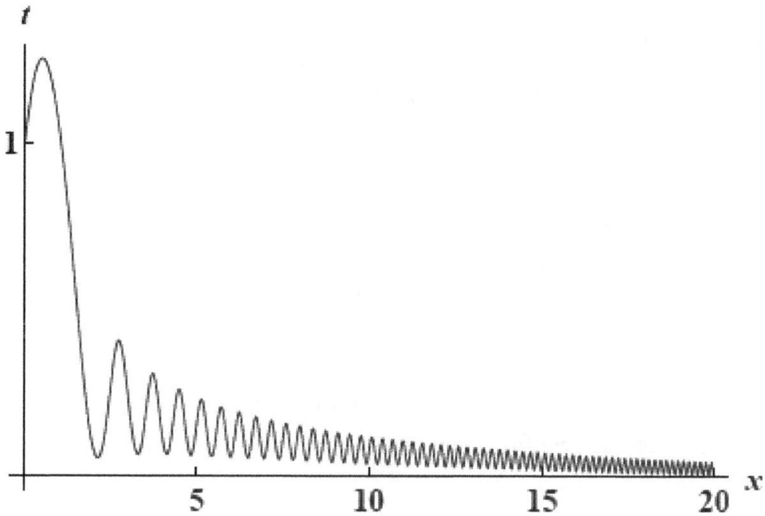

FIGURE 2.11: Numerical Solution of ODE by MATHEMATICA

2.4 Methods for Solving First Order Linear Differential Equations

2.4.1 Method of Separation of Variables

Definition 40. A first order differential equation of the form

$$\frac{dy}{dx} = g(x)h(y),$$

where $g(x)$ and $h(y)$ are functions of x and y only is called **separable** or said to have **separable variables.**

Procedure for solving separable differential equations

(i) If $h(y) = 1$, then

$$\begin{aligned} \frac{dy}{dx} &= g(x) \\ \text{or dy} &= g(x)dx. \end{aligned}$$

Integrating both sides we get

$$\int dy = \int g(x)d(x) + c$$

$$\text{or} \quad \text{y} = \int \text{g(x)d(x)} + \text{c}$$

where c is the constant of integration.
We can write

$$y = G(x) + c$$

where $G(x)$ is an anti-derivative (indefinite integral) of $g(x)$.

(ii) Let $\frac{dy}{dx} = f(x, y)$, where $f(x, y) = g(x)h(y)$ that is $f(x, y)$ can be written as the product of the two functions, one function of variable x only and other variable only y.

$$\frac{dy}{dx} = g(x)h(y)$$

can be written as

$$\frac{1}{h(y)}dy = g(x)dx.$$

By integrating both sides we get

$$\int \rho(y)dy = \int g(x)dx + C$$

where $\rho(y) = \frac{1}{h(y)}$ or $H(y) = G(x) + C$, where $H(y)$ and $G(x)$ are anti-derivatives of $p(y) = \frac{1}{h(y)}$ and $g(x)$, respectively.

Example 94. Solve the differential equation

$$y' = y/x.$$

Solution: Here $g(x) = \frac{1}{x}, h(y) = y$ and $p(y) = \frac{1}{y}$ and $G(x) = \ln x$ and $H(y) = \ln y$.

$$\begin{aligned} \text{Hence } H(y) &= G(x) + C \\ \text{or } \ln y &= \ln x + \ln c \\ \ln y - \ln x &= \ln c \quad \text{(See Appendix A for properties of} \ln x) \\ \frac{y}{x} &= c \\ y &= cx. \end{aligned}$$

Example 95. Solve the initial value problem

$$\frac{dy}{dx} = -\frac{x}{y}, \quad y(4) = 3.$$

Solution: $g(x) = x,\ h(y) = -1/y,\ p(y) = -y$

$$\begin{aligned} H(y) &= G(x) + c \\ -\frac{1}{2}y^2 &= \frac{1}{2}x^2 + c \\ y^2 &= -x^2 - 2c_1^2 \end{aligned}$$

where $c_1^2 = -2c$. By the initial value condition,

$$\begin{aligned} 16 + 9 &= c_1^2 \\ \text{or} \quad \mathrm{c}_1 &= \pm 5 \\ \text{or} \quad \mathrm{x}^2 + \mathrm{y}^2 &= 25. \end{aligned}$$

Thus the initial value problem determines

$$x^2 + y^2 = 25.$$

Example 96. Solve the differential equation

$$\frac{dy}{dx} = \cos 7x.$$

Solution: $dy = \cos 7x dx$. Integrating both sides, we get

$$\int dy = \int \cos 7x dx + c$$

$$y = \frac{\sin 7x}{7} + c.$$

2.4.2 Linear Equations

Definition 41. A first order differential equation of the form

$$a_1(x)\frac{dy}{dx} + a_0(x)y = g(x)$$

is called a **linear equation.**

If $a_1(x) \neq 0$, we can write this differential equation in the form

$$\frac{dy}{dx} + P(x)y = f(x) \tag{2.34}$$

where $P(x) = \dfrac{a_0(x)}{a_1(x)}, \quad f(x) = \dfrac{g(x)}{a_1(x)}.$
(2.34) is called the standard form of a linear differential equation of the first order.

Definition 42. $e^{\int P(x)dx}$ is called the **integrating factor** of the standard form of a linear differential Equation (2.34).

Procedure of Solution (2.34):
Step 1: Write the given equation in the standard form (2.34) if it is not already in this form.
Step 2: Identify $P(x)$ and compute the integrating factor $I(x) = e^{\int P(x)dx}$.
Step 3: Multiply the standard form by $I(x)$.
Step 4: The solution is

$$y.I(x) = \int f(x).I(x)dx + c. \tag{2.35}$$

Example 97. Find the general solution of the following differential equations:
(a) $\frac{dy}{dx} = 9y$ (b) $x\frac{dy}{dx} + 2y = 3$ (c) $x\frac{dy}{dx} + (3x+1)y = e^{-3x}$

Solution: (a) $\frac{dy}{dx} - 9y = 0.$

$$P(x) = -9.$$

Integrating function $= I(x) = e^{\int 9dx} = e^{-9x}$,

$$y.e^{-9x} = \int 0.e^{-9x}dx + c$$

or $\quad y = ce^{9x}, \quad \infty < x < \infty.$

(b) $\frac{dy}{dx} + \frac{2}{x}y = \frac{3}{x}.$
Integrating factor $= I(x) = e^{\int P(x)dx}$,

$$\text{where P(x)} = \frac{2}{\text{x}}$$

$$I(x) = e^{\int \frac{2}{x} dx} = x^2.$$

Solution is given by

$$y.I(x) = \int f(x).I(x)dx + c$$

where $I(x) = x^2, \ f(x) = \frac{3}{x}$. Thus

$$\begin{aligned} yx^2 &= \int \frac{3}{x}x^2dx + c \\ &= \int 3xdx + c = \frac{3}{2}x^2 + c \\ \text{or} \qquad y &= \frac{3}{2} + \frac{c}{x^2}, \qquad 0 < x < \infty. \end{aligned}$$

(c) Standard form is

$$\frac{dy}{dx} = \left(3+\frac{1}{x}\right)y = \frac{e^{-3x}}{x}$$

$$P(x) = 3+\frac{1}{x}, \quad f(x) = \frac{e^{-3x}}{x}.$$

Integrating factor $= I(x) = e^{\int P(x)dx}$

$$= e^{\int\left(3+\frac{1}{x}\right)dx} = xe^{3x}$$

$$y.xe^{3x} = \int \frac{e^{-3x}}{x}.xe^{3x}dx + c$$

$$= \int e^0 dx + c = x + c$$

or $\quad y = e^{-3x} + \frac{c}{x}e^{-3x}$, for $0 < x < \infty$.

One can find a particular solution of Equation (2.34) by a procedure known as variation of parameter (for higher order linear differential equations, this method is discussed in Section 2.5.7). The basic idea is to find a function u so that $y_p = u(x)e^{-\int P(x)dx}$ is a solution of (2.34). Substituting y_p into (2.34) yields

$$u\left[\frac{dy_1}{dx} + P(x)y_1\right] + y_1\frac{du}{dx} = f(x),$$

where $y_1 = e^{-\int P(x)dx}$, so that

$$y_1\frac{du}{dx} = f(x) \quad \text{as} \quad \left[\frac{dy_1}{dx} + P(x)y_1\right] = 0.$$

This gives us

$$du = \frac{f(x)}{y_1(x)}dx \quad \text{and} \quad u = \int \frac{f(x)}{y_1(x)}dx = \int f(x)e^{\int P(x)dx}.$$

Therefore, $y_p = e^{-\int P(x)dx}\int e^{\int P(x)dx}f(x)dx$.

$y_c = ce^{-\int P(x)dx}$ is a general solution of $\frac{dy}{dx} + P(x)y = 0$.

Hence

$$y = y_c + y_p = ce^{-P(x)dx} + e^{-\int P(x)dx}\int e^{\int P(x)dx}f(x)dx. \tag{2.36}$$

If (2.34) has a solution then it must be of the form (2.36).

2.4.3 Exact Equations

We consider here a special kind of non-separable differential equation called an **exact differential equation**. We recall that the total differential of a function of two variables $U(x,y)$ is given by

$$dU = \frac{\partial U}{\partial x}dx + \frac{\partial U}{\partial y}dy. \tag{2.37}$$

Definition 43. The first order differential equation

$$M(x,y)dx + N(x,y)dy = 0 \tag{2.38}$$

is called an exact differential equation if the left hand side of (2.38) is the total differential of some function $U(x,y)$.

Remark 22. (a) It is clear that a differential equation of the form (2.38) is exact if there is a function of two variables $U(x,y)$ such that

$$\begin{aligned} dU &= \frac{\partial U}{\partial x}dx + \frac{\partial U}{\partial y}dy \\ \text{or} \quad \frac{\partial \mathrm{U}}{\partial \mathrm{x}} &= M(x,y), \qquad \frac{\partial U}{\partial y} = N(x,y). \end{aligned}$$

(b) Let $M(x,y)$ and $N(x,y)$ be continuous and have continuous first derivatives in a rectangular region R defined by $a < x < b, \quad c < y < d$. Then a necessary and sufficient condition that $M(x,y)dx + N(x,y)dy$ is an exact differential is that

$$\frac{\partial M}{\partial y} = \frac{\partial N}{\partial x} \quad \text{or} \quad \mathrm{M_y} = \mathrm{N_x}. \tag{2.39}$$

(c) A linear differential equation of the first order can be made exact by multiplying with the integrating factor,

$$I(x) = e^{\int \frac{My - Nx}{N} dx}$$

$$I(y) = e^{\int \frac{My - Nx}{M} dy}.$$

Procedure of Solution 2.38:

Step 1: Check whether a differential equation written in the form (2.38) satisfies (2.39).

Step 2: If Equation (2.39) is satisfied then there exists a function f for which

$$\frac{\partial f}{\partial x} = M(x,y). \tag{2.40}$$

Integrating (2.40) with respect to x, while holding y constant, we get

$$f(x,y) = \int M(x,y)dx + g(y) \tag{2.41}$$

where the arbitrary function $g(y)$ is a constant of integration.

Step 3: Differentiate (2.41) with respect to y and assume $\dfrac{\partial f}{\partial y} = N(x,y)$ and we get

$$\frac{\partial f}{\partial y} = \frac{\partial}{\partial y}\int M(x,y)dx + g'(y) = N(x,y).$$

or

$$g'(y) = N(x,y) - \frac{\partial}{\partial y}\int M(x,y)dx. \tag{2.42}$$

Step 4: Integrate (2.40) with respect to y and substitute this value in (2.41) to obtain $f(x,y) = c$, the solution of the given equation.

Remark 23. (a) Right hand side of (2.42) is independent of variable x, because

$$\begin{aligned}\frac{\partial}{\partial x}\left[N(x,y) - \frac{\partial}{\partial y}\int M(x,y)dx\right] &= \frac{\partial N}{\partial x} - \frac{\partial}{\partial y}\left(\frac{\partial}{\partial x}\int M(x,y)dx\right)\\ &= \frac{\partial N}{\partial x} - \frac{\partial M}{\partial y} = 0.\end{aligned}$$

(b) We could just start the above mentioned procedure with the assumption that

$$\frac{\partial f}{\partial y} = N(x,y)$$

By integrating $N(x,y)$ with respect to y and differentiating the resultant expression, we would find the analogues of (2.41) and (2.42) to be, respectively,

$$f(x,y) = \int N(x,y)dy + h(x) \quad \text{and}$$

$$h(x) = M(x,y) - \frac{\partial}{\partial x}\int N(x,y)dy.$$

Example 98. Check whether $x^2y^3dx + x^3y^2dy = 0$ is exact.

Solution: In view of Remark 22(b) we must check whether $\dfrac{\partial M}{\partial y} = \dfrac{\partial N}{\partial x}$, where

$$M(x,y) = x^2y^3,\ N(x,y) = x^3y^2$$

$$\frac{\partial M}{\partial y} = 3x^2y^2, \qquad \frac{\partial N}{\partial x} = 3x^2y^2.$$

This shows that $\dfrac{\partial M}{\partial y} = 3x^2y^2 = \dfrac{\partial N}{\partial x}$. Hence the given equation is exact.

Example 99. Determine whether the following differential equations are exact. If they are exact solve them by the procedure given in this section.

(a) $(2x-1)dx+(3y+7)dy$.

(b) $(2x+y)dx-(x+6y)dy=0$.

(c) $(3x^2y+e^y)dx+(x^3+xe^y-2y)dy=0$.

Solution: (a) $M(x,y)=2x-1, \quad N(x,y)=3y+7$.

$$\frac{\partial M}{\partial y}=0, \qquad \frac{\partial N}{\partial x}=0.$$

Thus $\dfrac{\partial M}{\partial y}=\dfrac{\partial N}{\partial x}$ and so the given equation is exact.
Apply procedure of (2.38) for finding solution.
Put $\dfrac{\partial f}{\partial x}=2x-1$. Integrating and choosing $h(y)$ as a constant of integration we get

$$\int \frac{\partial f}{\partial x}=f(x,y)=x^2-x+h(y),$$

$h'(y)=N(x,y)=3y+7$, and by integrating with respect to y we obtain

$$h(y)=\frac{3}{2}y^2+7y.$$

The solution is

$$f(x,y)=x^2-x+\frac{3}{2}y^2+7y=c.$$

(b) It is not exact as

$$M(x,y)=2x+y, \quad N(x,y)=-x-6y$$

$$\text{and} \quad \frac{\partial M}{\partial y}=1\neq\frac{\partial N}{\partial x}=-1.$$

(c) $M(x,y)=3x^2y+e^y, \quad N(x,y)=x^3+xe^y-2y$,

$$\frac{\partial M}{\partial y}=3x^2+e^y$$

$$\frac{\partial N}{\partial x}=3x^2+e^y.$$

Thus $\dfrac{\partial M}{\partial y}=\dfrac{\partial N}{\partial x}$, the equation is exact.

Applying procedure for solving 2.38 and let $\dfrac{\partial f}{\partial x}=3x^2y+e^y$. Integrating with respect to x, we obtain

$$f(x,y)=x^3y+xe^y+g(y)$$

where $g(y)$ is a constant of integration.
Differentiating with respect to y we obtain

$$\frac{\partial f}{\partial y} = x^3 + xe^y + g'(y).$$

This gives $N(x, y) = \dfrac{\partial f}{\partial y} = x^3 + xe^y + g(y)$

$$\text{or} \quad g'(y) = -2y$$

$$\text{or} \quad g(y) = -y^2.$$

Substituting this value of $g(y)$ we get

$$f(x, y) = x^3 y + xe^y - y^2 = c.$$

Thus $x^3 y + xe^y - y^2 - c$ is the solution of the given differential equation.

2.5 Methods for Solving Higher Order Differential Equations

In this section we discuss the solution of differential equations of order two or more. In the following subsections underlying theory and certain important methods are presented. The main goal is to find general solutions of higher order linear differential equations.

2.5.1 Initial Value and Boundary Value Problems

Initial value Problem: For a linear differential equation the following problem is called an nth order **initial value problem.**
Find a solution of the differential equation

$$a_n(x)\frac{d^n y}{dx^n} + a_{n-1}(x)\frac{d^{n-1} y}{dx^{n-1}} + \cdots + a_0(x)y = g(x) \tag{2.43}$$

subject to

$$y(x_0) = y_0,\ y'(x_0) = y_1, \ldots, y^{(n-1)}(x_0) = y_{n-1}. \tag{2.44}$$

Conditions given by (2.44) are called n initial conditions. The following theorem proves existence and uniqueness of solutions of initial value problems.

Theorem 16. *(Existence and Uniqueness of Solutions).*
Let $a_n(x), a_{n-1}(x), \ldots, a_1(x), a_0(x)$ and $g(x)$ be continuous on an interval I, and let $a_n(x) \neq 0$ for every x in I. If $x \neq x_0$ is any point in I, then there exists a unique solution $y(x)$ of the initial value problems (2.54) and (2.44) on the interval I.

Example 100. The initial value problem

$$6\frac{d^3y}{dx^3} + 10\frac{d^2y}{dx^2} - 8y = 0$$

$$y(1) = 0, \quad y'(1) = 0, \quad y''(1) = 0$$

possesses the trivial solution $y = 0$. Since the third order equation is linear with constant coefficients, it follows that all conditions of Theorem 16 are satisfied. Hence $y = 0$ is the only solution on any interval containing $x = 1$.

Example 101. Check whether the function $y = 3e^{2x} + e^{-2x} - 3x$ is a solution of the initial value problem

$$\frac{d^2y}{dx^2} - 4y - 12x = 0$$

$$y(0) = 4, \quad y'(0) = 1.$$

Here $a_2(x) = 1 \neq 0$, $a_1(x) = -4 \neq 0$ for every interval containing $x = 0$, $g(x) = 12x$. $a_2(x), a_1(x)$ and $g(x)$ are continuous on any interval I containing $x = 0$. $y = 3e^{2x} + e^{-2x} - 3x$ is a solution of the initial value problem on any interval I containing $x = 0$ by Theorem 16. It is also a unique solution by the same theorem.

Boundary Value Problem

A boundary value problem consists of solving a linear differential equation of order two or greater in which dependent variable y or its derivatives are specified at different points. A typical boundary value problem (BVP) is to solve the linear differential equation of order two.

$$a_2(x)\frac{d^2y}{dx^2} + a_1(x)\frac{dy}{dx} + a_0(x)y = g(x) \tag{2.45}$$

subject to

$$y(a) = y_0, \quad y(b) = y_1. \tag{2.46}$$

Conditions in (2.46) are called boundary conditions.

Remark 24. (i) A solution of BVP [(2.45) and (2.46)] is a function $\phi(x)$ satisfying the differential Equation (2.45) on some interval I, containing a and b, whose graph passes through the two points (a, y_0) and (b, y_1).
(ii) For a second order differential equation, other pairs of boundary conditions could be

$$y'(a) = y_0, \quad y(b) = y_1,$$

$$y(a) = y_0, \quad y'(b) = y_1,$$

$$y'(a) = y_0, \quad y'(b) = y_1.$$

Example 102. Show that the following boundary value problem

$$\frac{d^2y}{dx^2} + 16y = 0$$

$$y(0) = 0, \quad y(\frac{\pi}{2}) = 0$$

has infinitely many solutions.

Solution: It can be checked that

$$y = c_1 \cos 4x + c_2 \sin 4x$$

is a solution of the equations

$$\frac{d^2y}{dx^2} + 16y = 0.$$

$$y(0) = 0 = c_1 \cos 4.0 + c_2 \sin 4.0 \quad \text{or} \quad c_1 = 0.$$

$y\left(\frac{\pi}{2}\right) = 0 = c_2 \sin 4\frac{\pi}{2}$ or $c_2 \sin 2\pi = 0$, for any choice c_2.
Hence the boundary value problem

$$\frac{d^2y}{dx^2} + 16y = 0$$

$$y(0) = 0, \quad y(\frac{\pi}{2}) = 0$$

has infinitely many solutions.

2.5.2 Homogeneous Equations

A linear nth-order differential equation of the form

$$a_n(x)\frac{d^ny}{dx^n} + a_{n-1}(x)\frac{d^{n-1}y}{dx^{n-1}} + \cdots + a_1(x)\frac{dy}{dx} + a_0(x)y = g(x) \qquad (2.47)$$

where $g(x) \neq 0$, ($g(x)$ is not identically zero) is called a **non-homogeneous** nth order differential equation. If $g(x) = 0$, then

$$a_n(x)\frac{d^ny}{dx^n} + a_{n-1}(x)\frac{d^{n-1}y}{dx^{n-1}} + \cdots + a_1(x)\frac{dy}{dx} + a_0(x)y = 0 \qquad (2.48)$$

is called a **homogeneous** linear differential equation of nth order. We shall see in the latter part of this chapter that in order to solve a non homogeneous differential Equation (2.47), we must be able to solve the associated homogeneous Equation (2.48). Here we discuss the general solution of (2.48). Throughout the discussion we assume that

- $a_i(x),\ i = 0, 1, 2, \ldots, n$ are continuous.

- $g(x) = 0$ as the case for homogeneous or continuous equations.
- $a_n(x) \neq 0$ for every x on which solution is considered.

Differential Operator

Let $\dfrac{dy}{dx} = Dy$. The symbol D is said to be a **differential operator** as it transforms a differentiable function into another function.

We can write $\dfrac{d^2y}{dx^2} = \dfrac{d}{dx}\left(\dfrac{dy}{dx}\right) = D(Dy) = D^2(y)$, that is acting (operating) twice on y.

Continuing this process we can write

$$\frac{d^n y}{dx^n} = D(D^{n-1}y) = D(D(D^{n-2})).$$

We define an nth order **differential operator** to be

$$L = a_n(x)D^n + a_{n-1}(x)D^{n-1} + \cdots + a_1(x)D + a_0(x). \tag{2.49}$$

By the two basic properties of differentiation, we have

$$D(\alpha f(x)) = \alpha D f(x),$$

where α is a constant, and

$$D\{f(x) + g(x)\} = Df(x) + Dg(x).$$

This means that the differential operator L possesses a linearity property; that is, L operating on a linear combination of two differentiable functions is the same as the linear combination of L operating on the individual functions. In symbol language, this means

$$L\{\alpha f(x) + \beta g(x)\} = \alpha L(f(x)) + \beta L(g(x)) \tag{2.50}$$

where α and β are constants. In view of (2.50) L is called a **linear operator**. Homogeneous Equation (2.48) can be expressed in terms of the D notion as $L(y) = 0$, where L is given by (2.49). So, (2.47) can be written as

$$L(y) = g(x).$$

Superposition Principle

The following theorem tells us that the sum or superposition of two or more solutions of (2.48) is also a solution of (2.48).

Theorem 17. *Let $y_1, y_2, y_3, \ldots, y_n$ be a solution of (2.48) on an interval I. Then the linear combination*

$$y = \alpha_1 y_1 + \alpha_2 y_2 + \cdots + \alpha_n y_n,$$

where α_i, $i = 1, 2, 3 \ldots, n$, are arbitrary constants, is also a solution of (2.48) on I.

Linear Dependence and Linear Independence

Definition 44. A set of functions $f_1(x), f_2(x), \ldots, f_n(x)$ is said to be **linearly independent** on an interval I if the only constants for which

$$\alpha_1 y f_1(x) + \alpha_2 f_2(x) + \cdots + \alpha_n f_n(x) = 0$$

for every x in the interval are $\alpha_1 = \alpha_2 = \cdots = \alpha_n = 0$.

A set of functions which is not **linearly independent** is called **linearly dependent**.

Remark 25. An equivalent formulation for linearly dependent set: A set of functions $f_1(x), f_2(x), \ldots, f_n(x)$ are **linearly dependent** on an interval I if there exist constants $\alpha_1, \alpha_2, \ldots, \alpha_n$ not all zero, such that

$$\alpha_1 f_1(x) + \alpha_2 f_2(x) + \cdots + \alpha_n f_n(x) = 0$$

for every x in the interval.
Let the set consist of two functions only, say $f_1(x)$ and $f_2(x)$. Therefore, assuming $\alpha_1 \neq 0, f_1(x) = -\dfrac{\alpha_2}{\alpha_1} f_2(x)$, that is, $f_1(x)$ is a constant multiple of $f_2(x)$.

Thus if the set of two functions is linearly dependent, then one must be a constant multiple of the other. Conversely, let $f_1(x) = \alpha_2 f_2(x)$ for some constant α_2. Then $-f_1(x) + \alpha_2 f_2(x) = 0$ for every x in the interval.

Hence the set of two functions is linearly dependent when neither function is a constant, namely $\alpha_1 = -1$ is not zero. We conclude that a set of two functions is linearly independent when neither function is a constant multiple of the other on the interval.

Example 103. *(i) Let* $f_1(x) = \sin 2x,\ f_2(x) = \sin x \cos x$. *The set of* $f_1(x)$ *and* $f_2(x)$ *is linearly dependent on* $(-\infty, \infty)$ *as* $f_1(x)$ *is a constant multiple of* $f_2(x) : f_1(x) = \sin 2x = 2 \sin x \cos x$ *on* $(-\infty, \infty)$.
(ii) Let $f_1(x) = e^x,\ f_2(x) = 5e^x$. *The set* $\{f_1(x), f_2(x)\}$ *is linearly dependent.*
(iii) Let $f_1(x) = 2 + x,\ f_2(x) = 2 + |x|$.
$\{f_1(x), f_2(x)\}$ *is linearly independent as* $f_1(x)$ *and* $f_2(x)$ *cannot be multiples of each other.*

Now we mention results characterizing linearly independent solutions of (2.48) in terms of ***Wronskian*** *determinants.*

Definition 45. Let each of the functions $f_1(x), f_2(x), \ldots, f_n(x)$ possess at least $n-1$ derivatives. The determinant

$$W(f_1, f_2, \ldots, f_n) = \begin{vmatrix} f_1 & f_2 & \cdot & \cdot & \cdot & \cdot & f_n \\ f_1' & f_2' & \cdot & \cdot & \cdot & \cdot & f_n' \\ \vdots & \vdots & \vdots & \vdots & \vdots & \vdots & \\ f_1^{(n-1)} & f_2^{(n-1)} & \cdot & \cdot & \cdot & \cdot & f_n^{(n-1)} \end{vmatrix} \tag{2.51}$$

where the primes denote derivatives is called the **Wronskian** of the functions.

Theorem 18. *Let $y_1, y_2, \dots, y_n$ be n solutions of (2.48) on an interval I. Then the set of solutions is linearly independent on I if and only if $W(y_1, y_2, \dots, y_n) \neq 0$ for every x in the interval.*

Definition 46. (Fundamental set of solutions). Any set $y_1, y_2, \dots, y_n$ of n linearly independent solution (2.48) on an interval I is said to be a fundamental set of solutions on the interval.

Theorem 19. *(Existence of a fundamental set) There exists a fundamental set of solutions of (2.48) on an interval I.*

Theorem 20. *(General solution) Let $y_1, y_2, \dots, y_n$ be a fundamental set of solutions of (2.48) on an interval I. Then*

$$y = \alpha_1 y_1(x) + \alpha_2 y_2(x) + \cdots + \alpha_n y_n(x)$$

where α_i, $i = 1, 2, \dots, n$ are arbitrary constants, is also called ***general solution*** *of (2.48).*

Remark 26. Theorem 20 states that for any solution $y(x)$ of (2.48) on an interval I, $c_1, c_2, \dots, c_n$ can be found such that

$$y(x) = c_1 y_1(x) + c_2 y_2(x) + \cdots + c_n y_n(x).$$

Example 104. The set consisting of e^{-3x} and e^{4x} is a fundamental set of solutions of the differential equation $y'' - y' - 12y = 0$ on $(-\infty, \infty)$. $y = e^{-3x}$ is a solution of the given differential equation, that is, $y'' - y' - 12y = 9e^{-3x} + 3e^{-3x} - 12e^{-3x} = 0$.
$y = e^{4x}$ is also a solution of the given differential equation, that is,

$$y'' - y' - 12y = 16e^{4x} - 4e^{4x} - 12e^{4x} = 0.$$

The set of $\{e^{-3x}, e^{4x}\}$ is linearly independent as $\dfrac{e^{-3x}}{e^{4x}} = e^{-7x}$ is a function and not a constant. In other words neither is a constant multiple of the other and $\{e^{-3x}, e^{4x}\}$ is independent. Also,

$$W(y_1, y_2) = \begin{vmatrix} e^{-3x} & e^{4x} \\ -3e^{-3x} & 4e^{4x} \end{vmatrix} = 4e^{x} + 3e^{x} = 7e^{x} \neq 0.$$

Therefore, $\{e^{-3x}, e^{4x}\}$ is a fundamental set of solutions on interval $(-\infty.\infty)$.

2.5.3 Non-homogeneous Equations

Any function y_p, free of arbitrary parameters, that satisfies (2.47) is said to be a **particular solution** or **particular integral** of the equation.

Theorem 21. *Let y_p be any particular solution of the non-homogeneous linear nth order differential Equation (2.47) on an interval I, and let $y_1, y_2, \ldots, y_n$ be a fundamental set of solutions of the associated homogeneous differential Equation (2.48) on I. Then the* **general solution** *of the equation on the interval I is*

$$y = c_1y_1(x) + c_2, y_2(x) + \cdots + c_ny_n(x) + y_p \tag{2.52}$$

where c_i, $i = 1, 2, \ldots, n$ are arbitrary constants.

The linear combination $y_c(x) = c_1y_1(x) + c_2, y_2(x) + \cdots + c_ny_n(x)$, which is the general solution of (2.48) is called a **complementary function** for non-homogeneous differential Equation (2.47). Thus, in order to solve (2.47) we first solve associated homogeneous linear differential Equation (2.48) and then find a particular solution of (2.47). The general solution of (2.48) is

$$\begin{aligned} y &= \text{complementary function} + \text{any particular solution} \\ &= y_c + y_p. \end{aligned} \tag{2.53}$$

Example 105. $y = c_1e^{2x} + c_2e^{5x} + 6e^x$ is the general solution of the non-homogeneous differential equation

$$y'' - 7y' + 10y = 24e^x \qquad \text{on } (-\infty, \infty).$$

Verification: We are required to check that $y_c(x) = c_1e^{2x} + c_2e^{5x}$ is the general solution of $y'' - 7y + 10y = 0$ and $y = 6e^x$ is a particular solution of

$$y'' - 7y' + 10y = 24e^x.$$

We have

$$y_c'(x) = 2c_1e^{2x} + 5c_2e^{5x}$$
$$y_c''(x) = 4c_1e^{2x} + 25c_2e^{5x}$$

$$\begin{aligned} y'' - 7y' + 10y &= (4c_1e^{2x} + 25c_2e^{5x}) - 7(2c_1e^{2x} + 5c_2e^{5x}) \\ &\quad +10c_1e^{2x} + 10c_2e^{5x} \\ &= (14c_1e^{2x} - 14c_1e^{2x}) + (35c_2e^{5x} - 35c_2e^{5x}) \\ &= 0. \end{aligned}$$

Thus, $y_c(x)$ is general solution of $y'' - 7y' + 10y = 0$. We also have

$$\begin{aligned} y' &= 6e^{2x} \\ y'' &= 6e^x, \text{ so} \\ y'' - 7y' + 10y &= 6e^x - 42e^x + 60e^x = 24e^x \end{aligned}$$

that is, $y = 6e^x$ is a particular solution of $y'' - 7y' + 10y = 24e^x$.

2.5.4 Reduction of Order

Let $a_2(x)y''+a_1(x)y'+a_0(x)y=0$ (2.54)
be a linear second order homogeneous differential equation. The main idea is to discuss procedure to reduce (2.54) to a linear first order differential equation.

Theorem 22. *If y_1 is a nontrivial solution of the second order differential Equation (2.54) then the substitution $y_2(x)=y_1(x)u(x)$, followed by the substitution $w(x)=u'(x)$ reduces it to a first order linear differential equation.*

Remark 27. (i) First order linear differential equation obtained in Theorem. 21 can be solved by computing an integrating factor $I(x)=e^{\int P(x)dx}$ (see Section 2.4.2).
(ii) This procedure holds also for higher order linear differential equation.

Example 106. Let y_1 be a solution of $y''-y=0$ on the interval $(-\infty,\infty)$. Use reduction of order to find a second solution y_2.

Verification: Let $y_2(x)=y_1(x)u(x)=u(x)e^x$. Differentiating this product function we get

$$y_2'(x) = ue^x+u'(x)e^x \tag{2.55}$$
$$y_2''(x) = ue^x+u'(x)e^x+u'(x)e^x+u''(x)e^x. \tag{2.56}$$

Therefore, $y_2''(x)-y_2(x)=e^x(u''+2u')=0$. Since $e^x\neq 0$, we get

$$u''+2u'=0.$$

By substituting $u'=w$ in this equation we get

$$w'+2w=0.$$

This is a linear first order differential equation. Applying integrating factor $e^{\int 2dx}=e^2x$, we can write

$$\frac{d}{dx}\left[e^{2x}w\right]=0.$$

By integrating we obtain

$$e^{2x}w=c_1 \quad \text{or} \quad w=u'=c_1e^{-2x}.$$

Integrating again with respect to x we get

$$u=\frac{-c_1}{2}e^{-2x}+c_2.$$

Thus

$$\begin{aligned} y_2(x) &= u(x)e^x=\left(\frac{-c_1}{2}e^{-2x}+c_2\right)e^x \\ &= \frac{-c_1}{2}e^{-x}+c_2e^x. \end{aligned}$$

By choosing $c_2 = 0$ and $c_1 = -2$, we get

$$y_2(x) = e^{-x}.$$

Since $W(y_1, y_2) = \begin{vmatrix} e^x & e^{-x} \\ e^x & -e^{-x} \end{vmatrix} = -e^0 - e^0 = -2 \neq 0$ for every $x \in (-\infty, \infty)$, the solutions are linearly independent in this interval.

2.5.5 Homogeneous Linear Equations with Constant Coefficients

We consider in this section equations of the type

$$a_n \frac{d^n y}{dx^n} + a_{n-1} \frac{d^{n-1} y}{dx^{n-1}} + \cdots + a_2 \frac{d^2 y}{dx^2} + a_0 y = 0 \tag{2.57}$$

where the coefficients $a_n, a_{n-1}, \ldots, a_2, a_1, a_0$ are real constants and $a_n \neq 0$. We focus mainly on case $n = 2$; similar discussion is possible for other higher numbers. It is interesting to note that all solutions of (2.57) for any n in general and $n = 2$ in particular are exponential functions or constructed from exponential functions.

Let us consider the special case $n = 2$ of (2.57) of the form

$$ay'' + by' + cy = 0. \tag{2.58}$$

If we try a solution of the form $y = e^{mx}$, then after substituting $y' = me^{mx}$ and $y'' = m^2 e^{mx}$ Equation (2.58) gives us

$$am^2 e^{mx} + bme^{mx} + ce^{mx} = 0$$

$$\text{or} \quad e^{mx}(am^2 + bm + c) = 0.$$

Since $e^{mx} \neq 0$ for all x,

$$am^2 + bm + c = 0, \tag{2.59}$$

(2.59) is called the auxiliary equation.
Equation (2.58) can be satisfied by roots of (2.59):

$$m_1 = \frac{-b + \sqrt{b^2 - 4ac}}{2a}$$

$$m_2 = \frac{-b - \sqrt{b^2 - 4ac}}{2a}.$$

We know that (i) m_1 and m_2 are real and distinct if $b^2 - 4ac > 0$. (ii) m_1 and m_2 are real and equal if $b^2 - 4ac = 0$. (iii) m_1 and m_2 are conjugate complex numbers if $b^2 - 4ac < 0$.

Case (i) Distinct Real Roots

Let m_1 and m_2 be two distinct real roots of (2.59). We find two solutions

$$y_1 = e^{m_1 x} \quad \text{and} \quad y_2 = e^{m_2 x}.$$

We can check that y_1 and y_2 are linearly independent on $(-\infty, \infty)$ and form a fundamental set;

$$y = c_1 e^{m_1 x} + c_2 e^{m_2 x} \tag{2.60}$$

is the general solution of (2.58).

Case (ii) Repeated Roots

If $m_1 = m_2$, we obtain only one exponential solution $y_1 = e^{m_1 x}$. A second solution is

$$\begin{aligned} y_2 &= e^{m_1 x} \int \frac{e^{2m_1 x}}{e^{2m_1 x}} dx \\ &= e^{m_1 x} \int dx = x e^{m_1 x}. \end{aligned}$$

In this equation we have used $-b/a = 2m_1$. The general solution in this case is

$$y = c_1 e^{m_1 x} + c_2 x e^{m_2 x}. \tag{2.61}$$

Case (iii) Conjugate Complex Roots

If $m_1$and m_2 are complex, then $m_1 = \alpha + i\beta$ and $m_2 = \alpha - i\beta$, where α and β are real and > 0. $y_1 = c_1 e^{(\alpha+i\beta)x}$ and $y_2 = c_2 e^{(\alpha-i\beta)x}$ are two linearly independent solutions.

Thus $y = y_1 + y_2 = c_1 e^{(\alpha+i\beta)x} + c_2 e^{(\alpha-i\beta)x}$ is the general solution of (2.58). y is in complex form. By applying Euler's formula

$$e^{i\theta} = \cos\theta + i\sin\theta$$

where θ is any real number, we write the general solution in real form. From this formula it follows that

$$\begin{aligned} e^{i\beta x} &= \cos\beta x + i\sin\beta x \\ \cos\beta &= \frac{e^{i\beta x} + e^{-i\beta x}}{2} \\ \sin\beta &= \frac{e^{i\beta x} - e^{-i\beta x}}{2i}. \end{aligned}$$

Since $y = c_1 e^{(\alpha+i\beta)x} + c_2 e^{(\alpha-i\beta)x}$ is a solution of (2.58) for every choice of c_1 and c_2, choices $c_1 = c_2 = 1$ and $c_1 = 1$ and $c_2 = -1$ give two solutions:

$$\begin{aligned} y_3 &= e^{(\alpha+i\beta)x} + e^{(\alpha-i\beta)x} \\ y_4 &= e^{(\alpha+i\beta)x} - e^{(\alpha-i\beta)x}. \end{aligned}$$

But $y_3 = e^{\alpha x}(e^{i\beta x} + e^{-i\beta x}) = 2e^{\alpha x}\cos\beta x$ and $y_4 = e^{\alpha x}(e^{i\beta x} - e^{-i\beta x}) = 2e^{\alpha x}\sin\beta x$.

By Theorem 16, $e^{\alpha x}\cos\beta x$ and $e^{\alpha x}\sin\beta x$ are real solutions of (2.58). Moreover, these solutions form a fundamental set on $(-\infty, \infty)$. Consequently the general solution is

$$y = c_1 e^{\alpha x}\cos\beta x + c_2 e^{\alpha x}\sin\beta x = e^{\alpha x}(c_1\cos\beta x + c_2\sin\beta x). \tag{2.62}$$

Example 107. Solve the following differential equations:

(i) $2y'' - 5y - 3y = 0$.

(ii) $y'' + 5y' - 6y = 0$.

(iii) $y'' + 8y' + 16y = 0$.

(iv) $y'' + 4y' + 7y = 0$.

Solution: of (i) The auxiliary equation is $2m^2 - 5m - 3 = 0$ which can be written as

$$(2m + 1)(m - 3) = 0.$$

Therefore, two roots are $m_1 = \frac{-1}{2}, \quad m_2 = 3$
The solution is of form (2.60), that is,

$$y = c_1 e^{\frac{-1}{2}} + c_2 e^{3x}.$$

(ii) The auxiliary equation is $m^2 + 5m - 6 = 0$. This can be written in the form

$$(m + 1)(m + 6) = 0.$$

Roots are $m_1 = 1, \; m_2 = -6$. Then the solution is of form (2.60), that is,

$$y = c_1 e^{x} + c_2 e^{-6x}.$$

(iii) The auxiliary equation is $m^2 + 8m + 16 = 0$.
Roots are $m_1 = m_2 = -4$. The solution is of form (2.61), that is,

$$y = c_1 e^{m_1 x} + c_2 x e^{m_1 x} = c_1 e^{-4x} + c_2 x e^{-4x}.$$

(iv) The auxiliary equation is $m^2 + 4m + 7 = 0$.
Roots m_1 and m_2 are given by $m_1 = -2 + i\sqrt{3}, \quad m_2 = -2 - i\sqrt{3}$.
The solution is of form (2.62), that is,

$$y = e^{-2x}(c_1 \cos \sqrt{3}x + c_2 \sin \sqrt{3}x).$$

Example 108. Solve the initial value problem

$$y'' + 3y' - 2y = 0.$$

$$y(0) = 1, \; y'(0) = 2.$$

Solution: The auxiliary equation is

$$m^2 + 3m + 2 = 0.$$

Roots are m_1 and m_2 are

$$m_1 = -1 \;\; and \;\; m_2 = -2.$$

Therefore, the solution is of form (2.60), that is,

$$y = c_1 e^{-x} + c_2 e^{-2x}.$$

To find c_1 and c_2 we use initial conditions

$$y(0) = 1 \text{ and } y'(0) = 2$$

$$y(0) = 1 = c_1 e^{-0} + c_2 e^{-0}$$

$$\text{or} \quad c_1 + c_2 = 1$$

$$y' = -c_1 e^{-x} - 2c_2 e^{-2x}$$

$$y'(0) = -c_1 e^{-0} - 2c_2 e^{-0} = -c_1 - 2C - 2 = 2$$

$$\text{or} \quad c_1 + 2c_2 = -2.$$

Thus $c_1 + c_2 = 1$ and $c_1 + 2c_2 = -2$.
This gives $c_2 = -3$ and $c_1 = 4$.
Therefore, $y = 4e^{-x} - 3e^{-2x}$.

Remark 28. In general, to solve an nth order differential Equation (2.57) we must solve an nth degree polynomial equation:

$$a_n m^n + a_{n-1} m^{n-1} + \cdots + a_2 m^2 + a_1 m + a_0 = 0.$$

If all roots (say $m_1, m_2, \ldots, m_n$) of this equation are real and distinct, then the general solution of (2.57) is

$$y = c_1 e^{m_1 x} + c_2 e^{m_2 x} + \cdots + c_n e^{m_n x}.$$

It is difficult to summarize other two cases because the roots of any auxiliary equation of degree greater than 2 can occur in many combinations.

2.5.6 Method of Undetermined Coefficients

To solve a homogeneous linear differential equation with constant coefficients

$$a_n \frac{d^n y}{dx^n} + a_{n-1} \frac{d^{n-1} y}{dx^{n-1}} + \cdots + a_2 \frac{d^2 y}{dx^2} + a_1 \frac{dy}{dx} + a_0 y = g(x), \tag{2.63}$$

one must find complementary function y_c, that is, the general solution of (2.57) (see Theorem 21) and a particular solution of (2.63). A process of finding a particular solution y_p of (2.63) is known as the method of undetermined coefficients. The underlying idea in this method is to guess about the form of y_p that is motivated by the form $g(x)$ in (2.63). The method is limited to those equations of the type (2.63), where $g(x)$ is of the following forms:

(i) $g(x)$ is constant.

(ii) $g(x)$ is a polynomial function (function of form)

$$g(x) = a_0 + a_1x + a_2x^2 + \cdots + a_nx^n.$$

(iii) $g(x) = e^{\alpha x}$, exponential function.

(iv) $g(x) = \sin\beta x$ or $\cos\beta x$ or finite sum and products of these functions.

It may be observed that this method is not applicable in cases where $g(x) = \ln x,\ g(x) = \dfrac{1}{x},\ g(x) = \tan x,\ g(x) = \sin^{-1} x$ etc.
The method is illustrated through the following examples.

Example 109. Find a general form of a particular solution y_p for the following equations

(a) $3y'' + 2y = 5e^{2x} + 2x^3$.

(b) $3y'' + 2y = x^2e^{-3x}$.

(c) $3y'' + 2y = 20\sin 2x$.

Solution: (a) The particular solution y_p will be of the form

$$y_p = Ae^{2x} + B + Cx + Dx^2 + Ex^3.$$

(b) The general form of y_p will be of the form

$$y_p = Ae^{-3x} + Bxe^{-3x} + Cx^2e^{-3x}.$$

(c) $y_p = A\sin 2x + B\cos 2x$.

Example 110. Find a particular solution y_p of differential equation

$$y'' - y' + y = 2\sin 3x.$$

Solution: A natural first guess for a particular solution would be $A\sin 3x$. Since successive differentiation of $\sin 3x$ produces $\cos 3x$ and $\sin 3x$, we are prompted instead to assume a particular solution that includes both of these terms:

$$y_p = A\cos 3x + B\sin 3x.$$

Differentiating y_p and substituting the results into the differential equation gives:

$$y_p' = -3A\sin 3x + 3B\cos 3x.$$

$$y_p'' = -9A\cos 3x - 9B\sin 3x.$$

$$\begin{aligned}
y_p'' - y_p' + y_p &= -9A\cos 3x - 9B\sin 3x + 3A\sin 3x - 3B\cos 3x \\
&\quad +A\cos 3x + B\sin 3x && (2.64)\\
&= 2\sin 3x. && (2.65)\\
y_p'' - y_p' + y_p &= (-8A-3B)\cos 3x + (3A-8B)\sin 3x && (2.66)\\
&= 2\sin 3x + 0\cos 3x. && (2.67)
\end{aligned}$$

Comparing coefficients of $\cos 3x$ and $\sin 3x$ we get

$$\begin{aligned}
-8A-3B &= 0. && (2.68)\\
3A-8B &= 2.
\end{aligned}$$

Solving for A and B we get

$$A = \frac{6}{73}, \quad \text{and} \quad B = \frac{-16}{73}.$$

Thus, a particular solution y_p is given by

$$y_p = \frac{6}{73}\cos 3x - \frac{-16}{73}\sin 3x.$$

Example 111. Find a particular solution of $y'' + 3y' + 2y = 5x^2$.

Solution: We guess that y_p is of the form

$$\begin{aligned}
y_p &= A + Bx + Cx^2 && (2.69)\\
y_p' &= B + 2Cx && (2.70)\\
y_p'' &= 2C && (2.71)\\
y_p'' + 3y_p' + 2y_p &= 2C + 3B + 6Cx + 2A + 2Bx + 2Cx^2 && (2.72)\\
&= 5x^2 && (2.73)\\
\text{or } (2C+3B+2A) &+ (6C+2B)x + 2Cx^2 = 0 + 0x + 5x^2. && (2.74)
\end{aligned}$$

This implies that

$$\begin{aligned}
2A + 3B + 2C &= 0 && (2.75)\\
2B + 6C &= 0 && (2.76)\\
2C &= 5. && (2.77)
\end{aligned}$$

Thus $C = \frac{5}{2}$, $B = \frac{-15}{2}$, $A = \frac{-3}{2}B - C = \frac{45}{4} - \frac{5}{2} = \frac{35}{4}$.
Therefore

$$y_p = \frac{35}{4} - \frac{15}{2}x + \frac{5}{2}x^2.$$

Example 112. Solve the differential equation $y'' - 10y' + 25y = 30x + 3$ by undetermined coefficients.

Solution: Step 1. Find the complementary function of $y'' - 10y' + 25y = 0$.
Step 2. Find y_p.

Step 1. The auxiliary equation is

$$m^2 - 10m + 25 = 0.$$

$$(m-5)^2 = 0.$$

$$m_1 = 5, \quad m_2 = 5.$$

Solution is of form (2.61), that is,

$$y = c_1 e^{5x} + c_2 x e^{5x}.$$

Step 2. Let $y_p = Ax + B$

$$y_p' = A.$$

$$y_p'' = 0.$$

$$0 - 10A + 25(Ax + B) = 30x + 3.$$

$$(-10A + 25B) + 25Ax = 30x + 3.$$

This implies

$$-10A + 25B = 3, \; 25A = 30.$$

Thus $A = \frac{6}{5}$

$$-10.\frac{6}{5} + 25B = 3 \;\text{ gives }\; \text{B} = \frac{3}{5}.$$

$$y_p = \frac{6}{5}x + \frac{3}{5}.$$

The general solution is

$$y = c_1 e^{5x} + c_2 x e^{5x} + \frac{6}{5}x + \frac{3}{5}.$$

Example 113. Solve the differential equation $y'' + 4y = 3\sin 2x$ by undetermined coefficients.

Solution: Step 1. Find the complementary function. Auxiliary equation is

$$m^2 + 4 = 0$$

$$m = \pm 2i.$$

Solution is of form (2.63), that is,

$$y_c(x) = e^{0x}(c_1 \cos 2x + c_2 \sin 2x).$$

Step 2. Find a particular solution y_p.

$$\begin{aligned}
y_p =&Ax\sin 2x + Bx\cos 2x\\
y_p' =&A\sin 2x + 2Ax\cos 2x + B\cos 2x - 2Bs\sin 2x\\
y_p'' =&2A\cos 2x + 2Ax\cos 2x - 4Ax\sin 2x - 2Bs\sin 2x\\
&- 2B\sin 2x - 4Bx\cos 2x\\
=&4A\cos 2x - 4Ax\sin 2x - 4B\sin 2x - 4Bx\cos 2x\\
y_p'' + 4y_p =&(4A\cos 2x - 4Ax\sin 2x - 4B\sin 2x - 4Bx\cos 2x)\\
&+ (4Ax\sin 2x + 4Bx\cos 2x) = 3\sin 2x
\end{aligned}$$

or $\quad 4\mathrm{A}\cos 2\mathrm{x} - 4B\sin 2x = 3\sin 2x.$

$$-4b = 3 \quad \text{or} \quad \mathrm{B} = \frac{-3}{4}$$

$$A = 0.$$

$$y_p = \frac{-3}{4}x\cos 2x$$

$$y = y_c + y_p = c_1\cos 2x + c_2\sin 2x - \frac{3}{4}x\cos 2x.$$

Undetermined Coefficients– Annihilator Approach: Differential Equation (2.63) can be written in terms of operators $D, D^2, D^3, \ldots, D^n$ as

$$Ly = g(x) \tag{2.78}$$

$$\text{where} \quad \mathrm{L} = \mathrm{a_n D^n + a_{n-1} D^{n-1}} + \cdots + \mathrm{a_1 D + a_0}. \tag{2.79}$$

L is said to be an **annihilator operator** of a function f if

$$L(f(x)) = 0$$

where $f(x)$ is sufficiently differentiable.
The differential operator D^n annihilates each of the functions

$$1, x, x^2, \ldots, x^{n-1}. \tag{2.80}$$

The differential operator $(D-\alpha)^n$, where α is a real constant, annihilates each of the functions

$$e^{\alpha x}, xe^{\alpha x}, x^2e^{\alpha x}, \ldots, x^{n-1}e^{\alpha x}. \tag{2.81}$$

Example 114. Find a differential operator that annihilates the function

$$4e^{2x} - 10xe^{2x}.$$

Solution: $n = 2$, $\alpha = 2$, $(D-2)^2$ is a differential operator which annihilates $4e^{2x} - 10xe^{2x}$, that is,

$$(D-2)^2(4e^{2x} - 10xe^{2x} = 0.$$

The differential operator $[D^2 - 2\alpha D + (\alpha^2+\beta^2)]^n$ annihilates each of functions.

$$e^{\alpha x}\cos\beta x, xe^{\alpha x}\cos\beta x, x^2e^{\alpha x}\cos\beta x, \ldots, x^{n-1}e^{\alpha x}\cos\beta x,$$
$$e^{\alpha x}\cos\beta x, xe^{\alpha x}\cos\beta x, x^2e^{\alpha x}\cos\beta x, \ldots, x^{n-1}e^{\alpha x}\cos\beta x. \quad (2.82)$$

Remark 29. (i) If L annihilates y_1 and y_2 then it also annihilates their linear combination, that is, their linear combination, $\alpha y_1 + \beta y_2$, where α and β are real numbers.
(ii) Let L_1 and L_2 be annihilator operator for y_1 and y_2 respectively. However $L_1(y_2) \neq 0$ and $L_2(y_1) \neq 0$. Then L_1L_2 annihilates $\alpha y_1 + \beta y_2$.

Steps for Solution:

(i) Find the complementary solution y_c of $L(y) = 0$.

(ii) Operate on both sides of $L(y) = g(x)$ with a differential operator L_1 that annihilates the function $g(x)$.

(iii) Find the general solution of the higher order homogeneous differential equation $L_1L_(y) = 0$.

(iv) Delete from the solution in step (iii) all those terms that are duplicate in the complementary solution y_c found in step (i). Form a linear combination y_p of the terms that remain. This is the form of a particular solution of $L(y) = g(x)$.

(v) Substitute y_p found in step (iv) into $L(y) = g(x)$. Match coefficients of the various functions on each side of the equality, and solve the resulting system of equations for the unknown coefficients in y_p.

(vi) With the particular solution found in step (v), form the general solution $y = y_c + y_p$ of the given differential equation.

Example 115. Solve $y'' + 3y' + 2y = 4x^2$ using undetermined coefficients.

Solution: Step 1. Solve the homogeneous equation

$$y'' + 3y' + 2y = 0.$$

The auxiliary equation is

$$m^2 + 3m = 2 = 0.$$

Roots of this equation are $m_1 = -1$ and $m_2 = -2$, and so complementary function is of form (2.60), that is,

$$y_c = c_1e^{-x} + c_2e^{-2x}.$$

Step 2. Now, since $4x^2$ is annihilated by the differential operator D^3, we find that

$$D^3(D^2 + 3D + 2)y = 4D^3x^2$$

is the same as

$$D^3(D^2 + 3D + 2)y = 0. \tag{2.83}$$

The auxiliary equation of the fifth order in (2.83),

$$m^3(m^2 + 3m + 2) = 0$$

$$\text{or} \quad m^3(m + 1)(m + 2) = 0,$$

has roots $m_1 = m_2 = m_3 = 0,\ m_4 = -1.$ and $m_5 = -2.$
Thus its general solution must be

$$y = c_1 + c_2x + c_3x^2 + \boxed{c_4e^{-x} + c_5e^{-2x}}. \tag{2.84}$$

The terms in the box in (2.84) constitute the complementary function of the given equation. We can very well argue that a particular solution y_p of the given equation should also satisfy (2.83). This means that the terms remaining (2.84) must be the basic form, of y_p :

$$y_p = A + Bx + Cx^2 \tag{2.85}$$

where, c_1, c_2, c_3 are replaced by A, B and C respectively. For (2.85) to be a particular solution of the given equation, it is necessary to find specific coefficients A, B and C.
Differentiating (2.85), we obtain

$$y_p' = B + 2Cx, \quad y_p'' = 2C.$$

Substituting these values into the given equation

$$y'' + 3y' + 2y = 4x^2,$$

we get

$$y_p'' + 3y_p' + 2y_p = 2C + 2B + 6Cx + 2A + 2B + 2Bx + 2Cx^2 = 4x^2$$

$$\text{or} \quad (2C + 2B + 2A) + (6C + 2B)x + 2Cx^2 = (\text{constant terms})0 + 0x + 4x^2.$$

Comparing constant terms, coefficients of x and x^2, we get

$$2C + 3B + 2a = 0, \quad 6C + 2B = 0, \text{ and } 2C = 4.$$

This implies $C = 2,\ B = -6$, and $A = 7$. Thus $y_p = 7 - 6x + 2x^2$.
Step 3. The general solution of the given equation is

$$y = y_c + y_p \text{ or } y = c_1e^{-x} + c_2e^{-2x} + 7 - 6x + 2x^2.$$

Example 116. Solve the differential equation

$$y'' - 9y = 54$$

by undetermined coefficient approach.

Solution: Applying D to the differential equation, we obtain $D(D^2-9)y = 0$.
The auxiliary equation is

$$m(m^2 - 9) = 0.$$

Roots are $m_1 = 0,\ m_2 = 3,\ m_3 = -3.$
The general solution is $y_c = c_1e^{3x} + c_2e^{-3x} + c_3$ and a particular solution is $y_p = A$.
Putting values of $y_p,\ , y_p',\ y_p''$ in the given differential equation we get

$$0 - 9A = 54 \quad \text{or} \quad A = -6.$$

Thus, the general solution is

$$y = c_1e^{3x} + c_2e^{-3x} - 6.$$

Example 117. Solve $y'' - 2y' + 5y = e^y \sin x$ using the undetermined coefficients–annihilator approach.

Solution: Applying $D^2 - 2D + 2$ to the differential equation we obtain

$$(D^2 - 2D + 2)(D^2 - 2D + 5)y = 0.$$

Then the general solution is

$$y = e^x(c_1 \cos 2x + c_2 \sin 2x) + e^x(c_3 \cos 2x + c_4 \sin 2x),$$

and

$$\begin{aligned}
y_p &= Ae^x \cos x + Be^x \sin x \\
y_p' &= Ae^x \cos x - Ae^x \sin x + Be^x \sin x + Be^x \cos x \\
y_p'' &= (Ae^x \cos x - Ae^x \sin x) - (Ae^x \sin x + Ae^x \cos x) \\
&+ (Be^x \sin x + Be^x \cos x) + (Be^x \cos x - Be^x \sin x).
\end{aligned}$$

Substituting $y_p,\ y_p',\ y_p''$ in the given equation we get:

$$(Ae^x \cos x - Ae^x \sin x) - (Ae^x \sin x + Ae^x \cos x) + (Be^x \sin x + Be^x \cos x)$$
$$+(Be^x \cos x - Be^x \sin x) - 2Ae^x \cos x - 2Ae^x \sin x + 2Be^x \sin x$$
$$+2Be^x \cos x + 5(Ae^x \cos x + Be^x \sin x) = e^x \sin x + 0.e^x \cos x + 0.\text{constant term}$$
$$\text{or} \quad 3Ae^x \cos x + 3Be^x \sin x = e^x \sin x.$$

Equating coefficients gives $A = 0$ and $B = \frac{1}{3}$.
The general solution is

$$y = e^x(c_1 \cos 2x + c_2 \sin 2x) + \frac{1}{3}e^x \sin x.$$

2.5.7 Method of Variation of Parameters

The method of variation of parameters described below is applied to solve a linear second order non-homogeneous differential equation of the form $a_2(x)y'' + a_1(x)y' + a_0(x)y = g(x)$ which can be written in the standard form

$$y'' + P(x)y' + Q(x)y = f(x). \tag{2.86}$$

In (2.86) we assume that $P(x)$, $Q(x)$ and $f(x)$ are continuous on some interval I. As we have seen earlier in Sections 2.5.3 through 2.5.5 there is no difficulty in finding the complementary function y_c of (2.86) when $P(x)$ and $Q(x)$ are constant functions.
Step 1. Find complementary function y_c of (2.86) of the form $y_c = c_1y_1 + c_2y_2$.
Step 2. Find Wronksian W of y_1 and y_2, that is,

$$W(y_1, y_2) = \begin{vmatrix} y_1 & y_2 \\ y_1' & y_2' \end{vmatrix}$$

Step 3. Write $W_1 = \begin{vmatrix} 0 & y_2 \\ f(x) & y_2' \end{vmatrix}$, $\quad W_2 = \begin{vmatrix} y_1 & 0 \\ y_1' & f(x) \end{vmatrix}$
and find u_1 and u_2 by integrating

$$u_1' = \frac{W_1}{W}, \quad u_2' = \frac{W_2}{W}.$$

Step 4. Find a particular solution which is of form $y_p = u_1y_1 + u_2y_2$.
Step 5. The general solution of the equation is $y = y_c + y_p$.

Example 118. Solve the differential equation $y'' - y = xe^x$ by applying the method of variation of parameters.

Solution: Corresponding homogeneous equation is $y'' - y = 0$.
The auxiliary equation is $m^2 - 1 = 0$.
Roots are $m_1 = 1, \ m_2 = -1$.
The complementary function is

$$\begin{aligned}
y &= c_1e^x + c_2e^{-x} \\
W &= \begin{vmatrix} e^x & e^{-x} \\ e^x & -e^{-x} \end{vmatrix} = -e^0 - e^{-0} = -2 \\
W_1 &= \begin{vmatrix} 0 & e^{-x} \\ xe^x & -e^{-x} \end{vmatrix} = -x \\
W_2 &= \begin{vmatrix} e^x & 0 \\ e^x & xe^x \end{vmatrix} = xe^{2x} \\
u_1' &= \frac{-x}{-2} = \frac{x}{2} \\
u_2' &= \frac{xe^{2x}}{-2} = \frac{-1}{2}xe^{2x}.
\end{aligned}$$

Integrating u_1' and u_2' we get

$$\begin{aligned} u_1 &= \frac{x^2}{4} \\ u_2 &= -(xe^{2x}/4) + (e^{2x}/8). \end{aligned}$$

The general solution is

$$\begin{aligned} y &= y_c + y_p \\ \text{where } \mathrm{y_c} &= c_1e^x + c_2e^{-x} \\ y_p &= u_1y_1 + u_2y_2 = \frac{x^4}{4}e^x - (xe^{2x}/4)e^{-x} + (e^{2x}/8)e^{-x}. \end{aligned}$$

Thus $y = c_1e^x + c_2e^{-x} + \dfrac{1}{4}x^4e^x - \dfrac{1}{4}xe^x + \dfrac{1}{8}e^x$.

Example 119. Apply the method of variation of parameters to solve the differential equation

$$y'' - y = \cosh x.$$

Solution: The auxiliary equation is

$$m^2 - 1 = 0, \quad \text{so} \quad m_1 = 1 \quad \text{and} \quad m_2 = -1$$

and complementary function

$$\begin{aligned} y_c &= c_1e^x + c_2e^{-x} = c_1y_1 + c_2y_2 \\ W(y_1, y_2) &= \begin{vmatrix} e^x & e^{-x} \\ e^x & -e^{-x} \end{vmatrix} = -2 \\ f(x) &= \cosh x = \frac{1}{2}(e^x + e^{-x}) \\ W_1 &= \begin{vmatrix} 0 & e^{-x} \\ \cosh x & -e^{-x} \end{vmatrix} = -e^{-x}\cosh x \\ W_2 &= \begin{vmatrix} e^x & 0 \\ e^x & \cosh x \end{vmatrix} = e^x \cosh x \\ u_1' &= \frac{W_1}{W} = \frac{1}{2}\cosh xe^{-x} = \frac{1}{4}(e^{-2x} + 1) \\ u_2' &= \frac{W_2}{W} = \frac{\cosh xe^x}{-2} = \frac{-1}{4}(1 + e^{2x}) \\ u_1 &= \frac{-1}{8}e^{-2x} + \frac{1}{4}x \\ u_2 &= \frac{-1}{4}x - \frac{1}{8}e^{2x} \\ y_p &= \left(-\frac{1}{8}e^{-2x} + \frac{1}{4}x\right)e^x + \left(\frac{-1}{4}x - \frac{1}{8}e^{2x}\right)e^{-x}. \end{aligned}$$

The general solution is

$$y = y_c + y_p.$$

2.5.8 Cauchy-Euler Equation

Second order equations of the form

$$ax^2\frac{d^2y}{dx^2} + bx\frac{dy}{dx} + cy = g(x) \tag{2.87}$$

where a, b and c are constants, $a \neq 0$, and $g(x)$ is continuous on a given interval are called *Cauchy-Euler equations.* By putting $y = x^m,\ y' = mx^{m-1},\ y'' = m(m-1)x^{m-2}$ in (2.87) we get

$$\begin{aligned} ax^2\frac{d^2y}{dx^2} + bx\frac{dy}{dx} + cy &= am(m-1)x^m + bmx^m + cx^m \\ &= (am(m-1) + bm + c)x^m. \end{aligned}$$

Thus $y = x^m$ is a solution of

$$ax^2\frac{d^2y}{dx^2} + bx\frac{dy}{dx} + cy = 0 \tag{2.88}$$

where m is solution of the auxiliary equation :

$$am(m-1) + bm + c = 0 \quad \text{or} \quad am^2 + (b-a)m + c = 0. \tag{2.89}$$

There are three different cases to be considered:

Case 1: Distinct Real Roots
Let m_1 and m_2 be real roots of (2.89) such that $m_1 \neq m_2$ such that $m_1 \neq m_2$. Then $y_1 = x^{m_1}$ and $y_2 = x^{m_2}$ form a fundamental set of solutions. Hence the general solution is

$$y = c_1x^{m_1} + c_2x^{m_2}. \tag{2.90}$$

Case 2: Repeated Roots
If the roots of (2.89) are repeated, that is, $m_1 = m_2$ then the general solution is of the form

$$y = c_1x_1^m + c_2x_1^m \ln x. \tag{2.91}$$

Case 3: Conjugate Complex Roots If the roots of (2.89) are conjugate pair $m_1 = \alpha + i\beta,\ m_2 = \alpha - i\beta$, where α and $\beta > 0$ are real, then a solution is

$$y = c_1e^{\alpha+i\beta} + c_2e^{\alpha+i\beta}.$$

This solution can be written in the real form as

$$y = x^\alpha[c_1\cos(\beta\ln x) + c_2\sin(\beta\ln x)]. \tag{2.92}$$

Verification: $xe^{i\beta} = (e^{lnx})^{i\beta} = e^{i\beta lnx}$ which, by Euler's formula, is the same as

$$x^{i\beta} = \cos(\beta \ln x) + i\sin(\beta \ln x).$$

Similarly,

$$x^{-i\beta} = \cos(\beta \ln x) - i\sin(\beta \ln x).$$

Adding and substracting the last two results yields

$$\begin{aligned} x^{i\beta} + x^{-i\beta} &= 2\cos(\beta \ln x) \quad \text{and} \\ x^{i\beta} - x^{-i\beta} &= 2i\sin(\beta \ln x), \quad \text{respectively.} \end{aligned}$$

Since $y = c_1 x^{\alpha + i\beta} + c_2 x^{\alpha - i\beta}$ is a solution for any values of the constants, we see, in turn, for $c_1 = c_2 = 1$ and $c_1 = 1,\ c_2 = -1$ that

$$y_1 = x^\alpha (x^{i\beta} + x^{-i\beta})$$

$$\text{and} \quad y_2 = x^\alpha (x^{i\beta} - x^{-i\beta})$$

$$\text{or} \quad y_1 = 2x^\alpha \cos(\beta \ln x)$$

$$\text{and} \quad y_2 = 2ix^\alpha \sin(\beta \ln x)$$

are also solutions.
Since the Wronskian for $x^\alpha \cos(\beta \ln x)$ and $x^\alpha \sin(\beta \ln x)$ is $\beta x^{2\alpha - 1} \neq 0,\ \beta > 0$, on the interval $(0, \infty)$, we conclude that

$$y_1 = x^\alpha \cos(\beta \ln x) \text{ and } y_2 = x^\alpha \sin(\beta \ln x)$$

constitute a fundamental set of real solutions of the differential equation. Hence we get the general solution in the real form

$$y = x^\alpha [c_1 \cos(\beta \ln x) + c_2 \sin(\beta \ln x)].$$

Remark 30. The method described above holds true for similar equations of order n.

Example 120. Solve the differential equations
(a) $x^2 y'' - 2y = 0$.
(b) $x^2 y'' - 3xy' - 2y = 0$.
(c) $x^2 y'' + xy' + y = 0$ subject to initial conditions $y(1) = 1,\ y'(1) = 2$.

Solution (a) The auxiliary equation is

$$m^2 - m - 2 = 0 \text{ or } (m+1)(m-2) = 0$$

$$\text{so} \quad m_1 = -1,\ m_2 = 2.$$

The general solution is

$$y = c_1 x^{-1} + c_2 x^2.$$

(b) The auxiliary equation is

$$m^2 - 4m - 2 = 0$$

$$\begin{aligned} m &= \frac{4 \pm \sqrt{16+8}}{2} \\ &= 2 \pm \frac{1}{2}\sqrt{24} \\ &= 2 \pm \sqrt{6}. \end{aligned}$$

The general solution is

$$y = c_1 x^{2+\sqrt{6}} + c_2 x^{2-\sqrt{6}}.$$

(c) The auxiliary equation is $m^2 + 1 = 0$ so that the general solution is given by

$$y = c_1 \cos(\ln x) + c_2 \sin(\ln x) \quad \text{and}$$

$$y' = -c_1 \frac{1}{x} \sin(\ln x) + c_2 \frac{1}{x} \cos(\ln x).$$

The initial conditions imply $c_1 = 1$ and $c_2 = 2$.
Thus, $y = \cos(\ln x) + 2\sin(\ln x)$.

2.6 Solution of Engineering Problems Modeled by Differential Equations

2.6.1 Population Dynamics

We have seen in Section 2.2.1 that the Equation (2.8) represents population growth or decay. In fact, it also represents many other situations such as radioactive decay and compound interest. The differential Equation (2.8),

$$\frac{dP}{dt} = \lambda P(t)$$

with $P(t) > 0$ and λ a constant (either negative or positive) is solved by the separable variable method

$$\int \frac{1}{P(t)} dP = \int \lambda dt,$$

$$\text{or} \quad \ln P(t) = \lambda t + c\ ,$$

where c is a constant of integration. This implies that

$$e^{\ln P(t)} = e^{\lambda t + c} = e^{\lambda t} e^c = A e^{\lambda t} ,$$

where $A = e^c$ is a new constant. It is also clear that $A = P(0) = P_0 =$ population at the time $t = 0$ (initial population). Thus the particular solution of the population growth model (2.8) is

$$P(t) = P(0)e^{\lambda t}. \tag{2.93}$$

Due to the presence of of the natural exponential function in the solution the population model (2.8)

$$\frac{dP}{dt} = \lambda P$$

is often called the **exponential** or **natural growth model**. Figure 2.12 shows a typical growth of $P(t)$ in the case of $\lambda > 0$ and the case $\lambda < 0$ is given in Figure 2.13.

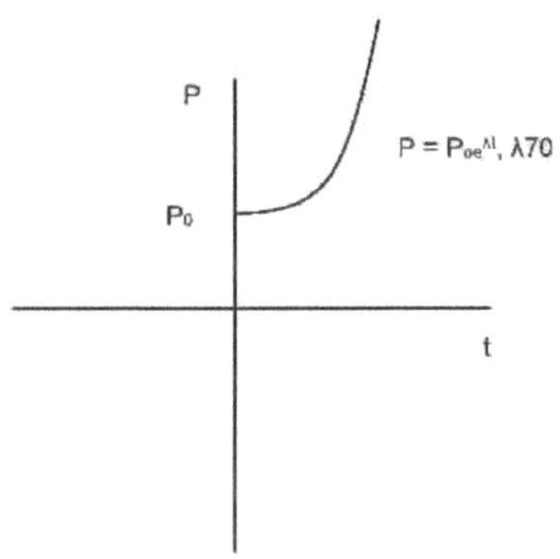

FIGURE 2.12: Natural Growth

It is clear that if population is given at any time t then initial population can be computed.

Example 121. The population of a city is known to increase at a rate proportional to the number of people present at a time t. If the population is doubled in 7 years, how long it will take to triple?

Solution: Let $P(t)$ denote the population at time t and let $P(0)$ denote the population at $t = 0$ (initial population). Then

$$\frac{dP(t)}{dt} = \lambda P(t) \qquad [\text{see } (2.8)].$$

Its solution is

$$P(t) = P(0)e^{\lambda t}$$

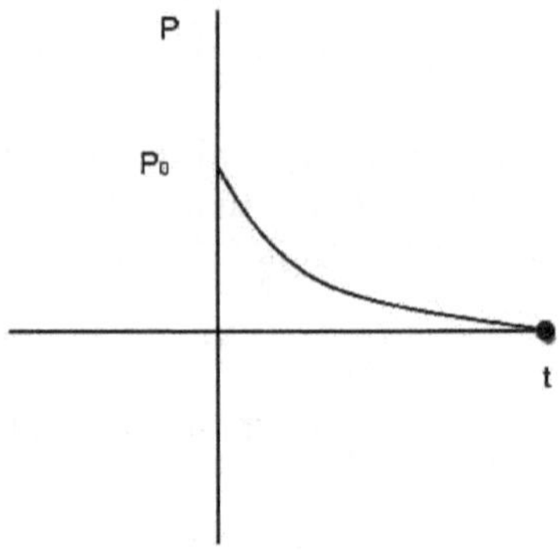

FIGURE 2.13: Natural Decay

as we saw earlier.

$$P(7) = 2P(0) = P(0)e^{\lambda - 7}$$

or

$$2 = e^{\lambda 7} \text{ or } \ln 2 = 7\lambda$$
$$\text{or } \lambda = \frac{\ln 2}{7}$$
$$3 = e^{\frac{\ln 2}{7}t} \text{ or } \ln 3 = \frac{\ln 2}{7}t$$
$$\text{or } t = 7\ln 3 - \ln 2 = \ln 3^7 - \ln 2$$
$$= 7\ln 3 - \ln 2 = 7\frac{\ln 3}{\ln 2}.$$

Example 122. According to an authentic source the total population of the world in 1999 was 6 billion persons. Assume that since then increase of about 212,000 persons per day according to Equation (2.8) : $\dfrac{dP(t)}{dt} = \lambda P(t)$.
(a) What is the annual growth rate λ?
(b) Find the world population at the middle of the 21st century.
(c) In how many years the will world population be 60 billion (grow by 10 times).

Solution: (a) $P_0 = P(0) = 6$ billion, $t = 0$ year corresponds to (mid) 1999. $P(t)$ is increasing by 212,000 or 0.000212 billion, persons per day at time $t = 0$ means that $P'(0) = 0.000212(365.25) \approx 0.07743$ billions per year. From the natural growth model

$$\frac{dP}{dt} = \lambda P$$

with $t = 0$, we get

$$\lambda = \frac{P'(0)}{P(0)} \approx \frac{0.07743}{6} \approx 0.0129.$$

Thus the world population grew at the rate of about 1.29% annually in 1999. This value of λ gives population function

$$P(t) = 6e^{0.0129t}.$$

(b) Here $t = 51$ and so

$$P(51) = 6e^{(0.0129)\times 51} \approx 11.58(\text{billion}),$$

that is, the world population in 2050 is 11.58 billion. Thus the world population at the given growth has doubled in 5 years.
(c) By given information $P(t) = 60$, $\lambda = 0.0129$ then by (2.93), $60 \approx 6e^{0.0129t}$ or

$$t = \frac{\ln 10}{0.0129} \approx 178\text{years}.$$

Thus, in 178 years or in 2177 the world population will be 10 times the population in mid 1999.

Example 123. Suppose the population of a city grows at a rate proportional to the population at time t. The initial population of 500 increases by 15% in 10 years. What will be the population in 30 years? How fast is the population growing at $t = 30$?

Solution: Let $P = P(t)$ be the solution at time t. $\frac{dP(t)}{dt} = \lambda P(t)$ and $P(t) = Ae^{\lambda t}$.
Since $P(0) = 500$, we have $A = 500$ and $P = 500e^{\lambda t}$.
Since 15% of 500 is 75, we get $P(0) = 500e^{10\lambda} = 575$.
Solving for λ we get $\lambda = \frac{1}{10} \ln \frac{575}{500} = \frac{1}{10} \ln 1.15$.
For $t = 30$, $P(30) = 500e^{\frac{1}{10} \ln 1.5 * 30} = 500e^{3 \ln 1.15} = 760\text{years}$.

$$P'(30) = \left.\frac{dP}{dt}\right|_{t=30} = \lambda P(30) = \frac{1}{10}(\ln 1.15)760 = 10.62 \text{ persons per year.}$$

Example 124. The population of an insect grows in a pond at a rate proportional to the number of insects present at time t. After 3 hours it is observed that 400 insects are present. After 10 hours 2000 insects are present. What was the initial number of insects?

Solution: Let $P = P(t)$ be insect population at time t and P_0 be the initial population of the insect.
We have $P = P_0 e^{\lambda t}$ by solving $\frac{dP}{dt} = \lambda P(t)$.
$P(3) = 400$, $P(10) = 2000$ give us $400 = P_0 e^{3\lambda}$ or $e^{\lambda} = (\frac{400}{P_0})^{\frac{1}{3}}$.
From $P(10) = 2000$, we get

$$2000 = P_0 e^{10\lambda} = P_0 (\frac{400}{P_0})^{\frac{10}{3}}$$

so $\dfrac{2000}{400^{\frac{10}{3}}} = (P_0)^{\frac{-7}{3}}$ and $P_0 = (\dfrac{2000}{400^{\frac{10}{3}}})^{\frac{-3}{7}} \approx 201.$

As discussed in Section 2.2.1 population growth is also modelled by Equation (2.9):

$$\frac{dP(t)}{dt} = aP(t) - bP^2(t).$$

We solve it by separation of variables. Decomposing the left hand side of $\dfrac{dP}{P(a-bP)} = dt$ into partial fractions and integrating yields

$$\begin{aligned}
\left(\frac{1}{a}\frac{1}{P} + \frac{b}{a(a-bP)}\right)dP &= dt \\
\frac{1}{a}\ln|P| - \frac{1}{a}\ln|a-bP| &= t+c \\
\ln\left|\frac{P}{a-bP}\right| &= at+ac \\
\frac{P}{a-bP} &= c_1e^{at} \\
P(t) &= ac_1e^{at} - bc_1P(t)e^{at} \\
P(t)(1+bc_1e^{at}) &= ac_1e^{at} \\
P(t) = \frac{ac_1e^{at}}{1+bc_1e^{at}} &= \frac{ac_1}{e^{-at}+bc_1}.
\end{aligned}$$

Let $P(0) = P_0$, $P_0 \neq \dfrac{a}{b}$, we find $c_1 = \dfrac{P_0}{(a-bP_0)}$.

After substitution and simplification

$$P(t) = \frac{aP_0}{bP_0 + (a-bP_0)e^{-at}}.$$

Example 125. A model for the population $P(t)$ in a suburb of a large city is given by the initial value problem

$$\frac{dP}{dt} = P(10^{-1} - 10^{-7}P),\ P(0) = 500,$$

where t is measured in months. What is the limiting value of the population? At what time will the population be equal to one-half of the limiting value?

Solution: From $\dfrac{dP}{dt} = P(\dfrac{1}{10} - \dfrac{1}{10^7}P)$ and $P(0) = 5000$ we obtain, by $P(t)$ after solution of Example 124, $a = \dfrac{1}{10}$, $b = \dfrac{1}{10^7}$

$$P = \frac{500}{(0.0005 + 0.0995e^{-.1t})}$$

so that

$$P \longrightarrow 1000,000 \ \text{ as } \ \mathrm{t} \longrightarrow \infty.$$

If $P(t) = 500,000$ then $t = 52.9$ months.

2.6.2 Radioactive Decay

Example 126. It is known that the radioactive isotope of lead, Pb−209, decays at a rate proportional to the amount present at time t and has a half life of 3.3 hours. If 1 gram of this isotope is present initially, how much time will it take for 90% of the lead to decay?

Solution: Let $A = A(t)$ be the amount of lead present at time t. By the discussion in Equation (2.10),

$$\frac{dA}{dt} = \lambda A \text{ and } A(0) = 1.$$

By solving this initial value problem we get

$$A = e^{\lambda t} \text{ at } t = 3.3 \text{ hrs},$$

$$A(3.3) = e^{(3.3)\lambda} = \frac{1}{2} \text{ implying } \lambda = \frac{1}{3.3}\ln\frac{1}{2}.$$

If 90% lead is decayed then 0.1 gram remains.

$$\begin{aligned} \text{Let } A(t) = 0.1 \text{ or } 0.1 &= e^{\lambda t} = e^{t(\frac{1}{3.3})\ln\frac{1}{2}} \\ \text{so } t &= \frac{3.3\ln 0.1}{\ln\frac{1}{2}} \approx 10.96 \text{ hrs.} \end{aligned}$$

Example 127. Assume 100 milligrams of a radioactive substance present at time $t = 0$ decreases 3% in 6 hours. If the rate of decay is proportional to the amount of the substance present at time t,
(a) Find the amount remaining after 24 hours.
(b) Determine the half life of the radioactive substance.

Solution: (a) Let $A = A(t)$ be the amount present at time t. From Equation (2.10) we get $\frac{dA}{dt} = \lambda A$, where $A(0)$ =100 grams. Solving this initial value problem we get $A = 100e^{\lambda t}$. Since the amount of the radioactive substance is 97grams after 6 hours, $97 = A(6) = e^{6\lambda}$ or $97 = 100e^{6\lambda}$ or $\lambda = \frac{1}{6}\ln 0.97$.
(b) When $A(t) = 50$ so $50 = 100e^{\lambda t}$ or $\lambda t = \ln\frac{1}{2}$ or $t = \frac{1}{\lambda}\ln\frac{1}{2}$ or

$$t = \frac{1}{\frac{1}{6}\ln 0.97}\ln\frac{1}{2} \approx 136.5 \text{ hours.}$$

2.6.3 Carbon Dating

The examples are related to finding the ages of artifacts, namely carbon dating of an artifact.

Example 128. Suppose a fossilized wood has 47% C−14. Assuming that wood died at the initial time ($t = 0$), compute the time T it would take for 1 gram of the radioactive carbon to decay this amount.

Solution: By Equation (2.10)

$$\frac{dA}{dt} = \lambda A$$

where λ is the half life C−14

$$A(t) = A_0 e^{\lambda t},$$

$A(t) =$ represents mass of C−14 at time t and $A(0) = A_0$ denotes the mass at the initial time. At time T

$$A(T) = A_0 e^{\lambda T}$$

or

$$\frac{A(T)}{A(0)} = e^{\lambda T}$$

or

$$\ln \frac{A(T)}{A_0} = \lambda T$$

or

$$\lambda = \frac{1}{T} \ln \frac{A(T)}{A_0}$$

or

$$A(t) = A_0 e^{\frac{1}{T}} \ln \frac{A(T)}{A_0}.$$

Half life of C−14 is 5600 years approximately. Hence

$$T = \frac{\ln 2}{5600} \ln(0.47).$$

Example 129. A fossilized bone is found to contain one thousandth the origin amount $C - 14$. Determine the age of the fossil.

Solution: $\frac{dA}{dt} = \lambda A(t)$, where λ is a constant of proportionality of decay (let $\lambda = -k$). The solution is $A(t) = A(0)e^{+\lambda t}$ by separation of variables. Half life of C−14 is approximately 5600 years

$$\begin{aligned} A(56000) &= \frac{A_0}{2} \\ \frac{1}{2} A_0 &= A_0 e^{5600\lambda} \\ \text{or} \quad \ln \frac{1}{2} &= 5600\lambda \\ \text{or} \quad \lambda &= -\frac{1}{5600} \ln 2 = -0.00012378 \end{aligned}$$

$$\begin{aligned}\frac{1}{1000}A_0 &= A_0e^{-0.00012378t}\\ t &= \frac{\ln 1000}{0.00012378} = 55800 \text{ years.}\end{aligned}$$

2.6.4 Newton's Law of Cooling

Example 130. A thermometer is removed from a room where temperature is 80^oF and is taken outside where the air temperature is 10^oF. After 2 minutes the thermometer shows the reading 40^oF. What is reading of the thermometer at $t = 3$ minutes? How long will it take for the thermometer to reach 20^oF?

Solution: This phenomenon is modelled by

$$\frac{dT}{dt} = \lambda(T - 10),$$

see Section 2.2.4. Its solution is $T = 10 + ce^{\lambda t}$.
If $T(0) = 80$ and $T(2) = 40$ then $80 = 10 + ce^0$, $40 = 10 + ce^{2\lambda}$ or $c = 70$ and $30 = 70e^{2\lambda}$ or $2\lambda = \ln\frac{30}{70}$ or $\lambda = \frac{1}{2}\ln\frac{3}{7}$.

$$T(3) = 10 + 70e^{3*\frac{1}{2}\ln\frac{3}{7}}.$$

If $T(t) = 20 = 10 + 70e^{\frac{1}{2}\ln\frac{3}{7}t}$

$$\begin{aligned}10 &= 70e^{\frac{1}{2}\ln\frac{3}{7}t}\\ \text{or}\quad \left(\frac{1}{2}\ln\frac{3}{7}\right)t &= \ln\frac{10}{70}\\ t &= \frac{1}{\frac{1}{2}\ln\frac{3}{7}}\ln\frac{1}{7}.\end{aligned}$$

Example 131. A 4 kg roast, initially at 60^oF is placed in a 375^oF oven at 6 p.m. At 7.15 p.m. the temperature $T(t)$ of the roast is 125^oF. Find the time when the roast will be at 150^oF.

Solution: We take $t = 0$ which corresponds to 6 p.m. We assume that at any instant the temperature $T(t)$ of the roast is uniform throughout. We have $T(t) < A = 375$, $T(0) = 60$, and T (at 7.15 p.m.) $= T(75) = 125$. Hence the Equation (2.11) takes the form

$$\begin{aligned}\frac{dT}{dt} &= \lambda(375 - T)\\ \int\frac{dT}{375 - T} &= \lambda\int dt\\ -\ln(375 - T) &= \lambda t + c\\ 375 - T &= Be^{-\lambda t}.\end{aligned}$$

Now $T(0) = 60$ implies $375 - 60 = B$ or $B = 315$. Thus $T(t) = 375 - 315e^{-\lambda t}$. We know that $T = 125$ when $t = 75$. Substituting in the preceding equation we get

$$\begin{aligned}
T(75) &= 125 = 375 - e^{-75\lambda} \\
315e^{-75\lambda} &= 250 \\
\frac{315}{250} &= e^{75\lambda} \\
\text{or} \quad \lambda &= \frac{1}{75}\ln\frac{315}{250}. \\
T(t) &= 375 - 315e^{-\lambda t} \\
150 &= 375 - 315e^{-\lambda t} \\
315 &= 225e^{\lambda t} \\
\text{or} \quad t &= \frac{1}{\lambda}\ln\frac{315}{225}.
\end{aligned}$$

Application of Newton's cooling law for determining time of death

A police officer discovers the body of a person presumably murdered. Using Newton's law we determine the time of death. The body is placed in a room that is kept at constant temperature 70°F. For some time after the death, the body will radiate the heat into the cooler room, causing the body's temperature to decrease assuming that the body temperature was normal, 98.6^{o}F at the time of death. From the body's current temperature the time of death can be computed.

According to Newton's cooling law the body will radiate heat energy into the room at a rate proportional to the difference in temperature between the body and the room. If $T(t)$ is the body temperature at time t, then for some constant of proportionality λ,

$$\begin{aligned}
\frac{dT}{dt} &= \lambda(T(t) - 70) \\
\int \frac{dT}{T-70} &= \lambda dt \\
\ln|T-70| &= \lambda t + A, \quad \text{where } A \text{ is consatnt} \\
|T-70| &= e^{\lambda t + A} = Be^{\lambda t}, \quad \text{where } B = e^A.
\end{aligned}$$

Then

$$T - 70 = Be^{\lambda t}$$
$$T(t) = 70 + Be^{\lambda t}.$$

Suppose the officer arrived at 8.30 p.m. and the body temperature was 94.4^{o}F; 8.30 p.m. is considered as $t = 0$ then

$$T(0) = 94.4 = 70 + Be^{\lambda 0} = 70 + B$$

$$\text{or}\quad \mathrm{B} = 24.4,$$

$$\text{giving}\quad \mathrm{T(t)} = 70 + 24.4\mathrm{e}^{\lambda \mathrm{t}}.$$

Suppose the officer makes another measurement of the body's temperature at 10 p.m. and recorded temperature is 89^0F. Taking $t = 90$ minutes (the temperature at 10 p.m.) and $T(90) = 89$,

$$89 = 70 + 24.4e^{e^{90\lambda}}$$

implying $\lambda = \dfrac{1}{90} \ln \dfrac{19}{24.4}$ Thus,

$$\begin{aligned}
T(t) &= 70 + 24.4e^{\frac{1}{90} \ln \frac{19}{24.4} t} \\
98.6 &= 70 + 24.4e^{\frac{t}{90} \ln \frac{19}{24.4}} \\
\frac{28.6}{24.4} &= e^{\frac{t}{90} \ln \frac{19}{24.4}} \\
\ln \frac{28.6}{24.4} &= \frac{t}{90} \ln \frac{19}{24.4} \\
t &= 90 \frac{1}{\ln \frac{19}{24.4}} \ln \frac{28.6}{24.4} \\
&= \text{approximately } -57.07 \text{ min.}
\end{aligned}$$

Therefore the time of death is 57.07 minutes before 8.30 p.m.

2.6.5 Spread of Diseases, Technologies and Rumor

Example 132. Suppose a student in a school of 1000 students is carrying a flue virus. Find a differential equation governing the number of people $N(t)$ who have contracted the flu if the rate at which the the disease spreads is proportional to the number of interactions between the number of students with flu and the number of students, who have not yet been exposed to it.

Solution: The number of students with the flu is N and the number not infected is $1000 - N$, so

$$\frac{dN}{dt} = KN(1000 - N).$$

Example 133. A technological innovation is introduced into a community with a fixed population of n people, say at time $t = 0$. Find a differential equation representing the number of people $N(t)$ who have adopted the innovation at time t and the number who have not adopted the innovation at time time t under the condition that the rate at which the innovation spreads through the community is jointly proportional to the number of people who have not adopted it.

Solution: The rate at which the technology innovation is adopted is proportional to the number of people N who have adopted it and the number of people M who have not adopted it. Thus $M + N = n + 1$ and so

$$\frac{dN}{dt} = \lambda N(n + 1 - N), \quad N((0) = 1.$$

Example 134. **(a)** Describe a model for the spread of a rumor and solve it. **(b)** Currently rumor is known to 300 students in a residential university of 45000 students. It will be known to 900 students after one week. Assuming logistic growth find the number of students who will know the rumor after 4 weeks.

Solution: **(a)** As seen in Section 2.2.5 the desired model is

$$\frac{dN}{dt} = \frac{\lambda}{M} N(M - N),$$

where $N(t)$ is number of students who know the rumor at time t, and M is the total number of students (population size). This equation can be written as

$$\frac{dN}{dt} = \mu N(M - N)$$

where $\dfrac{\lambda}{M}$ is constant

$$\frac{dN}{N(M-N)} = \mu dt$$

$$\int \frac{dN}{N(M-N)} = \mu \int dt + c$$

$$\frac{1}{M} \ln |\frac{N}{M-N}| = \mu t + c$$

so $\ln |\dfrac{N}{M-N}| = M\mu t + Mc$

$$\ln \frac{N}{M-N} = \ln |\frac{N}{M-N}|$$

as $N > 0$ and $M - N > 0$.

In exponential form

$$\begin{aligned}
\frac{N}{M-N} &= e^{M\mu t + Mc} = Ae^{M\mu t}, \quad A = e^{Mc} \\
N &= (M - N)Ae^{M\mu t} \\
N(1 + Ae^{M\mu t}) &= MAe^{M\mu t} \\
N &= \frac{MAe^{M\mu t}}{Ae^{M\mu t} + 1}
\end{aligned}$$

$$N = \frac{M}{1+\dfrac{1}{Ae^{M\mu t}}} = \frac{M}{1+\frac{1}{A}e^{-\mu Mt}} = \frac{M}{1+be^{-\mu Mt}} = \frac{M}{1+be^{-ct}}.$$

(b)

$$M = 45000.$$

For

$$t=0,\ N=300 \Rightarrow 300 = \frac{45000}{1+b} \Rightarrow b = 149$$

$$t=4,\ N=900.$$

Thus $N = \dfrac{45000}{1+149e^{-ct}}$.

For $t=1, N=900, 900 = \dfrac{45000}{1+149e^{-c}}$

$$1+49e^{-c} = \frac{45000}{900} = 50 \text{ or } e^{-c} = \frac{49}{149}, so$$

$$N = \frac{45000}{1+149(\frac{49}{149})t}.$$

For $t=4, N = \dfrac{45000}{1+149(\dfrac{49}{149})^2} \approx 16000$ students who know the rumor after 4 weeks.

2.6.6 Series Circuit

Example 135. A 12 volt battery is connected to a series circuit in which the inductance is $\frac{1}{2}$ henry and the resistance 15 ohms. Find the current I if the initial current is zero.

Solution: By the model of Section 2.2.6, we have

$$\frac{1}{2}\frac{dI}{dt} + 12i = 12,$$

$$I(0) = 0.$$

It is an initial value problem having solution

$$I(t) = 1 + Ae^{-24t}.$$

Now $I(0) = 0 = 1 + A \Longrightarrow A = -1$. Therefore

$$I(t) = 1 - e^{-24t}.$$

2.6.7 Draining Tank

Example 136. A tank in the form of a right circular cylinder is leaking water through a circular hole in its bottom. Find the height h of water in the tank at any time t if the initial height of the water is 5 ft.

Solution: As discussed in Section 2.6.7, $h(t)$ is the solution of the differential equation

$$\frac{dh}{dt} = -\frac{B}{A}\sqrt{2gh} \ \text{(see Equation (2.19))}$$

where A is the cross-sectional area of the cylinder and B is the cross-sectional area of the hole at the base of the container. This equation can be written as

$$\begin{aligned} \frac{dh}{\sqrt{h}} &= -\frac{B}{A}\sqrt{2g}dt \\ \frac{dh}{\sqrt{h}} &= Cdt, \text{ where } C = -\frac{B}{A} \\ \text{or} \quad \int h^{\frac{-1}{2}}dt &= C\int dt. \\ 2h^{\frac{1}{2}} &= Ct + C_1. \end{aligned}$$

For $t = 0,\ h = 5^{\frac{1}{2}}$ implying $C_1 = 2\sqrt{5}$. Thus, $h(t) = \left(\frac{ct + 2\sqrt{5}}{2}\right)^2$.

2.6.8 Spring and Mass

Example 137. A 16 lb weight stretches a spring 4 ft. Assuming that a damping force numerically equal to two times the instantaneous velocity acts on the system, find the equation of motion if the weight is released from the equilibrium position with an upward velocity of 6 ft/s.

Solution: By Hooke's law $16 = 4k$ or k =4lb/ft and

$$w = mg \text{ gives } m = \frac{16}{32} = \frac{1}{2}.$$

The differential equation of motion is

$$\frac{1}{2}\frac{d^2x}{dt^2} = -4x - 2\frac{dx}{dt}$$

$$\text{or} \quad \frac{d^2x}{dt^2} + 4\frac{dx}{dt} + 8x = 0, \quad \text{(see Section 2.2.8)}$$

The auxiliary equation is

$$m^2 + 4m + 8 = 0$$

$$m = \frac{-4 \pm \sqrt{16-32}}{2}$$

$$m_1 = -2+2i, \quad m_2 = -2-2i$$

and $\alpha = 2, \ \beta = 2,$ so solution is

$$x(t) = e^{2t}(c_1 \cos 2t + c_2 \sin 2t).$$

Example 138. A 16 lb weight is attached to a spring whose spring constant is 8 lb/ft. What is the period of simple harmonic motion?

Solution: $\dfrac{d^2x}{dt^2} + w^2x = 0$

$$k = 8, \ m = \frac{16}{32} = \frac{1}{2}, \ w^2 = \frac{8}{\frac{1}{2}} = 16.$$

The differential equation of motion is

$$\frac{1}{2}\frac{d^2x}{dt^2} + 8x = 0$$

$$\frac{d^2x}{dt} + 16x = 0.$$

The auxiliary equation is

$$m^2 + 16 = 0.$$

$$m = \pm 4i.$$

Therefore,

$$x = c_1 \cos 4kt + c_2 \sin 4kt.$$

$$\text{Period of oscillation} = \frac{2\pi}{w} = \frac{2\pi}{4} = \frac{\pi}{2}.$$

$$\text{Frequency} = \frac{1}{T} = \frac{4}{2\pi} = \frac{2}{\pi}.$$

See Section 2.2.8.

2.6.9 Mixture of Salt and Payment of Loan

Example 139. A tank contains 200 liters of fluid in which 30 grams of salt is dissolved. Pure water is pumped into the tank at a rate of 4 liter/min and the well mixed solution is pumped out at the same rate. Find the number $A(t)$ of grams of salt in the tank at time t.

Solution: From Equation (2.23), we have

$$\frac{dA}{dt} = 0 - \frac{A}{50}$$

$$\text{implying } A = ce^{\frac{-t}{50}}$$

$$A(0) = 30 - ce^0 \text{ then } c = 30$$

$$\text{Thus } A(t) = 30e^{\frac{-t}{50}}.$$

Example 140. A tank is filled to capacity with 500 gallons of water. Brine containing 2 pounds of salt per gallon is pumped into the tank at a rate 5 gal/min. The well mixed solution is pumped out at a faster rate of 10 gal/min. (*a*) Find the number $A(t)$ of pounds of salt in the tank at the time t. (b) when is the tank empty?

Solution: From Equation (2.23) and by given data in this problem we have

$$\frac{dA}{dt} = 10 - \frac{10A}{500 - (10-5)t} = 10 - \frac{2A}{100 - t}$$

$$\text{or} \quad \frac{dA}{dt} + \frac{2A}{100-t} = 10.$$

By integrating we have

$$A(t) = 1000 - 10t + c(100 - t)^2.$$

If $A(0) = 0$ then $c = -\frac{1}{10}$. The tank is empty in 100 minutes.

Example 141. Assume that a lake marked as A has a volume of 480 km^3 and that its rate of inflow from lake marked as B in 350 km^3 and outflow to lake marked as C is the same. Let us assume further that at the time $t = 0$ (years), the pollutant concentration of lake A caused by past industrial pollution that ceased to exist now is five times that of lake B. If the outflow henceforth is perfectly mixed lake water, how long will it take to reduce the pollution concentration in lake A to twice that of lake B?

Solution: It is modelled by Equation (2.23), where $n = 350$ km^3, $m = 480$ km^3, $p = c$ (pollutant concentration of lake B), $A_0 = A(0) = 5$cm, that is,

$$\frac{dA}{dt} = 350c - \frac{350}{480}A$$

$$\text{or} \quad \frac{dA}{dt} + \frac{35}{48}A = 350c$$

$$\frac{dA}{dt} + qA = k$$

where $q = \frac{35}{48}, k = 350c$ and integrating factor $I = e^{qt}$. By the method discussed in Section 2.4.2, we find that

$$\begin{aligned} A(t) &= e^{-qt}[A_o + \int_0^t ke^{qt}dt] \\ &= e^{-qt}[A_o + \frac{k}{q}(e^{qt} - 1)] \\ &= e^{\frac{-nt}{m}}[5cm + \frac{nc}{\frac{n}{m}}(e^{\frac{nt}{m}} - 1)] \\ A(t) &= cm + 4cme^{\frac{-nt}{m}}. \end{aligned}$$

Now, we find t such that $A(t) = 2$ cm, that is,

$$cm + 4cme^{\frac{-nt}{m}} = 2 \text{ cm}$$

$$\text{or } 4 = e^{\frac{nt}{m}}$$

$$\frac{nt}{m} = \ln 4 \text{ or } t = \frac{m \ln 4}{n} = \frac{480}{350} ln4 \simeq 1.901 \text{ (years)}$$

Example 142. Suppose a person needs loan of \$ 20.000 from a bank. The bank is willing to provide loan with an annual interest of 8%. What monthly payment is required to the bank if the borrower wants to pay in four years?

Solution: Let $c(t)$ be the balance due on the loan at any time t. Suppose that A is measured in dollars and t in years. Then $\frac{dA}{dt}$ has the unit of dollars per year. The balance on the loan is affected by two factors namely: accumulation of interest tends to increase $A(t)$ and the payments tend to reduce it. By Equation (2.23) we get

$$\frac{dA}{dt} = \gamma A - 12\mu$$

where γ is the annual interest rate, μ is the monthly payment rate and μ is multiplicative by 12 so that all the terms will have units of dollars per year. The initial condition is $A(0) = A_0$, where A_0 is the amount of the loan.

Here $\gamma = 0.08, A_0 = 20,000$, thus $\frac{dA}{dt} - 0.08A = -12\mu$. The integrating factor is $e^{-0.08t}$ and by Section 2.4.2, $A = 150\mu + ce^{0.08t}$. Since $A(0) = 20,000, c = 20,000 - 150\mu$ we have

$$A(t) = 20,000e^{0.08t} - 150\mu(e^{0.08t} - 1).$$

In order to find monthly payment to pay off the loan in four years we set $A(t) = 0$ for $t = 4$ in the previous equation to get

$$\mu = \frac{20,000}{150} \frac{e^{0.32}}{e^{0.32} - 1} = \$486.88.$$

2.7 Laplace Transform for Linear Differential Equations

2.7.1 Introduction to Laplace Transform

Pierre Simon de Laplace, a French Mathematician (1749-1827) was essentially interested in describing nature using mathematics. The main goal of this section is to present those results of Laplace which are used to find solutions of differential and integral equations.

Definition and Fundamental Properties of Laplace Transform

The Laplace transform is considered an extension of the idea of the indefinite integral transform :

$$I = \{f(t)\} = \int_0^x f(t)dt.$$

It is defined as follows:

Definition 47. The **Laplace transform** of $f(t)$, denoted by $L\{f(t)\}$ is defined by

$$F(s) = L\{f(t)\} = \int_0^\infty e^{-st} f(t)dt, \tag{2.94}$$

where s is a real number called a parameter of the transform.

Remark 31. (a) Laplace transform takes a function $f(t)$ into a function $F(s)$ of the parameter s.

(b) We present functions of t by lower case letters, f, g and h, and their respective Laplace transforms by corresponding capital letters F, G and H. Thus we write

$$L\{f(t)\} = F(s)$$

$$\text{or} \quad F(s) = \int_0^\infty e^{-st} f(t)dt.$$

(c) The defining equation for the Laplace transform is an improper integral, which is defined as

$$\int_0^\infty e^{-st} f(t)dt = \lim_{T\to\infty} \int_0^T e^{-st} f(t)dt.$$

Thus, the existence of the Laplace transform of f depends upon the existence of the limit.

(d) A Laplace transform is rarely computed by referring directly to the definition and integrating. In practice we use tables of Laplace transforms of commonly used functions; see for example Table 2.2 later in this section.

In Section 2.7.2 we will develop methods that are used to find the Laplace transforms of a shifted or translated function, step functions, pulses and various other functions that often arise in applications.

(e) We shall verify that the Laplace transform is linear, that is,

$$L(f+g) = L(f) + L(g)$$

$$L(\lambda f) = \lambda L(f) = \lambda F.$$

Example 143. Show that

(i) $L\{f(t)\} = \frac{1}{s}$ where, $s > 0,\ f(t) = 1.$

(ii) $L\{f(t)\} = \frac{1}{s^2}$, where $s > 0$ and $f(t) = t.$

(iii) $L\{f(t)\} = \frac{n!}{s^{n+1}}$, $s > 0$, where $f(t) = t^n.$

(iv) $L\{(f(t)\} = \frac{1}{s^2+1}$, where $f(t) = \sin t.$

(v) $L\{(f(t)\} = \frac{s}{s^2+4}$, where $f(t) = \cos 2t.$

(vi) $L\{e^{at}\} = \frac{1}{s-a}$, $s > a$, and a is any real number.

(vii) $L\{e^{3t+1}\} = \frac{e}{s-3}$, $s > 3.$

Solution:(i) By Definition 47, we have

$$\begin{aligned} L\{1\} &= \int_0^\infty e^{-st}(1)dt = \lim_{T\to\infty}\int_0^T e^{-st}dt \\ &= \lim_{T\to\infty}\left[e^{-st}\left(\frac{-1}{s}\right)\right]_0^T \\ &= \lim_{T\to\infty}\frac{1}{s}\left[-e^{-sT}+e^{-s.0}\right] \\ &= \frac{1}{s}\left[-\lim e^{-st}+1\right] \\ &= \frac{1}{s}[0+1] = \frac{1}{s}, \quad s > 0. \end{aligned}$$

(ii)

$$\begin{aligned} L\{t\} &= \int_0^\infty e^{-st}tdt \qquad \text{(by Definition 47)} \\ &= \lim_{T\to\infty}\int_0^\infty e^{-st}tdt \\ &= \lim_{T\to\infty}\left[\frac{-1}{s}e^{-st}t\right]_0^T + \frac{1}{s}\int_0^\infty e^{-st}dt \end{aligned}$$

by applying integration by parts. Since the first term is zero and the second is

$$L\{t\} = \frac{1}{s^2},$$

by part (i).
Therefore $L\{t\} = \frac{1}{s^2}$.
(iii) By Definition 47, we have

$$L(t^n) = \int_0^\infty e^{-st} t^n dt.$$

By applying the formula for integration by parts n times we conclude that

$$L(t^n) = \frac{n!}{s^{n+1}},$$

$$\int_0^\infty e^{-st} t^n dt = \left[-\frac{t^n e^{-st}}{s} \right]_0^\infty + \frac{n}{s} \int_0^\infty e^{-st} t^{n-1} dt.$$

The first term on the right side is equal to zero for $n > 0$ and $s > 0$, so

$$L(t^n) = \int_0^\infty e^{-st} t^n dt = \frac{n}{s} \int_0^\infty e^{-st} t^{n-1} dt = \frac{n}{s} L(t^{n-1}).$$

Replacing n with $n-1$ in this equation, we get

$$L(t^{n-1}) = \frac{(n-1)}{s} L(t^{n-2}).$$

Combining values of $L(t^n)$ and $L(t^{n-1})$ one can write

$$L(t^n) = \frac{n(n-1)}{s^2} L(t^{n-2}).$$

Continuing in this way one gets

$$L(t^n) = \frac{n(n-1)(n-2)\ldots 3.2.1}{s^n} L(t^0).$$

Since $L(t^0) = L(1) = \frac{1}{s}$ by part (i), we obtain

$$L(t^n) = \frac{n!}{s^{n+1}},$$

where $n! = n(n-1)(n-2)\ldots 3.2.1$.
(iv) $L(\sin t) = \int_0^\infty e^{-st} \sin t dt$, by Definition 47,

$$\text{or } L(\sin t) = \lim_{T \to \infty} \int_0^T e^{-st} \sin t dt.$$

$$\begin{aligned}
\text{Let } I &= \lim_{T\to\infty}\int_0^T e^{-st}\sin t dt \\
&= \left[-\frac{1}{s}e^{-st}\sin t\right]_0^\infty + \frac{1}{s}\int_0^T e^{-st}\cos t dt \\
&= \frac{-1}{s}e^{-sT}\sin T + \left[-\frac{1}{s^2}e^{-st}\cos t\right]_0^\infty - \int_0^T\left(-\frac{1}{s^2}\right)(-\sin t)e^{-st}dt \\
&= \frac{-1}{s}e^{-sT}\sin T - \frac{1}{s^2}e^{-sT}\cos T + \frac{1}{s^2} - \frac{1}{s^2}\int_0^T e^{-st}\sin t dt \\
&= \frac{-1}{s}e^{-sT}\sin T - \frac{1}{s^2}e^{-sT}\cos T + \frac{1}{s^2} - \frac{1}{s^2}I.
\end{aligned}$$

Bringing $-\frac{1}{s^2}I$ on the left hand side we get

$$\left(1+\frac{1}{s^2}\right)I = \frac{-1}{s}e^{-sT}\sin T - \frac{1}{s^2}e^{-sT}\cos T + \frac{1}{s^2}.$$

By taking the limit as $T \to \infty$ in this equation, we get

$$\left(1+\frac{1}{s^2}\right)I = \frac{1}{s^2}$$

$$\text{or I} = \frac{\mathrm{s}^2}{\mathrm{s}^2+1}\frac{1}{\mathrm{s}^2} = \frac{1}{\mathrm{s}^2+1}.$$

(v) By Definition 47,

$$\begin{aligned}
L\{\cos 2t\} &= \int_0^\infty \cos 2te^{-st}dt \\
&= \left[e^{-st}(-s\cos 2t + 2\sin 2t)\right]_0^\infty \\
&= \lim_{m\to\infty}\frac{e^{-sm}(-s\cos 2m + 2\sin 2m)}{s^2+4} + \frac{s}{s^2+4} \\
&= 0 + \frac{s}{s^2+4} \\
&= \frac{s}{s^2+4}.
\end{aligned}$$

(vi) By Definition 47,

$$\begin{aligned}
L\{e^{at}\} &= \int_0^\infty e^{-st}t^{at}dt, \\
&= \lim_{T\to\infty}\int_0^T e^{(a-s)t}dt \\
&= \lim_{T\to\infty}\left[\frac{1}{a-s}e^{(a-s)t}\right]_0^T
\end{aligned}$$

$$= \lim_{T\to\infty}\left[\frac{1}{a-s} - \frac{1}{a-s}\right]$$

$$= -\frac{1}{a-s} = \frac{1}{s-a}, \quad \text{provided } a-d<0 \text{ or } s>a.$$

Thus the Laplace transform of e^{at} is

$$F(s) = L(e^{at}) = \frac{1}{s-a} \text{ if s > a.}$$

(vii) By Definition 47,

$$\begin{aligned} L\{e^{3t+1}\} &= \int_0^\infty e^{3t+1}e^{-st}dt \\ &= e\int_0^\infty e^{(3-s)t}dt \\ &= e\left[e^{(3-s)}\right]_0^\infty \\ &= e\left[\lim_{m\to\infty}\frac{e^{(3-s)m}}{3-s} - \frac{1}{3-s}\right] \\ &= \frac{e}{s-3}, \quad s>3. \end{aligned}$$

It may be observed that for $s \le 0$, $L(1)$ **does not exist.**
If $s<0$, then the exponent of e is positive for $t>0$. Therefore

$$\begin{aligned} \lim_{T\to\infty} e^{-st}dt &= \lim_{T\to\infty}\left[\frac{e^{-st}}{-s}\right]_0^T \\ &= \lim_{T\to\infty}\left[\frac{-1}{s}e^{-sT} + \frac{1}{s}\right] \\ &= \infty. \end{aligned}$$

which means the integral diverges.
Let $s=0$, then the integral becomes

$$\lim_{T\to\infty} = \lim_{T\to\infty}[t]_0^T = -\lim_{T\to\infty} T = \infty.$$

Theorem 23. *Let $f_1(t)$ and $f_2(t)$ have Laplace transforms and let c_1 and c_2 be constants, then*

(i) $L\{f_1(t)+f_2(t)\} = L\{f_1(t)\} + L\{f_2(t)\}$,

(ii) $L\{c_1f_1(t)\} = c_1L\{f_1(t)\}$ *and*
$L\{c_2f_2(t)\} = c_2L\{f_2(t)\}$, *equivalently,*

(iii) $L\{c_1f_1(t)+c_2f_2(t)\} = c_1L\{f_1(t)\} + c_2L\{f_2(t)\}$.

Proof. (iii) $LHS = L\{c_1 f_1(t) + c_2 f_2(t)\}$

$$\begin{aligned}
&= \int_0^\infty e^{-st}\left[c_1 f_1(t) + c_2 f_2(t)\right] dt \\
&= \int_0^\infty \left[c_1 e^{-st} f_1(t) + c_2 e^{-st} f_2(t)\right] dt \\
&= \int_0^\infty c_1 e^{-st} f_1(t) dt + c_2 e^{-st} f_2(t) dt \\
&= c_1 L\{f_1(t)\} + c_2 L\{f_2(t)\}, \quad \text{by using properties of integrals.}
\end{aligned}$$

□

Definition 48. A function f is said to be of *exponential order* if there exist real numbers a, M and t_0 such that

$$|f(t)| \leq Me^{at} \quad \text{for } \mathrm{t} > \mathrm{t_0}.$$

Example 144. Check whether the following functions are of exponential order.

(a) $f(t) = t^2$ (b) $f(t) = e^t$

(c) $f(t) = \sin t$ (d) e^{t^2}.

Solution (a) Let a be any constant > 0; then

$$\lim_{t\to\infty} |f(t)|e^{-at} = \lim_{t\to\infty} |t|^2 e^{-at}$$

$$= \lim_{t\to\infty} \frac{2t}{ae^{at}} = \lim \frac{2}{a^2 e^{at}} = 0,$$

where the last two limits are obtained by using L'Hopital's rule. Therefore there exists a positive constant M such that

$$|f(t)|e^{-at} \leq M \quad \text{or} \quad \mathrm{f(t) \leq Me^{at}},$$

that is $f(t) = t^2$ is of exponential order.
(b) Let a be any constant > 0; then

$$\begin{aligned}
\lim_{t\to\infty} |f(t)|e^{-at} &= \lim_{t\to\infty} e^t e^{-at} \\
&= \lim_{t\to\infty} e^{(1-a)t} \to 0 \quad \text{for } \mathrm{a} > 1.
\end{aligned}$$

Therefore we can find a and $M > 0$ such that $|f(t)| \leq Me^{at}$.
(c) $\lim_{t\to\infty} |f(t)|e^{-at} \leq \lim_{t\to\infty} e^{-at} \to 0$ for $a > 0$ implying there exists a and $M > 0$ such that $|f(t)| \leq Me^{at}$.
(d) $f(t) = e^{t^2}$ is not of exponential order since its graph grows faster than any positive linear power of e for $t > a > 0$.

Now we prove the following basic existence theorem for the Laplace transform of a function f.

TABLE 2.2: Laplace Transforms

Sr. No.	f(t)	L{f(t)}	Sr. No.	f(t)	L{f(t)}
1.	1	$\frac{1}{s}$	24.	$\sin kt - kt\cos kt$	$\frac{2k^3}{(s^2+k^2)^2}$
2.	t	$\frac{1}{s^2}$	25.	$t\sinh kt$	$\frac{2ks}{(s^2-k^2)^2}$
3.	t^n	$\frac{n!}{s^{n+1}}$	26.	$t\cosh kt$	$\frac{s^2+k^2}{(s^2-k^2)^2}$
4.	$t^{-1/2}$	$\sqrt{\frac{\pi}{s}}$	27.	$\frac{e^{at}-e^{bt}}{a-b}$	$\frac{1}{(s-a)(s-b)}$
5.	$t^{1/2}$	$\frac{\sqrt{\pi}}{2s^{3/2}}$	28.	$\frac{e^{at}-e^{bt}}{a-b}$	$\frac{s}{(s-a)(s-b)}$
6.	$\sin kt$	$\frac{s}{s^2+k^2}$	29.	$1-\cos kt$	$\frac{k^2}{s(s^2+k^2)}$
7.	$\cos kt$	$\frac{k}{s^2+k^2}$	30.	$\frac{e^{at}-e^{bt}}{t}$	$\ln\frac{s-a}{s-b}$
8.	$\sin^2 kt$	$\frac{2k^2}{s(s^2+4k^2)}$	31.	$H(t-a)=u_a(t)$	$\frac{e^{-sa}}{s},\quad s>0$
9.	$\cos^2 kt$	$\frac{s^2+2k^2}{s(s^2+4k^2)}$	32.	$\delta(t)$	1
10.	e^{at}	$\frac{1}{s-a}$	33.	$\delta(t-t_0)$	$e_0^{-st},\quad t>0$
11.	$\sinh kt$	$\frac{k}{s^2-k^2}$	34.	$e^{at}f(t)$	$F(s-a)$
12.	$\cosh t$	$\frac{s}{s^2-k^2}$	35.	$f(t-a)H(t-a)$	$e^{-as}F(s)$

TABLE 2.2: Cont.

Sr. No.	f(t)	L{f(t)}	Sr. No.	f(t)	L{f(t)}
13.	$\sinh^2 kt$	$\dfrac{2k^2}{s(s^2-4k^2)}$	36.	$f^{(n)}(t)$	$s^nF(s)-s^{n-1}f(0)\cdots-f^{(n-1)}(0)$
14.	$\cosh^2 kt$	$\dfrac{s^2-2k^2}{s(s^2-4k^2)}$	37.	$t^nf(t)$	$(-1)^n\dfrac{d^n}{ds^n}F(s)$
15.	te^{at}	$\dfrac{1}{(s-a)^2}$	38.	$\int_0^t f(u)g(t-u)du$	$F(s)G(s)$
16.	t^ne^{at}	$\dfrac{n!}{(s-a)^{n+1}}$ (n a positive integer)	39.	$\dfrac{\sin at}{t}$	$arc\tan\left(\dfrac{a}{s}\right)$
17.	$e^{at}\sin kt$	$\dfrac{k}{(s-a)^2+k^2}$	40.	$\dfrac{1}{\sqrt{\pi t}}e^{-a^2/4t}$	$\dfrac{e^{-a\sqrt{s}}}{\sqrt{s}}$
18.	$e^{at}\cos kt$	$\dfrac{s-a}{(s-a)^2+k^2}$	41.	$f(at)$	$\dfrac{1}{a}L\{f\}\left(\dfrac{s}{a}\right)$
19.	$e^{at}\sinh kt$	$\dfrac{k}{(s-a)^2-k^2}$	42.	$\int_0^t f(u)du$	$\dfrac{1}{s}L\{f\}(s)$
20.	$e^{at}\cosh kt$	$\dfrac{s-a}{(s-a)^2-k^2}$	43.	$\dfrac{f(t)}{t}$	$\int_s^\infty L\{f\}(x)dx$
21.	$t\sin kt$	$\dfrac{2ks}{(s^2+k^2)^2}$	44.	$e^{at}f(t)$	$L\{f\}(s-a)$
22.	$t\cos kt$	$\dfrac{s^2-k^2}{(s^2+k^2)^2}$	45.	$f(t+T)=f(t), T>0$	$\dfrac{1}{1+e^{-sT}}\int_0^T e^{-st}f(t)dt$
23.	$\sin kt+kt\cos kt$	$\dfrac{2ks^2}{(s^2+k^2)^2}$	46.	$\dfrac{\sin at}{t}$	$arc\tan\left(\dfrac{a}{s}\right),\ \ s>0$

Theorem 24. *Let f be a positive continuous function of exponential order defined on $[0,\infty)$, then its Laplace transform exists for parameters greater than some constant a.*

Proof. *Since the function f is of exponential order, we know that there are constants t_0, a and M such that*

$$\begin{aligned} |f(t)| &\leq Me^{at} \text{ for } t > t_0 \\ \text{or } e^{-at}|f(t)| &\leq M \text{ for } t > t_0 \\ \text{or } |e^{-at}f(t)| &\leq M \text{ for } t > t_0 \\ \text{as } |e^{-at}f(t)| &= |e^{-at}||f(t)| = e^{-at}|f(t)|. \end{aligned}$$

It may be noted that e^{-at} is always positive. Multiplying by e^{-st+at}, we have

$$|e^{-st}f(t)| \leq Me^{-st}e^{at}.$$

Hence

$$\begin{aligned} \int_0^\infty |e^{-st}f(t)|dt &\leq \int_0^\infty Me^{-(s-a)}dt \\ &= \left[\frac{-Me^{-(s-a)}}{s-a}\right]_0^\infty \\ &= -\lim_{t\to\infty}\frac{Me^{-(s-a)t}}{s-a} + \frac{M}{s-a}. \end{aligned}$$

Since the first term is zero for $s > a$, we have

$$\int_0^\infty |e^{-st+at}dt \leq \frac{M}{s-a}, \quad s > a$$

which implies the existence of the improper integral defining the Laplace transform of f and completes the proof. □

Corollary: Let f be a piecewise continuous function of exponential order defined on $[0,\infty)$. Then $\lim_{s\to\infty} F(s) = L(f(t)) = 0$.

Proof.

$$\begin{aligned} |L(f(t))| &= \left|\int_0^\infty e^{-st}f(t)dt\right| \\ &\leq \int_0^\infty |e^{-st}f(t)|dt \\ &\leq \frac{M}{s-a}, \quad s > a \text{ as seen in the proof of Theorem 24.} \end{aligned}$$

Taking the limit as $s \to \infty$ we get

$$\lim_{s\to\infty} F(s) = \lim_{s\to\infty} |L(f(t))| = 0.$$

□

2.7.2 Translation Theorems

The following are called shifting theorems.

Theorem 25. *(The First Shifting Theorem):*
Let $L\{f(t)\} = F(s)$. Then

$$L\{e^{at}f(t)\} = F(s-a).$$

Proof. *By definition of the Laplace Transform, we write*

$$\begin{aligned} L\{e^{at}f(t)\} &= \int_0^\infty e^{-st}(e^{at}f(t))dt \\ &= \int_0^\infty e^{-(s-a)t}f(t)dt \\ &= F(s-a). \end{aligned}$$

□

Theorem 26. *(The Second Shifting Theorem):*
Let $e^{at}f(t) = F(s)$. Then

$$L\{H(t-a)f(t-a)\} = e^{-as}F(s)$$

where H is a Heaviside function defined as

$$H(t) = \begin{cases} 0 & \text{if t} < 0 \\ 1 & \text{if t} \geq 0 \end{cases}.$$

Proof. $L\{H(t-a)f(t-a)\} = \int_0^\infty e^{-st}H(t-a)f(t-a)dt$

$$= \int_a^\infty e^{-st}f(t-a)dt$$

because $H(t-a) = 0$ for $t < a$ and $H(t-a) = 1$ for $t \geq a$. Now let $u = t - a$ in the last integral. We get

$$\begin{aligned} L\{H(t-a)f(t-a)\} &= \int_0^\infty e^{-s(a+u)t}f(u)du \\ &= e^{-as}\int_0^\infty e^{-st}f(u)du \\ &= e^{-as}L\{f(u)\} = e^{-as}F(s). \end{aligned}$$

□

Example 145. Apply the first shifting theorem to find

(a) $L\{e^{3t}\sin t\}$.

(b) $L\{e^{-t}g(t)\}$, where

$$g(t) = \begin{cases} 2t & 0 \le t < 3 \\ -1 & t \ge 3 \end{cases}.$$

(c) $L^{-1}\left\{\dfrac{4}{s^2 - 4s + 20}\right\}$.

Solution: (a) Since $L\{\sin t\} = \dfrac{1}{s^2+1}$, it follows that by Theorem 25

$$L\{e^{3t}\sin t\} = \frac{1}{(s-3)^2+1}.$$

(b) By Theorem 25

$$L\{e^{-t}g(t)\} = F(s+a)$$

where $L\{g(t)\} = F(s)$.

$$\begin{aligned}
F(s) &= \int_0^\infty e^{-st}g(t)dt \\
&= \int_0^3 e^{-st}g(t)dt + \int_3^\infty e^{-st}g(t)dt \\
&= \int_0^3 e^{-st}2tdt - \int_3^\infty e^{-st}dt \\
&= 2\left[e^{-st}\left(-\frac{1}{s}\right)t\right]_0^3 - 2\int_0^3 e^{-st}\left(-\frac{1}{s}\right)dt - \left[-\frac{1}{s}e^{-st}\right]_3^\infty \\
&= \frac{2}{s^2} - \frac{2e^{-3s}}{s^2} - \frac{7e^{-3s}}{s}
\end{aligned}$$

Therefore $L\{e^{-t}g(t)\} = \dfrac{2}{(s+1)^2} - \dfrac{2e^{-3(s+1)}}{(s+1)^2} - \dfrac{7e^{-3(s+1)}}{s+1}$.

(c) We have $\dfrac{4}{s^2+4s+20} = \dfrac{4}{(s+2)^2+6}$.

$$F(s+2) = \frac{4}{(s+2)^2+16}.$$

This means we should choose

$$F(s) = \frac{4}{s^2+16}.$$

By the first shifting Theorem 25,

$$\begin{aligned}
L\{e^{-2t}\sin 4t\} &= F(s-(-2)) = F(s+2) \\
&= \frac{4}{(s+2)^2+16}
\end{aligned}$$

and therefore

$$L^{-1}\left\{\frac{4}{(s+2)^2+16}\right\} = e^{-2t}\sin 4t.$$

Example 146. Compute $L^{-1}\left\{\frac{se^{-3s}}{s^2+4}\right\}$.

Solution: By Theorem 26

$$\begin{aligned} L\{H(t-a)f(t-a)\} &= e^{-as}F(s) \\ \text{or}\quad \mathrm{H(t-a)f(t-a)} &= L^{-1}\{e^{-as}F(s)\} \\ F(s) &= \frac{s}{s^2+4} \\ L^{-1}\{F(s)\} &= L^{-1}\left(\frac{s}{s^2+4}\right) \quad \text{implies that} \quad f(t) = \cos 2t. \end{aligned}$$

Therefore,

$$L^{-1}\left\{\frac{se^{-3s}}{s^2+4}\right\} = H(t-3)\cos 2(t-3).$$

2.7.3 Inverse Laplace Transform

In the previous section we have seen the method for finding the Laplace transform. In this section we discuss the method for reversing the process of the previous section and we reconstruct a function $f(t)$ whose Laplace transform $F(s)$ is given.

Definition 49. Let $f(t)$ be a function such that $L\{f(t)\} = F(s)$, then $f(t)$ is called the inverse Laplace transform. The inverse Laplace transform is designated as L^{-1} and we write

$$f(t) = L^{-1}\{F(s)\}.$$

In order to find an inverse transform we must be familiar with the formulas for finding Laplace Transforms, see Table 2.2. One should learn to use this table in reverse. However in general the given Laplace transform will not be in the form that allows direct use of the table, so the given $F(s)$ have to be algebraically manipulated in a form that can be found in the table. The most relevant result for this purpose is the linearity property of the inverse Laplace transform which states that

$$L^{-1}\{c_1F_1(s) + c_2F_2(s)\} = c_1L^{-1}\{F_1(s)\} + c_2L^{-1}\{F_2(s)\}$$

where c_1 and c_2 are constants.

The proof of this result follows from definition of the inverse Laplace transform and the corresponding linearity of the Laplace transform.

Example 147. Find

(i) $L^{-1}\left\{\dfrac{1}{s+2}\right\}$.

(ii) $L^{-1}\left\{\dfrac{1}{s^2+2}\right\}$.

(iii) $L^{-1}\left\{\dfrac{-2s+6}{(s^2+4)}\right\}$.

(iv) $L^{-1}\left\{\dfrac{s+5}{(s^2-2s-3}\right\}$.

(v) $L^{-1}\left\{\dfrac{s+1}{(s^2-4s)}\right\}$.

(vi) $L^{-1}\left\{\dfrac{2s-6}{s^2+9}\right\}$.

(vii) $L^{-1}\left\{\dfrac{2s-5}{(s-1)^3(s^2+4)}\right\}$.

Solution: (i) From Table 2.2, let $L(e^{at}) = \dfrac{1}{s-a}$. Choosing $a = -2$ we get $L(e^{-2t}) = \dfrac{1}{s+2}$ and consequently by the definition of the inverse Laplace transform

$$L^{-1}\left\{\frac{1}{s+2}\right\} = e^{-2t}$$

(ii) By Table 2.2, for $k = \sqrt{2}$ and the linearity of the Inverse Laplace Transform we get

$$\begin{aligned} L^{-1}\left\{\frac{1}{s^2+2}\right\} &= \frac{1}{\sqrt{2}}L^{-1}\left\{\frac{\sqrt{2}}{s^2+2}\right\} \\ &= \frac{1}{\sqrt{2}}\sin\sqrt{2}t. \end{aligned}$$

(iii) As
$$\begin{aligned} L^{-1}\left\{\frac{-2s+6}{(s^2+4)}\right\} &= L^{-1}\left\{\frac{-2s}{(s^2+4)}+\frac{6}{(s^2+4)}\right\} \\ &= -2L^{-1}\left\{\frac{s}{(s^2+4)}\right\}+6L^{-1}\left\{\frac{1}{(s^2+4)}\right\}. \end{aligned}$$

By linearity of the inverse transform,

$$\begin{aligned} &= -2\cos 2t+\frac{6}{2}L^{-1}\left\{\frac{2}{(s^2+4)}\right\} \\ &= -2\cos 2t + 3\sin 2t \qquad \text{by Table 2.2 (items 6 and 7).} \end{aligned}$$

(iv) Since $s^2 - 2s - 3 = (s-3)(s+1)$ we get

$$\frac{s+5}{s^2-2s-3} = \frac{s+5}{(s-3)(s+1)} = \frac{A}{s-3} + \frac{s-3}{s+1}$$

where constants A and B are to be determined.

$$\begin{aligned}\frac{A}{s-3} + \frac{s-3}{s+1} &= \frac{A(s+1)+(s-3)B}{(s-3)(s+1)} \\ &= \frac{(A+B)s+(A-3B)}{(s-3)(s+1)}.\end{aligned}$$

We can write

$$\frac{s+5}{(s-3)(s+1)} = \frac{(A+B)s+(A-3B)}{(s-3)(s+1)}.$$

This implies that

$$s+5 = (A+B)s + (A-3B).$$

This gives $A+B=1$ and $A-3b=5$.
Subtracting second from first we get $B=-1$. Putting this value in the first equation we get $A=2$. Therefore, we have

$$\begin{aligned}L^{-1}\left\{\frac{s+5}{(s^2-2s-3}\right\} &= L^{-1}\left\{\frac{2}{s-3} + \frac{-1}{s+1}\right\} \\ &= 2L^{-1}\left\{\frac{1}{s-3}\right\} - L^{-1}\left\{\frac{1}{s+1}\right\}, \\ &\quad \text{using linearity of } L^{-1}\end{aligned}$$

By Table 2.2 (item 10, for $a=3$ and $a=-1$) we get

$L^{-1}\left\{\frac{1}{s-3}\right\} = e^{3t}$ and $L^{-1}\left\{\frac{1}{s+1}\right\} = e^{-t}$.

Hence $L^{-1}\left\{\frac{s+5}{(s^2-2s-3)}\right\}h = 2e^{3t} - e^{-t}$.

(v) $L^{-1}\left\{\frac{s+1}{(s^2-4s)}\right\} = L^{-1}\left\{-\frac{1}{4}\frac{1}{s} + \frac{5}{4}\frac{1}{s-4}\right\}$

$$= -\frac{1}{4}L^{-1}\left\{\frac{1}{s}\right\} + \frac{5}{4}\left\{\frac{1}{s-4}\right\}$$

$$= -\frac{1}{4} + \frac{5}{4}e^{4t}$$

(by Table 2.2, items 1 and 10).

(vi) $L^{-1}\left\{\frac{2s-6}{s^2+9}\right\} = L^{-1}\left\{2.\frac{s}{s^2+9} - 2.\frac{3}{s^2+9}\right\}$

$$= 2\cos 3t - 2\sin 3t$$

(vii) Given $F(s) = \dfrac{2s-5}{(s-1)^3(s^2+4)}$, decompose $F(s)$ into partial fractions as follows

$$\frac{2s-5}{(s-1)^3(s^2+4)} = \frac{A}{s-1} + \frac{B}{(s-1)^2} + \frac{C}{(s-1)^3} + \frac{Ds+E}{s^2+4}.$$

By comparison, we find that the coefficients are

$$A = \frac{-17}{125},\quad B = \frac{16}{25},\quad C = \frac{-3}{5},\quad D = \frac{17}{125},\quad E = \frac{-63}{125}.$$

Using shifting property, (Theorem 27) and the Laplace transforms of 1, t, t^2, $\sin 2t$ and $\cos 2t$, we have

$$\begin{aligned} L^{-1}\{F(s)\} &= -\frac{17}{125}e^t + \frac{16}{25}te^t - \frac{3}{10}t^2e^t + \frac{17}{125}\cos 2t - \frac{63}{250}\sin 2t \\ &= \frac{1}{250}(-34e^t + 160te^t - 75t^2e^t + 34\cos 2t - 63\sin 2t). \end{aligned}$$

2.7.4 Step and Impulse Functions

In engineering literature one often encounters functions that are discontinuous (off or on) or quite large in very small intervals. An important function representing such a situation is called the Heaviside function or unit step function. Oliver Heaviside was an electrical engineer who introduced this function. He lived from 1850 to 1925 and developed methods to solve differential equations by applying Laplace transforms.

Definition 50. The Heaviside function denoted by H is defined by

$$H(t) = \begin{cases} 0, & t < 0 \\ 1, & t \geq 0 \end{cases}$$

Remark 32. $H(t-a)$ or $u_a(t)$ is another notation for the Heaviside function defined by

$$H(t-a) = \begin{cases} 0, & 0 \leq t < a \\ 1, & t \geq a \end{cases}.$$

The graph of the Heaviside function $y = H(t-3)$ is shown in Figure 2.14.

Remark 33. (i) It may be observed that $H(t-a)f(t-a)$ is simply a translation of a function $f(t)$, that is,

$$H(t-a)f(t-a) = \begin{cases} 0, & 0 \leq t < a \\ f(t-a), & t \geq a \end{cases}.$$

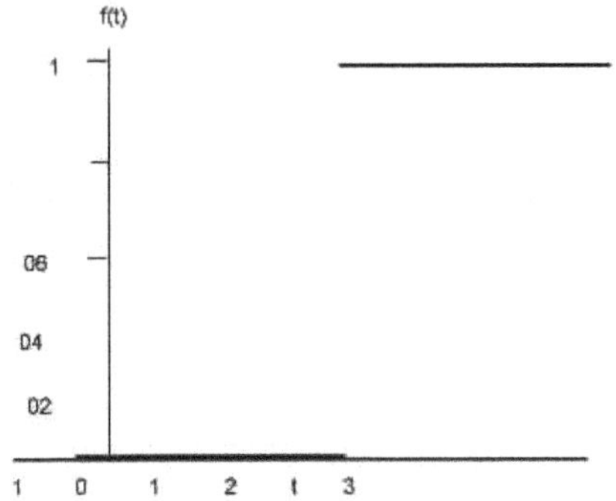

FIGURE 2.14: Heavyside Function $H(t-3)$

(ii) An external force acting on a mechanical system or a voltage impressed on a circuit turned off after a period of time can be represented by the Heaviside function.

Example 148. Find the Laplace transform of the following functions
(i) $H(t-8)$.
(ii) $t + H(t-2)(t-2)^2$.
(iii) $t-(t-1)H(t-1)$.

Solution: (i) $L\{H(t-8)\}(s) = \int_0^{\infty} H(t-8)e^{-st}ds$

$$
\begin{aligned}
&= \int_0^8 H(t-8)e^{-}sds + \int_8^{\infty} H(t-8)e^{-s}ds \\
&= \int_8^{\infty} e^{-s}ds = \frac{e^{-8s}}{s}.
\end{aligned}
$$

(ii) $L\{t + H(t-2)(t-2)^2\} = L\{t\} + L\{H(t-2)(t-2)^2\}$

$$
\begin{aligned}
&= \frac{1}{s^2} + \int_0^2 H(t-2)(t-2)^2 e^{-st}dt \\
&\qquad + \int_2^{\infty} H(t-2)(t-2)^2 e^{-st}dt \\
&= \frac{1}{s^2} + 0 + \int_2^{\infty} (t-2)^2 e^{-st}dt \\
&= \frac{1}{s^2} + e^{-2s}(L\{t^2\})(s) \\
&= \frac{1}{s^2} + e^{-2s}\frac{2}{s^3} \\
&= \frac{s + e^{-2s}}{s^3}.
\end{aligned}
$$

(iii) Using translation property (Theorem 26), we have

$$F(s) = -\frac{1}{s^2} - \frac{e^{-s}}{s^2}.$$

Example 149. Find the inverse Laplace transform of the function

$$F(s) = \frac{1}{s^4} + \frac{e^{-5s}}{s^4}.$$

Solution: $f(t) = L^{-1}\{F(s)\} = L^{-1}\{\frac{1}{s^4}\} + L^{-1}\{\frac{e^{-5s}}{s^4}\}$

$$= \frac{t^3}{3!} + H(t-5)\frac{(t-5)^3}{3!}, \quad \text{applying Theorem 26.}$$

Unit Impluse: Mechanical systems are often impacted by external forces of large magnitude that act only for very short times. One example is electromagnetic force acting within an electrical circuit. A vibrating airplane wing may be struck by lightning. A spring may receive a sharp blow from a mass. A baseball, golf ball, or tennis ball can soar into the air when struck violently by a bat, club, or racket. Such impacts are often called *generalized functions* or *distributions.*

Note, however, that these impacts are not functions in the classical sense (designated Dirac delta functions to honor P.A.M. Dirac (1902-1984, recipient of the Nobel Prize in physics in 1933). Dirac supervised another Nobel laureate, Abdus Salam who won the physics prize in 1979 and founded and led UNESCOs International Centre of Theoretical Physics in Trieste, Italy.

The Dirac delta function introduced in 1930 contradicted the well known Lebesgue integral theory used by engineers and physicists for 20 years. In 1950, Laurent Schwartz, a French mathematician, developed a rigorous mathematical theory of generalized functions. Known as the distribution theory, it provided logical foundations for delta function techniques. Every piecewise continuous function is a distribution. However, the delta function defined below is an example of a distribution that is *not* a function in the classical sense.

$$\delta_\epsilon = \frac{1}{2}[H(t) - H(t-\epsilon)]$$

For any positive number ϵ, the impulse $\delta\epsilon$ is defined by:

$$\begin{aligned} \delta(t) &= \lim_{\epsilon \to 0^+} \delta_\epsilon(t) \\ \text{and } \delta(\text{t} - \text{a}) &= \lim_{\epsilon \to 0^+} \delta_\epsilon(t-a), \quad a > 0 \end{aligned}$$

($\epsilon \to 0^+$ means ϵ tends to zero from the right hand side).

Definition 51. (a) The Dirac delta function concentrated at $t = 0$, denoted by $\delta(t)$ is defined as

$$\delta(t) = 0, \quad t \neq a$$

$$\int_{-\infty}^{\infty} \delta(t) = 1.$$

(b) The Dirac delta function, concentrated at a point $t = a$, denoted by $\delta(t-a)$ is defined as

$$\delta(t-a) = 0, \quad t \neq 0$$

$$\int_{-\infty}^{\infty} \delta(t-a) = 1.$$

Example 150. Find the Laplace transforms of the following functions: (a) Dirac delta function $\delta(t-a)$.
(b) $\delta(t-\pi)$.
(c) $\sin t.H_{2\pi}(t) + \delta(t - \frac{\pi}{2})$.

Solution: (a) $L\{\delta(t-a)\} = L\{\lim_{\epsilon \to 0^+} \delta_\epsilon(t-a)\}$

$$\begin{aligned}
&= \lim_{\epsilon \to 0^+} \frac{1}{\epsilon}\left[\int_0^{\infty} \{H(t-a) - H(t-a-\epsilon)\}\right] dt \\
&= \lim_{\epsilon \to 0^+} \frac{1}{\epsilon}\left[\frac{1}{s}e^{-as} - \frac{1}{s}e^{-(a+\epsilon)s}\right] \\
&= \frac{\left[\frac{1}{s}e^{-as} - \lim_{\epsilon \to 0^+} \frac{1}{s}e^{-(a+\epsilon)s}\right]}{\epsilon} \\
&= \lim_{\epsilon \to 0^+} \frac{e^{-as}(1 - e^{-\epsilon s})}{\epsilon s} = e^{-as}
\end{aligned}$$

(as $\lim_{\epsilon \to 0^+} \dfrac{1 - e^{-\epsilon s}}{\epsilon s} = 1$). If $a = 0$, $L\{\delta(t)\} = e^{-0.s} = 1$.
(b) We know $L\{\delta(t-a)\} = e^{-sa}$ and so $L\{\delta(t-\pi)\} = e^{-s\pi}$.

(c) Since $\sin t$ is periodic with period 2π, we have

$$\sin t H_{2\pi}(t) - \delta\left(t - \frac{\pi}{2}\right) = \sin(t - 2\pi) H_{2\pi}(t) - \delta\left(t - \frac{\pi}{2}\right).$$

Therefore, by the translation property Theorem 26 and by part (a) we get desired result.

2.7.5 Some Additional Properties

The Laplace transforms of the derivatives of a differentiable function exist under appropriate conditions. In this section we discuss results that are quite useful in solving differential equations. For solving second order differential equations we need to evaluate the Laplace transform of $\frac{dy}{dt}$ and $\frac{d^2y}{dt^2}$.

Let $f(t)$ be differentiable for $t \geq 0$ and let its derivative $f'(t)$ be continuous. By applying the formula for integration by parts we find that

$$L\{f'(t)\} = sF(s) - f(0). \tag{2.95}$$

Verification: By definition

$$\begin{aligned} L\{f'(t)\} &= \int_0^\infty e^{-st} f'(t)dt \\ &= \left[e^{-st} f'(t)\right] + s\int_0^\infty e^{-st} f(t)dt \\ &= -f(0) + sL\{f(t)\} \\ &= sF(s) - f(0). \end{aligned}$$

Here we have used the fact that

$$\lim_{t\to\infty} e^{-st} f(t) = 0.$$

Similarly for a twice differentiable function $f(t)$ such that $f'(t)$ is continuous we can prove that

$$L\{f'(t)\} = s^2 F(s) - sf(0) - f'(0). \tag{2.96}$$

In fact we can prove the following theorem by repeatedly applying integration by parts.

Theorem 27. *Let $f, f', \ldots, f^{(n-1)}$ be continuous on $[0, \infty)$ and of exponential order and if $f^{(n)}$ (t) be piecewise continuous on $[0, \infty)$, then*

$$L\{f^{(n-1)}(t)\} = s^n F(s) - s^{n-1} f(0) - s^{n-2} f'(0) \cdots - f^{n-1}(0). \tag{2.97}$$

Theorem 27 can be used to generate a formula for the Laplace transform of the indefinite integral of a function f. We then have the following theorem.

Theorem 28. *Let f be piecewise continuous and of exponential order for $t \geq 0$, then*

$$L\left\{\int_0^t f(u)du\right\} = \frac{1}{s}L\{f(t)\} = \frac{1}{s}L\{F(s)\}.$$

Proof. *Let $g(t) = \int_0^t f(u)du$. Then $g'(t) = f(t)$ and $g(0) = 0$. Furthermore, $g(t)$ is of exponential order. By Theorem 27*

$$\begin{aligned} L\{g'(t)\} &= sL\{g(t)\} - g(0) \\ \text{or } \mathrm{L\{f(t)\}} &= sL\{\int_0^t f(u)du\} \\ L\{\int_0^t f(u)du\} &= \frac{1}{s}L\{f(t)\}. \end{aligned}$$

□

Example 151. (a) Using the Laplace transform of f'', find $L\{\sin kt\}$.
(b) Show that $L^{-1}\left\{\dfrac{1}{s}F(s)\right\} = \int_0^t f(u)du.$

Solution: (a) Let $f(t) = \sin kt$, then $f'(t) = k\cos kt$, $f'(t) = -k^2 \sin kt$, $f(0) = 0$ and $f'(0) = k$. Therefore

$$\begin{aligned} L\{f''(t)\} &= s^2F(s) - sF(0) - f'(0) \\ \text{or } L\{f''(t)\} &= s^2F(s) - k \\ \text{or } L\{-k^2 \sin kt\} &= s^2F(s) - k \end{aligned}$$

where $F(s) = L\{f(t)\} = L\{\sin kt\}$

$$L\{-k^2 \sin kt\} = s^2F(s) - k = s^2L\{\sin kt\} - k.$$

Solving for $L\{\sin kt\}$ we get

$$L\{\sin kt\} = \frac{k}{s^2 + k^2}$$

$$(-k^2L\{\sin kt\} = s^2L\{\sin kt\} - k \quad \text{or} \quad L\{\sin kt\}(s^2 + k^2) = k).$$

(b) By Theorem 28,

$$L\{\int_0^t f(u)ds\} = \frac{1}{s}F(s).$$

This implies that $\int_0^t f(u)ds = L^{-1}\left\{\dfrac{1}{s}F(s)\right\}$.

Derivative of Laplace Transform

Theorem 29. *Let $f(t)$ be piecewise continuous and of exponential order over each finite interval, and let*

$$L\{f(t)\} = F(s).$$

Then $F(s)$ is differentiable and

$$F'(s) = L\{-tf(t)\}.$$

Proof. Suppose that $|f(t)| \leq Me^{at}, t > 0$ and take any $s_0 > a$. Then consider

$$\frac{\partial}{\partial s}\left[e^{-st}f(t)\right] = -tf(t)e^{-st}.$$

Choose $\epsilon > 0$ such that $s_0 > a + \epsilon$. Then we have $|t|, e^{\epsilon t}$ for all t large enough since in fact

$$\lim_{t \longrightarrow \infty} \left(\frac{s}{e^{at}}\right) = 0.$$

Thus $|tf(t)| \le Me^{(a+\epsilon)t}$, *for all large* t *and we find that* $tf(t)$ *is also of exponential order and*

$$\int_0^\infty tf(t)e^{-st}dt$$

exists by Theorem 24, that is, the integral converges uniformly. Hence $F(s)$ *is differentiable at* s_0 *and*

$$F(s_0) = \int_0^\infty \frac{\partial}{\partial s}\left[e^{-st}f(t)dt\right]dt \quad \text{at s} = \text{s}_0.$$

Therefore

$$\begin{aligned} F'(s) &= -\int_0^\infty tf(t)e^{-st}dt \\ &= L\{-tf(t)\} \qquad \textit{for all } s > a. \end{aligned}$$

□

Remark 34. It can be checked that $F(s) = L\{f(t)\}$ and $n = 1, 2, 3, \ldots$; then

$$L\{t^n f(t)\} = (-1)^n \frac{d^n}{ds^n} F(s).$$

Definition 52. (Convolution). Let f and g be piecewise continuous functions for $t \ge 0$. Then the *convolution* of f and g denoted by $f * g$, is defined by the integral

$$\begin{aligned} (f * g)(t) &= \int_0^t f(u)g(t-u)du \\ &= \int_0^t g(u)f(t-u)du \\ &= (g * f)(t). \end{aligned}$$

Theorem 30. *(Convolution theorem). Let* f *and* g *be piecewise continuous and of exponential order for* $t \ge 0$*; then the Laplace transform of* $f * g$ *is given by the product of the Laplace transform of* f *and the Laplace transform of* g*. That is*

$$L\{f * g\} = F(s)G(s).$$

Proof. *Let* $F = L\{f\}$ *and* $G = L\{g\}$*. Then*

$$\begin{aligned} F(s)G(s) &= F(s)\int_0^\infty e^{-st}g(t)dt \\ &= \int_0^\infty F(s)e^{-su}g(u)du \end{aligned}$$

in which we changed variables of integration from t to u and brought F(s) inside the integral. Let us recall that

$$e^{-su}F(s) = L\{H(t-u)f(t-u)\}$$

where $F(s) = L\{f(t)\}$ *and H(.) is the Heaviside function, see Theorem 26. Substitute this into the integral for F(s)G(s) to get*

$$F(s)G(s) = \int_0^\infty L\{H(t-u)f(t-u)\}g(u)du. \tag{2.98}$$

But, from the definition of the Laplace transform,

$$L\{H(t-u)f(t-u)\} = \int_0^\infty e^{-st}\{H(t-u)f(t-u)\}dt.$$

Substituting this into (2.98) we get

$$\begin{aligned} F(s)G(s) &= \int_0^\infty \left[\int_0^\infty e^{-st}H(t-u)f(t-u)dt\right] g(u)du \\ &= \int_0^\infty \int_0^\infty e^{-st}g(u)H(t-u)f(t-u)dtdu. \end{aligned}$$

Let us recall that

$$H(t-u) = \begin{cases} 0, & \text{if } 0 \le t < u \\ 1, & \text{if } t \ge u \end{cases}.$$

Therefore,

$$F(s)G(s) = \int_0^\infty \int_0^\infty e^{-st}g(u)H(t-u)f(t-u)dtdu.$$

Figure 2.15 show $t-s$ *plane.*

The Laplace integral is over the shaded region, consisting of points satisfying

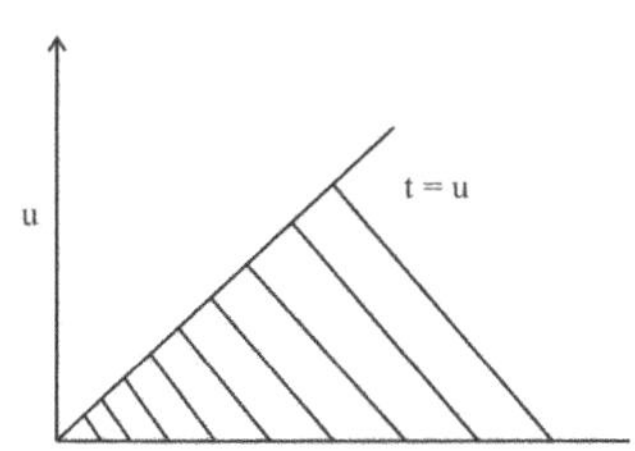

FIGURE 2.15: $t-S$ Plane

$0 \leq u \leq t < \infty$. *Reversing the order of integration gives us*

$$\begin{aligned} F(s)G(s) &= \int_0^\infty \int_0^\infty e^{-st} g(u) f(t-u)\,dt\,du \\ &= \int_0^\infty e^{-st} \left[\int_0^\infty g(u) f(t-u)\,du \right] dt \\ &= \int_0^\infty e^{-st} (f * g)(t)\,dt \\ &= L\{f * g\}. \end{aligned}$$

□

It follows immediately from Theorem 30 that:

Theorem 31. *Let* $L^{-1}(F) = f,\ L^{-1}(G) = g$. *Then*

$$L^{-1}\{FG\} = f * g.$$

Example 152. (a) Evaluate $L^{-1}\left\{\dfrac{1}{(s^2+25)^2}\right\}$.
(b) Let $f(t) = e^{-t}$ and $g(t) = \sin t$. Show that
(i) $f * g = g * f$ and (ii) $L\{f\}L\{g\} = (L\{f * g\})(s)$.

Solution: Let $F(s)G(s) = \dfrac{1}{s^2+5^2}$, so that

$$f(t) = g(t) = \frac{1}{5} L^{-1}\left\{\frac{1}{s^2+5^2}\right\} = \frac{1}{5}\sin 5t.$$

By the convolution theorem

$$\begin{aligned} L^{-1}\left\{\frac{1}{(s^2+5^2)^2}\right\} &= \frac{1}{25}\int_0^t \sin ku \sin k(t-u)\,du \\ &= \frac{1}{2.25}\int_0^t [\cos k(2u-t) - \cos 5t]\,du \\ &= \frac{1}{50}\left[\frac{1}{2.5}\sin k(2u-t) - u\cos kt\right]_0^t \\ &= \frac{\sin 5t - 5t\cos 5t}{2(5)^3}. \end{aligned}$$

(b) (i) $(f * g)(t) = \displaystyle\int_0^t f(t-x)g(x)\,dx$

$(f * g)(t) = \displaystyle\int_0^t f(t-x)g(x)\,dx$

$(f * g)(t) = \displaystyle\int_0^t f(t-x)g(x)\,dx$

$$
\begin{aligned}
&= \int_0^t e^{-(t-x)} \sin x dx \\
&= e^{-t} \int_0^t e^x \sin x dx \\
&= e^{-t}\left[\frac{e^x}{2}(\sin x - \cos x)\right]_0^t + \frac{1}{2}e^{-t} \\
&= \frac{1}{2}(\sin t - \cos t) + \frac{1}{2}e^{-t}.
\end{aligned}
$$

On other hand,

$$
\begin{aligned}
(g * f)(t) &= \int_0^t g(t-x) f(x) dx \\
&= \int_0^t \sin(t-x) e^x dx \\
&= \frac{1}{2}(\sin t - \cos t) + \frac{1}{2}e^{-t}.
\end{aligned}
$$

So, $f * g = g * f$.
(b) (ii) We know that $L\{e^{-at}\} = \dfrac{1}{s+1}$, and $L\{\sin t\} = \dfrac{1}{s^2+1}$.
Thus $F(s)G(s) = \dfrac{1}{s+1}\dfrac{1}{s^2+1}$, or

$$
\begin{aligned}
L^{-1}\{F(s)G(s)\} &= L^{-1}\left\{\frac{1}{s+1}.\frac{1}{s^2+1}\right\} \\
&= L^{-1}\left\{\frac{1}{2}\frac{1}{s+1} + \frac{1}{2}\frac{-s+1}{s^2+1}\right\} \\
&= \frac{1}{2}L^{-1}\left\{\frac{1}{s+1}\right\} - \frac{1}{2}L^{-1}\left\{\frac{s}{s^2+1}\right\} + \frac{1}{2}L^{-1}\left\{\frac{1}{s^2+1}\right\} \\
&= e^{-t} - \frac{1}{2}\cos t + \frac{1}{2}\sin t \\
&= (f * g)(t)
\end{aligned}
$$

or $F(s)G(s) = L\{(f * g)(t)$.

2.7.6 Application to Differential and Integral Equations

In this section we discuss applications of the Laplace transform and related methods in finding solutions of differential equations with initial conditions and integral equations. In view of discussion in Section 2.7.4, the most important feature of the Laplace method in that the initial value given in the problem is naturally incorporated into the solution process through Theorem 27 and Equations (2.95) and (2.96). One advantage of this method is that we need not find the general solution first, then solve for the constant to satisfy the initial condition.

General procedure of Laplace method for solving initial value problems.

Essentially a Laplace transform converts an initial value problem to an algebraic problem, incorporating initial conditions into the algebraic manipulations. There are three basic steps:

(i) Find the Laplace transforms of both sides of the given differential equation, making use of the linearity property of the transforms.

(ii) Solve the transformed equation for the Laplace transform of the solution function.

(iii) Find the inverse transform of the expression $F(s)$ found in step (ii).

Example 153. Apply the Laplace transform to solve the initial value problem

(i) $y' + 2y = 0,\ y(0) = 1.$

(ii) $\dfrac{dy}{dx} - y = 1,\ y(0) = 0.$

Solution : Given $y' + 2y = 0$, taking the Laplace transforms of both sides of this equation yields

$$L\{y' + 2y\} = L\{0\}$$

$$\text{or}\quad \mathrm{L}\{\mathrm{y}'\} + 2\mathrm{L}\{\mathrm{y}\} = \mathrm{L}\{0\}$$

(by linearity of the Laplace transform).

Let $L\{y(t)\} = Y(s)$ and by applying Equation (2.95), the previous equation takes the form

$$sY(s) - 1 + 2Y(s) = 0.$$

Solving for $Y(s)$ we have

$$Y(s) = \frac{1}{s+2}.$$

The function $y(t)$ is then found by taking the inverse transform of this equation. Thus

$$y(t) = L^{-1}\left(\frac{1}{s+2}\right) = e^{-2t} \quad \text{(see Table 2.2).}$$

(b) Let $L\{y\} = Y(s)$. Taking the Laplace transform of the initial value problem we have

$$sY(s) - y(0) - Y(y) = \frac{1}{s}.$$

Solving for $Y(s)$ we obtain

$$Y(s) = \frac{1}{s} + \frac{1}{s-1}.$$

Thus $y = -1 + e^t$.

Example 154. Apply the Laplace transform to solve the initial value problem

(i) $2\dfrac{dy}{dx} = -y,\ y(0) = -3.$

(ii) $y' - 4y = 1,\ y(0) = 1.$

Solution : The Laplace transform of the initial value problem gives us

$$2sY(s) - 2y(0) + Y(s) = 0.$$

Solving for $Y(s)$ we get

$$Y(s) = \frac{6}{2s+1} = \frac{3}{s+\frac{1}{2}}.$$

Taking the inverse Laplace transform of this equation we obtain

$$y(t) = 3e^{-t/2}$$

as the solution of the given initial value problem.
Let $L\{y(t)\} = Y(s)$. Taking the Laplace transform of the differential equation, using the linearity of L and Equation (2.95) we get

$$\begin{aligned} L\{y' - 4y\} &= L\{1\} \\ \text{or}\quad \mathrm{L\{y'\} - 4L\{y\}} &= \frac{1}{s} \\ \text{or}\quad \mathrm{sY(s) - y(0) - 4Y(s)} &= \frac{1}{s} \\ \text{or}\quad \mathrm{sY(s) - 1 - 4Y(s)} &= \frac{1}{s} \\ \text{or}\quad \mathrm{Y(s)(s-4)} &= 1 + \frac{1}{s} \\ \text{or}\quad \mathrm{Y(s)} &= \frac{1}{s-4} + \frac{1}{s(s-4)}. \end{aligned}$$

Taking the inverse Laplace transform of this equation we have

$$L^{-1}\{Y(s)\} = L^{-1}\left\{\frac{1}{s-4} + \frac{1}{s(s-4)}\right\}.$$

By the linearity of L^{-1}, we get

$$L^{-1}\{Y(s)\} = L^{-1}\left\{\frac{1}{s-4}\right\} + L^{-1}\left\{\frac{1}{s(s-4)}\right\}.$$

By Table 2.2

$$\begin{aligned}
L^{-1}\left\{\frac{1}{s-4}\right\} &= e^{4t}\\
L^{-1}\left\{\frac{1}{s(s-4)}\right\} &= \frac{1}{4}L^{-1}\left\{\frac{1}{s-4}-\frac{1}{s}\right\}\\
&= \frac{1}{4}L^{-1}\left\{\frac{1}{s-4}\right\}-\frac{1}{4}L^{-1}\left\{\frac{1}{s}\right\}\\
&= \frac{1}{4}e^{4t}-\frac{1}{4}.
\end{aligned}$$

Thus

$$\begin{aligned}
y(t) &= e^{4t}+\frac{1}{4}e^{at}-\frac{1}{4}\\
&= \frac{5}{4}e^{at}-\frac{1}{4}
\end{aligned}$$

is the solution of the given initial value problem.

Example 155. (i) $y' + 6y = e^{4t},\ y(0) = 2.$

(ii) $y' + 7y = H(t-2),\ y(0) = 3.$

(iii) $y'' - 2y' + y = 3\delta(t-2),\ y(0) = 0,\ y'(0) = 1.$

Solution : The Laplace transform of the initial value problem yields

$$sY(s) - y(0) + 6Y(s) = \frac{1}{s-4}.$$

Solving for $Y(s)$ we get

$$Y(s) = \frac{1}{(s-4)(s+6)} + \frac{2}{s+6} = \frac{1}{10}.\frac{1}{s-4} + \frac{19}{10}.\frac{1}{s+6}.$$

The inverse Laplace transform gives the solution of the initial value problem

$$y(t) = \frac{1}{10}e^{4t} + \frac{19}{10}e^{-6t}.$$

(ii) Taking Laplace transforms of both sides of the equation and using the property of the Laplace transform of higher order derivative, we get

$$sY(s) - 3 + 7Y(s) = e^{-2s}\frac{1}{s}.$$

Solving for $Y(s)$, we get

$$\begin{aligned}
Y(s) &= e^{-2s}\frac{1}{s(s+7)} + \frac{3}{s+7}\\
&= \frac{1}{7}e^{-2s}\frac{1}{s} - \frac{1}{7}e^{-2s}\frac{1}{s+7} + \frac{3}{s+7}.
\end{aligned}$$

Using inverse Laplace transform we obtain

$$y(t) = \frac{1}{7}H(t-2) - \frac{1}{7}e^{-7(t-2)}H(t-2) + 3e^{-7t}.$$

(iii) Taking Laplace transforms of the given equation and using the properties of higher order derivatives, we obtain

$$s^2Y(s) - 2 - 2sY(s) + Y(s) = 3e^{-2s}.$$

Solving for $Y(s)$ we get

$$Y(s) = \frac{2}{s^2-2s+1} + 3e^{-2s}\frac{1}{s^2-2s+1} = \frac{1}{(s-1)^2} + 3e^{-2s}\frac{1}{(s-1)^2}.$$

Taking the inverse Laplace transform we obtain

$$y(t) = te^t + 3(t-2)e^{t-2}H(t-2).$$

Example 156. Solve $y'' + 4y' + 3y = e^t, \quad y(0) = 0, \ y'(0) = 2.$

Solution : Take the Laplace transform of the given differential equation to get

$$L\{y''\} + L\{4y'\} + L\{3y\} = L\{e^t\}.$$

Using Equations (2.95) and (2.96) and applying initial conditions we get

$$\begin{aligned} L\{y''\} &= s^2Y(s) - sy(0) - y'(0) = s^2Y(s) - 2 \\ \text{and } \mathrm{L\{y'\}} &= sY(s) - y(0) = sY(s). \end{aligned}$$

Therefore,

$$s^2Y(s) - 2 + 4sY(s) + 3Y(s) = \frac{1}{s-1}.$$

Solving this for $Y(s)$, we get

$$Y(s) = \frac{2s-1}{(s-1)(s^2+4s+3)}.$$

Let $\dfrac{2s-1}{(s-1)(s^2+4s+3)} = \dfrac{A}{s-1} + \dfrac{B}{s+1} + \dfrac{C}{s+3}.$

The equation can hold only if,for all s,

$$A(s+1)(s+3) + B(s-1)(s+3) + C(s-1)(s+1) = 2s-1.$$

Now choose values of s to simplify the task of determining A, B and C.

Let $s = 1$ to get $8A = 1$, so $A = \frac{1}{8}$.

Let $s = -1$ to get $-4B = -3$, so $B = \frac{3}{4}$.

Choose $s = -3$ to get $8C = -7$, so $C = -\frac{7}{8}$.

Then

$$Y(s) = \frac{1}{8}.\frac{1}{s-1} + \frac{3}{4}.\frac{1}{s+1} - \frac{7}{8}.\frac{1}{s+3}.$$

By Table 2.2 we find that

$$L^{-1}\{Y(s)\} = \frac{1}{8}L^{-1}\left(\frac{1}{s-1}\right) + \frac{3}{4}L^{-1}\left(\frac{1}{s+1}\right) - \frac{7}{8}L^{-1}\left(\frac{1}{s+3}\right)$$

$$y(t) = \frac{1}{8}e^{t} + \frac{3}{4}e^{-t} - \frac{7}{8}e^{-3t}.$$

This is the solution of the given initial value problem.

Example 157. Find $L\{f(t) * g(t)\}$, where $f(t) = e^{-t}$ and $g(t) = \sin 2t$.

Solution: By Theorem 30, we have

$$\begin{aligned} L\{f(t) * g(t)\} &= F(s)G(s) \\ \text{where F(s)} &= \int_0^\infty e^{-st}e^{-t}dt = \frac{1}{s+1} \qquad \text{(by Table 2.2)} \\ G(s) &= \left(\frac{2}{s^2+4}\right) \qquad \text{(by Table 2.2).} \end{aligned}$$

Thus

$$\begin{aligned} L\{f(t) * g(t)\} &= \left(\frac{1}{s+1}\right)\left(\frac{2}{s^2+4}\right) \\ &= \frac{2}{(s+1)9s^2+4}. \end{aligned}$$

Example 158. Evaluate $L\left\{\int_0^t e^u \sin(t-u)du\right\}$.

Solution : $\int_0^t e^u \sin(t-u)du = f(t) * g(t)$, where $f(t) = e^t$ and $g(t) = \sin t$ (by definition of convolution). By Theorem 30 we get

$$\begin{aligned} L\{f(t) * g(t)\} &= L\{f(t)\}L\{g(t)\} \\ &= \frac{1}{s+1}\frac{1}{s^2+1} \qquad \text{(by Table 2.2)} \\ &= \frac{1}{(s-1)(s^2+1)}. \end{aligned}$$

An equation involving an unknown function $f(t)$, known functions $g(t)$ and $h(t)$ and integral of f and g is called a *Volterra integral equation* for $f(t)$:

$$f(t) = g(t) + \int_0^t f(u)h(t-u)du.$$

Example 159. Solve the following Volterra integral equation for $f(t)$:

$$f(t) = 3t^2 - e^{-t} - \int_0^t f(u)e^{t-u}du$$

Solution : We identify $h(t-u) = e^{t-u}$ so that $h(t) = e^t$.
Using Laplace transforms of both sides, we have

$$L\{f(t)\} = L\{3t^2\} - L\{e^{-t}\} - L\left\{\int_0^t f(u)e^{t-u}du\right\}$$

$$L\{f(t)\} = F(s),\ L\{3t^2\} = 3L\{t^2\} = 3\frac{2}{s^3}$$

$$\begin{aligned} L\{e^{-t}\} &= \frac{1}{s+1} \\ L\left\{\int_0^t f(u)e^{t-u}du\right\} &= L\{f(t) * h(t)\} \\ &= L\{f(t)\}L\{h(t)\}. \end{aligned}$$

By Definition 52 and Theorem 30

$$\begin{aligned} L\left\{\int_0^t f(u)e^{t-u}du\right\} &= L\{f(t)\}L\{h(t)\} \\ &= F(s).\frac{1}{s-1}. \end{aligned}$$

Therefore,

$$\begin{aligned} F(s) &= \frac{6}{s^3} - \frac{1}{s+1} - \frac{1}{s+1}F(s) \\ F(s)\left[1 + \frac{1}{s-1}\right] &= \frac{-s^3+6s+6}{s^3(s+1)} \\ F(s)\left[\frac{s}{s-1}\right] &= \frac{-s^3+6s+6}{s^3(s+1)} \\ F(s) &= \frac{6}{s^3} - \frac{6}{s^4} + \frac{1}{s} - \frac{2}{s+1} \end{aligned}$$

by carrying out the partial fraction decomposition. The inverse problem then gives

$$\begin{aligned} f(t) &= 3L^{-1}\left\{\frac{2!}{s^3}\right\} - L^{-1}\left\{\frac{3!}{s^4}\right\} + L^{-1}\left\{\frac{1}{s}\right\} - 2L^{-1}\left\{\frac{1}{s+1}\right\}. \\ &= 3t^2 - t^3 + 1 - 2e^{-t} \end{aligned}$$

Example 160. Find $f(t)$ such that

$$f(t) = 2t^2 - e^{-t} - \int_0^t f(t-u)e^{-u}du.$$

Solution : It is clear that

$$f(t) * g(t) = \int_0^t f(t-u)e^{-u}du$$

and by Theorem 30

$$\begin{aligned} L\{f(t) * g(t)\} &= L\{f(t)\}L\{g(t)\} \\ &= F(s)\frac{1}{s+1}. \end{aligned}$$

By using the Laplace transforms of both sides of the integral equation we get

$$\begin{aligned} L\{f(t)\} &= L\{2t^2\} + F(s)\frac{1}{s+1} \\ F(s) &= 2.\frac{2}{s^3} + \frac{1}{s+1}F(s) \\ F(s)\left[\frac{s}{s+1}\right] &= \frac{4}{s^3} \\ F(s) &= \frac{4(s+1)}{s^4} \\ &= \frac{4}{s^3} + \frac{4}{s^4}. \end{aligned}$$

With the inverse Laplace transform we get

$$f(t) = 2L^{-1}\left\{\frac{2}{s^3}\right\} + \frac{2}{3}L^{-1}\left\{\frac{3!}{s^4}\right\} = 2t^2 + \frac{2}{3}t^3.$$

Example 161. Find the function $f(t)$ if

(i) $f(t) = t + \int_0^t f(u)\sin(t-u)du.$

(ii) $f(t) = 4t - 3\int_0^t f(x)\sin(t-x)dx.$

Solution : We can identify the integral as

$$f(s) * h(t), \quad \text{where } h(t) = \sin t.$$

Taking the Laplace transform of the both sides of the integral equation we get

$$L\{f(t)\} = L\{t\} + L\{f(t) * h(t)\}.$$

By Theorem 30,

$$\begin{aligned} L\{f(t) * h(t)\} &= L\{f(t)\}L\{h(t)\} \\ &= F(s)\frac{1}{s^2+1}. \end{aligned}$$

Thus

$$\begin{aligned} F(s) &= \frac{1}{s^2} + F(s)\frac{1}{s^2+1} \\ \text{or} \quad \mathrm{F(s)}\frac{\mathrm{s}^2}{\mathrm{s}^2+1} &= \frac{1}{s^2} \\ \text{or} \quad \mathrm{F(s)} &= \frac{s^2+1}{s^4} = \frac{1}{s^2} + \frac{1}{s^4}. \end{aligned}$$

Taking the inverse Laplace transform of this equation we get

$$f(t) = t + \frac{1}{6}t^3.$$

(ii) Let $F = L\{f\}$. The equation $f(t) = 4t - 3\int_0^t f(x)\sin(t-x)dx$ can be written as

$$f(t) - 4t = -3(f(t) * \sin t)$$

Applying Laplace transforms to both sides of this equation and using the convolution Theorem 30, we get

$$F(s) = 8\frac{s^2+1}{s^2(s^2+4)} = 6\frac{1}{s^2+4} + 2.\frac{1}{s^2}.$$

Taking the inverse Laplace transform of this equation we get

$$f(t) = 3\sin 2t + 2t.$$

Example 162. Solve $y'' - 6y' + 9y = t^2e^{3t}$.

Solution : Applying the Laplace transform to both sides and using linearity we get

$$L\{y'') - 6L\{y'\} + 9L\{y\} = L\{t^2e^{3t}\}$$

$$\text{or} \quad \mathrm{sY(s) - sY(0) - y'(0) - 6[sY(s) - y(0)] + 9Y(s)} = \frac{2}{\mathrm{(s-3)^2}}$$

where $Y(s) = L\{y(t)\}$ (Using Laplace transform of derivatives).

$$\begin{aligned} (s^2 - 6s + 9)Y(s) &= 2s + 5 + \frac{2}{(s-3)^2} \\ Y(s) &= \frac{2s+5}{(s-3)^2} + \frac{2}{(s-3)^5} \\ Y(s) &= \frac{2}{s-3} + \frac{11}{(s-3)^2} + \frac{2}{(s-3)^5} \end{aligned}$$

(decomposing right hand side into partial fractions). Taking the inverse Laplace transform of this equation we get

$$y(t) = 2L^{-1}\left\{\frac{1}{s-3}\right\} + 11L^{-1}\left\{\frac{1}{(s-3)^2}\right\} + 2L^{-1}\left\{\frac{1}{(s-3)^5}\right\}$$

$$\text{or} \quad \mathrm{y(t) = 2e^{3t} + 11te^{3t}} + \frac{1}{12}\mathrm{t^4e^{3t}}.$$

Example 163. Solve $y'' + y = 4\delta(t - 2\pi)$, $y(0) = 1$ and $y'(0) = 0$.

Solution : Taking the Laplace transform to both sides, we get

$$\begin{aligned} L\{y''\} + L\{y\} &= 4L\{\delta(t - 2\pi)\} \\ s^2Y(s) - sY(0) - y'(0) + Y(s) &= 4e^{-2\pi s} \\ (s^2 + 1)Y(s) - s &= 4e^{-2\pi s} \\ Y(s) &= \frac{s}{s^2 + 1} + 4\frac{e^{-2\pi s}}{s^2 + 1}. \end{aligned}$$

Taking the inverse Laplace transform of this equation we get

$$y(t) = \cos t + 4\sin(t - 2\pi)H(t - 2\pi).$$

Since $\sin(t - 2\pi) = \sin t$, the solution can be written as

$$y(t) = \left\{ \begin{array}{ll} \cos t, & 0 \le t \le 2\pi \\ \cos t + 4\sin t, & t \ge 2\pi \end{array} \right\}.$$

Example 164. Solve the initial value problem $y'' - 5y' + 6y = H(t-1), \ y(0) = 0, \ y'(0) = 1$.

Solution: Applying the Laplace transform to both sides, we get

$$s^2L\{y\} - sY(0)0 - y'(0) - 5[sL\{y\} - y(0)] + 6L\{y\} = \frac{e^{-s}}{s}.$$

Solving for $L\{y\}$ and taking the inverse Laplace transform we get

$$(s^2 - 5s + 6)Y(s) = \frac{e^{-s}}{s} + 1$$

$$\text{or} \quad \mathrm{Y(s)} = \frac{\mathrm{e^{-s}}}{\mathrm{s}} + 1 + \frac{1}{\mathrm{(s-2)(s-3)}}$$

$$L^{-1}\left\{\frac{1}{6} - \frac{1}{2(s-2)} + \frac{1}{3(s-3)}\right\} = \frac{1}{6} - \frac{1}{2}e^{2t} + \frac{1}{3}e^{3t}$$

(by decomposition into partial fractions)

$$y = \left[\frac{1}{6} - \frac{1}{2}e^{2(t-1)} + \frac{1}{3}e^{3(t-1)}\right] H(t-1) + e^{3t} - e^{2t}.$$

Example 165. Solve the initial value problem $y'' + 4y' = \sin tH(t - 2\pi), \ y(0) = 1, \ y'(0) = 0$.

Example 166. Solve the initial value problem $y'' + 4y = \sin 3t, \ y(0) = y'(0) = 0$.

2.8 Series Solution of Differential Equations

We have seen in Section 2.5 that we can solve linear differential equations of order two or more with constant coefficients. The Cauchy-Euler equation is an exception. In fact most linear differential equations of higher order with variable coefficients cannot be solved in terms of elementary functions. The usual strategy for solving such equations is to assume a solution in the form of an infinite series and proceed in a manner similar to the method of undetermined coefficients (Subsection 2.5.6). Since these series solutions often turn out to be power series, it is appropriate to summarize properties of power series in the first subsection essential for proper understanding of discussion on Bessel's and Legendre's equations.

2.8.1 Review of Properties of Power Series

A **power series** in $(x-a)$ is an infinite series of the form

$$c_0 + c_1(x-a) + c_2(x-a)^2 + \cdots = \sum_{n=0}^{\infty} c_n(x-a)^n. \tag{2.99}$$

Series of (2.99) is also called a **power series centered** at a. The power series centered at $a=0$ is often referred as the **power series**, that is, the series $\sum_{n=0}^{N} c_n x^n$. A power series centered at a is called **convergent** at a specified value of x if its sequence of partial sums $S_N(x) = \sum_{n=0}^{\infty} c_n(x-a)^n$, that is, $\{S_N(x)\}$ is convergent. In other words the limit of $\{S_N(x)\}$ exists. If the limit does not exist the power series is called *divergent*. The set of points x, at which the power series is convergent, is called the **interval of convergence** of the power series.

For $R>0$, a power series $\sum_{n=0}^{\infty} c_n(x-a)^n$ converges if $|x-a|<R$ and diverges if $|x-a|>R$. If the series converges only at a then $R=0$, and if it converges for all x then $R=\infty$, $|x-a|<R$ is equivalent to $a-R<x<a+R$. A power series may or may not converge at the end points $a-R$ and $a+R$ of this interval. R is called the **radius of convergence**.

A power series is called **absolutely convergent** if the series $\sum_{n=0}^{\infty} |c_n(x-a)|^n$ converges. A power series converges absolutely within its interval of convergence. By the ratio test, a power series centered at a, (2.99) is absolutely convergent if $L = |x-a| \lim_{n=\infty} \left|\dfrac{c_{n+1}}{c_n}\right|$ is less than 1, that is, $L<1$, the series diverges if $L>1$; and test fails if $L=1$. A power series defines a function $f(x) = \sum_{n=0}^{\infty} c_n(x-a)^n$ whose domain is the interval of convergence of series. If the radius of convergence $R>0$, then f is continuous, differentiable and integrable on the interval $(a-R, a+R)$. Moreover $f'(x)$ and $\int f(x)dx$ can be found by term-by-term differentiation and integration. Convergence at an end point may be lost by differentiation or gained through integration.

$$\begin{aligned}\text{Let y} &= \sum_{n=0}^{\infty} c_n x^n \\ y' &= \sum_{n=0}^{\infty} n c_n x^{n-1} \\ y'' &= \sum_{n=0}^{\infty} n(n-1) c_n x^{n-2}.\end{aligned}$$

We observe that the first term in y' and first two terms in y'' are zero. Keeping this in mind we can write

$$\begin{aligned} y' &= \sum_{n=1}^{\infty} n c_n x^{n-1}. \\ y'' &= \sum_{n=2}^{\infty} n(n-1) c_n x^{n-2}. \end{aligned} \tag{2.100}$$

Identity property: If $\sum_{n=0}^{\infty} c_n(x-a)^n = 0,\ R > 0$ for all x in the interval of convergence, then $c_n = 0$ for all n.

Analytic at a point: A function f is analytic at a point a if it can be represented by a power series in $x - a$ with a positive or infinite radius of convergence. A power series where $c_n = \dfrac{f^{(n)}(a)}{n!}$, that is, the series of the type $\sum c_n \dfrac{f^{(n)}(a)}{n!}(x-a)^n$, is called the **Taylor series**. If $a = 0$ the Taylor series is called a **Maclaurin series**. In calculus $e^x,\ \cos x,\ \sin x, \ln(x-1)$ can be written in the form of a power series, more precisely in the form of a **Maclaurin series**. For example

$$\begin{aligned} e^x &= 1 + x + \frac{x^2}{2!} + \ldots \\ \sin x &= x - \frac{x^3}{3!} + \frac{x^5}{5!} \\ \cos x &= 1 - \frac{x^2}{2!} + \frac{x^4}{4!} - \frac{x^6}{6!} \end{aligned}$$

for $|x| \leq \infty$.

Arithmetic of power series: Two power series can be combined through operation of addition, multiplication and division. The procedures for power series are similar to those by which two polynomial are added, multiplied and divided. For example :

$$\begin{aligned}
e^x \sin x &= \left(1 + x + \frac{x^2}{2} + \frac{x^3}{6} + \frac{x^4}{24} + \dots\right)\left(x - \frac{x^3}{6} + \frac{x^5}{120} - \frac{x^7}{5040}\right) \\
&= (1)x + x^2 + \left(-\frac{1}{6} + \frac{1}{2}\right)x^3 + \left(-\frac{1}{6} + \frac{1}{6}\right)x^4 + \dots \\
&\dots \left(\frac{1}{120} - \frac{1}{12} + \frac{1}{24}\right)x^5 + \dots \\
&= x + x^2 + \frac{x^3}{3} - \frac{x^5}{30}.
\end{aligned}$$

Since the power series for e^x and $\sin x$ converges for $|x| < \infty$, the product series converges on the same interval.

Shifting summation index: In order to discuss power series solutions of differential equation it is advisable to combine two or more summations as a single summation.

Example 167. Express $\sum_{n=2}^{\infty} n(n-1)c_n x^{n-2} + \sum_{n=0}^{\infty} c_n x^{n+1}$ as one power series.

Solution: In order to add the two given series, it is necessary that both summation indices start with the same number and the powers of x in each series be such that if one series starts with a multiple of x to the first power, we want the other series to start with same power. In this problem the first series starts with the x^1. By writing the first term of the first series outside the summation notation,

$$\sum_{n=2}^{\infty} n(n-1)c_n x^{n-2} + \sum_{n=0}^{\infty} c_n x^{n+1} = 2.1c{-}2x^0 + \sum_{n=3}^{\infty} n(n-1)c_n x^{n-2} + \sum_{n=0}^{\infty} c_n x^{n+1}.$$

Both series on the right hand side start with same power of x, namely x^1. Let $k = n-2$ or $k = n+1$ respectively in the first and second series. Then the right hand side becomes

$$2c_2 + \sum_{k=1}^{\infty} (k+2)(k+1)c_{k+2}x^k + \sum_{k=1}^{\infty} c_{k-1}x^k. \tag{2.101}$$

Keep in mind that the value of the summation index is important, not the summation index which is a dummy variable, say $k = n-1$ or $k = n+1$. Now we are in position to add series in () term by term and we have

$$\sum_{n=2}^{\infty} n(n-1)c_n x^{n-2} + \sum_{n=0}^{\infty} c_n x^{n+1} = 2c_2 + \sum_{k=1}^{\infty} [(k+2)(k+1)c_{k+2}x + c_{k_1}]x^k.$$

2.8.2 Solution about Ordinary Point

We look for a power series solution of a linear second order differential equation about a special point:

$$a_2(x)\frac{d^2y}{dx^2} + a_1(x)\frac{dy}{dx} + a_0(x)y = 0 \tag{2.102}$$

where $a_2(x) \neq 0$.
This can be put into standard form

$$\frac{d^2y}{dx^2} + \frac{a_1(x)}{a_2(x)}\frac{dy}{dx} + \frac{a_0(x)}{a_2(x)}y = 0$$

$$\text{or} \quad \frac{d^2y}{dx^2} + P(x)\frac{dy}{dx} + Q(x)y = 0. \tag{2.103}$$

Note that x_0 is said to be an **ordinary point** of the differential Equation (2.102) if $P(x)$ and $Q(x)$ represented by a power series. A point that is not ordinary is called a **singular point**.

A solution of the form $y = \sum_{n=0}^{\infty} c_n(x - x_0)^n$ is said to be a **solution about the ordinary point** x_0.

Remark 35. It has been proved that if $x = x_0$ is an ordinary point of (2.102) then there exist two linearly independent solutions in the form of a power series centered at x_0, that is, $y = \sum_{n=0}^{\infty} c_n(x - x_0)^n$. A series solution converges at least on some interval defined by $|x - X_0| < R$, where R is the distance from x_0 to the closest singular point.

Power series solution about ordinary point:

Let $y = \sum_{n=0}^{\infty} c_n x^n$ and substitute values of y, $\dfrac{dy}{dx} = y'$, $\dfrac{d^2y}{dx^2} = y''$ in (2.103). Combine series as in Example 167, and then equate all coefficients to the right hand side of the equation to determine the coefficients c_n. We illustrate the method by the following examples. We also see through these examples how the single assumption that $y = \sum_{n=0}^{\infty} c_n x^n$ leads to two sets of coefficients, so we have two distinct power series $y_1(x)$ and $y_2(x)$ both expanded about the ordinary point $x = 0$. The general solution of the differential equation is $y = C_1 y_1(x) + C_2 y_2(x)$, and it can be shown that $C_1 = c_0$ and $C_2 = c_1$.

The differential equation $\dfrac{d^2y}{dx^2} + xy = 0$ is known as Airy's equation and is used in the studies of the diffraction of light, diffraction of radio waves around the surface of the earth, aerodynamics and other physical phenomena. We discuss here power series solution of this equation around its ordinary point $x = 0$.

Example 168. Write the general solution of Airy's equation $y' + xy = 0$.

Solution: In view of the remark, two power series solutions centered at 0 and convergent for $|x| < \infty$ exist. By substituting $y = \sum_{n=0}^{\infty} c_n x^n$, $y'' = \sum_{n=2}^{\infty} n(n-1)c_n x^{n-2}$ into Airy's differential equation we get

$$\begin{aligned} y'' + xy &= \sum_{n=2}^{\infty} n(n-1)c_n x^{n-2} + x\sum_{n=0}^{\infty} c_n x^n, \\ &= \sum_{n=2}^{\infty} n(n-1)c_n x^{n-2} + \sum_{n=0}^{\infty} c_n x^{n+1}. \end{aligned} \tag{2.104}$$

As seen in the solution of Example 167, (2.104) can be written as

$$y'' + xy = 2c_2 + \sum_{k=1}^{\infty}[(k+1(k+2)c_{k+2} + c_{k-1}]x^k = 0. \tag{2.105}$$

Since (2.105) is identically zero, it is necessary that coefficients of each power of x be set equal to zero, that is,

$$\begin{aligned} 2c_2 &= 0 \quad \text{(coefficients of } x^0\text{) and} \\ (k+1) &= (k+2)c_{k+2} + c_{k-1} = 0, \; k = 1, 2, 3, \ldots. \end{aligned} \tag{2.106}$$

The above holds in view of the identity property. It is clear that $c_2 = 0$. The expression in (2.106) is called a *recurrence relation* and it determines the c_k in such a manner that we can choose a certain subset of the set of coefficients to be non-zero. Since $(k+1)(k+2) \neq 0$ for all values of k, we can solve (2.106) for c_{k+2} in terms of c_{k-1}:

$$c_{k+2} = \frac{-c_{k-1}}{(k+1)(k+2)}, \quad k = 1, 2, 3, \ldots \tag{2.107}$$

$$\begin{aligned} \text{For } k &= 1, \; c_3 = -\frac{c_0}{2.3}. \\ \text{For } k &= 2, \; c_4 = -\frac{c_1}{3.4}. \\ \text{For } k &= 3, \; c_5 = -\frac{c_2}{4.5} = 0 \text{ as } c_2 = 0. \\ \text{For } k &= 4, \; c_6 = -\frac{c_3}{5.6} = \frac{1}{2.3.5.6}c_0. \end{aligned}$$

$$\begin{aligned} \text{For } k &= 5, \; c_7 = -\frac{c_4}{6.7} = -\frac{1}{3.4.6.7}c_1. \\ \text{For } k &= 6, \; c_8 = -\frac{c_5}{7.8} = 0 \text{ as } c_5 = 0. \\ \text{For } k &= 7, \; c_9 = -\frac{c_6}{8.9} = -\frac{1}{2.3.5.6.8.9}c_0. \\ \text{For } k &= 8, \; c_{10} = -\frac{c_7}{9.10} = -\frac{1}{3.4.6.7.8.10}c_1. \\ \text{For } k &= 9, \; c_{11} = -\frac{c_8}{10.11} = 0 \text{ as } c_8 = 0 \text{ and so on.} \end{aligned}$$

Substituting the coefficients just obtained into

$$y_n = \sum_{n=1}^{\infty} = c_0+c_1x+c_2x^2+c_3x^3+c_4x^4+c_5x^5+c_6x^6+c_7x^7+c_8x^8+c_9x^9+c_{10}x^{10}\ldots$$

we get

$$\begin{aligned} y &= c_0 + c_1x + 0 - \frac{c_0}{2.3}x^3 - \frac{c_1}{3.4}x^4 + 0 + \frac{c_0}{2.3.5.6}x^6 \\ &+ \frac{c_1}{3.4.6.7}x^7 + 0 - -\frac{c_0}{2.3.5.6.8.9}x^9 - \frac{c_1}{3.4.6.7.8.10}x^{10} + 0 + \ldots \end{aligned}$$

After grouping the terms containing c_0 and the terms containing c_1, we obtain

$$y = c_0y_1(x) + c_1y_2(x),$$

$$\begin{aligned} \text{where } \mathrm{y_1(x)} &= 1 - \frac{1}{2.3}x^3 + \frac{1}{2.3.5.6}x^6 - \frac{1}{2.3.5.6.8.9}x^9 + \ldots \\ &= 1 + \sum_{k=1}^{\infty} \frac{(-1)^k}{2.3\ldots(3k-1)(3k)}x^{3k} \\ y_2(x) &= x - \frac{1}{3.4}x^4 + \frac{1}{3.4.6.7}x^7 - \frac{1}{3.4.6.7.9.10}x^10 + \ldots \\ &= x + \sum_{k=1}^{\infty} \frac{(-1)^k}{3.4\ldots(3k)(3k+1)}x^{3k+1}. \end{aligned}$$

Since the recursive use of (2.107) leaves c_0 and c_1 completely undetermined, they can be chosen arbitrarily. $y = c_0y_1(x) + c_1y_2(x)$ is the general solution of Airy's equation.

Example 169. Find two power series solutions of the differential equation $y'' - xy = 0$ about the ordinary point $x = 0$.

Solution: Substituting $y = \sum_{n=0}^{\infty} c_nx^n$ into the differential equation we get

$$\begin{aligned} y'' - xy &= \sum_{n=2}^{\infty} n(n-1)c_nx^{n-2} - \sum_{n=0}^{\infty} c_nx^{n+1} \\ &= \sum_{k=0}^{\infty}(k+2)(k+1)c_{k+2}x^k - \sum_{k=1}^{\infty} c_{k-1}x^k \\ &= 2c_2 + \sum_{k=1}^{\infty}[(k+2)(k+1)c_{k+2} - c_{k-1}]x^k. \end{aligned}$$

Thus $c_2 = 0$,

$$(k+2)(k+1)c_{k+2} - c_{k-1} = 0$$

and

$$c_{k+2} = \frac{1}{(k+2)(k+1)} c_{k-1}, \quad k = 1, 2, 3, \ldots$$

Choosing $c_0 = 1$ and $c_1 = 0$ we find

$$c_3 = \frac{1}{6}, \quad c_4 = c_5 = 0, \quad c_6 = \frac{1}{180}$$

and so on. For $c_0 = 1$ and $c_1 = 0$ we obtain

$$c_3 = 0, \quad c_4 = \frac{1}{12}, \quad c_5 = c_6 = 0, \quad c_7 = \frac{1}{504}$$

and so on. Thus two solutions are

$$y_1 = 1 + \frac{1}{6}x^3 + \frac{1}{180}x^6 + \ldots \text{and}$$

$$y_2 = x + \frac{1}{12}x^4 + \frac{1}{504}x^7 + \ldots$$

2.8.3 Solution about Regular Singular Points: The Method of Frobenius

A singular point x_0 of (2.102) is called a *regular singular point* of this equation if the function $p(x) = (x - x_0)P(x)$ and $q(x) = (x - x_0)^2 Q(x)$ are both analytic at x_0. A singular point that is not regular is said to be an *irregular singular point* of the equation. This means that one or both of the functions $p(x) = (x - x_0)P(x)$ and $q(x) = (x - x_0)^2 Q(x)$ fails to be analytic at x_0.

In order to solve a differential equation given by (2.102) about a regular singular point we employ the following theorem due to Frobenius.

Theorem 32. *(Frobenius theorem) If $x = x_0$ is a regular singular point of the differential Equation (2.102), then there exists at least one solution of the form*

$$y = (x - x_0)^r \sum_{n=0}^{\infty} c_n (x - x_0)^n = \sum_{n=0}^{\infty} c_n (x - x_0)^{n+r}$$

where r is constant to be determined. The series will converge at least on some interval $0 < x - x_0 < R$.

Method of Frobenius: Fining a series solution about a regular singular point x_0, is similar to the method of previous subsection in which we substitute $y = \sum_{n=0}^{\infty} c_n (x - x_0)^{n+r}$ into the given differential equation and determine the unknown coefficients c_n by a recurrence relation. However we have an additional task in this procedure. Before determining coefficients we must find unknown exponent r and equate to 0 the coefficients of the lowest power of x. This is called an **indicial equation** and determines the value(s) of the index r.

If r is found to be number that is not a non-negative integer, then the corresponding solution $y = \sum_{n=0}^{\infty} c_n(x - x_0)^{n+r}$ is not a power series. For the sake of simplicity we assume that the regular singular point is $x = 0$.

Example 170. Apply the method of frobenius to solve the differential equation $2xy'' + 3y' - y = 9$ about the regular singular point $x = 0$.

Solution: Let us assume that the solution is of form

$$\begin{aligned} y &= \sum_{n=0}^{\infty} c_n x^{n+r}, \quad \text{then} \\ y' &= \sum_{n=0}^{\infty} c_n(n+r)x^{n+r-1} \text{ and} \\ y'' &= \sum_{n=0}^{\infty} c_n(n+r)(n+r-1)x^{n+r-2}. \end{aligned}$$

Substituting these values of $y,\ y',\ y''$ into $2xy'' + 3y' - y = 0$, we get

$$2\sum_{n=0}^{\infty} c_n(n+r)(n+r-1)x^{n+r-2} + 3\sum_{n=0}^{\infty} c_n(n+r)x^{n+r-1} - \sum_{n=0}^{\infty} c_n x^{n+r} = 0.$$

Shifting the index in the third series and combining the first two yields

$$\sum_{n=0}^{\infty} c_n(n+r)(2n+2r+1)x^{n+r-1} - \sum_{n=0}^{\infty} c_{n-1}x^{n+r-1} = 0.$$

Writing the term corresponding to $n = 0$ and combining the terms for $n \geq l$ into one series,

$$c_0 r(2r+1)x^{r-1} \sum_{n=1}^{\infty} [c_n(n+r)(2n+2r+1) - c_{n-1}]x^{n+r-1} = 0.$$

Equating the coefficients of x^{r-1} to zero yields the indicial equation

$$c_0 r(2r+1) = 0.$$

Since $c_0 \neq 0$, either $r = 0$ or $r = -\dfrac{1}{2}$.
Hence two linearly independent solutions of the given differential equation have the forms

$$y_1 = F_0(x) = \sum_{n=0}^{\infty} c_n x^n \quad \text{and}$$

$$y_2 = F_{-1/2}(x) = x^{-1/2} \sum_{n=0}^{\infty} c_n^* x^n.$$

Since $c_n(n+r)(2n+2r+1)-c_{n-1}=0$ for all b $n \geq 1$, we have the following information on the coefficients for the two series:

(i) c_0 is arbitrary, and for $n \geq 1$, $c_n = \dfrac{1}{n(2n+1)}c_{n-1}$.

(ii) c_0^* is arbitrary, and for $n \geq 1$, $c_n^* = \dfrac{1}{n(2n-1)}c_{n-1}^*$.

Iteration of the formula for c_0, yields

$$n=1, c_1 = \frac{1}{1.3}c_0 = \frac{2}{1.2.3}c_0 = \frac{2c_0}{3!}.$$

$$n=2, c_2 = \frac{1}{2.5}c_1 = \frac{1}{2.3.5}c_0 = \frac{2^2c_0}{5!}.$$

$$n=3, c_3 = \frac{1}{3.7}c_2 = \frac{1}{3.7}\frac{2^2c_0}{5!} = \frac{2^3c_0}{7!}.$$

Each term of c_n was multiplied by $\frac{2}{2}$ to make the denominator $(2n+1)!$. The general form of c_n is then

$$c_n = \frac{2^n c_0}{(2n+1)!}.$$

Similarly, the general form of c_n^* is found to be $c_n^* = \dfrac{2^n c_0}{(2n)!}$. The two solution are

$$y_1 = c_0 \sum_{n=0}^{\infty} \frac{2^n}{(2n+1)!}x^n \quad \text{and} \quad y_2 = c_0^* x^{-1/2} \sum_{n=0}^{\infty} \frac{2^n}{(2n)!}x^n.$$

y_2 is not a power series.

Example 171. Apply the method of Frobenius to obtain two linearly independent series solutions of the differential equation

$$2xy'' - y' + 2y = 0$$

about the regular singular point $x=0$ of the differential equation.

Solution: Substituting $y = \sum_{n=0}^{\infty} c_n x^{n+r}$,

$$y' = \sum_{n=0}^{\infty} c_n(n+r)x^{n+r-1} \quad \text{and}$$

$$y'' = \sum_{n=0}^{\infty} c_n(n+r)(n+r-1)x^{n+r-2}$$

into the differential equation and collecting terms, we obtain

$$\begin{aligned} 2xy'' - y' + 2y &= (2r^2-3r)c_0x^{r-1} \\ &+ \sum_{k=1}^{\infty}[2(k+r-1)(k+r)c_k - (k+r)c_k + 2c_{k-1}]x^{k+r-1} = 0, \end{aligned}$$

which implies that

$$2r^2 - 3r = r(2r-3) = 0$$

and

$$(k+r)(2k+2r-3)c_k + 2c_{k-1} = 0.$$

The indicial roots are $r = 0$ and $r = \frac{3}{2}$. For $r = 0$ the recurrence relation is

$$c_k = -\frac{2c_{k-1}}{k(2k-3)}, \quad k = 1, 2, 3, \ldots$$

and

$$c_1 = 2c_0, \quad c_2 = -2c_0, \quad c_3 = \frac{4}{9}c_0.$$

For $r = \frac{3}{2}$ the recurrence relation is

$$c_k = -\frac{2c_{k-1}}{k(2k+3)}, \quad k = 1, 2, 3, \ldots$$

and

$$c_1 = -\frac{2}{5}c_0, \quad c_2 = \frac{2}{35}c_0, \quad c_3 = -\frac{4}{945}c_0.$$

The general solution is

$$\begin{aligned} y &= C_1(1 + 2x - 2x^2 + \frac{4}{9}x^3 + \ldots) \\ &+ C_2 x^{3/2}(1 - \frac{2}{5} + \frac{2}{35}x^2 - \frac{4}{945}x^3 + \ldots). \end{aligned}$$

2.8.4 Bessel's Equation

$$x^2y'' + xy' + (x^2 - v^2)y = 0 \tag{2.108}$$

is called Bessel's equation.

Solution of Bessel's equation:

Because $x = 0$ is a regular singular point of Bessel's equation we know that there exists at least one solution of the form $y = \sum_{n=0}^{\infty} c_n x^{n+r}$. Substituting the last expression into (2.108) gives

$$x^2y'' + xy' + (x^2 - v^2)y = \sum_{n=0}^{\infty} c_n(n+r)(n+r-1)x^{n+r}$$

$$+\sum_{n=0}^{\infty} c_n(n+r)x^{n+r} + \sum_{n=0}^{\infty} c_n x^{n+r+2} - v^2 \sum_{n=0}^{\infty} c_n x^{n+r}$$

$$= c_0(r^2 - r + r - v^2)x^r + x^r \sum_{n=0}^{\infty} c_n x^{n+2}$$

$$+x^r \sum_{n=1}^{\infty} c_n[(n+r)(n+r-1) + (n+r) - v^2]x^n$$

$$= c_0(r^2 - v^2)x^r + x^r \sum_{n=1}^{\infty} c_n[(n+r)^2 - v^2]x^n$$

$$+x^r \sum_{n=0}^{\infty} c_n x^{n+2}. \tag{2.109}$$

From (2.109) we see that the indicial equation is $r^2 - v^2 = 0$, so the indicial roots are $r_1 = v$ and $r_2 = -v$. When $r_1 = v$, (2.109) becomes

$$\begin{aligned} - \quad & x^v \sum_{n=1}^{\infty} c_n n(n+2v)x^n + x^v \sum_{n=0}^{\infty} c_n x^{n+2} \\ = \quad & x^v \left[(1+2v)c_1 x + \sum_{n=2}^{\infty} c_n n(n+2v)x^n + \sum_{n=0}^{\infty} c_n x^{n+2}\right] \\ = \quad & x^v \left[(1+2v)c_1 x + \sum_{k=0}^{\infty}[(k+2)(k+2+2v)c_{k+2} + \sum_{n=0}^{\infty} c_n x^{k+2}\right] = 0. \end{aligned}$$

Therefore by the usual argument we can write $(1+2v)c_1 = 0$ and

$$(k+2)(k+2+2v)c_{k+2} + c_k = 0$$

$$\text{or} \quad c_{k+2} = \frac{-c_k}{(k+2)(k+2+2v)}, \quad k = 0, 1, 2, 3, \ldots \tag{2.110}$$

The choice $c_1 = 0$ in (2.110) implies $c_3 = c_5 = c_7 = \cdots = 0$, so for $k = 0, 2, 4, 6 \ldots$. We find after letting $k + 2 = 2n$, $n = 1, 2, 3, \ldots$ that

$$c_{2n} = -\frac{c_{2n-2}}{2^2 n(n+v)}. \tag{2.111}$$

Thus, $c_2 = -\dfrac{c_0}{2^2.1(1+v)}$

$$\begin{aligned} c_4 &= -\frac{c_2}{2^2.2(2+v)} = -\frac{c_0}{2^4.2.1(1+v)(2+v)} \\ c_6 &= -\frac{c_4}{2^2.3(3+v)} = -\frac{c_0}{2^6.1.2.3(1+v)(2+v)(3+v)} \\ &\vdots \\ &\vdots \\ &\vdots \\ &\vdots \\ c_{2n} &= -\frac{(-1)^n c_0}{2^{2n} n!(1+v)(2+v)\ldots(n+v)}, \quad n = 1, 2, 3, \ldots \end{aligned} \tag{2.112}$$

It is standard practice to choose c_0 to have a specific value namely.

$$c_0 = \frac{1}{2^v \Gamma(1+v)}$$

where $\Gamma(1+v)$ is the gamma function (see Appendix A). Since this latter function possesses the convenient property $\Gamma(1+\alpha) = \alpha\Gamma(\alpha)$, we can reduce the indicated product in the denominator of (2.112) to one term. For example:

$$\begin{aligned}\Gamma(1+v+1) &= (1+v)\Gamma(1+v).\\ \Gamma(1+v+2) &= (2+v)\Gamma(2+v) = (2+v)(1+v)\Gamma(1+v).\end{aligned}$$

Hence we can write (2.112) as

$$\begin{aligned}c_{2n} &= \frac{(-1)^n}{2^{2n+v} n!(1+v)(2+v)\ldots(n+v)\Gamma(1+v)}\\ &= \frac{(-1)^n}{2^{2n+v} n!\Gamma(1+v+n)}, \quad \text{for } n = 0, 1, 2, \ldots\end{aligned}$$

Bessel function of first kind

Using coefficients c_{2n} just obtained and $r = v$, a series solution of (2.108) is $y = \sum_{n=0}^{\infty} c_{2n} x^{2n+v}$. The solution is usually denoted by $J_v(x)$:

$$J_v(x) = \sum_{n=0}^{\infty} \frac{(-1)^n}{n!\Gamma(1+v+n)} \left(\frac{x}{2}\right)^{2n+v}. \tag{2.113}$$

If $v \geq 0$, the series converges at least on the interval $[0, \infty)$. Also, for the second exponent $r_2 = -v$ we obtain, in exactly the same manner,

$$J_{-v}(x) = \sum_{n=0}^{\infty} \frac{(-1)^n}{n!\Gamma(1-v+n)} \left(\frac{x}{2}\right)^{2n-v}. \tag{2.114}$$

The functions $J_v(x)$ and $J_{-v}(x)$ are called **Bessel functions of the first kind** of order v and $-v$, respectively. Depending on the values of v, (2.114) may contain negative powers of x and hence converge on $(0, \infty)$.

2.8.5 Legendre's Equation

$$(1-x^2)y'' - 2xy' + n(n+1)y = 0. \tag{2.115}$$

is known as **Legendre's equation.**

Solution of Legendre's equation: Since $x = 0$ is an ordinary point of the equation, we substitute the power series $y = \sum_{n=0}^{\infty} c_n x^n$, shift summation indices, and combine series to get

$$\begin{aligned}(1-x^2)y'' - 2xy' + n(n+1)y &= [n(n+1)c_0 + 2c_2] + [(n-1)(n+2)c_1 + 6c_3]x\\ &+ \sum_{j=2}^{\infty} [(j+2)(j+1)c_{j+2} + (n-j)(n+j+1)c_j]x^j = 0,\end{aligned}$$

which implies that

$$\begin{aligned} n(n+1)c_0 + 2c_2 &= 0 \\ (n-1)(n+2)c_1 + 6c_3 &= 0 \\ (j+2)(j+1)c_{j+2} + (n-j)(n+j+1)c_j &= 0. \end{aligned}$$

$$\begin{aligned} c_2 &= -\frac{n(n+1)}{2!} \\ c_3 &= -\frac{(n-1)(n+2)}{3!} \\ c_{j+2} &= -\frac{(n-j)(n+j+1)}{(j+2)(j+1)}c_j, \quad j = 2, 3, 4, \ldots \end{aligned} \tag{2.116}$$

If we let j take on the values 2,3,4,..., the recurrence relation (2.116) yields

$$\begin{aligned} c_4 &= -\frac{(n-2)(n+3)}{4.3}c_2 \\ &= \frac{(n-2)n(n+1)(n+3)}{4!}c_0 \end{aligned}$$

$$\begin{aligned} c_5 &= -\frac{(n-3)(n+4)}{5.4}c_3 \\ &= -\frac{(n-3)(n-1)(n+2)(n+4)}{5!}c_1 \\ c_6 &= -\frac{(n-4)(n+5)}{6.5}c_4 \\ &= -\frac{(n-4)(n-2)n(n+1)(n+3)(n+5)}{5!}c_1 \\ c_7 &= -\frac{(n-5)(n+6)}{7.6}c_5 \\ &= -\frac{(n-5)(n-3)(n-1)(n+2)(n+4)(n+6)}{7!}c_1 \end{aligned}$$

and so on. Thus for at least $|x| < 1$ we obtain two linearly independent power series solutions:

$$\begin{aligned} y_1(x) &= c_0\left[1 - \frac{n(n+1)}{2}x^2 + \frac{(n-2)n(n+1)(n+3)}{4!}x^4 \right. \\ &\quad \left. - \frac{(n-4)(n-2)n(n+1)(n+3)(n+5)}{6!}x^6 + \ldots\right]. \end{aligned} \tag{2.117}$$

$$\begin{aligned} y_2(x) &= c_1\left[x - \frac{(n-1)(n+2)}{3!}x^3 + \frac{(n-3)(n-1)(n+2)(n+4)}{5!}x^5 \right. \\ &\quad \left. - \frac{(n-5)(n-3)(n-1)(n+2)(n+4)(n+6)}{7!}x^7 + \ldots\right]. \end{aligned}$$

If n is an even integer, the first series terminates, whereas $y_2(x)$ is an infinite series. For example, if $n = 4$, then

$$\begin{aligned} y_1(x) &= c_0 \left[1 - \frac{4.5}{2!}x^2 + \frac{2.4.5.7}{4!}x^4\right] \\ &= c_0 \left[1 - 10x^2 + \frac{35}{3}x^4\right]. \end{aligned}$$

Similarly, when n is an odd integer, the series for $y_2(x)$ terminates with x^n, that is, when n is a non-negative integer, we obtain an nth degree polynomial solution of Legendre's equation.

Since we know that a constant multiple of a solution of Legendre's equation is also a solution, it is traditional to choose specific values for c_0 or c_1, depending on whether n is an even or odd positive integer, respectively. For $n = 0$ we choose $c_0 = 1$, and for $n = 2, 4, 6 \ldots$

$$c_0 = (-1)^{n/2} \frac{1.3\ldots(n-1)}{2.4\ldots n};$$

whereas for $n = 1$ we choose $c_1 = 1$, and for $n = 3, 5, 7, \ldots,$

$$c_1 = (-1)^{(n-1)/2} \frac{1.3\ldots n}{2.4\ldots(n-1)}.$$

For example, when $n = 4$, we have

$$\begin{aligned} y_1(x) &= (-1)^{4/2}\frac{1.3}{2.4}\left[1 - 10x^2 + \frac{35}{3}x^4\right] \\ &= \frac{1}{8}(35x^4 - 30x^2 + 3). \end{aligned}$$

These specific nth degree polynomial solutions are called **Legendre polynomials** and are denoted by $P_n(x)$. From the series for $y_1(x)$ and $y_2(x)$ and from the above choices of c_0 and c_1 we find that the first several Legendre polynomials are

$$\begin{aligned} P_0(x) &= 1 \\ P_1(x) &= x \\ P_2(x) &= \frac{1}{2}(3x^2 - 1) \\ P_3(x) &= \frac{1}{2}(5x^3 - 3x) \\ P_4(x) &= \frac{1}{8}(35x^4 - 30x^2 + 3) \\ P_5(x) &= \frac{1}{8}(63x^5 - 70x^3 + 15x). \end{aligned} \qquad (2.118)$$

Remember, $P_0(x)$, $P_1(x)$, $P_2(x), \ldots$ are, in turn, particular solutions of the differential equations

$$\begin{aligned} n &= 0 : (1-x^2)y'' - 2xy' = 0 \\ n &= 1 : (1-x^2)y'' - 2xy' + 2y = 0 \\ n &= 2 : (1-x^2)y'' - 2xy' + 6y = 0 \\ n &= 3 : (1-x^2)y'' - 2xy' + 12y = 0. \end{aligned}$$

2.9 Exercises

2.1. Classify the given differential equations by order, and tell whether it is linear or non linear.

(a) $y' + 2xy = x^2$

(b) $y'(y + x) = 6$

(c) $y \cos y = y''$

(d) $y^2 \sin y = y''$

(e) $y'' - 4y' + 3y = x^4$

(f) $y'' = e^{zy}$

2.2. State whether the following differential equations are linear or non-linear. Write the order of each equation.

(a) $(1-x^2)y'' - 6xy' + 9y = \sin x$

(b) $\dfrac{xd^3y}{dx^3} - 2\left(\dfrac{dy}{dx}\right)^2 + y = 0$

(c) $yy' + 2y = 2 + x^2$

(d) $\dfrac{d^2y}{dx^2} + 9y = \sin y$

(e) $\dfrac{dy}{dx} = \left(1 + \left(\dfrac{d^2y}{dx^2}\right)^2\right)^{1/2}$

(f) $\dfrac{d^2r}{dt^2} = -\dfrac{k}{r^2}$

Verify that in Exercises 2.3 to 2.8 the indicated function is a solution of the given differential equation. In some cases assume an appropriate interval.

2.3. $2y' + y = 0$; $y = e^{-x/2}$.

2.4. $y'' + 16y = 0;\ y = \sin 4x$.

2.5. $x^2 dy + 2xy dx = 0:\ y = -1/x^2$.

2.6. $y''' - 3y'' + 3y' - y = 0;\ y = x^2 e^x$.

2.7. $y = y - 1;\ y = e^x + 1$.

2.8. $y'' + 9y = 8 \sin x;\ y = \sin x + \lambda_1 \cos 3x + \lambda_2 \sin 3x$.

In Exercises 2.9 through 2.14 determine a region of the xy plane for which the given differential equation would have a unique solution through a point (x_0, y_0) in the region.

2.9. $\dfrac{dy}{dx} = \sqrt{xy}$

2.10. $\dfrac{dy}{dx} = \dfrac{y}{x}$

2.11. $\dfrac{dy}{dx} = y + x$

2.12. $y' = \dfrac{y^2}{x^2 + y^2}$

2.13. $\dfrac{dy}{dx} = x^2 \cos y$

2.14. $\dfrac{dy}{dx} = \dfrac{y + x}{y - x}$

2.15. Solve the following differential equations by the separation of variables method:

(i) $dx + e^{3x} dy$

(ii) $\dfrac{dy}{dx} = (1 + x)^3$

(iii) $\dfrac{dy}{dx} = e^{3x - 2y}$

(iv) $\dfrac{dP}{dt} = P - P^2$

(v) $\dfrac{dN}{dt} + N = N + e^{t+2}$

(vi) $\dfrac{dy}{dx} = \dfrac{y^2 - 1}{x^2 - 1}$

2.16. Solve the following linear differential equations :

(i) $(x^2 - 9)\dfrac{dy}{dx} + xy = 0$

(ii) $x\dfrac{dy}{dx} + 2y = 3$

(iii) $\cos x\dfrac{dy}{dx} + (\sin x)y = 1$

2.17. Solve the following initial value problems:

(i) $y\dfrac{dy}{dx} - x = 2y^2, \quad y(1) = 5$

(ii) $(x+1)\dfrac{dy}{dx} + y = \ln x, \quad y(1) = 10$

2.18. Determine whether the given differential equation is exact. If it is exact, solve it.

(i) $(2x + y)dx - (x + 6y)dy = 0$

(ii) $(5x + 4y)dx + (4x - 8y^3)dy = 0$

(iii) $x\dfrac{dy}{dx} = 2xe^x - y + 6x^2$

2.19. Find the general solutions of the second order differential equations below:

(i) $y'' + 8y' + 16y = 0$

(ii) $y'' - 4y' + 5y = 0$

(iii) $2y'' - 3y' + 4y = 0$

2.20. Solve the given initial value problems:

(i) $y'' + 16y = 0, \quad y(0) = 2, \; y'(0) = -2$

(ii) $4y'' - 4y' - 3y = 0, \quad y(0) = 1, \; y'(0) = 5$

2.21. Solve the following initial value problems:

(i) $\dfrac{d^2y}{dx^2} + 4y = -2$

$y\left(\dfrac{\pi}{8}\right) = \dfrac{1}{2}, \quad \left.\dfrac{dy}{dx}\right|_{x=\pi/8} = 2$

(ii) $5\dfrac{d^2y}{dx^2} + 4\dfrac{dy}{dx} = -6x$

$y(0) = 0, \quad \left.\dfrac{dy}{dx}\right|_{x=0} = -10$

2.22. Solve each differential equation below by variation of parameters.

(i) $y'' + y = \sin x$

(ii) $y'' + 3y' + 2y = \dfrac{1}{1 + e^x}$

2.23. Solve the following differential equations with or without initial values.

(i) $x^2\dfrac{d^2y}{dx^2} - 7x\dfrac{dy}{dx} + 41y = 0$

(ii) $x\dfrac{d^2y}{dx^2} - 3\dfrac{dy}{dx} = 0$

(iii) $x^2\dfrac{d^2y}{dx^2} + 3x\dfrac{dy}{dx} = 0$

$y(1) = 0, \quad \left.\dfrac{dy}{dx}\right|_{x=1} = 4$

2.24. Derive a population growth model that takes deaths into account.

2.25. A drug is infused into a patient's blood system at a constant rate of r grams per second. Simultaneously the drug is removed at a rate proportional to the amount $x(t)$ of the drug present at any time t. Determine a differential equation governing the amount $x(t)$.

2.26. Find the relation between doubling and tripling times for a population.

2.27. In an archaeological wooden specimen, only 25% of original radiocarbon 12 is present. Write a mathematical model, the solution of which will give time of its manufacturing.

2.28. Write a mathematical model whose solution will provide the rate of interest compounded continuously if a bank's rate of interest is 10% per annum.

2.29. The number of field mice in a certain pasture is given by the function $200-10t$, where t is measured in years. Determine a differential equation governing a population of owls that feed on the mice if the rate at which the owl population grows is proportional to the difference between the number of owls at the time t and the number of field mice at time t.

2.30. Let a dog start running to pursue a rabbit at time t_0 when the dog sights the rabbit. Determine a differential Equation (mathematical model) whose solution will give the path of pursuit assuming that the rabbit runs in a straight line at a constant speed so that its line of sight is always directed at the rabbit.

2.31. To save money the manager of a manufacturing firm decides to eliminate the advertising budget. In the absence of advertising, the sales manager finds that sale in Indian rupees, decline at a rate that is directly proportional to the volume of sales. Write a differential equation that describes the rates of declining sales.

2.32. Suppose you deposited 10,000 Indian rupees in a bank account at an interest rate of 5% compounded continuously. Write a mathematical model in terms of a differential equation whose solution will give the amount of money in your account after a year and a half.

2.33. Bacteria grown in a culture increase at a rate proportional to the number present. If the number of bacteria doubles every 2 hours, write a mathematical equation describing this situation by which you can find the population of bacteria (number) after 10 hours and 10 days.

2.34. The growth rate of a population of bacteria is directly proportional to the population. If the number of bacteria in a culture grows from 100 to 400 in 24 hours, write the initial value problem to determine the population after 12 hours.

2.35. A culture initially has P_0 number of bacteria. At $t = 1$ hour, the number of bacteria is measured to be $3/2P_0$. If the rate of growth is proportional to the number of bacteria $P(t)$ present at time t, determine the time necessary for the number of bacteria to triple.

2.36. Solve the logistic differential equation:

$$\frac{dN}{dt} = r_0 \left(1 - \frac{N}{k}\right) N,$$

$$t \geq 0, \quad N(0) = N_0.$$

2.37. Insects in a tank increase at a rate proportional to the number present. If the number increases from 50,000 to 100,000 in 1 hour, how many insects are present at the end of 2 hours?

2.38. It was estimated that the earth's human population in 1961 was 3.060 billion. Assume the population in 1996 using a model of population growth (2.8), check this number with the actual population of the earth based on available from authentic sources.

2.39. A breeder reactor converts relatively stable uranium 238 into the isotope plutonium 239. After 30 years 0.022% of the initial amount N_0 of plutonium has disintegrated. Find the half life of this isotope if the rate of disintegration is proportional to amount remaining.

2.40. The radioactive Pb–209 isotope of lead decays at a rate proportional to the amount present at time t and has a half life of 4 hours. If 1 gram of lead is present initially, how long will it take for 80% of the lead to decay?

2.41. In the 1950 excavation at Nippur in Babylonia, charcoal from a roof beam gave a count of 4.09 disintegrations per minute per gram. Living wood gave 6.68 disintegrations. Assuming that this charcoal was formed during Hammurabi's reign, find an estimate for the likely time Hammurabi's succession.

2.42. Suppose a large mixing tank initially holds 300 gallons of water in which 50 pounds of salt has been dissolved. Pure water is pumped into the tank at a rate of 3 gal/min, and when the solution is well stirred it is pumped out at the same rate. Write a differential equation for the amount $A(t)$ of salt in the tank at any time t.

2.43. A spherical rain drop evaporates at a rate proportional to its surface area. Write a differential equation for its volume V as a function of time.

2.44. A chemical A in a solution breaks down to form chemical B at a rate proportional to the concentration of unconverted A. Half of A is converted in 20 minutes. Write a differential equation describing this physical situation.

2.45. A tank with a capacity of 600 liters initially contains 200 liters of pure water. A solution containing 3 kilograms of salt per liter is allowed to run into the tank at a rate of 16 liters per minute. The mixture is then removed at a rate of 12 liters per minute. Find the expression for the number of kilograms of salt in the tank at any time t.

2.46. A large tank is filled with 600 liters of pure water. Brine containing 2 kilograms of salt per liter is pumped into the tank at a rate of 5 liters per minute. The well mixed solution is pumped out at the same rate. Find the number $P(t)$ of kilograms of salt in the tank at time t. What is the concentration of the solution in the tank at $t = 10$ minutes?

2.47. A 250 liter tank contains 100 liters of pure water. Brine containing 4 kilograms of salt per liter flows into the tank at a rate of 5 liters per minute. If the well stirred mixture flows out at a rate of 3 liters per hour, find the concentration of salt in the tank at the instant it is filled to the top.

2.48. A thermometer reading $100°F$ is placed in a pan of oil maintained at $10°F$. What is the temperature of the thermometer when $t = 20$ seconds, if its temperature is $60°F$ when $t = 8$ seconds?

2.49. A thermometer is removed from a room where the air temperature is $60°\ F$ and is taken outside, where the temperature is $55°F$. After 1 minute the thermometer reads $50°F$. What is the reading of the thermometer at $t = 2$ minutes? How long will it take for the thermometer to reach $20°F$.

2.50. Water is heated at $120°F$. It is then removed from the burner and kept in a room of $30°F$ temperature. Assuming that there is no change in the temperature of the room and the temperature of the hot water is $110°F$ after 3 minutes:

   (a) Find the temperature of water after 6 minutes.

(b) Find the duration in which water will cool down to the room temperature.

2.51. The diagram in Figure 2.3 represents an electric circuit in which voltage of V volts is applied to a resistance of R ohms and an inductance of L henrys connected in series.When the switch is closed, a current will vary with time, and it can be shown that a mathematical model for this circuit is the first order differential equation

$$L\frac{dI}{dt} + IR = V.$$

Verify that the current in the circuit is given by

$$I = \frac{V}{R}(1 - e^{-Rt/L}).$$

2.52. When an object at room temperature is placed in an oven whose temperature is $400°C$, the temperature of the object will increase with time, approaching the temperature of the oven. It is known that the temperature Q of the object is related to the time through the differential equation

$$\frac{dQ}{dt} = \alpha(Q - 400).$$

Verify that the temperature of the object is given by $Q = 400 + cc^{\alpha t}$, where c and α are constants.

2.53. A car leaves at 11:30 am and arrives at a heart research center in a city at 3 pm. Its driver started from rest and steadily increased his speed, as indicated on his speedometer, to the extent that when he reached the destination he was driving at a speed of 60 kilometer per hour. Write a mathematical model in terms of differential equation which may help to determine the distance to location from where the car started.

2.54. Find Laplace transforms of the functions:

(a) $2\sinh t - 4$

(b) $t^2 - 3t + 5$

(c) $4t\sin 2t$

(d) $t - \cos(5t)$

(e) $(t + 4)^2$

(f) $3e^{-t} + \sin 6t$

(g) $t^3 - 3t + \cos 4t$

(h) $-3\cos 2t + 5\sin 4t$

(i) te^{4t}

(j) t^2e^{-2t}

(k) $e^t \cos t$

(l) $f(t) = \begin{cases} t, & 0 \le t < 1 \\ 1, & t \ge 1 \end{cases}$

(m) $t(t-2)e^{3t}$

(n) $t^3 - \sinh 2t$

(o) $e^{-2t} + 4e^t$

(p) $f(t) = \begin{cases} 1, & 0 \le t < 2 \\ 2, & t \ge 2 \end{cases}$

(q) $g(t) = \begin{cases} \sin t, & 0 \le t \le \pi \\ 0, & t \ge \pi \end{cases}$

(r) $f(t) = \begin{cases} e^t, & 0 \le t < 2 \\ 0, & t \ge 2 \end{cases}$

(s) $L(t) = \begin{cases} 1, & 0 \le t < 2 \\ 2, & 2 \le t < 4 \\ 0, & t \ge 4 \end{cases}$

2.55. Show that t^n, where n is a positive integer is of exponential order. Show that the following functions are of exponential order.

(a) $t^{1/2}$

(b) $\sinh t$

(c) $t \sin 2t$

2.56. Evaluate $L^{-1}\left\{\dfrac{s^2}{(s+1)3}\right\}$.

2.57. Evaluate $L^{-1}\left\{\dfrac{s}{s^2+6s+13}\right\}$.

2.58. Use the first shifting theorem to find the Laplace transforms of the following functions.

(i) $e^t(\cos 2t - 3\sin 5t)$.

(ii) $e^{-2t}\cos 4t$.

2.59. Use the first shifting theorem to solve the initial value problems.

$$y'' - 6y' + 9y = t^2e^{3t}.$$

$$y(0) = 2, \quad y'(0) = 17.$$

2.60. Solve $f(t) = t - \int_0^t (t-u)f(u)du$.

2.61. Solve $f(t) + 2\int_0^t f(u)\cos(t-u)du = 4e^t + \sin t$.

2.62. Solve $f(t) + \int_0^t f(u)du = 1$.

2.63. Solve $f(t) = 1 + t - \frac{8}{3}\int_0^t (u-t)^3 f(u)du$.

2.64. Solve $y'' + y = 4\delta(t-2\pi)$, $y(0) = 1$, $y'(0) = 0$.

2.65. Solve $y_1'' + 10y_1' - 4y_2 = 0$, $-4y_1 + y_2'' + 4y_2 = 0$ subject to initial conditions $y_1(0) = 0$, $y_1'(0) = 1$, $y_2(0) = 0$, $y_2'(0) = -1$.

2.66. Solve the initial value problem $y' = 3y = 13\sin 2t$, $y(0) = 6$.

2.67. Solve the initial value problem $y''' + 2y'' - y' - 2y = \sin 3t$, $y(0) = 0$, $y'(0) = 0$, $y'(0) = 0$, $y''(0) = 1$.

2.68. Solve the initial value problems:

(a) $y'' + 4y = \sin t H(t-2\pi)$, $y(0) = 1$, $y'(0) = 0$.

(b) $y'' - 5y' + 6y = H(t-1)$, $y(0) = 0$, $y'(0) = 1$.

2.69. Let $F(s) = L\{f(t)\}$ and $n = 1, 2, 3, \ldots$, then prove that

$$L\{t^n f(t)\} = (-1)^n \frac{d^n}{ds^n} F(s).$$

Power series

2.70. Write $e^x \cos x$ in the form of a power series. Examine whether this power series is convergent.

Solution about ordinary point

Find the general solution of the following differential equations about an ordinary point in terms of two power series.

2.71. $y'' - (1+x)y = 0$.

2.72. $y'' + x^2 y'$.

2.73. $y'' + y = e^x$.

2.74. $(x^2 - 1)y' + y = 0$.

2.75. $y'' - (x+1)y' - y = 0$.

Use the power series method to solve the following initial value problems

2.76. $(x-1)y'' - xy' + y = 0$, $y(0) = -2$, $y'(0) = 6$.

2.77. $(x^2+1)y'' + 2xy' = 0$, $y(0) = 0$, $y'(0) = 1$.

Solution about regular singular point: method of Frobenius
Use the method of Frobenius to solve the following differential equations.

2.78. $xy'' - xy' + y = 0.$

2.79. $y'' + \frac{3}{x}y' - 2y = 0.$

2.80. $xy'' + y' + y = 0.$

2.81. $xy'' + y' + xy = 0.$

Bessel's equation
Find the general solution of the following equations.

2.82. $x^2y'' + xy' + (x^2 - 1)y = 0.$

2.83. $xy'' + xy' + xy = 0.$

2.84. Verify that $y = x^n Jn(x)$ is a particular solution of $xy''+(1-2n)y'+xy = 0,\ x > 0.$

Legendre's equation
Solve the following equations.

2.85. $(1 - x^2)y'' - 2xy' = 0.$

2.86. $(1-x^2)y''-2xy'+12y = 0$ subject to initial conditions $y(0) = 0,\ y'(0) = 1.$

2.10 Suggestion for Further Reading

In this chapter we have presented some well known methods of finding analytic solutions of linear differential equations of first and second orders with and without initial and boundary conditions. Solutions of second order differential equations applying Laplace transform as well as series solutions methods with variable coefficients are discussed. Modeling of engineering problems in terms of ordinary differential equations is explained. A natural question arises as to what related areas to this chapter could be studied independently after completing this chapter.

Section 2.6 models under simple assumptions are considered. One may take up the study of problems more realistic in nature, for example, problems from references [3 through 5, 7, 11 through 15]. Ordinary differential equations are also used in modeling of epidemics. One may pursue the study of this field from references [2, 12].

In Section 2.6.3 we have discussed Newton's cooling law and its formulation as a differential equation. Newton's cooling law is purely a surface principle; it involves a boundary condition, and it leads to a surface temperature that depends only on time. In extended bodies, interior temperature typically varies not only in time but also from place to place. Principles that govern interior temperature were first explained by Joseph Fourier (1768-1830) at the beginning of the 19th century. His theory is explained in Chapter 4. Fourier's method hinges in the relationship between heat and temperature and on the principle of conservation of energy. The equation governing distribution of temperature is known as the heat equation. It is discussed in Chapter 5. However, we recommend books by Cannon [6] and Groestch [8, 118-126, 9] for lucid introduction. It may be pointed out here that the heat equation has been used to model option pricing– important concept of financial engineering which fetched a Nobel Prize for economics in 1997. Roos [12] provides updated information on applications of ordinary differential equations to population dynamics which is quite useful for planning in different branches of engineering. MATLAB programs are introduced. A reading of the book may inculcate interest among engineering students and even faculty members for applying differential equations in engineering problems.

As mentioned above, ordinary differential equations are used for modeling epidemics; see for example a recent article by Beckely et al. [2]. References [3, 5, 7, 13, 14] provide valuable information of application of ordinary differential equations in diverse fields. We solve direct problems related to differential equations, that is, we have found solutions of differential equations in diverse fields and under certain conditions.

An inverse problem could be to find boundary or initial conditions or parameters if the solution of a given differential equation is known. A simple account of inverse problems associated with ordinary differential equations is given by Groetsch [8, 9]. An introduction of inverse problems and their applications to engineering problems is given in Chapter 9. The books by Krisch [10], Bank and Kunish [1] and Groetsch [9] provide slightly advanced level discussions on inverse problems in engineering fields.

It may be pointed out to the readers that solutions of inverse problems have been important in the development of mathematics and different areas of sciences and technology. That fact alone would justify the introduction of inverse problems in the early years of undergraduate programs. But more important is the development of a habit of inverse thinking. If students consider only the direct problem, they are not looking at the problem from all angles and will fail to see two thirds of real world problems. The authors have been benefited from references [3, 4, 11, 13 through 15] while writing this chapter.

Bibliography

[1] H. T. Bank and K. Kunish, *Estimation Techniques for Distributed Parameter Systems*, Birkhäuser, Boston, 1989.

[2] R. Beckley et al., *Modeling Epidemics with Differential Equations*, www.tnstate.edu/math.

[3] P. Blanchard, L. D. Robert and R. H. Glem, *Ordinary Differential Equations*, Richard Stratten, 2012.

[4] W. E. Boyce, R. C. DiPrima, *Elementary Differential Equations and Boundary Value Problems*, Ninth Edition, Wiley, 2010.

[5] J. R. Brenan, W. E. Boyce, *Differential Equations: An Introduction to Modern Methods and Applications*, John Wiley & Sons Inc, 2007.

[6] J. R. Cannon, *The One–Dimensional Heat Equation*, Addison Wesley, 1998.

[7] C. H. Edwards and D. E. Penney, *Differential Equations: Computing and Modeling*, Fourth Edition, Pearson Prentice Hall, 2008.

[8] C. W. Groestch, *Inverse Problems*, Mathematical Association of America, 1999.

[9] C. W. Groestch, *Inverse Problems in Mathematical Sciences*, Verlag Vieweg, Braunschweig, 1993.

[10] A. Krisch, *An Introduction to the Mathematical Theory of Inverse Problems*, Springer, New York, 1996.

[11] B. J. Rice and J. D. Strange, *Ordinary Differential Equations with Applications*, Third Edition, Brooks/Cole Publishing Company, 1994.

[12] A. M. De Roos, *Modeling Population Dynamics*, Amsterdam. *(staff.science.uva-nl/aroos/doenloads/pdf_sreader/syllabus.pdf)*, 2013.

[13] A. H. Siddiqi and P. Manchanda, *A First Course of Differential Equations with Applications*, Macmillan, India Ltd., 2006.

[14] W. Xie, *Differential Equations for Engineers*, Cambridge University Press, 2010.

[15] D. G. Zill, *A First Course in Differential Equations with Modeling Applications*, Seventh Edition, Brooks/Cole, 2001.

Chapter 3

Vector Calculus

3.1 Introduction

In the calculus course (Appendix B) we study properties of functions defined on R (the line), R^2 (the plane) or R^3 (the space) with values in R, which are called real valued or scalar functions, or scalar fields. Here, we study the calculus of functions taking values in R^2 or R^3, instead of R. Those functions are called vector valued functions or vector fields. In Section 3.2 the concept of a vector will be introduced along with its basic algebraic properties. Vector fields and their continuity and differentiation properties are discussed in Section 3.3 along with the notions of gradient, divergence and curl. Moreover, we explain how curves and surfaces are described by such functions. Integrals of vector fields are introduced in Section 3.4, first the line integral for scalar and vector fields, and then the surface integrals for scalar fields.

In Section 3.5, we present three fundamental theorems of vector calculus, namely the Green-Ostrogradski theorem, the Gauss divergence theorem and the theorem of Stokes. Section 3.6 is devoted to certain applications of vector calculus to science and engineering, with an emphasis on problems from mechanics.

The concept of a vector can be traced to the development of affine and analytic geometry in the 17th century and, later, the invention of complex numbers and quaternions. Vector calculus was developed mainly during the second half of the 19th century; contributions stem from, among others, Sir William Rowan Hamilton (1805-1865), William Kingdon Clifford (1845-1879), James Clerk Maxewell (1831-1879), Hermann Gunter Grassman (1809-1877) and Josiah Willard Gibbs (1839-1903). However, the fundamental results mentioned above were already discovered by George Green (1793-1841), Carl Friedrich Gauss (1777-1855), Gabriel Stokes (1819-1903) and Mikhail Ostrogradski (1801-1862).

Vector calculus now serves as a basic mathematical tool in all areas of science and engineering where mechanical, electromagnetic and thermodynamics forces determine the behaviours of solids, fluids, electric conductors, semiconductors and magnetic materials. Many ideas of this chapter are based on references [7, 8]: 7 is out of print.

3.2 Vectors

Vectors and their basic properties have been introduced in Section 3.1. In this section we discuss the concepts of magnitude of a vector and angle between two vectors. The magnitude of a vector is called the **norm**. The angle between two vectors is defined with the help of the inner product of two vectors.

Definition 53. The **inner product** of two vectors $x = (x_1, x_2, x_3)$ and $y = (y_1, y_2, y_3)$ denoted by $x.y$ or (x, y) is defined by $x.y = x_1y_1 + x_2y_2 + x_3y_3$.

Position vector (2-space and 3-space)
The vector shown in Figure 3.1b with initial point the origin O and terminal $P(x_1, x_2)$ is called the position vector of the point P and is written

$$\overrightarrow{OP} = (x_1, x_2).$$

Figure 3.2 depicts the position velocity in 3-space with the origin O and terminal point $P(x_1, x_2, x_3)$ of the point and is written (x_1, x_2, x_3). Addition

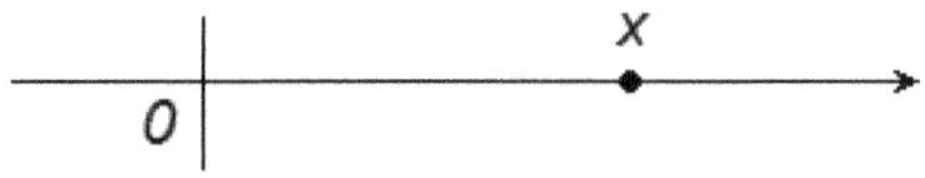

FIGURE 3.1a: Points on Line

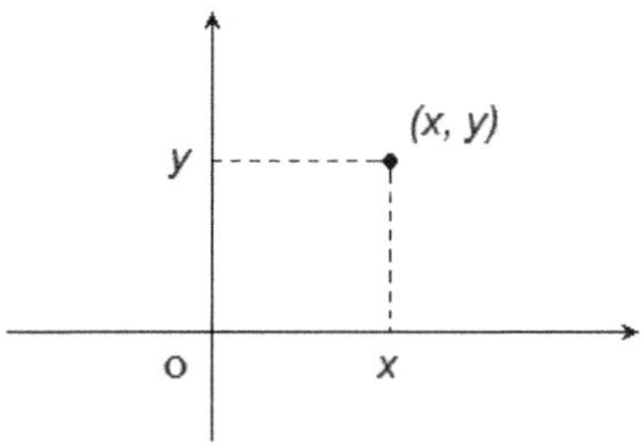

FIGURE 3.1b: Points in Plane

and scalar multiplication have the property that

$$x + y = x + y, \quad (x + y) + z = x + (y + z)$$

$$\alpha(x + y) = \alpha x + \alpha y, \quad (\alpha + \beta)x = \alpha x + \beta x, \quad (\alpha\beta)x = \alpha(\beta x),$$

hold for all vectors x,y and z and all scalars α and β. This follows from the corresponding properties of ordinary addition and multiplication. We know from elementary geometry that the length (or magnitude) of the vector $x = (x_1, x_2, x_3)$ is given by the expression

$$\sqrt{x_1^2 + x_2^2 + x_3^2}.$$

Definition 54. The norm has the following properties, which are analogous to the properties of the absolute value of a real number.

$$\| x \| \geq 0, \quad \text{and} \quad \| x \| = 0 \text{ if and only if x} = 0, \tag{3.1}$$

$$\| \alpha x \| = |\alpha| \, \| x \| \text{ for scalars } \alpha, \tag{3.2}$$

$$\| x + y \| \leq \| x \| + \| y \| . \tag{3.3}$$

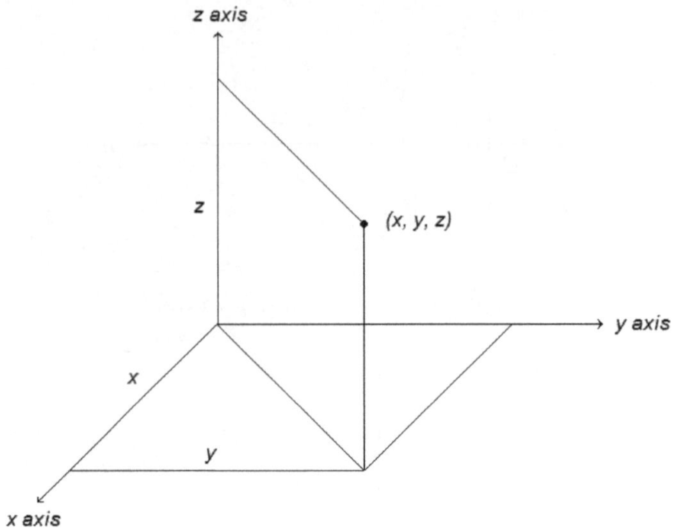

FIGURE 3.2: Points in R^3

Properties (3.1) and (3.2) can be checked immediately from the definition. It will be seen below that property (3.3) is a consequence of the Schwarz inequality. Setting $\alpha = -1$ in property (3.2), we obtain

$$\| -x \| = \| x \| \tag{3.4}$$

that holds for every vector x.

The $\| x + y \| \leq \| x \| + \| y \|$ is called the **triangle inequality**. Consider the triangle formed by the points $A = 0,\ B = x$ and $C = x + y$. The triangle inequality states that the length of the side connecting A and C cannot be larger than the sum of the length of the other two sides.

Example 172. A vector x with norm $\| x \| = 1$ is called a **unit vector**. If x is a non-zero vector, then

$$e = \frac{x}{\| x \|}$$

is a unit vector. The unit vectors

$$i = (1, 0, 0),\ j = (0, 1, 0),\ k = (0, 0, 1)$$

(they obviously satisfy $\| i \| = \| j \| = \| k \| = 1$) are also called **unit coordinate vectors** or **standard unit vectors**, see Figure 3.3. Every vector can be expressed as a linear combination of the unit coordinate vectors. For $x = (x_1, x_2, x_3)$, we have

$$x = x_1 i + x_2 j + x_3 k.$$

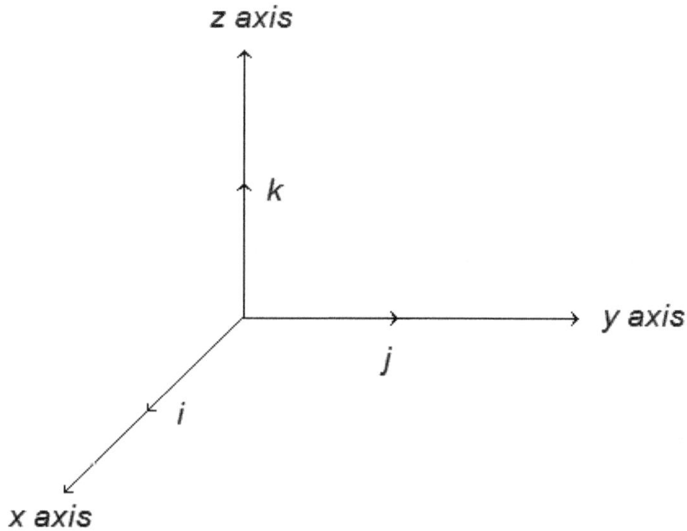

FIGURE 3.3: Unit Vectors in R^3

Indeed, from the rules of addition and scalar multiplication we see that

$$\begin{aligned} x &= (x_1, 0, 0) + (0, x_2, 0) + (0, 0, x_3) \\ &= x_1(1, 0, 0) + x_2(0, 1, 0) + x_3(0, 0, 1) \\ &= x_1 i + x_2 j + x_3 k. \end{aligned}$$

Example 173. Let us consider a vector $x = x_1 i + x_2 j + x_3 k$. If $x_3 = 0$, we have $x = x_1 i + x_2 j$, so x lies in the plane spanned by the first two unit coordinate vectors. We may identify this plane with the standard xy plane spanned by unit coordinate vectors (1,0) and (0,1), which we continue to denote by i and j. Every vector $x = (x_1, x_2)$ in the plane can be expressed as a linear combination

$$x = x_1(1, 0) + x_2(0, 1) = x_1 i + x_2 j$$

of the unit coordinate vectors.

Example 174. (a) Find the magnitude of the vector $5i + 3k$.
(b) Show that $x = (\frac{-2}{7}, \frac{3}{7}, \frac{6}{7})$ is a unit vector.

Solution: (a) Magnitude $= \| 5i + 3k \| = \sqrt{25 + 9} = \sqrt{34}$.

(b) $\| x \| = \sqrt{\frac{-2}{7}^2 + \frac{3}{7}^2 + \frac{5}{7}^2} = \sqrt{(\frac{4}{49} + \frac{9}{49} + \frac{36}{49})} = 1.$

3.2.1 Scalar Product

Definition 55. The scalar product and the norm are related by

$$x.x = \| x \|^2 = x_1^2 + x_2^2 + x_3^2, \quad \| x \| = \sqrt{x.x}, \tag{3.5}$$

which holds for every vector $x = (x_1, x_2), x_3$. The scalar product is **commutative**, that is,

$$x.y = y.x$$

holds for all vectors x and y. Moreover, it is **distributive** in the sense that

$$x.(y + z) = x.y + x.z \tag{3.6}$$

$$(x + y).z = x.z + y.z \tag{3.7}$$

$$(\alpha x).y = \alpha(x.y) = x.(\alpha y) \tag{3.8}$$

hold for all vectors x, y and z and all scalars α. In particular, setting $\alpha = 0$ we have for every vector x

$$x.0 = 0 = 0.x.$$

All these rules follow immediately from the definition of the scalar product.

Geometric interpretation of the scalar product.
We look at the triangle in Figure 3.4 with vertices A, B and C which has a right angle at B. Setting $x = B - A$, $y = C - B$, the theorem of Pythagoras tells us that

$$\| x + y \|^2 = \| x \|^2 + \| y \|^2 . \tag{3.9}$$

Using (3.5) and the properties of the scalar product we compute

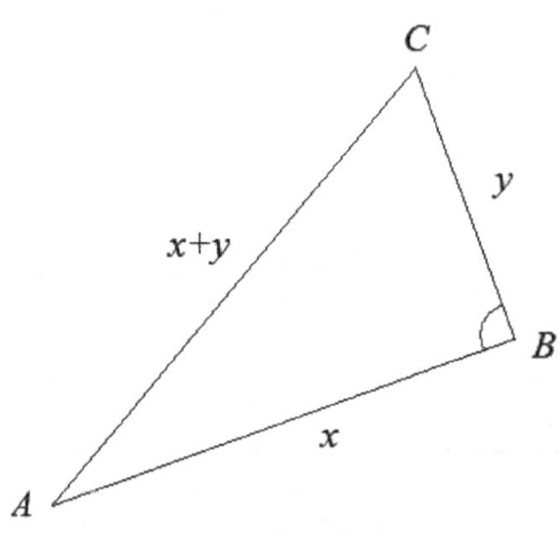

FIGURE 3.4: Geometric Interpretation of Pythagoras Theorem

$$\| x + y \|^2 = (x + y).(x + y) = x.x + x.y + y.x + y.y = \| x \|^2 + 2x.y + \| y \|^2 .$$

Comparing with (3.9) we see that we must have $x.y = 0$.

Definition 56. Two vectors x and y are called **orthogonal** or **perpendicular** if $x.y = 0$, and we write $x \perp y$ in this case.

Example 175. Examine whether the vectors $x = (2, 1, 1)$ and $y = (1, 1, -3)$ are orthogonal.

Solution: We have $x.y = 2.1 + 1.1 + 1.(-3) = 0$. This implies $x \perp y$.
Let us now look at the triangle with vertices A, B and C in Figure 3.4 setting $x = B - A$ and $y = C - B$. we want to determine the point D on the side BC such that the line AD becomes perpendicular to BC. We have $D - B = \alpha y$ for some scalar α, and $A - D = x - \alpha y$. In order to determine α, we exploit the orthogonality relation

$$0 = \alpha y.(x - \alpha y) = \alpha y.x - \alpha^2 y.y$$

and solving for α we obtain

$$\alpha = \frac{x.y}{y.y}. \tag{3.10}$$

Definition 57. Let x and y be vectors with $y \neq 0$. The projection of x on y, denoted by $P_y(x)$, is defined by

$$P_y(x) = \frac{x.y}{y.y} y. \tag{3.11}$$

The length of the projection is given by

$$\| P_y(x) \| = \frac{|x.y|}{\| y \|}, \tag{3.12}$$

since

$$\| P_y(x) \| = \| \frac{x.y}{y.y} y \| = |\frac{x.y}{y.y}| \| y \| = \frac{|x.y|}{\| y \|^2} \| y \| = \frac{|x.y|}{\| y \|}.$$

In the special case y is a unit vector, $\| y \| = 1$ and (3.11) becomes

$$P_y(x) = (x.y)y. \tag{3.13}$$

In this manner, setting

$$x_I = P_y(x), \quad x_{II} = x - P_y(x), \tag{3.14}$$

we can decompose an arbitrary vector x into a sum $x = x_I + x_{II}$ of a vector x_I parallel to y and a vector x_{II} perpendicular to y, provided y is nonzero. Indeed (3.14) is the only way to get a decomposition with those properties, so the decomposition is unique. This appears natural when we recall the geometric construction above; we will not write a formal proof here. One may check directly, however, that the vectors x_I and x_{II} are orthogonal. We have

$$x_I.x_{II} = P_y(x).(x - P_y(x)) = \frac{x.y}{y.y} y.(x - \frac{x.y}{y.y} y)$$

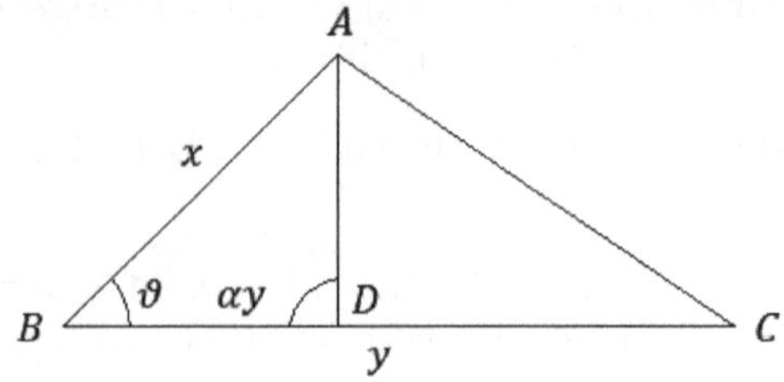

FIGURE 3.5: Projection $P_y(x) = \alpha y$

$$= \frac{x.y}{y.y}(y.x) - \frac{(x.y)^2}{(y.y)^2}(y.y) = 0.$$

We now relate the projection vector to the angle θ in Figure3.5. There θ is an acute angle (less than 90 degrees), and we have by (3.12)

$$\cos\theta = \frac{\| P_y(x) \|}{\| x \|} = \frac{|x.y|}{\| y \|\| x \|}.$$

If the angle θ is larger than 90 degrees, then the cosine and the scalar product $x.y$ become negative. (The reader is urged to draw a picture similar to Figure 3.5 for this situation.) The general formula relating the **angle θ between x and y** to the **scalar product** is

$$\cos\theta = \frac{x.y}{\| x \|\| y \|}. \tag{3.15}$$

Example 176. Find the angle between the vectors x = (2,3,2) and y = (1,2,-1).

Solution: We determine the angle from formula (3.15) above. We have $x.y = 2.1 + 3.2 + 2.(-1) = 6$ and

$$\| x \| = \sqrt{4+9+14} = \sqrt{17}, \quad \| y \| = \sqrt{1+2^2+(-1)^2} = \sqrt{6}.$$

Thus

$$\cos\theta = \frac{x.y}{\| x \|\| y \|} = \frac{6}{\sqrt{17}\sqrt{6}} = \frac{1}{17}\sqrt{102} \approx \frac{10.1}{17} \approx 0.594,$$

and $\theta \approx 0.935$ radians, which is about 54 degrees. Since the cosine ranges between -1 and 1, we immediately get from (3.15) the **Schwarz inequality** (or **Cauchy-Schwarz** or **Cauchy-Bunyakovski-Schwarz** inequality)

$$|x.y| \leq \| x \|\| y \|, \tag{3.16}$$

which is valid for any two vectors x and y. (Let us remark that one can prove this inequality directly from the properties of the scalar product, without recourse to the geometric construction used above.) Using the Schwarz inequality, we may now prove the triangle inequality. Indeed, for two vectors x and y we have

$$\begin{aligned} \| x+y \|^2 &= (x+y).(x+y) = x.x+y.x+x.y+y.y \\ &\leq \| x \|^2 + 2|x.y| + \| y \|^2 \\ &\leq \| x \|^2 + 2 \| x \| \| x \| + \| y \|^2 \quad \text{by Schwarz's inequality} \\ &= (\| x \| + \| y \|)^2. \end{aligned}$$

Taking the square roots of both sides we get

$$\| x+y \| \leq \| x \| + \| y \| .$$

Example 177. Verify triangle inequality for the vectors
(i) $(3,-1,2)$ and $(6,4,8)$
(ii) $(0,0,1)$ and $(1,1,0)$.

Solution: (i) $x+y=(9,3,10)$, $\quad \| x \| = \sqrt{9^2+3^2+10^2} = \sqrt{200}$

$$\| x \| = \sqrt{9+1+4} = \sqrt{14}, \quad \| y \| = \sqrt{36+16+64},$$

clearly $\sqrt{200} < \sqrt{14} + \sqrt{16}$.
(ii) $x+y=(1,1,1)$, $\quad \| x+y \| = \sqrt{3}$, $\quad \| x \| = 1, \| y \| = 1$

$$\| x+y \| = \sqrt{3} < 1+1 \quad \text{or} \quad \sqrt{3} < 2.$$

3.2.2 Cross Product

Definition 58. Let $x = x_1 i + x_2 j + x_3 k$ and $y = y_1 i + y_2 j + y_3 k$. The **cross product** (or **vector product**) of x and y is denoted by xy and defined as

$$x \times y = (x_2 y_3 - x_3 y_2) i + (x_3 y_1 - x_1 y_3) j + (x_1 y_2 - x_2 y_1) k. \tag{3.17}$$

Thus, the cross product of two vectors yields a vector.

The formula (3.17) for $x \times y$ is related to determinants as follows.

$$x \times y = \begin{vmatrix} i & j & k \\ x_1 & x_2 & x_3 \\ y_1 & y_2 & y_3 \end{vmatrix} = \begin{vmatrix} x_2 & x_3 \\ y_2 & y_3 \end{vmatrix} i - \begin{vmatrix} x_1 & x_3 \\ y_1 & y_3 \end{vmatrix} j + \begin{vmatrix} x_1 & x_2 \\ y_1 & y_2 \end{vmatrix} k.$$

Actually, the expression between the two equality signs is not a true determinant, since the first line consists of vectors, not of numbers, but it is a convenient way to memorize (3.17). The standard rule for computing the three 2×2 determinants on the right hand side gives (3.17). Another way to memorize (3.17) is to observe the cyclic behaviours of the indices.

Geometric interpretation of cross product

The vector $x \times y$ is geometrically related to the vectors x and y as follows. Whenever the vectors x and y are not parallel, they form the sides of a parallelogram, see Figure 3.6. The side AB is formed by the vector x and has

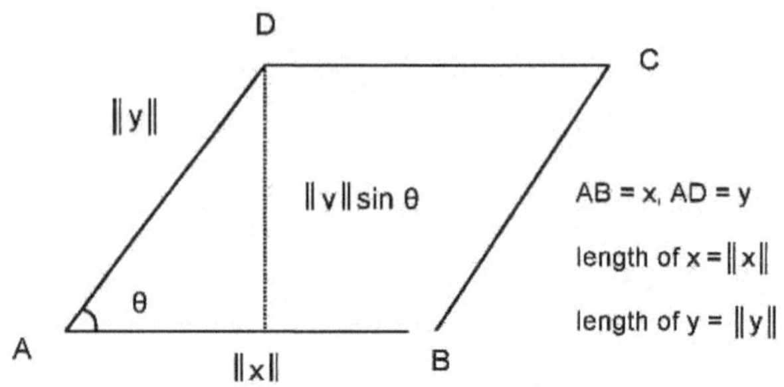

FIGURE 3.6: Geometric Interpretation of Cross Product

length $\| x \|$, AD is formed by y with length $\| y \|$. As we know, the area of a parallelogram with base b and height h is given by bh; here $b = \| x \|$ and $h = \| y \| \sin\theta$. It turns out (see the discussion of Lagrange identity below) that the magnitude of $x \times y$ equals the area of that parallelogram, so

$$\| xy \| = \| x \| \| y \| \sin\theta. \tag{3.18}$$

Moreover, $x \times y$ is orthogonal to both x and y. Finally, from the two remaining possibilities the direction of $x \times y$ is chosen according to the so-called **right-hand rule**: Point the index finger in the direction of **x** and the middle finger in the direction of **y**. The thumb then points in the direction of $x \times y$. Below and in the exercises we will show that these geometric properties follow from Definition 3.2.6, if the coordinate system has a right hand orientation (index finger points to i, middle finger points to j, thumb points to k.)

Algebraic properties of cross product

In contrast to the scalar product, the cross product is **anticommutative**, that is,

$$\| y \| \times \| x \| = - \| x \| \times \| y \| \tag{3.19}$$

holds for all vectors x and y. As a consequence, for all vectors x we have

$$x \times x = 0. \tag{3.20}$$

On the other hand, the cross product shares some properties of the scalar product. The **distributive** laws

$$x \times (y + z) = x \times y + x \times z \tag{3.21}$$

$$(x + y) \times z = x \times z + y \times z \tag{3.22}$$

$$(\alpha x) \times y = \alpha(x \times y) = x \times (\times y) \tag{3.23}$$

hold for all vectors x, y and z and all scalars α. In particular, setting $\alpha = 0$ we have for every vector x

$$x \times 0 = 0 = 0 \times x.$$

The properties (3.19) through (3.21) can be verified directly from the definition of the cross product.

Example 178. (a) Calculate $x \times y$ where $x = (1, -1, 3)$ and $y = (2, 1, -1)$.
(b) Show that $i \times j = k$, $j \times k = i$ and $k \times i = j$.
(c) Compute $i \times (i \times j)$ and $(i \times i) \times j$.
(d) Compute $x \times y$ where $x = i - j$ and $y = i + k$.

Solution:(a) We have

$$x \times y = \begin{vmatrix} i & j & k \\ 1 & -1 & 3 \\ 2 & 1 & -1 \end{vmatrix} = \begin{vmatrix} -1 & 3 \\ 1 & -1 \end{vmatrix} i - \begin{vmatrix} 1 & 3 \\ 2 & -1 \end{vmatrix} j + \begin{vmatrix} 1 & -1 \\ 2 & 1 \end{vmatrix} k = -2i + 7j + 3k.$$

(b) We have $i = (1, 0, 0)$ and $j = (0, 1, 0)$. We compute

$$i \times j = \begin{vmatrix} i & j & k \\ 1 & 0 & 0 \\ 0 & 1 & 0 \end{vmatrix} = \begin{vmatrix} 0 & 0 \\ 1 & 0 \end{vmatrix} i - \begin{vmatrix} 1 & 0 \\ 0 & 0 \end{vmatrix} j + \begin{vmatrix} 1 & 0 \\ 0 & 1 \end{vmatrix} k$$

so we have shown that $i \times j = k$. Similarly we can prove the other two identities.
(c) We obtain

$$i \times (i \times j) = i \times k = -j, \quad (i \times i) \times j = 0 \times j = 0.$$

(d) We calculate as before

$$x \times y = \begin{vmatrix} i & j & k \\ 1 & -1 & 0 \\ 1 & 0 & 1 \end{vmatrix} = \begin{vmatrix} -1 & 0 \\ 0 & 1 \end{vmatrix} i - \begin{vmatrix} 1 & 0 \\ 0 & 1 \end{vmatrix} j + \begin{vmatrix} 1 & -1 \\ 1 & 0 \end{vmatrix} = -i + j + k.$$

Alternatively we may use part (b) above with properties (3.19) through (3.21) to compute

$$x \times y = (i - j) \times (i + k) = i \times i + i \times k - j \times i - j \times k = -j + k - i.$$

We note that, as we see from part (c) above, that

$$(x \times y) \times z = x \times (y \times z).$$

Thus, the cross product is **not associative**. We come back to formula (3.18), $\| x \times y \| = \| x \| \| y \| \sin\theta$. From formula (3.15) we see that

$$\| x \|^2 \| y \|^2 - (xy)^2 = \| x \|^2 \| y \|^2 \left(1 - \cos^2\theta\right) = \| x \|^2 \| y \|^2 \sin\theta,$$

so (3.18) is equivalent to the **Lagrange identity**

$$\| x \times y \|^2 = \| x \|^2 \| y \|^2 - (x.y)^2. \tag{3.24}$$

Lagrange's identity in turn is a special case ($z = x, \quad w = y$) of the more general formula

$$(x \times y).(z \times w) = (x.z)(y.w) - (y.z)(x.w). \tag{3.25}$$

Therefore, formula (3.18) for the magnitude of the cross product follows from (3.25).

Scalar triple product.
If x, y and z are vectors, the expression $(x \times y).z$ is called the **scalar triple product** of x, y and z. Its absolute value $|(x \times y).z|$ represents the volume of the parallelogram with edges x, y and z.

Example 179. Show that the scalar triple product satisfies

$$(x \times y).z = \begin{vmatrix} x_1 & x_2 & x_3 \\ y_1 & y_2 & y_3 \\ z_1 & z_2 & z_3 \end{vmatrix}.$$

Solution:

$$\begin{aligned} (x \times y).z &= \left(\begin{vmatrix} x_2 & x_3 \\ y_2 & y_3 \end{vmatrix} i - \begin{vmatrix} x_1 & x_3 \\ y_1 & y_3 \end{vmatrix} j + \begin{vmatrix} x_1 & x_2 \\ y_1 & y_2 \end{vmatrix} k \right).(z_1 i + z_2 j + z_3 k)) \\ &= z_1 \begin{vmatrix} x_2 & x_3 \\ y_2 & y_3 \end{vmatrix} - z_2 \begin{vmatrix} x_1 & x_3 \\ y_1 & y_3 \end{vmatrix} + z_3 \begin{vmatrix} x_1 & x_2 \\ y_1 & y_2 \end{vmatrix}. \end{aligned}$$

The latter expression is the expansion of the 3×3 determinant

$$\begin{vmatrix} x_1 & x_2 & x_3 \\ y_1 & y_2 & y_3 \\ z_1 & z_2 & z_3 \end{vmatrix}.$$

Scalar and vector products often appear in physics and engineering. Work is the scalar product of force and displacement. Torque and angular momentum are the cross products of force and displacement or force and linear momentum. Maxwell's equations, which provide the foundation of electromagnetic theory, involve both scalar and cross products of electrical and magnetic variables.

Example 180. Let $x = 2i-3j+4k$ and $y = -i+2j+5k$. Find (i) $(2x).(x-2y)$ (ii) $(\frac{x.y}{y.y}y)$.

Solution:(i)
$$\begin{aligned}(2x).(x-2y) &= (4i-6j+8k)(2i-3j+4k+2i-4j-10k)\\ &= (4i-6j+8k)(4i-7j-6k)\\ &= 16+42-48 = 10.\end{aligned}$$

(ii)
$$\begin{aligned}(\frac{x.y}{y.y}y) &= \frac{(2i-3j+4k)(-i+2j+5k)}{(-i+2j+5k)(-i+2j+5k)}(-i+2j+5k)\\ &= \frac{(-2-6+20)}{1+9+25}(-i+2j+5k)\\ &= \frac{2}{5}(-i+2j+5k)\\ &= \frac{-2}{5}i+\frac{4}{5}j+2k.\end{aligned}$$

Example 181. Examine whether the vectors $i-j+k$ and $-4i+3j+8k$ are orthogonal?

Solution: $(i-j+k).(-4i+3j+8k) = -4-3+8 = 1 \neq 0$ so given vectors are not orthogonal.

Example 182. Find the value of α such that vectors $\alpha i+\frac{1}{2}j+\alpha k$ and $-3i+4j+\alpha$ are orthogonal.

Solution: If these two vectors are orthogonal then their dot product (inner product) is 0, that is,

$$\begin{aligned}(\alpha i+\frac{1}{2}j+\alpha k)(-3i+4j+\alpha k) &= 0\\ \text{or} \quad -3\alpha+\frac{1}{2}.4+\alpha^2 &= 0\\ \text{or} \quad \alpha^2-3\alpha+2 &= 0\\ \text{or} \quad (\alpha-1)(\alpha-2) &= 0.\end{aligned}$$

Thus given vectors are orthogonal if $\alpha = 1$ or 2.

Example 183. A sled is pulled horizontally over ice by a rope attached to its front. A 10 lb force acting at angle of 60° with the horizontal moves the sled 50 ft. Find the work done.

Solution: Magnitude of the force $F = \| F \| = 20, \quad \theta = 60^0$

$$\text{and} \quad \| d \| = 100.$$

Then work done $= W = \| F \| \| d \| \cos\theta = 20\ (100)\ \frac{1}{2} = 1000$ ft lb.

Example 184. Verify Lagrange's identity

$$\| x \times y \|^2 = \| x \|^2 \| y \|^2 - (x.y)^2$$

for vectors $x = (1, -1, 1)$ and $y = (2, 3, -1)$.

Solution:

$$\| x \times y \| = \begin{vmatrix} i & j & k \\ 1 & -1 & 1 \\ 2 & 3 & -1 \end{vmatrix} = i(1-3) - j(-1-2) + k(3+2)$$

$$= -2i + 3j + 5k$$

Left Hand Side $= \| x \times y \|^2 = 4 + 9 + 25 = 38$

Right Hand Side $= \| x \|^2 \| y \|^2 - (x.y)^2 = (1+1+1)(4+9+1) - (2-3-1)^2$

$$= 3(14) - (-2)^2 = 42 - 4 = 38$$

Left Hand Side = Right Hand Side.

Hence the identity is verified.

Example 185. (a) Find the area of the triangle determined by the points having coordinates $P_1(1,1,1)$, $P_2(1,2,1)$, $P_3(1,1,2)$.
(b) Find the volume of the parallelepiped for which the given vectors are three edges $x = 3i + j + k, \quad y = i + 4j + k, \quad z = i + j + k$.

Example 186. Prove Lagrange's identity:

$$\| x \times y \|^2 = \| x \|^2 \| y \|^2 - (x.y)^2.$$

Solution: Let $x = (x_1, x_2, x_3)$ and $y = (y_1, y_2, y_3)$. Then

Left Hand Side $= \| x \times y \|^2 = (x_2y_3 - x_3y_2)^2 + (x_1y_3 - x_3y_1)^2 + (x_1y_2 - x_2y_1)^2$

$$= x_2^2y_3^2 + x_3^2y_2^2 - 2x_2y_2x_3y_3 + x_1^2y_3^2 + x_3^2y_1^2 - 2x_1y_1x_3y_3 + x_1^2y_2^2 + x_2^2y_1^2 - 2x_1x_2y_1y_2.$$

Right Hand Side $= \| x \|^2 \| y \|^2 - (x.y)^2$

$$= (x_1^2 + {x_2}^2 + x_3^2)(y_1^2 + y_2^2 + y_3^2) - (x_1y_1 + x_2y_2 + x_3y_3)^2$$

$$= x_1^2y_1^2 + x_2^2y_1^2 + x_3^2y_1^2 + x_1^2y_2^2 + x_2^2y_2^2 + x_3^2y_2^2 + x_1^2y_3^2 + x_2^2y_3^2 + x_3^2y_3^2 - x_1^2y_1^2 - x_2^2y_2^2 - x_3^2y_3^2$$
$$-2x_1y_1x_2y_2 - 2x_1x_3y_1y_3 - 2x_2x_3y_2y_3.$$

This implies

$$\| x \times y \|^2 = \| x \|^2 \| y \|^2 - (x.y)^2.$$

3.3 Differential Calculus of Vector Fields

Let us begin with a specific situation. Let f_1, f_2, f_3 be real valued functions defined on some interval I; then for each $t \in 2$ I we can form the vector

$$\mathbf{F}(t) = f_1(t)i + f_2(t)j + f_3(t)k. \tag{3.26}$$

All those points (that is, the points in the range of F) form a curve in 3-space; if we think of t as a time variable, we may imagine an object moving along this curve, passing the point $\mathbf{F}(t)$ at time t. The functions f_1, f_2, f_3 are called the **component functions**, or simply the **components** of $\mathbf{F}$.
In general, let us consider a function $\mathbf{F}$ defined on some set S. Whenever the values of $\mathbf{F}$ are vectors, the function $\mathbf{F}$ is called a **vector field** (or a **vector valued function**, or a **vector function**). Its domain S may be an interval as above, or it may be a subset of the plane R^2 or the space R^3. In the latter case, the vector field has the form

$$F(x, y, z) = f_1(x, y, z)i + f_2(x, y, z)j + f_3(x, y, z)k. \tag{3.27}$$

As an example, $\mathbf{F}$ may be the gravity field, which to every point $P = (x, y, z)$ in space associates the vector $\mathbf{F}(x, y, z)$ representing the gravity force at P. The same situation occurs with the electric field and the magnetic field. Moreover, the flow of a fluid gives rise to a velocity field which associates with each point $P = (x, y, z)$ the velocity vector of the fluid at this point. The form (3.27) describes the situation when the velocity at a point P does not change with time (this is called stationary flow); for nonstationary flow the vector field F and its components f_i depend on time, too,

$$\mathbf{F}(x, y, z, t) = f_1(x, y, z, t)i + f_2(x, y, z, t)j + f_3(x, y, z, t)k, \tag{3.28}$$

so in this case the domain S of $\mathbf{F}$ is a subset of 4-space. Many more examples of vector fields will appear in the following sections. In the two-dimensional situation, (3.27) simplifies to $\mathbf{F}(x, y) = f_1(x, y)i + f_2(x, y)j$. In this case, the values of $\mathbf{F}$ are vectors in R_2 instead of R_3. As in Section 3.2 above, for the argument vector (x, y, z) we may also write (x_1, x_2, x_3), or x in compact notation. The value of $\mathbf{F}$ at $\mathbf{x}$ is then simply denoted by $\mathbf{F(x)}$.

3.3.1 Curves

In this subsection we consider vector fields which are defined on an interval I of the real line, as in (3.26). Such vector fields are commonly called **curves**, since one imagines the points $\mathbf{F}(t)$ to form a curve in space as t varies in I. $\mathbf{F}(t) = f(t)i + g(t)j$, is called curves in parametric form in 2-space.

Definition 59. The vector parametric equation of a straight line in 3-space passing through the points having coordinates (x_1, x_2, x_3) and (y_1, y_2, y_3) is given by

$$\gamma = \gamma_2 + tx, \quad -\infty < t < \infty \tag{3.29}$$

where $x = (y_1 - x_1, y_2 - x_2, y_3 - x_3)$, $\gamma_2 = (y_1, y_2, y_3)$, $\gamma_1 = (x_1, x_2, x_3)$ and $\gamma = z = (z_1, z_2, z_3)$. $x = (y_1 - x_1, y_2 - x_2, y_3 - x_3)$ is called a direction vector of the line. The scalar parametric form of the straight line (3.29) is

$$z_1 = y_1 + t(y_1 - x_1), \quad z_2 = y_2 + t(y_2 - x_2), \quad z_3 = y_3 + t(y_3 - x_3). \tag{3.30}$$

Definition 60. (Planes in 3-space). Let $P_0 = (x_1, x_2, x_3)$ be a point in R^3 with position vector $\gamma_0 = x_1 i + x_2 j + x_3 k$. If $n = a_1 i + a_2 j + a_3 k$ is any given non-zero vector then there is exactly one plane (flat surface) passing through the point P and its equation is

$$n.(\gamma - \gamma_0) = 0 \tag{3.31}$$

where $P = (p_1, p_2, p_3)$ has position vector γ. Thus the equation of the plane having non-zero normal vector $n = a_1 i + a_2 j + a_3 k$ and passing through the point $P_0 = (x_1 i, x_2 j, x_3 k)$ with position vector γ_0, has equation $n.(\gamma - \gamma_0) = 0$. Its scalar form is

$$a_1(p_1 - x_1) + a_2(p_2 - x_2) + a_3(p_3 - x_3) = 0.$$

Example 187. (a) The equation $2p_1 - 3p_2 - 4p_3 = 0$ represents a plane that passes through the origin and is normal to the vector $(2, -3, -4) = 2i - 3j - 4k$.
(b) Find an equation of the plane that passes through the three points $P = (1, 1, 0)$, $Q = (0, 2, 1)$ *and* $R = (3, 2, -1)$.

Solution: We are required to find the normal vector n to the plane in order to write the equation of the plane (3.31). Such a vector will be perpendicular to the vectors $\overrightarrow{PQ} = -i + j + k$ and $\overrightarrow{PR} = 2i + j - k$. Therefore, by a property, of the cross product (The vector cross product is orthogonal to each vector):

$$n = \overrightarrow{PQ} \times \overrightarrow{PR} == -2i + j - 3k$$

$$n.(\gamma - \gamma_0) = (-2i + 3j - 3k).(p_1 - 1, p_2 - 1, p_3)$$

where $(p_1 - 1, p_2 - 1, p_3) = (\gamma - \gamma_0)$ using (1,1,0) and (p_1, p_2, p_3) as the position vector of γ.

3.3.2 Distances

Definition 61. (a) (Distance from point to plane)
Let $P_0 = (x_1, x_2, x_3)$ be a point in the plane P having equation $a_1 p_1 + a_2 p_2 + a_3 p_3 = d$. Then the distance between P_0 and the plane, denoted by s, is given by

$$s = \frac{|a_1 x_1 + a_2 x_2 + a_3 x_3 - d|}{(a_1^2 + a_2^2 + a_3^2)^{\frac{1}{2}}}.$$

(b) (Distance between point and straight line)
Let $P_0(x_1, x_2, x_3)$ be a point and R be a straight line through a point P_1 having position vector γ_1 and parallel to a non-zero vector v. Then the distance between P_0 and the line, denoted by s, is given by

$$s = \frac{\| (\gamma_0 - \gamma_1) \times v \|}{\| v \|}.$$

Example 188. (a) Find the distance from $(1, -1, 2)$ to the plane $x - y - z = 9$.
(b) Find the distance from $(1, 0, -1)$ to the line $\gamma = i + (1 + 3t)j - (3 - 4t)k$ parallel to $v = 3j + 4k$ passing through $P_1 = (1, 1, -3)$.

Solution: (a) $s = \dfrac{|1 + 1 - 2 - 9|}{\sqrt{1 + 1 + 1}} = \dfrac{9}{\sqrt{3}}$.

(b) $s = \dfrac{|((1 - 1)i + (0 - 1)j) + (2 + 3)k \times (3j + 4k)|}{\sqrt{3^2 + 4^2}} = \dfrac{|(-j + 5k) \times (3j + k)|}{5}$.

Example 189. (a) $\gamma(t) = 2\cos ti + 2\sin tj + tk, \quad t \geq 0$ is a vector equation of circular helix.
(b) The parametric equation of the vector function $\gamma(t) = 2\cos ti + 2\sin tj + 5k$ is $x = 2\cos t, \quad y = 2\sin t, \quad z = 5$. Its geometrical representation is given in Figure 3.7.
(c) The vector equation of the circle in 2-space with radius α and center (0,0) is

$$\begin{aligned} \gamma(t) &= (\alpha \cos t, \alpha \sin t) \\ &= \alpha \cos ti + \alpha \sin tj. \end{aligned}$$

(d) Graph of the curve traced by the following vector functions (vector fields)
(i) $F(t) = \cos ti + j + \sin t + k, \quad t \geq 0$.
(ii) $F(t) = 4i + 2\cos tj + 3\sin tk$.
(iii) $z = x^2 + y^2, \quad y = x, \quad x = t$.

Definition 62. Let F be a vector field defined on an interval I, let $t_0 \in I$ be given. We say that the vector L is the limit of F as t tends to t_0, and we write

$$\lim_{t \to t_0} F(t) = L,$$

if $\lim_{t \to t_0} \| F(t) - L \| = 0$. We say that F is continuous at t_0 if

$$\lim_{t \to t_0} F(t) = F(t_0).$$

We say that F is continuous on I if it is continuous at every point $t_0 \in I$. The limit process can be carried out by components.

Theorem 33. *Let $F(t) = f_1(t)i + f_2(t)j + f_3(t)k$ and $L = l1i + l2j + l3k$. Then $F(t) = L$ if and only if*

$$\lim_{t\to t_0} f_1(t) = l_1, \quad \lim_{t\to t_0} f_2(t) = l_2, \quad \lim_{t\to t_0} f_3(t) = l_3. \tag{3.32}$$

F is continuous at t_0 if and only if all f_i are continuous at t_0. F is continuous on I if and only if all f_i are continuous on I.

Verification: Since $\| F(t) - L \| = \sqrt{\sum (f_i(t) - l_i)^2}$ and therefore $0 \leq |f_i(t) - l_i| \leq \| F(t) - L \|$ for all i, the properties of limits and the Sandwich theorem imply the first assertion. The other assertions are then consequences of Definition 10.

Remark 36. If $\lim_{t\to t_0} F(t) = L$ then $\lim_{t\to t_0} \| F(t) \| = \| L \|$. As for scalar functions, the converse of this result is wrong, namely $\lim_{t\to t_0} \| F(t) \| = \| L \|$ does not imply $\lim_{t\to t_0} F(t) = L$.

Verification: By the reverse triangle inequality (see Example 9.2.3(d))

$$0 \leq | \; \| F(t) \| - \| L \| \; | \leq \| F(t) - L \| \; .$$

It follows from the Sandwich theorem that if

$$\lim_{t\to t_0} \| F(t) - L \| = 0, \text{ then } \lim_{t\to t_0} (\| F(t) \| - \| L \|) = 0.$$

For the converse choose $F(t) = x$ and $L = -x$, where x is any fixed non-zero vector. We have $\lim_{t\to t_0} \| F(t) \| = \| x \| = \| L \|$, but $\lim_{t\to t_0} F(t) = x \neq -x = L$. The usual properties of the limit can be extended for limits of vector fields. Let $F(t) \to L$, $G(t) \to M$ and $\alpha(t) \to A$ as $t \to t_0$. Then

$$\lim_{t\to t_0} (F(t) + G(t) = \lim_{t\to t_0} F(t) + \lim_{t\to t_0} G(t) = L + M, \tag{3.33}$$

in particular, if β is a scalar and x is a vector, and moreover

$$\lim_{t\to t_0} (F(t).G(t) = (\lim F(t)).(\lim G(t)) = L.M, \tag{3.34}$$

$$\lim_{t\to t_0} (F(t) \times G(t)) = L \times M. \tag{3.35}$$

Definition 63. (Derivative of vector field) We define the derivative of F at t_0 by

$$F'(t_0) = \lim_{h\to 0} \frac{F(t_0 + h) - F(t_0)}{h}$$

if this limit exists. In that case, F is called **differentiable** at t_0. Alternatively, $F'(t_0)$ is also denoted as $\frac{dF}{dt}(t_0)$. Geometrically, $F'(t_0)$ points along the tangent of the curve in the point $F(t_0)$, see Figure 3.7. Differentiation can be

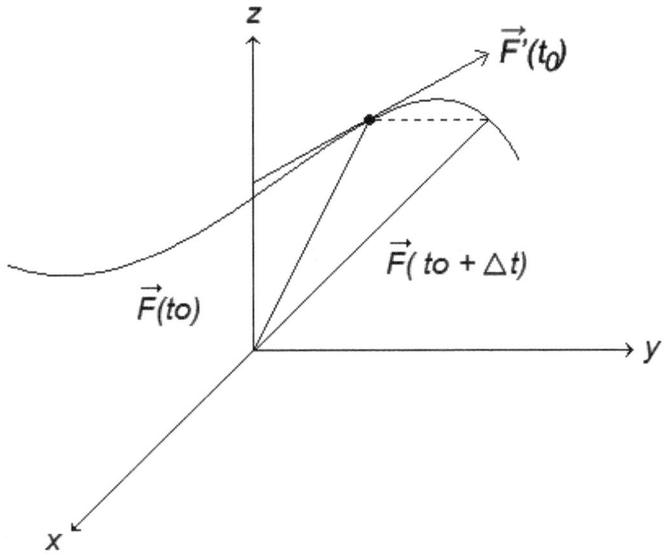

FIGURE 3.7: Geometric Interpretation of Derivative of Vector Field

carried out component by component, that is, if $F(t) = f_1(t)i + f_2(t)j + f_3(t)k$ is differentiable at t, then

$$F'(t) = f_1'i + f_2'j + f_3'k. \tag{3.36}$$

Indeed, applying rules (3.7) and (3.9) we see that

$$F'(t) = \lim_{h\to 0} \frac{F(t+h) - F(t)}{h}$$

$$= \lim_{h\to 0} [\frac{f_1(t+h) - f_1(t)}{h} i + \frac{f_2(t+h) - f_2(t)}{h} j + \frac{f_3(t+h) - f_2(t)}{h} k]$$

$$= [\lim_{h\to 0} \frac{f_1(t+h) - f_1(t)}{h}]i + [\lim_{h\to 0} \frac{f_2(t+h) - f_2(t)}{h}]j + [\lim_{h\to 0} \frac{f_3(t+h) - f_3(t)}{h}]k$$

$$= f_1'(t)i + f_2'(t)j + f_3'(t)k.$$

Differentiation properties of vector fields. We have already seen in calculus courses that we may define, for example, the sum $f + g$ of two functions by $(f + g)(x) = f(x) + g(x)$. In this manner, operations performed on function values give rise to operations on the functions themselves. The same applies to vector fields. Let F and G be two vector fields defined on an interval I, and let α, β be scalars. We define

$$(F+G)(t) = F(t) + G(t), \quad (\alpha F)(t) = \alpha F(t), \tag{3.37}$$

$$\text{thus} \quad (\alpha F + \beta G)(t) = \alpha F(t) + \beta G(t), \tag{3.38}$$

$$(F.G)(t) = F(t).G(t), \tag{3.39}$$

$$(F \times G)(t) = F(t) \times G(t). \tag{3.40}$$

In addition, when v is a scalar function defined on I, we set

$$(vF)(t) = v(t)Ft. \tag{3.41}$$

We also consider the composition

$$(F \circ u)(t) = F(u(t))$$

for a scalar function u whose range is contained in I. The following rules hold for differentiation.

$$(F+G)'(t) = F'(t) + G'(t) \tag{3.42}$$

$$(\alpha F)'(t) = \alpha F'(t) \quad \text{for constants } \alpha, \tag{3.43}$$

$$(vF)'(t) = v(t)F'(t) + v'(t)F(t), \tag{3.44}$$

$$(F.G)'(t) = F(t).G'(t) + F'(t).G(t), \tag{3.45}$$

and the chain rule for vector functions

$$(F \circ u)'(t) = F'(u(t))u'(t) = u'(t)F'(u(t)). \tag{3.46}$$

The formulas (3.42) through (3.46) are consequences of the definitions, the componentwise formula (3.36) and the corresponding differentiation rules for scalar functions . They can also be written in Leibniz notation, for example

$$\frac{d}{dt}(F+G) = \frac{dF}{dt} + \frac{dG}{dt},$$

$$\frac{d}{dt(F.G)} = (F.\frac{dG}{dt}) + (\frac{dF}{dt}.G),$$

$$\frac{d}{dt}(F \times G) = (F \times \frac{dG}{dt}) + (\frac{dF}{dt} \times G).$$

Example 190. Let $F(t) = 2t^2 i - 3j$, $G(t) = i + tj + t^2 k$, $u(t) = \frac{1}{3}t^3$. Verify (3.45)- (3.46) for these functions.

Solution: We first compute the derivatives $F'(t) = 4ti$, $G'(t) = j + 2tk$, $u'(t) = t^2$. We verify (3.45). We have $(FG)(t) = (2t^2i - 3j)(i + tj + t^2k) = 2t^2 - 3t$, therefore the left hand side becomes $(FG)'(t) = 4t - 3$. For the right hand side we get

$$F(t)G'(t) + F'(t)G(t) = (2t^2i - 3j)(j + 2tk) + 4ti(i + tj + t^2k) = -3 + 4t,$$

so indeed both sides are equal. We now consider (3.40). We have

$$(F \times G)(t) = (2t^2i - 3j) \times (i + tj + t^2k) = -3t^2i - 2t^4j + (2t^3 + 3)k,$$

so the left hand side equals

$$(F \times G)'(t) = -6ti - 8t^3j + 6t^2k.$$

We compute the right hand side

$$\begin{aligned} F(t) \times G'(t) + F(t) \times G(t) &= (2t^2i - 3j) \times (j + 2tk) + 4ti \times (i + tj + t^2k) \\ &= (2t^2k - 4t^3j - 6ti) + (4t^2k - 4t^3j) \\ &= -6ti - 8t^3j + 6t^2k, \end{aligned}$$

therefore both sides are equal. Finally, we investigate (3.46). We have

$$(F \circ u)(t) = 2(\frac{1}{3}t^3)^2i - 3j, \quad (F \circ u)'(t) = 2.\frac{1}{9}.6t^5i = \frac{4}{3}t^5i$$

on the left hand side, and

$$F'(u(t))u'(t) = [4u(t)i]u'(t) = [4(\frac{1}{3}t^3)i]t^2 = \frac{4}{3}t^5i$$

on the right hand side.

Example 191. (a) Let $F(t) = ti + \sqrt{t+1}j - e^tk$. Find $F'(t)$, $F'(0)$, $F''(1)$and $F(t)F'(t)$.
(b) Find $F''(t)$ if $F(t) = t\sin ti + e^{-t}j + tk$.

Solution: (a) We have $f_1(t) = t, f_2(t) = \sqrt{t+1}$ and $f_3(t) = -e^t$. We get

$$F'(t) = f_1'(t)i + f_2'(t)j + f_3'(t)k = i + \frac{1}{2\sqrt{t+1}}j - e^tk,$$

$$F'(0) = i + \frac{1}{2}j - k, \quad as \quad e^0 = 1, \quad \sqrt{t+1} = 1 \quad for \quad t = 0,$$

$$F''(t) = 0i - \frac{1}{4}\frac{1}{(t+1)^{\frac{3}{2}}}j - e^tk,$$

$$F''(0) = -\frac{1}{4}j - e^0k = -\frac{1}{4}j - k,$$

$$F(t).F'(t) = (ti + \sqrt{t+1}j - e^t k).(i + \frac{1}{2\sqrt{t+1}}j - e^t k) = t + \frac{1}{2} + e^{2t}.$$

(b) We have $f_1(t) = t\sin t, f_2(t) = e^{-t}$ *and* $f_3(t) = t$. We get

$$F'(t) = f_1'(t)i + f_2'(t)j + f_3'(t)k = (t\cos t + \sin t)i - e^{-t}j + k,$$

$$F''(t) = (-t\sin t + \cos t + \cos t)i + e^{-t}j + 0k = (2\cos t - t\sin t)i + e^{-t}j.$$

Vector Fields in Several Dimensions

In this section we introduce the concepts of gradient, divergence and curl. These are based on the notion of partial derivatives which we have introduced in Section. Since all kinds of derivatives of a function f at a certain point x are defined as limits involving function values at nearby points, it is most convenient if the domain of f always includes nearby points. This is made precise in the following definition.

Definition 64. (Open set) A subset D of R^n is called **open** if for every $x \in D$ there exists a ball B with center x which is contained in D. Thus, if $x \in D$ and if the ball B in question has radius r, then every point z whose distance from x is smaller than r must also belong to D. For example, the interior $D = (x, y, z) : 0 < x, y, z < 1$ of a cube is open, while if we include its boundary, the corresponding set $R = (x, y, z) : 0 \leq x, y, z \leq 1$ is not open.

Let us note that the definition of an open set in the plane is a special case of Definition 1.3.6.

Gradient.

Let f be a scalar function of three variables. The **gradient** of f at a point (x, y, z) is defined as the vector

$$\nabla f(x,y,z) = \frac{\partial f}{\partial x}(x,y,z)i + \frac{\partial f}{\partial y}(x,y,z)j + \frac{\partial f}{\partial z}(x,y,z)k. \tag{3.47}$$

If f and its partial derivatives are defined on some open subset D of R^3, the gradient thus becomes a vector field $\nabla f : D \to R^3$ whose component functions are just the partial derivatives $\frac{\partial f}{\partial x}, \frac{\partial f}{\partial y}, \frac{\partial f}{\partial z}$. According to the notation of vectors, we may also write

$$\nabla f(x,y,z) = (\frac{\partial f}{\partial x}(x,y,z), \frac{\partial f}{\partial y}(x,y,z), \frac{\partial f}{\partial z}(x,y,z),).$$

For $f(x, y, z) = x^2 y \sin z$, we have

$$\frac{\partial f}{\partial x}(x,y,z) = 2xyz\sin z, \quad \frac{\partial f}{\partial y}(x,y,z) = x^2\sin z, \quad \frac{\partial f}{\partial z}(x,y,z) = x^2 y\cos z,$$

and thus $\nabla f(x,y,z) = 2xy\sin zi + x^2\sin zj + x^2y\cos zk$, or $\nabla f(x,y,z) = (2xy\sin z, x^2\sin zj, x^2y\cos z)$. For a function f of two variables (x,y),

$$\nabla f(x,y) = \frac{\partial f}{\partial x}(x,y)i + \frac{\partial f}{\partial y}(x,y)j.$$

In dimensions, we define

$$\nabla f(x) = \Sigma \partial_i f(x) e_i, \tag{3.48}$$

where $x = (x_1, ...x_n), \partial_i f = \frac{\partial f}{\partial x_i}$ and e_i is the unit vector. The gradient obeys the rules

$$\nabla(f+g)(x) = \nabla f(x) + \nabla g(x), \tag{3.49}$$

$$\nabla(\alpha f)(x) = \alpha \nabla f(x), \text{ if } \alpha \text{ is a scalar}, \tag{3.50}$$

$$\nabla(fg)(x) = f(x)\nabla g(x) + g(x)\nabla f(x). \tag{3.51}$$

Those rules can be verified componentwise by the corresponding rules for partial derivatives.

Linearization. In the case of a function f of a single variable we have, as a consequence of the definition of the derivative,

$$f(x+h) = f(x) + f'(x)h + r(h) \tag{3.52}$$

for some remainder term $r(h)$ with $\frac{r(h)}{h} \to 0$ as $h \to 0$. For functions of several variables, the gradient plays the corresponding role.

Definition 65. Let f be a real valued function of n variables defined in an open subset D of R^n. We say that f is differentiable at a point $x \in D$, if all partial derivatives of f exist at x and

$$\lim_{h\to 0} \frac{|f(x+h) - f(x) - \nabla f(x).h|}{\parallel h \parallel} = 0. \tag{3.53}$$

As a consequence, for a differentiable function f we obtain the analogue of (3.52) in several dimensions as

$$f(x+h) = f(x) + \nabla f(x)h + r(h), \tag{3.54}$$

where $r(h)$ is a remainder term with $\frac{r(h)}{\parallel h \parallel} \to 0$. For this reason, the function g defined by

$$g(z) = f(x) + \nabla f(x)(z-x)$$

is called the **linearization** of f at x. Note that the difference

$$g(x+h) - f(x+h) = r(h)$$

converges to 0 faster than $\| h \|$. The notation

$$r(h) = o(\| h \|)$$

(read as $r(h)$ is small o of $\| h \|$) is a common way of stating that $\frac{r(h)}{\|h\|}$ *as* $h \to 0$. It would be very inconvenient if, whenever we want to apply linearization, we had to check the validity of (3.53) explicitly. Usually this is not necessary because of the following theorem.

Theorem 34. *Let f be a real valued function of n variables which is continuously differentiable in an open subset D of R^n. Then f is differentiable at every point $x \in D$.*

This means that we only have to check whether all partial derivatives of f are continuous in D, which is often obviously the directional derivative. We present its definition in the general case of n dimensions.

Definition 66. (Directional derivative) Let f be a real valued function of n variables. For each vector $u \in R^n$, the limit

$$f'_u(x) = \lim_{t \to 0} \frac{f(x+tu) - f(x)}{t},$$

if it exists, is called the **directional derivative** of f at x in the direction u. Instead of $f'_u(x)$, we also write $\partial_u f(x)$.

Remark 37. (a) When $u = e_i$ is the ith canonical unit vector, the directional derivative f'_u becomes the partial derivative $\partial_i f$. In particular, for a function f of three variables (x, y, z), we obtain

$$\frac{\partial f}{\partial x}(x, y, z) = f'_i(x, y, z) = \partial_x f(x, y, z), \quad i = (1, 0, 0),$$

and similarly

$$\frac{\partial f}{\partial y} = f'_j = \partial_y f \;\; and \;\; \frac{\partial f}{\partial z} = f'_k = \partial_z f.$$

(b) As we know already the partial derivatives $\frac{\partial f}{\partial x}, \frac{\partial f}{\partial y}$ and $\frac{\partial f}{\partial z}$ give the rates of change of f in the directions i, j and k respectively. Analogously, if $\| u \|$ is a unit vector, the directional derivative $f'_u(x)$ gives the rate of change of f in the direction u.
(c) The geometrical interpretation of the directional derivative is essentially the same as that of the partial derivative, except that we now look at tangents to the graph of f in arbitrary directions u, not only in the direction of the coordinate axes. The following theorem gives the connection between the gradient of f at x and its directional derivative at x.

Theorem 35. *If f is differentiable at x, then f has a directional derivative at x in every direction u, and*

$$f'_u(x) = \partial_u f(x) = \nabla f(x).u. \tag{3.55}$$

In particular, all partial derivatives $\partial_i f$ exist.

Proof. Let $u \in R^n$ be given, assume $u \neq 0$ (otherwise, the assertion is trivially satisfied). In order to verify the definition of the directional derivative, we consider the identity below valid for all $t \neq 0$,

$$|\frac{f(x+tu)-f(x)}{t} - \nabla f(x).u| = \frac{f(x+tu)-f(x)-t\nabla f(x).u}{|t|}$$

$$= \frac{f(x+tu)-f(x)-t\nabla f(x).u}{\| tu \| . \| u \|}.$$

Since f is differentiable at x, the fraction on the right hand side converges to 0, see Definition 65. Therefore, the left hand side too converges to 0, which yields the assertion. □

Theorem 21 yields an important geometric property of the gradient. According to (3.41), the angle θ between $\nabla f(x)$ and any vector u satisfies

$$\nabla f(x)u = \| f(x) \| \| u \| \cos\theta.$$

If moreover u is a unit vector and $\nabla f(x)$ is non-zero, we get from (3.55) that

$$f'_u(x) = \| f(x) \| \cos\theta.$$

Since $-1 \leq \cos\theta \leq 1$, we have

$$- \| \nabla f(x) \| \leq f'_u(x) \leq \| \nabla f(x) \| .$$

In particular, if u points in the direction of $\nabla f(x)$, then $\theta = 0$ and $\cos\theta = 1$, so

$$f'_u(x) = \| \nabla f(x) \| \tag{3.56}$$

where, if u points in the direction of $-\nabla f(x)$, then $\theta = \Pi$ and $\cos\theta = -1$, so

$$f'_u(x) = - \| \nabla f(x) \| . \tag{3.57}$$

Since the directional derivative gives the rate of change of the function in that direction, it follows that the function f increases most rapidly in the direction of the gradient and decreases most rapidly in the opposite direction.

Example 192. Let $f(x, y, z) = xyz$. Compute $\nabla f(1, 1, 1)$ and determine the maximum and minimum rates of change of f at (1,1,1).

Solution: We have

$$\frac{\partial f}{\partial x}(x,y,z) = yz, \quad \frac{\partial f}{\partial y}(x,y,z) = xz, \quad \frac{\partial f}{\partial z}(x,y,z) = xy.$$

Therefore

$$\nabla f(x,y,z) = yzi + xzj + xyk = (yz, xz, xy),$$

and hence $\nabla f(x,y,z) = i + j + k = (1,1,1)$. According to (3.56) and (3.57), the maximum and minimum rates of changes are $\nabla f(1,1,1) = \sqrt{3}$ and $-\nabla f(1,1,1) = -\sqrt{3}$ respectively.

Example 193. Calculate the directional derivative of $\phi(x,y,z) = 8xy^2 - xz$ at an arbitrary point (x,y,z) in the direction of the vector

$$u = \frac{1}{\sqrt{3}}i + \frac{1}{\sqrt{3}}j + \frac{1}{\sqrt{3}}k = (\frac{1}{\sqrt{3}}, \frac{1}{\sqrt{3}}, \frac{1}{\sqrt{3}}).$$

Solution: According to Theorem 21, we have

$$\partial_u \varphi(x,y,z) = \varphi'_u = \nabla \varphi(x,y,z).u.$$

We compute

$$\frac{\partial \varphi}{\partial x}(x,y,z) = 8y^2 - z, \quad \frac{\partial \varphi}{\partial y}(x,y,z) = 16xy \quad \frac{\partial \varphi}{\partial z}(x,y,z) = -x,$$

hence $\nabla \varphi(x,y,z) = (8y^2 - z, 16xy, -x)$ and

$$\partial_u \varphi(x,y,z) = (8y^2 - z, 16xy, -x).(\frac{1}{\sqrt{3}}, \frac{1}{\sqrt{3}}, \frac{1}{\sqrt{3}}) = \frac{8y^2 + 16xy - x}{\sqrt{3}}.$$

Example 194. Let the temperature at a point (x,y) on a metallic plate in the xy-plane be given by $T(x,y) = \dfrac{xy}{1+x^2+y^2}$ degrees Celsius. (a) Find the rate of change of temperature at (1,1) in the direction of $u = 2i - j$.
(b) An insect at (1,1) wants to walk in the direction in which the temperature drops most rapidly. Find a unit vector in that direction.

Solution: (a) We have

$$\frac{\partial T}{\partial x}(x,y) = \frac{y(1-x^2+y^2)}{(1+x^2+y^2)^2}, \quad \frac{\partial T}{\partial y}(x,y) = \frac{x(1+x^2-y^2)}{(1+x^2+y^2)^2}$$

so at $(x,y) = (1,1)$ we have

$$\nabla T(1,1) = \frac{i}{9} + \frac{j}{9} = (\frac{1}{9}, \frac{1}{9}).$$

The directional derivative gives the correct rate of change only after we normalize the vector u to unit length, so we set $e = \frac{u}{\| u \|} = \frac{(2i - j)}{\sqrt{5}}$ and compute the rate of change as

$$\partial_e T(1,1) = \nabla T(1,1).e = (\frac{i}{9} + \frac{j}{9}).\frac{(2i - j)}{\sqrt{5}} = \frac{1}{9\sqrt{5}}.$$

(b) The temperature drops most rapidly in the direction of of $-\nabla T(1,1)$. Using part (a), we compute a unit vector e in that direction as

$$-\nabla T(1,1) = -\frac{1}{9}i - \frac{1}{9}j, \quad \| -\nabla T(1,1) = \frac{\sqrt{2}}{9} \|,$$

so

$$e = \frac{-\nabla T(1,1)}{\| -\nabla T(1,1) \|} = -\frac{1}{2}i - -\frac{1}{2}j.$$

Example 195. Let the temperature at each point of a metal plate be given by the function

$$T(xy) = e^x \cos y + e^y \cos x.$$

(a) In what direction does the temperature increase most rapidly at the point (0,0)? What is the rate of increase?
(b) In what direction does the temperature decrease most rapidly at (0,0)?

Solution: We first compute

$$\nabla T(x,y) = \frac{\partial T}{\partial x}(x,y)i + \frac{\partial T}{\partial y}(x,y)j$$

$$= (e^x \cos y - e^y \sin x)i + e^y(\cos x - e^x \sin y)j.$$

(a) At (0,0) the temperature increases most rapidly in the direction of the gradient $\nabla T(0,0) = i + j$. The rate of increase is obtained as $\| \nabla T(0,0) \| = \| i + j \| = \sqrt{2}$.
(b) The temperature decreases most rapidly in the direction of $-\nabla T(0,0) = -i - j$.
We repeat that the gradient vector $\nabla f(x)$ tells us the direction of the steepest climb of f at the point x, and its length,$\nabla f(x)$, gives the steepness.

Divergence and rotation.
Here we present the definition and some properties of the differential expressions termed divergence and curl or rotation.

Definition 67. Let $F(x,y,z) = f_1(x,y,z)i + f_2(x,y,z)j + f_3(x,y,z)k$ be a vector field with components f_1, f_2, f_3 defined on some open subset D of R^3 which we assume to possess partial derivatives.
(a) The divergence of F, denoted by $div\mathbf{F}$ or $\nabla.F$ (read as nabla dot $\mathbf{F}$), is defined as

$$divF = \frac{\partial f_1}{\partial x} + \frac{\partial f_2}{\partial y} + \frac{\partial f_3}{\partial z}. \tag{3.58}$$

It is a scalar field with domain D.
(b) The curl of F, denoted by $curl\mathbf{F}$ or $\nabla \times \mathbf{F}$ (read as nabla cross $\mathbf{F}$), is defined as

$$curlF = (\frac{\partial f_3}{\partial y} - \frac{\partial f_2}{\partial z})i + (\frac{\partial f_1}{\partial z} - \frac{\partial f_3}{\partial x})j + (\frac{\partial f_2}{\partial x} - \frac{\partial f_1}{\partial y})k. \tag{3.59}$$

It is a vector field with domain D. In analogy to the definition of a 3×3 determinant, the curl can be expressed symbolically as a traditional way to memorize the definition of the curl. Alternatively, the notion rotation is used instead of curl, and $rotF$ is used instead of $curlF$.

In contrast to the gradient, the meanings of divergence and curl cannot be explained from this definition in a direct and intuitive manner. The notion of divergence is best understood in context with the divergence theorem of Gauss (Theorem 41) and will be explained in Subsection 3.6.3. Likewise, the theorem of Stokes (Theorem 44) helps to clarify the meaning of the curl, see Subsection 3.6.5. We can see that for a rotating body, angular velocity is proportional to the curl of the tangential velocity. Compute $\nabla . F$ (the divergence) and $\nabla \times F$ (the curl) of the vector field

$$F(x, y, z) = 2xyi + xe^y j + zk.$$

Solution: The components of F are $f_1(x, y, z) = 2xy$, $f_2 = xe^y$, $f_3 = 2z$. Therefore

$$(div\mathbf{F})(X, Y, Z) = (\frac{\partial f_1}{\partial x} + \frac{\partial f_2}{\partial y} + \frac{\partial f_3}{\partial z})(x, y, z) = 2y + xe^y + 2.$$

To evaluate

$$\mathbf{curl\ F} = (\frac{\partial f_3}{\partial y} - \frac{\partial f_2}{\partial z})i + (\frac{\partial f_1}{\partial z} - \frac{\partial f_3}{\partial x})j + (\frac{\partial f_2}{\partial x} - \frac{\partial f_1}{\partial y})k,$$

we compute

$$(\frac{\partial f_3}{\partial y} - \frac{\partial f_2}{\partial z})(x, y, z) = 0 - 0 = 0, \quad (\frac{\partial f_1}{\partial z} - \frac{\partial f_3}{\partial x}) = 0 - 0 = 0,$$

$$(\frac{\partial f_2}{\partial x} - \frac{\partial f_1}{\partial y}) = e^y - 2x,$$

and obtain

$$curlF(x, y, z) = 0i + 0j + (e^y - 2x)k = (e^y - 2x)k.$$

Example 196. (i) Let φ be a scalar field which possesses continuous first and second partial derivatives. Show that $\nabla \times (\nabla\varphi) = 0$ or $\mathbf{curl}(\nabla\varphi) = 0$, that is, the curl of the gradient of φ is identically equal to the zero vector field.

(ii) Let $F = f_1 i + f_2 j + f_3 k$ be a vector field which has continuous first and second partial derivatives. Show that

$$div\ curl F = 0 \quad or \quad \nabla(\nabla \times F) = 0.$$

This means that the divergence of the curl of F is identically equal to the zero scalar field.

Solution: (i) In order to compute the curl of $\nabla\varphi$, we apply (3.61) to the vector field $F = \nabla\varphi$, that is, $f_1 = \dfrac{\partial\varphi}{\partial x}, \quad f_2 = \dfrac{\partial\varphi}{\partial y}, \quad f_3 = \dfrac{\partial\varphi}{\partial z}$. Inserting those expressions into (3.63) we obtain

$$\nabla \times (\nabla\varphi) = (\frac{\partial^2\varphi}{\partial y\partial z} - \frac{\partial^2\varphi}{\partial z\partial y})i + (\frac{\partial^2\varphi}{\partial z\partial x} - \frac{\partial^2\varphi}{\partial x\partial z})j + (\frac{\partial^2\varphi}{\partial x\partial y} - \frac{\partial^2\varphi}{\partial y\partial x})k.$$

We know that we can interchange the sequence of partial derivatives if the second partial derivatives are continuous. Therefore

$$\frac{\partial^2\varphi}{\partial y\partial z} = \frac{\partial^2\varphi}{\partial z\partial y}, \quad \frac{\partial^2\varphi}{\partial z\partial x} = \frac{\partial^2\varphi}{\partial x\partial z}, \quad \frac{\partial^2\varphi}{\partial x\partial y} = \frac{\partial^2\varphi}{\partial y\partial x}$$

so all components of the vector field $\nabla \times (\nabla\varphi)$ are zero, and thus identically equal to the zero vector field.
(ii) By the definition

$$\nabla \times F = (\frac{\partial f_3}{\partial y} - \frac{\partial f_2}{\partial z})i + \frac{\partial f_1}{\partial z} - \frac{\partial f_3}{\partial x})j + \frac{\partial f_2}{\partial x} - \frac{\partial f_1}{\partial y})k.$$

Forming the divergence on both sides we get

$$div(\nabla \times F) = \frac{\partial}{\partial x}[(\frac{\partial f_3}{\partial y} - \frac{\partial f_2}{\partial z})] + \frac{\partial}{\partial y}[(\frac{\partial f_1}{\partial z} - \frac{\partial f_3}{\partial x})] + \frac{\partial}{\partial z}[(\frac{\partial f_2}{\partial x} - \frac{\partial f_1}{\partial y})]$$

$$= \frac{\partial^2 f_3}{\partial x\partial y} - \frac{\partial^2 f_2}{\partial x\partial z} + \frac{\partial^2 f_1}{\partial y\partial z} - \frac{\partial^2 f_3}{\partial y\partial x} + \frac{\partial^2 f_2}{\partial z\partial x} - \frac{\partial^2 f_1}{\partial z\partial y}$$

$$= [\frac{\partial^2 f_3}{\partial x\partial y} - \frac{\partial^2 f_3}{\partial y\partial x}] + [\frac{\partial^2 f_2}{\partial z\partial x} - \frac{\partial^2 f_2}{\partial x\partial z}] + [\frac{\partial^2 f_1}{\partial y\partial z} - \frac{\partial^2 f_1}{\partial y\partial z} - \frac{\partial^2 f_1}{\partial z\partial y}]$$

because for the same reason as in (i),

$$\frac{\partial^2 f_3}{\partial x\partial y} = \frac{\partial^2 f_3}{\partial y\partial x}, \quad \frac{\partial^2 f_2}{\partial z\partial x} = \frac{\partial^2 f_2}{\partial x\partial z}, \quad \frac{\partial^2 f_1}{\partial y\partial z} = \frac{\partial^2 f_1}{\partial z\partial y}.$$

Example 197. Consider the vector field $r(x, y, z) = xi + yj + zk$ (the identity mapping on R^3). We let $r = \parallel r \parallel$, that is, we also consider the scalar field r given by $r(x, y, z) = \parallel r(x, y, z) \parallel$.
(a) Let n be any integer (positive or negative). Prove that $\nabla(r^n) = nr^{n-2}r$.
(b) Let φ be a real valued function of one variable. Prove that $curl(\varphi(r)r = 0)$.

Solution: (a) We have $r(x, y, z) = \| r(x, y, z) \| = (x^2 + y^2 + z^2)^{\frac{1}{2}}$, so $r^n(x, y, z) = (x^2 + y^2 + z^2)^{\frac{n}{2}}$. This gives

$$\frac{\partial(r^n)}{\partial x}(x, y, z) = \frac{n}{2}(x^2 + y^2 + z^2)^{\frac{n}{2}-1}.2x = nxr^{n-2}(x, y, z),$$

and similarly

$$\frac{\partial(r^n)}{\partial y} = nyr^{n-2}, \quad \frac{\partial(r^n)}{\partial z} = nzr^{n-2}.$$

Thus, in abbreviated notation,

$$\nabla r^n = \frac{\partial r^n}{\partial x}i + \frac{\partial r^n}{\partial y}j + \frac{\partial r^n}{\partial z}k = nxr^{n-2}i + nyr^{n-2}j + nzr^{n-2}k$$

$$= nr^{n-2}(xi + yj + zk) = nr^{n-2}r.$$

(b) Recall the formula for the curl of a vector field F,

$$curlF = (\frac{\partial f_3}{\partial y} - \frac{\partial f_2}{\partial z})i + (\frac{\partial f_1}{\partial z} - \frac{\partial f_3}{\partial x})j + (\frac{\partial f_2}{\partial x} - \frac{\partial f_1}{\partial y})k.$$

In the present case (again we use abbreviated notation) $F = \varphi(r)r = \varphi(r)xi + \varphi(r)yj + \varphi(r)zk$, that is,

$$f_1 = \varphi(r)x, \quad f_2 = \varphi(r)y, \quad f_3 = \varphi(r)z,$$

and $r(x, y, z) = (x^2 + y^2 + z^2)^{\frac{1}{2}}$. Hence, using the chain rule,

$$\frac{\partial f_3}{\partial y} - \frac{\partial f_2}{\partial z} = z\varphi'(r)\frac{\partial r}{\partial y} - y\varphi'(r)\frac{\partial r}{\partial z}$$

$$= (\varphi'^2 + y^2 + z^2)^{\frac{-1}{2}} - yz(x^2 + y^2 + z^2)^{\frac{-1}{2}} = 0.$$

Similarly we obtain

$$\frac{\partial f_1}{\partial z} - \frac{\partial f_3}{\partial x} = 0, \quad \frac{\partial f_2}{\partial x} - \frac{\partial f_1}{\partial y} = 0.$$

This proves that $curl(\varphi(r)r) = 0$.

Finally we note that divergence and curl satisfy the property of linearity, that is,

$$div(F + G) = divF + divG, \quad div(\alpha F) = \alpha div + G,$$

$$curl(F + G) = curlF + curlG, \quad curl(\alpha F) = \alpha curlF.$$

Definition 68. A parametrization $r : D \to \Sigma$ is called regular, if it is continuously differentiable on the interior of D, and if the vectors $\partial_u r(u, v)$ and $\partial_v r(u, v)$ are not parallel at all interior points (u, v) of D. A surface Σ is called smooth if it has a regular parametrization.

Example 198. Consider the vertical cylinder with height H and a circular base of radius R.
Find the plane tangent to the side surface at the point $P_0 = ((\frac{1}{\sqrt{2}})R, (\frac{1}{\sqrt{2}})R, 1)$.

Solution: The parametrization $r(u, v) = R\cos ui + R\sin uj + vk$ yields $P_0 = r(u_0, v_0)$ with $u_0 = v_0 = \frac{\Pi}{4}$, as well as

$$\partial_u r(u, v) = -R\sin ui + R\cos uj, \quad \partial_v r(u, v) = k.$$

The tangent plane for Σ at P_0 is therefore spanned by the vectors

$$\partial_u r(\frac{\Pi}{4}, \frac{\Pi}{4}) = -\frac{1}{\sqrt{2}}i + \frac{1}{\sqrt{2}}j, \quad \partial_v r(\frac{\Pi}{4}, \frac{\Pi}{4}) = k.$$

Let P_0 be a point on a surface Σ. Any vector which is perpendicular to the tangent plane for Σ at P_0 is called a normal vector (or simply a normal) for Σ at P_0, or a vector normal to Σ at P_0. Note that any scalar multiple of a normal vector is again a normal vector. A normal vector of length 1 is called a unit normal. Unit normals are commonly denoted by n. Since the vector product $x \times y$ of two vectors is perpendicular to both vectors x and y, we see that the vector

$$N(P_0) = \partial_u r(u_0, v_0) \times \partial_v r(u_0, v_0)$$

is a normal vector for Σ at $P_0 = r(u_0, v_0)$. Let us return to the special situation where Σ is given as the graph of a (continuously differentiable) function S,

$$r(x, y) = (x, y, S(x, y)) = xi + yj + S(x, y)k, \quad (x, y) \in D. \tag{3.60}$$

In this case the vectors

$$\partial_x r(x_0, y_0) = i + \partial_x S(x_0, y_0)k, \partial_y r(x_0, y_0) = j + \partial_y S(x_0, y_0)k, \tag{3.61}$$

span the tangent plane at $P_0 = (x_0, y_0, S(x_0, y_0))$. Note that the parametrization e.g., (3.60) is regular, since the vectors in (3.61) are not parallel. The vector

$$N(P_0) = \partial_x r(x_0, y_0) \times \partial_y r(x_0, y_0) = -\partial_x S(x_0, y_0) - \partial_y S(x_0, y_0)j + k \tag{3.62}$$

is normal to Σ at P_0. A corresponding unit normal is given by

$$n = \frac{1}{\nu}(\partial_x r(x_0, y_0)) = -\partial_x S(x_0, y_0)i - \partial_y S(x_0, y_0)j + k), \tag{3.63}$$

where

$$\nu = \sqrt{1 + (\partial_x S(x_0, y_0))^2 + (\partial_y S(x_0, y_0))^2}.$$

As we know from geometry, any plane in the space R^3 can be described by an equation involving a vector perpendicular to it. In our case this means that the tangent plane to Σ at P_0 is equal to the set of all points (x, y, z) which satisfy the equation

$$-\partial_x S(x_0, y_0)(x - x_0) - \partial_y S(x_0, y_0)(y - y_0) + (z - z_0) = 0, \qquad (3.64)$$

or alternatively

$$z - z_0 = \partial_x S(x_0, y_0)(x - x_0) + \partial_y S(x_0, y_0)(y - y_0). \qquad (3.65)$$

As an alternative to the parametric description (3.64), we can describe a surface as a level set in the form

$$\Sigma = (x, y, z) : f(x, y, z) = c, \qquad (3.66)$$

3.4 Integration in Vector Fields

In this section we discuss integrals of vector fields over curves and surfaces. Since they describe important aspects of processes taking place in space and time, they constitute another basic tool for solving problems of science and engineering, see also Section 3.6 . There are two subsections of this section dealing respectively with line integrals and surface integrals.

3.4.1 Line Integrals

Suppose a curve C in R^3 is given by the parametric equations

$$x = x(t), \quad y = y(t), \quad z = z(t), \quad \text{for } \ a \le t \le b. \qquad (3.67)$$

The resulting vector function $r : [a, b] \to R^3$ with

$$r(t) = (x(t), y(t), z(t)) = x(t)i + y(t)j + z(t)k, \quad t \in [a, b] \qquad (3.68)$$

is called a **parametrization** of the curve C. We will think of C not only as a geometric locus of points $(x(t, y(t), z(t))$ but also as having a direction or an orientation induced by its parametrization. The $r(a) = (x(a), y(a), z(a))$ is called the **initial point** while $r(b) = (x(b), y(b), z(b))$ is called the **terminal** (or **end**) **point**. A curve is called **closed** if $r(a) = r(b)$, that is, if initial and terminal points coincide. In Figures 3.8a and 3.8b we display curves as the range of r, that is, as the set of all points $r(t) = (x(t), y(t), z(t))$ as t varies over the parameter interval. A curve C is called continuous if its parametrization r is a continuous function (or equivalently, if its components in (3.67) are continuous functions) on the parameter interval $[a, b]$. C is called differentiable if r is differentiable, and C is called smooth if the derivative r' is a continuous function and the vectors $r'(t)$ are not zero for all t. A continuous curve C is called piecewise smooth, if the parameter interval $[a, b]$ admits a partition $a = t_0 < \cdots\cdots < t_n = b$ such that C is smooth on every interval $[t_{j-1}, t_j$ of this partition.

Line Integrals of Scalar Fields

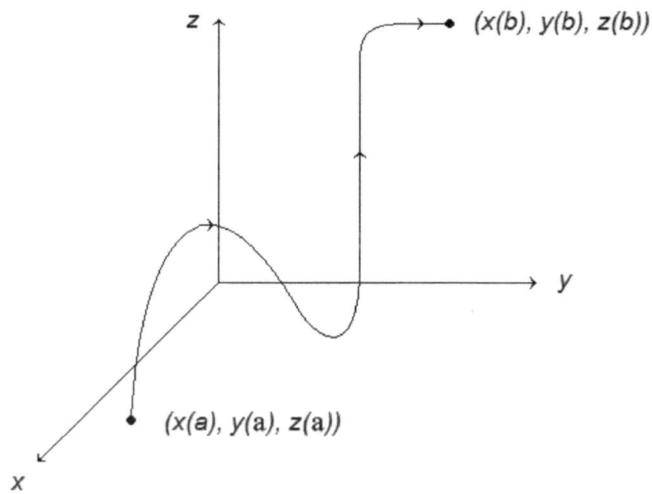

FIGURE 3.8a: Curve Starting at $t = 0$

Let us first consider a straight rod of length L which we think of as a one dimensional object with mass density ρ. If L is measured in centimeters and ρ in grams per centimeter, its total mass will be ρL grams. Now let us consider a curved rod as a curve C in space, whose points are given by the parametrization $r(t) = (x(t), y(t), z(t))$ for $t \in [a, b]$, and assume that its mass density varies along the rod as a function of (x, y, z). Then the total mass of the rod is given by the integral

$$\int_a^b \rho(rt) \parallel r'(t) \parallel dt \tag{3.69}$$

Indeed, a limit passage like the one expanded in calculus (based on an approximation of the rod consisting of straight pieces) shows that (3.69) is the correct formula to compute the total mass of the rod. This motivates the following definition.

Definition 69. (Line integral of scalar field) Let C be a smooth curve given by (3.68) and let f be a continuous scalar field whose domain contains

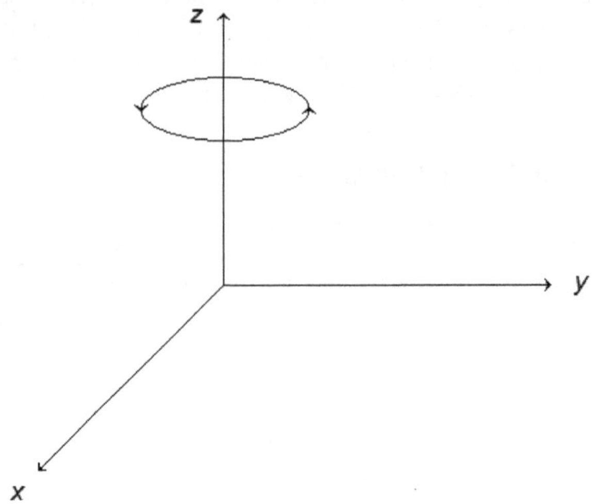

FIGURE 3.8b: Close Curve Directed from $t = 0$ to $t = 2\pi$

the curve C. Then the line integral of f over C is denoted by $\int_C f ds$:

$$\int_C f \ ds = \int_a^b f(r(t)) \parallel r'(t) \parallel dt. \tag{3.70}$$

Example 199. Evaluate the line integral

$$\int_C (x+y)ds,$$

where C is given by $x(t) = y(t) = t, \ \ z(t) = t^2 \ \ for \ \ 0 \le t \le 2.$

Solution: Since $r(t) = (x(t), y(t), z(t)) = (t, t, t^2)$, we get

$$r'(t) = (x'(t), y'(t), z'(t)) = (1, 1, 2t),$$

$$\parallel r'(t) \parallel = \sqrt{x'^2} + \sqrt{y'^2} + \sqrt{z'^2} = \sqrt{1+1+(2t)^2} = \sqrt{2+4t^2}.$$

For $f(x, y, z) = x + y \ \ we \ get \ \ f(x(t), y(t), z(t)) = f(t, t, t^2) = t + t = 2t$, so

$$\int_C (x+y)ds = \int_0^2 2t\sqrt{2+4t^2}dt = (\frac{1}{6}(2+4t^2)^{\frac{3}{2}})_0^2 = \frac{26\sqrt{2}}{3}.$$

In the special case $f = 1$, the line integral (3.70) equals the length of the curve C. Therefore, the function

$$s(t) = \int_0^t \sqrt{x'^2 + y'^2 + z'^2} d\tau = \int_o^t \parallel r'(\tau) \parallel \tag{3.71}$$

yields the length of C from the initial point $r(a)$ up to the point $r(t)$ and is called the arc length of the curve C; compare for the corresponding situation in two dimensions. From (3.71) we see that $s'(t) =\| r'(t) \|$. This explains the notation $\int_C f ds$ where ds stands for $s'(t)dt$ or $\| r'(t) \| dt$. Accordingly, ds is termed the **line element**.

Line Integrals of Vector Fields

We consider a vector field $F$with components f_1, f_2 and f_3,

$$F(x, y, z) = f_1(x, y, z)i + f_2(x, y, z)j + f_3(x, y, z)k, \tag{3.72}$$

whose domain contains the curve C, that is, the points $r(t)$ belong to the domain of F for all $t \in [a, b]$.

Definition 70. (Line integral of vector field) Let C be a smooth curve given by (3.68) and let F be a continuous vector field of the form (3.72) whose domain contains the curve C. Then the line integral of F over C is denoted by $\int_C F.dr$ and defined by

$$\int_C F.dr = \int_a^b F(r(t)).r'(t)dt. \tag{3.73}$$

When C is closed, one also writes

$$\oint_C F.dr$$

instead of $\int_C F.dr$. In this case, the value of the line integral is also called the circulation of F around C.

We give some explanations concerning the integral on the right hand side of (3.73). The integrand is real valued function of the single variable t, obtained as the scalar product of the vector $F(r(t))$ (the value of the vector field at the curve point $r'(t)$) and the vector (a vector tangent to the curve at that point). If we expand the scalar product into components according to its definition, we obtain, since $r'(t) = (x'(t), y'(t), z'(t))$,

$$\begin{aligned}\int_C F.dr &= \int_a^b f_1(r(t))x'(t) + f_2(r(t))y'(t) + f_3(r(t))z'(t) \\ &= \int_a^b f_1(x(t), y(t), z(t))x'(t) + f_2(x(t), y(t), z(t))y'(t) \\ &\quad + f_3(x(t), y(t), z(t))z'(t). \end{aligned} \tag{3.74}$$

One may ask why the line integral $\int_C F.dr$ is defined in this way. In fact, it expresses certain quantities which are important in applications, for example the mechanical work done by a force field F, moving a point mass along the curve C from the initial to the terminal point. This will elaborated in more detail in Subsection 3.6.2.

Example 200. Let the curve C be given by $r(t) = (\cos t)i - (\sin t)j + tk$ on the interval $0 \leq t \leq \pi$. Evaluate the integral $\int_C F.dr$ where $F(x, y, z) = yi - xj + 2k$.

Solution: We use formula (3.73) or (3.74). In order to form $F(r(t))$, we have to insert $\cos t$ for x, -$\sin t$ for y and t for z. This gives $F(r(t)) = (-\sin t, -\cos t, 2)$. Moreover, $r'(t) = (-\sin t, -\cos t, 1)$. Computing the scalar product of those two vectors and integrating, we get

$$\int_C F.dr = \int_0^\pi F(r(t)).r'(t)dt = \int_0^\pi (\sin^2 t + \cos^2 t + 2)dt = \pi + 2\pi = 3\pi.$$

The line integral $\int F.dr$ is also frequently denoted as

$$\int_C f_1(x, y, z)dx + f_2(x, y, z)dy + f_3(x, y, z)dz, \tag{3.75}$$

or in shorter form as

$$\int_C f_1 dx + f_2 dy + f_3 dz.$$

The definition of those expressions is, of course, the same as above in (3.73) or (3.74). If we formally replace x by $x(t)$ and dx by $x'(t)dt$ (and analogously for y and z) in (3.75), we arrive at (3.74).

Example 201. Evaluate $\int xydx - y\cos xdy$, where C is given by $x(t) = t^2$ and $y(t) = t$ on the interval $-2 \leq t \leq 5$.

Solution: We have $x'(t) = 2t$ and $y'(t) = 1$, therefore

$$\int_C xydx - y\cos xdy = \int_- 2^5[xyx' - y(\cos x)y'](t)dt$$

$$= \int_- 2^5 t^2 t 2t - t\cos t^2 dt = [2\frac{t^5}{5}]_- 2^5 - \frac{1}{2}[\sin t^2]_- 2^5$$

$$2(625 + \frac{32}{5}) - \frac{1}{2}(\sin 25 - \sin 4)$$

$$\frac{6314}{5} - \frac{1}{2}(\sin 25 - \sin 4).$$

Example 202. Find the circulation of the field $F(x, y) = (-y + x)i + xj$ around the circle $x^2 + y^2 = 1$.

Solution: The parametric equation of the given circle is

$$r(t) = \cos ti + \sin tj, \quad 0 \leq t \leq 2\Pi.$$

We have

$$F(r(t)) = (-\sin t + \cos t)i + \cos tj, \quad r'(t) = -\sin ti + \cos tj,$$

$$F(r(t))r'^2(t) - \sin t \cos t + \cos^2(t) + \cos^2(t) = 1 - \sin t \cos t.$$

Therefore, the circulation of F around the circle (denoted as C) becomes

$$\oint_C F.r = \int_0^2 \Pi F(r(t)).r'(t)dt = \int_0^2 \Pi(1 - \sin t \cos t)dt = [t - \frac{\sin^2 t}{2}]_0^2 \Pi.$$

The result of the following example will be used later in the proof of the theorem of Green and Ostrogradski.

Example 203. Let C be the graph of a function $y = h(x)$, where $h : [a,b] \to R$ is differentiable and let F be a vector field whose second component is zero, that is, $F(x,y) = (f(x,y), 0)$ for some continuous real valued function f. Show that

$$\int_C F.dr = \int_a^b f(x, h(x))dx. \tag{3.76}$$

Solution: We parameterize the curve C as $r(t) = (t, h(t))$. Then $r'(t) = (1, h'(t))$ and

$$\int_C F.dr = \int_a^b f(r(t)).1 + 0.h'(t)dt = \int_a^b f(t, h(t))dt,$$

so (3.76) is proved. (Recall that it does not determine whether the integration variable is denoted by t or by x.)
The following properties hold for the line integral. Let F and G be continuous vector fields defined on some region in space containing a curve C, which is given as the range of some vector function r as above. Then

$$\int_C (F + G).dr = \int_C F.dr + \int G.dr. \tag{3.77}$$

$$\int_C \alpha F.dr = \alpha \int_C F.dr, \quad \alpha \text{ being any scalar.} \tag{3.78}$$

This means that the line integral is linear with respect to the vector fields, that is, the line integral of the sum of two vector fields (the scalar multiple of a vector field) is equal to the sum of the line integrals (the scalar multiple of the line integral). These formulas are a direct consequence of the definition of the line integral, using the corresponding properties of the ordinary integral. For piecewise smooth curves, the line integral is evaluated on each piece separately, as stated in the following definition.

Definition 71. (Line integral, piecewise smooth curves) Let C be a piecewise smooth curve given by (3.68) and let F be a continuous vector field of the form (3.72) whose domain contains the curve C. Then the line integral of F over C is defined by

$$\int_C F.dr = \sum_1^n \int_t i - 1_i^t F(r(t)).r'(t)dt, \tag{3.79}$$

where $a = t_0 < .....t_n = b$ is a partition of $[a, b]$ such that C is smooth on every interval $[t_{i-1}, t_i]$.

Path Independence, Potential Functions and Conservative Fields

Definition 72. Let F be a vector field defined on an open region D in R^3. If $F = \nabla\psi$ for some scalar function ψ defined on D, then ψ is called a **potential function** (or simply potential) for F in D, and the vector field F is called conservative.

We will explain in Remark 46 why the term "conservative" is used here.

Theorem 36. *Let F be a continuous vector field defined on an open region D in R^3 which possesses a potential function in D. Let C be a smooth curve in D with initial point A and terminal point B. Then*

$$\int F.dr = \psi(B) - \psi(A). \tag{3.80}$$

The proof will be given in Appendix C. Let us take a look at formula (3.80). Given the potential ψ , the value of the right hand side depends only on the points A and B. This means in particular that the value of the line integral on the left hand side does not change if we replace C by a different curve $\tilde{C}$, as long as $\tilde{C}$ has the same initial and terminal points as C. This property is called path independence. Theorem 36 thus states that for conservative vector fields, line integrals are path independent. We present another related definition.

Definition 73. A continuous vector field F defined on an open region D in R^3 is called circulation free on D if we have

$$\oint_C F.dr = 0 \tag{3.81}$$

for every closed curve C in D. When the curve C is closed, its initial and terminal points coincide, and the right hand side of (3.80) becomes zero. Therefore, a conservative field is circulation free by Theorem 36. Actually, the converse is true, too, so the following theorem holds.

Theorem 37. *(Closed loop property of conservative fields) Let F be a continuous vector field defined on an open region D in R^3. Assume that D is connected, that is, any two points in C can be connected by a smooth curve. The following statements are equivalent:*

1. ***F** is conservative on D.*

2. ***F** is circulation free in D.*

The proof will be given in Appendix C. Another way of stating Theorem 37 would be to say that a vector field F possesses a potential in D if and only if the line integral over any closed curve in D is zero.

Example 204. Let $\psi(x, y, z) = xyz$, let C be a smooth curve with initial point $(-1, 6, 18)$ and terminal point $(2, 12, -8)$. Find $\int_C F.dr$, where $F = \nabla\psi$.

Solution: Setting $A = (-1, 6, 18)$ and $B = (2, 12, -8)$ in Theorem 43, we get

$$\int_C F.dr = \psi(B) - \psi(A) = 2.12.(-8) - (-1).6.18 = -192 + 108 = 84.$$

3.4.2 Surface Integrals

From Subsection 3.3.3 we know how a surface Σ in space is described by a parametrization $r : D \to \Sigma$ defined on some domain D of the plane. Our first task is to compute the area of the surface. As an introduction to the general formula, we consider the parallelogram with corners 0, x, y and $x + y$, where x and y are two nonparallel vectors in space. We know that its area equals $\| x \times y \|$. We interpret this expression in terms of the parametrization

$$r(u, v) = ux + vy, \quad \Sigma = r(D), \quad D = [0, 1] \times [0, 1].$$

Indeed, since

$$\partial_u r(u, v) = x, \quad \partial_v r(u, v) = y,$$

are constant as functions of u and v, we have

$$\| x \times y \| = \int_0^1 \int_0^1 \| x \times y \| \, dudv = \int\int_D \| \partial_u r(u, v) \times \partial_v r(u, v) \| \, dA,$$

so we have expressed the area in terms of a two-dimensional integral. If we just want to determine the area of a parallelogram, there is no need for such a complicated formula, but for a general surface, this is the way to go.

Definition 74. (Area of surface) Let Σ be a smooth surface given as $\Sigma = r(D)$ with a regular parametrization r. The area of Σ is defined as

$$A(\Sigma) = \int\int_D \| \partial_u r(u, v) \times \partial_v r(u, v) \| \, dA. \tag{3.82}$$

Since r is regular, the integral on the right hand side is well-defined as the limit of the corresponding Riemannian sums. Those sums can be interpreted as the area of small parallelograms which approximate corresponding portions of Σ. (We do not carry out the details, which in fact would be rather cumbersome to do). Thus, (3.82) is a natural definition of the area of an arbitrary curved surface. The most important special case arises when Σ is the graph of a function S.

Theorem 38. *Let Σ be a smooth surface given as $z = S(x, y)$, where S is a continuously differentiable function defined on some domain D of the plane. Then we have*

$$A(\Sigma) = \int\int_D \sqrt{1 + \partial_x S(x, y)^2 \partial_y S(x, y)^2} dA. \tag{3.83}$$

Verification: From Subsection 3.3.3 we know that with $r(x, y) = xi + yj + S(x, y)k$ we obtain

$$\partial_x r(x, y) \times r(x, y) = -\partial_x S(x, y)i - \partial_y S(x, y)j + k.$$

Taking the length of this vector, we see that (3.83) is indeed a special case of (3.82). Formulas (3.82) and (3.83) are analogous to the formula for the length of a curve as the integral over the length of tangent vectors. Similarly, the definition of the surface integral of a scalar function f defined on Σ is analogous to that of a line integral as given in Definition 34.

Definition 75. (Surface integral) Let Σ be a smooth surface, given as $\Sigma = r(D)$ with a regular parametrization r. Let f be a scalar continuous function defined on Σ. Then the **surface integral** of f over the surface Σ is denoted by $\int\int_\Sigma f(x, y, z)d\sigma$ or simply $\int\int_\Sigma f d\sigma$:

$$\int\int_\Sigma f d\sigma = \int\int_\Sigma f(x, y, z)d\sigma = \int\int_D f(r(u, v)) \parallel \partial_u r(u, v) \times \partial_v r(u, v) \parallel dA. \tag{3.84}$$

For the case where Σ is the graph of a function S, we obtain the following formula for the surface integral from (3.84) in the same manner as we obtained the area in Theorem 38 from Definition 74.

Theorem 39. *Let Σ be a smooth surface given as $z = S(x, y)$, where S is a continuously differentiable function defined on some domain D of the (x, y)-plane. Let f be a scalar continuous function defined on Σ. Then we have*

$$\int\int_\Sigma f d\sigma = \int\int_D f(x, y, S(x, y))\sqrt{1 + (\frac{\partial S}{\partial x})^2(x, y) + (\frac{\partial S}{\partial y})^2(x, y)}dA. \tag{3.85}$$

Remark 38. (i) In the special case $f = 1$ we obtain the area of the surface, $A(\Sigma) = \int\int_\Sigma 1 d\sigma$, as we can see from the definitions.
(ii) For a given surface Σ, it is possible to choose different parametrization r of it. One can prove that this does not change the value of the area resp. of the surface integral.
(iii) Let a surface Σ consist of smooth components $\Sigma_1, \Sigma_2, ...\Sigma_n$ which are mutually disjoint or intersect each other only in a set of zero area, for example along a curve. Such a surface is called **piecewise smooth.** The surface integral over Σ is then defined as

$$\int\int_\Sigma f d\sigma = \int\int_{\Sigma_1} f d\sigma + ... + \int\int_{\Sigma_n} f d\sigma. \tag{3.86}$$

For example, the boundary Σ of a cube is not smooth (the normal vector jumps across its edges), but it is piecewise smooth, since it consists of six square pieces which are smooth.
(iv) The formal expression $d\sigma = \sqrt{1 + \partial_x S(x, y)^2 + \partial_y S(x, y)^2}dA$. is often called the **surface element**.

Example 205. Evaluate the surface integral $\int\int f d\sigma$, where $f(x, y, z) = y^2$ and Σ is the part of the plane $z = x$ given by $0 \leq x \leq 2, \quad 0 \leq y \leq 4$.

Solution: We have $S(x, y) = x$ and $S : D \to \Sigma$ with $D = [0, 2] \times [0, 4]$, so

$$\sqrt{1 + (\frac{\partial S}{\partial x})^2(x, y) + (\frac{\partial S}{\partial x})^2(x, y)} = \sqrt{1 + 1^2 + 0^2} = \sqrt{2},$$

and therefore

$$\int\int_{\Sigma} f d\sigma = \int\int_{D} y^2\sqrt{2} dA = \int_0^2 \int_0^4 y^2\sqrt{2} dy dx$$
$$= \sqrt{2}\int_0^2 dx. \int_0^4 y^2 dy = \frac{128}{3}\sqrt{2}.$$

Example 206. Evaluate the surface integral $\int\int_{\Sigma} f d\sigma$ where $f(x, y, z) = z^2$, and Σ is the portion of the boundary surface of the vertical cone $z = \sqrt{x^2 + y^2}$ which lies between the planes $z = 1$ and $z = 2$.

Solution: We present two different approaches. For the first one, we use a parametrization adapted to the cone, namely

$$r(u, v) = u \cos vi + u \sin vj + uk,$$

so $\Sigma = r(D)$ with $D = (u, v) : 1 \leq u \leq 2, 0 \leq v \leq 2\Pi$. In order to use Definition 75 we compute

$$\partial_u r(u, v) = \cos vi + \sin vj + k, \quad \partial_v r(u, v) = -u \sin vi + u \cos vj.$$

We get

$$(\partial_u r \times \partial_v)r(u, v) = (0 - u\cos v)i - (-u \sin v + 0)j + (u\cos^2 v + u \sin^2 v)k$$
$$= -u \cos vi + u \sin vj + uk$$

and moreover

$$\| \partial_u r \times \partial_v)r(u, v) \| = \sqrt{u^2\cos^2 v + u^2 \sin^2 v + u^2} = \sqrt{2u^2} = u\sqrt{2}.$$

Inserting this into (3.84), we get, since $f(r(u, v)) = u^2$,

$$\int\int_{\Sigma} f d\sigma = \int\int_{D} u^2.u\sqrt{2} dA = \sqrt{2}\int_0^2 \Pi \int_1^2 u^3 du dv = \frac{15}{2}\Pi\sqrt{2}.$$

For the second approach, we use Theorem 39 with $S(x, y) = \sqrt{x^2 + y^2}$ and the annular domain $D = (x, y) : 1 \leq \sqrt{x^2 + y^2} \leq 2$. We have $f(x, y, S(x, y)) = S(x, y)^2 = x^2 + y^2$, so

$$\int\int f d\sigma = \int\int (x^2 + y^2)\sqrt{1 + (\frac{\partial S}{\partial x})^2(x, y) + \frac{\partial S}{\partial y})^2(x, y) dA}.$$

The partial derivatives of S are

$$\frac{\partial S}{\partial x}(x,y) = \frac{x}{\sqrt{x^2+y^2}}, \quad \frac{\partial S}{\partial y}(x,y) = \frac{y}{\sqrt{x^2+y^2}},$$

so

$$\sqrt{1+(\frac{\partial S}{\partial x})^2(x,y)} = \sqrt{1+\frac{x^2}{x^2+y^2+\frac{y^2}{x^2+y^2}}} = \sqrt{2}.$$

This yields

$$\int\int_D f d\sigma = \int\int_D (x^2+y^2)\sqrt{2}dA.$$

Since D is an annular region, it is best to evaluate this two-dimensional integral in polar coordinates. Using well known results of polar coordinates, we get with $G = (r,\theta) : 1 \le r \le 2, \quad 0 \le \theta \le 2\Pi$ that

$$\int\int_\Sigma f d\sigma = \int\int_D (x^2+y^2)\sqrt{2}dA = \sqrt{2}\int\int_G (r^2\cos^\theta + r^2\sin^2\theta)rdA$$

$= \sqrt{2}\int\int r^3 dr d\theta$ which is the same as above.

Example 207. Evaluate the surface integral $\int\int f d\sigma$, where $f(x,y,z) = x^2$ and Σ is the upper half of the sphere $x^2+y^2+z^2 = a^2$.

Solution: First we apply Theorem 39. We have $S(x,y) = \sqrt{a^2-x^2-y^2}$ and

$$\frac{\partial S}{\partial x}(x,y) = -\frac{x}{\sqrt{a^2-x^2-y^2}}, \quad \frac{\partial S}{\partial y}(x,y) = -\frac{y}{\sqrt{a^2-x^2-y^2}}.$$

The integration domain becomes the circular disk $D = (x,y) : 0 \le x^2+y^2 \le a^2$, and therefore

$$\int\int f d\sigma = \int\int_D x^2 \frac{a}{\sqrt{a^2-x^2-y^2}} dA.$$

Again, this two-dimensional integral is best evaluated in polar coordinates, and Theorem 39 yields

$$\int\int_\Sigma f d\sigma = \int_0^2 \Pi \int_0^a r^2\cos^2\theta . \frac{a}{\sqrt{a^2-r^2}} r dr d\theta$$

$$a. \int_0^2 \Pi \cos^2\theta d\theta . \int_0^a \frac{r^3}{\sqrt{a^2-r^2}} dr.$$

The rightmost integral can be evaluated with the substitution $r = a\sin t$, $dr = a\cos t dt$, and (we omit the details)

$$\int_0^a \frac{r^3}{\sqrt{a^2-r^2}} dr = \int_0^{\frac{\Pi}{2}} a^3\sin^3 t dt = \frac{2}{3}a^3.$$

We finally obtain

$$\int\int_{\Sigma} f d\sigma = a \int_0^2 \Pi \cos^2 \theta d\theta . \frac{2}{3} a^3 = \frac{2}{3} \Pi a^4 .$$

An alternative solution arises from the use of spherical coordinates,

$$r(u, v) = a \sin u \cos v i + a \sin u \sin v j + a \cos u k .$$

Here $\Sigma = r(D)$ with $D = (u, v) : 0 \leq u \leq \frac{\Pi}{2}, \quad 0 \leq v \leq 2\Pi$. A straightforward but somewhat lengthy computation yields

$$\| (\partial_u r \times \partial_v r)(u, v) \| = a^2 \sin u .$$

Since $f(x, y, z) = x^2$, we have $f(r(u, v)) = a^2 \sin^2 u \cos^2 v$, and from (3.84) we get

$$\int\int f \int d\sigma = \int\int_D a^2 \sin^2 u \cos^2 v . a^2 \sin u dA$$

$$= a^4 \int_0^{\frac{\Pi}{2}} \sin^3 u du . \int_0^2 \Pi \cos^2 v dv = \frac{2}{3} \Pi a^4$$

as before.

3.5 Fundamental Theorems of Vector Calculus

We discuss here three very important results of vector calculus. The first result bears the names Green and Ostrogradski, in honor of the British scientist George Green and the Ukrainian mathematician Mikhail Ostrogradski. This theorem establishes a relationship between a double integral over a domain in a plane and a line integral over the boundary of that domain. The second result which we present here is known as the divergence theorem of Gauss in honor of the German mathematician and scientist Carl Friedrich Gauss. This theorem relates a volume (triple) integral over a region in space to a surface integral over the boundary surface of that region.

The third theorem is the Stokes theorem, named for Sir George Gabriel Stokes, an Irish mathematician who worked at Cambridge University. This theorem connects a surface integral to a line integral over the boundary curve of the surface. Thus, all those theorems relate integrals over objects of different (adjacent) dimensions. In some sense they can be viewed as extensions of the fundamental theorem of calculus; recall that the latter relates an integral over an interval (a one-dimensional object) to function values on the boundary (a zero-dimensional object). There are numerous applications of this theorem in science and engineering and some of them will be described in the next section.

3.5.1 Theorem of Green and Ostrogradski

Let C be a piecewise smooth curve in the plane R^2 parameterized as $r(t) = (x(t), y(t))$ for $a \leq t \leq b$. Let C be closed, that is, $r(b) = r(a)$ holds for the initial and terminal points. The curve C is called positively oriented if $r(t)$ traverses C counterclockwise (anticlockwise) as t varies from a to b. If $r(t)$ traverses C clockwise, then C is said to be negatively oriented. The parametrization $r(t) = (\cos t, \sin t)$ traverses the unit circle C anticlockwise as t varies from 0 to 2Π, and hence C is positively oriented by r.

On the other hand, the parametrization $r(t) = (-\cos t, \sin t)$ defines a negatively oriented curve C (whose graph is again the unit circle), since it traverses C clockwise. A non-closed curve C in the plane is called simple if it does not intersect itself, that is, if it has a parametrization $r : [a, b] \to R^2$ which is one-to-one, so $r(t_1) = r(t_2)$ can hold only if $t_1 = t_2$. If we imagine the graph of a curve as a train track, the track does not cross itself. A closed curve is called simple if it does not intersect itself except for the initial and final point, that is, $r(t_1) = r(t_2)$ holds for $t_1 \leq t_2$ only if $t_1 = a$ and $t_2 = b$. Recall that by $\oint F.dr$ we denote the line integral of a vector field $F = (f, g)$ over a closed curve C, and also write it as

$$\oint F.dr = \oint f(x, y)dx + g(x, y)dy.$$

Theorem 40. ***(Green-Ostrogradski)*** *Let D be a bounded domain in the plane whose boundary C is a closed, simple and positively oriented curve. Let $F = (f, g)$ be a vector field whose components are continuously differentiable. Then*

$$\oint f(x, y)dx + g(x, y)dy = \int\int_D (\frac{\partial g}{\partial x} - \frac{\partial f}{\partial y})dA. \tag{3.87}$$

The proof of this theorem will be given in Appendix C.

Remark 39. (i) The theorem relates a line integral, which is one-dimensional, to an integral over a two-dimensional region.
(ii) The theorem is used for theoretical purposes as well as for the computation of specific integrals. In particular it may happen that one of the integrals is much easier to calculate directly than the other one.

Example 208. (a) A particle moves counterclockwise once around the triangle D with vertices (0,0), (4,0) and (1,6), under the influence of the force $F(x, y) = xyi + xj$. Calculate the work done by this force, if the units of length and force are meters and Newtons, respectively.
(b) Evaluate $\oint F.dr$, where $F(x, y) = (x^2 - y)i + (\cos^2 y - e^{3y} + 4x)j$ and C is (the boundary of) any square with sides of length 5. Assume C is oriented counterclockwise.

Solution: (a) Let us denote by C the curve formed by the three sides of the triangle. The total work done by the force F equals the line integral $\oint F.dr$. We have $F = (f, g)$ with $f(x, y) = xy, \;\; g(x, y) = x$, so $\frac{\partial f}{\partial y} = x$ and $\frac{\partial g}{\partial x} = 1$. From (40) we know that

$$\oint F.dr = \int\int_D (\frac{\partial g}{\partial x} - \frac{\partial f}{\partial y}(x, y))dA = \int\int_D (1 - x)dA.$$

We compute the two-dimensional integral over the triangular region D as a double integral,

$$\int\int_D (1-x)dA = \int_0^1 \int_0^6 x(1-x)dydx + \int_1^4 \int_0^8 -2x(1-x)dydx$$
$$= \int_0^1 6x(1-x)dx + \int_1^4 (8-2x)(1-x)dx = -8.$$

Therefore, the work done equals -8 Newton meter.
(b) We have $F = (f, g)$ with $f(x, y) = x^2 - y$, $g(x, y) = \cos^2 y - e^{3y} + 4x$, so $\frac{\partial f}{\partial y} = -1$ and $\frac{\partial g}{\partial x} = 4$. The two-dimensional region D is a square of side length 5. We calculate

$$\int F.dr = \int\int_D (\frac{\partial g}{\partial x} - \frac{\partial f}{\partial y}(x, y))dA = \int\int_D 4 - (-1)dA = 5\int\int_D dA.$$

Since $\int\int dA.$ equals the area of the square, which is equal to 25, we obtain $\oint Fdr = 125$.

Example 209. (a) Use the Green-Ostrogradski theorem to evaluate the line integral $\oint y^2dx + x^2dy$, where C is the boundary of the square with vertices (0,0), (1,0), (0,1) and (1,1), oriented counterclockwise. Check the answer by evaluating the line integral directly.
(b) Do the same for $\oint xydx + (y+x)dy$, where C is the unit circle $x^2+y^2 = 1$, oriented counterclockwise.
(c) Verify the validity of the Green-Ostrogradski theorem for the vector field $F = 2yi - xj$ and the curve C taken as the circle of radius 4 with center (1,3).

Solution: (a) Here $f(x, y) = y^2$, $g(x, y) = x^2$, so $\frac{\partial f}{\partial y} = 2y$ and $\frac{\partial g}{\partial x} = 2x$, moreover $D = [0, 1] \times [0, 1]$. Using (3.87) we get

$$\begin{aligned}\oint_C f(x, y)dx + g(x, y)dy &= \int\int_D (\frac{\partial g}{\partial x} - \frac{\partial f}{\partial y})(x, y)dA = \int\int_D 2x - 2ydydx \\ &= \int_0^1 \int_0^1 2x - 2ydydx = 1 - 1 = 0. \qquad (3.88)\end{aligned}$$

For the line integral, we have to evaluate $\int y(t)^2x'(t) + x(t)2y'(t)dt$ separately along all four sides of the square, traversed by a suitable parametrization $r(t) = (x(t), y(t))$. Along the side $x = 0$ we have $y(t) = 0 = x'(t)$, and along the side $y = 0$ we have $x(t) = 0 = y'(t)$, so the corresponding line integrals are zero. The side $x = 1$ is parameterized by $r(t) = (x(t), y(t)) = (1, t)$ for $0 \leq t \leq 1$, so the line integral becomes

$$\int_0^1 t^2.0 + 1^2.1dt = 1.$$

Analogously, the line integral along $y = 1$ yields the value -1, so the overall integral gives $0 + 0 + 1 - 1 = 0$, which is the same as the result in (3.88).
(b) We have $f(x, y) = xy$, $g(x, y) = y + x$, so $\dfrac{\partial g}{\partial x} = x$ and $\dfrac{\partial g}{\partial x} = 1$, moreover $D = (x, y) : 0 \leq x^2 + y^2 \leq 1$ is the unit disk. Thus with $F = (f, g)$ we obtain

$$\oint_C F.dr = \int\int_D (\frac{\partial g}{\partial x} - \frac{\partial f}{\partial y})(x, y)dA = \int\int_D 1 - xdA.$$

Using polar coordinates we get, according to Theorem 40,

$$\int\int_D 1 - xdA = \int_0^2 \Pi \int_0^1 (1 - r\cos\theta)rdrd\theta = \int_0^2 \Pi(\frac{1}{2} - \frac{1}{3\cos\theta})d\theta = \Pi.$$

In order to compute $\oint_C Fdr$ directly, we parameterize C by $r(t) = (x(t), y(t))$ with $x(t) = \cos t$, $y(t) = \sin t$, where $0 \leq t \leq 2\Pi$. We obtain

$$\oint_C xydx + (y + x)dy = \int_0^2 \Pi[xyx' + (y + x)y'](t)dt$$

$$= \int_0^2 \Pi[\cos t \sin t(-\sin t) + (\sin t + \cos t)\ cost]dt$$

$$= -\int_0^2 \Pi \cos t \sin^2 tdt + \int_0^2 \Pi \sin t \cos tdt + \int_0^2 \Pi \cos^2 tdt$$

$$= 0 + 0 + \int_0^2 \Pi \cos^2(t)dt = \Pi.$$

Recall that $\int_0^2 \Pi \cos^2 tdt = \Pi$.
(c) We have $F = (f, g)$ with $f(x, y) = 2y$, $g(x, y) = -x$, so $\frac{\partial f}{\partial y} = 2$, $\frac{\partial g}{\partial x} = -1$, and D is the disk of radius 4 with center $(1, 3)$. We get ($A(D)$ denotes the area of D).

$$\oint_C Fdr = \int\int_D (\frac{\partial g}{\partial x} - \frac{\partial f}{\partial y})(x, y)dA = \int\int_D -3dA$$

$$= -3.A(D) = -3.4^2\Pi = -48\Pi.$$

To evaluate $\oint_C Fdr$ directly, we parameterize C by $r(t) = (x(t), y(t))$ with $x(t) = 1 + 4\cos t$, $y(t) = 3 + 4\sin t$, where $0 \leq t \leq 2\Pi$. We obtain

$$\oint_C 2ydx - xdy = \int_0^2 \Pi[2yx' - xy'](t)dt$$

$$= \int_0^2 \Pi 2(3 + 4\sin t)(-4\sin t) - (1 + 4\cos t)4\cos t dt$$

$$= -24\sin t - 32\sin^2 t - 4\cos t - 16\cos^2 t dt$$

$$= 0 - 32\Pi - 0 - 16\Pi = -48\Pi.$$

Recall that $\int_0^2 \Pi sin^2 tdt = \int_0^2 \Pi cos^2 tdt = \Pi$.

3.5.2 Divergence Theorem of Gauss

We consider the following situation in the space R^3. Let D be an open set which is bounded by a surface Σ such that D lies completely inside Σ. For example, the open unit ball $x : x \in R^3, \| x \| < 1$ is bounded by and lies completely inside the unit sphere $x : x \in R^3, \| x \| = 1$; the same is true for the (full) cube whose boundary surface consists of its six sides. In the case of the ball the boundary surface (the sphere) is smooth and in the case of the cube the boundary surface is piecewise smooth. At every point x of a smooth surface, normal vectors can be defined (see Subsection 3.3.3).

In the present situation, we can distinguish between **outer normals** pointing away from D and **inner normals** pointing into D (compare Figure 3.9). Thus, while we have two unit normals (normals whose length equals 1) at each point $x \in \Sigma$, there is exactly one outer unit normal which we denote by $n(x)$. In this way we obtain a vector field n which is defined on the boundary surface Σ. As a consequence of our definition of a smooth surface (Definition 32), the field n is continuous. When the boundary surface consists of smooth pieces (as in the case of the cube), the outer unit normal field n is continuous within each piece, but has jumps across their connecting boundaries (the edges and corners, in the case of the cube) where it is not defined.

Theorem 41. *Let D be a domain in R^3 with boundary Σ and outer unit normal field n as described above. Suppose that $\boldsymbol{F}$ is a vector field with values in R^3 whose components are continuous and have continuous first partial derivatives in D and up to the boundary Σ. Then*

$$\int\int_\Sigma Fnd\sigma = \int\int\int_D div\boldsymbol{F}dV. \tag{3.89}$$

Remark 40. (i) A proof will be given in Appendix C.
(ii) The theorem relates an integral over a two-dimensional surface to an integral over a three-dimensional volume. The integral on the left side is a

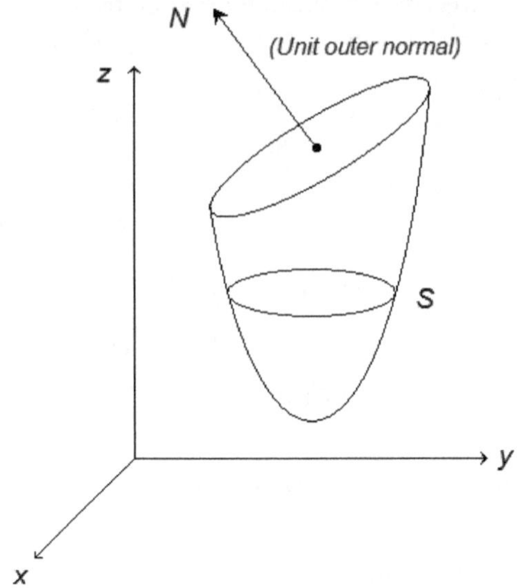

FIGURE 3.9: Unit Outwards Normal

surface integral (see Definition 74), the integrand being the scalar function $\mathbf{Fn} = f_1 n_1 + f_2 n_2 + f_3 n_3$, where f_j and n_j are the component functions of the vector fields F and n. The integral on the right side is a volume integral, which we treat in Calculus, of the scalar function $divF = \frac{\partial f_1}{\partial x} + \frac{\partial f_2}{\partial y} + \frac{\partial f_3}{\partial z}$.
(iii) Its interpretation and some applications will be discussed in Section 3.6. Indeed, the divergence theorem is a fundamental tool in the analysis of partial differential equations and of the phenomena described by them. Some identities useful for that purpose will be presented below, following the examples.
(iv) Equation (3.89) is sometimes helpful when one wants to compute surface or volume integrals, because the evaluation of one side may be more convenient than the evaluation of the other side.

Example 210. For each of the following data calculate the right hand side and left side of Equation (3.89), whichever is convenient. (i) $F(x, y, z) = xi + yj - zk$, Σ is the sphere of radius 4 centered at (1,1,1).
(ii)$F(x, y, z) = x^3 i + y^3 j + z^3 k$, Σ is the sphere of radius 1 with center at the origin.
(iii)$F(x, y, z) = x^2 i + y^2 j + z^2 k$, Σ is the rectangular box bounded by the coordinate planes $x = 0$, $y = 0$, $z = 0$ and the planes $x = 6$, $y = 2$ and $z = 7$.

Solution: In all cases, D denotes the region enclosed by Σ. Recall that

$$div\mathbf{F} = \frac{\partial f_1}{\partial x} + \frac{\partial f_2}{\partial y} + \frac{\partial f_3}{\partial z}$$

where $\mathbf{F} = f_1 i + f_2 j + f_3 k$.
(i) We have $f_1(x,y,z) = x, \quad f_2(x,y,z) = y, \quad f_3(x,y,z) = -z$, so

$$div\mathbf{F}(x,y,z) = 1 + 1 - 1 = 1.$$

We evaluate the right hand side of (3.89) as (here, vol(D) denotes the volume of D, a ball of radius 4).

$$\int\int\int_D div\mathbf{F}dV = \int\int\int_D dV = vol(D) = \frac{4}{3}\Pi.4^3 = \frac{256}{3}\Pi.$$

(ii) Here $f_1(x,y,z) = x^3, \quad f_2(x,y,z) = y^3, \quad f_3(x,y,z) = z^3$. So

$$\frac{\partial f_1}{\partial x}(x,y,z) = 3x^2, \quad \frac{\partial f_2}{\partial y}(x,y,z) = 3y^2, \frac{\partial f_3}{\partial z}(x,y,z) = 3z^2,$$

$$divF(x,y,z) = 3x^2 + 3y^2 + 3z^2.$$

The right hand side of (3.89) becomes

$$\int\int\int_D divFdV = 3\int\int\int_D (x^2 + y^2 + z^2)dV$$

Using spherical coordinates, we have already evaluated this integral in Example 8.7.4, so

$$\int\int\int_D divFdV = 3.\frac{4}{5}\Pi = \frac{12}{5}\Pi.$$

(iii) $F(x,y,z) = x^2 i + y^2 j + z^2 k$, so

$$divF(x,y,z) = 2x + 2y + 2z.$$

Again we evaluate the right and side of (3.89), this time with Frobenius theorem applied to the rectangular box,

$$\int\int\int_D div\mathbf{F}dV = 2\int_0^7\int_0^2\int_0^6 x + y + zdxdydz$$

$$= 2\int_0^7\int_0^2 [\frac{1}{2}x^2 + xy + xz]_{x=0}^{x=6}dydz = \int_0^7\int_0^2 36 + 12y + 12zdydz$$

$$= \int_0^7 [36y + 6y^2 + 12z]_{y=0}^{y=2}dz = \int_0^7 72 + 24 + 24zdz$$

$$= [96z + 12z^2]_{z=0}^{z=7} = 672 + 12.49 = 1260.$$

Example 211. Verify the statement of Gauss divergence theorem for the vector field $F(x,y,z) = xi + yj + zk$ and the cone whose interior is given by $D = (x,y,z) : 0 < z < 1, \quad 0 < x^2 + y^2 < z^2$.

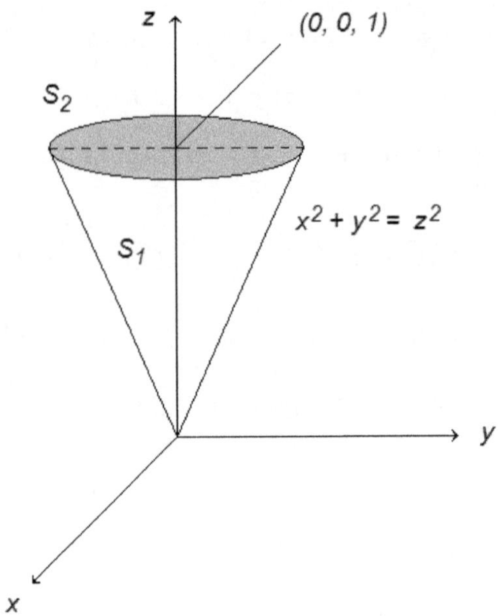

FIGURE 3.10: Boundary Surface

Solution: The boundary surface Σ of D consists of two parts, see Figure 3.10. Its lateral part Σ_1 is described by $z = S(x,y)$ with $S(x, \sqrt{x^2+y^2}$ for $x, y \in [-1,1]$ and its upper part Σ_2 is the flat circular disk described by $x^2+y^2 \leq 1$ lying in the plane $z = 1$. The surface integral on the left hand side of (3.89) thus decomposes into two parts,

$$\int\int_{\Sigma} F.nd\sigma = \int\int_{\Sigma_1} F.n_1 d\sigma + \int\int_{\Sigma_2} F.n_2 d\sigma$$

where n_1 and n_2 are the vector fields of outer unit normals to Σ_1 and Σ_2. According to (3.70),

$$N_1(x,y,z) = -\frac{\partial S}{\partial x}(x,y)i - \frac{\partial S}{\partial y}(x,y)j - k = -\frac{x}{z}i - \frac{y}{z}j + k$$

is a normal vector to Σ_1 in $(xy, z))$, where $z = \sqrt{x^2+y^2}$. The corresponding outer unit normal n_1 is then given by

$$n_1(x,y,z) = \frac{1}{\sqrt{2}}(\frac{x}{z}i + \frac{y}{z})j - k.$$

Moreover, $n_2 = k$ as Σ_2 is a flat horizontal surface. For the given vector field

$F(x, y, z) = xi + yj + zk$ we now compute

$$\int\int_{\Sigma} F.nd\sigma = \int\int_{\Sigma_1} F.n_1 d\sigma + \int\int_{\Sigma_2} F.n_2 d\sigma$$

$$= \int\int_{\Sigma_1} (xi + yj + zk).\frac{1}{\sqrt{2}}(\frac{x}{z}i + \frac{y}{z}j - k)d\sigma + \int\int_{\Sigma_1} (xi + yj + zk).kd\sigma$$

$$= \frac{1}{\sqrt{2}}\int\int_{\Sigma_1}(\frac{x^2}{z} + \frac{y^2}{z} - z) + \int\int_{\Sigma_2} zd\sigma = 0 + \int\int_{\Sigma_3} 1d\sigma$$

$$= \Pi,$$

as $z^2 = x^2 + y^2$ on Σ_1, $z = 1$ on Σ_2 and the area of the disk Σ_2 equals Π. The right hand side of (3.89) becomes

$$\int\int\int_D divFdV = \int\int\int_D 1 + 1 + 1dV = 3\int\int\int_D 1dV = 3vol(D).$$

Since the volume of the cone of height 1 and radius 1 is equal to $\frac{\Pi}{3}$, the right hand side, too, is equal to Π.
We now present Green's identities. Here Δ denotes the Laplace operator,

$$\Delta f = \frac{\partial^2 f}{\partial x^2} + \frac{\partial^2 f}{\partial y^2} + \frac{\partial^2 f}{\partial z^2}.$$

Theorem 42. ***(Green's first identity)*** *D, Σ and n as in Theorem 41. Let f and g be continuous scalar fields whose first and second partial derivatives are continuous in D and up to the boundary Σ. Then*

$$\int\int_{\Sigma} f\nabla g.nd\sigma = \int\int\int_D (f\Delta g + \nabla f.\nabla g)dV. \tag{3.90}$$

One can derive Green's first identity from the divergence theorem by choosing $F = f\nabla g$ in (3.89).

Theorem 43. ***(Green's second identity)*** *Let D, Σ, n, f and g as in Theorem 42. Then*

$$\int\int_{\Sigma}(f\nabla g - g\nabla f)nd\sigma = \int\int\int_D (f\Delta g - g\Delta f)dV. \tag{3.91}$$

Green's second identity is obtained from Green's first identity by interchanging the roles of f and g, and then subtracting the resulting formula from the original one. As a particular case of Green's identities, setting $f = 1$ we get

$$\int\int_{\Sigma} \nabla g.nd\sigma = \int\int\int_D \Delta gdV. \tag{3.92}$$

3.5.3 Theorem of Stokes

Let $F(x,y,z) = f_1(x,y,z)i + f_2(x,y,z)j + f_3(x,y,z)k$ be a vector field whose component functions are continuously differentiable, Σ be a smooth surface bounded by a piecewise smooth closed curve C, n be a vector field of unit normals to Σ whose orientation fits the orientation of C (this will be described below). Stokes theorem relates the line integral of F over C to a surface integral involving $curlF$ over Σ by the formula

$$\oint_C F dr = \int\int_\Sigma (curlF) n d\sigma. \tag{3.93}$$

Line integrals of vector fields have been studied in Subsection 3.4.1 and surface integrals in Subsection 3.4.2. The integrand of the surface integral on the right side is a scalar field, obtained as the scalar product of the vector fields $curlF$ and n. From Definition 27 we recall that the notation as a vector product or as a determinant

$$curlF = \nabla \times F = \begin{vmatrix} i & j & k \\ \frac{\partial}{\partial x} & \frac{\partial}{\partial y} & \frac{\partial}{\partial z} \\ f_1 & f_2 & f_3 \end{vmatrix} \tag{3.94}$$

is a convenient way to memorize the component definition of the rotation $curlF$ of the vector field F.
We assume that the surface Σ is given as $z = S(x,y)$ with a function S defined on the corresponding domain D in the xy-plane, and we assume that S and its first and second partial derivatives are continuous. We have seen in (3.63) that

$$n = \frac{1}{\nu}(-\partial_x S i - -\partial_y S j + k), \quad \nu = \sqrt{1 + (\partial_x S)^2 + (\partial_y S)^2} \tag{3.95}$$

defines a unit normal in the point $(x, y, S(x,y))$ of Σ, if the right hand side is evaluated at $(x,y) \in D$. We fix a suitable orientation of the boundary curve C of Σ as follows. Let τ be the boundary of D, positively oriented by a parametrization $q : [a,b] \to R^2$. Then

$$r(t) = q_1(t)i + q_2(t)j + S(q_1(t), q_2(t))k$$

defines a parametrization $r : [a,b] \to R^3$ of C which orients C as required.

Theorem 44. ***(Stokes)*** *Under the assumptions above we have*

$$\oint_C F dr = \int\int_\Sigma (curlF) n d\sigma. \tag{3.96}$$

Remark 41. (i) A proof will be given in Appendix C.
(ii) The theorem also holds for more general surfaces, that is, surfaces which cannot be described as the graph of a function S.
(iii) The theorem relates a one-dimensional integral (the line integral) to a two-dimensional integral (the surface integral).

Example 212. Verify the Stokes theorem by evaluating both sides of (3.96) for the following data.
(a) $F(x, y, z) = (x - y)i + (y - z)j + (z - x)k$, the surface Σ is the portion of the plane $x + y + z = 1$ which lies in the first octant.
(b) $F(x, y, z) = (y - z)i + (z + x)j + (y - x)k$, the surface Σ is the portion of the paraboloid $z = 9 - x^2 - y^2$ which lies above the xy-plane.

Solution: (a) The surface Σ forms a planar triangular region in space whose vertices are the unit vectors i, j and k and the boundary curve C is the triangle connecting those points and thus is piecewise smooth. The corresponding domain D in the xy-plane is the triangular region enclosed by $x+y = 1$,$x = 0$ and $y = 0$. Thus, if we go through the corners of C in the sequence $i \to j \to k \to i$, the orientation corresponds to a positive orientation of τ, the boundary of D. We parameterize the first piece $C_1 : i \to j$ with $r(t) = (1-t)i + tj,\ \ t \in [0, 1]$. For $F(x, y, z) = (x - y)i + (y - z)j + (z - x)k$, the line integral becomes

$$\int_{C_1} Fdr = \int_0^1 F(r(t))r'(t)dt$$

$$= \int_0^1 ((1 - 2t)i + tj + (t - 1)k)(-i + j)dt$$

$$= \int_0^1 (3t - 1)dt = \frac{1}{2}.$$

The other two pieces of C are parameterized by $r(t) = (1 - t)j + tk$ and $r(t) = (1 - t)k + tj,\ \ t \in [0, 1]$, respectively. An analogous computation as above shows that

$$\int_{C_2} F.dr = \int_{C_3} F.dr = \frac{1}{2}$$

so

$$\oint_C Fdr = \frac{1}{2} + \frac{1}{2} + \frac{1}{2} = \frac{3}{2}.$$

The parametrization S of Σ is given by $S(x, y) = 1-x-y$, so $\partial_x S = \partial_Y S = -1$ is constant. The unit normal according to (3.63) is

$$n = \frac{1}{\sqrt{3}}(i + j + k).$$

Moreover, $curlF = i+j+k$. Thus, both n and $curlF$ are constant on Σ .Since $\sqrt{1 + (\partial_x S)^2 + (\partial_y S)^2} = \sqrt{3}$, the surface integral becomes

$$\int\int_\Sigma (curlF).nd\sigma = \int\int_\Sigma \sqrt{3}d\sigma = \int\int_D \sqrt{3}\sqrt{3}dA = \frac{3}{2},$$

because the area of the triangular region D equals $\frac{1}{2}$.
(b) The boundary curve C is the circle lying in the xy-plane with radius 3 and

center at the origin and it can be parameterized as $r(t) = 3\cos ti + 3\sin tj$ for $0 \leq t \leq 2\Pi$. We have

$$\oint_C Fdr = \int_0^2 \Pi F(r(t))r'(t)dt = \int_0^2 \Pi 9(-\sin^2 t + \cos^2 t)dt = 0.$$

Since the line integral equals 0, it does not matter which orientation we choose. Moreover, we compute that $curlF = 0$, so

$$\int\int_\Sigma (curlF)nd\sigma = \int\int_\Sigma 0d\sigma = 0.$$

Hence, we have verified the Stokes theorem for both sets of data.

Remark 42. A vector field F defined on some open set G of space R^3 is called irrotational in G if $curl\mathbf{F} = 0$ at all points of G. This terminology is motivated by the Stokes theorem, since the circulation $\oint_C Fdr$ equals zero for all closed curves which arise as a boundary of a surface Σ domain G. If all closed curves in G arise as boundaries, we then can conclude from Theorem 45 that $curl\mathbf{F} = 0$ in G implies that F is conservative on G. This is the case, for example, when G is the whole space R^3 or when G is a ball.

On the other hand, if $curl\mathbf{F}(x, y, z)$ is not zero at some point (x, y, z), then for small disks around this point whose normal is parallel to $curl\mathbf{F}(x, y, z)$, we have $\oint_C Fdr \neq 0$ and hence F cannot be conservative. However, note that for some types of regions G there may exist closed curves C which are not boundaries of such a surface Σ and have non-zero circulation (and hence, F is not circulation free and not conservative on G according to Definitions 44 and 37, even though we might have $curl\mathbf{F} = 0$ on G). Consider, for example, the vector field

$$F(x, y, z) = -\frac{y}{x^2 + y^2}i + \frac{x}{x^2 + y^2}j.$$

Its domain G is the whole space R^3 except the z-axis. It satisfies $curl\mathbf{F} = 0$ on G, but $\oint_C Fdr = 2\Pi \neq 0$ for the circle C defined by $x^2 + y^2 = 1$. Indeed, if Σ is a surface with boundary C, it must intersect the z-axis, so it cannot be contained in G.

Example 213. For the following data, evaluate $\oint_C Fdr$ or $\int\int_\Sigma (curlF).nd\sigma$, whichever is easier.
(a) $F = yx^2 i - xy^2 j + z^2 k$, the surface Σ is the hemisphere $x^2 + y^2 + z^2 = 4, z \geq 0$.
(b) $F = zi + xj + y^2 k$, the surface Σ is the cone $z = x^2 + y^2$ for $0 \leq z \leq 4$.

Solution: (a) The boundary curve C of Σ can be described by $r(t) = 2\cos ti + 2\sin tj$ for $0 \leq t \leq 2\Pi$, so along C we have

$$F(r(t))r'(t) = -16\cos^2 t\sin^2 t - 16\cos^2 t\sin^2 t = -8\sin^2(2t) = -4(1 - \cos 4t),$$

and therefore

$$\oint_C Fdr = \int_0^2 \Pi - 4(1 - \cos 4t)dt = -8\Pi.$$

(b) The parametric equation of C is $r(t) = 4\sin ti + 4\cos tj + 4k$ for $0 \leq t \leq 2\Pi$, so

$$F(r(t)).r'^2 t,$$

$$\oint_C F.dr = \int_0^2 \Pi 16\cos t - 16\sin^2 t = -16\Pi,$$

Note that the parametrization $z = S(x, y) = \sqrt{x^2 + y^2}$ is not differentiable at 0; we remark that the Stokes theorem remains valid in this particular case.

3.6 Applications of Vector Calculus to Engineering Problems

A major part of science and engineering deals with the analysis of forces such as the force of water on a dam or of air on a wing, stresses within buildings and bridges, electric and magnetic forces in the power industry and in computer hardware, and so on. These forces vary with position, time and other circumstances. Vector calculus provides basic tools for analyzing and understanding these situations.

In the first subsection we exhibit relationships between elements of vector calculus and physical concepts like velocity, acceleration, momentum, angular momentum and temperature. The second subsection presents applications of line integrals. The third subsection is concerned with applications of surface integrals, in particular with the flux of a vector field across a surface, which leads to a better understanding of the notion of divergence. The fourth subsection is devoted to further applications of the divergence theorem of Gauss, namely Archimedes' principle, mass balance and heat flow. As an application of the theorem of Stokes, we interpret the notion of curl in the fifth subsection. The final subsection contains some basic examples of fluid flow, with an application to hurricane modelling.

3.6.1 Elements of Vector Calculus and Physical World

We present here some physical phenomena modeled by vector fields. According to Newton's law of gravitation, two objects with masses m and M attract each other with a force F of magnitude

$$\| F \| = \frac{GmM}{r^2} \tag{3.97}$$

where r is the distance between the two objects (treated as point masses), and G is the gravitational constant ($G = 6.673\ m^3/(kgsec^2)$). Assume that the object with mass M is located at the origin in 3-space, r is the position vector of the object of mass m and $r = \| r \|$, and the force $F(r)$ exerted by the object of mass M on the object of mass m points in the direction of the unit vector $\frac{-r}{\|r\|}$. Thus, from (3.97)

$$F(r) = -\frac{GmM^{2}}{\| r \|} \frac{r}{\| r \|} = -\frac{GmM^{3}}{\| r \|} r, \tag{3.98}$$

or, in Cartesian coordinates,

$$F(x, y, z) = -\frac{GmM}{(x^2 + y^2 + z^2)}^{\frac{3}{2}} (xi + yj + zk).$$

This defines a vector field whose domain D equals the whole space R^3 except the origin. It describes the gravitational force of a point mass M located at the origin, as a function of the position of the point mass m.

Electric Force Field

Coulomb's law states that the electrostatic force exerted by one charged particle on another is directly proportional to the product of the charges and inversely proportional to the square of the distance between them. Let two particles of charge Q and q be located at the origin of R^3 and at the position given by a vector r, respectively. Then the force $F(r)$ that the particle of charge Q exerts on the particle of charge q equals

$$F(r) = \frac{qQ}{4\Pi\epsilon_0 \| r \|^3} r, \tag{3.99}$$

where ϵ_0 is a positive constant (called the permittivity constant or the dielectric constant). Note that the force is repellent (directed outward) if Q and q have the same sign, and attractive otherwise. Formula (3.99) defines the vector field F of the electrostatic force generated by a the point charge Q at the origin, as a function of the position of the point charge q. If we divide by q we obtain the electrostatic force per unit charge, which is called the **electric field**.

$$E = \frac{F}{q}. \tag{3.100}$$

Because of their form (3.97) and (3.99), both Newton's law and Coulomb's law are instances of what is termed an inverse square law. Recall that, for a scalar vector field $\psi = \psi(x, y, z)$, the gradient is defined as

$$\nabla\psi = \frac{\partial\psi}{\partial x} i + \frac{\partial\psi}{\partial y} j + \frac{\partial\psi}{\partial z} k$$

and is called a potential of the vector field F if $F = \nabla\psi$. In this case, F is called a gradient field.

Gravitational and Electric Potential

The gravitational force field (3.98) possesses the potential (called gravitational potential)

$$\psi(r) = \frac{GmM}{\| r \|}. \tag{3.101}$$

Indeed, we have computed in Example 1.97(a), setting $n = -1$ where $F = \nabla\psi$. Analogously, the electric field (3.100) possesses the potential $-\psi$ (called electric potential, the minus sign is conventional) with

$$\psi(r) = \frac{Q}{4\Pi\epsilon_0 \| r \|}. \tag{3.102}$$

We have seen in Section 3.3 that at each point in a gradient field F where the gradient is non-zero, the latter points in the direction in which the rate of increase of the corresponding potential is maximal, and that moreover the gradient is perpendicular to the tangent plane of the level surface $\psi(x, y, z) = c$ through that point.

Vector fields F with domain and range in the plane R^2 can be represented graphically by drawing the vectors $F(r)$ at some points r of the plane. We give some examples.

(i) Setting $r(x, y) = xi + yj$ as above, we consider the vector field $F(r) = r$ as well as

$$F(r) = \frac{r}{\| r \|} = \frac{x}{\sqrt{x^2 + y^2}}i + \frac{y}{\sqrt{x^2 + y^2}}j.$$

These vector fields are shown in Figures 3.11a and 3.11b respectively.

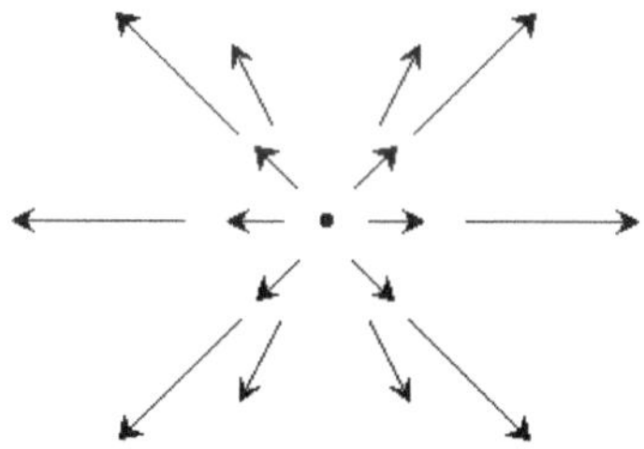

FIGURE 3.11a: Vector Field

(ii) The vector field defined as $F(r) = F(x, y) = -yi + xj$ is called the spin field or rotation field or turning field. $F(r)$ is perpendicular to r, since

$$F(r)r = -yx + xy = 0.$$

Moreover, we have $\| F(r) \| = \sqrt{x^2 + y^2} = \| r \|$. The vector fields F and $\frac{F}{\|F\|}$ are respectively shown in Figures 3.12a and 3.12b.

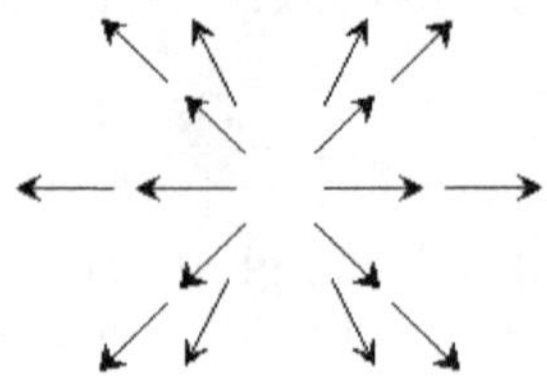

FIGURE 3.11b: Vector Field

Velocity and Acceleration Let a particle be moving along a path having position $r(t) = x(t)i + y(t)j + z(t)k$ as t varies from a to b. Its velocity field v also has the domain $[a, b]$ and is defined as

$$v(t) = r'(t) = x'(t)i + y'(t)j + z'(t)k. \tag{3.103}$$

The speed of the particle is defined as the magnitude of the velocity and it therefore is equal to

$$\| v(t) \| = (x'(t)^2 + y'(t)^2 + z'(t)^2)^{\frac{1}{2}} = \| r'(t) \| . \tag{3.104}$$

Let $s(t)$ denote the total distance traveled up to time t, that is,

$$s(t) = \int \| F'(\tau) \| \, d\tau, \text{ see (3.101)}$$

. Since $s'(t) = \|r'(t)\|$, we see from (3.104) that the speed of the particle is equal to $s'(t)$. The **acceleration a** of the particle is defined as the rate of change of the velocity with respect to time,

$$a(t) = v'(t) = r''(t) = x''(t)i + y''(t)j + z''(t)k.$$

Momentum The momentum p of an object is defined as the mass of the object times its velocity,

$$p(t) = mv(t) = mr'(t).$$

By Newton's law, its time derivative

$$p'(t) = mr''(t) = ma(t)$$

equals the total (or net) force acting upon the object. We see therefore that $p'(t) = 0$ if the net force is zero at time t.

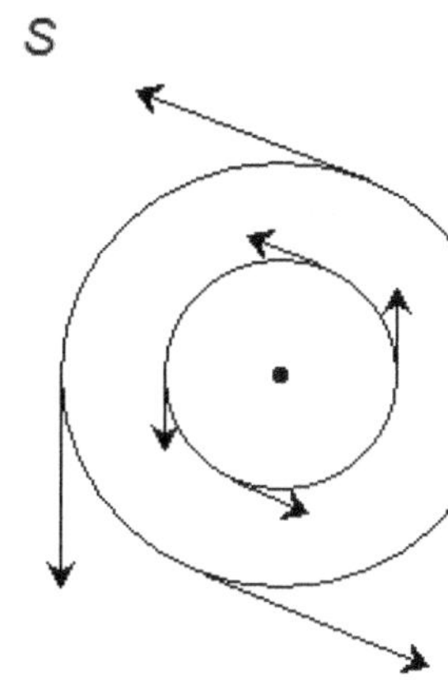

FIGURE 3.12a: Spin Field F

Remark 43. As a consequence, the momentum p of an object stays constant during time intervals in which no force is applied to it. This statement is called the law of conservation of momentum. Indeed conservation laws (which assert that certain quantities stay constant under certain conditions) are basic ingredients of the sciences, since in particular when analyzing situations with several (or many) changing quantities it can be very helpful to identify quantities which are not changing. Instead of saying "quantity X stays constant" one also says "quantity X is invariant" or "quantity X remains invariant".

Angular Momentum If an object of mass m has velocity $v(t)$ at position $r(t)$ at time t, its angular momentum $L(t)$ w.r.t. the origin is defined by the equation

$$L(t) = r(t) \times p(t) = r(t) \times mv(t) = r(t) \times mr'(t).$$

Note that the angular momentum $L(t)$ is perpendicular to $r(t)$ and $v(t)$. According to (3.18), its magnitude is given by

$$\| L(t) \| = \| r(t) \| \| mv(t) \| \sin \theta(t),$$

where $\theta(t)$ is the angle between $r(t)$ and $v(t)$.

Example 214. A particle in the plane moves along the circle with radius 2 centered at the origin in such a way that its x- and y-coordinates are given by $x(t) = 2\cos t$, $y(t) = 2\sin t$.
(a) Find the velocity, the speed, and the acceleration of the particle at an arbitrary time t.

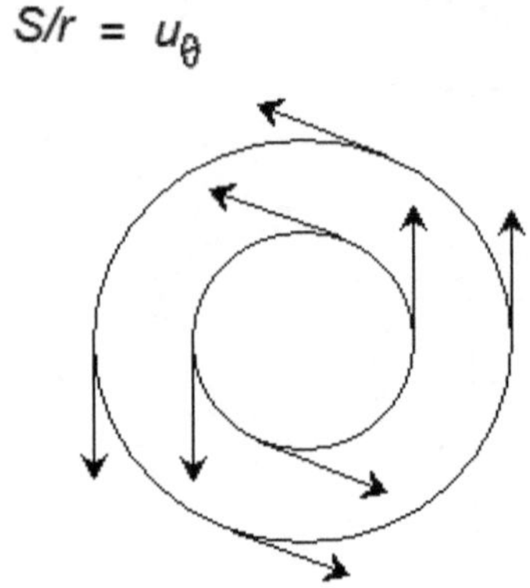

FIGURE 3.12b: Spin Field $\frac{F}{\|F\|}$

(b) Sketch the path of the particle and show the position and velocity vectors at $t = \frac{\Pi}{4}$.

Solution: (a) The position is described by the vector function

$$r(t) = 2\cos ti + 2\sin tj.$$

Its velocity and speed at time t are therefore

$$v(t) = r'(t) = -2\sin ti + 2\cos tj,$$

$$\| v(t) \| = \sqrt{(-2\sin t)^2 + (2\cos t)^2} = \sqrt{4(\sin^2 t + \cos^2 t)} = 2.$$

Its acceleration at time t is $a(t) = r''(t) = -2\cos ti - 2\sin tj$. At time $t = \frac{\Pi}{4}$, we have

$$r(\frac{\Pi}{4}) = 2\cos\frac{\Pi}{4}i + 2\sin\frac{\Pi}{4}j = \sqrt{2}i + \sqrt{2}j,$$

$$v(\frac{\Pi}{4}) = r'(\frac{\Pi}{4}) = -2\sin\frac{\Pi}{4}i + 2\cos\frac{\Pi}{4}j = -\sqrt{2}i + \sqrt{2}j.$$

Example 215. A particle of charge q moving in a magnetic field B is subject to the so-called **Lorentz force**:

$$F = \frac{q}{c}v \times B \tag{3.105}$$

where c is the speed of light and v is the velocity of the particle. Assume that the magnetic field is constant and vertically oriented, $B(t) = B_0 k$ with $B_0 \neq 0$. Find the path

$$r(t) = x(t)i + y(t)j + z(t)k$$

of the particle, given its initial position $r(0) = r_0$ and velocity $v(0) = v_0$, as well as its mass m.

Solution: By Newton's law and (3.105) we have

$$mv'(t) = mr''(t) = F(t) = \frac{q}{c}v(t) \times B(t).$$

Since $B(t) = B_0 k$, we get

$$v'(t) = \lambda v(t) \times k, \quad where \lambda = \frac{qB_0}{mc}. \tag{3.106}$$

Written in components $v(t) = v_1(t)i + v_2(t)j + v_3(t)k$ becomes

$$v_1'(t)i + v_2'(t)j + v_3'(t)k = \lambda[v_2(t)i.v_1(t)j].$$

This implies that

$$v_1'(t) = \lambda v_2(t), \quad v_2'(t) = -\lambda v_1(t), \quad v_3'(t) = 0. \tag{3.107}$$

Since $v_3'(t) = 0$ for all t, v_3 is a constant function, say $v_3(t) = C$. From the first two equations of (3.107) we get

$$v_1''(t) = \lambda v_2'^2 v_1(t),$$

or

$$v_1''(t) + \lambda^2 v_1(t) = o. \tag{3.108}$$

A solution of (3.108) is given by

$$v_1(t) = Asin(\lambda t + \varphi),$$

because

$$v_1'(t) = \lambda A \cos(\lambda t + \varphi), \quad v_1''^2(t) = \lambda A \sin(\lambda t + \varphi).$$

Since $v_1'(t) = \lambda v_2(t)$, we get

$$v_2(t) = \frac{v_1'(t)}{\lambda} = A \cos(\lambda t + \varphi).$$

Therefore

$$r'(t) = v(t) = A \sin(\lambda t + \varphi)i + A \cos(\lambda t + \varphi)j + Ck.$$

A final integration with respect to t gives us

$$r(t) = [-\frac{A}{\lambda} \cos(\lambda t + \varphi) + K_1]i + [\frac{A}{\lambda} \sin(\lambda t + \varphi) + K_2]j + [Ct + K_3]k, \tag{3.109}$$

where K_1, K_2, K_3 are constants of integration. All six constants A, φ, C, K_1,$K_2$and K_3 can be evaluated from the six initial conditions given by the vector equations $r(0) = r_0$ and $v(0) = v_0$, but we omit this computation here. Instead, we observe that the path of the particle is a circular helix with axis parallel to B, that is, parallel to k. One can see this from equation (3.109): The z-component of r varies linearly with t, while the x- and y-components represent uniform circular motion with angular velocity λ and radius $|\frac{A_1}{\lambda}|$ around the center (K_1, K_2). Thus, charged particles spiral around the magnetic field lines. Qualitatively, this behavior still occurs even if the magnetic field lines are curved, as in the case of the earth's magnetic field. Charged particles trapped by the earth's magnetic field spiral around the magnetic field lines that run from pole to pole.
A heat-seeking particle is located at the point (2,3) on a flat metal sheet whose temperature at a point (x, y) is

$$T(x, y) = 10 - 8x^2 - 2y^2.$$

Example 216. Find an equation for the trajectory of the particle if it moves continuously in the direction of maximum temperature increase.

Solution: Let the trajectory (a curve) be parameterized by $r(t) = (x(t), y(t))$ with $r(0) = (2, 3)$. Since the direction of maximum temperature increase at any point $(x, y$ is given by $\nabla T(x, y)$, the velocity vector $v(t)$ of the particle at time t points in the direction of the gradient at its current position $r(t)$. Thus there is a scalar μ that may depend on t such that

$$v(t) = \mu(t)\nabla(x(t), y(t)).$$

Since $v(t) = r'(t)$ and $\nabla T(x, y) = (-16x, -4y)$, we get

$$x'(t)i + y'(t)j = \mu(t)(-16x(t)i - 4y(t)j),$$

or, equating components,

$$x'(t) = -16\mu(t)x(t), \quad y'(t) = -4\mu(t)y(t). \tag{3.110}$$

Let $M = M(t)$ be any antiderivative of μ. We may check that

$$x(t) = e^{-16M(t)}x_0, \quad y(t) = e^{-4M(t)}y_0$$

are solutions of (3.110) with initial values $x(0) = x_0$ and $y(0) = y_0$. Using the initial values $x_0 = 2$ and $y_0 = 3$ we see that the trajectory $(x(t), y(t))$ satisfies, for all values of t, the equation

$$y = \frac{3}{\sqrt[4]{2}}\sqrt[4]{x}. \tag{3.111}$$

The graph of the trajectory and Cantor plot of the temperature function are shown in Figure 3.13.

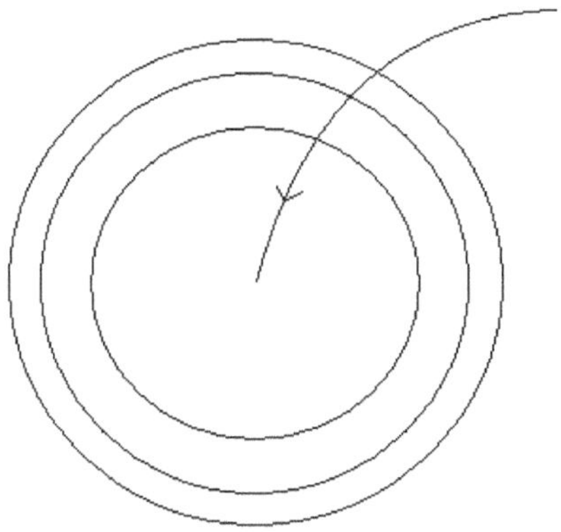

FIGURE 3.13: Trajectory and Cantor Plot

Remark 44. The preceding example exhibits dynamics controlled by the gradient of some scalar field. This is called gradient flow. There are numerous applications where such a situation arises.

Example 217. Assume that a certain distribution of electric charges in the plane produces the electric potential $(x, y) = e^{-.2x}\cos(2y)$.
(i) Find the electric field vector $E = -\nabla r$ at $(\frac{\Pi}{4}, 0)$.
(ii) Find the direction in which the potential decreases most rapidly at this point.

Solution:(i) We have$\nabla(x, y) = -2e^{-2x}\cos 2yi - 2e^{-2x}\sin 2yj$,

$$E(x, y) = -\nabla\psi(\frac{\Pi}{4}, 0) = -2e^{-\frac{\Pi}{2}}j.$$

(ii) At $(\frac{\Pi}{4}, 0)$, decreases most rapidly in the direction of $-\nabla\psi(\frac{\Pi}{4}, 0)$ which was computed in (i).
Suppose a rigid object rotates with constant angular speed around an axis through the origin with direction $a \in R^3$. Let $r = xi + yj + zk$ be any point of space. We decompose $r = r_\perp + \alpha a$ into a radial component $r_\perp$ and a component αa parallel to a, where α is a scalar. Since any material point of the object which passes through a fixed space point r will have the same velocity vector, we can associate with this motion a velocity field $v = v(x, y, z)$. We see that v is perpendicular to $r_\perp$ as well as to the rotation axis a and the speed $\| v \|$ is proportional to $r_\perp = \| r \| \sin\theta$. Therefore, there exists a unique vector ω which is parallel to a such that

$$v = \omega \times r_\perp = \omega \times r.$$

Angular velocity as the curl of linear velocity: This vector ω is called the angular velocity of the object; its length $\| \omega \|$ equals the angular speed. Let $\omega = Ai + Bj + Ck$. Then

$$v(x, y, z) = (Ai + Bj + Ck) \times (xi + yj + zk)$$
$$= (Bz - Cy)i + (Cx - Az)j + (Ay - Bx)k.$$

We compute

$$curl(v) = \begin{vmatrix} i & j & k \\ \partial_x & \partial_y & \partial_z \\ B_z - C_y & C_x - A_z & A_y - B_x \end{vmatrix} = 2Ai + 2Bj + 2Ck$$
$$= 2\omega. \tag{3.112}$$

Thus the angular velocity of a uniformly rotating body equals one-half the curl of the linear velocity, as the latter is called in this context to emphasize its different character.

3.6.2 Applications of Line Integrals

Line integrals have been introduced in Section 3.4.1. Here we present some applications in the area of mechanics.

Line Integrals of Scalar Fields

Let us consider a bent wire as a (one-dimensional) smooth curve $r(t) = x(t)i + y(t)j + z(t)k$ in space, where $r : [a, b] \to R^3$. We assume that the distribution of its mass is described by a continuous density function ρ (in units of mass per unit length) defined on the set $C = r([a, b])$ of the curve points. We have already mentioned in Subsection 3.4.1 that its total mass M can be expressed as the line integral

$$M = \int_C \rho ds = \int_a^b \rho(r(t)) \| r'(t) \| dt.$$

Its center of mass is another quantity of interest in mechanics. We define the coordinates of the so-called first moment of ρ as the line integrals

$$M_x = \int_C x\rho ds, \quad M_y = \int_C y\rho ds, \quad M_z = \int_C z\rho ds,$$

that is,

$$M_x = \int x(t)\rho(r(t)) \| r'(t) \| dt, \qquad M_y = \ldots, \qquad M_z = \ldots.$$

The coordinates of the center of mass are then given by

$$\overline{x} = \frac{M_x}{M}, \quad \overline{y} = \frac{M_y}{M}, \quad \overline{z} = \frac{M_z}{M}.$$

Let us now assume that the wire rotates around an axis. Its moment of inertia with respect to this axis is given by the line integral

$$I = \int_C d^2 \rho ds,$$

and $d(x, y, z)$ denotes the distance of the point (x, y, z) from the axis. In particular, we obtain the moment of inertia of the wire with respect to the x-axis as

$$Ix = \int_C (y^2 + z^2) \rho ds.$$

Line Integrals of Vector Fields

Let C be the straight line connecting two points r_0 and r_1 in space. We know from elementary mechanics that if a mass is moved from r_0 to r_1 under the influence of a constant force F, the work W done by the force is given by

$$W = F.(r_1 - r_0) = \| F \| \| r_1 - r_0 \| \cos \alpha,$$

where α is the angle between the vectors F and $r_1 - r_0$. Let us now parameterize C as $r(t) = r_0 + t(r_1 - r_0)$ with $r : [0, 1] \to R^3$. Since $r'(t) = r_1 - r_0$, we have

$$F.(r_1 - r_0) = \int_0^1 F.r'(t) dt.$$

The latter integral is nothing more than the line integral of Definition 70, so we can express the work W as the line integral.

$$W = \int_C F.dr. \tag{3.113}$$

This line integral also yields the correct value of the total work done by an arbitrary force field Let $F(x, y, z) = f_1(x, y, z)i + f_2(x, y, z)j + f_3(x, y, z)k$ on a mass which has moved along an arbitrary curve C from its initial point $r(a)$ to its final point $r(b)$, if C is given by $r = x(t)i + y(t)j + z(t)k$ with $r : [a, b] \to R^3$. This can be seen if we approximate C by a curve consisting of pieces of straight lines and pass to the limit. The procedure is analogous to the one used to compute the length of a curve, we will not carry out the details.

Example 218. Let $F(x, y, z) = i - yj + xyzk$ be a force field. Calculate the work done when moving a particle from $(0, 0, 0)$ to $(1, -1, 1)$ along the curve $x = t, \quad y = -t^2, \quad z = t, \quad 0 \leq t \leq 1.$

Solution: The work done is equal to the line integral $\int_C F.dr$. We compute

$$\int_C F.dr = \int_C dx - ydy + xyzdz = \int_0^1 [1x' - yy' + xyzz'](t) dt$$

$$= \int_0^1 1 + t^2(-2t) - t^4 dt = [t - \frac{1}{2}t^4 - \frac{1}{5}t^5]_0^1$$

$$= 1 - \frac{1}{2} - \frac{1}{5} = \frac{3}{10.}$$

Example 219. Find the work done by $F(x,y,z) = x^2 i - 2yzj + zk$ in moving an object along the line segment from (1,1,1) to (4,4,4).

Solution: First parameterize the line segment as $x = y = z = 1 + 3t$ with $0 \leq t \leq 1$, so $x'(t) = y'(t) = z'(t) = 3$. The work done by F moving the object along C is given by

$$W = \int_C F.dr = \int_0^1 [x^2 x' - 2yy' + zz'](t)dt$$

$$= \int_0^1 [(1+3t)^2 - 2(1+3t)^2 + (1+3t)].3dt$$

$$[\frac{(1+3t)^2}{2} - \frac{(1+3t)^3}{3}]_0^1 = -\frac{27}{2}.$$

In Definition 72 we have called a vector field F conservative if it possesses a potential function, that is, if $F = \nabla\psi$ for some scalar field. The following remark explains this terminology.

Remark 45. (Conservation of mechanical energy)
Let an object of mass m move in space according to Newton's law $F = ma = mr''$. Assume that the force field F has a potential function ψ , so $F = \nabla\psi$. In mechanics, one considers the potential energy $-\psi$ and the kinetic energy $-\frac{1}{2}m \parallel v \parallel^2 2$, where $v = r'$. The total mechanical energy $E(t)$ of the object at time t is given as the sum of its kinetic and its potential energy at that time, so

$$E(t) = \frac{1}{2}mr'(t).r'(t) - \psi(r(t)). \tag{3.114}$$

Let us compute its time derivative $E'(t)$. Using the product rule and the chain rule we get

$$E'(t) = \frac{1}{2}mr''(t).r'(t) + \frac{1}{2}mr'(t) - \nabla\psi(r(t)).r'(t)$$

$$= [mr''(t) - \nabla\psi(r(t))].r'(t).$$

Since $mr''(t) = F(r(t)) = \nabla(r(t))$, we conclude that $E'(t) = 0$. This means that the total energy remains constant (or, is conserved) along the trajectory of the object.

Remark 46. *As we have seen in Section 3.6.1, the gravitational field possesses a potential and hence it is conservative.*

3.6.3 Applications of Surface Integrals

Mass on Surface

Let Σ be a piecewise smooth bounded surface. We already know from Subsection 3.4.2 that its area is given by the surface integral of the constant 1 over Σ,

$$A(\Sigma) = \int\int_{\Sigma} 1 d\sigma.$$

Assume now that the surface carries mass whose distribution is described by a continuous density function $\rho = \rho(x, y, z)$ (in units of mass per unit area) defined on Σ. The total mass carried by the surface is then given by

$$M = \int\int_{\Sigma} \rho d\sigma.$$

Flux of a Vector Field across Surface Consider the motion of a fluid (a liquid or a gas) through a pipe with rectangular cross-section Σ_0 of area A_0. Let Σ be a (hypothetical) rectangular surface obtained by intersecting the pipe with a plane tilted by an angle α, see Figure 3.14.

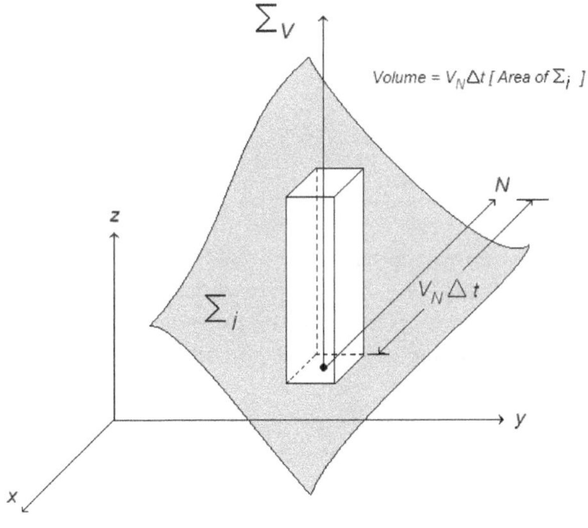

FIGURE 3.14: Flux

Assume for the moment that the fluid moves with constant velocity vector v during a certain time interval $[0, T]$. At time T, those particles of the fluid which have passed through Σ during the interval $[0, T]$ occupy the region D. The volume of D is given by

$$vol(D) = \parallel v \parallel TA_0 = \parallel v \parallel TA \cos\alpha = TAv.n, \tag{3.115}$$

where n is a unit normal to Σ, and A is the area of Σ. Thus, the number $Av.n$ gives the rate (per unit time) of fluid volume passing through Σ. The sign of this number depends on the direction of the flow (which is the direction of the velocity vector v) as well as on the orientation of the normal n.
In general, the velocity field $v = v(x, y, z, t)$ is not constant, but depends on space and time, and the surface Σ does not lie in a plane. If Σ, however, consists of plane parts Σ_j with unit normals n^j and area A_j on which, moreover, v is a constant vector v^j. By the above considerations the rate of fluid volume passing through all of Σ is equal to $\Sigma_j A_j v^j.n^j$. For the general case, the surface integral can be viewed as the result of a limit procedure of approximating Σ and $v.n$ in such a piecewise fashion. Thus, the rate of fluid volume passing through Σ at time t is equal to

$$\int\int_\Sigma v(.,t).nd\sigma. \tag{3.116}$$

Here, the notation $v(.,t)$ indicates that the surface integral $\int\int_\Sigma fd\sigma$ has to be evaluated with the integrand $f(x, y, z) = v(x, y, z, t).n(x, y, z)$ for that value of the time t for which we want to determine the rate. The integral in (3.116) is called the flux of the vector field v across the surface Σ. In this context, the vector field v is called the corresponding flux density. Since the rate of a quantity is nothing else than its time derivative, we obtain the total volume of the fluid passing through Σ during some time interval $[t_1, t_2]$ as the integral of (3.116) with respect to time,

$$\int_{t_1}^{t_2}\int\int_\Sigma v(.,t).nd\sigma dt. \tag{3.117}$$

If we want to determine the rate of fluid mass (instead of volume) passing through Σ, we have to replace the vector field v by the vector field ρv, where ρ is the mass density (mass per unit volume). Indeed, in the situation described above by Figure 3.14, the total mass of the fluid occupying the region D equals $\rho vol(D) = TA\rho v.n$, and the subsequent considerations yield the rate of fluid mass passing through Σ as

$$\int\int_\Sigma \rho(.,t)v(.,t).nd\sigma. \tag{3.118}$$

Note that (3.118) also applies to situations where ρ is not constant, that is, if the fluid in question is compressible. This typically occurs when the fluid is a gas.
Accordingly, for an arbitrary vector field $F = F(x, y, z, t)$ with values in R^3 one defines the flux of F across the surface Σ as the surface integral

$$\int\int_\Sigma F(.,t).nd\sigma, \quad or \quad \int\int_\Sigma F.nd\sigma. \tag{3.119}$$

The latter if F is stationary (that is, F does not depend on t) or if one just wishes to keep the notation simple. In this context, the vector field F useful is called the flux density.

Let the surface Σ enclose a region D in space, and let n be the field of outer unit normals on Σ. At points of Σ where $F.n > 0(< 0)$, the field vector F points outward (inward). Thus in this case, the (3.119) yields the outward flux across Σ.

Interpretation of Divergence

We have just seen that the surface integral $\int\int_\Sigma F.nd\sigma$ over the boundary Σ of a region D in space can be understood as the outward flux of the vector field F. In combination with the divergence theorem, this yields an interpretation of the divergence $divF$ at a point $P = (x, y, z)$ as follows. Assume that F is continuously differentiable and let D be any region which includes P. If D is small, then $divF(x, y, z)$ is close to the average value of $divF$ over D,

$$divF(x, y, z) \approx \frac{1}{volD} \int\int\int_D divF dV. \tag{3.120}$$

Actually, when $divF$ is continuous as we have assumed, we can prove that this average value tends to $divF(x, y, z)$ in the limit when D shrinks to the single point P. From the divergence theorem (39) we obtain

$$\frac{1}{vol(D)} \int\int\int_D divF dV = \frac{1}{vol(D)} \int\int_\Sigma F.nd\sigma, \tag{3.121}$$

where Σ is the boundary of D. Putting together (3.120) and (3.121) we see that the number $divF(x, y, z)$ yields the outward flux per unit volume near the point $P = (x, y, z)$, in the limit of small volumes. In particular, if $divF(x, y, z) < 0$ the quantity described by F tends to accumulate near P, whereas if $divF(x, y, z) < 0$ the quantity tends to move away from P. In the first case P is sometimes called a sink, and in the second case it is called a source.

It is instructive to look at this situation in more detail. Let D be a rectangular box with P as one of its corner points, see Figure 3.15. Let us look at the front face II and the back face I in the figure. On face II, the unit outer normal is constant and equal to the vector i. Consequently, the outward flux across face II is

$$\int\int_I IF.nd\sigma = \int\int_I IF.id\sigma = \int\int_I If_1 d\sigma,$$

where $F = f_1 i + f_2 j + f_3 k$. On face I, the unit outer normal is $-i$, so the sign is reversed, and we obtain

$$\int\int_I IF.nd\sigma + \int\int_I F.nd\sigma = \int\int_I If_1 d\sigma - \int\int_I f_1 d\sigma. \tag{3.122}$$

For the other four faces we get corresponding expressions involving the components f_2 and f_3 of F. Now let us apply the divergence theorem to the vector

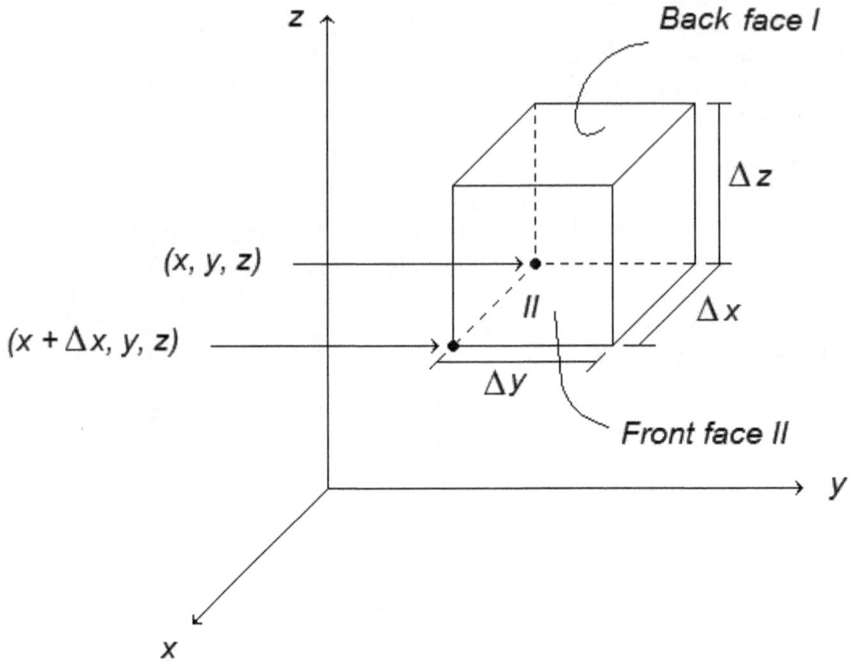

FIGURE 3.15: Interpretation of Divergence Theorem

field $\tilde{F} = f_1 i$, instead of F. We have $div\tilde{F} = \dfrac{\partial f_1}{\partial x}$ and therefore

$$\int\int\int_D \frac{\partial f_1}{\partial x} dV = \int\int\int_D div\tilde{F} dV = \int\int_\Sigma \tilde{F}.n dV = \int\int_I I f_1 d\sigma - \int\int_I f_1 d\sigma \tag{3.123}$$

since the contributions of the other four faces are zero for $\tilde{F}$. Therefore,

$$\frac{\partial f_1}{\partial x}(x,y,z) \approx \frac{1}{volD}\int\int\int_D \frac{\partial f_1}{\partial x} dV = \frac{1}{volD}[\int\int_I I f_1 d\sigma - \int\int_I f_1 d\sigma] \tag{3.124}$$

for small values of Δx, Δy and Δz. Thus, the number $\dfrac{\partial f_1}{\partial x}$ expresses the difference of outward flux across face II and inward flux across face I, per unit volume. The partial derivatives $\dfrac{\partial f_2}{\partial y}$ and $\dfrac{\partial f_3}{\partial z}$ can be interpreted in an analogous manner.

3.6.4 Applications of Gauss Divergence Theorem

Archimedes Principle

We use the divergence theorem to prove the validity of the Archimedes principle. This principle states that the buoyant force, exerted by a fluid on a solid object immersed in it, is equal to the weight of the fluid displaced by the object. This forms the basis for a ship to float or sink in the ocean. The issue is the balance between the force of gravity and the buoyant force. Consider a solid object immersed in a fluid which occupies a region D bounded by a piecewise smooth surface Σ with unit outer normal field n. Assume that the fluid is at rest, has constant density ρ and is acted upon by gravity in the downward direction, with constant gravity acceleration g. In mechanics, as a consequence of the balance of momentum, the pressure $p = p(x, y, z)$ within the fluid satisfies

$$p(x, y, z) = p_0.\rho g z, \tag{3.125}$$

if p has a constant value p_0 in the horizontal plane $z = 0$ (for example, the atmospheric pressure at the ocean surface). We see that the pressure increases linearly with depth $-z$ (the vertical coordinate is directed upward). The total buoyant force on the object is given by the vertical component f_3^{tot} of the total force $F^{tot} = f_1^{tot} i + f_2^{tot} j + f_3^{tot} k$ exerted by the fluid on the object. (This is a single vector, not a vector field.) On the surface Σ, the fluid pressure acts as a force density (force per unit area) of magnitude p and direction $-n$. Since $-pn.k$ is the vertical component of this force density, the total buoyant force is obtained as the surface integral

$$f_3^{tot} = \int\int_{\Sigma} -pn.k d\sigma = \int\int_{\Sigma} (\rho g z - p_0) n.k d\sigma. \tag{3.126}$$

As an illustration, if Σ_0 would be a flat horizontal piece, having area A_0, of the lower part of Σ at depth $d = -z$ (so the force acts from below with $n = -k$), the buoyant force on this piece would be equal to $(\rho g d + p_0)A_0$.

In order to apply the divergence theorem to (3.126), we define the vector field $G = -pk$. From (3.125) we see that

$$(divG)(x, y, z) = -\frac{\partial p}{\partial z}(x, y, z) = \rho g \tag{3.127}$$

is constant, so the total buoyant force becomes

$$f_3^{tot} = \int\int_{\Sigma} -pn.k d\sigma = \int\int_{\Sigma} -pk.n d\sigma = \int\int_{\Sigma} G.n d\sigma$$

$$\int\int\int_{D} (divG) dV = \int\int\int_{D} \rho g dV = \rho g vol(D). \tag{3.128}$$

Since $m = \rho vol(D)$ is the mass of the fluid displaced by the object, its weight mg (the force exerted by gravity on the fluid mass) equals the buoyant force acting on the object. In the same manner, one can show that $f_1^{tot} = f_2^{tot} = 0$, that is, the total force on the object is indeed vertically upward. To see this, one replaces k by i or j in the derivation prior to (3.127). Note that the field G in this case satisfies $divG = 0$, because $\frac{\partial p}{\partial x} = \frac{\partial p}{\partial y} = 0$. Thus we have derived the principle of Archimedes from the balance of momentum which underlies Equation (3.125) for the pressure.

Mass Balance

Consider a region G through which a fluid moves which has mass density $\rho = \rho(x, y, z, t)$ and velocity $v = v(x, y, z, t)$. Let D be any subregion of G with boundary surface Σ and unit outer normal field n. At any given time t, the total mass $M(t)$ contained in D is given by the volume integral of the mass density with respect to the spatial coordinates (x,y,z),

$$M(t) = \int\int\int_D \rho(., t)dV. \tag{3.129}$$

On the other hand, when discussing the flux of a vector field in Subsection 3.6.3, we have seen in (3.117) that the rate of fluid mass passing through the surface Σ in the outward direction is given by

$$\int\int_\Sigma \rho(., t)v(., t).nd\sigma. \tag{3.130}$$

In continuum mechanics, the principle of mass conservation states that the rate of change $M'(t)$ of the total mass $M(t)$ in D equals the rate of mass passing through the boundary of D (that is, the only way for the total mass to change is that part of it leaves or enters D through the boundary). In view of (3.129) and (3.130) this means that

$$\frac{d}{dt}\int\int\int_D \rho(., t)dV = M'(t) = -\int\int_\Sigma \rho(., t)v(., t).nd\sigma \tag{3.131}$$

This is the equation of **mass balance**. If we interchange the time derivative with the volume integral but it can also be done in several dimensions) and (3.131) becomes

$$\int\int\int_\Sigma \frac{\partial \rho}{\partial t}(., t)dV = -\int\int_D \rho(., t)v(., t).nd\sigma. \tag{3.132}$$

By the Gauss divergence theorem,

$$\int\int_\Sigma (\rho v)(., t).nd\sigma = \int\int\int_D div(\rho v)(., t)dV. \tag{3.133}$$

We combine the two previous equations and merge all components under one integral sign to obtain

$$\int\int\int_D \frac{\partial \rho}{\partial t}(., t) + div(\rho v)(., t)dV = 0. \tag{3.134}$$

This is valid for an arbitrary subregion D of G for which the divergence theorem can be applied. A theorem of analysis (which we do not discuss in this book) says that this is only possible if the integrand is equal to the zero function, that is, if

$$\frac{\partial \rho}{\partial t} + div(\rho v) = 0 \tag{3.135}$$

holds at all points (x, y, z) of G and for all times t considered. If the density ρ is constant, we have $\frac{\partial \rho}{\partial t} = 0$ and $div(\rho v) = \rho div v$, so (3.135) becomes

$$div v = 0, \tag{3.136}$$

that is, the velocity field is divergence free.
Equation (3.135), called the continuity equation, is one of the fundamental balance equations of continuum mechanics, used explicitly or implicitly in almost all computer simulations of the behaviors of fluids.

Heat Flow
In this subsection we show that the heat equation in Example 8.4.15, can be obtained as a consequence of the principle of energy conservation, using the divergence theorem. The derivation proceeds along similar lines as in the previous subsection, so our exposition will be shorter. The total amount of heat contained in some region D at a time t is given by

$$Q(t) = \int\int\int_D (\rho u)(., t) dV, \tag{3.137}$$

where u denotes the specific inner energy (that is, energy per mass). Denoting further its flow by q, energy conservation tells us that

$$\frac{d}{dt}\int\int\int_D (\rho u)(., t) dV = Q'(t) = -\int\int_\Sigma q(., t).n d\sigma. \tag{3.138}$$

The divergence theorem yields

$$\int\int_\Sigma q(., t).n d\sigma = \int\int\int_D (div q)(., t) dV.$$

As in the preceding subsection we can show that the previous two equations imply that

$$\frac{\partial}{\partial t}(\rho u) + div q = 0 \tag{3.139}$$

holds at all points (x, y, z) and times t considered. We assume that $u = cT + \alpha$, where T is the temperature, and the specific heat capacity c as well as the number α and the mass density ρ are constant. Fourier's law of heat conduction states that

$$q = -\kappa \nabla T, \tag{3.140}$$

where κ is a constant, called the heat conductivity. Since $div(\nabla T) = \triangle T$, the Laplace operator, we obtain from (3.139)

$$\frac{\partial T}{\partial t} = \frac{\kappa}{c\rho} \Delta T \tag{3.141}$$

is the heat equation. The steady state case occurs when T does not change with time. In this case $\frac{\partial T}{\partial t} = 0$ and (3.141) becomes the Laplace equation

$$\triangle T = 0. \tag{3.142}$$

Both preceding equations are bases for the modelling and simulation of heat conduction and flow in many kinds of engineering problems.

3.6.5 Application of Stokes Theorem

As an application of Stokes theorem, we present an interpretation of the curl of a vector field by measuring its rotation. Let $v = v(x, y, z)$ be the velocity field of a fluid and P_0 be any point in the fluid. Suppose Σ_r is a disk of radius r with center P_0, with unit normal vector n and boundary circle C_r as indicated in Figure 3.16. Because the disk is flat, the normal vector n is constant on Σ_r. By Stokes theorem

$$\oint_{C_r} v.dr = \int\int_{\Sigma_r} (curlv).nd\sigma. \tag{3.143}$$

If the parametrization r of C_r is chosen such that $\parallel r'(t) \parallel = 1$ along the curve, then $v(r(t)).r'(t)$ is the tangential component of the velocity, and the line integral $\oint v.dr$ gives the circulation of the fluid around C_r. Assuming v to be continuously differentiable for small r we have

$$(curlv)(P_0).n \approx \frac{1}{\Pi r^2} \int\int_{\Sigma_r} (curlv).nd\sigma,$$

since the right hand side represents the average of $(curlv).n$ over Σ_r, and the area of Σ_r equals Πr^2. In fact, since $curlv$ is continuous and Σ_r shrinks to the point P_0 as $r \to 0$,

$$(curlv)(P_0).n = lim \frac{1}{\Pi r^2} \int\int_{\Sigma_r} (curlv).nd\sigma. \tag{3.144}$$

Putting (3.143) and (3.144) together we arrive at

$$(curlv)(P_0).n = lim \frac{1}{\Pi r^2} \oint_{C_r} v.dr. \tag{3.145}$$

Thus, the number $(curlv)(P_0).n$ represents the circulation of v per unit area in the plane normal to n near P_0. In this manner, the curl of the velocity serves as a local model for the rotation of the fluid.

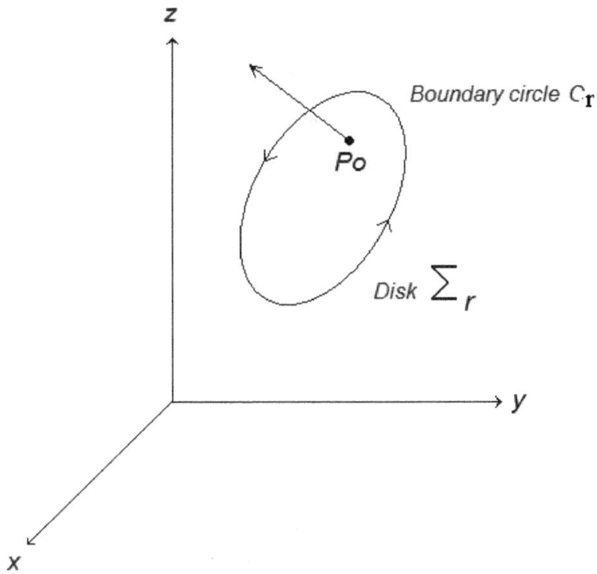

FIGURE 3.16: Interpretation of Curl

3.6.6 Example of Planar Fluid Flow

We begin with two basic situations, sink flow and vortex flow, in a situation made as simple as possible. Consider an incompressible fluid (its density ρ is constant) whose flow takes place in the plane and is stationary, that is, its velocity field $v = v(x, y)$ does not depend on time. It is also assumed that the fluid is inviscid, that is, its internal friction (the viscosity) is zero.

Sink flow. Imagine a hole at the origin (the sink) where the fluid leaves the plane and that (i) the fluid flows toward the origin, that is, the velocity vector at every point (x, y) is directed toward the origin, (ii) the flow is radially symmetric, that is, the speed of the fluid is the same at all points of every given circle centered at the origin.
Conditions (i) and (ii) above are modeled by a velocity field of the form

$$v(x, y) = \beta(r)(xi + yj), \quad r = \sqrt{x^2 + y^2}, \tag{3.146}$$

with a function $\beta = \beta(r) < 0$ yet to be fixed. We have seen in Section 3.6.4 that we must have $divv = 0$ in order to satisfy the principle of mass conservation. Using the chain rule we compute

$$\frac{\partial v_1}{\partial x} = \beta r + x\beta'(r)\frac{\partial r}{\partial x} = \beta(r) + \beta'(r)\frac{x^2}{\sqrt{x^2 + y^2}}.$$

In the same manner we obtain

$$\frac{\partial v_2}{\partial y} = \beta r + \beta'(r)\frac{y^2}{\sqrt{x^2+y^2}}.$$

Since in two dimensions

$$divv = \frac{\partial v_1}{\partial x} + \frac{\partial v_2}{\partial y},$$

we get

$$divv = 2\beta(r) + \beta'(r)\frac{x^2+y^2}{\sqrt{x^2+y^2}} = 2\beta(r) + r\beta'(r).$$

The condition

$$divv = 0$$

leads to

$$\beta(r) = \frac{c}{r^2}$$

for some constant c which must be negative by the above. Setting $c = -\frac{q}{2\Pi}$ (q is called the sink strength), (3.147) becomes

$$v(x,y) = -\frac{q}{2\Pi(x^2+y^2)}(xi+yj). \tag{3.147}$$

Since it follows that $\| v(x,y) \| = \frac{q}{2\Pi r}$, the sink flow has the further property that (iii) the speed of the fluid at any point $P = (x,y)$ is inversely proportional to the distance of P from the origin; in particular, it tends to $+\infty$ as that distance tends to zero.

Vortex flow. Here the fluid flows along concentric circles around the origin in the counterclockwise direction, that is,

(i) The velocity vector $v(x,y)$ at a point (x,y) is tangent to the circle centered at the origin which passes through (x,y),

(ii) $v(x,y)$ points in the counterclockwise direction.

(iii) Moreover, the speed $\| v \|$ is constant along those circles, and

(iv) The speed is inversely proportional along any such circle to the radius r of the latter, and hence it tends to $+1$ as r tends to 0. The vector field

$$v(x,y) = \frac{k}{2\Pi(x^2+y^2)}(-yi+xj) \tag{3.148}$$

possesses those four properties. (The constant $k > 0$ is called the vortex strength.) Indeed, by (3.148) we have $\| v(x,y) \| = \frac{k}{2\Pi x^2+y^2} = \frac{k}{2\Pi r}$. Moreover, $v(x,y).(xi+yj) = 0$, so $v(x,y)$ is perpendicular to the radius vector of the circle, and by drawing a picture one sees that the direction is counterclockwise.

Streamlines and stream functions. The paths followed by the fluid particles in a fluid flow are called the streamlines of the flow. If the streamlines can be represented as the level curves of some function $\psi = \psi(x, y)$, then ψ is called the stream function of the flow. In this case, every particle path must satisfy $(x(t), y(t)) = c$ for a suitable constant c. By the chain rule,

$$\frac{\partial\psi}{\partial x}(x(t), y(t))x\dot{t} + \frac{\partial\psi}{\partial y}(x(t), y(t))y\dot{t} = 0$$

must hold along particle paths. Therefore, ψ and the velocity field v of the flow are related by

$$\nabla\psi(x, y).v(x, y) = 0 \tag{3.149}$$

at all points (x, y) in the domain of the flow; the velocity vectors are tangent, while the vectors $\nabla\psi(x, y)$ are perpendicular to the streamlines.

Combined sink and vortex flow. Here we have the velocity field

$$v(x, y) = \frac{1}{2\Pi\sqrt{x^2 + y^2}}[(-qx - ky)i + (-qy + kx)j]. \tag{3.150}$$

The particles in this flow rotate while moving toward the sink, so we expect that they spiral inward. In order to find the streamlines, we will compute its stream function, using (3.149). It is convenient to do this in polar coordinates, using the vectors introduced in (3.60) as

$$e_r = \cos\theta i + \sin\theta j, \quad e_\theta = -\sin\theta i + \cos\theta j.$$

Setting $x = r\cos\theta,\ \ y = r\sin\theta$, the velocity field in polar coordinates becomes

$$v(r, \theta) = \frac{1}{2\Pi r^2}[(-qr\cos\theta - kr\sin\theta)i] + (-rr\sin\theta + kr\cos\theta)i]$$

and $\nabla\psi$ transforms according to (3.61) into

$$\nabla\psi = \frac{\partial\psi}{\partial r}e_r + \frac{1}{r}\frac{\partial\psi}{\partial\theta}e_\theta.$$

Condition (3.149) for the stream function now becomes in polar coordinates

$$0 = (\frac{\partial\psi}{\partial r}(r, \theta)e_r + \frac{1}{r}\frac{\partial\psi}{\partial\theta(r, \theta)}e_\theta).v(r, \theta) = \frac{1}{2\Pi r}(-q\frac{\partial\psi}{\partial r}(r, \theta) + \frac{k}{r}\frac{1}{r}\frac{\partial\psi}{\partial\theta(r, \theta)}). \tag{3.151}$$

We see that (3.151) is satisfied if we choose ψ such that

$$\frac{\partial\psi}{\partial r}(r, \theta) = \frac{k}{r}, \quad \frac{\partial\psi}{\partial\theta}(r, \theta) = q. \tag{3.152}$$

This is indeed possible and we set

$$\psi(r, \theta) = k\ln r \ + \ q\theta. \tag{3.153}$$

We want to compute r as a function of θ for the streamlines $\psi = c$. From $k \ln r + q\theta = c$ we get

$$\ln r = \frac{1}{k}(c - q\theta), \quad r = e^{\frac{c-q\theta}{k}} = e^{\frac{c}{k}}.e^{\frac{-q\theta}{k}}.$$

Since c is an arbitrary constant, we may replace $e^{\frac{c}{k}}$ by c and finally obtain that

$$r = ce^{\frac{-q\theta}{k}}, \quad c > 0 \tag{3.154}$$

holds along the streamlines. Thus, the spirals are determined by the value of $\frac{q}{k}$.

Modelling of hurricane. Let us assume that the preceding flow model can be used for a hurricane and that only a single measurement of velocity is available.

Example 220. Find the strength of the parameters k and q of the flow model (3.150) for a hurricane from the report that at 20 km distance from the eye the wind velocity has a component of 15 km/hr towards the eye and a counterclockwise tangential component of 45 km/hr. Estimate the size of the hurricane by finding a radius beyond which the wind speed is less than 5 km/hr.

Solution: The velocity component toward the eye is equal to the speed of the sink part of the flow. We have already seen that the latter is given by $q/(2\Pi r)$, so at $r = 20$ we have

$$15 = \frac{q}{2\Pi.20}, \quad so \quad q = 600\Pi$$

with unit 1/hr. The tangential velocity component is equal to the speed of the vortex part of the flow, and we have already seen that the latter is given by $k/(2\Pi r)$, so at $r = 20$ we have

$$45 = \frac{q}{2\Pi.20}, \quad so \quad q = 1800\Pi,$$

again with unit 1/hr. To estimate the size, we determine r from the condition that $\| v \| = 5km/h$. Since tangential and inward velocity are perpendicular, we have

$$5 = \| v \| = \frac{1}{r}\sqrt{(\frac{q}{2\Pi})^2 + (\frac{k}{2})^2}$$

Therefore

$$r = \frac{1}{5}\sqrt{300^2 + 900^2} = \frac{100}{5}\sqrt{9+81} = 60\sqrt{10} \approx 189.7 \text{ km}.$$

3.7 Exercises

3.1. Let x, y, z be vectors of three-dimensional space, let λ be a scalar. Show that

(i) $(x + y) + z = x + (y + z)$
(ii) $\lambda(x + y) = \lambda x + \lambda y$
(iii) $x.(y + z) = x.y + x.z$

(iv) $\lambda(x.y) = (\lambda x).y = x.(\lambda y)$
(v) $x.(x \times y) = 0$, that is, $x \times y$ is perpendicular to x.
(vi) $y > (x \times y) = 0$, that is, $x \times y$ is perpendicular to y.
(vii) $x \times (y \mid z) = x \times y + x \times z$
(viii) $x \times (y \times z) = (x.z)y - (x.y)z$
(ix) $x.(y \times z) = y.(z \times x) = z.(x \times y)$
(x) $x \times x = 0$

3.2. The goal of this exercise is to derive the identity

$$(x \times y).(z \times w) = (x.z)(y.w) - (y.z)(x.w), ...(*)$$

for arbitrary vectors x, y, z and w of three-dimensional space.

(i) Show that $(*)$ holds for $x = y = k$ and arbitrary vectors z and w.

(ii) Show that $(*)$ holds for $x = k$, $y = \alpha i + \beta j$ and arbitrary vectors z and w, where α and β are arbitrary scalars.

(iii) Show that $(*)$ holds for $x = k$ and arbitrary vectors y, z and w.

(iv) Show that $(*)$ holds for for arbitrary vectors x, y, z and w.

3.3. The line in R^2 that passes through the point $r_0 = (x_0, y_0)$ and is parallel to the non-zero vector $v = (a, b) = ai + bj$ has parametric equations

$$x(t) = x_0 + at, \quad y(t) = y_0 + bt,$$

or, in vector form $r(t) = (x(t), y(t))$,

$$r(t) = r_0 + tv.$$

In R^3, the vector form is the same, but now $r(t) = (x(t), y(t), z(t)), r_0 = (x_0, y_0, z_0)$ and $v = (a, b, c)$ and the component equations are

$$x(t) = x_0 + at, \quad y(t) = y_0 + bt, \quad z(t) = z_0 + ct.$$

(a) Find the parametric equations of the line

(i) passing through (4 2) and parallel to $v = (-1, 5)$,

(ii) passing through (1,2,-3) and parallel to $v = (4, 5, -7)$.

(b) Find parametric equations for the line whose vector equation is given as

(i) $r(t) = 2i - 3j + ti - 4tj$,

(ii) $r(t) = -i + 0j + 2k - ti + 3tj$.

3.4. The equation of the plane passing through a point $r_0 = (x_0, y_0, z_0)$ in R^3 and perpendicular to a vector $N = (a, b, c)$, called a normal for the plane, is given by

$$a(x - x_0) + b(y - y_0) + c(z - z_0) = 0,$$

that is, the plane consists of the points $r = (x, y, z)$ satisfying this equation. Its vector form is

$$N.(r - r_0) = 0.$$

Find an equation of the plane that passes through the point $r_0 = (2, 6, 1)$ having the vector $N = (1, 4, 2)$ as a normal.

3.5. Let F and G be two vector fields defined on an interval with values in R^3. Prove that

(a) $\frac{d}{dt}[F.G] = F.\frac{dG}{dt} + \frac{dF}{dt}.G.$

(b) $\frac{d}{dt}[F \times G] = F \times \frac{dG}{dt} + \frac{dF}{dt} \times G.$

(c) Let $r = r(t)$ be a vector valued function with values in R^3 such that $\| r(t) \| = c$ for some constant c. Show that $r(t).r'(t) = r(t).\frac{dr}{dt}(t) = 0$, that is, $r(t)$ is orthogonal to $\frac{dr}{dt}(t) = r'(t)$, for all t.

3.6. (a) Verify explicitly that $div(\mathbf{curlF}) = 0$, where

(i) $F = \sinh(x - z)i + 2yj + (z - y^2)k$,

(ii) $F = x^2 i + y^2 j + z^2 k$.

(b) Verify explicitly that $curl(\nabla\varphi) = 0$, where

(i) $\varphi(x, y, z) = -2x^3 y z^2$,

(ii) $\varphi(x, y, z) = e^{x+y+z}$.

3.7. (a) Let F and G be two vector fields with domain in R^3 and values in R^3. Prove that

$$div(F \times G) = G.(\nabla \times F) - F.(\nabla \times G).$$

(b) Let $\varphi = \varphi(x, y, z)$ and $\psi = \psi(x, y, z)$ be scalar fields. Prove that $div(\nabla\varphi \times \nabla\psi) = 0$.

3.8. Let C be a constant vector and $r(x, y, z) = xi + yj + zk$. Prove that
(a) $\nabla(C.r) = C$,
(b) $div(r - C) = 3$,
(c) $curl(r - C) = 0$.

3.9. Evaluate $\int_C F.dr$, where $F(x, y, z) = i - xj + k$ and C is parameterized by $r(t) = \cos t i - \sin t j + tk$ for $0 \leq t \leq \Pi$.

3.10. Evaluate the surface integral $\int\int_\Sigma f d\sigma$, where
(a) $f(x, y, z) = y^2, \Sigma$ is the part of the cone $z = \sqrt{x^2 + y^2}$ lying in the first octant and between the plane $z = 2$ and $z = 4$.
(b) $f(x, y, z) = xyz, \Sigma$ is the part of the surface $z = 1 + y^2$ for $0 \leq x \leq 1, \ \ 0 \leq y \leq 1$.

3.11. A particle moves once counterclockwise around the circle of radius 12 centered at the origin under the influence of the force $F(x, y, z) = (e^x - y + x \cosh x)i + (y^{\frac{3}{2}} + x)j$. Calculate the work done.

3.12. Apply the Green-Ostrogradski theorem to evaluate the line integrals $\int_C F.dr$ for the following data. The curves are oriented counterclockwise.
(a) $F = x^2yi - xy^2j$, C is the boundary of the region defined by $x^2 + y^2 \leq 4$, $x \geq 0$ and $y \geq 0$.
(b) $F = (e^{\sin x} - y)i + (\sinh(y^3) - 4x)j$, C is the circle of radius 4 with center $(-8, 0)$.
(c) $F = (x^2 + y^2)i + (x^2 - y^2)j$, C is the ellipse $4x^2 + y^2 = 10$.

3.13. Let D be a region bounded by the surface Σ. Evaluate $\int\int_\Sigma F.nd\sigma$ or $\int\int\int_D div(F)dV$ for the following data, whichever is more convenient.
(a) $F = 4xi - 6yj + k$, Σ is the surface of the solid cylinder $x^2 + y^2 \leq 4$, $0 \leq z \leq 6$ (the surface includes both caps of the cylinder).
(b) $F = 2yzi - 4xzj + xyk, \Sigma$is the sphere of radius 6 with center $(-1, 3, 1)$.

3.14. Find the value of $\int\int_\Sigma F.nd\sigma$ for the data $F = 3xyi + z^2k, \Sigma$ the sphere of radius 1 centered at the origin.

3.15. Let Σ be a smooth surface enclosing some region D and C be a constant vector. Show that $\int\int_\Sigma C.nd\sigma = 0$.

3.16. Evaluate $\int\int_\Sigma (curlF).nd\sigma$, where $F = xyi + yzj + xyk$ and Σ is the part of the plane $2x + 4y + z = 8$ in the first octant.

3.17. Calculate the circulation of $F = (x - y)i + x^2yj + xzak$ counterclockwise around the circle $x^2 + y^2 = 1$, where a is a positive constant.

3.18. Examine whether the following vector fields are conservative or not.
(a) $F = \cosh(x+y)(i+j-k)$.
(b) $F = 2xi - 2yj + 2zk$.
(c) $F = i - 2j + k$.

3.19. Let Σ be the portion of the paraboloid $z = 1 - x^2 - y^2$ for which $z \geq 0$, and let C be the circle $x^2 + y^2 = 1$ that forms the boundary of Σ. Verify Stokes'theorem for the vector field

$$F = (x^2y - z^2)i + (y^3 - x)j + (2x + 3z - 1)k$$

by evaluating the line integral as well as the surface integral.

3.8 Suggestion for Further Reading

As we know vector calculus is the backbone of two fundamental equations, one set governing electrodynamics named Maxwell equations and other related to fluid dynamics called Navier-Stokes equations which are expressed in terms of gradient, Laplacian, curl and divergence terms. Thus everything involving electricity, magnetism and fluid (or air) flow derives from gradient, divergence, curl and Laplacian operators. Different vector and differentiation operators are also encountered in many different Hamiltonians for Schrodinger wave equations.

A more practical and useful situations are antenna/scattering problems where we are required to find the vector and scalar potential due to a current distribution and its application to find electromagnetic fields of an object. Concepts of *grad*, *curl*, *div*, are required to determine fields from the potentials.

Problems involving surface or volume integrals can be transformed to a lower dimensional integral using the results of this chapter. Around 150 years ago James Clerk Maxwell first wrote what are now known as the four Maxwell equations which explain magnetic and electrical fields are essentially two sides of the same coin. The equations contain the mechanism by which electromagnetic radiation exists how all light, all X-rays, microwaves, radio waves and gamma rays arise as solutions to Maxwell equations.

Maxwell Equations

$$\nabla . D = \rho v$$

$$\nabla . B = 0$$

$$\nabla \times E = -\frac{\partial B}{\partial t}$$

$$\nabla \times H = \frac{\partial D}{\partial t} + J$$

where ρ, v, J, H, B, E, D respectively represent electric charge density, electric current Π density, magnetizing field, magnetic field, electric field, and displacement field. Monk [4], Neunzert and Siddiqi [5], Li and Huang [3], and references therein provide interesting reading material on Maxwell equations and applications. The Navier-Stokes equations named after Claude-Louis Navier and George Gabriel Stokes describe the motions of fluid substances. The vector form of Navier-Stokes equations for incompressible flow of Newton fluids is:

$$\rho(\frac{\partial v}{\partial t} + v.\nabla v) = -\nabla p + \mu \nabla^2 v + f$$

where f represents other body forces such as gravity and ρ, μ, v, p, respectively denote density, viscosity, velocity, and pressure. Navier-Stokes equations are very important as they describe many physical problems. They may be used to model the weather, ocean current, water flow in a pipe and air flow around a wing. They help to design aircraft and cars, the studies of blood flow, the design of power stations, analysis of pollution. Coupled with Maxwell equations, they can be used to model and study magnetohydrodynamics.

The Navier-Stokes equations pose challenging problems purely in mathematical sense, as shown by a U.S.$1 million prize offered by the Clay Mathematics Institute for proving or disproving existence of smooth solution in three dimensions. Books by Doering and Gibbon [1], Galdi [2], and Temam [6] provide detailed account; interested readers should consult them to gain deeper knowledge about the Navier-Stokes equations.

Bibliography

[1] C. R. Doering, J. D. Gibbon, *Applied Analysis of the Navier-Stokes Equations*, Cambridge University Press, 1995.

[2] G. P. Galdi, *An Introduction to Mathematical Theory of the Navier Stokes Equations*, Second Edition, 2011, Springer.

[3] J. Li, Y. Huang, *Time-Domain Finite Element Methods for Maxwell Equations in Metamaterials*, Series in Computational Mathematics, Springer, 2013.

[4] P. Monk, *Finite Element Methods for Maxwell Equations*, Oxford Science Publications, 2004.

[5] H. Neunzert, A.H. Siddiqi, *Topics in Industrial Mathematics: Case Studies*, Kluwer Academic Publisher, 2000. Springer electronic version 2014.

[6] R. Temam, *Navier-Stokes Equations: Theory and Numerical Analysis*, AMS Chelsea Publishing, 2000.

[7] A. H. Siddiqi, P. Manchanda, M. Brokate, *Calculus with Applications*, IK International, 2011.

[8] D. G. Zill and W. S. Wright, *Advanced Engineering Mathematics*, Jones and Bartlett Publications, 2012.

Chapter 4

Fourier Methods and Integral Transforms

4.1 Introduction

Fourier methods or Fourier analysis constitute a branch of mathematics developed formally some 150 years after Newton's and Leibniz' calculus and heavily depend on integral and differential calculus. Jean Baptiste Joseph Fourier was born in 1768 in Auxerre, a town between Paris and Dijon. He became fascinated by mathematics at the age of 13. After the French Revolution, Fourier taught in Paris, then accompanied Napoleon to Egypt and served as permanent secretary of the Institute of Egypt. He wrote a book on Egypt and in certain quarters he is famous more as an Egyptologist than for his contributions to mathematics and physics. In the world of science he is famous for, among other things, the ideas he set forth in a memoir in 1807 and published in 1822 in his book in French entitled *The Analytic Theory of Heat.*

Fourier analysis shows that we can represent periodic functions, even very jagged and irregular-looking ones, in form of a finite or infinite sum of sine and cosine functions, called **Fourier series**. Non-periodic functions can be treated with the **Fourier transform**. Fourier showed how these mathematical tools can be used to study natural phenomena such as heat diffusion, making it possible to solve equations that had until then remained intractable. Under the action of the Fourier transform, derivatives are transformed into multiplications, thus turning differential equations into equations containing algebraic expressions. In this way, many important differential equations are transformed into equations which are much easier to study and solve.

If a phenomenon is described by a function of time or space, Fourier analysis tells us, loosely speaking, how much of each frequency this phenomenon contains. In many cases, this frequency information is not simply a mathematical trick to make calculations easier, but corresponds to relevant properties of the phenomenon under study.

During the second half of the 20th century, Fourier analysis was refined in various directions in order to make it easier and more efficient to transmit, compress and analyze information or to separate information from surrounding noise. One of those techniques, the wavelet transform invented during the 1980s, will be presented in Chapter 10.

Fourier analysis and its descendants have been applied to a broad range of areas of science and engineering, for example, telecommunications and signal processing, physics, imaging in the biomedical sciences (EEG, ECG, CAT, MRI, NMR), meteorology, oceanography, seismology, economics and finance. A few of them will be touched in this chapter. Interested readers may find details of some of these applications in Prestini [4] and Cartwright [2]. There are also many interactions within mathematics in areas as diverse as statistics and number theory.

For several reasons, we have chosen not to present proofs (detailed or sketched) for the majority of results of this chapter, and we refer the reader to the literature on the subject.

Finally, we want to acknowledge the contributions of Dr. A. K. Verma to this chapter, whose program we used to illustrate the partial sums of several of the Fourier series discussed in this Chapter.

4.2 Orthonormal Systems and Fourier Series

4.2.1 Orthonormal Systems

In Chapter 3 we introduced the concept of orthogonal (or perpendicular) vectors. Namely, two vectors $\mathbf{u}$ and $\mathbf{v}$ are called orthogonal if their scalar product (dot product) $\mathbf{u} \cdot \mathbf{v}$ equals zero. Recall also that $\mathbf{u} \cdot \mathbf{u} = \|\mathbf{u}\|^2 > 0$ if $\mathbf{u} \neq \mathbf{0}$. We now extend this concept in two respects. We define orthogonality for systems of more than two vectors, and we define orthogonality for functions instead of vectors.

Throughout this chapter, the functions considered will be piecewise continuous. Such functions are formed by putting together continuous functions defined on separate intervals, for example the sign function or the greatest integer function as considered in Calculus Books, or the "sawtooth" function defined by

$$f(x) = x - k\,, \quad \text{if } k - \frac{1}{2} \leq x < k + \frac{1}{2}\,. \tag{4.1}$$

The formal definition runs as follows. A function f defined on an interval $I = [a, b]$ is called **piecewise continuous**, if we can decompose I into finitely many subintervals $I_k = [x_{k-1}, x_k]$ such that f is continuous in the interior of I_k and the one-sided limits $\lim_{x \to x_{k-1}+} f(x)$ and $\lim_{x \to x_k-} f(x)$ exist for all such subintervals. If the domain of f is unbounded, we require this property to hold for every bounded interval in the domain. As a consequence, we can integrate piecewise continuous functions, and

$$\int_a^b f(x)\,dx = \sum_k \int_{x_{k-1}}^{x_k} f(x)\,dx\,.$$

Definition 76. (Orthogonal functions)
Let f_1, f_2 be piecewise continuous real-valued functions defined on an interval $[a, b]$. Their **scalar product** (or **inner product**) is denoted by $\langle f_1, f_2 \rangle$ and defined by

$$\langle f_1, f_2 \rangle = \int_a^b f_1(x) f_2(x)\,dx\,. \tag{4.2}$$

The functions f_1 and f_2 are said to be **orthogonal** on $[\boldsymbol{a}, \boldsymbol{b}]$ if

$$\langle f_1, f_2 \rangle = \int_a^b f_1(x) f_2(x)\,dx = 0\,. \tag{4.3}$$

Example 221. (a) The functions $f_1(x) = e^x$ and $f_2(x) = \sin x$ are orthogonal on the interval $[\pi/4, 5\pi/4]$.

(b) The functions $f_1(x) = x$ and $f_2(x) = \cos 2x$ are orthogonal on the interval $[-\pi/2, \pi/2]$.

In analogy to the case of vectors, for functions $f : [a, b] \to \mathbb{R}$ we define

$$\|f\| = \sqrt{\langle f, f \rangle} = \sqrt{\int_a^b f(x)^2\,dx} \tag{4.4}$$

and call it the **norm** of f. (We might also call $\|f\|$ the length of f, although its geometric meaning is no longer obvious.) We then have

$$\|f\|^2 = \langle f, f \rangle = \int_a^b f(x)^2\,dx\,.$$

As in the case of vectors, we have $\|0\| = 0$ for the zero function. On the other hand, a piecewise continuous function f satisfies $\|f\| = 0$ only if f is the zero function; in other words, if f is nonzero, we must have $\|f\| > 0$.

If two nonzero functions f_1 and f_2 are orthogonal, we say that the set $\{f_1, f_2\}$ formed by those two functions is an orthogonal system. If we have more than two functions, we require that the functions are mutually (or pairwise) orthogonal in the sense of the following definition.

Definition 77. (Orthogonal system, orthonormal system)
A system (or set) consisting of finitely or infinitely many nonzero functions $f_1, f_2, \dots$ is said to be **orthogonal on the interval** $[a, b]$ if

$$\langle f_m, f_n \rangle = \int_a^b f_m(x) f_n(x) = 0\,,$$

whenever $m \neq n$.
If moreover

$$\langle f_n, f_n \rangle = 1 = \|f_n\|\,, \quad \text{for all } n,$$

the system is called **orthonormal on** $[a, b]$.

Remark 47. An orthogonal system $\{f_1, f_2, \dots\}$ can be made into an orthonormal system by replacing each function f_n with its scalar multiple

$$\frac{f_n}{\|f_n\|}\,.$$

Example 222. (a) The set $\{1, \cos x, \cos 2x, \dots, \cos nx\}$ is an orthogonal system on $[-\pi, \pi]$.

(b) The set $\left\{\frac{1}{\sqrt{2\pi}}, \frac{\cos x}{\sqrt{\pi}}, \cdots, \frac{\cos nx}{\sqrt{\pi}}\right\}$ is an orthonormal system on $[-\pi, \pi]$.

(c) Examine whether the following systems are orthonormal or not on the intervals indicated.

(i) $\{\sin x, \sin 3x, \sin 5x, \cdots\}$, $I = [0, \pi/2]$,
(ii) $\{\sin nx\} : n = 1, 2, 3, \dots\}$, $I = [0, \pi]$,
(iii) $\{1, \frac{\cos n\pi x}{l} : n = 1, 2, 3, \dots\}$, $I = [0, l]$,
(iv) $\{1, \frac{\cos n\pi x}{l}, \frac{\sin m\pi x}{l}\} : n, m = 1, 2, 3, \dots\}$, $I = [-l, l]$.

Solution: We discuss here part (c).
(i) We set $f_n(x) = \sin(2n+1)x$. For $m \neq n$, we obtain

$$\begin{aligned}\langle f_n, f_m \rangle &= \int_0^{\pi/2} \sin(2n+1)x \sin(2m+1)x \, dx \\ &= \frac{1}{2}\int_0^{\pi/2} [\cos 2(n-m)x - \cos 2(n+m+1)x] \, dx\end{aligned}$$

(using the trigonometric identity $2\sin Ax \sin Bx = \cos(A-B)x - \cos(A+B)x$)

$$\begin{aligned}&= \frac{1}{2}\Big[\frac{\sin 2(n-m)x}{2(n-m)}\Big]_0^{\pi/2} - \frac{1}{2}\Big[\frac{\sin 2(n+m+1)x}{2(n+m+1)}\Big]_0^{\pi/2}\\ &= 0 - 0 = 0\,.\end{aligned}$$

For $n = m$ we get

$$\int_0^{\pi/2} \sin^2(2n+1)x\,dx = \int_0^{\pi/2} \Big(\frac{1}{2} - \frac{1}{2}\cos 2(2n+1)x\Big)\,dx$$

(using the identity $2\sin^2 Ax = 1 - \cos 2Ax$)

$$\begin{aligned}&= \frac{1}{2}\int_0^{\pi/2} dx - \frac{1}{2}\int_0^{\pi/2} \cos 2(2n+1)x\,dx\\ &= \Big[\frac{1}{2}x\Big]_0^{\pi/2} - \frac{1}{2}\Big[\frac{1}{2(2n+1)}\sin 2(2n+1)x\Big]_0^{\pi/2}\\ &= \frac{\pi}{4}\,.\end{aligned}$$

Hence the given system is orthogonal but not orthonormal. However, if we multiply each function by $2/\sqrt{\pi}$, the new system obtained in this way is orthonormal, that is,

$$\Big\{\frac{2\sin x}{\sqrt{\pi}}, \frac{2\sin 3x}{\sqrt{\pi}}, 2\frac{\sin 5x}{\sqrt{\pi}}, \ldots, \frac{2\sin(2n+1)x}{\sqrt{\pi}}, \cdots,\Big\}$$

is an orthonormal system on $[0, \pi/2]$.

(ii) Set $f_n(x) = \sin nx$. For $m \neq n$ we have

$$\langle f_n, f_m\rangle = \int_0^{\pi} \sin nx \sin mx\,dx = \frac{1}{2}\int_0^{\pi} \Big[\cos(n-m)x - \cos(n+m)x\Big]\,dx$$

(by the trigonometric identity $2\sin Ax \sin Bx = \cos(A-B)x - \cos(A+B)x$)

$$\begin{aligned}&= \frac{1}{2}\int_0^{\pi} [\cos(n-m)x]\,dx - \frac{1}{2}\int_0^{\pi} [\cos(n+m)x]\,dx\\ &= \frac{1}{2}\Big[\frac{\sin(n-m)x}{(n-m)}\Big]_0^{\pi} - \frac{1}{2}\Big[\frac{\sin(n+m)x}{(n+m)}\Big]_0^{\pi} = 0\,.\end{aligned}$$

For $m = n$ we have

$$\begin{aligned}\langle f_n, f_n\rangle &= \int_0^{\pi} \sin^2 nx\,dx = \int_0^{\pi} \Big(\frac{1}{2} - \frac{1}{2}\cos 2nx\Big)\,dx\\ &= \frac{1}{2}\pi - \frac{1}{4n}\Big[\sin 2nx\Big]_0^{\pi} = \frac{\pi}{2}\,,\end{aligned}$$

where again we used $2\sin^2 Ax = 1-\cos 2Ax$. Hence $\{\sin nx\}$ is an orthogonal, but not an orthonormal system on $[0,\pi]$. But the system $\left\{\dfrac{\sqrt{2}\sin nx}{\sqrt{\pi}}\right\}$ is an orthonormal system on $[0,\pi]$.

(iii) For $m \neq n$ we have

$$\int_0^l \cos\frac{n\pi x}{l}\cos\frac{m\pi x}{l}\,dx = \frac{1}{2}\int_0^l \left[\cos\frac{(n-m)\pi x}{l} + \cos\frac{(n+m)\pi x}{l}\right] dx$$

(by the trigonometric identity $2\cos Ax\cos Bx = \cos(A-B)x + \cos(A+B)x$)

$$\begin{aligned}
&= \frac{1}{2}\int_0^l \cos\frac{(n-m)\pi x}{l}\,dx + \frac{1}{2}\int_0^l \cos\frac{(n+m)\pi x}{l}\,dx \\
&= \frac{l}{2(n-m)\pi}\left[\sin\frac{(n-m)\pi x}{l}\right]_0^l + \frac{l}{2(n+m)\pi}\left[\sin\frac{(n+m)\pi x}{l}\right]_0^l \\
&= 0 + 0 = 0\,.
\end{aligned}$$

For $m = n$ we have

$$\begin{aligned}
\int_0^l \cos^2\frac{n\pi x}{l}\,dx &= \int_0^l \left(\frac{1}{2} + \frac{1}{2}\cos\frac{2n\pi x}{l}\right) dx = \left[\frac{1}{2}\,x\right]_0^l + \frac{l}{4n\pi}\left[\sin\frac{2n\pi x}{l}\right]_0^l \\
&= \frac{l}{2}\,.
\end{aligned}$$

Moreover,

$$\int_0^l 1\cdot\cos\frac{n\pi x}{l}\,dx = \left[\frac{l}{n\pi}\sin\frac{n\pi x}{l}\right]_0^l = 0\,, \qquad \int_0^l 1^2 dx = l\,.$$

The given system is orthogonal but not orthonormal. However, the system $\left\{\dfrac{1}{\sqrt{l}}, \dfrac{\sqrt{2}}{\sqrt{l}}\dfrac{\cos n\pi x}{l}\right\}$ is orthonormal.

(iv) We know already from part (iii) that

$$\int_{-l}^l \cos\frac{n\pi x}{l}\cos\frac{m\pi x}{l}\,dx = 2\int_0^l \cos\frac{n\pi x}{l}\cos\frac{m\pi x}{l}\,dx = 0\,.$$

Concerning the sine, for $m \neq n$ we get

$$\int_{-l}^l \sin\frac{n\pi x}{l}\sin\frac{m\pi x}{l}\,dx = 2\int_0^l \sin\frac{n\pi x}{l}\sin\frac{m\pi x}{l}\,dx$$

(by the identity $2\sin Ax\sin Bx = \cos(A-B)x - \cos(A+B)x$)

$$\begin{aligned}
&= \int_0^l \left[\cos\frac{(n-m)\pi x}{l} - \cos\frac{(n+m)\pi x}{l}\right] dx \\
&= \left[\sin\frac{(n-m)\pi x}{l}\cdot\frac{l}{n-m}\right]_0^l - \left[\sin\frac{(n+m)\pi x}{l}\cdot\frac{l}{n+m}\right]_0^l \\
&= 0 - 0 = 0\,.
\end{aligned}$$

Using the same identities as in the previous computations one derives that

$$\int_{-l}^{l} \sin^2 \frac{n\pi x}{l}\, dx = l\,, \quad \int_{-l}^{l} \cos^2 \frac{n\pi x}{l}\, dx = l\,,$$
$$\int_{-l}^{l} \sin \frac{n\pi x}{l} \cos \frac{m\pi x}{l}\, dx = 0\,, \quad \text{for } n \neq m,$$

moreover

$$\int_{-l}^{l} 1 \cdot \cos \frac{n\pi x}{l}\, dx = 0\,, \quad \int_{-l}^{l} 1 \cdot \sin \frac{n\pi x}{l}\, dx = 0\,, \quad \int_{-l}^{l} 1^2\, dx = 2l\,.$$

The given system is orthogonal but not orthonormal. However, the system $\left\{ \frac{1}{\sqrt{l}}, \frac{1}{\sqrt{l}} \frac{\cos n\pi x}{l}, \frac{1}{\sqrt{l}} \frac{\sin n\pi x}{l} \right\}$ is orthonormal.

Example 223. (a) Haar systems The Haar system $\{h_n\}$ is defined as follows:
Set $h_0 = 1$. For $n,\ k \in N$ with $0 \leq k < 2^n$ define h_n on [0,1] by

$$h_{2^n+k}(x) = \begin{cases} 2^{n/2}, & \text{if } x \in I(2k, n+1) \\ -2^{n/2}, & \text{if } x \in I(2k+1, n+1) \\ 0, & \text{otherwise} \end{cases}$$

$$h_{2^n+k}(x+1) = h_{2^n+k}(x).$$

Each Haar function is continuous from the right and the Haar system $\{h_n\}$ is orthonormal on [0,1).
(b) Walsh systems Let N denote the set of non-negative integers, P denote the set of positive integers, R denote the set of real numbers, C denote the set of complex numbers and Q denote the set of rationals in the unit interval [0,1). Thus each dyadic element of Q has the form $\frac{p}{2^n}$ for some $p, n \in N,\ 0 \leq p < 2^n$.

During the discussion we will use the notation $x = (x_\alpha, \alpha \in A)$ to represent a collection x indexed by a set. Let v be a function defined on [0,1] by

$$r(x) = \begin{cases} 1, & 0 \leq x < 1/2 \\ -1, & 1/2 \leq x < 1 \end{cases}, \tag{4.5}$$

$r(x+1) = r(x)$ (extended to R by periodicity 1).
The Rademacher system $r = (r_n,\ n \in N)$ is defined by

$$r_n(x) = r(2^n x), \quad x \in R, n \in N. \tag{4.6}$$

Given $n \in N$ it is possible to write

$$\sum_{k=0}^{\infty} n_k 2^k \tag{4.7}$$

where $n_k = 0$ or 1 for $k \in N$. The expression will be called the binary expansion of n and the numbers n_k will be called the binary coefficients of n.

There are three types of Walsh functions, namely the original Walsh system, Walsh-Kaczmarz system and Walsh-Paley system. We discuss here the Walsh-Paley system invented in 1932 and we will call it the Walsh system. This contains the Rademacher system as a special case and enjoys many properties analogous to the classical trigonometric, Sturm-Liouville and Legendre systems. Let $n \in N$ have binary coefficients, n_k, $k \in N$, then

$$w_n = \prod_{k=0}^{\infty} (r_k)^{n_k}. \tag{4.8}$$

We observe that the product is always finite because $n_k = 0$ for k sufficiently large and by definition $w_0 = 1$ and $w_{2^n} = r_n$ for $n \in N$. It is also clear that the Walsh system is closed under the finite products and every Walsh function is piecewise constant with finitely many jump discontinuities on [0,1), and takes values +1 and −1. It may be noted that the Walsh system is complete and orthonormal. Every function $f(x)$ which is periodic with period 1 and Lebesgue integrable on [0,1] can be expanded in a Walsh-Fourier series.

$$f(x) \sim \sum_{n=0}^{\infty} c_n w^n(x),$$

where $c_n = \int_0^1 f(x) w_n(x) dx, \ n = 0, 1, 2.....$
Many properties of Walsh-Fourier series are analogous to trignometric Fourier series, for example, the Dirichlet kernels for the Walsh system and trigonometric system have the same order of growth. The theory of Walsh-Fourier series is quite rich. It has been shown that Walsh analysis is a special case of the harmonic analysis on compact Abelian groups.

Remark 48. The systems named after Rademacher, Haar and Walsh are other well-known orthonormal systems.

A function f is said to be a **linear combination** of the functions f_1, $f_2, \ldots, f_n$ if

$$f = \alpha_1 f_1 + \alpha_2 f_2 + \ldots + \alpha_n f_n$$

holds for suitably chosen scalars $\alpha_1, \alpha_2, \ldots, \alpha_n$.

Definition 78. (Linear dependence and independence)
A system $\{f_1, f_2, \cdots, f_n\}$ of functions is said to be **linearly independent** if $\alpha_1 f_1 + \alpha_2 f_2 + \alpha_3 f_3 + \cdots + \alpha_n f_n = 0$ implies that $\alpha_1 = \alpha_2 = \alpha_3 = \cdots = \alpha_n = 0$. The system is called **linearly dependent** if it is not linearly independent. In other words, at least one element of the system is a linear combination of the remaining $n-1$ elements. An infinite system $\{f_1, f_2, \ldots\}$ of functions is said to be linearly independent, if every finite subset taken from it forms a linearly independent system in the sense above, and it is said to be linearly dependent if this is not the case.

Remark 49. Every orthogonal (and, hence, every orthonormal) system is linearly independent. However, the converse is not true, so a system may be linearly independent without being orthogonal.

4.2.2 Fourier Series

A series of the form

$$\frac{1}{2}a_0 + (a_1 \cos x + b_1 \sin x) + (a_2 \cos 2x + b_2 \sin 2x) + \cdots$$

that is,

$$\frac{1}{2}a_0 + \sum_{n=1}^{\infty}(a_n \cos nx + b_n \sin nx) \tag{4.9}$$

is called an (infinite) **trigonometric series.**

Let f be a piecewise continuous function defined on the interval $[-\pi, \pi]$. The numbers

$$a_0 = \frac{1}{\pi}\int_{-\pi}^{\pi} f(x)\, dx\,, \tag{4.10}$$

$$a_n = \frac{1}{\pi}\int_{-\pi}^{\pi} f(x) \cos nx\, dx\,, \quad b_n = \frac{1}{\pi}\int_{-\pi}^{\pi} f(x) \sin nx\, dx\,, \tag{4.11}$$

where $n = 1, 2, 3, \ldots$, are called the **Fourier coefficients** (more precisely, the Fourier cosine resp. sine coefficients) of f on the interval $[-\pi, \pi]$, and the series (4.9) with the coefficients from (4.10) and (4.11) is called the **Fourier series** of f on this interval. Note that $a_0/2$ is just the average of the function f over $[-\pi, \pi]$.

The Fourier coefficients arise when we want to represent a function $f : [-\pi, \pi] \to \mathbb{R}$ as a trigonometric series,

$$f(x) = \frac{1}{2}a_0 + \sum_{n=1}^{\infty}(a_n \cos nx + b_n \sin nx)\,, \tag{4.12}$$

for the following reason. For an arbitrary trigonometric series, we define the partial sums

$$s_N(x) = \frac{1}{2}a_0 + \sum_{n=1}^{N}(a_n \cos nx + b_n \sin nx)\,. \tag{4.13}$$

(We remind the reader that we have studied series and their partial sums in Chapter 5.) Now let us compute, for $N \geq n \geq 1$, using the orthogonality of the system $\{\cos nx, \sin nx : n = 1, 2, \dots\}$ on $[-\pi, \pi]$,

$$\begin{aligned}\int_{-\pi}^{\pi} s_N(x)\cos nx\,dx &= \frac{1}{2}a_0\int_{-\pi}^{\pi}\cos nx\,dx + \sum_{k=1}^{N} a_k\int_{-\pi}^{\pi}\cos nx\cos kx\,dx \\ &\quad + \sum_{k=1}^{N} b_k\int_{-\pi}^{\pi}\cos nx\sin kx\,dx \qquad (4.14)\\ &= 0 + a_n\pi + 0 = \pi a_n\,.\end{aligned}$$

If f can be represented as in (4.12), that is, if

$$f(x) = \lim_{N\to\infty} s_N(x)\,,$$

and if moreover

$$\int_{-\pi}^{\pi} f(x)\cos nx\,dx = \lim_{N\to\infty}\int_{-\pi}^{\pi} s_N(x)\cos nx\,dx\,,$$

then we see from (4.14) that

$$a_n = \frac{1}{\pi}\int_{-\pi}^{\pi} f(x)\cos nx\,dx\,.$$

Similarly, we get the corresponding formula for b_n by using $\sin nx$ instead of $\cos nx$ in (4.14).

Instead of the interval $[-\pi, \pi]$ we may consider an interval $[-l, l]$, where $l > 0$ is an arbitrary number.

Definition 79. (Fourier series)
Let f be a piecewise continuous function defined on the interval $[-l, l]$, where $l > 0$. Then the Fourier series of f on $[-l, l]$ is given by

$$\frac{a_0}{2} + \sum_{n=1}^{\infty}\left(a_n\cos\frac{n\pi x}{l} + b_n\sin\frac{n\pi x}{l}\right), \quad x \in [-l, l]\,,$$

where

$$a_0 = \frac{1}{l}\int_{-l}^{l} f(x)\,dx\,, \qquad (4.15)$$

$$a_n = \frac{1}{l}\int_{-l}^{l} f(x)\cos\frac{n\pi x}{l}\,dx\,, \quad b_n = \frac{1}{l}\int_{-l}^{l} f(x)\sin\frac{n\pi x}{l}\,dx\,. \qquad (4.16)$$

If the Fourier series at x converges to $f(x)$, we write as usual

$$f(x) = \frac{a_0}{2} + \sum_{n=1}^{\infty}\left(a_n\cos\frac{n\pi x}{l} + b_n\sin\frac{n\pi x}{l}\right). \qquad (4.17)$$

The following theorem gives conditions under which (4.17) holds, that is, when the Fourier series of f on $[-l,l]$ converges to f.

Theorem 45. *(Dirichlet convergence theorem)*
Let f and f' be piecewise continuous on the interval $[-l,l]$, let $x \in (-l,l)$.

(i) If f is continuous at x, the Fourier series of f at x converges to $f(x)$.

(ii) If f is discontinuous at x, the Fourier series of f at x converges to the mean value

$$\frac{f(x+)+f(x-)}{2},$$

where $f(x+)$ and $f(x-)$ denote the right- resp. left-hand limit of f at x.

Note that in fact (i) can be viewed as a special case of (ii), since $f(x) = f(x+) = f(x-)$ whenever f is continuous at x.

Example 224. Expand the following functions in Fourier series.

(a) $$f(x) = \begin{cases} 0\,, & -\pi < x < 0\,, \\ \pi - x\,, & 0 \le x < \pi\,, \end{cases} \qquad \text{on } [-\pi,\pi],$$

(b) $$f(x) = \begin{cases} 0\,, & -\pi < x < 0\,, \\ 1\,, & 0 \le x < \pi\,, \end{cases} \qquad \text{on } [-\pi,\pi],$$

(c) $$f(x) = \begin{cases} 0\,, & -1 < x < 0\,, \\ x\,, & 0 \le x < 1\,, \end{cases} \qquad \text{on } [-1,1],$$

(d) $$f(x) = \begin{cases} 0\,, & -\pi < x < 0\,, \\ \sin x\,, & 0 \le x < \pi\,, \end{cases} \qquad \text{on } [-\pi,\pi],$$

(e) $$f(x) = \begin{cases} -\frac{\pi}{4}\,, & -\pi < x < 0\,, \\ 0\,, & x = 0\,, \\ \frac{\pi}{4}\,, & 0 \le x < \pi\,, \end{cases} \qquad \text{on } [-\pi,\pi],$$

(f) $$f(x) = \begin{cases} -1\,, & -4 < x < 0\,, \\ 1\,, & 0 \le x < 4\,, \end{cases} \qquad \text{on } [-4,4],$$

(g) $$f(x) = e^x\,, \qquad \text{on } [-\pi,\pi].$$

Solution: (a) Here we have $l = \pi$. We get

$$
\begin{aligned}
a_0 &= \frac{1}{\pi}\int_{-\pi}^{\pi} f(x)\,dx = \frac{1}{\pi}\Big[\int_{-\pi}^{0} f(x)\,dx + \int_{0}^{\pi} f(x)\,dx\Big] \\
&= \frac{1}{\pi}\Big[\int_{-\pi}^{0} 0\,dx + \int_{0}^{\pi} (\pi - x)\,dx\Big] = \frac{1}{\pi}\Big[\pi x - \frac{x^2}{2}\Big]_0^{\pi} = \frac{\pi}{2}. \\
a_n &= \frac{1}{\pi}\int_{-\pi}^{\pi} f(x)\cos nx\,dx = \frac{1}{\pi}\Big[\int_{-\pi}^{0} 0\,dx + \int_{0}^{\pi} (\pi - x)\cos nx\,dx\Big] \\
&= \frac{1}{\pi}\Big[(\pi - x)\frac{\sin nx}{n}\Big]_0^{\pi} + \frac{1}{n}\int_0^{\pi} \sin nx dx\Big] \\
&= \Big[-\frac{1}{n\pi}\frac{\cos nx}{n}\Big]_0^{\pi} = \frac{-\cos n\pi + 1}{n^2\pi} = \frac{1-(-1)^n}{n^2\pi}.
\end{aligned}
$$

Similarly we can calculate that

$$
\begin{aligned}
b_n &= \frac{1}{\pi}\int_{-\pi}^{\pi} f(x)\sin nx\,dx = \frac{1}{\pi}\Big[\int_{-\pi}^{0} f(x)\sin nx\,dx + \int_{0}^{\pi} f(x)\sin nx\,dx\Big] \\
&= \frac{1}{\pi}\int_0^{\pi} (\pi - x)\sin nx\,dx = \frac{1}{n}.
\end{aligned}
$$

Since the given function is continuous except at $x = 0$, the Fourier series converges. See the graphs of the function and of a partial sum of its Fourier series in Figures 4.1a and 4.1b respectively.

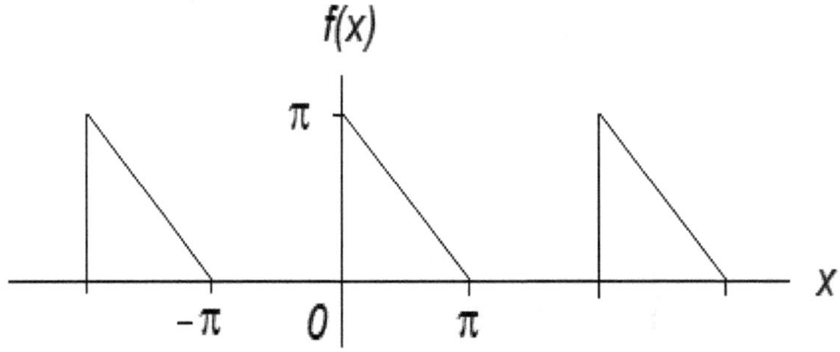

FIGURE 4.1a: Function

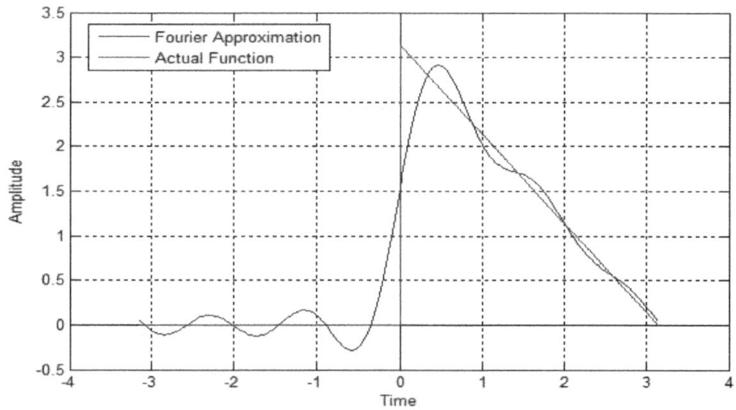

FIGURE 4.1b: Fourier Approximation Plot for 5 Harmonics

(b) We have

$$\begin{aligned}
a_0 &= \frac{1}{\pi}\int_{-\pi}^{\pi} f(x)\,dx = \frac{1}{\pi}\Big[\int_{-\pi}^{0} f(x)\,dx + \int_{0}^{\pi} f(x)\,dx\Big] = 0 + \frac{1}{\pi}\int_0^{\pi} 1\,dx \\
&= 1\,, \\
a_n &= \frac{1}{\pi}\int_{-\pi}^{\pi} f(x)\cos nx\,dx = \frac{1}{\pi}\Big[\int_{-\pi}^{0} f(x)\cos nx\,dx + \int_{0}^{\pi} f(x)\cos nx\,dx\Big] \\
&= 0 + \int_0^{\pi} 1\cos nx\,dx = \Big[\frac{\sin nx}{n}\Big]_0^{\pi} = 0\,, \\
b_n &= \frac{1}{\pi}\int_{-\pi}^{\pi} f(x)\sin nx\,dx = \frac{1}{\pi}\Big[\int_{-\pi}^{0} f(x)\sin nx\,dx + \int_{0}^{\pi} f(x)\sin nx\,dx\Big] \\
&= 0 + \frac{1}{\pi}\int_0^{\pi} \sin nx\,dx = \frac{1}{\pi}\Big[-\frac{\cos nx}{n}\Big]_0^{\pi} \\
&= \frac{1}{\pi}\Big[-\frac{\cos n\pi}{n} + \frac{1}{n}\Big] \\
&= -\frac{(-1)^n}{n\pi} + \frac{1}{n\pi}\,, \qquad \text{as } \cos n\pi = (-1)^n, \\
&= \frac{1}{n\pi}(1 - (-1)^n)\,.
\end{aligned}$$

Since the given function is continuous except at $x = 0$, the Fourier series converges to the given function for $x \neq 0$, and we have

$$f(x) = \frac{1}{2} + \frac{1}{\pi}\sum_{n=1}^{\infty} \frac{1-(-1)^n}{n} \sin nx\,.$$

For the graphs of f and of a partial sum of its Fourier series see Figures 4.2a and 4.2b respectively.

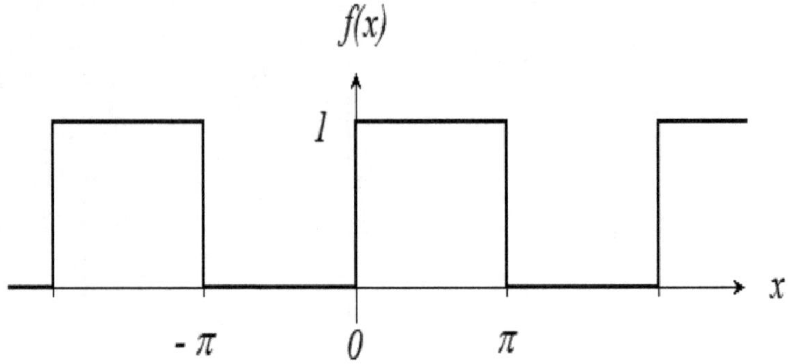

FIGURE 4.2a: Function

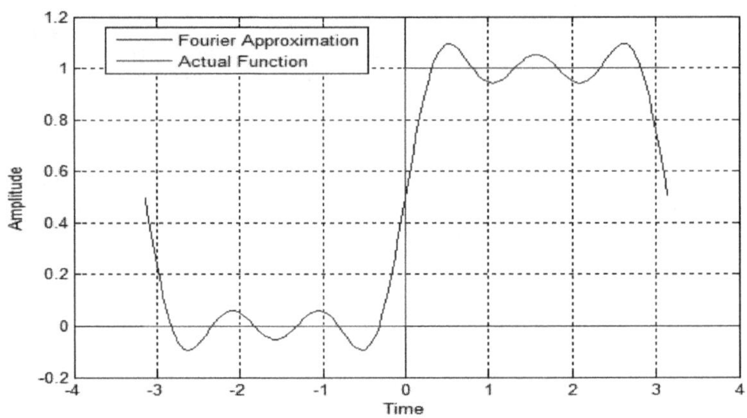

FIGURE 4.2b: Fourier Approximation Plot for 5 Harmonics

(c) Here $l = 1$. The given function is continuous also at $x = 0$, hence its

Fourier series converges everywhere. We get

$$a_0 = \int_{-1}^{1} f(x)\,dx = \frac{1}{\pi}\Big[\int_{-1}^{0} f(x)\,dx + \int_{0}^{1} f(x)\,dx\Big] = \Big[\int_{-1}^{0} 0\,dx + \int_{0}^{1} x\,dx\Big]$$
$$= 0 + \int_{0}^{1} x\,dx = \Big[\frac{1}{2}x^2\Big]_0^1 = \Big[\frac{1}{2}\cdot 1^2 - 0\Big] = \frac{1}{2},$$
$$a_n = \frac{1}{1}\int_{-1}^{1} f(x)\cos nx\,dx = \int_{-1}^{0} \cos n\pi x\,dx + \int_{0}^{1} f(x)\cos n\pi x\,dx$$
$$= \int_{-1}^{0} 0\cos n\pi x\,dx + \int_{0}^{1} x\cos n\pi x\,dx$$
$$= 0 + \Big[\frac{x\sin n\pi x}{n}\pi\Big]_0^1 - \int_{0}^{1} \frac{\sin n\pi x}{n\pi}\,dx$$
$$= \Big[\frac{\sin n\pi}{n\pi} - 0\Big] + \Big[\frac{\cos n\pi x}{\pi^2 n^2}\Big]_0^1 = 0 + \Big[\frac{\cos n\pi}{n^2\pi^2} - \frac{\cos n\pi 0}{\pi^2 n^2}\Big]$$
$$= \frac{(-1)^n}{\pi^2 n^2} - \frac{1}{n^2\pi^2} = \frac{1}{n^2\pi^2}[(-1)^n - 1]\,.$$

Similarly

$$b_n = \int_{-1}^{1} f(x)\sin n\pi x\,dx = \int_{-1}^{0} f(x)\sin n\pi x\,dx + \int_{0}^{1} f(x)\sin n\pi x\,dx$$
$$= 0 + \int_{0}^{1} x\sin n\pi x\,dx = \frac{(-1)^{n+1}}{n\pi},$$

through integration by parts. See Figures 4.3a and 4.3b for the graphs of the function and of a partial sum of its Fourier series.

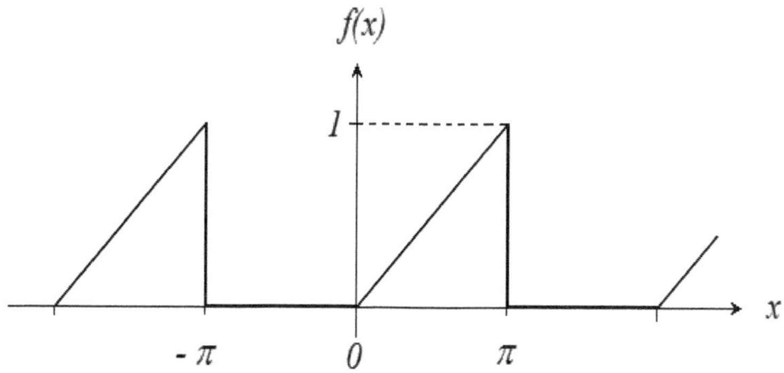

FIGURE 4.3a: Function

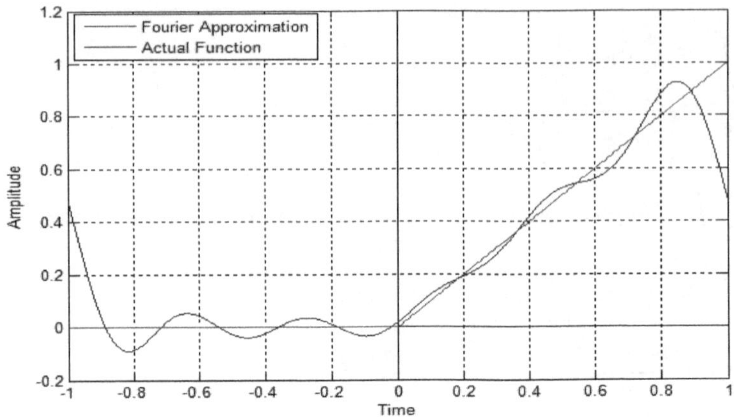

FIGURE 4.3b: Fourier Approximation Plot for 5 Harmonics

(d) We have

$$\begin{aligned} a_0 &= \frac{1}{\pi}\int_{-\pi}^{\pi} f(x)\,dx = \frac{1}{\pi}\int_{-\pi}^{0} f(x)\,dx + \frac{1}{\pi}\int_{0}^{\pi} f(x)\,dx \\ &= 0 + \frac{1}{\pi}\int_{0}^{\pi} \sin x\,dx = \frac{1}{\pi}[-\cos x]_0^{\pi} = \frac{2}{\pi}\,. \end{aligned}$$

$$\begin{aligned} a_n &= \frac{1}{\pi}\int_{-\pi}^{\pi} f(x)\cos nx\,dx = \Big[\int_{-\pi}^{0} f(x)\cos nx\,dx + \int_{0}^{\pi} f(x)\cos nx\,dx\Big] \\ &= \frac{1}{\pi}\int_{0}^{\pi} \sin x \cos nx\,dx \\ &= \frac{1}{2\pi}\int_{0}^{\pi} [\sin(n+1)x + \sin(1-n)x]dx \end{aligned}$$

(by the identity $2\sin Ax\cos Bx = \sin(A+B)x + \sin(A-B)x$)

$$\begin{aligned} &= \frac{1}{2\pi}\Big[\frac{-\cos(n+1)x}{n+1}\Big]_0^{\pi} + \frac{1}{2\pi}\Big[\frac{-\cos(n-1)x}{1-n}\Big]_0^{\pi} \\ &= \frac{1}{2\pi}\Big[\frac{-\cos(n+1)\pi}{n+1} + \frac{1}{n+1}\Big] + \frac{1}{2\pi}\Big[\frac{-\cos(1-n)\pi}{1-n} + \frac{1}{1-n}\Big] \\ &= \frac{1+(-1)^n}{\pi(1-n^2)}, \quad \text{for } n = 2, 3, 4, \ldots \end{aligned}$$

$$a_1 = \frac{1}{2\pi}\int_0^{\pi} \sin 2x\,dx = 0\,.$$

$$\begin{aligned}
b_n &= \frac{1}{\pi}\int_{-\pi}^{\pi} f(x)\sin nx\,dx = \frac{1}{\pi}\int_0^{\pi} \sin x \sin nx\,dx \\
&= \frac{1}{2\pi}\int_0^{\pi} \cos(1-n)x - \cos(1+n)x]\,dx \\
&= 0\,, \quad \text{for } n = 2, 3, 4, \ldots
\end{aligned}$$

$$b_1 = \frac{1}{2\pi}\int_0^{\pi} (1-\cos 2x)\,dx = \frac{1}{2}\,.$$

Thus,

$$f(x) = \frac{1}{\pi} + \frac{1}{2}\sin x + \sum_{n=2}^{\infty} \frac{1+(-1)^n}{\pi(1-n^2)}\cos nx$$

is the Fourier series for f. See the graph of the function in Figure 4.4a and the graph of a partial sum of its Fourier series in Figure 4.4b.

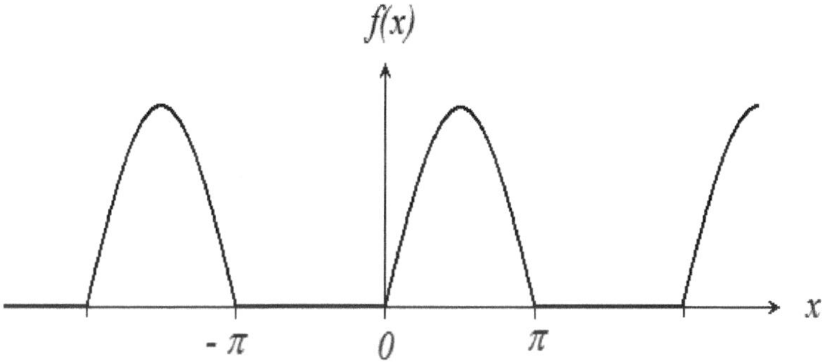

FIGURE 4.4a: Function

(e) We have

$$\begin{aligned}
a_0 &= \frac{1}{\pi}\int_{-\pi}^{\pi} f(x)\,dx = \frac{1}{\pi}\int_{-\pi}^{0} f(x)\,dx + \frac{1}{\pi}\int_{-\pi}^{\pi} f(x)\,dx \\
&= \frac{1}{\pi}\int_{-\pi}^{0} -\frac{\pi}{4}\,dx + \frac{1}{\pi}\int_0^{\pi} \frac{\pi}{4}\,dx - \frac{\pi}{4} + \frac{\pi}{4} = 0\,.
\end{aligned}$$

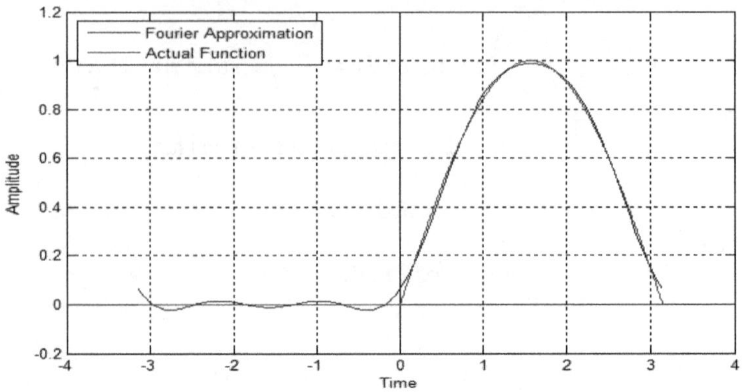

FIGURE 4.4b: Fourier Approximation Plot for 5 Harmonics

$$\begin{aligned}
a_n &= \frac{1}{\pi}\int_{-\pi}^{\pi} f(x)\cos nx\,dx \\
&= \frac{1}{\pi}\int_{-\pi}^{0} f(x)\cos nx\,dx + \frac{1}{\pi}\int_{0}^{\pi} f(x)\cos nx\,dx \\
&= -\frac{1}{\pi}\frac{\pi}{4}\int_{-\pi}^{0} \cos nx\,dx + \frac{1}{\pi}\frac{\pi}{4}\int_{0}^{\pi} \cos nx\,dx \\
&= -\frac{1}{4}\left[\frac{\sin nx}{n}\right]_{-\pi}^{0} + \frac{1}{4}\left[\frac{\sin nx}{n}\right]_{0}^{\pi} = 0\,.
\end{aligned}$$

$$\begin{aligned}
b_n &= \frac{1}{\pi}\int_{-\pi}^{\pi} f(x)\sin nx\,dx \\
&= \frac{1}{\pi}\int_{-\pi}^{0} f(x)\sin nx\,dx + \frac{1}{\pi}\int_{0}^{\pi} f(x)\sin nx\,dx \\
&= \frac{1}{\pi}\int_{-\pi}^{0} -\frac{\pi}{4}\sin nx\,dx + \frac{1}{\pi}\int_{0}^{\pi} \frac{\pi}{4}\sin nx\,dx \\
&= \frac{1}{4}\left[\frac{\cos nx}{n}\right]_{-\pi}^{0} - \frac{1}{4}\left[\frac{\cos nx}{n}\right]_{0}^{\pi} \\
&= \frac{1}{4n} - \frac{\cos n\pi}{4n} - \frac{1}{4n}\cos n\pi + \frac{1}{4n} = \frac{2}{4n} - \frac{2(-1)^n}{4n}\,.
\end{aligned}$$

Thus, the Fourier series for f is

$$f(x) = \sum_{n=1}^{\infty} \frac{1}{2n-1}\sin(2n-1)x\,.$$

The graph of this function and of a partial sum of its Fourier series are given in Figures 4.5a and 4.5b respectively.

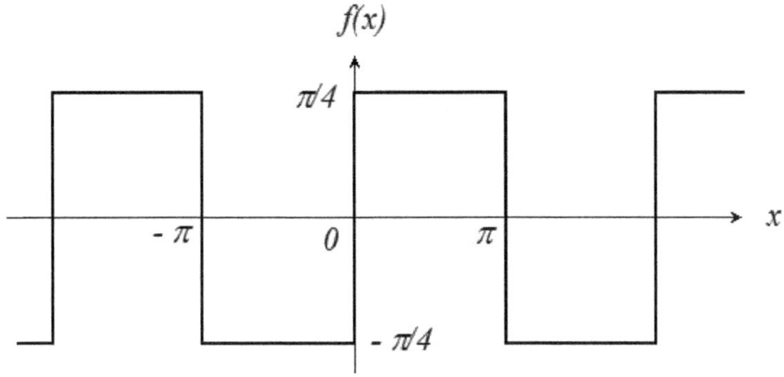

FIGURE 4.5a: Function

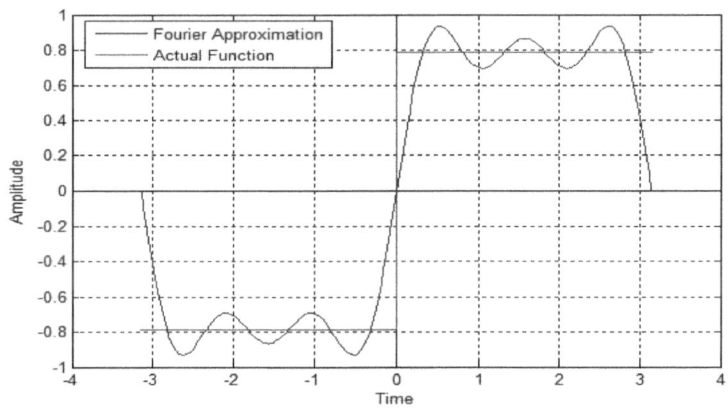

FIGURE 4.5b: Fourier Approximation Plot for 5 Harmonics

(f) We have

$$a_0 = \frac{1}{4}\int_{-4}^{4} f(x)\,dx = \frac{1}{4}\int_{-4}^{0} f(x)\,dx + \frac{1}{4}\int_{0}^{4} f(x)\,dx = \frac{1}{4}[-x]_{-4}^{0} + \frac{1}{4}[x]_0^4$$
$$= -1 + 1 = 0\,.$$
$$a_n = -\frac{1}{4}\int_{-4}^{0} \cos\frac{n\pi x}{4}\,dx + \frac{1}{4}\int_{0}^{4} \cos\frac{n\pi x}{4}\,dx$$
$$= -\frac{1}{4}\Big[\frac{4}{n\pi}\sin\frac{n\pi x}{4}\Big]_{-4}^{0} + \frac{1}{4}\Big[\frac{4}{n\pi}\sin\frac{n\pi x}{4}\Big]_{0}^{4} = 0\,.$$

$$b_n = \frac{1}{4}\int_{-4}^{0} f(x)\sin\frac{n\pi x}{4}\,dx + \frac{1}{4}\int_{0}^{4} f(x)\sin\frac{n\pi x}{4}\,dx$$
$$= -\frac{1}{4}\int_{-4}^{0} \sin\frac{n\pi x}{4}\,dx + \frac{1}{4}\int_{0}^{4} \sin\frac{n\pi x}{4}\,dx = \frac{1}{4}\Big[\frac{4}{n\pi}\cos\frac{n\pi x}{4}\Big]_{-4}^{0}$$
$$-\frac{1}{4}\Big[\frac{4}{n\pi}\cos\frac{n\pi x}{4}\Big]_{0}^{4} = \frac{1}{n\pi}[1-(-1)^n] - \frac{1}{n\pi}[(-1)^n - 1]$$
$$= \frac{2}{n\pi} - \frac{2(-1)^n}{n\pi} = \frac{4}{\pi}\frac{1}{2n-1}\,.$$

Therefore, the Fourier series of f is

$$f(x) = \frac{4}{\pi}\sum_{n=1}^{\infty} \frac{1}{2n-1}\sin\Big[\frac{(2n-1)\pi x}{4}\Big]\,.$$

The graph of this function and of a partial sum of its Fourier series are given in Figures 4.6a and 4.6b.

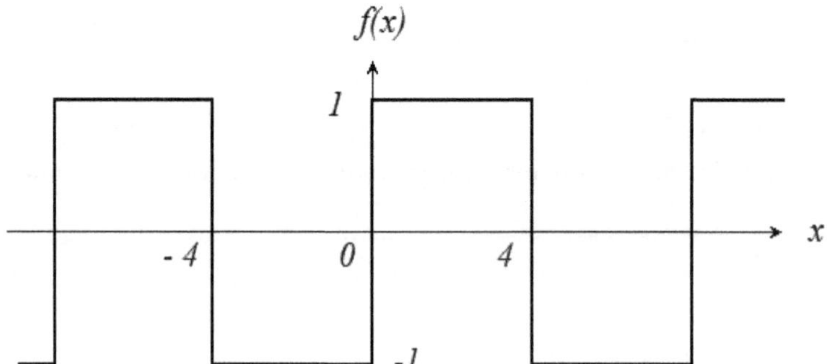

FIGURE 4.6a: Function

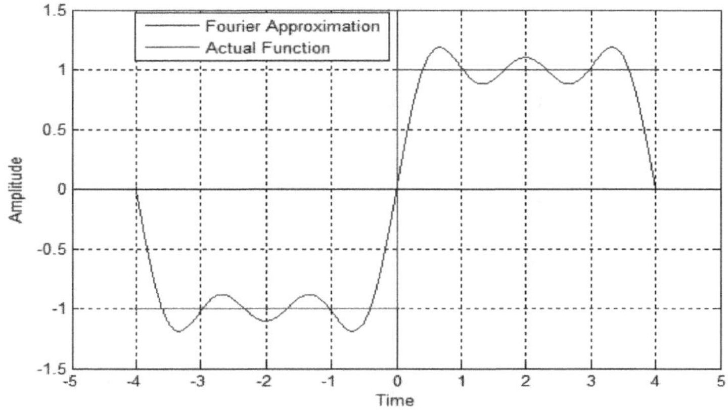

FIGURE 4.6b: Fourier Approximation Plot for 5 Harmonics

(g) $a_0 = \dfrac{1}{\pi}\displaystyle\int_{-\pi}^{\pi} f(x)\,dx = \dfrac{1}{\pi}\int_{-\pi}^{\pi} e^x\,dx = \dfrac{1}{\pi}(e^{\pi} - e^{-\pi})$,

$$a_n = \frac{1}{\pi}\int_{-\pi}^{\pi} f(x)\cos nx\,dx = \frac{1}{\pi}\int_{-\pi}^{\pi} e^x \cos nx\,dx = \frac{(-1)^n(e^{\pi} - e^{-\pi})}{\pi(1+n^2)},$$

$$b_n = \frac{1}{\pi}\int_{-\pi}^{\pi} f(x)\sin nx\,dx = \frac{1}{\pi}\int_{-\pi}^{\pi} e^x \sin nx\,dx = \frac{(-1)^n(e^{-\pi} - e^{\pi})}{\pi(1+n^2)}.$$

The Fourier series of f is

$$f(x) = \frac{e^{\pi} - e^{-\pi}}{2\pi} + \sum_{n=1}^{\infty}\Big[\frac{(-1)^n(e^{\pi} - e^{-\pi})}{\pi(1+n^2)}\cos nx + \frac{(-1)^n(e^{-\pi} - e^{\pi})}{\pi(1+n^2)}\sin nx\Big].$$

See Figures 4.7a and 4.7b for the graph of the function and of a partial sum of its Fourier series.

4.2.3 Further Properties of Fourier Series

Periodic Extension

Let us recall the notion of a periodic function given in Chapter 1. A real function of a single variable is called periodic with period p if $f(x+p) = f(x)$ for all x. For example, 8π is a period of the sine function as $\sin(x+8\pi) = \sin x$ for all x. The smallest value of p for which $f(x+p) = f(x)$ holds for all x is called the fundamental period. For example, $p = 2\pi$ is the fundamental period of the sine function as 2π is the smallest value of p which satisfies

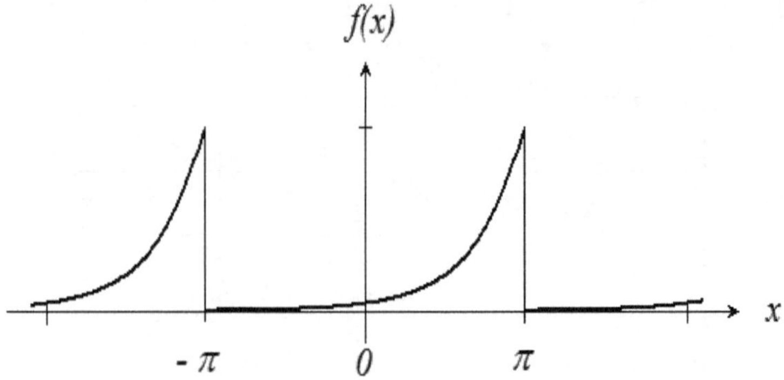

FIGURE 4.7a: Function

$f(x+p) = f(x)$ for all x. Let us point out that often "period" is defined to be the fundamental period.

Let f be an arbitrary function defined on $(-l, l)$. Its Fourier series (if convergent)

$$\frac{1}{2}a_0 + \sum_{n=1}^{\infty}\left(a_n \cos\frac{n\pi x}{l} + b_n \sin\frac{n\pi x}{l}\right)$$

is a periodic function of x of period $p = 2l$ and thus (if convergent to f) not only represents f on $(-l, l)$, but also gives the periodic extension of f on the real line.

Approximation by Partial Sums

Let

$$s_N(x) = \frac{1}{2}a_0 + \sum_{n=1}^{N}\left(a_n \cos nx + b_n \sin nx\right)$$

denote the Nth partial sum of the Fourier series of a function f defined on $[-\pi, \pi]$. One may ask how well s_N approximates f. We present without proof the following two results. If f is continuous on $[-\pi, \pi]$ and $f(-\pi) = f(\pi)$, then

$$|f(x) - s_N(x)| \leq \frac{1}{\sqrt{N}}\frac{1}{\sqrt{\pi}}\sqrt{\int_{-\pi}^{\pi}|f'(t)|^2 dt}, \quad \text{for all } x.$$

If f is continuous, piecewise differentiable and satisfies $f(-\pi) = f(\pi)$, then

$$|f(x) - s_N(x)| \leq \sum_{n=N}^{\infty}\left(|a_n| + |b_n|\right), \quad \text{for all } x.$$

Gibbs Phenomenon

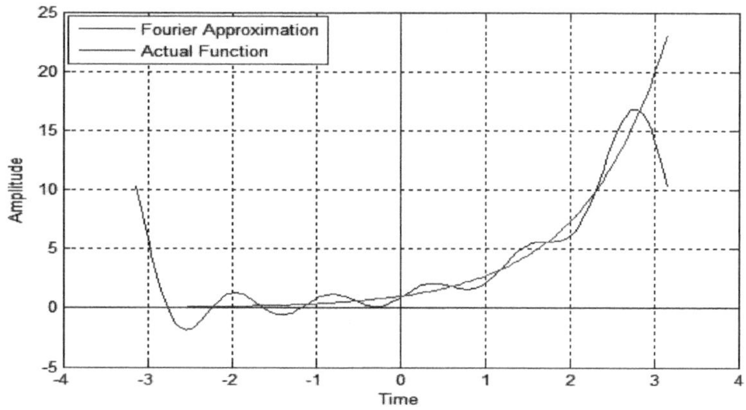

FIGURE 4.7b: Fourier Approximation Plot for 5 Harmonics

If we examine the graphs of the partial sums of the Fourier series in Example 222, we observe that all of them are overshooting the true values of the function f near its point of discontinuity. In fact, this phenomenon always occurs when we approximate a discontinuous function with Fourier series. It is known as the **Gibbs phenomenon**, in honor of Josiah William Gibbs, a mathematical physicist working at Yale, who analyzed it prior to 1900, after it had been discovered by Henry Wilbraham in 1848. One can show that the overshooting amounts to approximately 9% of the size of the jump, that is, of the difference $|f(x+) - f(x-)|$ of the one-sided limits at the discontinuity point x. The main point is that the amount of overshooting of the partial sums s_N does not decrease when N tends to infinity.

As an example, let us consider the Fourier series of the step function

$$f(x) = \begin{cases} -1\,, & \text{if } x \in [-\pi, 0), \\ 0\,, & \text{if } x = 0, \\ 1\,, & \text{if } x \in (0, \pi], \end{cases}$$

extended to a 2π-periodic function on $\mathbb{R}$. Proceeding along the lines of the solution of Example 224(e) we obtain

$$f(x) = \frac{4}{\pi} \sum_{x=1}^{\infty} \frac{1}{2n-1} \sin(2n-1)x\,.$$

The graph of s_{15} is given in Figure 4.8, and the graph of s_{100} is given in Figure 4.9. We observe the overshooting of the partial sum around the points where f is discontinuous.

Complex Form of Fourier Series

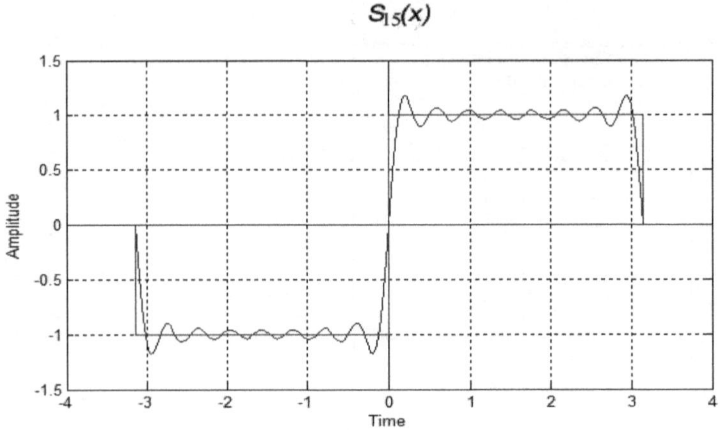

FIGURE 4.8: Fourier Approximation Plot for 15 Harmonics for Square Function

For this paragraph, we assume the reader to have some basic familiarity with complex numbers. The sine and cosine are related to the complex exponential function by

$$e^{i\theta} = \cos\theta + i\sin\theta\,, \quad e^{-i\theta} = \cos\theta - i\sin\theta\,,$$
$$\cos\theta = \frac{e^{i\theta} + e^{-i\theta}}{2}\,, \quad \sin\theta = \frac{e^{i\theta} - e^{-i\theta}}{2i}\,,$$

where θ is any real number. For a real-valued function f with domain $(-\pi, \pi)$, we define the **complex Fourier coefficients** by

$$c_n = \frac{1}{2\pi}\int_{-\pi}^{\pi} f(x)e^{-inx}\,dx\,, \tag{4.18}$$

where n is any integer $0, \pm 1, \pm 2, \ldots$. Since

$$\begin{aligned}
\int_{-\pi}^{\pi} f(x)e^{-inx}\,dx &= \int_{-\pi}^{\pi} f(x)(\cos nx - i\sin nx)\,dx \\
&= \int_{-\pi}^{\pi} f(x)\cos nx\,dx - i\int_{-\pi}^{\pi} f(x)\sin nx\,dx \\
&= \pi a_n - i\pi b_n\,, \quad n > 0\,,
\end{aligned}$$

and, analogously,

$$\int_{-\pi}^{\pi} f(x)e^{inx}\,dx = \pi a_n + i\pi b_n\,, \quad n > 0\,,$$

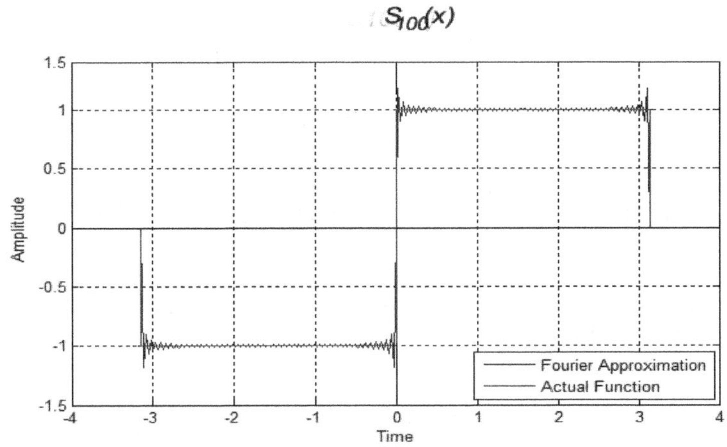

FIGURE 4.9: Fourier Approximation Plot for 100 Harmonics for Square Function

we see that the complex Fourier coefficients are related to the Fourier sine and cosine coefficients by

$$\begin{aligned} c_n &= \frac{1}{2}(a_n - ib_n)\,, \quad c_{-n} = \frac{1}{2}(a_n + ib_n)\,, \\ a_n &= c_n + c_{-n}\,, \quad b_n = i(c_n - c_{-n})\,, \end{aligned} \tag{4.19}$$

for any $n > 0$. Moreover,

$$c_0 = \frac{1}{2\pi}\int_{-\pi}^{\pi} f(x)\,dx = \frac{a_0}{2}\,. \tag{4.20}$$

From (4.19) we obtain

$$\begin{aligned} c_n e^{inx} + c_{-n} e^{-inx} &= c_n(\cos nx + i\sin nx) + c_{-n}(\cos nx - i\sin nx) \\ &= a_n \cos nx + b_n \sin nx\,. \end{aligned}$$

Therefore, the partial sums s_N of the Fourier series can also be represented as

$$s_N(x) = \frac{1}{2}a_0 + \sum_{n=1}^{N}(a_n \cos nx + b_n \sin nx) = \sum_{n=-N}^{N} c_n e^{inx}\,,$$

and the Fourier series of f can be written in complex form as

$$\sum_{n=-\infty}^{\infty} c_n e^{inx}\,.$$

Over an interval $(-l, l)$, $l > 0$, the complex form of the Fourier series is defined as

$$\sum_{n=-\infty}^{\infty} c_n e^{\frac{n\pi x}{l}}, \quad \text{where} \quad c_n = \frac{1}{2l}\int_{-l}^{l} f(x) e^{-\frac{in\pi x}{l}}\, dx\,, \quad n = 0, \pm 1, \pm 2, \pm 3, \ldots$$

It can be verified that the set of functions $\{\frac{1}{\sqrt{2\pi}} e^{inx}\}_{n\in Z}$ is an orthonormal set.

Sine and Cosine Fourier Series

We consider Fourier series of even and odd functions. Recall that if g is an even function on $(-\pi, \pi)$, then

$$g(-x) = g(x) \quad \text{for } x \in [0, \pi), \qquad \int_{-\pi}^{0} g(x)\, dx = \int_{0}^{\pi} g(x)\, dx\,,$$

and if h is an odd function on $(-\pi, \pi)$, then

$$h(-x) = h(x) \quad \text{for } x \in [0, \pi), \qquad \int_{-\pi}^{0} h(x)\, dx = -\int_{0}^{\pi} h(x)\, dx\,.$$

Thus, if f is even on $(-\pi, \pi)$, then its Fourier coefficients satisfy

$$\begin{aligned}
a_0 &= \frac{1}{\pi}\int_{-\pi}^{\pi} f(x)\, dx = \frac{1}{\pi}\int_{-\pi}^{0} f(x)\, dx + \frac{1}{\pi}\int_{0}^{\pi} f(x)\, dx \\
&= \frac{2}{\pi}\int_{0}^{\pi} f(x)\, dx\,, \\
a_n &= \frac{1}{\pi}\int_{-\pi}^{\pi} f(x)\cos nx\, dx = \frac{1}{\pi}\int_{-\pi}^{0} f(x)\cos nx\, dx + \frac{1}{\pi}\int_{0}^{\pi} f(x)\cos nx\, dx \\
&= \frac{2}{\pi}\int_{0}^{\pi} f(x)\cos nx\, dx\,,
\end{aligned}$$

since the function $g(x) = f(x)\cos nx$ is even, and

$$\begin{aligned}
b_n &= \frac{1}{\pi}\int_{-\pi}^{\pi} f(x)\sin nx\, dx = \frac{1}{\pi}\int_{-\pi}^{0} f(x)\sin nx\, dx + \frac{1}{\pi}\int_{0}^{\pi} f(x)\sin nx\, dx \\
&= 0\,,
\end{aligned}$$

since the function $h(x) = f(x)\sin nx$ is odd. Consequently, the Fourier series of an even function f is a cosine series,

$$\frac{a_0}{2} + \sum_{n=1}^{\infty} a_n \cos nx\,,$$

where the coefficients a_n are given above.

Similarly we find that the Fourier series of an odd function f contains only sine terms, hence it is a sine series

$$\sum_{n=1}^{\infty} b_n \sin nx\,, \quad \text{where} \quad b_n = \frac{2}{\pi}\int_0^{\pi} f(x)\sin nx\,dx\,.$$

Indeed, in this case $g(x) = f(x)\cos nx$ is an odd function, while $h(x) = f(x)\sin nx$ is an even function, so in particular

$$\begin{aligned} a_0 &= \frac{1}{\pi}\int_{-\pi}^{\pi} f(x)\,dx = -\frac{1}{\pi}\int_0^{\pi} f(x)\,dx + \frac{1}{\pi}\int_0^{\pi} f(x)\,dx = 0\,,\\ a_n &= \frac{1}{\pi}\int_{-\pi}^{\pi} f(x)\cos nx\,dx = \frac{1}{\pi}\int_{-\pi}^{0} f(x)\cos nx\,dx + \frac{1}{\pi}\int_0^{\pi} f(x)\cos nx\,dx\\ &= -\frac{1}{\pi}\int_0^{\pi} f(x)\cos nx\,dx + \frac{1}{\pi}\int_0^{\pi} f(x)\cos nx\,dx = 0\,. \end{aligned}$$

As a consequence, if we know that f is odd or even on $(-\pi,\pi)$, we need to compute only the b_k's if f is an odd function, the a_k's if f is an even function.

Example 225. Find the Fourier series of the function $f(x) = x^2$, $x \in (-\pi,\pi)$.

Solution: Since $f(-x) = (-x)^2 = x^2 = f(x)$, f is even and so $b_n = 0$ for $n = 1, 2, 3, \ldots$ We compute

$$a_0 = \frac{2}{\pi}\int_0^{\pi} f(x)\,dx = \frac{2}{\pi}\int_0^{\pi} x^2\,dx = \frac{2}{\pi}\cdot\frac{1}{3}\big[x^3\big]_0^{\pi} = \frac{2}{3}\pi^2\,,$$

and

$$\begin{aligned} a_n &= \frac{2}{\pi}\int_0^{\pi} f(x)\cos nx\,dx = \frac{2}{\pi}\int_0^{\pi} x^2\cos nx\,dx\\ &= \frac{2}{\pi}\left[\frac{x^2}{n}\sin nx\right]_0^{\pi} - \frac{4}{n\pi}\int_0^{\pi} x\sin nx\,dx\\ &= 0 + \frac{4}{n^2\pi}\big[x\cos nx\big]_0^{\pi} - \frac{4}{n^2\pi}\int_0^{\pi}\cos nx\,dx\\ &= \frac{4}{n^2}(-1)^n - \frac{4}{n^3\pi}\big[\sin nx\big]_0^{\pi} = \frac{4}{n^2}(-1)^n\,. \end{aligned}$$

Thus the Fourier series of f is given as the cosine series

$$\frac{1}{3}\pi^2 + 4\sum_{n=1}^{\infty}\frac{(-1)^n}{n^2}\cos nx\,.$$

Phase Angle Form and Frequency Spectrum

Let f be a periodic function defined on the real line which has the fundamental period l, that is, $f(x+l) = f(x)$ for all x, and l is the smallest number satisfying this condition. We define $\omega = 2\pi/l$ as the frequency corresponding to l. Let

$$\frac{1}{2}a_0 + \sum_{n=1}^{\infty} a_n \cos(n\omega x) + b_n \sin(n\omega x) \tag{4.21}$$

be the Fourier series of f on $[-l/2, l/2]$. The series

$$\frac{1}{2}a_0 + \sum_{n=1}^{\infty} d_n \cos(n\omega x + \delta_n) \tag{4.22}$$

is called the **phase angle form** of the Fourier series. Indeed, if two pairs (a, b) and (d, δ) of numbers are related by

$$\begin{aligned} a &= d\cos\delta\,, & b &= -d\sin\delta\,, \\ d &= \sqrt{a^2+b^2}\,, & \delta &= \arctan\left(-\frac{b}{a}\right), \end{aligned} \tag{4.23}$$

then

$$a\cos(n\omega x) + b\sin(n\omega x) = d\cos(n\omega x + \delta)$$

holds for all x, as a consequence of the trigonometric identities, so the series (4.21) and (4.22) correspond term by term. The phase angle form of the Fourier series is also called its **harmonic form**. It represents a periodic function as a superposition of cosine waves. The term $\cos(n\omega x + \delta_n)$ is the nth **harmonic**, d_n is the nth **harmonic amplitude** and δ_n is the nth **phase angle** of f. Note that the harmonic amplitude satisfies

$$d_n = 2|c_n|\,, \tag{4.24}$$

where $c_n = (a_n - ib_n)/2$ is the nth complex Fourier coefficient of f as introduced earlier.

The **amplitude spectrum** or **frequency spectrum**(Figure 4.10) of the periodic function f is a plot of $|c_n| = d_n/2$ on the vertical versus n along the horizontal axis. The **phase spectrum** of f is a plot of the points (n, δ_n) for $n = 0, 1, 2, \ldots$, where $\delta_n = \arctan(-b_n/a_n)$ is the nth phase angle of f.

Example 226. Find the complex Fourier series of f on the given intervals. Furthermore find the frequency spectrum of the function.

(a)

$$f(x) = \begin{cases} 0\,, & -\pi < x < 0\,, \\ x\,, & 0 \le x < \pi\,, \end{cases} \qquad \text{on } [-\pi, \pi].$$

(b)

$$f(x) = \begin{cases} -1\,, & -2 < x < 0\,, \\ 1\,, & 0 \le x < 2\,, \end{cases} \qquad \text{on } [-2, 2].$$

(c)

$$f(x) = \begin{cases} \cos x\,, & 0 < x < \pi/2\,, \\ 0\,, & \pi/2 \le x < \pi\,, \end{cases} \qquad \text{on } [0, \pi].$$

Solution: (a) We have

$$\begin{aligned} c_n &= \frac{1}{2\pi}\int_{-\pi}^{\pi} f(x)e^{-inx}\,dx = \frac{1}{2\pi}\Big[\int_{-\pi}^{0} f(x)e^{-inx}\,dx + \int_{0}^{\pi} f(x)e^{-inx}\,dx\Big] \\ &= \frac{1}{2\pi}\Big[0 + \int_0^{\pi} xe^{-inx}\,dx\Big] = \frac{1}{2\pi}\Big[-xe^{-inx}\cdot\frac{1}{in}\Big]_0^{\pi} - \frac{1}{2\pi}\int_0^{\pi} -e^{-inx}\frac{1}{in}\,dx \end{aligned}$$

$$\begin{aligned} &= \frac{1}{2}\frac{i}{n}e^{-in\pi} - \frac{1}{2\pi}\Big[e^{-inx}\frac{1}{in}\frac{1}{in}\Big]_0^{\pi} \\ &= \frac{1}{2}\frac{i}{n}e^{-in\pi} + \frac{1}{2\pi n^2}e^{-in\pi} - \frac{1}{2\pi n^2} \\ &= \frac{1 + in\pi}{2\pi n^2}e^{-in\pi} - \frac{1}{2n^2\pi}\,, \quad \text{for } n \neq 0, \\ c_0 &= \frac{1}{2\pi}\int_0^{\pi} x\,dx = \frac{\pi}{4}\,. \end{aligned}$$

The complex Fourier representation of f becomes

$$f(x) = \frac{\pi}{4} + \frac{1}{2\pi}\sum_{n=-\infty, n\neq 0}^{\infty} \frac{1}{n^2}[(1 + in\pi)e^{-in\pi} - 1]e^{inx}\,.$$

Here we have used the formulas

$$e^{in\pi} = (-1)^n = e^{-in\pi}\,, \quad e^{-2\pi in} = 1\,, \quad e^{-in\pi/2} = (-i)^n)\,.$$

(b) We have

$$\begin{aligned} c_n &= \frac{1}{4}\int_{-2}^{2} f(x)e^{-in\pi x/2}\,dx = \frac{1}{4}\Big[\int_{-2}^{0}(-1)e^{-in\pi x/2}\,dx + \int_0^2 e^{-in\pi x/2}\,dx\Big] \\ &= \frac{i}{2n\pi}[-1 + e^{in\pi} + e^{-in\pi} - 1] = \frac{1}{2n\pi}[-2 + (-1)^n + (-1)^n] \\ &= \frac{1 - (-1)^n}{n\pi i}\,, \quad \text{for } n \neq 0, \\ c_0 &= \frac{1}{4}\int_{-2}^{2} f(x)\,dx = 0\,. \end{aligned}$$

Thus

$$f(x) = \sum_{n=-\infty}^{\infty} \frac{1 - (-1)^n}{n\pi i}e^{in\pi x/2}\,.$$

The fundamental period is equal to 4 so $\omega = 2\pi/4 = \pi/2$, $c_0 = 0$ and $|c_n| =$

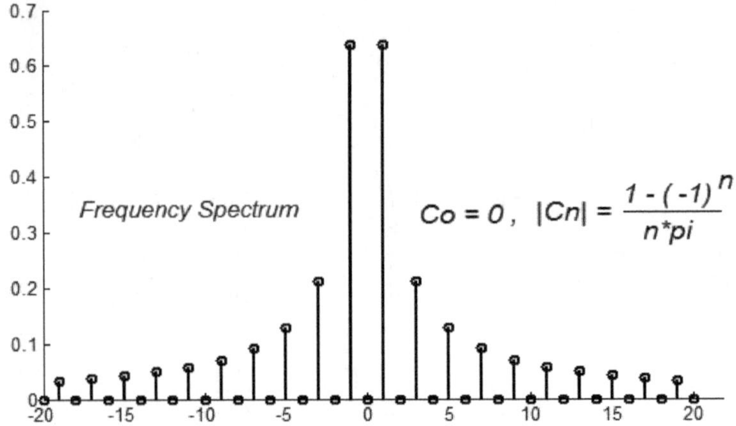

FIGURE 4.10: Frequency Spectrum

$(1-(-1)^n)/n\pi$.

(c) Using $\cos x = (e^{ix} - e^{-ix})/2$ we get

$$c_n = \frac{1}{\pi}\int_0^{\pi} f(x)e^{-2inx}\,dx = \frac{1}{\pi}\int_0^{\pi/2} \cos x e^{-2inx}\,dx$$
$$= \frac{1}{\pi}\int_0^{\pi/2} \frac{1}{2}(e^{ix} - e^{-ix})e^{-2inx}\,dx = \frac{1}{2\pi}\int_0^{\pi/2} (e^{(1-2n)ix} - e^{-(1+2n)ix})\,dx$$
$$= \frac{1}{2\pi}\left[\frac{1}{i(1-2n)}e^{(1-2n)ix} + \frac{1}{i(1+2n)}e^{(1+2n)ix}\right]_0^{\pi/2} = \frac{2ne^{-in\pi}+i}{\pi(1-4n^2)}.$$

4.3 Fourier Transform

The Fourier transform is a mathematical procedure that breaks a function into the frequencies that compose it, as a prism breaks light into colors. It transforms a function f into a new function, $\hat{f}$ or $\mathcal{F}[f]$ (read as "f hat" or "script f") which is called the Fourier transform of f. Depending on the context, the argument of f is a time variable or a spatial variable. The argument of $\hat{f}$ usually has the meaning of a frequency.

A function and its Fourier transform are two faces of the same information. The function exhibits the time (or spatial) information and hides the information about frequencies. The Fourier transform displays information about frequencies and hides the time (or spatial) information. Nevertheless the function and its Fourier transform both contain all the information about the function. One can compute the transform from the original function as well as reconstruct the function from its transform, that is, one can invert the transform.

In the previous section, we have studied the decomposition of a function into its Fourier series, which is a periodic function. This works well for functions defined on a bounded interval, as we can always think of them as periodically extended to the whole real line. In contrast to that, the Fourier transform acts on arbitrary (non-periodic) functions.

4.3.1 Basic Properties of Fourier Transform

When dealing with the Fourier transform, one constantly encounters integrals over the whole real line $(-\infty, \infty)$, that is, improper integrals. In order to simplify the exposition in this section, we call a function f defined on $\mathbb{R} = (-\infty, \infty)$ **integrable on** $\mathbb{R}$ resp. **square integrable on** $\mathbb{R}$, if the improper integral

$$\int_{-\infty}^{\infty} |f(x)|\, dx\,, \quad \text{resp.} \quad \int_{-\infty}^{\infty} |f(x)|^2\, dx$$

converges.

For the definition of the Fourier transform, several variants are in common use, see Remark 57 below. We choose the following one.

Definition 80. (Fourier transform)
Let the function f be defined on $\mathbb{R}$ and assume that it is integrable on $\mathbb{R}$. The function $\hat{f}$, defined on $\mathbb{R}$ by

$$\hat{f}(\xi) = \int_{-\infty}^{\infty} f(t)e^{-i\xi t}\, dt\,, \tag{4.25}$$

is called the **Fourier transform** of f.

In signal analysis, t is understood as the time variable and ξ is understood as the frequency variable, see Section 4.6 below.

Remark 50. (i) According to (4.25), integrals of complex valued functions are involved in the definition of the Fourier transform. They are defined as

$$\begin{aligned}\int_{-\infty}^{\infty} f(t)e^{-i\xi t} dt &= \int_{-\infty}^{\infty} f(t)(\cos(\xi t) - i\sin(\xi t))\, dt \\ &= \int_{-\infty}^{\infty} f(t)\cos(\xi t)\, dt - i\int_{-\infty}^{\infty} f(t)\sin(\xi t)\, dt\,,\end{aligned}$$

that is, the real and imaginary parts of the integrand are evaluated separately and yield the real and imaginary part of the integral, which is a complex number (in this case, the number $\hat{f}(\xi)$).

(ii) The integrand in (4.25) satisfies

$$|f(t)e^{-i\xi t}| = |f(t)| \cdot |e^{-i\xi t}| = |f(t)|\,,$$

since $|e^{ix}| = 1$ for all real numbers x. Therefore and since f is assumed to be integrable on $\mathbb{R}$, the improper integral in (4.25) is defined. One then infers from the properties of parameter-dependent integrals that the Fourier transform $\hat{f}$ is a continuous function. In addition, one can prove that $\hat{f}(\xi)$ tends to zero as ξ tends to $\pm\infty$. (The latter result is called the Riemann-Lebesgue lemma.)

(iii) As it stands, the requirement that f has to be integrable on the whole line is rather restrictive. For example, Definition 80 does not cover the case when f is a constant function. Indeed, the Fourier transform of the constant 1 is defined, but it is no longer a function defined on $\mathbb{R}$, but a more general mathematical object (although it is called the Dirac function). This, however, is outside the scope of this book.

Example 227. Find the Fourier transforms of the following functions.

(a) $f(t) = e^{-|t|}$.

(b) $f(t) = \begin{cases} 0, & t < 0 \\ e^{-t}, & t \geq 0 \end{cases}$.

(c) Let a and k be positive numbers, let

$$f(t) = \begin{cases} k, & -a \leq t < a \\ 0, & \text{otherwise} \end{cases}.$$

Solution: (a) For $f(t) = e^{-|t|}$ we get

$$\begin{aligned}
\hat{f}(\xi) &= \int_{-\infty}^{\infty} e^{-|t|-it\xi}\,dt = \int_{-\infty}^{0} e^{-|t|-it\xi}\,dt + \int_{0}^{\infty} e^{-|t|-it\xi}\,dt \\
&= \int_{-\infty}^{0} e^{t}e^{-i\xi t}dt + \int_{0}^{\infty} e^{-t}e^{-i\xi t}\,dt \\
&= \left[\frac{1}{1-i\xi}e^{(1-i\xi)t}\right]_{t=-\infty}^{t=0} + \left[\frac{-1}{1+i\xi}e^{-(1+i\xi)t}\right]_{t=0}^{t=\infty} \\
&= \frac{1}{1-i\xi} + \frac{1}{1+i\xi} = \frac{2}{1^2+\xi^2} = \frac{2}{1+\xi^2}\,.
\end{aligned}$$

(b) Let

$$H(t) = \begin{cases} 1\,, & t \geq 0 \\ 0\,, & t < 0 \end{cases}$$

be the Heaviside function (see Chapter 2). Then $f(t) = H(t)e^{-t}$ and

$$\hat{f}(\xi) = \int_{-\infty}^{\infty} f(t)e^{-i\xi t}\,dt = \int_{-\infty}^{\infty} H(t)e^{-t}e^{-i\xi t}\,dt = \int_{0}^{\infty} e^{-t}e^{-i\xi t}\,dt$$
$$= \int_{0}^{\infty} e^{-(1+i\xi)t}dt = \Big[-\frac{1}{1+i\xi}e^{-(1+i\xi)t}\Big]_{t=0}^{t=\infty} = \frac{1}{1+i\xi}\,.$$

(c) We obtain

$$\hat{f}(\xi) = \int_{-\infty}^{\infty} f(t)e^{-i\xi t}\,dt = \int_{-a}^{a} ke^{-i\xi t}\,dt = \Big[\frac{-k}{i\xi}\,e^{-i\xi t}\Big]_{t=-a}^{t=a}$$
$$= -\frac{k}{i\xi}\Big[e^{-i\xi a} - e^{i\xi a}\Big] = \frac{2k}{\xi}\sin(a\xi)\,,$$

since

$$\sin(a\xi) = \frac{e^{i\xi a} - e^{-i\xi a}}{2i}\,.$$

Remark 51. **(i)** For a given function f, the Fourier transform (if defined) yields a new function $\hat{f}$. We may thus view the Fourier transform as a mapping whose domain and range are certain sets of functions. Such a mapping is commonly called an **operator**. Let us denote it by $\mathcal{F}$, so

$$\mathcal{F}(f) = \hat{f}\,. \tag{4.26}$$

From Definition 80 we see that $\mathcal{F}$ is linear, that is,

$$\mathcal{F}(\alpha f + \beta g) = \alpha\mathcal{F}(f) + \beta\mathcal{F}(g)$$

holds for functions f, g and scalars α, β.
(ii) Instead of $\mathcal{F}(f) = \hat{f}$ one often writes

$$\mathcal{F}[f(t)] = \hat{f}(\xi)\,.$$

Although this is, in a strict sense, mathematically not correct (it confuses the functions f and $\hat{f}$ with their function values $f(t)$ and $\hat{f}(\xi)$), it leads to a concise way of writing formulas. In this notation, the result of Example 227 (a) becomes

$$\mathcal{F}[e^{-|t|}] = \frac{2}{1+\xi^2}\,.$$

In the next two theorems, we state some important properties of the Fourier transform.

Theorem 46. *Let f be a function which is integrable on $\mathbb{R}$.*
(a) Time shift. *Let t_0 be a real number, let $g(t) = f(t - t_0)$. Then*

$$\hat{g}(\xi) = e^{-i\xi t_0}\hat{f}(\xi)\,, \quad \xi \in \mathbb{R}\,. \tag{4.27}$$

This means that the Fourier transform of the translated function f equals the Fourier transform of f multiplied by a factor. In shorter notation, without introducing the function g explicitly, (4.27) becomes

$$\mathcal{F}[f(t-t_0)] = e^{-i\xi t_0}\hat{f}(\xi)\,.$$

(b) Frequency shift. *Let ξ_0 be a real number, let $g(t) = e^{i\xi_0 t}f(t)$. Then*

$$\hat{g}(\xi) = \hat{f}(\xi - \xi_0)\,, \quad \xi \in \mathbb{R}\,, \tag{4.28}$$

or, in shorter notation,

$$\mathcal{F}[e^{i\xi_0 t}f(t)] = \hat{f}(\xi-\xi_0)\,.$$

(c) Scaling or dilation. *Let a be a real number with $a \neq 0$, let $g(t) = f(at)$. Then*

$$\hat{g}(\xi) = \frac{1}{|a|}\hat{f}\Big(\frac{\xi}{a}\Big)\,, \quad \xi \in \mathbb{R}\,. \tag{4.29}$$

This states that the Fourier transform of the scaled function is obtained by replacing ξ by ξ/a in the Fourier transform of the original function and dividing by the magnitude of the scaling factor.

The formulas in the theorem above are obtained from properties of the integral. For example, (4.27) results from the computation

$$\begin{aligned}\hat{g}(\xi) &= \int_{-\infty}^{\infty} f(t-t_0)e^{-i\xi t}\,dt = e^{-i\xi t_0}\int_{-\infty}^{\infty} f(t-t_0)e^{-i\xi(t-t_0)}\,dt \\ &= e^{-i\xi t_0}\int_{-\infty}^{\infty} f(s)e^{-i\xi s}\,ds = e^{-i\xi t_0}\hat{f}(\xi)\,,\end{aligned}$$

and (4.29) from the computation

$$\hat{g}(\xi) = \int_{-\infty}^{\infty} f(at)e^{-i\xi t}\,dt = \int_{-\infty}^{\infty} f(u)e^{-iu\xi/a}\frac{1}{|a|}\,du = \frac{1}{|a|}\hat{f}\Big(\frac{\xi}{a}\Big)\,.$$

Remark 52. The properties given in Theorem 46 can also be written in operator form. For example, the time shift can be expressed by the **translation operator** T_{t_0} which maps a function f to its translate $T_{t_0}f$ defined by $(T_{t_0}f)(t) = f(t-t_0)$. Equation (4.27) then takes the form

$$\widehat{T_{t_0}f}(\xi) = e^{-\xi t_0}\hat{f}(\xi)\,.$$

Theorem 47. ***(a)*** *Suppose that f is continuous, f' is piecewise continuous and both f and f' are integrable on $\mathbb{R}$. Then*

$$\widehat{f'}(\xi) = \widehat{\mathcal{F}[f']}(\xi) = i\xi\hat{f}(\xi)\,. \tag{4.30}$$

(b) *Suppose that f satisfies $\int_{-\infty}^{\infty}|f(t)|\,dt < \infty$ and $\int_{-\infty}^{\infty}|tf(t)|\,dt < \infty$. Then (see Remark 51 for the notation)*

$$\mathcal{F}[tf(t)] = i\frac{d}{d\xi}\hat{f}(\xi)\,. \tag{4.31}$$

Remark 53. If we apply Theorem 47 to derivatives of f, we obtain the formulas

$$\widehat{f^{(n)}}(\xi) = \mathcal{F}[f^{(n)}](\xi) = (i\xi)^n \hat{f}(\xi)\,, \tag{4.32}$$

$$\mathcal{F}[t^n f(t)] = i^n \frac{d^n}{d\xi^n} \hat{f}(\xi)\,, \tag{4.33}$$

provided f and its derivatives up to order n satisfy the corresponding assumptions. In particular, for $n = 2$ we have

$$\mathcal{F}[t^2 f(t)] = -\frac{d^2}{d\xi^2} \hat{f}(\xi)\,. \tag{4.34}$$

Theorem 48. *(Plancherel's identity)*
If f is integrable as well as square integrable on $\mathbb{R}$, and if the same holds for g, then

$$\int_{-\infty}^{\infty} \hat{f}(\xi)\overline{\hat{g}(\xi)}\, d\xi = 2\pi \int_{-\infty}^{\infty} f(t)\overline{g(t)}\, dt\,. \tag{4.35}$$

(Here $\overline{c}$ denotes the complex conjugate of the complex number c.)

Setting $g = f$ in the preceding theorem, we obtain the following.

Theorem 49. *(Parseval's identity)*
If f and f^2 are integrable on $\mathbb{R}$, then

$$\int_{-\infty}^{\infty} |\hat{f}(\xi)|^2\, d\xi = 2\pi \int_{-\infty}^{\infty} |f(t)|^2 dt\,. \tag{4.36}$$

If we interpret f as a signal, the norm

$$\|f\| = \left(\int_{-\infty}^{\infty} |f(t)|^2 dt\right)^{1/2}$$

represents the energy of the signal.

Inverse Fourier Transform

In the beginning of this section, we defined the Fourier transform $\hat{f}$ of a function f. We can reverse this procedure if we know $\hat{f}$ and we can obtain f according to the following result.

Theorem 50. *Suppose that f is continuous and that f and $\hat{f}$ are integrable on $\mathbb{R}$. Then*

$$f(t) = \frac{1}{2\pi} \int_{-\infty}^{\infty} \hat{f}(\xi) e^{i\xi t}\, d\xi \tag{4.37}$$

holds for all $t \in \mathbb{R}$.

In abstract terms, the right hand side of (4.37) defines the inverse $\mathcal{F}^{-1}$ of the Fourier transform $\mathcal{F}$. It is called the **inverse Fourier transform**

$$(\mathcal{F}^{-1}[g])(t) = \frac{1}{2\pi}\int_{-\infty}^{\infty} g(\xi)e^{i\xi t}\,d\xi\,. \tag{4.38}$$

Indeed, we see that $\mathcal{F}^{-1}[\mathcal{F}[f]] = f$.

Remark 54. The following interpretation of Theorem 50 is fundamental for many applications of the Fourier transform. Consider t as a time variable. For fixed ξ, the values $e^{i\xi t}$ traverse along the unit circle at constant speed. Since $\xi = 2\pi$ corresponds to the completion of one cycle in one unit of time, the $\xi/2\pi$ gives the number of cycles per unit time, which is called the frequency. Its unit is Hertz if t is measured in seconds. The ξ is called the angular frequency and it gives the number of radians traversed per unit time. Seen in this light, formula (4.37) is a decomposition of the original function f into a weighted sum of oscillations in form of an integral. The weight of the angular frequency ξ is given by the value $\hat{f}(\xi)$ of the Fourier transform of f.

Remark 55. If f is twice differentiable and if f, f' and f'' are integrable on $\mathbb{R}$, then $\hat{f}$ is integrable on $\mathbb{R}$, so in this case we can apply Theorem 50.

For piecewise continuous functions, one has the following result.

Theorem 51. *Let f and f' be piecewise continuous and assume that f is integrable on $\mathbb{R}$. Then*

$$\lim_{R\to\infty}\frac{1}{2\pi}\int_{-R}^{R}\hat{f}(\xi)e^{i\xi t}\,d\xi = \frac{1}{2}(f(t+) - f(t-)) \tag{4.39}$$

holds for all $t \in \mathbb{R}$.

Remark 56. For a function h, the limit

$$\lim_{R\to\infty}\int_{-R}^{R} h(x)\,dx\,,$$

if it exists, is called the **principal value** (or **Cauchy principal value**) of $\int_{-\infty}^{\infty} h(x)\,dx$. Thus, under the assumptions of Theorem 51 we also obtain the formula

$$f(t) = \frac{1}{2\pi}\int_{-\infty}^{\infty}\hat{f}(\xi)e^{i\xi t}\,d\xi$$

at points t where f is continuous, provided we interpret the integral as its principal value.

Remark 57. If one wants the frequency variable ξ to denote ordinary frequency instead of angular frequency, one defines the Fourier transform by

$$\hat{f}(\xi) = \int_{-\infty}^{\infty} f(t)e^{-2\pi i t\xi}\,dt\,.$$

The inverse formula then becomes

$$f(t) = \int_{-\infty}^{\infty} \hat{f}(\xi) e^{2\pi i t \xi} \, d\xi \, .$$

In this case, the frequency ξ is measured in Hertz (cycles per second), if t is measured in seconds. If one keeps the angular frequency, but wants a more symmetric relation between the transform and its inverse, one uses

$$\hat{f}(\xi) = \frac{1}{\sqrt{2\pi}} \int_{-\infty}^{\infty} f(t) e^{-it\xi} \, dt \, , \quad f(t) = \frac{1}{\sqrt{2\pi}} \int_{-\infty}^{\infty} \hat{f}(\xi) e^{it\xi} \, d\xi \, .$$

Less common is an interchange of the sign in the exponent,

$$\hat{f}(\xi) = \int_{-\infty}^{\infty} f(t) e^{it\xi} \, dt \, , \quad f(t) = \frac{1}{2\pi} \int_{-\infty}^{\infty} f(t) e^{-it\xi} \, d\xi \, .$$

It is also possible to mix these variants. Therefore, when dealing with the Fourier transform, one has to make sure which convention is used.

Localization and Uncertainty Principle

In this subsection we explain the fact that a function f and its Fourier transform $\hat{f}$ cannot both be concentrated on a small interval.

First, consider the dilation $g(t) = f(at)$. For $a > 1$, g represents a compression of f around $t = 0$ by the factor a. On the other hand, Theorem 46(c) says that

$$\hat{g}(\xi) = \frac{1}{|a|} \hat{f}\Big(\frac{\xi}{a}\Big) \, ,$$

that is, we have to stretch $\hat{f}$ by a factor a to obtain $\hat{g}$. If $a < 1$, g is a stretched version of f while $\hat{g}$ is a compressed version of $\hat{f}$.

Second, assume that $\hat{f}(\xi) = 0$ outside some interval $[-l, l]$. We say in this case that f has **bandwidth** $\boldsymbol{l}$. From the Fourier inversion formula which becomes

$$f(t) = \frac{1}{2\pi} \int_{-l}^{l} \hat{f}(\xi) e^{i\xi t} \, d\xi$$

one concludes, with the aid of a result of complex function theory which we cannot present here, that f can be zero only at isolated points, so it spreads out to infinity and in particular cannot have the property that $f(t) = 0$ outside some interval $[-M, M]$.

Third, one can quantify this phenomenon. Consider the expression

$$\Delta f = \Big(\int_{-\infty}^{\infty} t^2 |f(t)|^2 \, dt \Big) \cdot \Big(\int_{-\infty}^{\infty} |f(t)|^2 \, dt \Big)^{-1} .$$

If the large values of f arise only at small values of t and f decays rapidly as t gets large, the numerator will be small in comparison with the denominator, so

Δf somehow measures the concentration (or localization) of f around $t = 0$. It can be proved that

$$(\Delta f) \cdot (\Delta \hat{f}) \geq \frac{1}{4},$$

that is, if Δf is small then $\Delta \hat{f}$ has to be large and vice versa. This is called the **uncertainty principle**.

4.3.2 Convolution

The convolution of two sequences $a = \{a_n\}$ and $b = \{b_n\}$, where n ranges over all integers, is defined as

$$(a * b)_k = \sum_{j=-\infty}^{\infty} a_j b_{k-j},$$

provided the infinite series converges. If a_j and b_j are non-zero only for $j \geq 0$, the convolution becomes the finite sum

$$(a * b)_k = \sum_{j=0}^{k} a_j b_{k-j}.$$

For example, for $k = 2$ and $k = 3$ we have

$$(a * b)_2 = \sum_{j=0}^{2} a_j b_{2-j} = (a_0 b_2 + a_1 b_1 + a_2 b_0)$$

$$(a * b)_3 = \sum_{j=0}^{3} a_j b_{3-j} = (a_0 b_3 + a_1 b_2 + a_2 b_1 + a_3 b_0).$$

For functions, convolution involves the integral instead of the sum.

Definition 81. Let f and g be two functions defined on the real line $\mathbb{R}$ and assume that f and g are integrable on $\mathbb{R}$. The **convolution** of f and g is denoted by $f * g$ and defined as

$$(f * g)(x) = \int_{-\infty}^{\infty} f(x - y) g(y)\, dy \tag{4.40}$$

(read as f star g, or f convolved with g).

One can prove that, under the stated assumptions, the improper integral in (4.40) indeed converges (we will not do it here), so that $f * g$ is integrable on $\mathbb{R}$, too.

Remark 58. (a) The convolution has the following properties:

(i) For all functions f, g, h as in Definition 81,

$$(f + g) * h = (f * h) + (g * h),$$

that is, $[(f + g) * h](x) = (f * h)(x) + (g * h)(x)$ for all $x \in \mathbb{R}$.

(ii) $(\lambda f) * g = \lambda(f * g)$ for functions f, g and scalars λ.

(iii) $f * (g + h) = (f * g) + (f * h)$.

(iv) $f * (\lambda g) = \lambda(f * g)$.

(v) $f * (g * h)] = (f * g) * h$.

(vi) $f * g = g * f$.

These properties tell us that the convolution $f * g$ is linear with respect to f and g separately, and that it is a commutative (property (vi)).

(c) The convolution can also be interpreted as a moving weight average of the function f, where the weighting is determined by the function g. In view of (a) (vi), $f * g$ can also be interpreted as a moving weight average of g, where the weighting is determined by f.

(d) If the function $f(x)$ has large oscillations, sharp peaks or discontinuities, then averaging about each point x will tend to decrease the oscillations, lower the peaks and smooth out the discontinuities. In view of all this, convolution acts as a smoothing operator. Let us mention two particular results in this direction.

(i) If $\sup_{t\in\mathbb{R}} |f(t)|$ and $\int\limits_{-\infty}^{\infty} |g(t)|\, dt$ are finite, then the function $f * g$ is continuous on $\mathbb{R}$.

(ii) If $\int_{-\infty}^{\infty} |f(t)|^2\, dt$ and $\int_{-\infty}^{\infty} |g(t)|^2\, dt$ are finite, then the function $f * g$ is continuous on $\mathbb{R}$.

(e) Convolutions arise as basic tools to describe input-output systems. Such a system transforms a time-dependent input function $u = u(t)$ into a time-dependent output function $w = w(t)$ according to

$$w(t) = (f * u)(t) = \int_{-\infty}^{\infty} f(t - s)u(s)\, ds\,. \tag{4.41}$$

In signal analysis, such an input-output system is called a **filter**. For example, an electrical circuit with input and output lines can be described in this way, and indeed much of mathematical systems theory has been developed in this context. A filter may serve various purposes such as letting through certain frequencies while blocking other ones and removing noise or blurring in pictures.

We may write the system (4.41) in operator form as

$$w = \mathcal{S}[u]\,.$$

The system is linear,

$$\mathcal{S}[\alpha u + \beta v] = \alpha \mathcal{S}[u] + \beta \mathcal{S}[v]\,,$$

and it is time-invariant, that is, if $\tilde{u}(t) = u(t-h)$ is a translate of u, then $\tilde{w} = \mathcal{S}[\tilde{u}]$ satisfies $\tilde{w}(t) = w(t-h)$, that is, $\tilde{w}$ is the corresponding translate of w. This means that the behavior of the system does not change when time passes.

We state some important properties of the convolution.

Theorem 52. *Suppose that f and g are integrable on $\mathbb{R}$. Then*

$$\int_{-\infty}^{\infty} |(f * g)(t)|\, dt \leq \int_{-\infty}^{\infty} |f(t)|\, dt \cdot \int_{-\infty}^{\infty} |g(t)|\, dt\,. \tag{4.42}$$

For convolution in the time domain, we have

$$\widehat{f * g}(\xi) = \hat{f}(\xi)\hat{g}(\xi)\,. \tag{4.43}$$

For convolution in the frequency domain, we have

$$\widehat{fg}(\xi) = \frac{1}{2\pi}(\hat{f} * \hat{g})(\xi)\,. \tag{4.44}$$

Thus, under the action of the Fourier transform or its inverse, multiplication becomes convolution and vice versa. This is a major reason why convolution plays a prominent role in the calculus of Fourier transforms.

4.3.3 Discrete Fourier Transform

In digital signal processing, signals are represented by sequences $\{x_n\}$, also written as $\{x(n)\}$, where n ranges over all integers. In other words, we consider functions x whose domain are the integers, instead of functions defined on the real numbers $\mathbb{R}$.

More specifically, let x be a periodic sequence with period N, that is, $x(n+N) = x(n)$ for all n. Any such sequence is completely specified by the values $x(0)$, $x(1)$, …, $x(N-1)$. The (N-point) **discrete Fourier transform** (DFT) of x, denoted by $\hat{x}$, is the N-periodic sequence defined by

$$\hat{x}(n) = \sum_{j=0}^{N-1} x(j) e^{-2\pi i j n/N}\,, \quad 0 \leq n \leq N-1\,, \tag{4.45}$$

and extended by periodicity to all integer values of n.

The following theorem yields the inverse of the discrete Fourier transform.

Theorem 53. *Let x be an N-periodic sequence $x(n)$ with DFT $\hat{x}$. Then*

$$x(j) = \frac{1}{N} \sum_{n=0}^{N-1} \hat{x}(n) e^{2\pi i n j/N}\,, \quad 0 \leq j < N-1\,. \tag{4.46}$$

We define the convolution of N-periodic sequences.

Definition 82. Let x and y be N-periodic sequences. The circular convolution of x and y is defined by

$$(x * y)(n) = \sum_{k=0}^{N-1} x(k)y(n-k)\,. \tag{4.47}$$

One may check immediately from (4.47) that $x * y$ is also N-periodic.

Theorem 54. *Let x and y be N-periodic sequences with DFT's $\hat{x}$ and $\hat{y}$. Then*

$$\widehat{x * y}(n) = \hat{x}(n)\hat{y}(n)\,, \tag{4.48}$$

where $\widehat{x * y}$ denotes the DFT of $x * y$.

If one computes the discrete Fourier transform of an N-periodic sequence directly from its definition, one needs N multiplications (and N additions) for each element $\hat{x}(n)$, thus N^2 multiplications for all elements $x(0),\ldots,x(N-1)$. However, due to specific properties of the factors $e^{-2\pi ijn/N}$, it is possible to compute the DFT with only $cN\log_2 N$ multiplications and additions, where c is a small constant.

An algorithm for this purpose was discovered by James Cooley and John Tukey, published in 1965, and is known as the **fast Fourier transform** (FFT). Its history goes back to Carl Friedrich Gauss. The algorithm is based on the recursive factorization of a particular matrix. When N is large, the speedup from N^2 to $cN\log_2 N$ is enormous. Indeed, the FFT is most widely used in computations involving the Fourier transform in all areas of science and technology.

4.4 Integral Transforms

It is well known that various functions can be defined and expressed as improper integrals of the form

$$F(y) = \int_{-\infty}^{\infty} K(x,y)f(x)dx.$$

A function defined in this manner is called an integral transform of f. The function $(K\cdot\cdot)$ appearing under the integral sign is called the kernal of the transform. We have already discussed in Section 2.7 and in Section 4.3, the Laplace transform and the Fourier transform. Here we introduce Fourier sine and Fourier cosine transforms and Hankel transform.

Integral transforms are used very extensively in both pure and applied mathematics, as well as in science and engineering. They are especially useful in solving certain boundary value problems, partial differential equations and some types of integral equations. An integral transform is a linear and invertible transformation, and a partial differential equation can be reduced to a system of algebraic equations by application of an integral transform. The algebraic problem is easy to solve for the transformed function and the function f can be recovered from F by some inversion formula.

1. The **Fourier cosine transform** $F_c\{f\} = \hat{f}_c$ of a function f is defined by
$$F_c\{f\}(w) = \hat{f}_c(w) = \int_0^\infty \cos(wx)f(x)dx.$$
2. The **Fourier sine transform** $F_s\{f\} = \hat{f}_s$ of a function f is defined by
$$F_s\{f\}(w) = \hat{f}_s(w) = \int_0^\infty \sin(wx)f(x)dx.$$
3. The **Hankel transform** $H_n\{f\}$ of order n of a function f is defined by
$$(H_n\{f\})(s) = \int_0^\infty xf(x)J_n(sx)dx.$$
where J_n is the Bessel function of order n of the first kind.

4.5 Sturm-Liouville Problems

We have seen that the trigonometric functions sine and cosine can be used to represent functions in the form of Fourier series expansions. Now we will generalize these ideas. The methods developed here will generally produce solutions of various boundary value problems in the form of infinite function series. Technical questions and issues, such as convergence, termwise differentiation and integration and uniqueness, will not be discussed in detail in this chapter. Interested readers may acquire these detail from advanced literature on these topics, such as the book by Folland [3].

4.5.1 Regular Sturm-Liouville Problems

In mathematical physics and other disciplines fairly large numbers of problems are defined in the form of boundary value problems involving second order ordinary differential equations. Therefore, let us consider the differential equation.

$$a(x)y''(x) + b(x)y'(x) + [c(x) + \lambda r(x)]y = 0, a < x \leq b \tag{4.49}$$

subject to some boundary conditions on a bounded interval $[a, b]$. We suppose that the real functions $a(x)$, $r(x)$ are continuous on the interval $[a, b]$, λ a parameter, and we suppose that $a(x)$ is not zero for every $x \in [a, b]$. It is much more convenient to rewrite the differential equation (4.49) in its equivalent form, the so-called Sturm-Liouville form

$$(p(x)y'(x))' + [q(x) + \lambda r(x)]y = 0,$$

where the real functions $p(x)$, $p'(x)$, $r(x)$ are continuous on $[a, b]$, and $p(x)$ and $r(x)$ are positive on $[a, b]$.

Remark 59. Any differential equation of the form (4.49) can be written in Sturm- Liouville form.
Indeed, first divide (4.49) by $a(x)$ to obtain

$$y''(x) + \frac{b(x)}{a(x)}y'(x) + \left[\frac{c(x)}{a(x)} + \lambda\frac{r(x)}{a(x)}\right] y - 0.$$

Definition 83. A regular Sturm-Liouville problem is a second order homogeneous linear differential equation of the form

$$p(x)y'(x) + [q(x) + \lambda r(x)]y(x) = 0, \quad a < x < b \tag{4.50}$$

where $p(x)$, $p'(x)$ and $r(x)$ are real continuous functions on a finite interval $[a, b]$ and $p(x)$, $r(x) > 0$ on $[a, b]$, together with the set of homogeneous boundary conditions of the form

$$\begin{aligned} \alpha_1 y(a) + \beta_1 y'(a) = 0, \\ \alpha_1 y(b) + \beta_1 y'(b) = 0, \end{aligned} \tag{4.51}$$

where α_1 and β_1 are constants. We regard λ as an undermined constant parameter.

Notation. Sometimes, we will denote by L the following linear differential operator.

$$L[y] = (p(x)y(x)) + q(x)y(x). \tag{4.52}$$

Using this operator the differential equation (4.50) can be expressed in the following form

$$L[y] = -ry. \tag{4.53}$$

We use the term linear differential operator because of the following important linear property of the operator L.

$$L[c_1y_1 + c_2y_2] = c_1L[y_1] + c_2L[y_2],$$

for any constants c_1 and c_2 and any differentiable functions y_1 and y_2.

Remark 60. For different values of the constants α's and β's in the boundary conditions (4.51) we have special types of boundary conditions.
For $\beta_1 = \beta_2 = 0$, (4.51) these are called Dirichlet boundary conditions.
For $\alpha_1 = \alpha_2 = 0$, (4.51) these are called Neumann boundary conditions. Other types of boundary conditions that are often encountered are periodic boundary conditions when

$$y(a) = y(b), \quad y'(a) = y'(b).$$

Our goal is to find all solutions of the Sturm-Liouville problem. It is clear that $y \equiv 0$ is the trivial solution of (4.50) for every λ, and satisfies the boundary conditions (4.51). However, we are interested here to find those parameters λ for which the Sturm-Liouville problem has non-trivial solutions y. Those values of λ and the corresponding solutions $y(x)$ have special names:

Definition 84. If $y(x) \neq 0$ is a solution of a regular Sturm-Liouville problem (4.50), (4.51) corresponding to some constant λ, then this solution is called an eigenfunction corresponding (associated) to the eigenvalue λ.

Let us take now an example.

Example 228. Find all eigenvalues and the corresponding eigenfunctions of the following problem.

$$y''(x) + \lambda y(x) = 0, \quad 0 < x < l$$

subject to be boundary conditions $y(0) = 0$ and $y(l) = 0$.

Solution: First we observe that this is a regular Sturm-Liouville problem. Indeed, the differential equation can be written in the form

$$(1.y'(x))' + [0 + 1.]y = 0,$$

so $p(x) = 1,\ q(x) = 0$ and $r(x) = 1$.
Case 1: $\lambda = 0$. is not an eigenvalue.
Case 2: $\lambda < 0$. It can be verified that this λ is not an eigenvalue.
Case 3: $\lambda > 0$. The eigenvalues for this problem are

$$\lambda_n = \frac{n^2\pi^2}{l^2} \; n = 1, 2, 3, \ldots.$$

The corresponding eigenfunctions are

$$y_n(x) = \sin\left(\frac{n\pi x}{l}\right), \quad n = 1, 2, 3, ...$$

It can be proved in general that every eigenvalue of the regular Sturm-Liouville problem is a real number.

Properties of the regular Sturm-Liouville problem are:

(a) There exist an infinite number of real eigenvalues that can be arranged in increasing order $\lambda_1 < \lambda_2 < \lambda_3 < \cdots < \lambda_n < \cdots$ such that $\lambda_n \to \infty$ as $n \to \infty$.

(b) For each eigenvalue there is only one eigenfunction (except for constant multiples).

(c) Eigenfunctions corresponding to different eigenvalues are linearly independent.

(d) The set of eigenfunctions corresponding to the set of eigenvalues is orthogonal with respect to the weight function $p(x)$ on the interval $[a, b]$.

4.6 Application of Fourier Methods to Signal Analysis

Signals are ubiquitous. They are synonymous functions. A square integrable function (Section 4.3.1) are also called a signal with finite energy. A signal is characterized by its energy defined as

$$\sqrt{\int_a^b |f(t)|^2 dt} = \|f\|.$$

The train's whistle and the blinking of a car's beam can be viewed as quantities varying in time which contain information; they are examples of time dependent signals. Early in human history, signals of smoke by day and of fire by night have been used to transmit information. In recent times, telegraph, telephone, radio, television and radar have been used as signal transmitters. A radio signal consists of a sine or cosine wave with radio frequency, called the carrier wave, which has been modulated with the information to be transmitted.

Signals can be divided into two categories, analog signals (functions defined on a continuum of numbers, for example an interval in $\mathbb{R}$) and digital signals, which are defined on a discrete set like the integers. In 1949, Claude E. Shannon of Bell Telephone Laboratories published a mathematical result now known as the Shannon sampling theorem. This result provided the foundation for digital signal processing. It tells us that if the range of frequencies of a signal measured in cycles per second does not exceed n, then the time-continuous signal can be reconstructed with complete accuracy by measuring its amplitude $2n$ times a second.

The study of signals is relevant not only in telecommunication, but also in telemetry, astronomy, oceanography, optics, crystallography, geophysics, bioengineering, bioinformatics and medicine.

Shannon sampling theorem

Let us assume that a signal f is **band-limited**, that is,

$$\hat{f}(\xi) = 0\,, \quad \text{for all } |\xi| > l, \tag{4.54}$$

holds for some $l > 0$, and the smallest such number l is called the bandwidth of the signal. This means that the total frequency content of the signal f lies in the band (or interval) $[-l, l]$. Moreover, let us assume that f is integrable and square integrable on $\mathbf{R}$, that is, the signal f has finite energy. Let it be recovered from its Fourier transform as

$$f(t) = \frac{1}{2\pi}\int_{-\infty}^{\infty} \hat{f}(\xi)e^{i\xi t}\,d\xi\,.$$

Because of (4.54),

$$f(t) = \frac{1}{2\pi}\int_{-l}^{l} \hat{f}(\xi)e^{i\xi t}\,d\xi\,. \tag{4.55}$$

We expand $\hat{f}$ on $[-l, l]$ in a complex Fourier series, so

$$\hat{f}(\xi) = \sum_{n=-\infty}^{\infty} c_n e^{n\pi i\xi/l}\,, \tag{4.56}$$

where

$$c_n = \frac{1}{2l}\int_{-l}^{l} \hat{f}(\xi)e^{-n\pi i\xi/l}\,d\xi\,. \tag{4.57}$$

We insert (4.56) in (4.55) and obtain

$$\begin{aligned} f(t) &= \frac{1}{2\pi}\int_{-l}^{l} \hat{f}(\xi)e^{i\xi t}\,d\xi = \frac{1}{2\pi}\int_{-l}^{l} \sum_{n=-\infty}^{\infty} c_n e^{n\pi i\xi/l}e^{i\xi t}\,d\xi \\ &= \frac{1}{2\pi}\sum_{n=-\infty}^{\infty} c_n \int_{-l}^{l} e^{n\pi i\xi/l}e^{i\xi t}\,d\xi\,. \end{aligned} \tag{4.58}$$

(Interchanging the integral with the sum is possible by a general property of Fourier series of functions of finite energy. Note that $\hat{f}$ has finite energy by Parseval's formula (4.36).) Next, we compare (4.55) for $t = -n\pi/l$ with (4.57) and see that

$$c_n = \frac{\pi}{l} f\Big(-\frac{n\pi}{l}\Big)\,. \tag{4.59}$$

We insert this value into (4.58) and compute

$$\begin{aligned} f(t) &= \frac{1}{2\pi}\sum_{n=-\infty}^{\infty} \frac{\pi}{l} f\Big(-\frac{n\pi}{l}\Big)\int_{-l}^{l} e^{n\pi i\xi/l}e^{i\xi t}\,d\xi \\ &= \frac{1}{2l}\sum_{n=-\infty}^{\infty} f\Big(\frac{n\pi}{l}\Big)\int_{-l}^{l} e^{-n\pi i\xi/l}e^{i\xi t}\,d\xi \end{aligned}$$

(we have replaced n by $-n$, as n ranges over all integers)

$$= \frac{1}{2l} \sum_{n=-\infty}^{\infty} f\left(\frac{n\pi}{l}\right) \int_{-l}^{l} e^{i\xi(t-n\pi/l)}\, d\xi$$

$$= \frac{1}{2l} \sum_{n=-\infty}^{\infty} f\left(\frac{n\pi}{l}\right) \frac{1}{i(t-n\pi/l)} \left[e^{i\xi(t-n\pi/l}\right]_{\xi=-l}^{\xi=l}$$

$$= \sum_{n=-\infty}^{\infty} f\left(\frac{n\pi}{l}\right) \frac{1}{lt-n\pi} \frac{1}{2i} \left[e^{i(lt-n\pi)} - e^{-i(lt-n\pi)}\right],$$

so finally we arrive at the Whittaker-Shannon interpolation formula

$$f(t) = \sum_{n=-\infty}^{\infty} f\left(\frac{n\pi}{l}\right) \frac{\sin(lt-n\pi)}{lt-n\pi}. \tag{4.60}$$

This is the main content of Shannon's sampling theorem which states that a function of bandwidth l can be completely recovered by (4.60) from its values at the points $n\pi/l$, where $n = 0, \pm 1, \pm 2, \ldots$. This result forms the basis for the conversion between analog and digital signals. If we convert an analog signal f of bandwidth l into a digital signal by evaluating it at times $t = 0, \pm\pi/l, \pm 2\pi/l$, we can convert it back to an analog signal without loss of information, at least in principle. Note that an exact evaluation of (4.60)) involves an infinite sum of values of f at arbitrarily large positive and negative times. Developing suitable approximations, both from the theoretical and practical standpoints, is one of the areas of signal analysis and signal processing.

4.7 Exercises

4.1. Discuss the relationship between linear independence and orthonormality. Can you convert an orthogonal system into an orthonormal system?

4.2. (a) Show that $f(x) = e^x$ and $g(x) = \sin x$ are orthogonal on the interval $\left[\frac{\pi}{4}, \frac{5\pi}{4}\right]$.

(b) Show that $\{\cos x, \cos 3x, \cos 5x, \ldots\}$ is an orthogonal set on $[0, \pi/2]$.

4.3. Expand the following functions into Fourier series on the given interval.

(a) $f(x) = x + \pi$, $-\pi < x < \pi$.

(b) $f(x) = e^{-8x}$ for $-4 \le x \le 4$.

(c) $f(x) = \begin{cases} 0, & -\pi < x < 0 \\ x^2, & 0 \le x < \pi \end{cases}$.

4.4. Show that

(a) $$\frac{\pi}{4} = 1 - \frac{1}{3} + \frac{1}{5} - \frac{1}{7} + \ldots \qquad \text{(using 10.8.3 (a))},$$

(b) $$\frac{\pi^2}{6} = 1 + \frac{1}{2^2} + \frac{1}{3^2} + \ldots \qquad \text{(using 10.8.3 (c))}.$$

4.5. Compare the graph of the function $f(x) = x^2$ with the 3rd, 6th, 10th and 13th partial sums of its Fourier series on the interval $[-2, 2]$.

4.6. Write the complex Fourier series of the following functions.

(a) $f(x) = \cos x$, $0 \leq x < 1$, and f has period 1.

(b) f has period 4, and $f(x) = \begin{cases} 0, & 0 \leq x < 1, \\ 1, & 1 \leq x < 4. \end{cases}$

4.7. Expand the following functions in an appropriate cosine and sine Fourier series.

(a) $f(x) = x^3$, $-\pi < x < \pi$.

(b) $f(x) = \begin{cases} x - 1, & -\pi < x < 0, \\ x + 1, & 0 \leq x < \pi. \end{cases}$

4.8. Let $f(x) = 4 \sin x$, $0 < x < \pi$, $f(x + \pi) = f(x)$. Sketch this function and its Fourier series. Find the frequency spectrum of f.

4.9. Let f be integrable on $[-l, l]$.

(a) Prove that

$$\frac{1}{2} a_0 + \sum_{n=1}^{\infty} (a_n^2 + b_n^2) \leq \frac{1}{l} \int_{-l}^{l} |f(x)|^2 \, dx \,,$$

where a_0, a_n, b_n are the Fourier coefficients of f.

(b) Prove Parseval's identity

$$\frac{1}{2} a_0 + \sum_{n=1}^{\infty} (a_n^2 + b_n^2) = \frac{1}{l} \int_{-l}^{l} |f(x)|^2 \, dx$$

if $\int_{-l}^{l} |f(x)|^2 \, dx$ is finite and a_0, a_n and b_n are its Fourier coefficients.

(c) Show that $a_n \to 0$ and $b_n \to 0$ as $n \to \infty$.

4.10. Find the Fourier transforms of the following functions

(a) $f(t) = te^{-|t|}$ for all real t.

(b) $f(t) = \begin{cases} \sin(\pi t), & -5 \le t \le 5 \\ 0, & |t| > 5 \end{cases}$

(c) $f(t) = \begin{cases} 1, & 0 \le t \le k \\ -1, & -k \le t < 0\,, \\ 0, & |t| > k \end{cases}$, where k is a positive constant.

4.11. Prove Theorem 46 (b): Let ξ_0 be a real number, let $g(t) = e^{i\xi_0 t} f(t)$, then $\hat{g}(\xi) = \hat{f}(\xi - \xi_0)$.

4.12. Show that

(a) For $g(t) = f(-t)$ show that $\hat{g}(\xi) = \hat{f}(-\xi)$.

(b) For $g(t) = \hat{f}(t)$ show that $\hat{g}(\xi) = 2\pi f(-\xi)$.

4.13. Prove that $(f * g)(x) = (g * f)(x)$ for all x.

4.14. Let $f(x) = e^{-ax}\chi_{(0,\infty)}(x)$ and $g(x) = e^{-bx}\chi_{(0,\infty)}(x)$, where

$$\chi_{(0,\infty)}(x) = \begin{cases} 1, & x \in (0,\infty) \\ 0, & x \notin (0,\infty) \end{cases}$$

Calculate $(f * g)(x)$.

4.15. Find $\hat{f}$ and $\hat{g}$ if

$$f(t) = \chi_{[-1/2,1/2]}(t) = \begin{cases} 1, & -1/2 \le t \le 1/2 \\ 0, & t < -1/2 \text{ or } t > 1/2 \end{cases}$$

4.16. Let f and g be square integrable. Show that the convolution $f * g$ is a continuous function on $\mathbb{R}$.

4.17. Let $x = x(n)$ and $y = y(n)$ be N-periodic signals. Prove that $(x*y)(n) = (y * x)(n)$.

4.8 Suggestion for Further Reading

For applications of Fourier Methods to sound, music, computers, X-ray crystallography, computerized tomography, nuclear magnetic resonance and radiostronomy, we refer to [2] and [4].

We suggest reading of relevant Chapter 3 of [1] and Chapter 12, Section 12.5 of [5] and [3] on the Sturm-Liovville problem. For acquiring a good knowledge of Haar system, Walsh system and wavelets constructed by Walsh systems, we suggest a forthcoming book on the theme wavelets constructed through Walsh functions by Farkov, Manchanda and Siddiqi Springer series due for publication in Dec 2017.

Bibliography

[1] K. Adzievski, A. H. Siddiqi, *Introduction to Partial Differential Equations for Scientists and Engineers Using Mathematica*, CRC Press, 2014.

[2] M. Cartwright, *Fourier Methods for Mathematicians, Scientists and Engineers*, Ellis Horwood Series in Mathematics Applications, New York, 1990.

[3] G. B. Folland, *Fourier Series and Its Applications*, Wadsworth, Pacific Grove, CA, 1992.

[4] E Prestini, *The Evolution of Applied Harmonic Analysis Models of the Real World*, Birkhäuser, 2004.

[5] D. G. Zill, Michael R. Cullen, *Advanced Engineering Mathematics*, Jones and Bartlett Publishers, 2012.

Chapter 5

Applied Partial Differential Equations

5.1 Introduction to Partial Differential Equations

Often real world problems are described by partial differential equations (PDEs) with or without boundary and initial value conditions. PDEs involve unknown functions of two or more variables and their partial derivatives. There is a wide class of PDEs which play a significant role in electromagnetic theory, fluid dynamics, traffic flow, medical imaging, financial engineering and many other disciplines. In this chapter we focus on basic ingredients of PDEs and applications of important classes such as the **wave equation**, **heat equation**, and **Laplace equation**.

Basic concepts and terminology

Let Ω be a nonempty subset of R^2 or in R^3 which is open and connected. It may be recalled that Ω is called open if at every point $P \in \Omega$, there is an open disc or ball (without circumference or surface) with center P which is subset of Ω. Note that Ω is called connected if any two points in Ω can be joined by a polygonal line which lies entirely in Ω. If $u = u(x, y, ...)$, $(x, y, ...) \in \Omega$ is a function of two or more variables, then the partial derivatives of u of the first order will be denoted by

$$\frac{\partial u}{\partial x} \text{ or } u_x, \ \frac{\partial u}{\partial y} \text{ or } u_y.$$

Similarly, for partial derivatives of the second order, we will use the notation

$$\frac{\partial^2 u}{\partial x^2}, \ \frac{\partial^2 u}{\partial y \partial x}, \ \frac{\partial^2 u}{\partial y^2}.$$

The class $C^k(\Omega)$, $k = 1, 2, ...$ is defined as the set of functions $u = u(x, y)$, $(x, y) \in \Omega$. All its partial derivatives up to the order of k are continuous functions in Ω. These concepts can be generalized for more variables.

A partial differential equation for a function $u = u(x, y, z, \dots)$ contains the function u and some of its partial derivatives:

$$F(x, y, u, u_x, u_y, u_{xx}, u_{xy}, u_{yy}, ...) = 0. \tag{5.1}$$

A solution of a partial differential equation of order k is a function $u = u(x, y, z, \dots) \in C^k(\Omega)$ which together with partial derivatives satisfies the PDE.

Examples of PDEs and their solutions

Example 229. Show that $u(x, y) = \ln(x^2 + y^2)$ is a solution of the PDE

$$u_{xx} + u_{yy} = 0$$

known as the Laplace equation.

Solution:

$$u_x = \frac{2x}{x^2 + y^2}, \ u_{xx} = \frac{-2x^2 + 2y^2}{(x^2 + y^2)^2}$$

$$u_y = \frac{2y}{x^2 + y^2}, \ u_{yy} = \frac{-2y^2 + 2x^2}{(x^2 + y^2)^2}.$$

It is clear that

$$u_{xx} + u_{yy} = 0.$$

Example 230. Show that $u(x, t) = \sin x \cos t$ is a solution of the PDE

$$u_{xx} = u_{tt}.$$

This type of PDE is known as a one-dimensional wave equation.

Solution:

$$u_x = \cos x \cos t,\ u_{xx} = -\sin x \cos t,$$

$$u_t = -\sin x \sin t,\ u_{tt} = -\sin x \cos t,$$

and it is clear that

$$u_{xx} = u_{tt}.$$

Example 231. Show that $u(x,t) = t^{-\frac{1}{2}}e^{-\frac{x^2}{t}}$ is a solution of the PDE

$$u_t = \frac{1}{4}u_{xx}.$$

This type of equation is known as a one-dimensional heat equation or diffusion equation.

Solution:

$$u_t = \frac{-1}{2}t^{\frac{-3}{2}}e^{-\frac{x^2}{t}} + t^{\frac{-5}{2}}x^2e^{-\frac{x^2}{t}}$$

$$u_x = -2xt^{\frac{-3}{2}}e^{-\frac{x^2}{t}}$$

$$u_{xx} = -2te^{-\frac{x^2}{t}} + 4t^{\frac{-5}{2}}x^2e^{-\frac{x^2}{t}}$$

so, clearly,

$$u_t = \frac{1}{4}u_{xx}.$$

Example 232. Find the general solution of the PDE

$$u_{xx}(x,y) = 0.$$

Solution: This equation means that $u_x(x,y)$ does not depend on x. Therefore, $u_x(x,y) = A(y)$, where $A(y)$ is an arbitrary function. Integrating the last equation with respect to x (keeping y constant), it follows that the general solution of the equation is given by

$$u(x,y) = A(y)x + B(y)$$

where $A(y)$ and $B(y)$ are arbitrary functions.

Example 233. Find the general solution of the PDE

$$u_{yx} + u_y = 0.$$

Solution: In order to get the general solution, we convert this PDE into an ordinary differential equation (ODE) by the substitution $w = w(x,y) = u_y$. Then the given equation becomes

$$w_x + w = 0.$$

The general solution of the above ODE is given by

$$w(x, y) = a(y)e^{-x}$$

where $a(y)$ is any differentiable function. Therefore,

$$u_y = a(y)e^{-x}.$$

Integrating this equation with respect to y (keeping x constant), the general solution of the given PDE is

$$u(x, y) = e^{-x}f(y) + g(x)$$

where f is any differentiable function so that $\dfrac{df}{dy} = a(y)$ and g is any arbitrary function.

Boundary and initial conditions

Boundaries are a set of constraints that describe the nature of unknown function $u(x, y)$ satisfying the PDE, on the boundary of the domain Ω, denoted by $\partial\Omega$. There are three important classes of boundary conditions:

1. Dirichlet conditions

$$u(x, y) = f(x, y) \text{ for } x, y \in \partial\Omega.$$

2. Neumann conditions

$$\frac{\partial u}{\partial n}(x, y) = h(x, y) \text{ for } x, y \in \partial\Omega$$

where $\dfrac{\partial u}{\partial n}$ denotes the derivative of u in the direction of the outward unit normal vector to $\partial\Omega$.

3. Robin mixed conditions

$$\alpha u(x, y) + \beta\frac{\partial u}{\partial n}(x, y) = l(x, y) \text{ for } x, y \in \partial\Omega$$

where α, β are nonzero constants or functions, and $l(x, y)$ is a function defined on $\partial\Omega$.

Initial conditions

When PDEs involve the time variable t, then we have to consider initial conditions called Cauchy conditions. These conditions specify the value of the unknown function and its higher order derivative at the initial time $t = t_0$.

If the functions $f(x, y)$, $h(x, y)$, and $l(x, y)$ are identically zero on their domain $\partial\Omega$, they are called homogeneous boundary conditions; otherwise they are called non-homogeneous conditions.

A differential equation is said to be linear if the function F is algebraically

linear in each of the variables $u, u_x, u_y, u_{xx}, u_{xy}, u_{yy}, \ldots$, and if the coefficients of u and its derivatives are functions only of the independent variables. An equation that is not linear is said to be non-linear; a non-linear equation is quasi-linear if it is linear in the highest order derivatives.
For example, an equation of the form

$$a(x, y)u_x + b(x, y)u_y + c(x, y)u(x, y) = 0$$

is a first order linear equation. The following equation

$$a(x, y, u)u_x + b(x, y, u)u_y + c(x, y, u)u(x, y) = 0$$

is a first order quasi-linear equation.

Classification of linear second order partial differential equations
The second order PDE

$$au_{xx} + bu_{xy} + cu_{yy} + du_x + eu_y + fu = 0$$

where a, b, c, d, e, f are real constants, is called

Elliptic if $b^2 - 4ac < 0$

Parabolic if $b^2 - 4ac = 0$

Hyperbolic if $b^2 - 4ac > 0$

Example 234. Classify the following PDEs
(i) $\frac{1}{4}u_{xx} = u_t$ (ii) $u_{xx} = u_{yy}$ (iii) $u_{xx} + u_{yy} = 0$ (iv) $u_{xx} - u_{xy} - 3u_{yy} = 0$.

Solution: (i) $\frac{1}{4}u_{xx} - u_t = 0$. $a = \frac{1}{4}$, $b = 0$, $c = 0$, $d = 0$, $e = -1$, and $f = 0$. $b^2 - 4ac = 0$. Then the PDE is parabolic.
(ii) $u_{xx} - u_{yy} = 0$. $a = 1$, $b = 0$, $c = -1$, $d = 0$, $e = f = 0$. $b^2 - 4ac = 4 > 0$ and the PDE is hyperbolic.
(iii) $u_{xx} + u_{yy} = 0$. $a = 1$, $b = 0$, $c = 1$, $d = 0$, $e = 0$, $f = 0$. $b^2 - 4ac = -4 < 0$. So the PDE is elliptic.
(iv) $u_{xx} - u_{xy} - 3u_{yy} = 0$. $a = 1$, $b = -1$, $c = -3$, $d = 0$, $e = 0$, $f = 0$. $b^2 - 4ac = 13 > 0$ and the equation is hyperbolic.

Some well-known linear PDEs

1. Laplace's equation

$$u_{xx} + u_{yy} = 0$$

2. Helmholtz's equation

$$-(u_{xx} + u_{yy}) = \lambda u$$

3. Heat (or diffusion) equation

$$u_t = a^2 u_{xx}$$

4. Wave equation

$$u_{tt} = c^2 u_{xx}$$

5. Schrodinger equation

$$iu_t + u_{xx} + u_{yy} = 0$$

6. Telegraph equation

$$u_{tt} + du_t - u_{xx} = 0$$

7. Airy's equation

$$u_t + u_{xxx} = 0$$

8. Beam equation

$$u_t + u_{xxxx} = 0$$

Well-known non-linear PDEs

1. Eikonal equation

$$\sqrt{(u_x)^2 + (u_y)^2} = 1$$

2. Non-linear Poisson equation

$$-(u_{xx} + u_{yy}) = f(u)$$

3. Scalar conservation law

$$u_t + \operatorname{div} F(u) = 0$$

4. Inviscid Burger's equation

$$u_t + uu_x = 0$$

5. Non-linear heat equation

$$u_t - u_{xx} = f(u)$$

6. Non-linear wave equation

$$u_{tt} - u_{xx} = f(u)$$

7. Korteweg-deVries (KDV) equation

$$u_t + uu_x + u_{xxx} = 0$$

8. Maxwell's equation

$$\operatorname{div} E = 0, \quad \operatorname{curl} E = \frac{-1}{c} H_t$$
$$\operatorname{div} H = 0, \quad \operatorname{curl} H = \frac{1}{c} E_t$$

where $E(x, y, z)$ and $H = H(x, y, z)$ represent electric and magnetic field in empty space, and c is the speed of the light.

9. Euler's equations for incompressible, inviscid flow

$$\begin{cases} u_t + u.\nabla u = -\nabla p \\ \operatorname{div} u = 0 \end{cases}$$

10. Navier-Stokes equation for incompressible, inviscid flow

$$\begin{cases} u_t + u.\nabla u - \Delta u = -\nabla p \\ \operatorname{div} u = 0 \end{cases}$$

5.2 Heat Equation

Consider the homogeneous heat equation

$$u_t = a^2 u_{xx}, \; 0 < x < l, \;\; t > 0 \tag{5.2}$$

which satisfies the initial condition

$$u(x, 0) = f(x), \; 0 < x < l, \tag{5.3}$$

and the homogeneous Dirichlet boundary conditions

$$u(0, t) = u(l, t) = 0, \; t > 0 \tag{5.4}$$

where $u(x, t)$ denotes the temperature distribution and k the thermal diffusivity.

The equation, in its simplest form, goes back to the beginning of the 19th century. Besides modelling temperature distribution it has been used to model the following physical phenomena:

- Diffusion of one material within another such as smoke particles in air.
- Chemical reactions, such as the Belousov-Zhabotinsky reaction which exhibits fascinating wave structure.
- Electrical activity in the membranes of living organisms, for example, the Hodgkin-Huxley model.

- Population dispersions; individuals move randomly and to avoid overcrowding.
- Pursuit and evasion in predator-prey systems.
- Pattern and evasion in predator-prey systems.
- Dispersion of pollutants in running streams.

More recently it has been used in financial mathematics or financial engineering for determining appropriate prices of options.

5.2.1 Solution Using Fourier Series

We will solve the above initial boundary value problem using the Fourier series method.
Let

$$u(x,t) = X(x)T(t) \tag{5.5}$$

where $X(x)$ and $T(t)$ are functions of single variables x and t, respectively. Differentiating (5.5) with respect to x and t, and substituting the partial derivatives in Equation (5.2), we obtain

$$\frac{X''(x)}{X(x)} = \frac{1T'(t)}{a^2T(t)}. \tag{5.6}$$

Equation (5.6) holds identically for every $0 < x < l$ and every $t > 0$. Since x and t are independent variables, Equation (5.6) is possible only if each function on both sides is equal to same constant λ :

$$\frac{X''(x)}{X(x)} = \frac{1T'(t)}{a^2T(t)} = \lambda.$$

This leads us to two ordinary differential equations:

$$X''(x) - \lambda X(x) = 0 \tag{5.7}$$

and

$$T'(t) - a^2\lambda T(t) = 0. \tag{5.8}$$

From the boundary conditions (7.75), we obtain

$$X(0)T(t) = X(l)T(t) = 0,\ t > 0.$$

This implies that

$$X(0) = X(l) = 0 \tag{5.9}$$

because $T(t) = 0$ for every $t > 0$ would lead to the trivial solution $u(x,t) = 0$. Solving the eigenvalue problem (5.7) and (5.9), as in Chapter 3, it follows that the eigenvalues λ_n and the corresponding eigenfunctions $X_n(x)$ are given by

$$\lambda_n = -\left(\frac{n\pi}{l}\right)^2,\ X_n(x) = \sin\frac{n\pi x}{l},\ n = 1, 2, \ldots. \tag{5.10}$$

The solution of the eigenvalue problem (5.8) corresponding to the above λ_n is given by

$$T_n(t) = c_n e^{-\frac{n^2\pi^2a^2}{l^2}t}, \quad n = 1, 2, ... \tag{5.11}$$

where c_n are constants to be determined. Therefore, we obtain a sequence of functions

$$u_n(x,t) = c_n e^{-\frac{n^2\pi^2a^2}{l^2}t} \sin\frac{n\pi x}{l}, \quad n = 1, 2, ... \tag{5.12}$$

each of which satisfies the heat equation (5.2) and the boundary conditions (7.75). Since the heat equation and the boundary conditions are linear and homogeneous, a function $u(x,t)$ of the form

$$u(x,t) = \sum_{n=1}^{\infty} u_n(x,t) = \sum_{n=1}^{\infty} c_n e^{-\frac{n^2\pi^2a^2}{l^2}t} \sin\frac{n\pi x}{l} \tag{5.13}$$

will also satisfy the heat equation and the boundary conditions. If we assume that the above series is convergent and the initial condition (5.3) is satisfied, we obtain

$$f(x) = u(x,0) = \sum_{n=1}^{\infty} c_n \sin\frac{n\pi x}{l}, \ 0 < x < l.$$

Section 4.2.2 this shows that the above is the sine Fourier series of $f(x)$. Therefore, the coefficients c_n are given by

$$c_n = \frac{2}{l}\int_0^l f(x) \sin\frac{n\pi x}{l} dx, \ n = 1, 2, \ldots. \tag{5.14}$$

5.2.2 Solution Using Fourier Transform

We discussed the Fourier transform in Chapter 4. In this section we study its application in solving initial value problems for the heat equation.

Notation. We denote by $U(\omega,t)$ the Fourier transform with respect to $u(x,t)$. Similarly, the Fourier transforms of $y(x,t)$ and $f(x,t)$ are respectively, $Y(\omega,t)$ and $F(\omega,t)$.

Consider the homogeneous heat equation with initial condition

$$\begin{cases} u_t = a^2 u_{xx}, & -\infty < x < \infty, \ t > 0 \\ u(x,0) = f(x), & -\infty < x < \infty \end{cases}. \tag{5.15}$$

Keeping the above notations in mind, the problem (5.15) can be written in the form of the following ordinary differential equation

$$\frac{dU}{dt} = -a^2\omega^2 U, \ U(\omega,0) = F(\omega)$$

whose solution is

$$U(\omega,t)=F(\omega)e^{-a^2\omega^2 t}.$$

Now, from the inverse Fourier transform discussed in Chapter 4, it follows that

$$u(x,t)=\frac{1}{2\pi}\int_{-\infty}^{\infty}U(\omega,t)e^{i\omega x}d\omega=\frac{1}{2\pi}\int_{-\infty}^{\infty}F(\omega)e^{-a^2\omega^2 t}e^{i\omega x}d\omega$$

$$=\frac{1}{2\pi}\int_{-\infty}^{\infty}\left(\int_{-\infty}^{\infty}f(\xi)e^{-i\omega^2\xi}\right)e^{-a^2\omega^2 t}e^{i\omega x}d\omega$$

$$=\frac{1}{2\pi}\int_{-\infty}^{\infty}\left(\int_{-\infty}^{\infty}e^{-a^2\omega^2 t}e^{i\omega(x-\xi)}d\omega\right)f(\xi)d\xi$$

$$=\frac{1}{\pi}\int_{-\infty}^{\infty}\left(\int_{0}^{\infty}e^{-a^2\omega^2 t}\cos\left(\omega(x-\xi)\right)d\omega\right)f(\xi)d\xi$$

$$=\frac{1}{2c\sqrt{\pi t}}\int_{-\infty}^{\infty}e^{-\frac{(x-\xi)^2}{4a^2t}}f(\xi)d\xi.$$

In the above we have used the following result

$$\int_{0}^{\infty}e^{-a^2\omega^2 t}\cos\left(\omega(x-\xi)\right)d\omega=\frac{1}{2c}\sqrt{\frac{\pi}{t}}.$$

This result can be derived by differentiating the function

$$g(\omega,\lambda)=\int_0^\infty e^{-c^2\omega^2 t}\cos(\omega\lambda)d\omega.$$

5.2.3 Solution Using Laplace Transform

Let us first introduce the following notations for Laplace transforms

$$\mathcal{L}u(x,t)=U(x,s),\ \mathcal{L}y(x,t)=Y(x,s),\ \mathcal{L}f(x,t)=F(x,s).$$

In this section we solve the initial value and boundary value problems for the heat equation using the Laplace transforms.

$$\begin{cases} u_t = a^2 u_{xx},\ 0 < x < 2,\ t > 0 \\ u(0,t) = 0,\ u(2,t) = 1,\ t > 0 \\ u(x,0) = 0,\ 0 < x < 2 \end{cases} . \tag{5.16}$$

Applying the Laplace transform to both sides of the heat equation and using the initial condition, we have

$$\frac{d^2U}{dx^2} - sU = 0.$$

The general solution of this ordinary differential equation is

$$U(x,s) = c_1 \cosh(\sqrt{s}x) + c_2 \sinh(\sqrt{s}x).$$

Using the given boundary conditions, we get

$$U(0,s) = \mathcal{L}(u(0,t)) = /,\ U(1,s) = \mathcal{L}(u(1,t)) = \mathcal{L}(\infty) = \frac{1}{s}.$$

The above boundary conditions for $U(x,s)$ imply

$$U(x,s) = \frac{1}{s}\frac{\sinh(\sqrt{s}x)}{\sinh(\sqrt{s})} = \frac{1}{s}\frac{e^{(x-1)\sqrt{s}} - e^{-(x+1)\sqrt{s}}}{1 - e^{-2\sqrt{s}}}.$$

Using the geometric series

$$\frac{1}{1 - e^{-2\sqrt{s}}} = \sum_{n=0}^{\infty} e^{-2n\sqrt{s}}$$

we obtain

$$U(x,s) = \sum_{n=0}^{\infty} \left[\frac{e^{-2(n+1-x)\sqrt{s}}}{s} - \frac{e^{-2(n+1+x)\sqrt{s}}}{s} \right].$$

Therefore, from the Laplace transform (Table 2.2) we get

$$u(x,t) = \sum_{n=0}^{\infty} \left[\mathcal{L}^{-\infty} \left(\frac{e^{-2(n+1-x)\sqrt{s}}}{s} \right) - \mathcal{L}^{-\infty} \left(\frac{e^{-2(n+1+x)\sqrt{s}}}{s} \right) \right]$$

$$= \sum_{n=0}^{\infty} \left[\operatorname{erf} \left(\frac{2n+1-x}{2\sqrt{t}} \right) - \operatorname{erf} \left(\frac{2n+1+x}{2\sqrt{t}} \right) \right]$$

where the error function $erf(.)$ is defined by

$$erf(x) = \frac{2}{\sqrt{\pi}} \int_0^{\infty} e^{-t^2}\, dt.$$

5.3 Wave Equation

Consider the initial and boundary value problem for the wave equation

$$u_{tt} = c^2 u_{xx},\ 0 < x < l,\ t > 0 \tag{5.17}$$

$$u(0,t) = 0,\ u(l,t) = 0,\ t > 0 \tag{5.18}$$

$$u(x,0) = f(x),\ u_t(x,0) = g(x),\ 0 < x < l \tag{5.19}$$

where $u(x,t)$ represents the displacement, for example of a vibrating from its equilibrium position, and c the wave speed.

Such equations have been used to model vibrating membrane, acoustic problems to determine velocity potential of the fluid flow through which sound can be transmitted, longitudinal vibrations of an elastic road or beam, and electric and magnetic fields.

5.3.1 Solution Using Fourier Series

We look for a non-trivial solution

$$u(x,t) = X(x)T(t) \tag{5.20}$$

where $X(x)$ and $T(t)$ are functions of single variables x and t, respectively. Differentiating (5.20) with respect to x and t, and substituting the partial derivatives in equation (5.17), we obtain

$$\frac{X''(x)}{X(x)} = \frac{1T''(t)}{c^2T(t)}. \tag{5.21}$$

Equation (5.21) holds identically for every $0 < x < l$. Notice that the left side of this equation is a function which depends only on x, while the right hand side is a function of only t.Since x and t are independent variables, this can happen only if each function in both sides of (5.21) is equal to the same constant λ.

$$\frac{X''(x)}{X(x)} = \frac{1T''(t)}{c^2T(t)} = \lambda.$$

From the last equation we obtain the two ordinary differential equations

$$X''(x) - \lambda X(x) = 0 \tag{5.22}$$

and

$$T''(t) - c^2\lambda T(t) = 0. \tag{5.23}$$

From the boundary conditions (5.18) it follows that

$$X(0)T(t) = X(l)T(t) = 0,\ t > 0.$$

This implies

$$X(0) = X(l) = 0$$

since $T(t) = 0$ would imply $u(x,t) = 0$. Solving the eigenvalue problem (5.22) with the above boundary conditions just as in Chapter 4, we find that the eigenvalues λ_n and corresponding eigenfunctions $X_n(x)$ are given by

$$\lambda_n = -\left(\frac{n\pi}{l}\right)^2, \quad X_n(x) = \sin\frac{n\pi x}{l}, \; n = 1, 2, \ldots.$$

The solution of the differential equation (5.23) corresponding to the above λ_n is given by

$$T_n(t) = a_n \cos\frac{n\pi ct}{l} + b_n \sin\frac{n\pi ct}{l}, \; n = 1, 2, ...$$

where a_n and b_n are constants to be determined. Therefore, we obtain a sequence of functions

$$u_n(x,t) = X_n(x)T_n(t)$$

or

$$u_n(x,t) = \left(a_n \cos\frac{n\pi ct}{l} + b_n \sin\frac{n\pi ct}{l}\right)\sin\frac{n\pi x}{l}, \; n = 1, 2, \ldots,$$

each of each satisfies the wave equation (5.17) and the boundary conditions (5.18). Since the wave equation and the boundary conditions are linear and homogeneous, a function $u(x,t)$ of the form

$$u(x,t) = \sum_{n=1}^{\infty} u_n(x,t) = \sum_{n=1}^{\infty}\left(a_n \cos\frac{n\pi ct}{l} + b_n \sin\frac{n\pi ct}{l}\right)\sin\frac{n\pi x}{l} \tag{5.24}$$

will also satisfy the wave equation and the boundary conditions. If we assume that the above series is convergent and that it can be differentiated term by term with respect to t, from (5.24) and the initial conditions (5.19), we obtain

$$f(x) = u(x,0) = \sum_{n=1}^{\infty} a_n \sin\frac{n\pi x}{l}, \; 0 < x < l \tag{5.25}$$

and

$$g(x) = u_t(x,0) = \sum_{n=1}^{\infty} \frac{n\pi c}{l} b_n \sin\frac{n\pi x}{l}, \; 0 < x < l. \tag{5.26}$$

Using the Fourier sine series (from Chapter 4) for the functions $f(x)$ and $g(x)$ from (5.25) and (5.26), we obtain

$$a_n = \frac{2}{l}\int_0^l f(x)\sin\frac{n\pi x}{l}dx, \quad b_n = \frac{2}{n\pi c}\int_0^l g(x)\sin\frac{n\pi x}{l}dx, \; n = 1, 2, \ldots.$$

A formal justification that the function $u(x,t)$ is the solution of the wave equation is given by Theorem 45 in Chapter 4.

5.3.2 Solution Using Fourier Transform

Let us recall notations of Fourier transforms and a few important properties

$$\mathcal{F}u(x,t) = U(x,\omega),\ \mathcal{F}f(x) = F(\omega),\ \mathcal{F}g(x) = G(\omega)$$

$$\mathcal{F}\left(u(x,t)\right) = U(x,\omega) = \int_{-\infty}^{\infty} u(x,t)e^{-i\omega x}dx$$

$$\mathcal{F}\left(u_x(x,t)\right) = i\omega U(\omega,t)$$

$$\mathcal{F}\left(u_{xx}(x,t)\right) = (i\omega)^2 U(\omega,t) = -\omega^2 U(\omega,t)$$

$$u(x,t) = \mathcal{F}^{-\infty}U(x,\omega) = \frac{1}{2\pi}\int_{-\infty}^{\infty} U(x,\omega)e^{i\omega x}d\omega$$

$$\mathcal{F}\left(u_t(x,t)\right) = \frac{d}{dt}\left(\mathcal{F}\left(u(x,t)\right)\right)$$

$$\mathcal{F}\left(u_{tt}(x,t)\right) = \frac{d^2}{dt^2}\left(\mathcal{F}\left(u(x,t)\right)\right) = U_{tt}(\omega,t).$$

Let us now solve the wave equation with initial conditions

$$u_{tt} = c^2 u_{xx}, \quad -\infty < x < \infty,\ t > 0 \tag{5.27}$$

$$u(x,0) = f(x),\ u_t(x,0) = g(x),\ t > 0 \tag{5.28}$$

using the Fourier transform. Applying the Fourier transform $\mathcal{F}$ to the wave equation, we get

$$\mathcal{F}\left(u_{tt}(x,t)\right) = c^2\mathcal{F}\left(u_{xx}(x,t)\right)$$

or the ordinary differential equation

$$\frac{d^2U(\omega,t)}{dt^2} + c^2\omega^2 U(\omega,t) = 0.$$

The general solution is

$$U(\omega,t) = C_1(\omega)\cos(\omega ct) + C_2(\omega)\sin(\omega ct).$$

Now,

$$C_1(\omega) = U(\omega,0) = F(\omega)$$

and

$$\omega c C_2(\omega) = U_t(\omega,0) = G(\omega).$$

so

$$C_2(\omega) = \frac{G(\omega)}{\omega c}.$$

Hence

$$U(\omega,t) = F(\omega)\cos(\omega ct) + \frac{1}{\omega c}G(\omega)\sin(\omega ct).$$

Applying the inverse Fourier transform, we get

$$u(x,t) = \frac{1}{2\pi}\int_{-\infty}^{\infty}\left[F(\omega)\cos(\omega ct) + \frac{1}{\omega c}G(\omega)\sin(\omega ct)\right]e^{i\omega x}\,d\omega.$$

5.3.3 Solution Using Laplace Transform

As usual, we use the following notations for the Laplace transform

$$\mathcal{L}u(x,t) = U(x,s),\ \mathcal{L}y(x,t) = Y(x,s),\ \mathcal{L}f(x,t) = F(x,s)$$

and

$$\begin{aligned}
\mathcal{L}\left(u_t(x,t)\right) &= sU(x,s) - u(x,0)\\
\mathcal{L}\left(u_{tt}(x,t)\right) &= s^2U(x,s) - su(x,0) - u_t(x,0)\\
\mathcal{L}\left(u_x(x,t)\right) &= \frac{d}{dx}\left(\mathcal{L}u(x,t)\right)\\
\mathcal{L}\left(u_{xx}(x,t)\right)\frac{d^2}{dx^2}\left(\mathcal{L}u(x,t)\right) &= \frac{d^2U(x,s)}{dx^2}.
\end{aligned}$$

Now, using the Laplace transform of both sides of the wave equation (5.17) and the initial conditions (5.19), we have

$$\frac{d^2U}{dx^2} - s^2U = -\cos\pi x$$

and the general solution of the above ordinary differential equation is

$$U(x,s) = c_1e^{sx} + c_2e^{-sx} + \frac{\cos\pi x}{s^2+\pi^2}.$$

Using the boundary condition for the function $u(x,t)$ we see that the function $U(x,s)$ satisfies the conditions

$$U(0,s) = \mathcal{L}u(0,t) = 0,\ U(1,s) = \mathcal{L}u(1,t) = 0.$$

These conditions for $U(x,s)$ imply $c_1 = c_2 = 0$ and so

$$U(x,s) = \frac{\cos\pi x}{s^2+\pi^2}.$$

Therefore

$$\begin{aligned}
u(x,t) &= \mathcal{L}^{-\infty}\left(U(x,s)\right) = \mathcal{L}^{-\infty}\left(\frac{\cos\pi x}{s^2+\pi^2}\right)\\
&= \cos\pi x\mathcal{L}^{-\infty}\left(\frac{1}{s^2+\pi^2}\right) = \frac{1}{\pi}\cos\pi x\sin\pi t.
\end{aligned}$$

5.4 Laplace Equation

$$\frac{\partial^2 \phi}{\partial x^2} + \frac{\partial^2 \phi}{\partial y^2} = 0. \tag{5.29}$$

The equation is also written as $\nabla^2 \phi = 0$, where $\nabla = \begin{bmatrix} \frac{\partial}{\partial x} \\ \frac{\partial}{\partial y} \end{bmatrix}$. This equation has no dependence on time, just on the spatial variables x, y. The Laplace's equation describes **steady state** situations such as:

(i) Temperature distributions

(ii) Stress distributions

(iii) Potential distributions (also called potential equations)

(iv) Flows, for example in a cylinder or around a corner

Stress analysis example: Dirichlet conditions
One steady state stress analysis problem, which satisfies the Laplace equation is a stretched elastic membrane on a rectangular form that has prescribed out-of-plane displacements along the boundaries.
To solve:

$$\frac{\partial^2}{\partial x^2} + \frac{\partial^2}{\partial y^2} = 0.$$

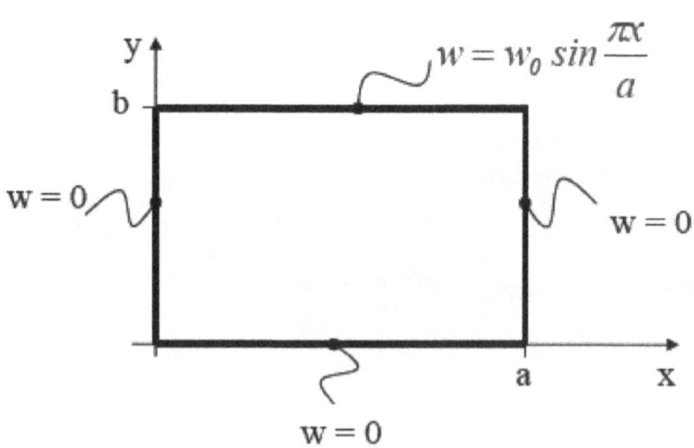

FIGURE 5.1: Laplace Equation with Dirichlet Equation

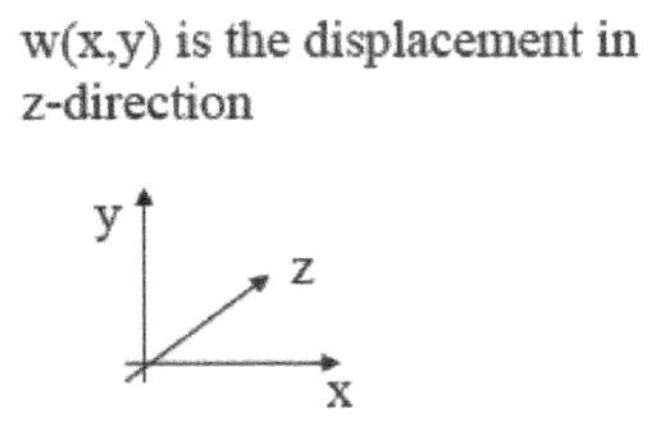

FIGURE 5.2: Displacement

Boundary conditions are:

$$\begin{aligned} w(0,y) &= 0, && \text{for } 0 \le y \le b. \\ w(x,0) &= 0, && \text{for } 0 \le x \le a. \\ w(a,y) &= 0, && \text{for } 0 \le y \le b. \\ w(x,b) &= w_0 \sin \frac{\pi}{x} a, && \text{for } 0 \le x \le a. \end{aligned}$$

Solution by separation of variables

Let $w(x,y) = X(x)Y(y)$ be solution of (5.29). Substituting it in (5.29), we get

$$\begin{aligned} X''Y + XY'' &= 0 \\ \frac{X''}{X} + \frac{Y''}{Y} &= 0 \\ \frac{X''}{X} = -\frac{Y''}{Y} &= k \end{aligned}$$

where k is a constant that is either equal to 0, or > 0, or < 0.

Case I: When $k = 0$, then

$$X(x) = (Ax + B), \quad Y(y) = (Cy + D)$$

$$w(0,y) = 0 \Rightarrow B = 0 \text{ or } C = D = 0.$$

If $C = D = 0$, then $Y(y) \equiv 0$, so $w(x,y) \equiv 0$.
Continuing with $B = 0$, we have

$$w(x,y) = Ax(Cy + D).$$

When $w(x,0) = 0 \Rightarrow ADx = 0$
Either $A = 0$ (so $w \equiv 0$) or $D = 0$.
Continuing with $w(x,y) = ACxy$,

$$\begin{aligned} w(a,y) = 0 &\Rightarrow ACay = 0 \\ A = 0 \text{ or } C = 0 &\Rightarrow w(x,y) \equiv 0. \end{aligned}$$

That is, the case $k = 0$ is not possible.

Case II: When $k > 0$,
Suppose that $k = \alpha^2$, so that

$$w(x, y) = (A \cosh \alpha x + B \sinh \alpha x)(A \cos \alpha y + B \sin \alpha y).$$

Recall that $\cosh 0 = 1$, $\sinh 0 = 0$.

$$\begin{aligned} w(0, y) = 0 \quad &\Rightarrow \quad A(C \cos \alpha y + D \sin \alpha y) = 0 \\ C = D = 0 \quad &\Rightarrow \quad w(x, y) \equiv 0. \end{aligned}$$

Continue with $A = 0 \Rightarrow w(x, y) = B \sinh \alpha x(C \cos \alpha y + D \sin \alpha y)$

$$\begin{aligned} w(x, 0) = 0 \quad &\Rightarrow \quad BC \sinh \alpha x = 0 \\ B = 0 \quad &\Rightarrow \quad w(x, y) \equiv 0. \end{aligned}$$

Continue with $C = 0 \Rightarrow w(x, y) = BD \sinh \alpha x \sin \alpha y$

$$\begin{aligned} w(a, y) = 0 \quad &\Rightarrow \quad BD \sinh \alpha a \sin \alpha y = 0 \\ \text{so either } B = 0 \text{ or } D = 0 \quad &\Rightarrow \quad w(x, y) \equiv 0. \end{aligned}$$

Again, we find that the case $k > 0$ is not possible.

Case III: When $k < 0$,
Suppose that $k = -\alpha^2$,

$$w(x, y) = (A \cos \alpha x + B \sin \alpha x)(C \cosh \alpha y + D \sinh \alpha y)$$

$$\begin{aligned} w(0, y) = 0 \quad &\Rightarrow \quad A(C \cosh \alpha y + D \sinh \alpha y) = 0 \\ C = D = 0 \quad &\Rightarrow \quad w(x, y) \equiv 0. \end{aligned}$$

Continue with $A = 0 \Rightarrow w(x, y) = B \sin \alpha x(C \cosh \alpha y + D \sinh \alpha y)$

$$\begin{aligned} w(x, 0) = 0 \quad &\Rightarrow \quad BC \sin \alpha x = 0 \\ B = 0 \quad &\Rightarrow \quad w(x, y) \equiv 0. \end{aligned}$$

Continue with $C = 0 \Rightarrow w(x, y) = BD \sin \alpha x \sinh \alpha y$

$$\begin{aligned} w(a, y) = 0 \quad &\Rightarrow \quad BD \sin \alpha a \sinh \alpha y = 0 \\ \text{so either } B = 0 \text{ or } D = 0 \quad &\Rightarrow \quad w(x, y) \equiv 0. \end{aligned}$$

$$\sin \alpha a = 0 \Rightarrow \alpha = n\frac{\pi}{a}$$

$$\Rightarrow w_n(x, y) = BD \sin n\frac{\pi}{a}x \sinh n\frac{\pi}{a}y.$$

Applying the first three boundary conditions, we have

$$w(x, y) = \sum_{n=1}^{\infty} K_n \sin \frac{n\pi x}{a} \sinh \frac{n\pi y}{a}.$$

The final boundary condition is:

$$w(x,b) = w_0 \sin\frac{\pi x}{a}$$

which gives

$$w_0 \sin\frac{\pi x}{a} = \sum_{n=1}^{\infty} K_n \sin\frac{n\pi x}{a} \sinh\frac{n\pi y}{a}.$$

We can see from this that n must take only one value, namely 1, so that $K_1 = \dfrac{w_0}{\sinh\frac{\pi b}{a}}$ and the final solution to the stress distribution is

$$w(x,y) = \frac{w_0}{\sinh\frac{\pi b}{a}} \sin\frac{\pi x}{a} \sinh\frac{\pi y}{a}.$$

5.4.1 Solution Using Fourier Series

Example 235. Consider the following Laplace equation with Dirichlet boundary value

$$\begin{cases} u_{xx} + u_{yy} = 0.\ 0 < x < a.\ 0 < y < b \\ u(x,0) = 0,\ 0 \leq x \leq a \\ u(0,y) = u(a,y) = 0,\ \ 0 \leq y \leq b \\ u(x,b) = f(x),\ 0 \leq x \leq a. \end{cases}$$

We solve this problem by separation of variables. Let $u(x,y) = X(x)Y(y)$ and substitute into Laplace's equation to obtain

$$X''Y + XY'' = 0$$

or

$$\frac{X}{X} = -\frac{Y''}{Y} = -\lambda$$

in which both $\dfrac{X''}{X}$ and $-\dfrac{Y''}{Y}$ must be constant because x and y are independent. Then

$$X'' + \lambda X = 0$$

and

$$Y'' - \lambda Y = 0.$$

From the boundary conditions

$$u(x,0) = X(x)Y(0) = 0,$$

hence $Y(0) = 0$ because $X(x)$ cannot be identically zero. Similarly,

$$X(0) = X(a) = 0.$$

The problem for X is the well-known eigenvalue problem

$$X'' + \lambda X = 0,\ X(0) = X(a) = 0$$

with eigenvalues

$$\lambda_n = \frac{n^2\pi^2}{a^2},\ n = 1, 2, \ldots$$

and eigenfunctions

$$X_n(x) = \sin\left(\frac{n\pi x}{a}\right),\ n = 1, 2, \ldots.$$

For any eigenvalue, the corresponding equation for y is

$$Y'' - \frac{n^2\pi^2}{a^2} Y = 0$$

with general solutions

$$Y_n = c_n e^{\frac{n\pi y}{a}} + d_n e^{-\frac{n\pi y}{a}},$$

but $Y(0) = 0$ so $d_n = -c_n$ and Y_n is of the form

$$Y_n = 2c_n \sinh\left(\frac{n\pi y}{a}\right),\ n = 1, 2, \ldots$$

and therefore,

$$\begin{aligned} u_n(x, y) &= X_n(x) Y_n(y) \\ &= b_n \sin\left(\frac{n\pi x}{a}\right) \sinh\left(\frac{n\pi y}{a}\right),\ n = 1, 2, \ldots \end{aligned}$$

which all satisfy the Laplace equation in the rectangle and the homogeneous boundary conditions on the lower and vertical sides. To find a solution satisfying the condition on the sides $y = b$, we use superposition principle

$$u(x, y) = \sum_{n=1}^{\infty} b_n \sin\left(\frac{n\pi x}{a}\right) \sinh\left(\frac{n\pi y}{a}\right). \tag{5.30}$$

Now, we need to choose b_n so that

$$u(x, b) = \sum_{n=1}^{\infty} b_n \sin\left(\frac{n\pi x}{a}\right) \sinh\left(\frac{n\pi b}{a}\right) = f(x).$$

Since this is a Fourier sine expansion of f on $[0, b]$, we have

$$b_n \sinh\left(\frac{n\pi b}{a}\right) = \frac{2}{a} \int_0^a f(x) \sin\left(\frac{n\pi x}{a}\right) dx$$

and thus

$$b_n = \frac{2}{a \sinh\left(\frac{n\pi b}{a}\right)} \int_0^a f(x) \sin\left(\frac{n\pi x}{a}\right) dx.$$

With this choice of b_n, Equation (5.30) gives the solution of this Dirichlet problem for a rectangle.

Example 236. Consider the following Laplace equation with mixed boundary conditions

$$\begin{cases} u_{xx} + u_{yy} = 0. \ 0 < x < a. \ 0 < y < b \\ u_x(0, y) = 0, \ u_x(a, y) = 0, \ 0 < y < a \\ u(x, 0) = 0, u(x, b) = f(x), \ 0 < x < a. \end{cases}$$

With $u(x, y) = X(x)Y(y)$, separation of variables leads to

$$\frac{X}{X} = -\frac{Y''}{Y} = -\lambda$$

$$X'' + \lambda X = 0$$

and

$$Y'' - \lambda Y = 0.$$

From the boundary conditions

$$u_x(0, y) = X'(0)Y(y) = 0$$

and

$$u_x(a, y) = X'(a)Y(y) = 0$$

we have

$$X'(0) = X'(a) = 0$$

because $Y(y)$ cannot be identically zero. Similarly, from

$$u(x, 0) = X(x)Y(0) = 0$$

we have

$$Y(0) = 0.$$

- **Case** $\lambda = 0$ $X'' = 0$. The solution of this ordinary differential equation (ODE) is $X(x) = c_1 + c_2 x$. The boundary value $X'(0) = 0$ then implies $c_2 = 0$, and so $X(x) = c_1$.

- **Case** $\lambda = \alpha^2 > 0$: ODE becomes

  $$X'' + \alpha^2 X = 0, \ X'(0) = X'(a) = 0.$$

  The general solution is

  $$X(x) = c_1 \cos \alpha x + c_2 \sin \alpha x.$$

Using the boundary condition $X'(0) = 0$, with

$$X'(x) = -c_1 \sin \alpha x + c_2 \cos \alpha x$$

we get $c_2 = 0$. The second boundary condition $X'(a) = 0$ yields $-c_1 \sin \alpha a = 0$. Because $\alpha > 0$, the last equation is satisfied when $\alpha a = n\pi$ or $\alpha = \dfrac{n\pi}{a}$, $n = 1, 2, \ldots$. The eigenvalues of ODE $(\ldots)$ are $\lambda_0 = 0$ and $\lambda_n = \dfrac{n^2\pi^2}{a^2}$, $n = 1, 2, \ldots$. By corresponding $\lambda_0 = 0$ with $n = 0$, the eigenfunctions of ODE are

$$X_0(x) = 1 \text{ and } X_n(x) = \cos \frac{n\pi x}{a},\ n = 1, 2, \ldots.$$

Let us now solve ODE subject to the boundary condition $Y(0) = 0$. First, $\lambda_0 = 0$ leads to $Y'' = 0$, and thus its solution is $Y(y) = c_3 + c_4 y$. But $Y(0) = 0$ implies $c_3 = 0$ so $Y(y) = c_4 y$. Second, for $\lambda_n = \dfrac{n^2\pi^2}{a^2}$, the ODE becomes

$$Y'' - \frac{n^2\pi^2}{a^2} Y = 0$$

and its general solution is

$$Y_n(y) = c_5 \cosh\left(\frac{n\pi y}{a}\right) + c_6 \sinh\left(\frac{n\pi y}{a}\right).$$

From this solution $Y(0) = 0$ implies $c_5 = 0$ so $Y = c_6 \sinh\left(\dfrac{n\pi y}{a}\right)$. Thus the products

$$u_n(x, y) = X_n(x) Y_n(y)$$

that satisfy the Laplace equation and the homogeneous boundary conditions are

$$A_0 y \text{ and } A_n \sinh \frac{n\pi y}{a} \cos \frac{n\pi x}{a}$$

where

$$A_0 = c_1 c_4.$$

The superposition principle yields

$$u(x, y) = A_0 y + \sum_{n=1}^{\infty} A_n \sinh \frac{n\pi y}{a} \cos \frac{n\pi x}{a}.$$

Finally, by substituting $y = b$ in the above Equation, we get

$$u(x, b) = f(x) = A_0 b + \sum_{n=1}^{\infty} A_n \sinh \frac{n\pi b}{a} \cos \frac{n\pi x}{a}$$

which is the cosine Fourier series of f. Therefore, by Chapter 4,

$$A_0 b = a_0/2 \text{ and } A_n \sinh \frac{n\pi b}{a} = a_n, n = 1, 2, \ldots.$$

It follows that

$$A_0 = \frac{1}{ab}\int_0^a f(x)dx$$

and

$$A_n \sinh\frac{n\pi b}{a} = \frac{2}{a}\int_0^a f(x)\cos\frac{n\pi x}{a}$$

or

$$A_n = \frac{2}{a\sinh\frac{n\pi b}{a}}\int_0^a f(x)\cos\frac{n\pi x}{a}.$$

5.4.2 Solution Using Fourier Transform

Example 237. Here we discuss solution of the Laplace equation using the Fourier transform for the problem

$$\begin{cases} u_{xx} + u_{yy} = 0, \; x > 0, \; -\infty < y < \infty \\ u(0, y) = f(y), \; \infty < y < \infty \\ \lim\limits_{x\to\infty} u(x, y) = 0, \; \infty < y < \infty. \end{cases}$$

Solution: Let

$$U(x, \omega) = \mathcal{F}(u(x, y)) = \int_{-\infty}^{\infty} u(x, y)e^{-i\omega y}dy$$

and

$$U(0, \omega) = \mathcal{F}(\{) = F(\omega).$$

If we apply the Fourier transform to the Laplace equation, then in view of the boundary conditions for the function $u(x, y)$ at $x = 0$, we get the ordinary differential equation

$$\frac{d^2U(x, \omega)}{dx^2} - \omega^2 U(x, \omega) = 0, \; x > 0.$$

The general solution of of the above equation is given by

$$U(x, \omega) = c_1(\omega)e^{|\omega|x} + c_2(\omega)e^{-|\omega|x}.$$

By using the condition $\lim\limits_{x\to\infty} u(x, y) = 0$, we obtain $c_1(\omega) = 0$. and from $U(0, \omega) = F(\omega)$, we have $c_2(\omega) = F(\omega)$. Thus

$$U(x, \omega) = F(\omega)e^{-|\omega|x}.$$

Now, by applying Table 2.2 for inverse Laplace transform, we obtain

$$\mathcal{F}^{-\infty}(e^{-|\omega|x})(y) = \frac{1}{\pi}\frac{x}{x^2+y^2}.$$

Therefore, using the convolution theorem for the Fourier transform, we get

$$u(x,y) = \frac{1}{\pi}\left(f * \frac{x}{x^2+y^2}\right)(x) = \frac{1}{\pi}\cdot\int_{-\infty}^{\infty} f(t)\frac{x}{x^2+(y-t)^2}dt$$

5.4.3 Solution Using Laplace transform

Example 238. Here we solve the Laplace equation, under appropriate boundary conditions, using the Laplace transform.

$$\begin{cases} u_{xx} + u_{yy} = 0,\ 0 < x < \pi,\ y > 0 \\ u(0,y) = u(\pi,y) = 0 \\ u(x,0) = 0,\ \ u_y(x,0) = -\sin \pi x. \end{cases}$$

We define the Laplace transform of $u(x,y)$ with respect to y by

$$\mathcal{L}(u(x,y)) = \int_0^{\infty} e^{-sy}u(x,y)dy = U(x,s).$$

The transforms of the derivatives are

$$\mathcal{L}(u_{yy}) = s^2U(x,s) - su(x,0) - u_y(x,0) \tag{5.31}$$

and

$$\mathcal{L}(u_{xx}) = \frac{d^2U}{dx^2}. \tag{5.32}$$

From (5.31) and (5.32) and the boundary conditions with respect to y, we get

$$\frac{d^2U}{dx^2} + s^2U = \sin \pi x. \tag{5.33}$$

The general solution of the corresponding homogeneous ordinary differential equation (5.33) is given by

$$U_c(x,s) = c_1 \cos sx + c_2 \sin sx.$$

The method of undeterminate coefficients gives a particular solution

$$U_p(x,s) = \frac{1}{s_2 - \pi_2}\sin \pi x,$$

then the general solution of the non-homogeneous ODE is

$$U(x,s) = U_c(x,s) + U_p(x,s) = c_1 \cos sx + c_2 \sin sx + \frac{1}{s_2 - \pi^2} \sin \pi x.$$

Now, using $\mathcal{L}(u(0,y)) = U(0,s) = 0$ and $\mathcal{L}(u(\pi,y)) = U(\pi,s) = 0$, we get $c_1 = 0$ and c_2 can be any real number. For sake of simplicity, we take $c_2 = 0$, which yields

$$U(x,s) = \frac{1}{s_2 - \pi^2} \sin \pi x$$

and, taking the inverse of both sides, we obtain

$$\begin{aligned} u(x,y) &= \mathcal{L}^{-\infty}\left(\frac{1}{s_2 - \pi_2} \sin \pi x\right) \\ &= \left(\frac{1}{\pi} \sin \pi x\right)\left(\mathcal{L}^{-\infty}\left(\frac{\pi}{s_2 - \pi^2}\right)\right) \\ &= \frac{1}{\pi} \sin \pi x \sinh \pi y. \end{aligned}$$

5.5 Poisson Equation

Poisson equation (non-homogeneous Laplace equation)
The Poisson equation is given by

$$\frac{\partial^2 w}{\partial x^2} + \frac{\partial^2 w}{\partial y^2} = f(x,y).$$

One encounters this equation while studying electrostatic potential in the presence of charge, gravitational potential in the presence of distributed matter, equilibrium displacement of a membrane under distributed forces, velocity potential for an inviscid, incompressible, irrotational homogeneous fluid in the presence of distributed sources of sinks, and state temperature in the presence of thermal sources or sinks. For solutions we refer to Adzievski and Siddiqi [1].

5.6 Simulation of Heat Equation, Wave Equation and Laplace Equation by MATLAB

Solution of one-dimensional heat equation

The reader is referred to www.Colorado.edu,http://www.colorado.edu/geography/class_homepages/geog_4023_s07/labs/html/PDE_lab.html#1

Problem : Periodic Heat Diffusion in Subsurface Rocks

```
dz = 0.25;
```
% each step depth is 1/4 meter
```
Nz = 400;
```
% Choose number of depth steps (at least 100 m)
```
Nt = 5000;
```
% Choose number of time steps
```
dt = (365*24*60*60)/Nt;
```
% Length of each time step in seconds

($\sim 6.3 * 10\wedge 3$ seconds, or ~ 105 minutes)
```
K = 2*10^-6;
```
% canonical K is $10e-6m\wedge 2/s$
```
T = 15*ones(Nz+1,Nt+1);
```
% Create temperature matrix with $Nz+1$ rows, and $Nt+1$ columns

% Initial guess is that T is 15 everywhere.
```
time = [0:12/Nt:12];
T(1,:) = 15-10*\sin(2*pi*time/12);  Nt = 5000;
```
% Choose number of time steps
```
dt = (365*24*60*60)/Nt;
```
% Length of each time step in seconds

($\sim 6.3 * 10\wedge 3$ seconds, or ~ 105 minutes)

K $= 2 * 10\wedge - 6$;

% canonical K is $10e-6m\wedge 2/s$
```
T = 15*ones(Nz+1,Nt+1);
```
% Create temperature matrix with $Nz+1$ rows and $Nt+1$ columns

% Initial guess is that T is 15 everywhere.
```
time = [0:12/Nt:12];
T(1,:) = 15-10*\sin(2*pi*time/12);     Nt = 5000;
```
% Choose number of time steps

```
dt = (365*24*60*60)/Nt;
```
% Length of each time step in seconds
($\sim 6.3 * 10$^3 seconds, or $\sim$ 105 minutes)
```
K =  2*10^ - 6;
```
% canonical K is $10e - 6m$^2/s
```
T = 15*ones(Nz+1,Nt+1);
```
% Create temperature matrix with $Nz + 1$ rows, and $Nt + 1$ columns
% Initial guess is that T is 15 everywhere.
```
time = [0:12/Nt:12];
T(1,:) = 15-10*sin(2*pi*time/12);     Nt = 5000;
```
% Choose number of time steps
```
dt = (365*24*60*60)/Nt;
```
% Length of each time step in seconds
($\sim 6.3 * 10$^3 seconds, or $\sim$ 105 minutes)
```
K =  2*10^ - 6;
```
% canonical K is $10e - 6m$^2/s
```
T = 15*ones(Nz+1,Nt+1);
```
% Create temperature matrix with $Nz + 1$ rows and $Nt + 1$ columns
% Initial guess is that T is 15 everywhere.
```
time = [0:12/Nt:12];
T(1,:) = 15-10*\sin(2*pi*time/12);
```
% Set surface temperature
```
maxiter = 500
for iter = 1:maxiter
Tlast = T;
```
% Save the last guess
```
 T(:,1) = Tlast(:,end);
```
% Initialize the temp at $t = 0$ to the last temp
```
Nt = 5000;
```
% Choose number of time steps
```
dt = (365*24*60*60)/Nt;
```
% Length of each time step in seconds
($\sim 6.3 * 10$^3 seconds, or $\sim$ 105 minutes)
```
K =  2 * 10^ - 6;
```
% canonical K is $10e - 6m$^2/s
```
T = 15*ones(Nz+1,Nt+1);
```
% Create temperature matrix with $Nz + 1$ rows, and $Nt + 1$ columns

% Initial guess is that T is 15 everywhere.

```
time = [0:12/Nt:12];
T(1,:) = 15-10*sin(2*pi*time/12);   Nt = 5000;
```

% Choose number of time steps

```
dt = (365*24*60*60)/Nt;
```

% Length of each time step in seconds

($\sim 6.3 * 10^\wedge 3$ seconds, or ~ 105 minutes)

K = $2 * 10^\wedge - 6$;

% canonical K is $10e - 6m^\wedge 2/s$

```
T = 15*ones(Nz+1,Nt+1);
```

% Create temperature matrix with $Nz + 1$ rows, and $Nt + 1$ columns

% Initial guess is that T is 15 everywhere.

```
time = [0:12/Nt:12];
T(1,:) = 15-10*\sin(2*pi*time/12);
```

% Set surface temperature

```
maxiter = 500
for iter = 1:maxiter
Tlast = T;
```

% Save last guess

```
T(:,1) = Tlast(:,end);
```

% Initialize temp at $t = 0$ to last temp

% Set surface temperature

```
maxiter = 500
for iter = 1:maxiter
Tlast = T;
```

% Save last guess

```
T(:,1) = Tlast(:,end);
```

% Initialize temp at $t = 0$ to last temp

% Set surface temperature

```
maxiter = 500
for iter = 1:maxiter
Tlast = T;
```

% Save last guess

```
T(:,1) = Tlast(:,end);
```

% Initialize temp at $t = 0$ to last temp

```
% Set surface temperature
maxiter = 500
for iter = 1:maxiter
Tlast = T;
% Save last guess
T(:,1) = Tlast(:,end);
% Initialize temp at t = 0 to last temp
% Set surface temperature
maxiter = 500
for iter = 1:maxiter
Tlast = T;
% Save last guess
T(:,1) - Tlast(:,end);
% Initialize temp at t = 0 to last temp
for i=2:Nt+1,
depth_2D = (T(1:end-2,i-1)-2*T(2:end-1,i-1)+T(3:end,i-1))/dz^2;
time_1D = K*depth_2D;
T(2:end-1,i) = time_1D*dt + T(2:end-1,i-1);
T(end,i) = T(end-1,i);
% Enforce bottom BC
end
err(iter) = max(abs(T(:)-Tlast(:)));
% Find difference between last two solutions
if err(iter)<1E-4
break;
% Stop if solutions very similar, we have convergence
end
end
if iter==maxiter;
warning('Convergence not reached')
end
figure(1)
plot(log(err)), title('Convergence plot')
figure(2)
imagesc([0 12],[0 100],T);
title('Temperature plot (imagesc)')
```

```
colorbar
figure(3)
depth = [0:dz:Nz*dz];
contourf(time,-depth,T);
title('Temperature plot (contourf)')
colorbar
```

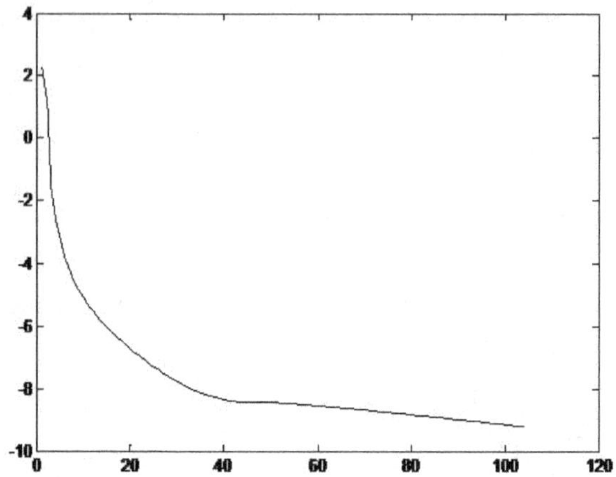

FIGURE 5.3: Convergence Plot of Heat Equation

Solution of two-dimensional heat equation in MATLAB:

```
clear; close all; clc
n =10;
% Grid has n − 2 interior points per dimension (overlapping)
x = linspace(0,1,n); dx = x(2)-x(1); y = x; dy = dx;
TOL = 1e-6;
T = zeros(n);
T(1,1:n) = 10;      % TOP
T(n,1:n) = 1;       % BOTTOM
T(1:n,1) = 1;       % LEFT
T(1:n,n) = 1;       % RIGHT
dt = dx^2/4;
error = 1; k = 0;
while error > TOL
```

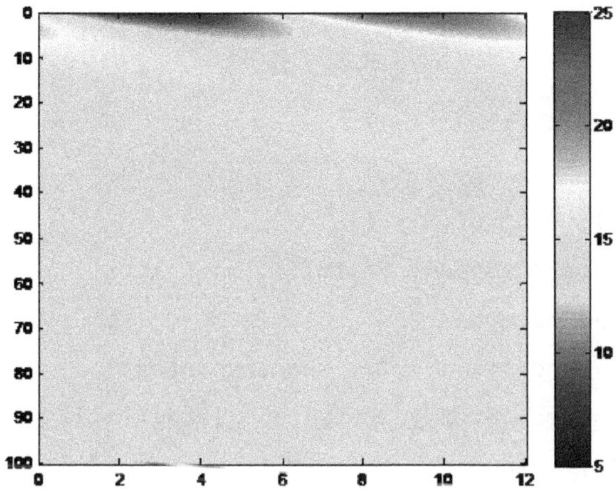

FIGURE 5.4: Temperature Plot (imagesc)

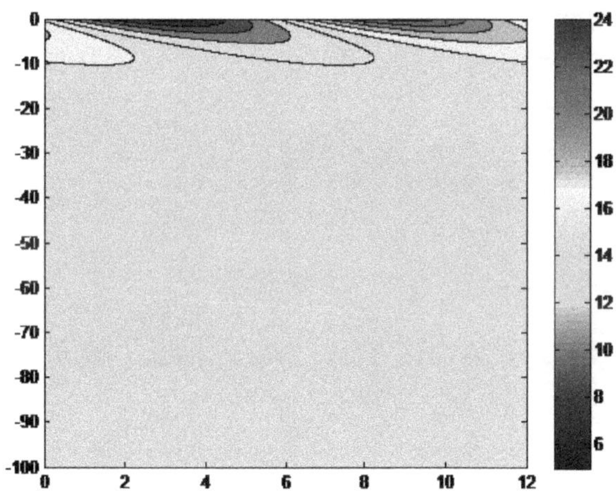

FIGURE 5.5: Temperature Plot (contourf)

```
k = k+1;
Told = T;
for i = 2:n-1
for j = 2:n-1
T(i,j) = dt*((Told(i+1,j)-2*Told(i,j)+Told(i-1,j))/dx^2...
```

```
    + (Told(i,j+1)-2*Told(i,j)+Told(i,j-1))/dy^2)...
   + Told(i,j); end
end
error = max(max(abs(Told-T)));
end
subplot(2,1,1), contour(x,y,T)
title('Temperature (Steady State)'), xlabel('x'),
   ylabel('y'), colorbar
subplot(2,1,2), pcolor(x,y,T), shading interp,
title('Temperature (Steady State)'), xlabel('x'),
   ylabel('y'), colorbar
```

```
clear; close all; clc;
c = 1;
n = 21;
x = linspace(0,1,n);
dx = 1/(n-1);
dt = dx;
u(:,1) = sin(pi*x);
u(1,2) = 0;
for i = 2:n-1
u(i,2) = 0.5*(dt^2*c^2*(u(i+1,1)-2*u(i,1)+u(i-1,1))/dx^2
     +2*u(i,1));
end
u(n,2) = 0;
error = 1; k = 1;
while k < 100
k = k+1;
u(1,k+1) = 0;
for i = 2:n-1
u(i,k+1) = dt^2*c^2*(u(i+1,k)-2*u(i,k)+u(i-1,k))/dx^2
    +2*u(i,k)-u(i,k-1);
```

```
end
u(n,k+1) = 0;
end
plot(x,u), xlabel('x'),ylabel('y')
```

Wave equation in one dimension

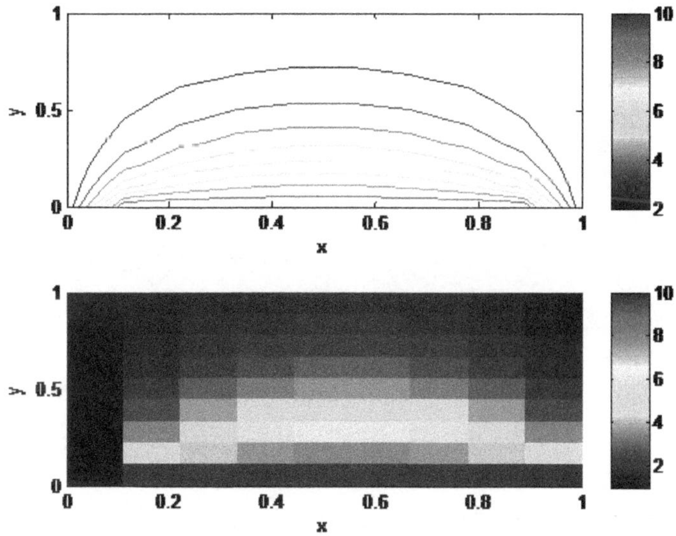

FIGURE 5.6: Temperature (Steady State)

```
clear; close all; clc;
for c = [1 -1];
cc = 0;
n = 261;
x = linspace(0,13,n);
u = zeros(n,1);
u(121:141) = sin(pi*x(121:141));
dx = x(2)-x(1)
dt = dx;
error = 1;
TOL = 1e-6;
k = 0;
```

```
while k < 110
uold = u;
k = k+1;
 for i = 2:n-1
     if c == 1, u(i) = dt*c*(uold(i+1)-uold(i))/dx+uold(i);
     end \%c = 1
if c == -1, u(i) = dt*c*(uold(i)-uold(i-1))/dx+uold(i);
       end \%c = -1
end
error = max(abs(u-uold));
   if mod(k,10)==0, cc = cc+1; out(cc,:) = u;
end
end
if c == 1
subplot(2,1,1),
for hh = 1:cc
plot(x,out(hh,:)+hh),
hold on,
end
u = zeros(n,1);
u(121:141) = sin(pi*x(121:141)); plot(x,u)
xlabel('u(x)'), ylabel('Time'),
title('Translation to the Left')
elseif c == -1
subplot(2,1,2)
for hh = 1:cc
plot(x,out(hh,:)+hh),hold on,
end
u = zeros(n,1);
u(121:141) = \sin(pi*x(121:141)); plot(x,u)
xlabel('u(x)'), ylabel('Time'),
title('Translation to the Right')
end
end
```

Wave equation in two dimensions

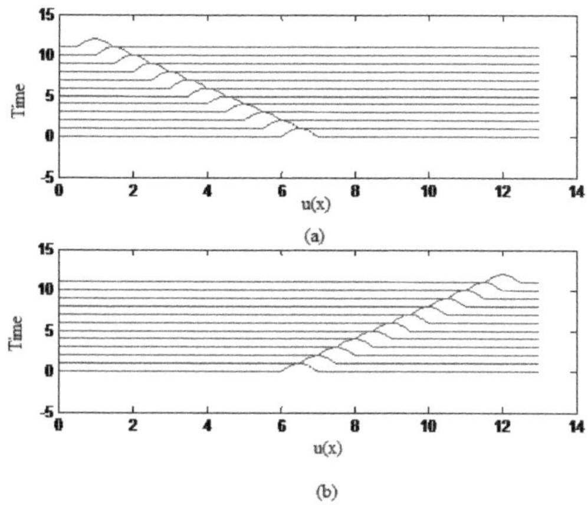

FIGURE 5.7: (a) Translation to Left (b) Translation to Right

```
clear; close all; clc;
c = 1;
n = 21;
x = linspace(0,1,n);
dx = 1/(n-1);
dt = dx;
u(:,1) = sin(pi*x);
u(1,2) = 0;
for i = 2:n-1
u(i,2) = 0.5*(dt^2*c^2*(u(i+1,1)-2*u(i,1)+u(i-1,1))/dx^2
        +2*u(i,1));
end
u(n,2) = 0;
error = 1; k = 1;
while k < 100
k = k+1;
u(1,k+1) = 0;
for i = 2:n-1
u(i,k+1) = dt^2*c^2*(u(i+1,k)-2*u(i,k)+u(i-1,k))/dx^2
       +2*u(i,k)-u(i,k-1);
end
u(n,k+1) = 0;
```

```
end
plot(x,u), xlabel('x'),ylabel('y')
```

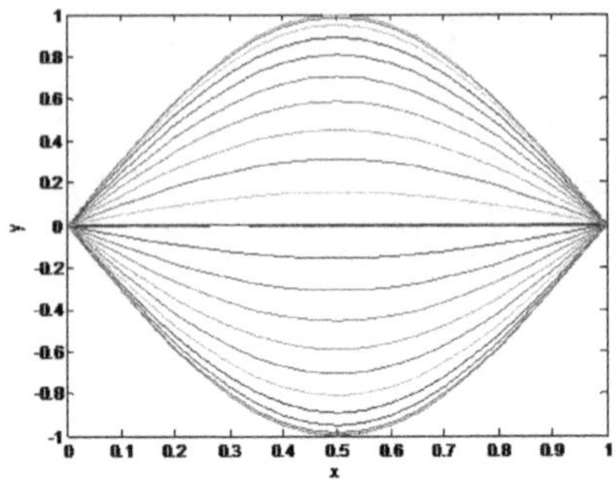

FIGURE 5.8: Solution of Laplace Equation in one dimension

Laplace equation in one dimensional with matlab dirichlet boundary condition

(References: https://math.stackexchange.com/questions/692532/laplace-equation-in-1d-with-matlab-dirichlet-boundary-condition)

MATLAB Program

% Solve equation $-u''(x) = f(x)$ with Dirichlet boundary condition

```
clear all
close all
N=2;
for j=1:10
a=0;
b=14;
N=2*N;
M(j)=N;
delta_x=(b-a)/N;
k=2;
A=sparse(N-1,N-1);
for i=1:N-1
if (i==1)
A(i,i)=2;
A(i,i+1)=-1;
```

```
elseif(i==N-1)
A(i,i-1)=-1;
A(i,i)=2;
else
A(i,i-1)=-1;
A(i,i+1)=-1;
A(i,i)=2;
end
end
A=A/((delta_x)^2);
for i=1:N-1
b(i)=functionf(i*delta_x,k)
end
u=A^(-1)*b';
for i=1:N+1
u_ex(i)=exact_solution((i-1)*delta_x,k);
end
for i=1:N+1
if (i==1)
u_dis(i)=0;
elseif(i==N+1)
u_dis(i)=0;
else
u_dis(i)=u(i-1,1);
end
end
for i=1:N+1
t(i)=(i-1)*delta_x;
end
norm_maxl2(j)=0;
for i=1:N+1
if (abs(u_dis(i)-u_ex(i)) > norm_maxl2(j))
norm_maxl2(j)=abs(u_dis(i)-u_ex(i));
end
end
norm_maxl2(j);
```

```
norm_l2(j)=0;
for i=1:N+1
norm_l2(j)=norm_l2(j)+(u_dis(i)-u_ex(i))^2*delta_x;
end
norm_l2(j)=(norm_l2(j))^1/2);
norm_maxh1(j)=0;
for i=1:N
if (abs(((u_dis(i+1)-u_ex(i+1))
       -(u_dis(i)-u_ex(i)))/delta_x) > norm_maxh1(j))
norm_maxh1(j)=abs(((u_dis(i+1)-u_ex(i+1))
       -(u_dis(i)-u_ex(i)))/delta_x);
end
end
norm_maxh1(j);
norm_h1(j)=0;
for i=1:N
norm_h1(j)=norm_h1(j)+(((u_dis(i+1)-u_ex(i+1))
      -(u_dis(i)-u_ex(i)))/delta_x)^2*delta_x;
end
norm_h1(j)=(norm_h1(j))^(1/2);
figure
hold on
plot(t,u_ex,'blue', t,u_dis,'red')
ylim([-0.02 0.12]);
hold off
end
plot(log(M), -log(norm_maxl2),'blue',log(M), -log(norm_l2), 'red',
     log(M), -log(norm_maxh1), 'cyan', log(M), -\log(norm_h1),
     'magenta', log(M), 2*log(M),'black')
function f=functionf(x,k);
if (k==1)
f=2;
end
if (k==2)
f=-6*x+12*x^2;
end
```

```
function u_ex=exact_solution(x,k);
if (k==1)
u_ex=x*(1-x);
end
if(k==2)
u_ex=x^3*(1-x);
end
```

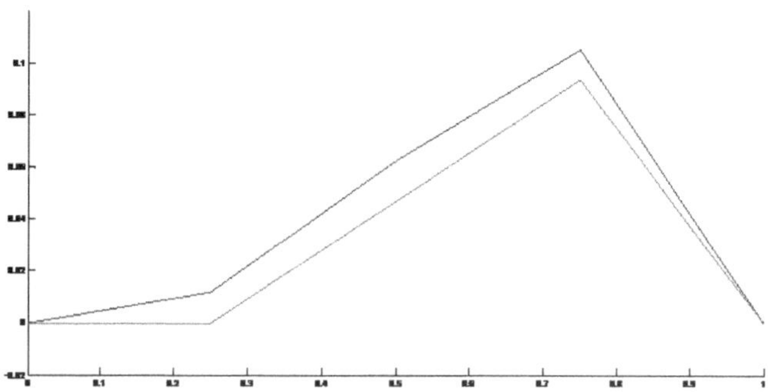

FIGURE 5.9: Solution of Laplace Equation in Two dimensions

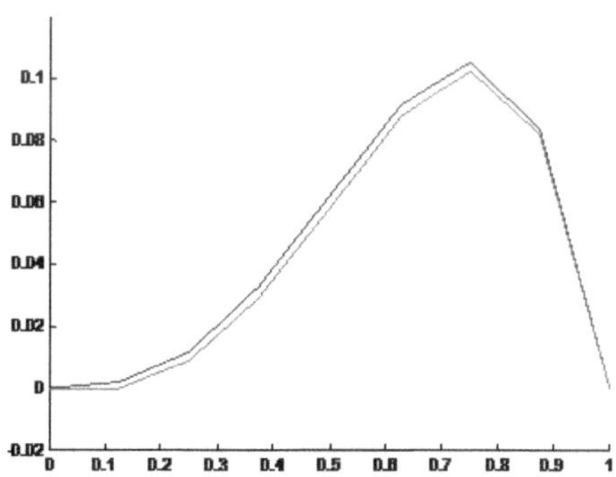

FIGURE 5.10: Solution of Laplace Equation in $1D$

Laplace equation in two dimensions

There are two parallel semi-infinite metal plates in the XZ plane. One plate is located at $y = -1m$ and extends from $x = 0m$ to $x = 4m$ and the other plate is located at $y = +1m$ and extends from $x = 0m$ to $x = 4m$. At $x = 0m$, an

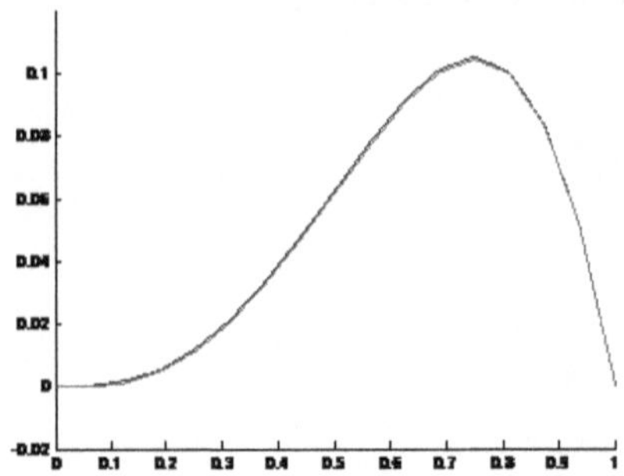

FIGURE 5.11: Solution

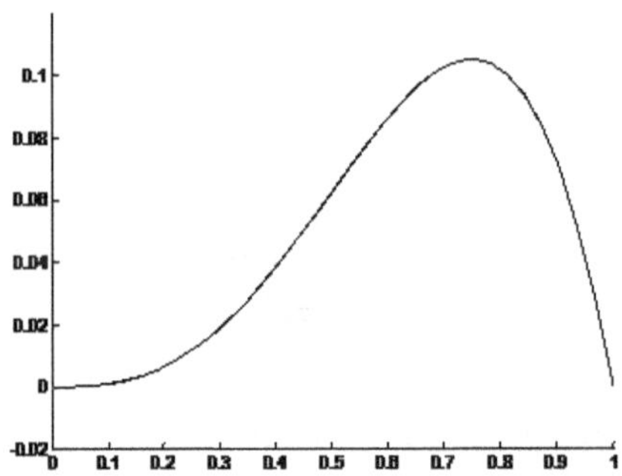

FIGURE 5.12: Solution

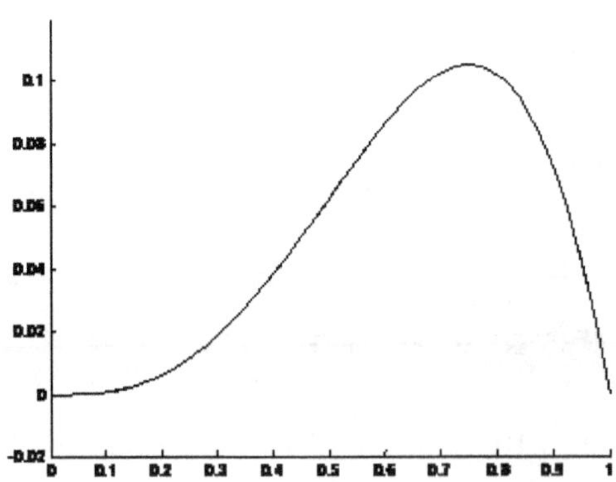

FIGURE 5.13: Solution

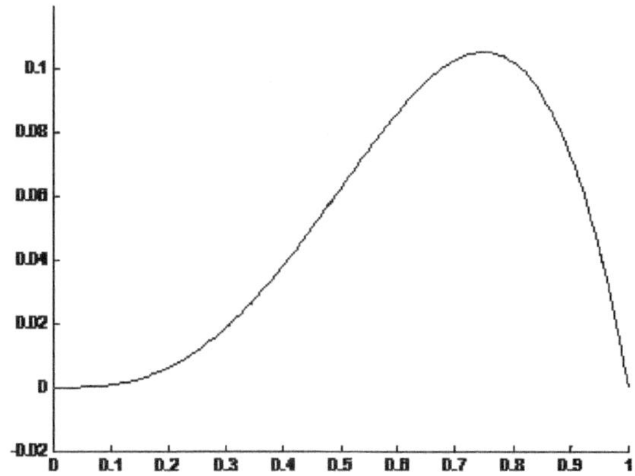

FIGURE 5.14: Solution

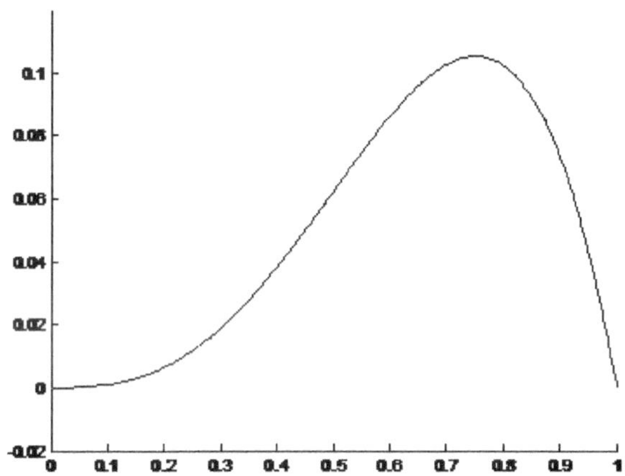

FIGURE 5.15: Solution

insulated strip connects the two semi-infinite plates and is held at a constant potential $100V$. Find the potential and the electric field in the region bounded by the plates.

Figures 5.17 plot the potential and show that the potential V has no local maximum or minima; all extremes are on the boundaries. The solution of the Laplace equation is the most featureless function possible, consistent with the boundary conditions: no hills, no valleys, only the smoothest surface.

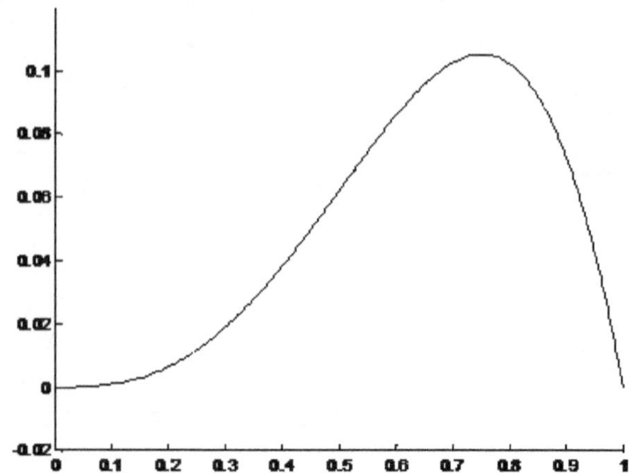

FIGURE 5.16: Solution

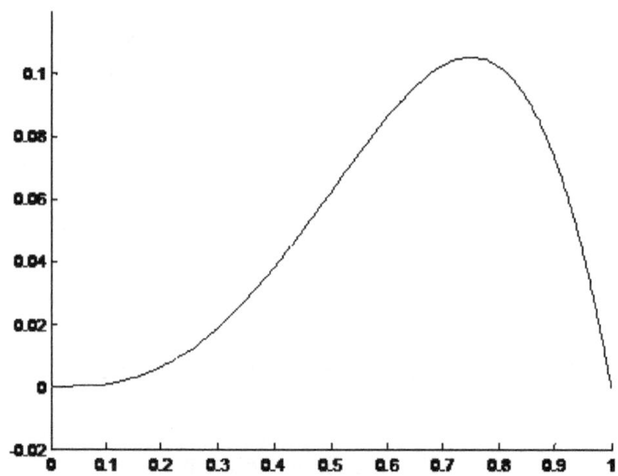

FIGURE 5.17: Solution

5.7 Solving Partial Differential Equation by MATHEMATICA

Solving Differential Equations Symbolically

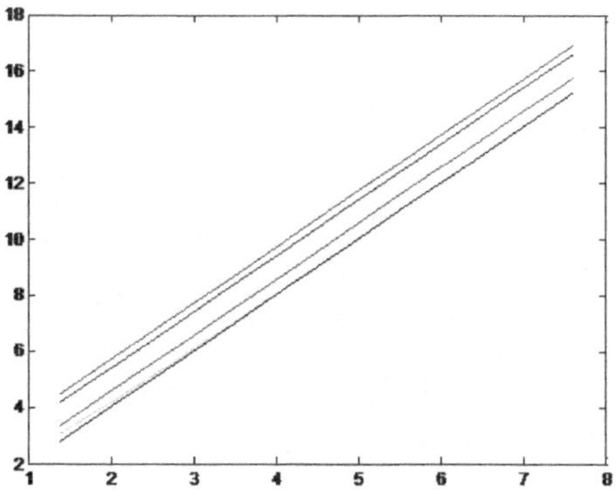

FIGURE 5.18: Simulation of Heat Equation by Mathematica

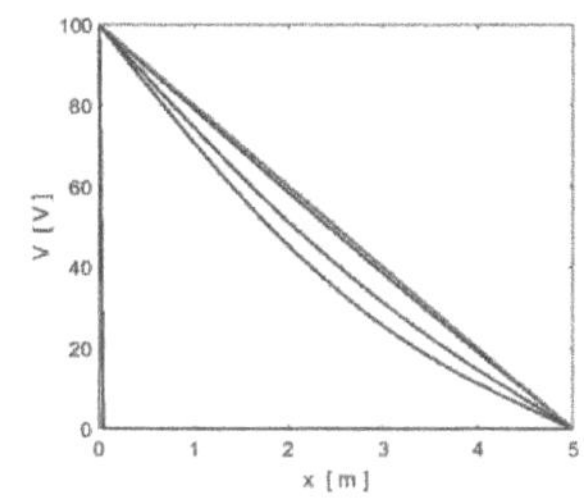

FIGURE 5.19: Solution of Heat Equation by Mathematica

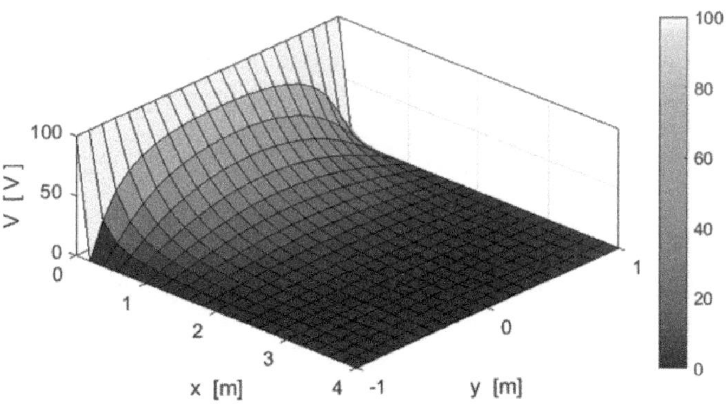

FIGURE 5.20: Solution of Heat Equation by Mathematica

Example 239. Find the general solution of the partial differential equation

$$2z_x(x,t) + 5z_t(x,t) = z(x,t) - 1.$$

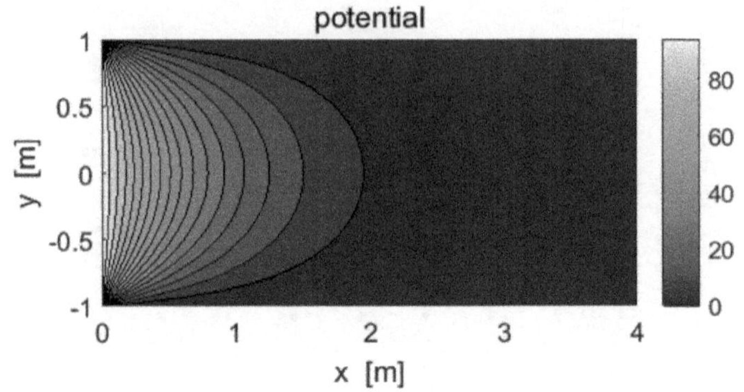

FIGURE 5.21: Potential V in region between plates

```
In[ ]:=DSolve[2D[z[x, t], x] + 5D[z[x, t], t] ==
    z[x, t] + 1, z, \{x, t\}].
Out[ ]:={{z ->({x, t} - > e^{1/2}c_1(1/2(2t-5x))-1)\}\}.
```

Example 240. Find the general solution of the second order partial differential equation

$$3x_{xx}(x,t) - 2z_{tt}(x,t) = 1.$$

```
In[ ]:=DSolve[3D[z[x, t], {x, 2}] - 2D[z[x, t], y, 2] == 1,
   z, {x, t}].
Out[ ]:= {{z -> ({x,t} -> (e^{x/2}c_1((1/2)(2t-5x))-1))}}.
```

```
In[ ]:={{z -> ({x,t} -> x c_1(t-sqrt{2/3}x)
      +c_2(t+\sqrt{2/3}x)+(x^2/6)\}\}.
```

Solving Differential Equations Symbolically

Example 241. Plot the solution of the initial value problem (partial differential equation)

$$u_t(x,t) = 9u_{xx}(x,t), \quad 0 < x < 5,\ 0 < t < 10$$
$$u(x,0) = 0, \quad 0 < x < 5$$
$$u(0,t) = \sin 2t, \quad u(5,t) = 0\ 0 < t < 10.$$

```
In[ ]:= NDSolve[{D[u[x,t],t]==9D[u[x,t],x,x],u[x,0]==x,
u[0,t]==sin[2t],u[5,t]==0},u,{t,0,10},{x,0,5}].
Out[ ]:={{u -> InterpolatingFunction [{{0,5},{0,10} }<>]}}.
```

We plot the solution with

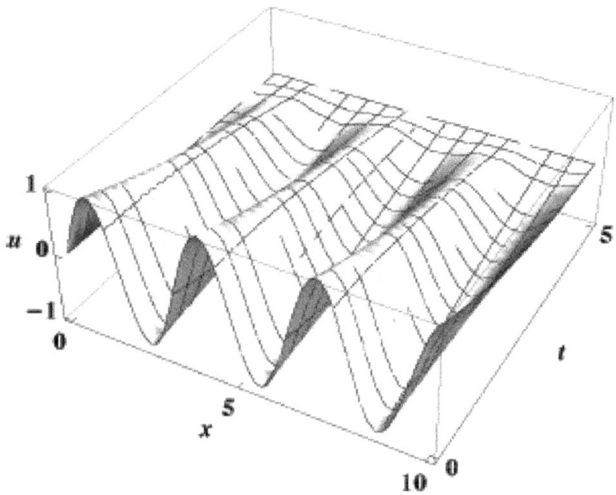

FIGURE 5.22: Simluation of Example 241

```
In[ ]:= Plot3D[Evaluate[u[x,t].%],{t,0,20},{x,o},
  PlotRange -> All].
```

5.8 Practical Applications in Physics and Mechanics

Damping

Air resistance and non-elastic effects in a string will contribute to reduce the amplitudes of the waves so that the motion dies out after some time. This damping effect can be modeled by a term but on left hand side of the equation

$$\varrho\frac{\partial^2 u}{\partial t^2}+b\frac{\partial u}{\partial t}=T\frac{\partial^2 u}{\partial x^2}.$$

The parameter b is normally determined from physical experiments.

External Forcing

It is easy to include an external force acting on the string. Say we have a vertical force $\tilde{f}_{ij}$ acting on mass m_i. This force affects the vertical component of Newton's law and gives rise to an extra term $\tilde{f}(x,t)$ on the right-hand side of (5.1). In this model we would add a term $\tilde{f}(x,t)$.

Waves on membrane

Elastic waves in rod

Consider an elastic rod subject to a hammer impact at the end. This experiment will give rise to an elastic deformation pulse that travels through the rod. A mathematical model for longitudinal waves along an elastic rod starts with the general equation for deformations and stresses in an elastic medium,

$$\varrho u_{tt} = \nabla.\delta + \varrho \mathbf{f}, \tag{5.34}$$

where ϱ is the density, **u** the displacement field, the stress tensor, and **f** body forces. The latter has normally no impact on elastic waves. For stationary deformation of an elastic rod, $\sigma_{xx} = Eu_x$, with all other stress components being zero. Moreover, $\mathbf{u} = u(x)\mathbf{i}$. The parameter is known as Youngs modulus. Assuming that this simple stress and deformation field, which is exact in the stationary case, is a good approximation in the transient case with wave motion. (5.34) simplifies to

$$\rho\frac{\partial^2 u}{\partial t^2} = \frac{\partial}{\partial x}\left(E\frac{\partial u}{\partial x}\right).$$

The associated boundary conditions are u or $\sigma_{xx} = Eu_x$, typically $u = 0$ for a clamped end and $\sigma_x x = 0$ for a free end.

Acoustic model for seismic waves

Seismic waves are used to infer properties of subsurface geological structures. The physical model is a heterogeneous elastic medium where sound is propagated by small elastic vibrations. The general mathematical model for deformations in an elastic medium is based on Newton's second law,

$$\rho u_{tt} = \nabla.\sigma + f, \rho f \tag{5.35}$$

and a constitutive law relating to **u**, often Hooke's generalized law,

$$\sigma = K\nabla.uI + G(\nabla u + (\nabla u)T - \frac{2}{3}\nabla.uI). \tag{5.36}$$

Here is the displacement field, is the stress tensor, **I** is the identity tensor, ϱ is the medium's density, f are body forces (such as gravity), **K** is the medium's bulk modulus and **G** is the shear modulus. All these quantities may vary in space, while u and will also show significant variations in time during wave motion.

The acoustic approximation to elastic waves arises from a basic assumption that the second term in Hookes law, representing the deformations that give rise to shear stresses, can be neglected. This assumption can be interpreted as approximating the geological medium by a fluid. Neglecting also the body forces f, (5.35) becomes

$$\rho u_{tt} = \nabla(K\nabla \cdot \mathbf{u}). \tag{5.37}$$

Introducing p as a pressure via

$$p = K\nabla \cdot \mathbf{u}, \tag{5.38}$$

and dividing (5.37) by ϱ, we get

$$u_{tt} = -\frac{1}{\varrho}\nabla p.$$

Taking the divergence of this equation, using $\nabla \cdot u = p/K$ from (5.38), gives the acoustic approximation to elastic waves:

$$p_{tt} = K\nabla \cdot \left(\frac{1}{\varrho}\nabla p\right). \tag{5.39}$$

This is a standard linear wave equation with variable coefficients. It is common to add a source term $s(x, y, z, t)$ to model the generation of sound waves:

$$p_{tt} = K\nabla \cdot \left(\frac{1}{\varrho}\nabla p\right) + s. \tag{5.40}$$

A common additional approximation is based on using the chain rule on the right hand side,

$$K\nabla \cdot \left(\frac{1}{\varrho}\nabla p\right) = \frac{K}{\varrho}\nabla^2 p + K\nabla \cdot \left(\frac{1}{\varrho}\right) \cdot \nabla p \approx \left(\frac{K}{\varrho}\right)\nabla^2 p \tag{5.41}$$

under the assumption that the relative spatial gradient $\nabla \varrho^1 = \varrho^2 \nabla \varrho$ is small. This approximation results in the simplified equation

$$p_{tt} = \left(\frac{K}{\varrho}\right)\nabla^2 p + s. \tag{5.42}$$

The acoustic approximations to seismic waves are used for sound waves in the ground, and the Earth's surface is then a boundary where p equals the atmospheric pressure p_o such that the boundary condition becomes $p = p_o$.

Anisotropy
Quite often in geological materials, the effective wave velocity $c = \sqrt{K/\varrho}$ is different in different spatial directions because geological layers are compacted such that the properties in the horizontal and vertical directions differ. With z as the vertical coordinate, we can introduce a vertical wave velocity c_z and a horizontal wave velocity c_h, and generalize (G) to

$$p_{tt} = c_z^2 p_z z + C_h^2(p_{xx} + p_{yy}) + s. \tag{5.43}$$

Sound waves in liquids and gases
Sound waves arise from pressure and density variations in fluids. The starting point of modeling sound waves involves the basic equations for a compressible fluid where we omit viscous (frictional) forces, body forces (gravity, for instance), and temperature effects:

$$p_+ + \nabla \cdot (pu) = 0 \qquad (5.44)$$
$$p_{ut} + pu.\nabla u = \nabla p \qquad (5.45)$$
$$p = p(p). \qquad (5.46)$$

These equations are often referred to as the Euler equations for the motion of a fluid. The parameters involved are the density ϱ, the velocity u, and the pressure p. Equation (5.45) reflects mass balance, (K) is Newton's second law for a fluid, with frictional and body forces omitted, and (M) is a constitutive law relating density to pressure by thermodynamics considerations. A typical model for (M) is the so-called isentropic relation, valid for adiabatic processes where there is no heat transfer:

$$\varrho = p_o \left(\frac{P}{p0}\right)^{1/r}$$
$$p = p_{o(\frac{p}{po})}{}^{1/r}.$$

p_o and p_0 are reference values for p and ϱ when the fluid is at rest, and γ is the ratio of specific heat at constant pressure and constant volume ($\gamma = 5/3$ for air). It is modeled by a standard linear wave $p_{tt} = c^2\nabla^2 p$ where $C = \sqrt{r}p_o/p_o$ is the speed of sound in the fluid.

5.9 Exercises

Separation of variables method for heat equation

In Exercises 5.1 through 5.7 use the separation of variables method to solve the heat equation

$$u_t = a^2 u_{xx},\ 0 < x < l,\ t > 0.$$

5.1. $a = \sqrt{3},\ l = \pi,\ u(0,t) = u(\pi,t) = 0,\ u(x,0) = x(\pi - x)$

5.2. $a = 1,\ l = 3,\ u_x(0,t) = u_x(3,t) = 0,\ u(x,0) = x$

5.3. $a = \sqrt{2},\ l = 4,\ u(0,t) = u_x(4,t) = 0,\ u(x,0) = x + 1$

5.4. $a = \sqrt{2},\ l = \pi,\ u(0,t) = u(\pi,t) = 0,\ u(x,0) = 20$

5.5. $a = \sqrt{2},\ l = 2,\ u(0,t) = u(2,t) = 0$, and $u(x,0) = \begin{cases} 20,\ 0 \le x < 1 \\ 0,\ 1 \le x < 2 \end{cases}$

Laplace transform method for heat equation

Use the Laplace transform method to solve the heat equation

$$u_t = a^2 u_{xx},\ 0 < x < l,\ t > 0$$

subject to:

5.6. $u_t = a^2 u_{xx}, 0 < x < 1, t > 0$, $u(x,0) = 0$, $u(0,t) = 1$, $u(1,t) = u_0$, u_0 is a constant.

5.7. $u_t = a^2 u_{xx}$, $0 < x < \infty$, $t > 0$, u_0 and u_1 are constants.

Fourier transform method for heat equation

Use the Fourier transform method to solve the indicated initial boundary value problems on the indicated intervals, subject to the given conditions:

5.8. $u_t = a^2 u_{xx}$, $-\infty < x < \infty$, $t > 0$, $u(x,0) = \mu(x)$.

5.9. $u_t = a^2 u_{xx}$, $-\infty < x < \infty$, $t > 0$, $u(x,0) = \begin{cases} 1, & |x| < 1 \\ 0, & |x| > 1 \end{cases}$.

5.10. $u_t = a^2 u_{xx}$, $u(x,0) = e^{-|x|}$.

5.11. $u_t = a^2 u_{xx} + f(x)$, $-\infty < x < \infty$, $t > 0$.

Use the Fourier transform method to solve the indicated heat equation, subject to the given initial and boundary value conditions.

$$u_t = a^2 u_{xx},\ 0 < x < \infty,\ t > 0.$$

5.12. $u_x(0,t) = \mu(t)$, and $u(x,0) = 0$

5.13. $u(x,0) = \begin{cases} 1, 0 < x < 1 \\ 0, \ \le 1 \le x < \infty \end{cases}$

5.14. $u(x,0) = 0$, $u_t(x,0) = 1$, $\lim_{x \to \infty} u(x,t) = 0$

5.15. $u_t = a^2 u_{xx} + f(x)$,$0 < x < \infty$, $t > 0$
$u(x,0) = 0$, $0 < x < \infty$, $u(0,t) = 0$

Separation of variables for wave equation

Using the method of separation of variables solve the wave equation

$$u_{tt}(x,t) = u_{xx}(x,t),\ 0 < x < \pi,\ t > 0,$$

subject to the Dirichlet boundary conditions

$$u(0,t) = u(\pi,t) = 0,\ \ t > 0,$$

and the following initial conditions.

5.16. $u(x,0) = 0,\ u_t(x,0) = 1,\ 0 < x < \pi.$

5.17. $u(x,0) = \pi x - x^2,\ u_t(x,0) = 0,\ 0 < x < \pi.$

5.18. $u(x,0) = \begin{cases} \frac{3}{2}x, 0 < x < \frac{2}{3}\pi \\ 3(\pi - x),\ \frac{2}{3}\pi < x < \pi \end{cases},\ u_t(x,0) = 0,\ 0 < x < \pi.$

5.19. $u(x,0) = \sin x,\ u_t(x,0) = \begin{cases} \frac{3}{2}x, & 0 < x < \frac{\pi}{4} \\ 1, & \frac{\pi}{4} < x < \frac{3\pi}{4} \\ 3(\pi - x), & \frac{3}{4}\pi < x < \pi \end{cases}.$

5.20. $u(x,0) = \begin{cases} x, & 0 < x < \frac{\pi}{2} \\ \pi - x, & \frac{\pi}{2} < x < \pi \end{cases},\ u_t(x,0) = 0,\ 0 < x < \pi.$

Fourier transform for wave equation

5.21. Use the Fourier transform to solve the following wave equations subject to the initial and boundary conditions.

$$u_{tt}(x,t) = c^2 u_{xx}(x,t),\ \ -\infty < x < \infty,\ t > 0.$$

$$u(x,0) = \frac{1}{1+x^2},\ u_t(x,0) = 0.$$

5.22. $u_{tt}(x,t) = c^2 u_{xx}(x,t),\ 0 < x < \infty,\ t > 0$
$u(x,0) = \frac{1}{1+x^2},\ u_t(x,0) = 0$
$u_x(0,t) = f(t),\ 0 < t < \infty$
(Hint: Use the Fourier cosine transform)

5.23. $u_{tt}(x,t) = c^2 u_{xx}(x,t) + f(x,t),\ 0 < x < \infty,\ t > 0$
$u(x,0) = 0, \quad u_t(x,0) = 0$
$u_x(0,t) = f(t),\ 0 < t < \infty$
(Hint: Use the Fourier sine transform)

Laplace transform for wave equation

Use the Laplace transform method to solve the indicated initial boundary value problems on the interval $(0,\infty)$, subject to the given conditions.

$$u_{tt}(x,t) = c^2 u_{xx}(x,t),\ x > 0,\ t > 0.$$

5.24. $u(0,t) = f(t),\ \lim\limits_{x\to\infty} u(x,t) = 0, t > 0,\ u(x,0) = 0,\ u_t(x,0) = 0,\ x > 0.$

5.25. $u(0,t) = 0,\ \lim\limits_{x\to\infty} u(x,t) = 0, t > 0,\ u(x,0) = xe^{-x},\ u_t(x,0) = 0,\ x > 0.$

5.26. $u(0,t) = \sin t,\ u(0,t) = 0, t > 0,\ u(x,0) = 0,\ u_t(x,0) = 0,\ x > 0.$

Use the Laplace transform to solve the initial boundary value problem for the wave equation on the interval $(0,1)$, subject to the given conditions.

5.27. $u(x,0) = \sin \pi x,\ u_t(x,0) = -\sin \pi x,\ u(0,t) = u(1,t) = 0.$

Separation of variables method for Laplace equation

Using the separation of variables method solve the following Laplace problem with the given data.

$$\begin{aligned} &u_{xx} + u_{yy} = 0,\ \ 0 < x < a,\ \ 0 < y < b \\ &u(x,0) = f_1(x),\ u(x,b) = f_2(x), 0 < x < a \\ &u(0,y) = g_1(y),\ u(a,y) = g_2(y), 0 < y < b. \end{aligned}$$

5.28. $a = b = 1,\ f_1(x) = 100,\ f_2(x) = 200,\ g_1(y) = g_2(y) = 0.$

5.29. $a = b = \pi,\ f_1(x) = 0,\ f_2(x) = 1,\ g_1(y) = g_2(y) = 1.$

5.30. $a = 1,\ b = 2,\ f_1(x) = 0,\ f_2(x) = x,\ g_1(y) = g_2(y) = 0.$

5.31. $a = 2,\ b = 1,\ f_1(x) = 100,\ f_2(x) = 0,\ g_1(y) = 0,\ g_2(y) = 100y(1-y).$

5.32. $a = 2,\ b = 1,\ f_1(x) = 100,\ f_2(x) = 0,\ g_1(y) = 0,\ g_2(y) = 100y(1-y).$

5.33. $a = b = 1$, $f_1(x) = 7\sin(7\pi x)$, $f_2 y(x) = \sin(\pi x)$, $g_1(y) = \sin 3\pi y$, $g_2(y) = 6\pi y$.

5.34. $a = b = 1$, $f_1(x) = 7\sin(7\pi x)$, $f_2(x) = \sin(\pi x)$, $g_1(y) = \sin 3\pi y$, $g_2(y) = 6\pi y$.

In the following problems, solve the Laplace equation in $[0, a] \times [0, b]$ subject to the given boundary conditions.

5.35. $u(x, 0) = f(x)$, $u(x, b) = 0$, $0 < x < a$; $u(0, y) = 0$, $u(a, y) = 0$, $0 < y < b$.

5.36. $u_y(x, 0) = 0$, $u_y(x, 1) = 0$, $0 < x < a$; $u(0, y) = 0$, $u(1, y) = 1 - y$, $0 < y < b$.

5.37. $u_x(0, y) = u(0, y)$, $u(0, y) = 1$, $0 < y < b$; $u(x, 0) = 0$, $u(x, \pi) = 0$, $0 < x < a$.

5.38. $u_x(0, y) = u(0, y)$, $u_x(0, y) = 0$, $0 < y < b$; $u(x, 0) = 0$, $u(x, b) = f(x)$, $0 < x < a$.

5.39. $u_x(0, y) = 0$, $u_x(a, y) = 1$, $0 < y < a$; $u(x, 0) = 1$, $u(x, b) = 1$, $0 < x < a$.

5.40. $u(x, 0) = 1$, $u(x, 1) = 4$, $0 < x < \pi$; $u(0, y) = 0$, $u(\pi, y) = 0$, $0 < y < 1$.

Fourier transform for Laplace equation

Use one of the Fourier transforms to solve the Laplace equation

$$u_{xx} + u_{yy} = 0,\ (x, y) \in D$$

on the indicated D, subject to the given boundary value conditions.

5.41. $D = \{(x, y) :\ 0 < x < \pi, 0 < y < \infty\}$; $u(0, y) = 0$, $u(\pi, y) = e^{-y}$ for $y > 0$ and $u_y(x, 0) = 0$ for $0 < x < \pi$.

5.42. $D = \{(x, y) :\ 0 < x < \infty, 0 < y < 2\}$; $u(0, y) = 0$, for $0 < y < 2$, $u(x, 0) = f(x)$, $u(x, 2) = 0$ for $0 < x < \pi$.

5.43. $D = \{(x, y) :\ 0 < x < \infty, 0 < y < \infty\}$; $u(0, y) = e^{-y}$, for $u(\pi, y) = e^{-y}$ for $0 < y < \infty$ and $u(x, 0) = e^{-x}$ for $0 < x < \infty$.

5.44. $D = \{(x, y) :\ -\infty < x < \infty, 0 < y < \infty\}$; $u(x, 0) = \begin{cases} 1, & |x| < 1 \\ 0, & |x| > 1 \end{cases}$.

5.45. $D = \{(x, y) :\ -\infty < x < \infty, 0 < y < \infty\}$; $u(x, 0) = \dfrac{1}{4 + x^2}$.

5.46. $D = \{(x, y) :\ -\infty < x < \infty, 0 < y < \infty\}$; $u(x, 0) = \cos x$.

5.10 Suggestion for Further Reading

Applications of MATHEMATICA to well known partial differential equations are explained well in Adzievski and Siddiqi [1]. As we saw in this chapter, partial differential equations constitute a major field of study in contemporary mathematics and also in other fields such as differential geometry and probability theory. Partial differential equations are also used widely in theoretical and applied aspects of physics, mechanics, engineering, chemistry, bioscience, medicine, meteorology, climatology, and economics. As discussed earlier, several numerical and analytical methods have been devised to solve various types of partial differential equations. A recent example is the finite volume method. Interested readers can obtain more information in Chapter 10 of a book by Le Dret and Lucquin [8] that provides interesting accounts of partial differential equation modeling, analysis and numerical approximation. References [2] through [7] are useful adjuncts for further study of the contents of this chapter.

Bibliography

[1] K. Adzievski, A. H. Siddiqi, *Introduction to Partial Differential Equations for Scientists and Engineers Using Mathematica*, CRC Press, 2014.

[2] N. Asmar, *Partial Differential Equations with Fourier Series and Boundary Value Problems*, Second Edition, Pearson, 1992.

[3] W. E. Boyce, R. C. Diprima, *Elementary Differential Equations and Boundary Value Problems*, Pearson, 2010.

[4] M. P. Coleman, *An Introduction to Partial Differential Equations with MATLAB*, Applied Mathematics and Nonlinear Science Series, Chapman & Hall, New York, 2005.

[5] G. H. Edwards, D. E. Penney, *Differential Equations, Computing and Modeling*, Pearson, 2008.

[6] L. C. Evans, *Partial Differential Equations: Graduate Studies in Mathematics*, American Mathematical Society, Providence, 2010.

[7] G. B. Folland, *Fourier Series and its Applications*, Wadsworth & Brooks, Pacific Grove CA, 1992.

[8] H. Le Dret, B. Lucquin, *Partial Differential Equations: Modeling, Analysis and Numerical Approximation*, Birkhäuser, 2016.

Chapter 6

Algorithmic Optimization

6.1 Basic Concepts

An optimization problem is characterized by its specific objective function that is to be maximized or minimized, depending upon the problem and, in the case of a constrained problem, a given set of constraints. Possible objective functions include expressions representing profits, costs, market share, portfolio risk and other calculations. Possible constraints include those that represent limited budgets or resources, non-negativity constraints on variables and conservation equations.

6.1.1 Gradient Method

In this section we will study the problem of minimizing a function of n variables. If f is a function from $\mathbb{R}^n$ to $\mathbb{R}$, the **gradient** of f at a point x is the vector $G = (G_1, ..., G_n)^T$ whose components are

$$G_i = G_i(x) = \frac{\partial f}{\partial x_i}, \ (1 \leq i \leq n). \tag{6.1}$$

The gradient is also denoted by ∇G or simply by $f'(x)$.

The **Hessian** of f at x is the matrix H whose entries are

$$H_{ij} = H_{ij}(x) = \frac{\partial^2 f}{\partial x_i x_j}. \tag{6.2}$$

With the gradient ad Hessian in hand, we can write the linear and quadratic terms in the Taylor expansion of f

$$f(x+h) = f(x) + G(x)^T h + \frac{1}{2} h^T H(x) h + \dots. \tag{6.3}$$

The gradient vector points in the direction of **steepest ascent** or (increase), and its negative points in the direction of **steepest decrease.**This important fact is proved by taking any unit vector u, and considering how the function changes locally at x in the direction of u. This can be seen by computing the directional derivative

$$\frac{d}{dt} f(x+tu) \mid_{t=0} . \tag{6.4}$$

By using the Taylor formula (6.3), with h replaced by tu, we obtain

$$f(x+h) = f(x) + tG(x)^T u + \frac{1}{2} t^2 u^T H(x) u + \dots \tag{6.5}$$

whence

$$\frac{d}{dt} f(x+tu) = G(x)^T u + tu^T H(x) u + \dots. \tag{6.6}$$

Using $t = 0$, we obtain

$$\frac{d}{dt} f(x + tu) \mid_{t=0} = G(x)^T u \tag{6.7}$$

as the rate of change of f at x in the direction of u. By the Cauchy-Schwarz inequality,

$$\begin{aligned} \|G(x)u\| &\leq \|G(x)\| \, \|u\| \\ &\leq \|G(x)\| \, . \end{aligned} \tag{6.8}$$

That is, this rate of change cannot exceed $\|G(x)\|$. On the other hand, we can attain this upper bound by taking u to be the unit vector in the direction of $G(x)$.

An obvious strategy for locating a minimum point of f is to start at any point x and determine the direction in which f decreases at the fastest rate, that is, in the direction of $-G(x)$. A line search can be carried out on $\{x - tG(x) : t > 0\}$; we start over by computing a new direction of **steepest descent**, and so on.

6.2 Analysis of Quadratic Functions

The problem consists of finding a vector x^*(called local minimizer) such that

$$f(x^*) = \min_x f(x) \text{ for } x \text{ in } F$$

where F is a subset of $\mathbb{R}^n$.

If f is continuously differentiable in the vicinity of x^*, a necessary and sufficient condition for x^* to be a strong local minimum are:

$$G(x^*)^T u = 0 \text{ for all } u \neq 0$$

and

$$H(x^*) \text{ positive-definite.}$$

A common strategy in minimization problem is to assume that our function is locally approximately a quadratic polynomial. A general quadratic function of n variables is of the form

$$f(x) = \frac{1}{2} x^T A x - b^T x + a. \tag{6.9}$$

Here, a is scalar and b is a constant vector having n components. The matrix A is $n \times n$, symmetric, and constant (constant entries).The right side of the equation displays (from left to right) the term of exact degree 2, the terms of exact degree 1, and the term of degree 0. In the case of two variables, the first terms are the three types x_1^2, x_2^2, and x_1x_2.

Because quadratic functions should be good local models for any given function that is twice differentiable, we will derive some basic facts about them. First, the full form of f is

$$f(x) = \frac{1}{2}\sum_{i=1}^{n}\sum_{j=1}^{n} x_i x_j A_{ij} - \sum_{i=1}^{n} b_i x_i + a \tag{6.10}$$

Next, we compute the gradient of f, which is by definition the vector having components

$$\frac{\partial f}{\partial x_k} = \frac{1}{2}\sum_{j=1}^{n} x_j A_{kj} + \frac{1}{2}\sum_{i=1}^{n} x_i A_{ik} - b_k \tag{6.11}$$

Because the matrix A is symmetric, this equation becomes

$$\frac{\partial f}{\partial x_k} = \sum_{j=1}^{n} A_{kj} x_j - b_k \tag{6.12}$$

Thus, the vector expression for the gradient is

$$G(x) = \nabla f(x) = Ax - b \tag{6.13}$$

To find the critical point of f, that is, the point where the derivative of f vanishes, we set $G(x) = 0$ and note that this occurs when $Ax = b$, or in other words, when

$$x = A^{-1}b \tag{6.14}$$

Example 242. Write the quadratic function

$$f(x) = 4x_1^2 - 2x_1x_2 + 3x_2^2 + 3x_1 - 2x_2 + 1$$

in the standard vector-matrix notation, and find the minimizer and the minimum value for the function.

Solution:

$$f(x) = (x_1, x_2)\begin{pmatrix} 4 & -1 \\ -1 & 3 \end{pmatrix}\begin{pmatrix} x_1 \\ x_2 \end{pmatrix} - 2\,(x_1, x_2)\begin{pmatrix} -\frac{3}{2} \\ 1 \end{pmatrix} + 1$$

whereby

$$A = \begin{pmatrix} 4 & -1 \\ -1 & 3 \end{pmatrix}, \; b = \begin{pmatrix} -\frac{3}{2} \\ 1 \end{pmatrix}$$

From (6.13) and (6.14), the minimizer can be found by solving the linear system

$$Ax = b$$

or

$$\begin{pmatrix} 4 & -1 \\ -1 & 3 \end{pmatrix} \begin{pmatrix} x_1 \\ x_2 \end{pmatrix} = \begin{pmatrix} -\frac{3}{2} \\ 1 \end{pmatrix}. \tag{6.15}$$

Applying the usual Gauss Elimination algorithm, only one row operation is required to replace the coefficient matrix in upper triangular form. This gives $x^* = \left(x_1^*, x_2^{*T}\right) = \left(-\dfrac{7}{22}, \dfrac{5}{22}\right)^T = (-0.31818, .22727)^T$. The minimal value for f is $f(x^*) = f\left(-\dfrac{7}{22}, \frac{5}{22}\right) = \dfrac{13}{44} \approx .29546$.

It is instructive to compare the algebraic solution with the minimization procedure learned in multi-variable calculus. The *critical points* of $f(x_1, x_2)$ are found by setting both partial derivatives equal to zero

$$\frac{\partial f}{\partial x_1} = 8x_1 - 2x_2 + 3 = 0$$
$$\frac{\partial f}{\partial x_2} = -2x_1 + 6x_2 - 2 = 0$$

or

$$4x_1 - x_2 = -\tfrac{3}{2}$$
$$-x_1 + 3x_2 = 1$$

This is precisely the *same linear system* we already constructed in (7.63). To check whether x^* is a (local) minimum, we need to analyze the *Hessian matrix*, which is the symmetric matrix of second order derivatives evaluated at $\left(x_1^*, x_2^{*T}\right) = \left(-\frac{7}{22}, \frac{5}{22}\right)^T$

$$H = \begin{pmatrix} \frac{\partial^2 f}{\partial x_1^2} & \frac{\partial^2 f}{\partial x_2 \partial x_1} \\ \frac{\partial^2 f}{\partial x_1 \partial x_2} & \frac{\partial^2 f}{\partial x_2^2} \end{pmatrix} = \begin{pmatrix} 8 & -2 \\ -2 & 6 \end{pmatrix} = 2A$$

If the Hessian matrix is positive definite-which we already in this case-then the critical point is indeed a (local) minimum.

Example 243. Minimize the quadratic function

$$f(x) = x_1^2 + 2x_1x_2 + x_1x_3 + 2x_2^2 + x_2x_3 + 2x_3^2 + 6x_2 - 7x_3 + 5$$

Solution: This has the matrix form with

$$A = \begin{pmatrix} 1 & 1 & \frac{1}{2} \\ 1 & 2 & \frac{1}{2} \\ \frac{1}{2} & \frac{1}{2} & 2 \end{pmatrix}, \; x = \begin{pmatrix} x_1 \\ x_2 \\ x_3 \end{pmatrix}, \; b = \begin{pmatrix} 0 \\ -3 \\ \frac{7}{2} \end{pmatrix}, \; c = 5$$

Gaussian Elimination produces the LDL^T factorization

$$A = \begin{pmatrix} 1 & 1 & \frac{1}{2} \\ 1 & 2 & \frac{1}{2} \\ \frac{1}{2} & \frac{1}{2} & 2 \end{pmatrix} = \begin{pmatrix} 1 & 0 & 0 \\ 1 & 1 & 0 \\ \frac{1}{2} & 0 & 1 \end{pmatrix} \begin{pmatrix} 1 & 0 & 0 \\ 0 & 1 & 0 \\ 0 & 0 & \frac{7}{4} \end{pmatrix} \begin{pmatrix} 1 & 1 & \frac{1}{2} \\ 0 & 1 & 0 \\ 0 & 0 & 1 \end{pmatrix}$$

The pivots, i.e., the diagonal entries of D, are all positive, and hence A is positive definite. Therefore, $f(x)$ has a unique minimizer $x^* = (x_1^*, x_2^*, x_3^*)$, which is found by solving the linear system $Ax = b$. The solution is then quickly obtained by forward and backward substitution:

$$x_1^* = 2,\ x_2^* = -3,\ x_3^* = 2$$

with

$$f(x^*) = f(2, -3, 2) = -11$$

6.2.1 Newton's Method

The general indirect method for determining x^* is to solve the system of equations $G(x) = 0$ by some numerical method to yield all stationary points. A popular method to achieve this is **Newton's method**. Since in general the system will be nonlinear, multiple stationary points are possible.These stationary points must then be further analyzed in order to determine whether or not they are local minima.

Assume x^* is a local minimum and x^k an approximate solution, with associated unknown error δ such that $x^* = x^k + \delta$. Then by applying Taylor's theorem and the first necessary condition for a minimum at x^* it follows that

$$0 = G(x^*) = G(x^k + \delta) = G(x^k) + H(x^k)\delta + O\left\|\delta\right\|^2$$

If x^k is a good approximation then the solution of the linear system

$$H(x^k)\delta + G(x^k) = 0$$

, obtained by ignoring the second order in δ above. A better approximation is therefore expected to be $x^{k+1} = x^k + \delta$ which leads to the Newton iterative scheme:Given an initial approximation x^0, compute

$$x^{k+1} = x^k - H^{-1}(x^k)G(x^k),\ k = 1, 2, ... \tag{6.16}$$

Newton's method applied to a quadratic problem

Consider the unconstrained problem

$$\text{Minimize } f(x) = \frac{1}{2}x^T Ax - b^T x + a \tag{6.17}$$

In this case the first iteration in (6.16) yields

$$x^1 = x^0 - A^{-1}(Ax^0 - b) = x^0 - x^0 + A^{-1}b = A^{-1}b$$

that is

$$x^1 = x^* = A^{-1}b$$

6.2.2 Line Search Descent Algorithm

Over the last four decades a number of powerful direct search methods have been developed for the unconstrained minimization of general functions.These algorithms require an initial estimate to the optimum point, denoted by x^0. With this estimate as starting point, the algorithm generates a sequence of estimates x^0, x^1, x^2,...by successively searching directly from each point in a direction of descent to determine the next point.The process is terminated if either no further progress is made, or if a point x^k is reached (for smooth functions) at which the necessary condition $G(x) = 0$ is sufficiently accurately satisfied, in which case $x^* = x^k$.

An important sub-class of direct search methods, specifically suitable for smooth functions, are the so-called line search descent methods. Basic to these methods is the selection of a descent direction u^{k+1} at each iterate x^k that ensures descent at x^* in the direction u^{k+1}, that is, it is required that the directional derivative in the direction u^{k+1} be negative:

$$\left.\frac{df(x^k)}{d\lambda}\right|_{u^{k+1}} = G(x^k)u^{k+1} < 0$$

General structure of a line search descent method

1. Given starting point x^0 and positive tolerances ϵ_1, ϵ_2 and ϵ_3, set $k = 1$.
2. Select a descent direction u^k
3. Perform a *one-dimensional line search* in direction u^k: i.e.

$$\min_{\lambda} F(\lambda) = \min_{\lambda} f(x^{k-1} + \lambda u^k)$$

to give minimizer λ_k.
4. Set $x^k = x^{k-1} + \lambda_k u^k$
5.Test for convergence:
if $\left\|x^k - x^{k-1}\right\| < \epsilon_1$, or $\left\|G(x^k)\right\| < \epsilon_2$, or $\left|f(x^k) - f(x^{k-1})\right| < \epsilon_3$
then STOP and $x^* \simeq x^k$
else go to Step 6.
6.Set $k = k + 1$ and go to Step 2.

6.2.3 The Method of Steepest Descent

Line search descent methods (see Section 6.1.1), that use the gradient vector $G(x)$ to determine the search direction for each iteration, are called first order methods because they employ first order partial derivatives of $f(x)$ to compute the search direction at the current iterate. The simplest and most famous of these methods is the method of steepest descent, first proposed by Cauchy in 1847.

In this method the direction of steepest descent is used as the search direction in the line search descent algorithm given above. The expression for the direction of steepest descent is derived below. At x'we seek the unit vector u such that for $F(\lambda) = f(x' + \lambda u)$, the directional derivative

$$\left.\frac{df(x)}{d\lambda}\right|_u = \frac{dF(0)}{d\lambda} = G(x)u$$

assumes a minimum value with respect to all possible choices for the unit vector u at x'. Clearly for the particular choice $u = -\frac{G(x')}{\|G(x')\|}$ the directional derivative at x' is given by $\frac{dF(0)}{d\lambda} = -G(x)\frac{G(x')}{\|G(x')\|} = -\|G(x')\| =$ least value.

Thus this particular choice for the unit vector corresponds to the direction of steepest descent. The search direction

$$u = -\frac{G(x')}{\|G(x')\|}$$

is called the normalized steepest descent direction at x.

Steepest descent algorithm

Given x°, do for iteration $i = 1, 2, ...$ until convergence:

1. $u^k = -\frac{G(x^{k-1})}{\|G(x^{k-1})\|}$
2. $x^k = x^{k-1} + \lambda_k u^k$ where λ_k is such that

$$F(\lambda_k) = f(x^{k-1} + \lambda_k u^k) = \min_{\lambda} f(x^{k-1} + \lambda_k u^k).$$

6.2.4 Conjugate Gradient Method

In spite of its local optimal descent property the method of steepest descent often performs poorly. There is, however, a class of first order fine search descent methods, known as conjugate gradient methods^for which it can be proved that a method from this class will converge exactly in a finite number of iterations when applied to a positive-definite quadratic function of the form

$$f(x) = \frac{1}{2}x^T Ax - b^T x + a$$

where A is a positive-definite nxn real symmetric matrix.

Mutually conjugate directions

Two non zero vectors u, v are defined to be orthogonal if the scalar product $u^T v = (u, v) = 0$. Two non zero vectors u, v are said to be A-conjugate if $u^T Av = (u, Av) = 0$.

The Fletcher-Reeves directions

These are defined as follows:

$$u^1 = -G\left(x^0\right)$$

and for $k = 1, 2, ..., n-1$

$$u^{k+1} = G\left(x^k\right) + \beta_k u^k$$

where $x^k = x^{k-1} + \lambda_k u^k$, and λ_k corresponds to the optimal descent step in iteration k, and

$$\beta_k = \frac{\left\|G(x^k)\right\|^2}{\left\|G(x^{k-1})\right\|^2} \tag{6.18}$$

Note that the above directions can be shown to be A-conjugate.

Fletcher-Reeves conjugate gradient algorithm for general functions

Given x^0 perform the following steps:

1. Compute $G(x^0)$ and set $u^1 = -G(x^0)$.
2. For $k = 1, 2, ..., n$ do:

2.1 $x^k = x^{k-1} + \lambda_k u^k$ where λ_k is such that

$$f(x^{k-1} + \lambda_k u^k) = \min_{\lambda} f(x^{k-1} + \lambda_k u^k)$$

2.2 Compute $G(x^k)$.

2.3 If convergence criteria satisfied, then STOP and $x^* \simeq x^k$, else go to step 2.4.

2.4 If $1 \leq k \leq n-1$, $u^{k+1} = -G(x^k) + \beta_k u^k$ with β_k given by (6.18).

3. Set $x^0 = x^k$ and go to step 2 (restart).

Example 244. Apply the Fletcher-Reeves method to minimize

$$f(x) = \frac{1}{2}x_1^2 + x_1 x_2 + x_2^2$$

with $x^0 = [10, -5]^T$.

Solution: *Iteration 1:*

$$G(x) = \begin{bmatrix} x_1 + x_2 \\ x_1 + 2x_2 \end{bmatrix}$$

and therefore

$$u^1 = -G(x^0) = \begin{bmatrix} -5 \\ 0 \end{bmatrix}$$

and

$$x^1 = x^\circ + \lambda u^1 = \begin{bmatrix} 10 - 5\lambda \\ -5 \end{bmatrix}$$

and

$$F(\lambda) = f(x^0 + \lambda u^1) = \frac{1}{2}(10 - 5\lambda)^2 + (10 - 5\lambda)(-5) + 25$$

For optimal descent

$$\frac{dF}{d\lambda}(\lambda) = \left.\frac{df(x)}{d\lambda}\right|_{u^1} = -5(10 - 5\lambda) + 25 = 0$$

This gives

$$\lambda_1 = 1, x^1 = \begin{bmatrix} 5 \\ -5 \end{bmatrix} \text{ and } G(x^1) = \begin{bmatrix} 0 \\ -5 \end{bmatrix}$$

Iteration 2:

$$u^2 = -G(x^1) + \frac{\|G(x^1)\|^2}{\|G(x^0)\|^2} u^1 = -\begin{bmatrix} 0 \\ -5 \end{bmatrix} + \frac{25}{25}\begin{bmatrix} -5 \\ 0 \end{bmatrix} = \begin{bmatrix} -5 \\ 5 \end{bmatrix}$$

$$x^2 = x^1 + \lambda u^2 = \begin{bmatrix} 5 \\ -5 \end{bmatrix} + \lambda \begin{bmatrix} -5 \\ 5 \end{bmatrix} = \begin{bmatrix} 5(1-\lambda) \\ -5(1-\lambda) \end{bmatrix}$$

and

$$F(\lambda) = f(x^1 + \lambda u^2) = \frac{1}{2}\left[25(1-\lambda)^2 - 50(1-\lambda)^2 + 50(1-\lambda)^2\right]$$

Again for optimal descent

$$\frac{dF}{d\lambda}(\lambda) = \left.\frac{df(x)}{d\lambda}\right|_{u^2} = -25(1-\lambda) = 0.$$

This gives

$$\lambda_2 = 1, x^2 = \begin{bmatrix} 0 \\ 0 \end{bmatrix} \text{ and } G(x^2) = 0$$

Therefore STOP.

6.3 Linear Programming

The term **linear programming (LP)** does not refer to the programming of a computer but to the programming of business or economic enterprises.

An explicit and technical meaning has been assigned to the term. It means finding the maximum value of a linear function of n variables over a convex polyhedral set in $\mathbb{R}^n$.

The goal of an LP problem is to maximize or minimize (*optimize*) a linear objective function f in n decisions variables $x_1, x_2, ..., x_n$ subject to a set of m constraints.

Let $c = (c_1, c_2, ..., c_n)^T \in \mathbb{R}^n$, $b = (b_1, ..., b_m)^T \in \mathbb{R}^m$, and $A = (a_{ij})$ is an $m \times n$ matrix. This is called a **standard maximization problem** if it has the form: Find the maximum of

$$f = c_1x_1 + c_2x_2 + ... + c_nx_n \tag{6.19}$$

subject to the constraints

$$\begin{aligned} a_{11}x_1 + a_{12}x_2 + ... + a_{1n}x_n &\leq b_1 \\ a_{21}x_1 + a_{22}x_2 + ... + a_{2n}x_n &\leq b_2 \\ &\vdots \\ a_{m1}x_1 + a_{m2}x_2 + ... + a_{mn}x_n &\leq b_m \end{aligned} \tag{6.20}$$

where the decision variables $x_1, x_2, ..., x_n$ are non-negative. In matrix form, find the maximum of

$$f = c^T x, x \in \mathbb{R}^n, \tag{6.21}$$

subject to the constraint

$$Ax \leq b, \text{and } x \geq 0$$

We recall that if $x = (x_1, ..., x_n)^T$, then the vector inequality $x \geq 0$ means that $x_i \geq 0$ for all $i = 1, 2, ..., n$. Likewise, the inequality $Ax \leq b$ means that

$$\sum_{j=1}^{n} a_{ij}x_j \leq b_i \text{ for all } i = 1, 2, ..., m \tag{6.22}$$

Next we describe some common terminology. **The feasible set** in our problem is the set

$$K = \{x \in \mathbb{R}^n : Ax \leq b, \ x \geq 0\}. \tag{6.23}$$

The *value of the problem* is the number

$$v = \sup \{c^T x : x \in K\}. \tag{6.24}$$

A **feasible point** is any element of K. A **solution** or **optimal feasible point** is any $x \in K$ such that $c^T x = v$. The function $x \to c^T x = \sum_{j=1}^{n} c_j x_j$ is the **objective function.** Since the problem is completely determined by the data A, b, and c, we refer to it as linear programming problem (A, b, c).

TABLE 6.1

	P_1	P_2	Time Constraints (hr)
M_1	1	2	≤ 40
M_2	1	2	≤ 30

Example 245. A company makes products P_1 and P_2 using two machines M_1 and M_2. Producing one unit of P_1 requires one hour on M_1 and M_2 or four hour in total. Producing one unit of P_2 requires two hour on each of M_1 and M_2. Due to the need for regular servicing, M_1 and M_2 can only run for 30 and 40 hours per week, respectively. The preceding data are organized in the Table 6.1.

The company expects net profits of \$20 and \$30, respectively, on each unit of P_1 and P_2 sold. Write down the LP problem How should the company schedule production in order to maximize net weekly profit?

Solution: The model translates directly into linear programming problem in standard form with $n = m = 2$ in (6.19) and (6.20).Let x_1 and x_2 represent, respectively, the number of units of products P_1 and P_2 produced per week. The company's objective is to maximize

$$f = 20x_1 + 30x_2 \text{ (net profit per week)}$$

such that

$$\begin{aligned} x_1 + 2x_2 &\leq 40 \text{ (hours, } M_1) \\ x_1 + x_2 &\leq 30 \quad \text{(hours, } M_1). \end{aligned}$$

6.4 Simplex Method

The simplex method was invented by G.B. Dantzig in 1949. His monograph (Dantzig 1963) is the classical reference. Most texts describe the simplex method as a sequence of pivots on a table of numbers called the simplex tableau.

We shall describe an algebraic step-by-step procedure, called the *simplex algorithm*, that can be applied to solve LP problems with any number of decisions and constraints. Let us first illustrate this algorithm by the following linear programming (LP) problem: Maximize the objective function

$$f = 6x_1 + 14x_2$$

TABLE 6.2

6	14	0	0	0	0
2	1	1	0	0	12
2	3	0	1	0	15
1	7	0	0	1	21
0	0	12	15	21	

subject to the constraint

$$\begin{cases} 2x_1 + x_2 \leq 12 \\ 2x_1 + 3x_2 \leq 15 \\ x_1 + 7x_2 \leq 21 \\ x_1 \geq 0,\ x_2 \geq 0 \end{cases}$$

In preparation for the simplex algorithm, we introduce **slack variables** and rewrite the problem as follows:

$$\text{Maximize: } F = 6x_1 + 14x_2 + 0x_3 + 0x_4 + 0x_5$$

subject to the constraints

$$\begin{cases} 2x_1 + x_2 + x_3 \qquad\qquad = 12 \\ 2x_1 + 3x_2 + \qquad\qquad + x_4 = 15 \\ x_1 + 7x_2 + ... + x_5 \qquad = 21 \\ x_1 \geq 0,\ x_2 \geq 0,\ x_3 \geq 0, x_4 \geq 0, x_5 \geq 0 \end{cases}$$

Hence, in matrix form

$$\text{Maximize: } F = (6, 14, 0, 0, 0,)^T$$

subject to the constraints

$$\left\{ \begin{bmatrix} 2 & 1 & 1 & 0 & 0 \\ 2 & 3 & 0 & 1 & 0 \\ 1 & 7 & 0 & 0 & 1 \end{bmatrix} x = \begin{bmatrix} 12 \\ 15 \\ 21 \end{bmatrix} \right.$$
$$x = (x_1,\ x_2,\ x_3, x_4, x_5)^T \geq 0$$

Our first vector will be $x = (0,\ 0, 12, 15, 21)^T$. All these data are summarized in the first tableau

Every step of the simplex method begins wit a tableau. The top row contains coefficients that pertain to the objective function F. The current value of $F(x) = c^T x$ is displayed in the top right corner. The next m rows in the tableau represent a system of equations embodying the quality constraints.

TABLE 6.3

cT	0	$F(x)$
A	I	b
x (nonbasic)	x (basic)	

It is worth reminding that elementary row operations can be performed on this system of equations without altering the set of solutions. The last row of the tableau contains the current x-vector. Notice that $F(x) = cTx$ ids easily computed using the top row and the bottom row. The preceding tableau id of the form

6.4.1 Tableau Rules

Each tableau that occurs in the simplex method must satisfy these five rules:

1. The x-vector must satisfy the equality constraint $Ax = b$.

2. The x-vector must satisfy the inequality constraint $x \geq 0$.

3. There are n components of x (called **nonbasic variables**) that are 0. The remaining m components are usually nonzero and are called **basic variables.**

4. In the matrix that defined the constraints, each basic variable occurs in only one row.

5. The objective function F must be expressed only in terms of nonbasic variables.

In each step, we examine the current tableau to see whether the value of $F(x)$ can be increased by allowing a nonbasic variable to become a basic variable In our example, we see that if we allow x_1 or x_2 to increase (and compensate by adjusting x_3, x_4, x_5), then the value of $F(x)$ will indeed increase. Because the coefficient 14 in F is greater than the coefficient 6, a unit increase in x_2 will increase F faster than a unit increase in x_1. Hence, we hold x_1 fixed at 0 and allow x_2 to increase as much as possible. These constraints apply:

$$0 \leq x_3 = 12 - x_2$$
$$0 \leq x_4 = 15 - 3x_2$$
$$0 \leq x_5 = 21 - 7x_2$$

These constraints tell us that

$$x_2 \leq 12 \qquad x_2 \leq 15 \qquad x_2 \leq 3$$

the most stringent of these is the inequality $x_2 \leq 3$, and therefore x_2 is allowed

TABLE 6.4

6	0	0	0	-2	42
$\frac{13}{7}$	0	0	1	$-\frac{1}{7}$	9
$\frac{11}{7}$	3	0	1	$-\frac{3}{7}$	6
1	7	0	0	1	21
0	3	9	6	0	

to increase to 3. The resulting values of x_3, x_4, and x_5 are obtained by the three given constraints. Hence, our new x-vector is

$$x = \begin{bmatrix}0 & 3 & 9 & 6 & 0\end{bmatrix}^T$$

The new basic variables are x_2, x_3, and x_4, and we must now determine the next tableau in accordance with the preceding five rules. In order to satisfy Rule **5**, we note that $x_2 = (21 - x_5)/7$. When this is substituted in F, we find a new form for the objective functions:

$$F = 6(x_1 + 14x_2 = 6x_1 + 14(21 - x_5)/7 = 6x_1 - 2x_5 + 42$$

To satisfy Rule **4,** Gaussian elimination steps (elementary row operations) are applied, using 7 as the pivot element. The purpose of this is to eliminate x_2 from all but one equation. After all this work has been carried out, Step 1 is finished abd Step 2 begins with the second tableau, which is

The situation presented now is similar to that of the beginning. The nonbasic variables are x_1 and x_5. Ant increase in x_5 will decrease $F(x)$, and so it is x_1 that is now allowed to become a basic variable. Hence, we hold x_5 fixed at 0 and allow x_1 to increase as much as possible. These constraints apply:

$$\begin{aligned} 0 &\le x_3 = 9 - \left(\tfrac{13}{7}\right) x_2 \\ 0 &\le x_4 = 6 - \left(\tfrac{11}{7}\right) x_2 \\ 0 &\le 7x_2 = 21 - x_1 \end{aligned}$$

These lead to

$$x_1 \le \frac{63}{13} \qquad x_1 \le \frac{42}{11} \qquad x_1 \le 21$$

The new basic variable x_1 is allowed to increase to only $42/11$, and new values x_3, x_4, and x_5 are computed from tableau or from the constraints equations immediately above. The new x-vector is

$$x = \begin{bmatrix}42/11 & 27/11 & 21/2 & 0 & 0\end{bmatrix}^T$$

The nonbasic variables are now x_4, and x_5. To satisfy Rule **5**, we use the substitution $x_1 = (7/11)(6 - x_4)$. then

$$\begin{aligned} F(x) &= 6x_1 - 2x_5 + 42 \\ &= (42/11)\,(6 - x_4) - 2x_5 + 42 \\ &- (42/11)\,x_4 - 2x_5 + 714/11 \end{aligned}$$

It is not necessary to complete the third tableau because both coefficients in F are negative. This means that the current x is a solution because neither of the the variables x_4, and x_5 can become basic variables without decreasing $F(x)$. Thus, in the original problem, the maximum value if $F(42/11, 27/11) = 630/11$.

Summary

On the basis of this example and the explanation, we can summarize the work to be done on any given tableau as follows:

1. If all coefficients in F (that is, the top row in the tableau) are ≤ 0, then the current x is the solution.

2. Select the nonbasic variable whose coefficient in F is positive and as large as possible. This becomes a new basic variable. Call it x_j.

3. Divide each b_i by the coefficient of the new basic variable in that row, a_{ij}. The value assigned to the new basic variable is the least of ratios. Thus, if b_k/a_{kj}, we set $x_j = b_k/a_{kj}$.

4. Using pivot element a_{kj}, create $0's$ in column j of A with Gaussian elimination steps.

6.5 Complementarity Problems

Complementarity is a new domain of applied mathematics, having deep relations with several aspects of fundamental mathematics and numerical analysis. Complementarity problems represent a wide class of mathematical models related to optimization, economics, engineering, mechanics, elasticity, fluid mechanics and game theory.

Complementarity problems arise naturally in the study of many phenomena in economics and engineering. A comprehensive and excellent treatment of applications of complementarity problems is provided in [8] and [14].

6.5.1 Problem Statement

A linear complementarity problem consists of finding a vector in a finite-dimensional real vector space that satisfies a certain system of inequalities.

More precisely, given a vector $q \in \mathbb{R}^n$ and a matrix $M \in \mathbb{R}^{n\times n}$, the linear complementarity problem (LCP), is to find a vector $z \in \mathbb{R}^n$ such that

$$\begin{cases} z \geq 0 \\ q + Mz \geq 0 \\ z^T(q + Mz) = 0 \end{cases} \tag{6.25}$$

or equivalently

$$\text{Find } w = (w_j) \in \mathbb{R}^n,\ z = (z_j) \in \mathbb{R}^n$$

satisfying

$$\begin{cases} w - Mz = q \\ w, z \geq 0 \\ w^T z = 0 \end{cases} \tag{6.26}$$

We denote the above LCP by the pair (q, M). The name comes from the third condition, the *complementarity condition* which requires that at least one variable in the pair (w_j, z_j) should be equal to 0 in the solution of the problem, for each $j = 1$ to n.

Example: The Obstacle Problem

The obstacle problem consists of finding the equilibrium position of an elastic membrane that is held at a fixed position on its boundary and which lies over an obstacle.

Consider stretching an elastic string fixed at the endpoint $(0, 0)$, and $(0, 4)$ over an obstacle defined by a function ψ (in this example, we use $\psi(x) = 1 - (x - 2.2)^2$. Notice that the position of the string will be defined by $\psi(x)$ for x between unknown points P and Q. , and that in the intervals $0 \leq x \leq P$, and $Q \leq x \leq 4$, the string lie along straight line segments connecting $(0, 0)$ to $(P, \psi(P))$ and $(Q, \psi(Q))$ to $(4, 0)$, respectively. If we represent the equilibrium position of the string by the function u, then u must satisfy the following conditions.

This representation of the problem is complicated by the presence of the free boundaries P and Q. The complementarity framework allows a simpler representation, which does not require free boundaries. First, note that since there is no downward force on the string, $u''(x) \leq 0$ for all x, except possibly at $x = P$ or $x = Q$ where u'' may be discontinuous. Also, note that $u(x) \geq \psi(x)$ everywhere. Finally, at each point x, either $u''(x) = 0$ or $u(x) = \psi(x)$. Thus, if we ignore momentarily the discontinuity of u'' at P and Q, we see that u must satisfy the conditions

$$\begin{cases} u(x) \geq \psi(x),\ 0 \leq x \leq 4 \\ u''(x) \leq 0 \\ (u(x) - \psi(x))u''(x) = 0 \\ u(0) = u(4) = 0. \end{cases} \tag{6.27}$$

This system can be solved numerically using a finite difference or finite element scheme. For example using a central scheme on a regular mesh with step size $h = 4/n$, u is approximated by the vector $u = (u_0, u_1, ..., u_n)$, where $u_i = \psi(x_i)$, $x_i = x_0 + ih$, $i = 0, ..., n$, and $x_0 = x_n = 0$. The above system is then approximated by

$$\begin{aligned}
&u_0 = u_n = 0 \\
&\frac{u_{i-1} - 2u_i + u_{i+1}}{h^2} \leq 0 \qquad i = 1, ..., n-1 \\
&u_i - \psi(x_i) \geq 0 \\
&(u_i - \psi(x_i)) \left(\frac{u_{i-1} - 2u_i + u_{i+1}}{h^2} \right) = 0.
\end{aligned}$$

Using the simple change of variables $z_i = u_i - \psi(x_i)$, this system is equivalent to the linear complementarity problem (q, M), where M is an $(n-1) \times (n-1)$ tridiagonal matrix and q is an $(n-1)$-vector defined by

$$M = \begin{bmatrix} 2 & -1 & 0 & 0 & \ldots & 0 \\ -1 & 2 & -1 & & & 0 \\ 0 & -1 & 2 & -1 & & 0 \\ . & & . & . & . & . \\ . & & & . & 2 & -1 \\ 0 & & & & -1 & 2 \end{bmatrix},$$

$$q = \begin{bmatrix} -2\psi(x_1) + \psi(x_2) \\ \psi(x_1) - 2\psi(x_2) + \psi(x_3) \\ \\ \\ \psi(x_{n-3}) - 2\psi(x_{n-2}) + \psi(x_{n-1}) \\ \psi(x_{n-2}) - 2\psi(x_{n-1}) \end{bmatrix}$$

The columns $z = (z_1, ..., z_{n-1})$ of this LCP then gives the discrete approximation to u at the interior points by the relation $u_i = z_i + \psi(x_i)$, $i = 1, ..., n-1$.

6.6 Variational Inequalities

Variational inequality theory was introduced by Hartman and Stampacchia (1966) as a tool for the study of partial differential equations with applications arising from mechanics. Such variational inequalities were *infinite-dimensional* rather than *finite-dimensional* as we describe in this section. The breakthrough in finite-dimensional theory of variational inequalities occurred in 1980 when Dafermos established that the traffic network equilibrium conditions of Smith (1979) contained a formulation of a variational inequality. This gave birth to a new methodlogy for the study of problems in economics, management science, operation research, and also in engineering with a focus on transportation. Variational inequality theory is a powerful tool for formulating a variety of equilibrium problems. It contains, as special cases, such well-known problems in mathematical programming as systems of non-linear equations, optimization problems, complementarity problems.

6.6.1 Variational Inequality Problem

The finite-dimensional variational inequality problem consists of seeking $x^* \in K \subset \mathbb{R}^n$, such that

$$F(x^*)^T.(x - x^*) \geq 0 \;\; \forall x \in K \tag{6.28}$$

or equivalently

$$\langle F(x^*)^T, (x - x^*)\rangle \geq 0 \;\; \forall x \in K \tag{6.29}$$

where K is a closed convex set, F is a given continuous function from K to $\mathbb{R}^n$, and $\langle ., . \rangle$ denotes the inner product in n dimensional Euclidean space.

In geometric terms, the variational inequality (6.28) states that $F(x^*)^T$ is orthogonal to the feasible set K at the point x^*. This formulation is particularly convenient because it allows for a unified treatment of equilibrium problems and optimization problems. Indeed, many mathematical problems can be formulated as variational inequality problems. Some examples are given below.

6.6.2 Systems of Equations

Many classical economic equilibrium problems have been formulated as systems of equations, since market clearing conditions necessarily equate the total supply with total demand. In terms of a variational inequality problem, the formulation of a system of equations is as follows.

Let $K = \mathbb{R}^n$ and let F from $\mathbb{R}^n$ into itself be a function. A vector $x^* \in \mathbb{R}^n$ solves VI$(F, \mathbb{R}^n)$ if and only if $F(\ x^*) = 0$.

Indeed, If $F(\ x^*) = 0$, then (6.29) holds with inequality. Conversely, if x^* satisfies (6.29), let $x = x^* - F(x^*)$. Then

$$\langle F(x^*)^T, (-F(x^*))\rangle \geq 0 \;\text{ or }\; - \|F(x^*)\|^2 \geq 0$$

and, therefore, $F(x^*) = 0$.

6.6.3 Optimization and Variational Inequalities

Both unconstrained and constrained optimization problems can be formulated as variational inequality problems. Next we show the relationship between an optimization problem and a variational inequality problem.

Proposition 1. *Let x^* be a solution to the optimization problem:*

$$\text{Minimize } f(x) \tag{6.30}$$

$$\text{subject to: } x \in K$$

where f is a continuously differentiable function and K is closed and convex. Then x^* is a solution of the variational inequality problem

$$\nabla f(x^*)^T.(x - x^*) \geq 0 \ \ \forall x \in K \tag{6.31}$$

Proof. Let $\phi(t) = f(x^* + t(x - x^*))$, for $t \to [0, 1]$. Since $\phi(t)$ achieves its minimum at $t = 0$, $0 \leq \phi'(0) = \nabla f(x^*)^T.(x - x^*)$, that is, x^* is a solution of (6.31). □

Proposition 2. *If $f(x)$ is a convex function and x^* is a solution to VI $(\nabla f, K)$, then x^* is a solution to the optimization problem (6.30).*

Proof. Since $f(x)$ is convex,

$$f(x) \geq f(x^*) + \nabla f(x^*)^T.(x - x^*), \ \forall x \in K$$

But $\nabla f(x^*)^T.(x - x^*) \geq 0$, since x^* is a solution to VI$(\nabla f, K)$. Therefore, from (6.31) one concludes that

$$f(x) \geq f(x^*), \ \forall x \in K$$

that is, x^* is a minimum point of the optimization problem (6.30). □

If the feasible set $K = \mathbb{R}^n$, then the unconstrained optimization problem is also a variational inequality problem. On the other hand, in the case where a certain symmetry holds, the variational inequality problem can be reformulated as an optimization problem.

6.7 Queuing Theory

Recall the last time that you had to wait at a supermarket checkout counter, wait a teller at your local bank, or wait to be served at a restaurant. In these and many other waiting line situations, the time is undesirable. Since adding more checkout clerks, more bank tellers, or more waiters is not always the most economical strategy for improving service, businesses are trying harder to understand their waiting line characteristics and find ways to keep waiting times within tolerable limits.

Indeed, for instance, there was a time when airline terminal passengers formed separate queues in front of check-in-counters. But now we see invariably only one line feeding into several counters. This is the result of the realization that a single line policy serves better for the passengers as well as the airline management.

Such a conclusion emanated from analyzing the mode by which a queue is formed and a service is provided. The analysis is based on building a mathematical model representing the process of arrival of passengers who join the queue, the rules by which they are allowed into service, and the time it takes to serve the passengers. In management science terminology a waiting line is referred to as a *queue*, and the body of knowledge dealing with waiting lines is known as *queuing theory*. In the early 1900s A.K. Erlang, a Danish telephone engineer, began a study of congestion and waiting completion of telephone calls. Since then, queuing theory has grown far more sophisticated and has been applied to wide variety of waiting line situations.

We identify the unit demanding service, whether it is human or otherwise, as the **customer**. The unit providing service is known as the *server*.This terminology of customers and servers is used in a generic sense regardless of the nature of the physical context. Some examples are given below:

(a) In communication systems, voice or data traffic queue up for lines for transmission. A simple example is the telephone exchange.

(b) In a manufacturing system with several work stations, units completing work in one station wait for access to the next.

(c) Vehicles requiring service wait for their turn in garage.

(d) Patients arrive at a doctor's clinic for treatment.

Numerous examples of this type are everyday occurrence. While analyzing them we can identify some basic elements of the systems. These are the following:

- *Input Process.* In most cases the arrival are products of external factors. Therefore, input is described in terms of random variables that can represent either the number arriving during a time interval or the time interval between successive arrivals.

- *Service mechanism.* The uncertainties involved in the service mechanism are the number of servers, the number of customers served at any time, and the duration and mode of service. Networks of queues consist of more than one server arranged in series and/or parallel. Random variables represent service times, and the number of servers, when appropriate.

- *System capacity.* The number of customers who wait at a time in a queueing system is an important factor for consideration. If the waiting room is large, one can assume that, for all piratical purposes, is infinite.

- *Queue discipline.* Servers follow rule in accepting customers for service. In this context, rules such as "first come, first served" (FCFS), "last come, first served" (LCFS), and "random selection for service" (RS) are self explanatory.

The identification of these elements leads to symbolically representing queueing systems with a variety of system elements. The basic representation widely used in queueing theory is made up of symbols representing three elements, input/service/number of servers. For instance, we use M for Poisson or exponential, D for deterministic (constant), E_k for the Erlang distribution with scale parameter k, and G for general (also GI for general independent).

- *Customer waiting time.* From a customer view, time spent in the queue and in the system are two characteristics of importance.

Let T_q and T be the time a customer spends in queue and in the system, respectively. We assume that the system operates according to a "first come, first served" (FCFS) queue discipline. With an FCFS queue discipline, the waiting time for service (T_q) of an arriving customer is the amount of time required to serve the customers already in the system. The total time T in the system is T_q plus the service time.

The ratio of arrival rate to service rate plays a significant role in measuring the performance of queueing systems.

$$\rho = \text{traffic intensity} = \frac{\text{arrival rate}}{\text{service rate}}.$$

6.7.1 Queue

M/M/1

The $M/M/1$ is the simplest of the queuing models used in practice.

1. The waiting line has a single channel.
2. The number of customers is finite.
3. The pattern of arrivals follows a Poisson process with mean arrival λ.
4. The service time follows an exponential probability distribution with mean service rate μ.
5. The queue discipline is first-come, first served (FCFS).

Let $Q(t)$ be the number of customers in the system, Q_q be the number of customers in the queue, excluding the one in service. Then, estimated number of customers in the system is given by:

$$L = E(Q) = \frac{\lambda}{\mu - \lambda} \tag{6.32}$$

and the estimated number of customers in the queue

$$L_q = E(Q_q) = \frac{\lambda^2}{\mu\,(\mu - \lambda)} \tag{6.33}$$

Now, let T_q and T be the time a customer spends in queue and in the system respectively. We assume that the system operates according to a "first come, first served" (FCFS) queue discipline. With an FCFS queue discipline, the waiting time for service (T_q) of an arriving customer is the amount of time required to serve the customers already in the system. The total time T in the system is T_q plus the service time. When there are n customers in the system, since service times are exponential with parameter μ, the total service time of n customers is Erlang with probability density

$$f_n(x) = e^{-\mu x} \frac{\mu^n x^{n-1}}{(n-1)!}.$$

Let $F_q(t) = P(T_q \leq t)$ be the distribution function of the waiting time T_q. Then, we have

$$F_q(t) = 1 - \rho e^{-\mu(1-\rho)t} \tag{6.34}$$

Let now $E(T_q) = W_q$ be the estimated waiting time in the the queue, then,

$$W_q = E(T_q) = \frac{\rho}{\mu\,(1-\rho)} = \frac{\lambda}{\mu\,(\mu - \lambda)}$$

And, since the total time in the system, T, is the sum of T_q and the service time, we get

$$W = E(T) = \frac{\lambda}{\mu\,(\mu - \lambda)} + \frac{1}{\mu} = \frac{1}{\mu - \lambda} \tag{6.35}$$

Combining the result (6.32) and (6.35), we note the relationship

$$L = \lambda W \tag{6.36}$$

A similar comparison between results (6.38) and (6.39) establishes

$$L_q = \lambda W_q \tag{6.37}$$

The formula (6.36) is known as *Little's law* in queueing literature.

Example 246. An airport has a single runway. Airplanes have been found to arrive at the rate of 15 per hour and it is estimated that each landing takes 3 minutes. Assuming a Poisson process for arrivals and an exponential distribution for landing times, we use the $M/M/1$ model to determine the following performance measures.

(a) Runway utilization:

$$\text{arrival rate} = 15/\text{hour } (\lambda).$$

$$\text{service rate } = \frac{60}{3}/\text{hour.}$$

$$\text{utilization} = \rho = \frac{\lambda}{\mu} = \frac{3}{4}.$$

(b) Expected number of airplanes waiting to land:

$$L_q = \frac{\rho^2}{1-\rho} = \frac{(0.75)^2}{0.25} = 2.25.$$

(c) Expected waiting time:

$$E(W_q) = \frac{\lambda}{\mu(\mu-\lambda)} = \frac{15}{20(20-15)} = \frac{3}{20} \text{ hour} = 9 \text{ minutes.}$$

(d) Probability that the waiting will be more than 5 minutes, 10 minutes, No waiting.

$$P(\text{no waiting}) = P(T_q = 0) = 1 - \rho = 0.25.$$

Since

$$P(T_q > t) = 1 - P(T_q \leq t) = \rho e^{-\mu(1-\rho)t}$$

then

$$P(T_q > 5 \text{ minutes}) = \frac{3}{4}e^{-20(1-\frac{3}{4})5/60}$$
$$= \frac{3}{4}e^{-\frac{25}{60}} = 0.4944$$

and

$$P(T_q > 10 \text{ minutes}) = \frac{3}{4}e^{-\frac{50}{60}} = 0.3259$$

(e) Expected number of landings in 20 minutes period $= \dfrac{15}{60} \times 20 = 5.$

6.7.2 Queue

M/M/s

The multiserver queue $M/M/s$ is the most used model. It aims at analyzing service stations with more than one server such as banks, checkout counters in stores, check-in counters in airports. The arrival of customers is assumed to follow a Poisson process, service times are assumed to have an exponential distribution, and the number of servers is s. These servers are assumed to provide service independently of each other. The arriving customers are assumed to form a single queue and the one at the head of the waiting line enters the service as soon as a server is free.

Let λ be the arrival rate and μ the service rate. Denoting by L and L_q the mean number of customers in the system and the number in the queue, respectively, and setting $\rho = \frac{\lambda}{s\mu}$ and $\alpha = \frac{\lambda}{\mu}$ following (6.32), we get

$$L = \alpha + \frac{\rho\alpha^s p_0}{s!(1-\rho)^2}$$

$$L_q = \frac{\rho\alpha^s p_0}{s!(1-\rho)^2} \tag{6.38}$$

where

$$p_0 = \left[\sum_{r=0}^{s-1} \frac{\alpha^r}{r!} + \frac{\alpha^s}{s!}\left(1 - \frac{\alpha}{s}\right)\right]^{-1}.$$

The estimated waiting time in the the queue, will be

$$W_q = E(T_q) = \frac{\alpha^s p_0}{s! s\mu(1-\rho)^2}. \tag{6.39}$$

Comparing (6.38) with (6.39), we have again Little's law formula

$$L_q = \lambda W_q.$$

Let T_q be the waiting time and $F_q(t) = P[T_q \leq t]$ be its distribution function. Then

$$F_q(t) = 1 - \frac{\alpha^s p_0}{s!(1-\rho)} e^{-s\mu(1-\rho)t}.$$

Example 247. In the airport problem of example, how would the performance measures change if there are two runways while assuming the same arrival and service rates?

(a) Runway utilization:

$$\begin{aligned} &\text{arrival rate} = 15/\text{hour } (\lambda), \\ &\text{service rate } = 20/\text{hour } (\mu), \\ &\text{number of servers} = 2(s) \\ &\text{utilization of each runway} = \rho = \frac{\lambda}{s\mu} = \frac{3}{8} \end{aligned}$$

(b) Expected number of airplanes waiting to land.

$$L_q = \frac{\rho\alpha^s p_0}{s!(1-\rho)^2}$$

with

$$p_0 = \left[\sum_{r=0}^{s-1} \frac{\alpha^r}{r!} + \frac{\alpha^s}{s!}\left(1 - \frac{\alpha}{s}\right)\right]^{-1}$$

$$= \left[1 + \frac{3}{4} + \frac{\left(\frac{3}{4}\right)^2}{2}\left(1 - \frac{3}{8}\right)^{-1}\right]^{-1} = 0.4545$$

$$L_q = \frac{\left[\left(\frac{3}{8}\right)\left(\frac{3}{4}\right)^2(0.4545)\right]}{2\left(\frac{5}{8}\right)^2} = 0.1227.$$

(c) Expected waiting time.

$$W_q = \frac{\alpha^s p_0}{s!s\mu(1-\rho)^2} = \frac{\left[\left(\frac{3}{4}\right)^2(0.4545)\right]}{2\times 2\times 20\left(1-\frac{3}{8}\right)^2}$$
$$= 0.00818 \text{ hour} = 0.49 \text{ minute}.$$

(d) Probability that the waiting time will be more than 5 minutes? 10 minutes? No waiting?

$$P(\text{no waiting}) = F_q(0) = 1 - \frac{\alpha^s p_0}{s!(1-\rho)}$$
$$= 1 - \frac{\left(\frac{3}{4}\right)^2(0.4545)}{2(1-3/8)}$$
$$= 0.7955.$$

$$P(T_q > t) = \frac{\alpha^s p_0}{s!(1-\rho)} e^{-s\mu(1-\rho)t}.$$

$$P(T_q > 5 \text{ minutes}) = \frac{\left(\frac{3}{4}\right)^2(0.4545)}{2(5/8)} e^{-2\left(\frac{1}{3}\right)\left(\frac{5}{8}\right)5}$$
$$= 0.1245.$$

$$P(T_q > 10 \text{ minutes}) = 0.0155.$$

Expected number of landings in a 20-minutes period $= \frac{15}{60} \times 20 = 5.$

6.8 Iterative Methods and Pre-conditioning

6.8.1 Norms of Vectors and Matrices

Definition 85. *A vector-norm on $\mathbb{R}^N$ is a function $\|.\|$, from $\mathbb{R}^N$ into $\mathbb{R}$ with the following properties:*

$$
\begin{aligned}
&(i)\ \ \|x\| \geq 0 \text{ for all } x \in \mathbb{R}^N \\
&(ii)\ \ \|x\| = 0 \text{ if and only if } x = 0 \\
&(iii)\ \ \|\alpha x\| \ = |\alpha|\,\|x\| \text{ for all } \alpha \in \mathbb{R} \text{ and } x \in \mathbb{R}^N \\
&(iv)\ \ \|x+y\| \leq \|x\| + \|y\|
\end{aligned}
$$

Definition 86. *The most used norms on $\mathbb{R}^N$ are the l_2 and l_∞ norms. For $x = (x_{1,x2}, ..., x_n)^t$ are defined by*

$$\|x\|_\infty = \max_{1\leq i\leq n} |x_i|, \quad \|x\|_1 = \sum_{i=1}^{n} |x_i|, \quad \|x\|_2 = \left\{\sum_{i=1}^{n} x_i^2\right\}^{1/2}$$

Example 248. The vector $x = (-1, 1, -2)$ in $\mathbb{R}^3$ has norms

$$
\begin{aligned}
\|x\|_2 &= \sqrt{(-1)^2 + (1)^2 + (-2)^2} = \sqrt{6} \\
\|x\|_\infty &= \max\{|-1|, |1|, |-2|\} = 2 \\
\|x\|_2 &= \sum_{i=1}^{n} |x_i| = |-1| + |1| + |-2| = 4
\end{aligned}
$$

Definition 87. *on the set of $n \times n$ matrices is a real-values function, $\|.\|$, defined on this set, satisfying for all $n \times n$ matrice A and B and all real numbers α :*

$$
\begin{aligned}
&\|A\| \geq 0. \\
&\|A\| = 0 \text{ if and only if } A = 0, \text{the matrix with } 0 \text{ entries.} \\
&\|A+B\| \leq \|A\| + \|B\|. \\
&\|AB\| \leq \|A\|\,\|B\|.
\end{aligned}
$$

The only norms we consider are those that are natural consequences of the vector norms l_2 and l_∞. Let $A = (a_{ij})$ be a These are $n \times n$ matrix, then the l_∞ norm of A is

$$\|A\|_\infty = \max_{1\leq i\leq n} \sum_{j=1}^{n} |a_{ij}|, \quad \|A\|_1 = \max_{1\leq j\leq n} \sum_{i=1}^{n} |a_{ij}|, \quad \|A\|_2 = \sqrt{\rho(A^T A)}$$

where $\rho\left(A^T A\right)$ is the largest eigenvalue of the positive and definite matrix $A^T A$. If A is a real symmetric and positive definite matrix, then $\|A\|_2 = |\lambda_{\max}| = \rho(A)$

Example 249. Compute the l_∞, l_1, and l_2 norms of

$$A = \begin{bmatrix} 1 & 1 & 0 \\ 1 & 2 & 1 \\ -1 & 1 & 2 \end{bmatrix}$$

Solution:

$$\sum_{j=1}^{3} |a_{1j}| = |1| + |1| = 2,$$

$$\sum_{j=1}^{3} |a_{2j}| = |1| + |2| + |1| = 4,$$

$$\sum_{j=1}^{3} |a_{3j}| = |-1| + |1| + |2| = 4$$

$$\|A\|_\infty = \max_{1\le i\le 3} \sum_{j=1}^{3} |a_{ij}| = 4$$

$$\sum_{i=1}^{3} |a_{i1}| = |1| + |1| + |-1| = 3,$$

$$\sum_{i=1}^{3} |a_{i2}| = |-1| + |2| + |2| = 4,$$

$$\sum_{i=1}^{3} |a_{i3}| = |1| + |2| = 3$$

$$\|A\|_1 = \max_{1\le j\le 3} \sum_{i=1}^{3} |a_{ij}| = 4$$

$$A^T A = \begin{bmatrix} 1 & 1 & -1 \\ 1 & 2 & 1 \\ 0 & 1 & 2 \end{bmatrix} \begin{bmatrix} 1 & 1 & 0 \\ 1 & 2 & 1 \\ -1 & 1 & 2 \end{bmatrix} = \begin{bmatrix} 3 & 2 & -1 \\ 2 & 6 & 4 \\ -1 & 4 & 5 \end{bmatrix}$$

To calculate $|\lambda_{\max}|$ we need the eigenvalues of $A^T A$. If

$$\det(A^T A - \lambda I) = 0$$

or

$$\det\begin{bmatrix} 3-\lambda & 2 & -1 \\ 2 & 6-\lambda & 4 \\ -1 & 4 & 5-\lambda \end{bmatrix}$$
$$= -\lambda^3 + 14\lambda^2 - 42\lambda = -\lambda(\lambda^2 - 14\lambda + 42) = 0$$

Then

$$\lambda = 0 \text{ or } \lambda = 7 \pm \sqrt{7},$$

so

$$\|A\|_2 = \sqrt{\rho\left(A^T A\right)} = \sqrt{\max\left\{0, 7-\sqrt{7}, 7+\sqrt{7}\right\}} = \sqrt{7+\sqrt{7}} \approx 3.108$$

6.8.2 General Iterative Method

In this section we describe the Jacobi and the Gauss-Seidel iterative methods, classic methods that date to the late eighteenth century. Iterative techniques are rarely used for solving linear systems of small dimension since the time required for sufficient accuracy exceeds that required for direct techniques such as Gaussian elimination. For large systems, however, these methods are efficient in terms of both computer storage and computation. System of this type arises frequently in circuit analysis and in the numerical solution of boundary-value problems and partial-differential equations.

An iterative technique to solve the $n \times n$ linear system

$$Ax = b \tag{6.40}$$

starts with an initial approximation $x^{(0)}$ to the solution x and generates a sequence of vectors $\left\{x^{(k)}\right\}_{k=0}^{\infty}$ that converges to x. Iterative technique involve a process that converts the system $Ax = b$ into an equivalent system of the form $x = Tx + c$ for some fixed matrix T and vector c. After the initial vector $x^{(0)}$ is selected, the sequence of approximate solution vectors is generated by computing

$$\mathbf{x}^{(k+1)} = T\mathbf{x}^{(k)} + c,\ k = 0, 1, 2, ... \tag{6.41}$$

Convergence analysis of iterative methods for both linear and nonlinear systems is based on the following important from mathematical analysis, known as the *Banach Fixed Point Theorem* or *Contraction Mapping Theorem* for a proof of this theorem we refer to [18] or [19].

Theorem 55. Let K be a closed and bounded subset of $\mathbb{R}^n$, equipped with a norm $\|.\|$. Suppose that $f : K \to K$ is a function such that there is a positive constant $q < 1$ with the property

$$\|f(x) - f(y)\| \leq q\,\|x - y\| \tag{6.42}$$

for all x, $y \in K$. Then, there is a unique point $x^* \in K$ such that $\mathbf{x}^* = f(\mathbf{x}^*)$. Moreover, the sequence $\{\mathbf{x}^{(k)}\}_{k\geq 0}$, defined recursively by

$$\mathbf{x}^{(k+1)} = f(\mathbf{x}^{(k)}),\ k = 0, 1, 2, ... \tag{6.43}$$

converges to $\mathbf{x}^*$ for any initial guess $\mathbf{x}^{(0)} \in K$.

Theorem 56. The following inequalities hold and describe the rate of convergence of the sequence $\{\mathbf{x}^{(k)}\}_{k\geq 0}$:

$$\left\|\mathbf{x}^{(k)} - \mathbf{x}^*\right\| \leq \frac{q}{1-q}\left\|\mathbf{x}^{(k)} - \mathbf{x}^{(k-1)}\right\|,\ k = 0, 1, 2, ... \tag{6.44}$$

or equivalently

$$\left\|\mathbf{x}^{(k)} - \mathbf{x}^*\right\| \leq \frac{q^k}{1-q}\left\|\mathbf{x}^{(1)} - \mathbf{x}^{(0)}\right\|,\ k = 0, 1, 2, ... \tag{6.45}$$

For a proof of this theorem we refer to [18] and [19].

Next we describe some iterative methods for solving a linear system.

6.8.3 Jacobi Iterative Method

We shall describe this method by the following example

Example 250. Use the Jacobi method to solve the linear system

$$\begin{aligned}
&(\mathrm{e}_1) 10x_1 - x_2 + 2x_3 = 6 \\
&(\mathrm{e}_2)\ -x_1 + 11x_2 - x_3 + 3x_4 = 25 \\
&(\mathrm{e}_3)\ 2x_1 - x_2 + 10x_3 - x_4 = -11 \\
&(\mathrm{e}_4)\ 3x_2 - x_3 + 8x_4 = 15
\end{aligned} \tag{6.46}$$

which has the unique solution $\mathbf{x} = (1, 2, -1, 1)^t$.

Solution: We first convert $A\mathbf{x} = b$ to the form solve equation (e_i) for x_i, for each $i = 1, 2, 3, 4$, to obtain

$$\begin{aligned}
x_1 &= \frac{1}{10}x_2 - \frac{1}{5}x_3 + \frac{3}{5} \\
x_2 &= \frac{1}{11}x_1 + \frac{1}{11}x_3 - \frac{3}{11}x_4 + \frac{25}{11} \\
x_3 &= -\frac{1}{5}x_1 + \frac{1}{10}x_2 + \frac{1}{10}x_4 = -11 \\
x_4 &= -\frac{3}{8}x_2 + \frac{1}{8}x_3 + \frac{15}{8}
\end{aligned}$$

TABLE 6.5

k	0	1	2	3	4	5
$x_1^{(k)}$	0.000	0.6000	1.0473	0.9326	1.0152	0.9890
$x_2^{(k)}$	0.000	2.2727	1.7159	2.053	1.9537	2.0114
$x_3^{(k)}$	0.000	−1.1000	−0.8052	−1.0493	−0.9681	−1.0103
$x_4^{(k)}$	0.000	1.8750	0.8852	1.1309	0.9739	1.0214
k	6	7	8	9	10	
$x_1^{(k)}$						
$x_2^{(k)}$						
$x_3^{(k)}$						
$x_4^{(k)}$	0.9944	1.0036	0.9989	1.0006	0.9998	

by solving for x_i in each row. That is, for an initial guess $\mathbf{x}^{(0)} = (0,0,0,0)^t$ we compute the iterate $x^{(1)}$ as follows

$$
\begin{aligned}
x_1^{(k)} &= \frac{1}{10}x_2^{(k-1)} - \frac{1}{5}x_3^{(k-1)} + \frac{3}{5}\\
x_2^{(k)} &= \frac{1}{11}x_1^{(k-1)} + \frac{1}{11}x_3^{(k-1)} - \frac{3}{11}x_4^{(k-1)} + \frac{25}{11}\\
x_3^{(1)} &= -\frac{1}{5}x_1^{(k-1)} + \frac{1}{10}x_2^{(k-1)} + \frac{1}{10}x_4^{(k-1)} - \frac{11}{10}\\
x_4^{(1)} &= -\frac{3}{8}x_2^{(k-1)} + \frac{1}{8}x_3^{(k-1)} + \frac{15}{8}
\end{aligned}
$$

Some iterates, $x^{(k)} = (x_1^{(k)}, x_2^{(k)}, x_3^{(k)}, x_4^{(k)})^t$, are displayed in Table the following table.

Solution: The decision to stop after ten iterations was based on the criterion

$$\frac{\left\|\mathbf{x}^{(10)} - \mathbf{x}^{(9)}\right\|_\infty}{\left\|\mathbf{x}^{(10)}\right\|_\infty} = \frac{8.0 \times 10^{-4}}{1.9998} < 10^{-3}$$

The Jacobi iterative method consists of solving the ith equation in $A\mathbf{x} = \mathbf{b}$ for x_i to obtain (provided $a_{ii} \neq 0$)

$$x_i = \sum_{\substack{j=1\\ j\neq i}}^{n}\left(-\frac{a_{ij}x_j}{a_{ii}}\right) + \frac{b_i}{a_{ii}}, \text{ for } i = 1,2,...,n \tag{6.47}$$

and generating each $x_i^{(k)}$ from components of $\mathbf{x}^{(k-1)}$ for $k \geq 1$ by

$$x_i^{(k)} = \sum_{\substack{j=1 \\ j \neq i}}^{n} \left(-\frac{a_{ij} x_j^{(k-1)}}{a_{ii}} \right) + \frac{b_i}{a_{ii}}, \text{ for } i = 1, 2, ..., n$$

6.8.4 The Gauss-Seidel Method

A possible improvement of Jacobi Method can be obtained by using the components, for $i > 1$, $x_1^{(k)}, ..., x_{i-1}^{(k)}$, already computed, to compute the component $x_i^{(k)}$. That is, we can use

$$x_i^{(k)} = \frac{-\sum_{j=1}^{i-1} (a_{ij} x_j^{(k)}) - \sum_{j=i+1}^{n} (a_{ij} x_j^{(k-1)}) + b_i}{a_{ii}}, \text{ for } i = 1, 2, ..., n \qquad (6.48)$$

Example 251. Use the Gauss-Seidel method to solve the linear system (6.46) solved in example 1 by the Jacobi method.

Solution: Let D be the diagonal matrix whose diagonal entries are those of A, $-L$ be the strictly lower-triangular part of A, $-U$ be the strictly upper triangular part of A. With this notation, the equationThe Gauss Seidel method applied to this system yields

Example 252.

$$\begin{aligned}
x_1^{(k)} &= \phantom{-\frac{1}{5}x_1^{(k)}} \quad \frac{1}{10} x_2^{(k-1)} - \frac{1}{5} x_3^{(k-1)} + \frac{3}{5} \\
x_2^{(k)} &= \frac{1}{11} x_1^{(k)} + \frac{1}{11} x_3^{(k-1)} - \frac{3}{11} x_4^{(k-1)} + \frac{25}{11} \\
x_3^{(1)} &= -\frac{1}{5} x_1^{(k)} + \frac{1}{10} x_2^{(k)} + \frac{1}{10} x_4^{(k-1)} - \frac{11}{10} \\
x_4^{(1)} &= -\frac{3}{8} x_2^{(k)} + \frac{1}{8} x_3^{(k)} + \frac{15}{8}
\end{aligned}$$

For $\mathbf{x}^{(0)} = (0, 0, 0, 0)^t$, the Gauss- Seidel method generates the iterates in Table

As

$$\frac{\left\| \mathbf{x}^{(5)} - \mathbf{x}^{(4)} \right\|_\infty}{\left\| \mathbf{x}^{(5)} \right\|_\infty} = \frac{0.0008}{2.000} = 4 \times 10^{-4}$$

$\mathbf{x}^{(5)}$ is accepted as a reasonable approximation to the solution. Note that Jacobi's method in example 1 required twice as many iterations for the same accuracy.

$$Ax = b$$

TABLE 6.6

k	0	1	2	3	4	5
$x_1^{(k)}$	0.000	0.6000	1.030	0.0065	1.0009	1.0001
$x_2^{(k)}$	0.000	2.3272	2.037	2.0036	2.0036	2.0000
$x_3^{(k)}$	0.000	−0.9893	−1.014	−1.0025	−1.0003	−1.0000
$x_4^{(k)}$	0.000	0.8789	0.9844	0.9983	0.9999	1.0000

becomes

$$(D - L - U)x = b$$

Then one can notice that the Jacobi iterative method is

$$\mathbf{x}^{(k+1)} = T\mathbf{x}^{(k)} + c \tag{6.49}$$

with

$$T = D^{-1}(L + U) \text{ and } c = D^{-1}b$$

and, concretely, in the following example.

$$T = \begin{bmatrix} 0 & \frac{1}{10} & -\frac{1}{5} & 0 \\ \frac{1}{11} & 0 & \frac{1}{11} & -\frac{3}{11} \\ -\frac{1}{5} & \frac{1}{10} & 0 & \frac{1}{10} \\ 0 & -\frac{3}{8} & \frac{1}{8} & 0 \end{bmatrix} \text{ and } c = \begin{bmatrix} \frac{3}{5} \\ \frac{25}{5} \\ -\frac{11}{10} \\ \frac{15}{8} \end{bmatrix}$$

while the Gauss-Seidel method is also an iterative method of the form (6.49) with

$$T = (D - L)^{-1} U \quad \text{and } c = (D - L)^{-1} b$$

Theorem 57. For any $\mathbf{x}^{(0)} \in \mathbb{R}^n$, the sequence $\{\mathbf{x}^{(k)}\}_{k \geq 1}$ defined in (6.41) converges to the unique solution of $x = Tx + c$ and, therefore, to the solution of system (6.40), if and only if the spectral radius of T, $\rho(T) < 1$.

6.9 Krylov Methods

Let A be a $n \times n$ matrix, and v a vector in $\mathbb{R}^m$. A Krylov subspace is a space of the form

$$K_m(A, v) = Span\left\{v, Av, A^2v, ..., A^{m-1}v\right\}$$

An important property of Krylov subspace is that K_m is the space of all vectors of $\mathbb{R}^n$ which can be written as $x = p(A)v$, where p is a polynomial not exceeding $m - 1$.

6.9.1 Arnoldi's Orthogonalization Method

Arnoldi's Method is an orthogonal projection method onto K_m for general non-Hermitian matrices. The procedure was introduced in 1951 as a mean of reducing a dense matrix into Hessenberg form.

Arnoldi's procedure is an algorithm for constructing an orthogonal basis for the Krylov subspace K_m. Below is one variant of this algorithm.

Algorithm

1. Choose a vector v_1 of norm 1 ($\|v_1\|_2 = 1$).
2. For $j = 1, 2, ..., m$ Do:
3. Compute $h_{ij} = (Av_j)$ for $i = 1, 2, ..., j$
4. Compute $wj = Avj - \sum hi_j v_j$
5. $h_{j+1}, j = \|w_j\|_2$
6. If $hj + 1, j = 0$, then stop
7. $vj_{+1} = w_{j/} h_{j+1,j}$
8. EndDo

At each step, the algorithm multiplies the previous Arnoldi vector v_j by A and then orthogonalizes the resulting vector w_j against all previous $v_i' s$ by a standard Gram-Schmidt procedure.

Proposition 3. Denote by V_m, the $n \times m$ matrix with column vectors $v_1, ..., v_m$, by $\bar{H}_m$, the $(m+1) \times m$ Hessenverg matrix whose nonzero entries h_{ij} are defined by Arnoldi's algorithm, and by H_m the matrix obtained from $\bar{H}_m$ by deleting its last row. Then the following relations hold:

$$\begin{aligned} AV_m &= V_{m+1}\bar{H}_m, \\ V_m^T AV_m &= H_m. \end{aligned} \tag{6.50}$$

6.9.2 Arnoldi's Method for Linear Systems

Consider the linear system

$$Ax = b.$$

Given an initial guess x_0 to this system, and the Krylov subspace

$$K_m(A, r_0) = \{r_0, Ar_0, A^2 r_0, ..., A^{m-1} r_0\}$$

where $r_0 = b - Ax_0$, the Arnoldi's method seeks an approximate solution x_m from the the affine subspace $x_0 + K_m$ of dimension m by imposing the so-called Galerkin condition

$$b - Ax_m \perp K_m.$$

If $v_1 = r_0/\left\|r_0\right\|_2$ in Arnoldi's method, and set $\beta = \left\|r_0\right\|_2$, then, by (6.50)

$$V_m^T r_0 = V_m^T(\beta v_1) = \beta e_1$$

which yields

$$x_m = x_0 + V_m y_m$$
$$y_m = H_m^{-1}(\beta e_1),\ e_1 = (1, 0, ..., 0).$$

Algorithm

1. Compute $r_0 = b - Ax_0$, $\beta = \left\|r_0\right\|_2$, and $v_1 = r_0/\beta$

2. Define the $m \times m$ matrix $H_m = (h_{ij})$; set $H_m = 0$

3.For $j = 1, 2, ..., m$ Do:

4. Compute $w_j = Avj$

5. For $i = 1, ..., j$ Do:

6. $\quad h_{ij} = (w_j, v_i)$

7. $\quad w_j = w_j - h_{ij} v_j$

8. EndDo

9. Compute $h_{j+1,j} = \left\|w_j\right\|_2$. If $h_{j+1,j} = 0$ set $m = j$ and GoTo 12

10. Compute $v_{j+1} = w_j/h_{j+1,j}$

11. EndDo

12. Compute $y_m = H_m^{-1}(\beta e_1)$ and $x_m = x_0 + V_m y_m$.

6.9.3 Conjugate Gradient Method

Consider the linear system

$$Ax = b \tag{6.51}$$

where A is symmetric and positive definite. Then it is easily verified that solving (6.51) is equivalent to minimizing the quadratic functional

$$\phi(v) = \frac{1}{2}(Av, v) - (b, v)$$

as the minimum value of ϕ is $-\frac{1}{2}(A^{-1}v, v) - (b, v)$,and it is attained for $x = A^{-1}b$. We also notice that

$$-\nabla\phi(x) = b - Ax = r$$

where r is known as the *residual vector* for x. Note that $r = 0$ if and only if x is the solution, and so the size of r measures, in a certain sense, how accurately x comes to solving the system. Moreover, the residual vector indicates the direction of steepest decrease in the quadratic function, and is thus a good choice of direction to head off in search the true minimizer.

The initial result is the *gradient descent algorithm,* in which each successive approximation u_k to the solution is obtained by going a certain distance in the residual direction:

$$x^{k+1} = x^k + \alpha_k r^k, \quad \text{where } r^k = b - Ax^k \tag{6.52}$$

The conjugate gradient method is one of the best iterative techniques for solving linear systems

$$Ax = b$$

with A, as an $n \times n$ Symmetric Positive Definite matrix. The method can be viewed as an orthogonal projection technique onto the Krylov subspace $K_m(r_0, A)$ where r_0 is the initial residual. In the conjugate gradient method the update of the iterate is obtained from (6.52) by replacing r^k with a new direction p^k that is not parallel to that of the gradient. These methods were originally proposed by Hestenes and Stielfel [20] (see also Hestenes [21]) as direct methods for symmetric definite linear systems, as (in the absence of round-off errors) they produce the exact solution in a finite number of steps. Nowadays, they are implemented as iterative schemes which are capable of producing very accurate results in a small number of iterations.

In the Conjugate Gradient (CG), method, the directions $\{p^k\}$ are A-*conjugate*, that is, they satisfy the orthogonality property $(p^j, Ap^m) = 0$ for $m \neq j$. In particular

$$(p^{k+1}, Ap^k) = 0 \quad \forall k = 0, 1, ...$$

The idea of the method is based on the following remark. Let $p^0, ..., p^m$ be linearly independent vectors, x^0 being an initial guess, and construct the solution x by successive approximation, with the *kth* iterate having the form

$$x^{k+1} = \alpha_1 p^1 + ... + \alpha_k p^k \quad \text{so that } x^{k+1} = x^k + \alpha_{k+1} p^{k+1}$$

where α_k are non-zero real numbers. The idea is not to try to specify the conjugate basis vectors in advance, but rather to successively construct them during the course of the algorithm. We begin, merely for convenience, with an initial guess $x^0 = 0$ for that solution. The residual vector $r^0 = b - Ax^0 = b$ indicates the direction of steepest decrease of $\phi(x)$ at x^0, and we update the original guess by moving in this direction, taking $p^1 = r^0 = b$ as our first conjugate direction. The next iterate is $x^1 = x^0 + \alpha_1 p^1$, and we choose the parameter α_1 so that the corresponding residual vector

$$r^1 = b - Ax^1 = r^0 - \alpha_1 p^1$$

is such that $\|r^1\|$ is as close to zero as possible. This occurs when r^1 is orthogonal to r^0, and so we require

$$\begin{aligned}0 = \left(r^0\right)^T r^1 &= \|r^0\|^2 - \alpha_1 \left(r^0\right)^T Ap^1 \\ &= \|r^0\|^2 - \alpha_1(r^0, p^1) \\ &= \|r^0\|^2 - \alpha_1(p^1, p^1).\end{aligned}$$

Therefore

$$\alpha_1 = \frac{\|r^0\|^2}{(p^1, p^1)}$$

and so

$$x^1 = x^0 + \frac{\|r^0\|^2}{(r^1, r^1)} p^1$$

is our new approximation to the solution. We can assume that $\alpha_1 \neq 0$, since otherwise the residual $r^0 = 0$. In this case, $x^0 = 0$ would be the exact solution of the system, and there would be no reason to continue the procedure.

Now let us set $p^2 = r^1 + s_1 p^1$, where the scalar factor s_1 is determined by the orthogonality requirement

$$0 = \left(p^1, p^2\right) = \left(r^1 + s_1 p^1, p^1\right) = \left(r^1, p^1\right) + s_1 \left(p^1, p^1\right)$$

so

$$s_1 = -\frac{\left(r^1, p^1\right)}{\left(p^1, p^1\right)}.$$

Now, using the orthogonality of r^0 and r^1, we get

$$\left(r^1, p^1\right) = \left(r^1\right)^T Ap^1 = \left(r^1\right)^T \left(\frac{r^0 - r^1}{\alpha_1}\right) = -\frac{1}{\alpha_1} \|r^1\|^2$$

while

$$\left(r^1, p^1\right) = \left(p^1\right)^T Ap^1 = \frac{1}{\alpha_1} \|r^0\|^2$$

Therefore, the second conjugate direction is

$$p^2 = r^1 + s_1 p^1, \text{ where } s_1 = \frac{\|r^1\|^2}{\|r^0\|^2}.$$

We then update

$$x^2 = x^1 + \alpha_2 p^2 = x^0 + \alpha_1 p^1 + \alpha_2 p^2 = \alpha_1 p^1 + \alpha_2 p^2$$

so as to make the corresponding residual vector

$$r^2 = b - Ax^2 = r^1 - \alpha_2 Ap^2$$

as small as possible, which is accomplished by requiring it to be orthogonal to r^1. Thus, using the A-orthogonality of p^1 and p^2

$$0 = \left(r^1\right)^T r^2 = \left\|r^1\right\|^2 - \alpha_2 \left(r^1\right)^T Ap^2 = \left\|r^1\right\|^2 - \alpha_2 \left(p^2, p^2\right)$$

and so

$$\alpha_2 = \frac{\left\|r^1\right\|^2}{\left(p^2, p^2\right)}$$

Continuing in this manner, at the kth stage, we have already constructed the conjugate vectors $p^1, ..., p^k$, and the solution approximation x^k as a suitable linear combination of them. The next conjugate direction given by

$$p^{k+1} = r^k + s_k p^k, \text{ where } s_k = \frac{\left\|r^k\right\|^2}{\left\|r^{k-1}\right\|^2}$$

results from the A-orthogonality requirement:$\left(p^i, p^k\right) = 0$ for $i < k$. The updated solution approximation

$$x^{k+1} = x^k + \alpha_{k+1} p^{k+1}, \text{ where } \alpha_{k+1} = \frac{\left\|r^k\right\|^2}{\left(p^{k+1}, p^{k+1}\right)}$$

is then specified so as to make the corresponding residual

$$r^{k+1} = b - Ax^k = r^k - \alpha_{k+1} p^{k+1}$$

as small possible, by requiring that it be orthogonal to r^k.

Therefore the **Conjugate Gradient method** can be formulated as follows.

Algorithm

1. Compute $r^0 = b - Ax^0$, and set $p^1 = r^0$. For $k = 1, 2, ...$, the *kth* iteration is computed as follows:

2. For $j = 0, 1, ...,$until convergence Do

3. $\alpha_{k+1} = \dfrac{\left\|r^k\right\|^2}{\left(p^{k+1}, p^{k+1}\right)}$

4. $x^{k+1} = x^k + \alpha_{k+1} p^{k+1}$

5. $r^{k+1} = r^k - \alpha_{k+1} p^{k+1}$

6. $s_k = \frac{\left\|r^k\right\|^2}{\left\|r^{k-1}\right\|^2}$

7. $p^{k+1} = r^k + s_k p^k$.

Observe that the algorithm does not require solving any linear systems: apart from multiplication of a matrix times a vector to evacuate Ap^k, all other operations are rapidly evaluated Euclidean dot products. Unlike Gaussian elimination, the method produces a sequence of successive approximations x^1,x^2,...to the solution x, and so the iteration can be stopped as soon as a desired solution accuracy is reached, which can be assessed by comparing how close the successive iterates are to each other. On the other hand, unlike purely iterative methods, *the conjugate gradient method does eventually terminate at exact solution*, because, there are at most n conjugate directions, forming an orthogonal basis of $\mathbb{R}^n$ for the inner product induced by A. Therefore, $x^n = \alpha_1 p^1 + ... + \alpha_n p^n = x$ must be the solution since its residual $r^n = b - Ax^n$ is orthogonal to all the conjugate basis vectors $p^1, ..., p^n$ and hence must be zero.

Example 253. Solve the linear system $Ax = b$ by the conjugate gradient method with

$$A = \left(\begin{bmatrix} 3 & -1 & 0 \\ -1 & 2 & 1 \\ 0 & 1 & 1 \end{bmatrix} \right), b = \begin{pmatrix} 1 \\ 2 \\ -1 \end{pmatrix}$$

exact solution is $x = (2, 5, -6)^T$.

Solution: Implement the conjugate gradient method, starting from the initial guess $x^0 = (0, 0, 0)^T$. The corresponding residual vector is $r^0 = b - Ax^0 = b = (1, 2, -1)^T$. The first conugate direction is $p^1 = r^0 = (1, 2, -1)^T$. Use the formula to obtain the updated approximation to the solution

$$x^1 = x^0 + \frac{\|r^0\|^2}{(r^1, r^1)} p^1 = \frac{6}{4} \begin{pmatrix} 1 \\ 2 \\ -1 \end{pmatrix} = \begin{pmatrix} \frac{3}{2} \\ 3 \\ -\frac{1}{2} \end{pmatrix}.$$

noting that $p^1, p^1) = (p^1)^T Ap^1 = 4$. In the next stage of the algorithm, we compute the residual $r^1 = b - Ax^1 = \left(-\frac{1}{2}, -1, -\frac{5}{2}\right)$. The conjugate direction is

$$p^2 = r^1 + \frac{\|r^1\|^2}{\|r^0\|^2} x^1 = \begin{pmatrix} -\frac{1}{2} \\ -1 \\ -\frac{5}{2} \end{pmatrix} + \frac{\frac{15}{2}}{6} \begin{pmatrix} 1 \\ 2 \\ -1 \end{pmatrix} = \begin{pmatrix} \frac{3}{4} \\ \frac{3}{2} \\ -\frac{15}{4} \end{pmatrix}$$

which, as designed, satisfies the conjugacy condition $(p^1, p^2) = (p^1)^T Ap^2 = 0$. Each entry of the ensuing approximation

$$x^2 = x^1 + \frac{\|r^1\|^2}{(p^2, p^2)} p^2 = \begin{pmatrix} \frac{3}{2} \\ 3 \\ -\frac{3}{2} \end{pmatrix} + \frac{\frac{15}{2}}{\frac{27}{4}} \begin{pmatrix} \frac{3}{4} \\ \frac{3}{2} \\ -\frac{15}{4} \end{pmatrix} = \begin{pmatrix} \frac{7}{3} \\ \frac{14}{3} \\ -\frac{17}{3} \end{pmatrix}$$

Since we are dealing with a 3×3 system, we will recover the exact solution by one more iteration of the algorithm.The new residual is $r^2 = b - Ax^2 =$

$(-\frac{4}{3},\frac{2}{3},0)^T$. the final conjugate direction is

$$p^3 = r^2 + \frac{\|r^2\|^2}{\|r^1\|^2}p^2 = \begin{pmatrix} -\frac{4}{3} \\ \frac{2}{3} \\ 0 \end{pmatrix} + \frac{\frac{20}{9}}{\frac{15}{2}} \begin{pmatrix} \frac{3}{4} \\ \frac{3}{2} \\ -\frac{15}{4} \end{pmatrix} = \begin{pmatrix} -\frac{10}{9} \\ \frac{10}{9} \\ -\frac{10}{9} \end{pmatrix}$$

which can be, easily checked, is conjugate to both p^1 and p^2. Thus, the solution is obtained from

$$x^3 = x^2 + \frac{\|r^2\|^2}{(p^3,p^3)}p^3 = \begin{pmatrix} \frac{7}{3} \\ \frac{14}{3} \\ -\frac{17}{3} \end{pmatrix} + \frac{\frac{20}{9}}{\frac{15}{2}} \begin{pmatrix} -\frac{10}{9} \\ \frac{10}{9} \\ -\frac{10}{9} \end{pmatrix} = \begin{pmatrix} 2 \\ 5 \\ -6 \end{pmatrix}$$

6.9.4 Preconditioned Conjugate Gradient Method

If the matrix A is ill conditioned, the conjugate gradient method is highly sensible to rounding errors. So, although the exact answer should be obtained in n steps, this is not always the case. The main benefit of the conjugate gradient method is as an iterative method applied to a better conditioned system. In this case an acceptable approximate solution is often obtained in about $\sqrt{n}$ steps.

To apply the conjugate gradient method to a better-conditioned system, we select a non-singular conditioning matrix C so that

$$\hat{A} = C^{-1}A\left(C^{-1}\right)^T$$

is better conditioned.

Consider the linear system

$$\hat{A}\hat{x} = \hat{b}$$

where

$$\hat{x} = C^T x \text{ and } \hat{b} = C^{-1}b.$$

Then

$$\hat{A}\hat{x} = C^{-1}A\left(C^{-1}\right)^T = C^{-1}Ax$$

which means, we can solve for $\hat{A}\hat{x} = \hat{b}$ for $\hat{x}$ and then obtain x by $\left(C^{-1}\right)^T$. Since

$$\hat{x} = C^T x^k$$

we have

$$\hat{r}^k = \hat{b} - \hat{A}x^k = C^{-1}b - C^{-1}A\left(C^{-1}\right)^T x^k = C^{-1}(b - Ax^k) = C^{-1}r^k.$$

Let $\hat{p}^k = C^T p^k$ and $w^k = C^{-1}r^k$. Then

$$\hat{s}_k = \frac{\left(\hat{r}^k, \hat{r}^k\right)}{\left(r^{k-1}, r^{k-1}\right)} = \frac{\left(C^{-1}r^k, C^{-1}r^k\right)}{\left(C^{-1}r^{k-1}, C^{-1}r^{k-1}\right)}$$

so

$$\hat{s}_k = \frac{(w^k, w^k)}{(w^{k-1}, w^{k-1})}.$$

Thus

$$\hat{\alpha}^k = \frac{(\hat{r}^{k-1}, \hat{r}^{k-1})}{(\hat{p}^k, \hat{p}^k)} = \frac{(C^{-1}r^{k-1}, C^{-1}r^{k-1})}{\left(C^T p^k, C^{-1}A\,(C^{-1})^T\, C^T p^k\right)} = \frac{(w^{k-1}, w^{k-1})}{(C^T p^k, C^{-1}Ap^k)}$$

and

$$\hat{\alpha}^k = \frac{(w^{k-1}, w^{k-1})}{(p^k, p^k)}.$$

Further

$$\hat{x}^k = \hat{x}^{k-1} + \hat{\alpha}^k \hat{p}^k$$

so

$$C^T \hat{x}^k = C^T \hat{x}^{k-1} + \hat{\alpha}^k C^T \hat{p}^k$$

and

$$x^k = x^{k-1} + \hat{\alpha}^k p^k.$$

Also,

$$\hat{r}^k = \hat{r}^{k-1} - \hat{\alpha}^k \hat{A} \hat{p}^k$$

so

$$C^{-1}r^k = C^{-1}r^{k-1} - \hat{\alpha}^k C^{-1} A \left(C^{-1}\right)^T \hat{p}^k$$

that is

$$r^k = r^{k-1} - \hat{\alpha} A p^k.$$

Finally

$$\hat{p}^{k+1} = \hat{r}^k + \hat{s}_k \hat{p}^k$$

and

$$C^T p^{k+1} = C^{-1} r^k + \hat{s}_k C^T p^k$$

yields

$$p^{k+1} = \left(C^{-1}\right)^T C^{-1} r^k + \hat{s}_k p^k$$

or

$$p^{k+1} = \left(C^{-1}\right)^T w^k + \hat{s}_k p^k.$$

Example 254. The next example illustrates the effect of preconditioning on a poorly conditioned matrix. The linear system $Ax = b$ with

$$A = \begin{pmatrix} 0.2 & 0.1 & 1 & 1 & 0 \\ 0.1 & 4 & -1 & 1 & -1 \\ 1 & -1 & 60 & 0 & -2 \\ 1 & 1 & 0 & 8 & 4 \\ 0 & -1 & -2 & 4 & 700 \end{pmatrix} \quad \text{and} \quad b = \begin{pmatrix} 1 \\ 2 \\ 3 \\ 4 \\ 5 \end{pmatrix}$$

has the solution

$$x^* = (7.859713071,\ 0.4229264082,\ -0.07359223606,\ -0.5406430164,\ 0.01062616286)^T.$$

Solution: The matrix A is symmetric and positive definite but is ill-conditioned with condition number $K_\infty(A) = \|A\|_\infty \left\|A^{-1}\right\|_\infty = 13961.71$. We will use tolerance 0.01 and compare the results obtained from the Jacobi, Gauss-Seidel iterative methods and from the conjugate gradient method with $C^{-1} = I$. Then we precondition by choosing C^{-1} as $D^{-1/2}$, the diagonal matrix whose diagonal entries are the reciprocal of the positive square roots of the diagonal entries of the positive definite matrix A. The results are presented in Table...The Conjugate method gives the most accurate approximation with the smallest number of iterations.

Method	**Number of Iterations**	$\left\|x^{(k)} - x^*\right\|_\infty$
Jacobi	49	0.00305834
Gauss-Seidel	15	0.02445559
Conjugate gradient	5	0.00629785
Conjugate gradient	4	
Preconditioned		0.00009312

The preconditioned conjugate gradient method is often used in the solution of linear systems in which the matrix is sparse and positive definite. These systems arise from the discretization by finite elements methods or finite difference methods of boundary value problems. The larger the system, the more impressive the conjugate gradient method becomes since it significantly reduces the number of iterations required. In these systems, the preconditioning matrix C is approximately equal to L in the Choleski factorization LL^t of A. Generally, small entries in A are ignored and Choleski's method is applied to obtain what is called an incomplete LL^t factorization of A. Thus $\left(C^{-1}\right)^T C^{-1} \approx A^{-1}$ and a good approximation is obtained.

6.9.5 Generalized Minimum Residual Method (GMRES)

Consider the linear system

$$Ax = b$$

Given an initial guess x_0 to this system, and the Krylov subspace, this method consists of minimizing the residual norm over all vectors in $x_0 + K_m$. Any vector x in $x_0 + K_m$ can be written as

$$x = x_0 + V_m y$$

where y is an vector in $\mathbb{R}^m$. Let

$$J(y) = \|b - Ax\|_2 = \|b - A(x_0 + V_m y)\|_2$$

So

$$\begin{aligned} b - Ax &= b - A(x_0 + V_m y) \\ &= r_0 - AV_m y \\ &= \beta v_1 - V_{m+1}\bar{H}_m y \\ &= V_{m+1}\left(\beta e_1 - \bar{H}_m y\right). \end{aligned}$$

Since the columns-vectors ofV_{m+1} are orthogonal, then

$$J(y) = \|b - A(x_0 + V_m y)\|_2 = \left\|\beta e_1 - \bar{H}_m y\right\|_2 \tag{6.53}$$

The GMRES approximation is the unique vector of $x_0 + K_m$ which minimizes the functional (6.53). If we denote by y_m the vector that minimizes (6.53), this approximation can be obtained as

$$x_m = x_0 + V_m y_m$$

Algorithm (GMRES)

1.Compute $r_0 = b - Ax_0,\ ,\ \beta = \|r_0\|_2$, and $v_1 = r_0/\beta$

2.Define $(m+1) \times n$ matrix $\bar{H}_m = (h_{ij})$. Set $\bar{H}_m = 0$.

3. For $j = 1, 2, ..., m$ Do:

4. Compute $w_j = Avj$

5. For $i = 1, ..., j$ Do:

6. $h_{ij} = (w_j, v_i)$

7. $w_j = w_j - h_{ij}v_j$

8. EndDo

9. Compute $h_{j+1,j} = \|w_j\|_2$. If $h_{j+1,j} = 0$ set $m = j$ and GoTo 12

10. Compute $v_{j+1} = w_j/h_{j+1,j}$

11. EndDo

12. Compute $y_m = H_m^{-1}(\beta e_1)$ and $x_m = x_0 + V_m y_m$

6.10 Multi-grid Methods

The classical iterative methods fail to be effective whenever the spectral radius of the iteration matrix is close to one. A Fourier analysis shows that large eigenvalues are associated with errors of low frequency. Therefore, the smooth component (low frequencies) of the error holds back convergence, whereas high frequency errors are rapidly damped.

The basic idea of the method is to change the grid, whenever the linear system at hands arises from a discretization of differential equations over a certain mesh. As a matter of fact, errors that are smooth on a grid of width h can be attacked on a coarser grid, while high frequency errors that are not visible on the coarse grid of width $2h$ can be solved on the fine grid.

6.10.1 Multi-grid Cycles

The basic multi-grid algorithm can be presented in an abstract fashion in the following form. We are concerned with a family of linear problems

$$A_j x_{(j)} = b_{(j)},\ j = 0, ..., m \tag{6.54}$$

where A_j is an invertible linear operator on a finite dimensional space V^j, with $\dim V^j < \dim V^{j+1}$, $j = 0, ..., m-1$, and $b_{(j)}$ is a given right hand side.The goal is to solve (6.54) for $j = m$. For this reason we omit the subscript m for the highest level,thus writing

$$V = V^m,\quad A = A_m,\quad b = b_{(m)},\quad x = x_{(m)}$$

For each j, problem (6.54) should be regarded as the discretization of a differential equation on a grid whose characteristic mesh spacing is h_j. If we assume that $h_{j-1} = 2h_j$ and set $h = h_m$, then problem (6.54) provides a family of discretizations with a refinement factor 2.

To solve the problem (6.54) we consider a basic iterative method

$$x_{(j)}^{k+1} = x_{(j)}^k + P_j(b_{(j)} - A_j x_{(j)}^k) \tag{6.55}$$

where P_j is a preconditioner for A_j. The associated error $e_{(j)}^k$ is transformed according to

$$e_{(j)}^{k+1} = B_j e_{(j)}^k, B_j = I - P_j A_j.$$

The multi-grid algorithm can be regarded as an acceleration of (6.55) for $j = m$, and can be formulated as follows.

1. *Pre-smoothing on fine grid.*

Given $x_{(j)}^0$, do n_1 times

$$x_{(j)}^k = x_{(j)}^{k-1} + P_j(b_{(j)} - A_j x_{(j)}^{k-1})\ ,\ k = 1, ..., n_1$$

2. *Coarse grid correction*
(i) Form the residual (or defect) $r_{(j)} = b_{(j)} - A_j x_{(j)}^{n_1}$;
(ii) Restrict the residual on the coarse grid setting $r_{(j-1)} = R_j^{j-1} r_{(j)}$, where

$$R_j^{j-1} : V^j \to V^{j-1}$$

is a restriction operator;
(iii) solve the the defect problem

$$A_{j-1} x_{(j-1)} = r_{(j-1)}$$

(iv) correct the solution in V^j by setting

$$\bar{x}_{(j)} = x_{(j)}^{n_1} + \Pi_{j-1}^{j} x_{(j-1)}$$

where

$$\Pi_{j-1}^{j} : V^j \to V^{j-1}$$

is a prolongation operator.
2. *Post-smoothing on the fine grid*
Set $\hat{x}_{(j)}^0 = \bar{x}_{(j)}$ and do n_2 times

$$\hat{x}_{(j)}^k = \hat{x}_{(j)}^{k-1} + P_j (b_{(j)} - A_j \hat{x}_{(j)}^{k-1}) \ , \ k = 1, ..., n_2$$

6.11 Exercises

Analysis of a Quadratic Function

6.1. Find the minimum value of the function $f(x_1, x_2, x_3) = x_1^2 + 2x_1x_2 + 3x_2^2 + 2x_2x_3 + x_3^2 - 2x_1 + 3x_3 + 2$. How do you know that your answer is really the global minimum ?

6.2 For the following quadratic functions, determine if there is a minimum. If so, find the minimizer and the minimum value for the function.

a. $x_1^2 - 2x_1x_2 + 4x_2^2 + x_1 - 1$

b. $3x_1^2 + 3x_1x_2 + 3x_2^2 - 2x_1 - 2x_2 + 4$

c. $x_1^2 + 5x_1x_2 + 3x_2^2 + 2x_1 - x_2$

6.3 For each matrix A, vector b,and scalar c, write out the quadratic function $f(x)$ given by (6.9). Then either find the minimizer x^* and minimum value $f(x^*)$, or explain why there is none.

a. $A = \begin{pmatrix} 4 & -12 \\ -12 & 45 \end{pmatrix}$, $b = \begin{pmatrix} -\frac{1}{2} \\ 2 \end{pmatrix}$, $c = 3$

b. $A = \begin{pmatrix} 3 & 2 \\ 2 & 1 \end{pmatrix}$, $b = \begin{pmatrix} 4 \\ 1 \end{pmatrix}$, $c = 0$

c. $A = \begin{pmatrix} 3 & -1 & 1 \\ -1 & 2 & -1 \\ 1 & -1 & 3 \end{pmatrix}$, $b = \begin{pmatrix} 4 \\ 1 \end{pmatrix}$, $c = -3$

d. $A = \begin{pmatrix} \\ \\ \\ \\ \end{pmatrix}$, $b = \begin{pmatrix} -1 \\ 2 \\ -3 \\ 4 \end{pmatrix}$, $c = 0$

LP and Simplex Algorithm

6.4. Write each Linear Programming as a standard maximization problem using matrix notation. Use the simplex algorithm to maximize the objective function f subject to the given constraints.

(a) $f = 4x_1 + 3x_2, \begin{cases} -x_1 + x_2 \leq 3 \\ 4x_1 - x_2 \leq 1 \end{cases}$.

(b) $f = 2x_1 + 3x_2, \begin{cases} x_1 - x_2 \leq 2 \\ 4x_1 - x_2 \leq 11 \\ x_1 + x_2 \leq 9 \end{cases}$.

(c) $f = 2x_1 + 2x_2 + x_3, \begin{cases} 4x_1 - 2x_2 + x_3 \leq 3 \\ x_1 + x_2 + x_3 \leq 6 \end{cases}$.

(d) $f = 2x_1 + x_2 + 2x_3, \begin{cases} 4x_1 + x_2 - x_3 \geq -3 \\ x_1 - x_2 + 2x_3 \leq 5 \\ 3x_1 + 2x_2 + x_3 \leq 10 \end{cases}$.

Jacobi and Gauss-Seidel Methods

6.5. Find the first two iterations of the Jacobi method for the following linear system, using $\mathbf{x}^{(0)} = 0$

a.

$$\begin{aligned} 3x_1 - x_2 + x_3 &= 1 \\ 3x_1 + 6x_2 + 2x_3 &= 0 \\ 3x_1 + 3x_2 + 7x_3 &= 4 \end{aligned}$$

b.

$$\begin{aligned} 10x_1 - x_2 \qquad &= 9 \\ -x_1 + 10x_2 - 2x_3 &= 0 \\ -2x_2 + 10x_3 &= 4 \end{aligned}$$

c.

$$\begin{aligned} 10x_1 + 5x_2 \qquad\qquad &= 6 \\ 5x_1 + 10x_2 - 4x_3 &= 25 \\ -4x_2 + 8x_3 - x_4 &= -11 \\ -x_3 + 5x_4 &= -11 \end{aligned}$$

6.6 Repeat Exercise 1 using Gauss-Seidel method.

Conjugate Gradient Method

6.7. Solve the following linear systems $Ax = b$, using the conjugate gradient method, keeping track of the residual vectors and solution approximation as you iterate

a.

$$A = \begin{bmatrix} 3 & -1 \\ -1 & 5 \end{bmatrix}, \quad b = \begin{bmatrix} 2 \\ 1 \end{bmatrix}$$

b.

$$A = \begin{bmatrix} 6 & 2 & 1 \\ 2 & 3 & -1 \\ 1 & -1 & 2 \end{bmatrix}, \quad b = \begin{bmatrix} 1 \\ 0 \\ -2 \end{bmatrix}$$

c.

$$A = \begin{bmatrix} 6 & -1 & -1 & 5 \\ -1 & 7 & 1 & -1 \\ -1 & 1 & 3 & -3 \\ 5 & -1 & -3 & 6 \end{bmatrix}, \quad b = \begin{bmatrix} 1 \\ 2 \\ 0 \\ -1 \end{bmatrix}$$

6.8. The 9×9 matrix

$$A = \begin{pmatrix} 4 & -1 & 0 & -1 & 0 & 0 & 0 & 0 & 0 \\ -1 & 4 & -1 & 0 & -1 & 0 & 0 & 0 & 0 \\ 0 & -1 & 4 & 0 & 0 & -1 & 0 & 0 & 0 \\ -1 & 0 & 0 & 4 & -1 & 0 & -1 & 0 & 0 \\ 0 & -1 & 0 & -1 & 4 & -1 & 0 & -1 & 0 \\ 0 & 0 & -1 & 0 & -1 & 4 & 0 & 0 & -1 \\ 0 & 0 & 0 & -1 & 0 & 0 & 4 & -1 & 0 \\ 0 & 0 & 0 & 0 & -1 & 0 & -1 & 4 & -1 \\ 0 & 0 & 0 & 0 & 0 & -1 & 0 & -1 & 4 \end{pmatrix}$$

arises in the finite difference (and finite element) discretization of the Poisson equation on a nine point square grid. Solve the linear system $Ax = b$ with tolerance 0.05, using

(a) Jacobi method

(b) Gauss-Seidel method

(c) Conjugate gradient method

(d) Use the preconditioned conjugate gradient method with $C^{-1} = D^{-1/2}$ and to

(e) Compare the results in (a), (b), (c) and (d).

6.12 Suggestion for Further Reading

1. Application of algorithmic optimization to industrial problems is covered in detail in [11] and [12].

2. Consult [6] and [7] regarding application of variation inequality to superconductivity.

3. Read Chapters 6 and 9 of [15] for advanced study of algorithmic optimization and variational inequality.

4. Many concepts discussed in the chapter are presented effectively in references [1] through [5] and [8] through [14].

Bibliography

[1] K. Atkinson, *Elementary Numerical Analysis*, New York, Wiley, 1985.

[2] K. Atkinson, H. Weimin, *Theoretical Numerical Analysis*, Third Edition, Springer, 2010.

[3] O. Axelson, *Iterative Solution Methods*, Cambridge University Press, 1994.

[4] U. N. Bhat, *An Introduction to Queueing Theory, Modeling and Analysis in Applications*, Springer, 2008.

[5] Z. Dostal, *Optimal Quadratic Programming Algorithms*, Springer, 2009.

[6] K. M. Furati, A. H. Siddiqi, Fast algorithm for the Bean critical model for superconductivity, *Numerical Functional Analysis and Optimization*, 26(2):177-192, 2005.

[7] K. M. Furati, A. H. Siddiqi, Quasi-variational inequality modeling problems in superconductivity, *Numerical Functional Analysis and Optimization*, 26(2): 193-204, 2005.

[8] P.T. Harker, J.-S. Pang, Finite-dimensional variational inequality and nonlinear complementarity problems: a survey of theory, algorithms and applications, *Math. Program.* 48:161-220 (1990).

[9] R. J. LeVeque, *Finite Difference Methods for Ordinary and Partial Differential Equations*. SIAM, 2007.

[10] R. J. LeVeque, *Finite Difference Methods for Ordinary and Partial Di erential Equations*. SIAM 2007.

[11] G. M. Lee, N. N. Tam, N. D. Yen, *Quadratic Programming and Variational Inequalities*. Springer, 2005.

[12] H. Neunzert and A. H. Siddiqi, *Topics is Industrial Mathematics-Case Studies and Related Mathematical Methods*, Kluwer Academic Publishers (Now Springer), 2000.

[13] H. Neunzert, D. P. Wolters (Eds) *Currents in Industrial Mathematics From Concepts to Research Education*, Springer, 2015.

[14] P. Qi Pan, *Linear Programming Computation*, Springer, 2014.

[15] Y. Saad, *Iterative Methods for Sparse Linear Systems*, PWS, Publishing Company, 1996.

[16] A. H. Siddiqi, *Functional Analysis with Applications*, Springer, Nature, 2017.

[17] Stephen C. Billupsa, Katta G. Murtyb, Complementarity problems, *Journal of Computational and Applied Mathematics* 124: 303-318, 2000.

[18] M. Rosenlicht, *Introduction to Analysis*, Dover Mathematics series, 1968.

[19] M. Reed and B. Simon, *Functional Analysis, (Methods of Modern Mathematics)*, Academic Press, 1980.

[20] M. R. Hestenes and E. L. Stiefel, Methods of Conjugate gradients for solving linear systems, *J. Research Nat. Bur. Standards Section B*, 49: 409-432, 1952.

[21] M. R. Hestenes, *Conjugate Direction Methods in Optimization, Application of Mathematics book series (SMAP),* Vol 12: 81-149, Springer-Verlag, 1980.

Chapter 7

Computational Numerical Methods in Engineering

7.1 Introduction to Numerical Differentiation

In a case in which an analytical expression of the solution is difficult to obtain or cannot be given, numerical solutions can be investigated. Many numerical methods are available in the literature and, among them, the very popular finite difference method. With this method, a particular differential equation is replaced with a difference equation, that is, a system of linear equations that can be solved by a computer.

The finite difference schemes which will be discussed in the next several sections are developed by approximating derivatives.

7.1.1 Introduction to Finite Difference Methods

We shall begin with a real valued function f of a single variable x. Let f be a sufficiently differentiable function. First let us recall the definition of the first derivative:

$$f'(x) = \lim_{h \to 0} \frac{f(x+h) - f(x)}{h}. \tag{7.1}$$

This means that

$$\frac{f(x+h) - f(x)}{h} \tag{7.2}$$

can be a good candidate for an approximation of $f'(x)$ for sufficiently small h. In order to determine how good this approximation is, we use the well-known Taylor formula.

7.1.2 Taylor's Formula

Let f have $n+1$ continuous derivatives in the interval $(x-l, x+l)$. Then for any h such that $x+h \in (x-l, x+l)$ there is a number $0 < \theta < 1$ such that

$$f(x+h) = f(x) + \frac{f'(x)}{1!}h + \frac{f''(x)}{2!}h^2 + \cdots + \frac{f'^{(n)}(x)}{n!}h^n + \frac{f'^{(n+1)}(x+\theta h)}{(n+1)!}h^{n+1}. \tag{7.3}$$

If the derivative $f'^{(n+1)}$ is bounded, the we can write

$$f(x+h) = f(x) + \frac{f'(x)}{1!}h + \frac{f''(x)}{2!}h^2 + \cdots + \frac{f'^{(n)}(x)}{n!}h^n + O(h^{n+1}). \tag{7.4}$$

The O has the following meaning. If $f(x)$ and $g(x)$ are two functions such that there exist constant $M > 0$ and $\delta > 0$ with the property

$$|f(x)| \leq M\,|g(x)| \;\text{ for }\; |x| < \delta \tag{7.5}$$

then we write

$$|f(x)| = O(|g(x)|). \tag{7.6}$$

If we take $n = 1$ in (7.3), we can obtain

$$f(x+h) = f(x) + f'(x)h + O(h^2) \tag{7.7}$$

which can be written as

$$f'(x) = \frac{f(x+h) - f(x)}{h} + O(h). \tag{7.8}$$

From the last expression we have one approximation of $f'(x)$:

$$f'(x) \approx \frac{f(x+h) - f(x)}{h}, \;\textbf{forward difference approximation}.$$

Replacing h by $-h$ in (7.7), we obtain

$$f'(x) = \frac{f(x) - f(x-h)}{h} + O(h). \tag{7.9}$$

From the last expression we have another approximation of $f'(x)$:

$$f'(x) \approx \frac{f(x) - f(x-h)}{h}, \quad \textbf{backward difference approximation.} \tag{7.10}$$

Replacing h by $-h$ in (7.4), we have

$$f(x-h) = f(x) - \frac{f'(x)}{1!}h + \cdots + (-1)^n \frac{f'^{(n)}(x)}{n!}h^n + O(h^{n+1}). \tag{7.11}$$

If we take $n = 1$ in (7.4) and (7.11) and subtract (7.11) from (7.4), we get

$$f'(x) = \frac{f(x+h) - f(x-h)}{2h} + O(h^2).$$

Therefore, we have another approximation of the first derivative

$$f'(x) \approx \frac{f(x+h) - f(x-h)}{2h}, \quad \textbf{central difference approximation.} \tag{7.12}$$

The truncation errors for the above approximations are given by

$$E_{for}(h) = \frac{f(x+h) - f(x)}{h} - f'(x) = O(h) \tag{7.13}$$

$$E_{back}(h) = \frac{f(x) - f(x-h)}{h} - f'(x) = O(h)$$

$$E_{cent}(h) = \frac{f(x+h) - f(x-h)}{h} - f'(x) = O(h^2)$$

Example 255. Let $f(x) = \ln x$. Approximate $f'(2)$ taking $h = 0.1$ and $h = 0.001$ in (a) backward difference approximation, (b) backward difference approximation and (c) central difference approximation.

Solution: We know that $f'(x) = \dfrac{1}{x}$ and the exact value of $f'(2)$ is 0.5.
(a) For $h = 0.1$ we have

$$f'(2) \approx \frac{f(2+0.1) - f(2)}{0.1} = \frac{\ln(2.1) - \ln(2)}{0.1} \approx 0.4879.$$

For $h = 0.001$

$$f'(2) \approx \frac{f(2+0.001) - f(2)}{0.001} = \frac{\ln(2.001) - \ln(2)}{0.001} \approx 0.49988.$$

(b) For $h = 0.1$ we have

$$f'(2) \approx \frac{f(2) - f(2-0.1)}{0.1} = \frac{\ln(2) - \ln(1.9)}{0.1} \approx 0.5129.$$

For $h = 0.001$

$$f'(2) \approx \frac{f(2) - f(2-0.001)}{0.001} = \frac{\ln(2) - \ln(1.999)}{0.001} \approx 0.50018.$$

(c) For $h = 0.1$ we have

$$f'(2) \approx \frac{f(2+0.1) - f(2-0.1)}{2(0.1)} = \frac{\ln(2.1) - \ln(1.9)}{2(0.1)} \approx 0.500417.$$

For $h = 0.001$ we have

$$f'(2) \approx \frac{f(2+0.1) - f(2-0.1)}{2(0.001)} = \frac{\ln(2.001) - \ln(1.999)}{2(0.001)} \approx 0.5000.$$

If we use the Taylor series (7.4) and (7.11), then we obtain the following approximations for the second derivative

$$f''(x) \approx \frac{f(x+2h) - 2f(x+h) + f(x)}{h^2}, \textbf{ forward approximation.} \quad (7.14)$$

$$f''(x) \approx \frac{f(x) - 2f(x-h) + f(x-2h)}{h^2}, \textbf{ backward approximation.} \quad (7.15)$$

$$f''(x) \approx \frac{f(x+h) - 2f(x) + f(x-h)}{h^2}, \textbf{ central approximation.} \quad (7.16)$$

The approximations (7.14) and (7.15) for the second derivative are accurate of first order, and the central difference approximation (7.16) for the second derivative is accurate of second order.

Example 256. Approximate $f''(2)$ for the function $f(x) = \ln x$ taking $h = 0.1$ with the central difference approximation.

Solution: We find $f''(x) = -\frac{1}{x}$ and so $f''(2) = -0.25$. With the central difference approximation for the second derivative,

$$f''(x) \approx \frac{f(x+h) - 2f(x) + f(x-h)}{h^2} \approx \frac{\ln 2.1 - 2\ln 2 + \ln 1.9}{(0.1)^2} \approx -0.25031.$$

7.1.3 Finite Differences for Function f of Two Variables

The approximation of the derivative of a function of a single variable can be extended to functions of several variables. For example, if $f(x, y)$ is a given function of two variables $(x, y) \in \Omega \subset \mathbb{R}^2$, for the first partial derivative we have

$$f_x(x,y) = \frac{f(x+h,y)-f(x,y)}{h} + O(h), \text{ forward difference.} \quad (7.17)$$

$$f_x(x,y) = \frac{f(x,y)-f(x-h,y)}{h} + O(h), \text{ backward difference.} \quad (7.18)$$

$$f_x(x,y) = \frac{f(x+h,y)-f(x-h,y)}{h} + O(h^2), \text{ central difference.} \quad (7.19)$$

$$f_y(x,y) = \frac{f(x,y+k)-f(x,y)}{k} + O(k), \text{ forward difference.} \quad (7.20)$$

$$f_y(x,y) = \frac{f(x,y)-f(x,y-k)}{k} + O(k), \text{ forward difference.} \quad (7.21)$$

$$f_y(x,y) = \frac{f(x,y+k)-f(x,y-k)}{2k} + O(k^2), \text{ backward difference.} \quad (7.22)$$

For the partial derivative of second order we have

$$f_{xx}(x,y) = \frac{f(x+2h,y) - 2f(x+h,y) + f(x,y)}{h} + O(h), \text{ central difference.} \tag{7.23}$$

$$f_{xx}(x,y) = \frac{f(x,y) - 2f(x-h,y) + f(x-2h,y)}{h} + O(h), \text{ forward difference.} \tag{7.24}$$

$$f_{xx}(x,y) = \frac{f(x,y) - 2f(x-h,y) + f(x-2h,y)}{h} + O(h^2), \text{ backward difference.} \tag{7.25}$$

$$f_{xx}(x,y) = \frac{f(x,y) - 2f(x-h,y) + f(x-2h,y)}{h} + O(h^2), \text{ central difference.} \tag{7.26}$$

For the second partial derivatives with respect to y, as well as for the mixed derivatives, we have similar expressions.

Application to ordinary differential equations

Euler's method

The object of the method is to obtain an approximation to the initial value problem (IVP):

$$\begin{cases} \dfrac{dy}{dt} = f(t,y), \ t \geq t_0 \\ y(t_0) = y_0 \end{cases} . \tag{7.27}$$

Starting from the data given at the initial time $t_0 = 0$, one should be able to march forward in time, computing approximation at successive times t_1, $t_2, \ldots$ We will use k to denote the time step, so $t_n = nk$ for $n \geq 0$. We will use Taylor's theorem to derive Euler's method. Suppose that $y(t)$, the unique solution to (7.27), is twice differentiable, so that for $k = 0, 1, 2, \ldots$

$$y(t_{n+1}) = y(t_n) + (t_{n+1} - t_n)y'(t_n) + \frac{(t_{n+1} - t_n)^2}{2} y''(\xi_n) \tag{7.28}$$

for some ξ_n in (t_n, t_{n+1}). Since $k = t_{n+1} - t_n$, we have

$$y(t_{n+1}) = y(t_n) + ky'(t_n) + \frac{k^2}{2} y''(\xi_n) \tag{7.29}$$

and since $y(t)$ satisfies the differential equation(7.27),

$$y(t_{n+1}) = y(t_n) + kf(t_n, y(t_n)) + \frac{k^2}{2} y''(\xi_n). \tag{7.30}$$

If we denote by w_n the approximation of $y(t_n)$, that is, $w_n \approx y(t_n)$, for each $n = 0, 1, \ldots,$. Now, by ignoring the remainder term, we obtain the Euler's method

$$y_{n+1} = y_n + kf(t_n, y_n) \text{ for each } n = 0, 1, \ldots \tag{7.31}$$

with an error estimate

$$|y(t_n) - y_n| = O(k). \tag{7.32}$$

TABLE 7.1

t_n	y_n	$y(t_n)$	$\|y_n - y(t_n)\|$
0.0	0.5000000	0.5000000	0.0000000
0.2	0.8000000	0.8292986	0.0292986
0.4	1.1520000	1.2140877	0.0620877
0.6	1.5520000	1.6489406	0.0985406
0.8	1.9884800	2.1272295	0.1387495
1.0	24.8657845	2.6408591	0.18268311
1.2	2.9498112	3.1799415	0.2301303
1.4	3.4517734	3.7324000	0.2801303
1.6	3.9501281	4.2834838	0.3333557
1.8	4.4281538	4.8151763	0.3870225
2.0	4.8657845	5.3054720	0.4396874

Example 257. Use Euler's method to approximate the solution to the initial value problem

$$\begin{cases} \dfrac{dy}{dt} = y - t^2 + 1, 0 \leq t \leq 2, \\ y(0) = 0.5 \end{cases} \tag{7.33}$$

with $k = 0.2$. The exact solution is $y(t) = (t+1)^2 - 0.5e^t$. Table 7.1 shows comparison between the approximate values at t_n and the exact values.

Note that the error grows slightly as the value of t increases. This controlled error growth is a consequence of the stability of Euler's method, which implies that the error is expected to grow no worse than it would in a linear manner.

Runge-Kutta Methods

Second order Runge Kutta method

The **second order** Runge-Kutta method is also known as the **improved Euler's method.** Starting from the initial point (t_0, y_0), we compute two slopes:

$$\begin{aligned} s_{n1} &= f(t_0, y_0) \\ s_{n2} &= f(t_0 + k, y_0 + ks_{n1}). \end{aligned} \tag{7.34}$$

With these slopes, we define the next values of the dependent variables to be

$$y_n = y_0 + k\frac{s_{n1} + s_{n2}}{2}. \tag{7.35}$$

An analysis using Taylor's expansion reveals an improvement in the estimate for the truncation error. For the second order Runge Kutta method, we have

$$|y(t_n) - y_n| = O(k^2). \tag{7.36}$$

TABLE 7.2

t_n	y_n	$y(t_n)$	$\lvert y_n - y(t_n)\rvert$
0.2	0.8292933	0.8292986	0.0000053
0.4	1.2140762	1.2140877	0.0000114
0.6	1.6489220	1.6489406	0.0000186
0.8	1.1272027	2.1272295	0.0000269
1.0	2.6408227	2.6408591	0.0000364
1.2	3.1798942	3.1799415	0.0000474
1.4	3.7323401	3.7324000	0.0000599
1.6	4.2834095	4.2834838	0.0000743
1.8	4.8150857	4.8151763	0.0000906
2.0	5.3053630	5.3054720	0.0001089

Fourth order Runge-Kutta method

The next method is also due to Runge and Kutta. It is probably the most commonly used solution algorithm. For most equations and systems it is suitably fast and accurate. For this method we use four slopes. Starting with the initial point (t_0, y_0), we compute

$$\begin{aligned} s_{n1} &= f(t_0, y_0) \\ s_{n2} &= f(t_{k-1} + \frac{k}{2}, y_{k-1} + \frac{k}{2}s_{n1}) \\ s_{n3} &= f(t_{k-1} + \frac{k}{2}, y_{k-1} + \frac{k}{2}s_{n2}) \\ s_{n4} &= f(t_{k-1} + k, y_{k-1} + ks_{n3}). \end{aligned} \tag{7.37}$$

With these slopes we define the next value of the dependent variable:

$$y_n = y_{\ n-1} + k\frac{s_{n1} + 2s_{n2} + 2s_{n3} + s_{n4}}{6}. \tag{7.38}$$

For the fourth order Runge-Kutta method, we have

$$|y(t_n) - y_n| = O(k^4). \tag{7.39}$$

Example 258. Use the Runge-Kutta method of order four to obtain approximation to the solution of the initial value problem

$$\begin{cases} \dfrac{dy}{dt} = y - t^2 + 1, 0 \le t \le 2, \\ y(0) = 0.5 \end{cases} \tag{7.40}$$

with $k = 0.2$. The exact solution is $y(t) = (t+1)^2 - 0.5e^t$. Table 7.2 shows comparison between the approximate values at t_n and the exact values.

Multistep methods

In previous sections we have discussed numerical procedures for approximating the solutions of the initial value problem in which the data at the point $t = t_n$ are used to compute an approximate value of the solution $y(t_{n+1})$ at the next point $t = t_{n+1}$. In other words, the computed value of y at any mesh point depends only on the data at the preceding mesh point.

These are called **one-step methods.** However, once approximate values of the solution y have been obtained at a few points beyond t_0, it is natural to ask whether we can make use of some of this information, rather than just the value at the last point, to compute the approximate value at the next point. More precisely, if y_1 at t_1, y_2 at $t_2, \ldots,$ y_n at t_n are known, how can we use this information to determine y_{n+1} at t_{n+1}? Methods that use information at more than the last mesh point are referred to as **multistep methods.** In this section we describe two types of multistep methods: Adams methods and backward differentiation formulas.

Adams methods

Recall that

$$y(t_{n+1}) - y(t_n) = \int_{t_n}^{t_{n+1}} y'(t)dt \tag{7.41}$$

where $y(t)$ is the solution of the initial value problem (7.27). The basic idea of an Adams method is to approximate $y'(t)$ by a polynomial $P_k(t)$ of degree k and to use the polynomial to evaluate the integral on the right side of (7.41). The coefficients in $P_k(t)$ are determined by using $k+1$ previously calculated data points. For example, suppose that we wish to use a first degree polynomial $P_1(t) = At + B$. Then we need only the two data points (t_n, y_n) and (t_{n-1}, y_{n-1}). Since P_1 is to be an approximation to y', we require that $P_1(t_n) = f(t_n, y_n)$ and that $P_1(t_{n-1}) = f(t_{n-1}, y_{n-1})$. Now, denoting $f(t_n, y_n)$ by f_n, A and B must satisfy the equations

$$\begin{aligned} At_n + B &= f_n \\ At_{n-1} + B &= f_{n-1}. \end{aligned} \tag{7.42}$$

Solving for A and B, we obtain

$$A = \frac{f_n - f_{n-1}}{k}, \qquad B = \frac{f_{n-1}t_n - f_n t_{n-1}}{k}. \tag{7.43}$$

Replacing $y'(t)$ by $P_1(t)$ and evaluating the integral in (7.41), we find that

$$y(t_{n+1}) - y(t_n) = \frac{A}{2}(t_{n+1}^2 - t_n^2) + B(t_{n+1} - t_n). \tag{7.44}$$

Finally, we replace $y(t_{n+1})$ and $y(t_n)$ by y_{n+1} and y_n, respectively, and carry out algebraic simplification to get

$$y_{n+1} = y_n + \frac{3}{2}kf_n - \frac{1}{2}kf_{n-1}. \tag{7.45}$$

Equation (7.45) is the **second order Adams-Bashforth formula**. It is an explicit formula for y_{n+1} in terms of y_n and y_{n-1} and has a local truncation error proportional to k^3.

Note that the first order Adams-Bashforth formula, based on the polynomial $P_0(t) = f_n$ of degree zero, is just the original Euler formula.

Adams-Bashforth method

More accurate Adams formula can be obtained by following the procedure outlined above, but using a higher degree polynomial and correspondingly more data points. For example, suppose that a polynomial $P_3(t)$ of degree three is used. The coefficients are determined from the four points (t_n, y_n), (t_{n-1}, y_{n-1}), (t_{n-2}, y_{n-2}), and (t_{n-3}, y_{n-3}). Substituting this polynomial for $y'(t)$ in Equation (7.41), evaluating the integral, and simplifying the result, we eventually obtain the fourth order Adams-Bashforth formula

$$y_{n+1} = y_n + (k/24)\left(55f_n - 59f_{n-1} + 37f_{n-2} - 9f_{n-3}\right). \tag{7.46}$$

Adams-Moulton method

A variation of the derivation of the Adams-Bashforth formula gives another set of formulas called the Adams-Moulton formulas. We use a first degree polynomial $\phi_1(t) = \alpha t + \beta$, but we determine the coefficients by using the points (t_n, y_n) and (t_{n+1}, y_{n+1}). Thus α and β must satisfy

$$\begin{aligned} \alpha t_n + \beta &= f_n \\ \alpha t_{n+1} + \beta &= f_{n+1} \end{aligned} \tag{7.47}$$

and it follows that

$$\alpha = \frac{f_{n+1} - f_n}{k}, \quad \beta = \frac{f_n t_{n+1} - f_{n+1} t_n}{k}. \tag{7.48}$$

Substituting $\phi_1(t)$ for $y'(t)$ in equation (7.41) and simplifying, we obtain

$$y_{n+1} = y_n + \frac{1}{2}kf_n + \frac{1}{2}kf(t_{n+1}, y_{n+1}) \tag{7.49}$$

which is the second degree Adams-Moulton formula. The presence of $f(t_{n+1}, y_{n+1})$ in the right hand side shows that the Adams-Moulton formula is implicit, rather than explicit, since the unknown y_{n+1} appears on both sides of the equation.

More accurate higher order formulas can be obtained by using an approximating polynomial of higher degree. The fourth order Adams-Moulton formula is

$$y_{n+1} = y_n + (k/24)\left(9f_{n+1} + 19f_n - 5f_{n-1} + f_{n-2}\right) \tag{7.50}$$

Observe that this is also an implicit formula because y_{n+1} appears in f_{n+1}.

Example 259. Use the Adams-Moulton-method with $k = 0.2$ to obtain an approximation to $y(0.8)$ for the solution of the initial value problem

$$y' = x + y - 1,\ y(0) = 1.$$

Solution: With a step of $k = 0.2$, $y(0.8)$ will be approximated by y_4. To start computation, we use the Runge-Kutta fourth method with $t_0 = 0$, $y_0 = 1$, and $k = 0.2$ to obtain

$$y_1 = 1.02140000, \ \ y_2 = 1.09181796, \ \ y_3 = 1.22210646.$$

Now with identifications $t_0 = 0$, $t_1 = 0.2$, $t_2 = 0.4$, $t_3 = 0.6$, and $f(x, y) = x + y - 1$, we find

$$\begin{aligned}
f_0 &= f(t_0, y_0) = 0 + 1 - 1 = 0 \\
f_1 &= f(t_1, y_1) = 0.2 + 1.02140000 - 1 = 0.22140000 \\
f_2 &= f(t_2, y_2) = 0.4 + 1.09181796 - 1 = 0.49181796 \\
f_3 &= f(t_3, y_3) = 0.6 + 1.22210646 - 1 = 0.82210646.
\end{aligned}$$

With these values, the Adams-Bashforth formula gives

$$\bar{y}_4 = y_3 + \frac{0.2}{24}(55f_3 - 59f_2 + 37f_1 - 9f_0).$$

And, as

$$f_4 = f(t_4, \bar{y}_4) = 0.8 + 1.42553535975 - 1 = 1.22535975,$$

the Adams-Moulton method yields

$$y_4 = y_3 + \frac{0.2}{24}(9f_4 + 19f_3 - 5f_2 + f_1) = 1.42552788.$$

The exact value of $y(0.8)$ is $y(0.8) = 1.42554093$.

7.1.4 Application to Partial Differential Equations

7.1.4.1 Poisson's Problem with Dirichlet Boundary Conditions

To illustrate how difference methods can be used to solve elliptic PDEs, we consider Poisson's equation

$$\begin{cases} -(u_{xx} + u_{yy}) = f(x, y) \ \forall (x, y) \in \Omega = (0, 1)^2, \\ u/\partial\Omega = g(x, y) \ \forall (x, y) \in \partial\Omega \end{cases} \tag{7.51}$$

where $\partial\Omega$ denotes the boundary of Ω. We assume that Ω is covered by a uniform grid

$$x_i = ih, \ y_j = jh, \ 0 \leq i, j \leq J, \ h = \frac{1}{J}. \tag{7.52}$$

The approximate solution at point (x_i, y_j) is denoted by u_{ij}. Using the second order central difference

$$\delta_x u_{ij} = \frac{u_{i+1j} - 2u_{ij} + u_{i-1j}}{h^2} \tag{7.53}$$

and

$$\delta_y u_{ij} = \frac{u_{ij+1} - 2u_{ij} + u_{ij-1}}{h^2} \tag{7.54}$$

to approximate the derivatives u_{xx} and u_{yy} in (...), respectively, we obtain the five point difference scheme

$$\frac{u_{i+1j} - 2u_{ij} + u_{i-1j}}{h^2} + \frac{u_{ij+1} - 2u_{ij} + u_{ij-1}}{h^2} = -f_{ij},\ 1 \le i, j \le J-1 \tag{7.55}$$

or

$$u_{i+1j} + u_{i-1j} + u_{ij+1} + u_{ij-1} - 4u_{ij} = -h^2 f_{ij}. \tag{7.56}$$

The boundary condition (7.50) is replaced by

$$u_{i0} = g(x_i, 0) = g_{i0},\ 0 \le i \le J \tag{7.57}$$

and the right hand side by

$$f_{ij} = f(x_i, y_j),\ 1 \le i, j \le J-1. \tag{7.58}$$

The local truncation error

$$\begin{aligned} T_{ij} &= \frac{1}{h^2}[u(x_{i+1}, y_j) + u(x_{i-1}, y_j) + u(x_i, y_{j+1}) + u(x_i, y_{j-1}) \\ &\quad -4u(x_i, y_j)] + f(x_i, y_j). \end{aligned} \tag{7.59}$$

Assuming that the derivatives of u are continuous up to order 4 in both x and y, we find, by Taylor expansion, that

$$T_{ij} = (u_{xx} + u_{yy})(x_i, y_j) + \frac{h^2}{12}\left(\frac{\partial^4 u}{\partial x^4} + \frac{\partial^4 u}{\partial y^4}\right)(x_i, y_j) + f(x_i, y_j) \tag{7.60}$$

from which we obtain

$$|T_{ij}| \le \frac{h^2}{12} \max_{(x,y)\in(0,1)^2} \left(\left|\frac{\partial^4 u}{\partial x^4}\right| + \left|\frac{\partial^4 u}{\partial y^4}\right|\right) \tag{7.61}$$

that is, the truncation error is $O(h^2)$.
Denote the $(J-1)^2$ dimensional vector

$$U = [u_{1,1}, \ldots, u_{J-1,1}; u_{1,2}, \ldots, u_{J-1,2}; \ldots; u_{1,J-1}, \ldots, u_{J-1,J-1}]^T \tag{7.62}$$

where T means the transpose. With this notation, the scheme (7.56) leads to a linear system

$$AU = F \tag{7.63}$$

where A is a matrix of order $(J-1)^2$ given by

$$\begin{bmatrix} B & -I & & & & \\ -I & B & -I & & & \\ & \cdot & \cdot & \cdot & & \\ & & \cdot & \cdot & \cdot & \\ & & & \cdot & \cdot & \cdot \\ & & & -I & B & -I \\ & & & & -I & B \end{bmatrix} \tag{7.64}$$

with I being the identity matrix of order $J-1$, and B being a matrix of order $J-1$ given by

$$A = \begin{bmatrix} 4 & -1 & & & & \\ -1 & 4 & -1 & & & \\ & \cdot & \cdot & \cdot & & \\ & & \cdot & \cdot & \cdot & \\ & & & \cdot & \cdot & \overline{}. \\ & & & -1 & 4 & -1 \\ & & & & -1 & 4 \end{bmatrix}.$$

Furthermore, F is a $(J-1)^2$ dimensional vector whose elements are

$$\begin{aligned} F_{1,1} &= h^2 f_{1,1} + g_{1,0} + g_{0,1}, \\ F_{i,1} &= h^2 f_{i,1} + g_{i,0},\ 2 \le i \le J-2, \\ F_{J-1,1} &= h^2 f_{J-1,1} + g_{J-1,0} + g_{J,1}, \\ F_{1,2} &= h^2 f_{1,2} + g_{0,2}, \\ F_{i,2} &= h^2 f_{i,2},\ 2 \le i \le J-2, \\ F_{J-1,2} &= h^2 f_{J-1,2} + g_{J,2}, \\ &\cdot \\ &\cdot \\ &\cdot \\ F_{1,J-1} &= h^2 f_{1,J-1} + g_{1,J} + g_{0,J-1} \\ F_{i,J-1} &= h^2 f_{i,J-1} + g_{i,J},\ 2 \le i \le J-2 \\ F_{J-1,J-1} &= h^2 f_{J-1,J-1} + g_{J-1,J} + g_{J,J-1}. \end{aligned} \tag{7.65}$$

7.1.4.2 Finite Difference Methods for Parabolic Problems

To illustrate how difference methods can be used to solve parabolic PDEs, we consider the heat equation model: find $u(x,t)$ such that

$$\begin{cases} u_t = c^2 u_{xx}, & \forall (t,x) \in (0,T) \times (0,l) \\ u(0,t) = u(1,t) = 0, & \forall t \in (0,T) \\ u(x,0) = f(x), & \forall x \in [0,l\,] \end{cases} \tag{7.66}$$

where T denotes the terminal time for the model.

To solve the problem (7.66), we first divide the physical domain $(0,T) \times (0,l)$ by $N \times J$ uniform grid points

$$t_n = n\triangle t,\ \triangle t = \frac{T}{N},\ n = 0, 1, \ldots, N. \tag{7.67}$$

$$x_j = j\triangle x,\ \triangle x = \frac{l}{j},\ j = 0, 1, \ldots, J. \tag{7.68}$$

We denote the approximate solution $u_j^n \approx u(x_j, t_n)$ at an arbitrary point (x_j, t_n). To obtain a finite difference scheme, we need to approximate the derivatives in (7.66) by some finite differences. Below we give two simple finite difference schemes for solving (7.66).

Explicit scheme

Substituting

$$u_t(x_j, t_n) \approx \frac{(u_j^{n+1} - u_j^n)}{\Delta t} \tag{7.69}$$

and

$$u_{xx}(x_j, t_n) \approx \frac{(u_{j+1}^n - 2u_j^n + u_{j-1}^n)}{h^2} \tag{7.70}$$

into (7.66), we obtain the explicit scheme for

$$\frac{u_j^{n+1} - u_j^n}{\Delta t} = \frac{(u_{j+1}^n - 2u_j^n + u_{j-1}^n)}{h^2} \tag{7.71}$$

$$u_j^{n+1} = u_j^n + \mu(u_{j+1}^n - 2u_j^n + u_{j-1}^n),\ 1 \le j \le J-1,\ 0 \le n \le N-1, \tag{7.72}$$

or

$$u_j^{n+1} = \mu u_{j-1}^n + (1 - 2\mu)\, u_j^n + u_{j+1}^n),\ 1 \le j \le J-1,\ 0 \le n \le N-1, \tag{7.73}$$

where we denote

$$\mu = \frac{\Delta t}{h^2}. \tag{7.74}$$

The boundary condition can be approximated directly as

$$u_0^n = u_J^n = 0,\ \ 0 \le n \le N-1, \tag{7.75}$$

and the initial condition can be approximated as

$$u_j^0 = f(jh),\ \ 0 \le j \le J. \tag{7.76}$$

Note that with the scheme (7.73), the approximate solution u_j^{n+1} at any interior point can be obtained by a simple marching in time.

Example 260. Consider the problem

$$\begin{cases} u_t = u_{xx}, & 0 < x < 1,\ 0 < t < 0.5 \\ u(0,t) = u(1,t) = 0, & 0 < t < 0.5 \\ u(x,0) = \sin \pi x, & 0 \le x \le 1 \end{cases}$$

with

$$f(x) = \begin{cases} 10x\ , & 0 \le x \le \frac{1}{2} \\ 10(1-x), & \frac{1}{2} < x < 1 \end{cases}.$$

TABLE 7.3

Time	0.20	0.40	0.60	0.80
0.00	0.5878	0.9511	0.9511	0.5878
0.10	0.2154	0.3486	0.3486	0.2154
0.20	0.0790	0.1278	0.1278	0.0790
0.30	0.0289	0.0468	0.0468	0.0289
0.40	0.0106	0.0172	0.0172	0.0106
0.50	0.0039	0.0063	0.0063	0.0039

TABLE 7.4

Exact	Approximation
$u(0.4, 4.005) = 0.5806$	$u_2^5 = 0.5758$
$u(0.6,0.06) = 0.5261$	$u_3^6 = 0.5208$
$u(0.2, 0.010) = 0.2191$	$u_1^{10} = 0.2154$
$u(0.8, 0.14) = 0.1476$	$u_4^{14} = 0.1442$

Using the explicit finite difference approximation, find a numerical solution of the problem at different locations and several time instants, for $h = 0.2$, $\triangle t = 0.01$, and compare the solution with with the exact solution

$$u(x,t) = e^{-\pi^2 t} \sin \pi x$$

Since we have homogeneous boundary conditions, the system can be written in matrix form

$$u^{n+1} = A^n u^0$$

where

$$u^{n+1} = \left(u_1^{n+1}, \ldots, u_{J-1}^{n+1}\right)^T$$

$$u^0 = (f(h), \ldots, f(h))^T$$

and the $(J-1) \times (J-1)$ matrix A is given by

$$\begin{pmatrix} 1-2\mu & \mu & 0 & 0 & 0 \\ \mu & 1-2\mu & \mu & & \\ 0 & \mu & . & . & 0 \\ 0 & 0 & \mu & 1-2\mu & \mu \\ 0 & 0 & \ldots 0 & \mu & 1-2\mu \end{pmatrix}$$

For $h = 0.1$, $\triangle t = 0.4$, $\mu = 0.4 < 0.5$. Solving the system at $t = 1.2$, $t = 4.8$, $t = 14.4$, we obtain approximations that are presented in Table 7.3.

We compare in Table 7.4 a sample of exact values with their corresponding approximations.

Implicit scheme

Similarly, by substituting

$$u_t(x_j, t_n) \approx \frac{(u_j^n - u_j^{n-1})}{\triangle t} \tag{7.77}$$

and

$$u_{xx}(x_j, t_n) \approx \frac{(u_{j+1}^n - 2u_j^n + u_{j-1}^n)}{(\triangle x)^2}. \tag{7.78}$$

into (7.66), another difference scheme can be constructed as

$$\frac{(u_j^n - u_j^{n-1})}{\triangle t} = \frac{(u_{j+1}^n - 2u_j^n + u_{j-1}^n)}{(\triangle x)^2}. \tag{7.79}$$

In which case, we obtain an implicit scheme

$$-\mu u_{j-1}^n + (1 + 2\mu)u_j^n - \mu u_{j+1}^n = u_j^{n-1},\ 1 \leq j \leq J - 1 \tag{7.80}$$

Note that we have to solve a linear system at each time step to obtain the approximate solution u_j^n at all interior points. That is why the scheme (7.80) is called implicit in order to distinguish it from the explicit scheme (7.73).

Crank-Nicolson scheme

A more rewarding scheme can be derived by averaging the explicit scheme (7.73) at the point (x_j, t_n) and the implicit scheme (7.80) at the point (x_j, t_{n+1}) to obtain

$$\frac{(u_j^{n+1} - u_j^n)}{\triangle t} = \frac{1}{2}\left[\frac{u_{j+1}^n - 2u_j^n + u_{j-1}^n}{h^2} + \frac{u_{j+1}^{n+1} - 2u_j^{n+1} + u_{j-1}^{n+1}}{h^2}\right] \tag{7.81}$$

or

$$\mu u_{j+1}^{n+1} - (2 + \mu)u_j^{n+1} + u_{j-1}^{n+1} = \mu u_{j+1}^n - (2 + \mu)u_j^n + u_{j-1}^n. \tag{7.82}$$

The advantage of the Crank-Nicolson scheme (Table 7.5) over the explicit one is that it doesn't depend on the parameter μ.

Example 261. Use the Crank-Nicolson method to approximate the solution of the problem

$$\begin{cases} u_t = \frac{1}{4}u_{xx},\ \ 0 < x < 2,\ 0 < t < 0.3 \\ u(0,t) = u(2,t) = 0,\ 0 < t < 0.3 \\ u(x,0) = \sin \pi x,\ \ 0 \leq x \leq 2 \end{cases}$$

with $h = 0.25$, $\triangle t = 0.01$, and $c = 0.25$. We obtain approximations that are presented in the Table 7.6.

Time $x = 0.25$ $x = 0.50$ $x = 0.75$ $x = 1.00$ $x = 1.25$ $x = 1.50$ $x = 1.75$.

TABLE 7.5

$t \setminus x$	0.25	0.50	0.75	1.00	1.25	1.50	1.75
0.00	0.701	1.000	0.7071	0.0000	−0.7071	−1.0000	−0.7071
0.05	0.6289	0.8894	0.6289	0.0000	−0.6289	−0.8894	−0.6289
0.10	0.5594	0.7911	0.5594	0.0000	−0.5594	−0.7911	−0.5594
0.15	0.4975	0.7036	0.4975	0.0000	−0.4975	−0.7036	−0.4975
0.20	0.4425	0.6258	0.4425	0.0000	−0.4425	−0.6258	−0.4425
0.25	0.3936	0.5567	0.3936	0.0000	−0.3936	−0.5567	−0.3936
0.3	0.3501	0.4951	0.3501	0.000	−0.3501	−0.4951	−0.3501

TABLE 7.6

Exact	Approximation
$u(0.75, 0.005) = 0.6250$	$u_3^5 = 0.6289$
$u(0.50, 0.20) = 0.5261$	$u_2^{20} = 0.5208$
$u(0.25, 0.10) = 0.5525$	$u_1^{10} = 0.5594$

7.1.4.3 Finite Difference Methods for Hyperbolic Problems

Explicit scheme

We consider the wave equation model: find $u(x,t)$ such that

$$u_{tt} = u_{xx}, \forall (t,x) \in (0,T) \times (0,l) \tag{7.83}$$

subject to the boundary conditions

$$u(0,t) = u(1,t) = 0, \ \forall t \in (0,T) \tag{7.84}$$

and the initial conditions

$$u(x,0) = f(x), \text{ and } u_t(x,0) = g(x) \ \forall x \in [0, l.\ ] \tag{7.85}$$

Let $u_j^n \approx u(x_j, t_n)$ be the approximation to the exact solution $u(x,t)$ at the point (x_j, t_n). Now, using the central differences for both time and space partial derivatives in the wave equation, we obtain

$$\frac{u_j^{n+1} - 2u_j^n + u_j^{n-1}}{(\triangle t)^2} = \frac{u_{j+1}^n - 2u_j^n + u_{j-1}^n}{h^2} \tag{7.86}$$

or

$$u_j^{n+1} = -u_j^{n-1} + 2(1-\delta^2)u_j^n + \delta^2(u_{j+1}^n + u_{j-1}^n), \ 1 \le j \le J-1 \tag{7.87}$$

where

$$\delta = \frac{\triangle t}{h}. \tag{7.88}$$

To start the approximation, we use Taylor's formula for $u(x,t)$ with respect to t to obtain

$$u(x_j,t_1) \approx u(x_j,0) + u_t(x_j,0)\triangle t + u_{tt}(x_j,0)\frac{(\triangle t)^2}{2}. \tag{7.89}$$

From the wave equation and the initial condition $u(x,0)=f(x)$ we have

$$u_{tt}(x_j,0) = u_{xx}(x_j,0) = f''(x_j). \tag{7.90}$$

Using the other initial condition $u_t(x_j,0)=g(x)$, Equation (7.89) becomes

$$u(x_j,t_1) \approx u(x_j,0) + g(x_j)\triangle t + f''(x_j)\frac{(\triangle t)^2}{2}. \tag{7.91}$$

Now if we use the central finite difference approximation for $f''(x_j)$, then (7.91) takes the form

$$u(x_j,t_1) \approx u(x_j,0) + g(x_j)\triangle t + (f(x_{j+1}) - 2f(x_j) + f(x_{j-1})\frac{(\triangle t)^2}{2h^2} \tag{7.92}$$

and since $f(x_j)=u(x_j,0)=u_{j,0}$ we have

$$u(x_j,t_1) \approx \left(1-\delta^2\right)u_{j,0} + \frac{\delta^2}{2}(u_{j-1,0}+u_{j+1,0}) + g(x_j)\triangle t. \tag{7.93}$$

The last approximation allows us to have the required first step

$$u_j^1 = \left(1-\delta^2\right)u_j^0 + \frac{\delta^2}{2}(u_{j-1}^0 + u_{j+1}^0) + g(x_j)\triangle t. \tag{7.94}$$

The boundary condition (7.84) can be approximated directly as

$$u_0^n = u_J^n = 0,\ n = 1,2,3,\ldots \tag{7.95}$$

and the initial condition (7.85) can be approximated as

$$u_j^0 = f(jh),\ 1 \le j \le J-1. \tag{7.96}$$

Writing this set of equations in matrix form gives

$$u^{n+1} = Au^n \tag{7.97}$$

where

$$\begin{pmatrix} 2(1-2\delta^2) & \delta^2 & 0 & 0 & 0 \\ \delta^2 & 2(1-2\delta^2) & \delta^2 & & \\ 0 & \delta^2 & . & . & 0 \\ 0 & 0 & \delta^2 & 2(1-2\delta^2) & \delta^2 \\ 0 & 0 & \ldots 0 & \delta^2 & 2(1-2\delta^2) \end{pmatrix}. \tag{7.98}$$

The technique defined by (7.87) and (7.94) is known as the explicit finite difference approximation of the wave equation.

Example 262. Consider the wave problem

$$\begin{cases} u_{tt} = 4u_{xx}, 0 < x < 1, \ t > 0 \\ u(0,t) = u(1,t) = 0, \ t > 0 \\ u(x,0) = \sin \pi x, \ 0 \le x \le 1 \text{ and } u_t(x,0) = 0, \ 0 \le x \le 1. \end{cases}$$

It is easily verified that the exact solution to this problem is

$$u(x,t) = \sin \pi x \cos 2\pi t.$$

The finite difference scheme applied to this problem with $h = 0.1$, $\triangle t = 0.05$, $\delta = 1$, $J = 10$, gives the values.

xi	0.0000

7.2 Finite Element in One Dimension

To have some idea of the finite element method, we consider an example on solving one-dimensional boundary value problems. Such examples show various aspects of the finite element method in the simple context of one-dimensional problems.

Let us consider a finite element method to solve the boundary value problem. Find $u(x)$ such that

$$\begin{cases} -u'' + \alpha u = f \\ u(0) = 0, \ u(1) = 0 \end{cases} \tag{7.99}$$

where $f \in L^2(0,1)$ is given. Let

$$V = H_0^1(0,1) = \{v \in H^1(0,1) \text{ such that } v(0) = v(1) = 0 \,\} \tag{7.100}$$

where

$$H^1(0,1) = \left\{ v \in L^2(0,1) \text{ such that } \frac{dv}{dx} \in L^2(0,1) \right\} \tag{7.101}$$

and $L^2(0,1)$ is the space of square integrable functions on $(0,1)$.

The weak formulation of the problem is

$$u \in V, \ \int_0^1 (u'v' + uv)dx = \int_0^1 fvdx \ \ \forall v \in V. \tag{7.102}$$

Let us develop a finite element method for the problem. For a natural number N, we partition the set $\bar{\Omega} = [0, 1]$ into N parts:

$$\bar{\Omega} = \cup_{i=1}^{N} K_i \tag{7.103}$$

where $K_i = [x_{i-1}, x_i]$, $1 \le i \le N$, are called the elements, and the x_i, $1 \le i \le N$, are called the nodes, $0 = x_0 < x_1 < \cdots < x_N = 1$. In this example we have Dirichlet conditions at x_0 and x_N. Denote $h_i = x_i - x_{i-1}$ and $h = \max_{1 \le i \le N} h_i$. The value h is called the mesh size or mesh parameter. We use continuous piecewise linear functions for the approximation, i.e., we choose

$$V_h = \left\{ v_h \in C(\bar{\Omega}) \text{ such that } v_h/_{K_i} \in \mathbb{P}_1(K_i), 1 \le i \le N, \ v_h(0) = v_h(1) = 0 \right\}.$$

For the basis functions of the space V_h, we introduce the "hat" functions associated with the nodes $x_1, \ldots, x_{N-1}$. For $i = 1, \ldots, N-1$, let

$$\phi_i(x) = \begin{cases} [c]l(x - x_{i-1})/h_i, & x_{i-1} \le x \le x_i \\ (x_{i+1} - x)/h_{i+1}, & x_i \le x \le x_{i+1} \\ 0, & \text{otherwise.} \end{cases} \tag{7.104}$$

These functions are continuous and piecewise linear. The first order weak derivatives of the basis functions are piecewise constants:

$$\phi_i'(x) = \begin{cases} [c]l1/h_i, & x_{i-1} \le x \le x_i \\ -1/h_{i+1}, & x_i \le x \le x_{i+1} \\ 0, & \text{otherwise.} \end{cases} \tag{7.105}$$

The corresponding finite element problem is

$$u_h \in V_h, \ \int_0^1 (u_h' v_h' + u_h v_h) dx = \int_0^1 f v_h dx \ \ \forall v_h \in V_h. \tag{7.106}$$

Write

$$u_h = \sum_{j=1}^{N-1} u_j \phi_j.$$

Note that $u_j = u_h(x_j)$, $1 \le j \le N-1$. We see that the finite element problem (7.106) is equivalent to the following linear system for the unknowns $u_1, \ldots, u_{N-1}$:

$$\sum_{j=1}^{N-1} u_j \int_0^1 (\phi_i' \phi_j' + \alpha \phi_i \phi_j) dx = \int_0^1 f \phi_i dx, \ 1 \le i \le N-1 \tag{7.107}$$

Let us find the coefficients matrix of the system (7.105) in the case of a uniform partition, $h_1 = \cdots = h_N$, The following formulas are useful for this purpose

$$\int_0^1 \phi_i'\phi_{i-1}'dx = -\frac{1}{h},\ 2 \le i \le N$$
$$\int_0^1 (\phi_i')^2\,dx = \frac{2}{h},\ 1 \le i \le N-1$$
$$\int_0^1 \phi_i\phi_{i-1}dx = \frac{h}{6},\ 2 \le i \le N$$
$$\int_0^1 (\phi_i)^2\,dx = \frac{2h}{3},\ 1 \le i \le N-1.$$

We see that in matrix and vector notation, in the case of a uniform partition, the finite element system (7.107) can be written as

$$Au = b \tag{7.108}$$

where

$$u = (u_1, \ldots, u_{N-1})^T \tag{7.109}$$

is the unknown vector, $a = \dfrac{2h}{3} + \dfrac{2}{h}$, $b = c = \dfrac{h}{6} - \dfrac{1}{h}$,

$$A = \begin{pmatrix} a & b & & & & & \\ c & a & b & & \cdot & & \\ & c & \cdot & \cdot & & & \\ & & \cdot & \cdot & \cdot & & \\ & & & \cdot & a & b & \\ & & & & c & a & b \\ & & & & & c & a \end{pmatrix} \tag{7.110}$$

is the stiffness matrix, and

$$b = \left(\int_0^1 f\phi_1 dx, \ldots, \int_0^1 f\phi_{N-1}dx\right)^T \tag{7.111}$$

is the load vector. The matrix A is sparse, thanks to the small supports of the basis functions. One distinguished feature of the finite element method is that the basis functions are constructed in such a way that their supports are as small as possible, so that the corresponding stiffness matrix is as sparse as possible.

Spectral method

The spectral element method represents a special case of Galerkin methods in which the finite dimensional space of test functions is made of continuous piecewise algebraic polynomials of high degree on each element of a partition of the computational domain. For ease of exposition, we will focus only on the one-dimensional problem: Find $u(x)$ such that:

$$\begin{cases} -u'' + u = f \\ u(0) = u(1) = 0. \end{cases} \tag{7.112}$$

Weighted Galerkin formulation

Let $\omega(x)$ be a weight function (positive and integrable on $(0,1)$. We define the weighted inner product in $L^2(0,1)$ by

$$(u,v)_\omega = \int_0^1 u(x)v(x)\omega(x)dx. \tag{7.113}$$

We consider the approximation of (7.112) by using Galerkin method in the polynomial space

$$X_N = \{\phi \in P_N : \phi(0) = \phi(1) \ = 0\}. \tag{7.114}$$

The Galerkin spectral formulation is

$$\left(u_N^{''}, v_N\right)_\omega + (u_N, v_N)_\omega = (f_N, v_N)_\omega \; \forall v \in X_N \tag{7.115}$$

where f_N is an appropriate polynomial approximation of f, which is usually taken to be the interpolation of f associated with the Gauss-type quadrature points. We shall see below that by choosing appropriate basis functions of X_N, we can reduce to a linear system with a sparse matrix that can be solved efficiently.

Given a set of basis functions $\{\phi_j\}_{j=1}^{N-2}$ of X_N, we denote

$$f_k = \int_0^1 f_N \phi_k dx, \;\; \mathbf{f} = (f_1, \ldots, f_{N-1})^T$$

,

$$u_N = \sum_{j=0}^{N-2} \bar{u}_j \phi_j, \qquad \mathbf{u} = (\bar{u}_0, \bar{u}_1, \ldots, \bar{u}_{N-2})^T$$

,

$$s_{kj} = -\int_0^1 \phi_j \phi_k, \quad m_{kj} = \int_0^1 \phi_j \phi_k dx$$

and

$$S = (s_{kj})_{0 \le k,j \le N-2}, \quad M = (m_{kj})_{0 \le k,j \le N-2}.$$

Taking $v_N = \phi_k, 0 \le k \le N-2$ in (7.115), we can see that (7.115) is equivalent to the following linear system

$$(S + \alpha M)\mathbf{u} = \mathbf{f}. \tag{7.116}$$

Next we determine the entries of the matrices S and M for two special cases: $\omega(x) \equiv 1$ and $\omega(x) \equiv \left(1 - x^2\right)^{-1/2}$.

We set $\omega(x) \equiv 1$ and $f_N = I_N f$, the Legendre interpolation polynomial of f with respect to the Legendre-Gauss-Lobatto points. Then problem (7.115) becomes

$$-\int_0^1 u_N'' v_N dx + \int_0^1 u_N v_N dx = \int_0^1 I_N f v_N dx \;\forall v_N \in X_N \tag{7.117}$$

which is called the **Legendre-Galerkin method.**

The linear system for (7.117) depends on the choice of the basis functions of X_N, which can be constructed with Legendre polynomials as

$$\phi_k(x) = L_k(x) + \alpha_k L_{k+1}(x) + \beta_k L_{k+2}(x), \; k \geq 0$$

where the constants α_k and β_k are uniquely determined by the boundary conditions of the continuous problem.

7.3 Exercises

Use the Runge-Kutta of order 4 (RK4) method with $k = 0.1$ to compute a four-decimal approximation to the following initial value problems:

7.1. $y' = 2x - 3y + 1$, $y(1) = 5$.

7.2. $y' = 4x - 2y$, $y(0) = 2$.

7.3. $y' = 1 + y^2$, $y(0) = 0$.

7.4. $y' = x^2 + y^2$, $y(0) = 1$.

7.5. $y' = e^{-y}$, $y(0) = 0$.

7.6. Use RK4 method with $k = 0.1$ to approximate $y(0.5)$ where $y(x)$ is the solution of the initial value problem $y' = (x + y - 1)^2$, $y(0) = 2$. Compare this approximate value with the actual value.

7.7. If the air resistance is proportional to the square of the instantaneous velocity, then the velocity ν of a mass m dropped from a given height h is determined from

$$m\frac{dv}{dt} = mg - \rho\nu^2, \; k > 0$$
$$\nu(0) = 0.$$

Let $\rho = 0.125$, $m = 5$ slags, and $g = 32 \; ft/s^2$.

(a) Use the RK4 method with $k = 1$ to approximate the velocity $\nu(5)$.

(b) Use a numerical solver to graph the solution of the IVP on the interval $[0, 6]$.

(c) Use separation of variables to solve the IVP and then find the actual value $\nu(5)$.

Use the **Adams-Moulton** method to approximate $y(1.0)$, where $y(x)$ is the solution of the following initial value problem

7.8. $y' = 1 + y^2$, $y(0) = 0$.

7.9. $y' = y + \cos x$, $y(0) = 1$.

7.10. $y' = (x - y)^2$, $y(0) = 0$.

7.11. $y' = xy + \sqrt{y}$, $y' = 1 + y^2$, $y(0) = 1$.

Laplace equation exercises

7.12. Use the numerical scheme (7.53) to solve the following boundary value problems:

(a)

$$\begin{aligned} -(u_{xx} + u_{yy}) &= 0,\ 0 < x < 2,\ 0 < y < 1 \\ u(0, y) = 0,\ u(2, y) &= 0,\ 0 < y < 1 \\ u(x, 0) = 100,\ u(x, 1) &= 0,\ 0 < x < 2 \\ \text{with mesh size } h &= \frac{1}{2}. \end{aligned}$$

(b)

$$\begin{aligned} -(u_{xx} + u_{yy}) &= 0,\ 0 < x < 2,\ 0 < y < 1 \\ u(0, y) = 0,\ u(1, y) &= 0,\ 0 < y < 1 \\ u(x, 0) = 0,\ u(x, 1) &= \sin \pi x,\ 0 < x < 1 \\ \text{with mesh size } h &= \frac{1}{2}. \end{aligned}$$

(c)

$$\begin{aligned} -(u_{xx} + u_{yy}) &= 0,\ 0 < x < 2,\ 0 < y < 1 \\ u(0, y) = 18y^2(1 - y),\ u(1, y) &= 0,\ 0 < y < 1 \\ u(x, 0) = 0,\ u(x, 1) &= 0,\ 0 < x < 1 \\ \text{with mesh size } h &= \frac{1}{3}. \end{aligned}$$

(d)

$$-(u_{xx} + u_{yy}) = -2,\ 0 < x < 2,\ 0 < y < 1$$
$$u(0, y) = \ ,\ u(\pi, y) = 1, 0 < y < 1$$
$$u(x, 0) =, ,\ 0 < x < \pi$$

with mesh size $h = \frac{1}{3}$.

Heat equation exercises

7.13. Use the explicit finite difference scheme (7.68 to 7.74) to approximate the solution of the initial boundary value problem

$$\begin{cases} u_t = u_{xx}, & 0 < x < 2,\ 0 < t < 1 \\ u(0,t) = u(2,t) = 0, & 0 \le t \le 1 \\ u(x,0) = f(x) \end{cases}$$

with

$$f(x) = \begin{cases} 1,\ 0 \le x \le 1 \\ 0,\ 1 \le x \le 2 \end{cases},\ h = 0.25 \text{ and } \triangle t = 0.025.$$

7.14 Solve exercise 7.12 by the Crank-Nicolson scheme (7.80) with

$$f(x) = \begin{cases} x, & 0 \le x \le 1 \\ 1 - x, & 1 \le x \le 2. \end{cases}$$

7.15 Solve problem 7.13 by the Crank-Nicolson Method with $h = 0.25$ and $\Delta t = 0 : 025$.

7.16 Solve the boundary value problems

$$u_t = c^2 u_{xx},\ u(x,t) \in D = \{0 < x < a,\ 0 < t < T\}$$

$$u(x,0) = x,\ u(x,T) = 2T,\ 0 < x < a,$$

$$u(0,t) = 2t,\ u(a,t) = a,\ 0 < t < T.$$

7.17 Use the explicit finite difference for the wave equation to approximate the solution of the initial boundary value problem

$$u_{tt} = c^2 u_{xx}; 0 < x < l;\ 0 < t < T$$
$$u(0;t) = u(l;t) = 0; 0 \le t \le T$$
$$u(x;0) = f(x), u_t(x;0) = 0; 0 \le x \le l$$

when

(a) $c = 1, l = 1, T = 1; f(x) = x(1 - x); h = 0 : 25, \Delta t = 0.1.$

(b) $c = 1, l = 2, T = 1; f(x) = e^{-16(x-1)^2}, h = 0.4, \Delta t = 0.1$

(c) Consider the boundary-value problem (BVP)

$$\begin{aligned} u_{tt} &= u_{xx}; 0 < x < 1;\ 0 < t < 0.5 \\ u(0;t) &= u(1;t) = 0;\ 0 \leq t \leq 0.5 \\ u(x,0) &= \sin \pi x,\ u_t(x;0) = 0;\ 0 \leq x \leq 1 \end{aligned}$$

(i) Use the separation method to solve the above (BVP).

(ii) Use the explicit scheme to approximate the solution of the problem with $h = 0 : 25$ and $\Delta t = 0.1$.

(iii) Compute the absolute error at each interior grid point.

7.4 Suggestion for Further Reading

Numerical Methods are used to solve problems in the oil industry and other fields. Interested readers may find References [6] and [18] helpful. We suggest to read references [5], [7], [9] and [19] to readers who want more knowledge of applications of mathematical methods, models and algorithms to various areas of science, engineering and technology. References [1] through [4], [8] and [10] through [17] are also useful for further learning.

Bibliography

[1] W. Ames, *Numerical Methods for Partial Differential Equations*, Third Edition, Academic Press, Boston, 1992.

[2] K. Atkinson, W. Hall, *Theoretical Numerical Analysis*, Springer, New York, 2010.

[3] A. Di Bucchianico, R. M. M. Martheij, M. A. Peletier (Eds.), *Progress in Industrial Mathematics at ECMI 2014*, Springer, 2016.

[4] J. V. Butcher, *Numerical Methods for Ordinary Differential Equations*, Wiley, Chichester, UK, 2003.

[5] C. S Desai, J. Kundu, *Introductory Finite Element Method*, CRC Press, Boca Raton, 2001.

[6] A. Iske, T. Randen (Eds.), *Mathematical Methods and Modelling in Hydrocarbon Exploration and Production*, Springer, 2005.

[7] C. Kanuto, M. Y. Hussaini, A. Quarteroni, T. A. Zang, *Spectral Methods in Fluid Dynamics*, Springer, New York, 1998.

[8] L. Lapidus, G. F. Pinder, *Numerical Solution of Partial Differential Equations in Science and Engineering*, Wiley-Interscience, New York, 1982.

[9] C. R. MacCluer, *Industrial Mathematics: Modelling in Industry, Science and Government*, Prentice Hall, 2000.

[10] P. Manchanda, K. Ahmad, A. H. Siddiqi (Eds), *Current Trends in Industrial and Applied Mathematics*, Anamya Publishers, 2002.

[11] K. W. Morton, D. F. Meyers, *Numerical Solution of Partial Differential Equations*, Cambridge University, Press, 1996.

[12] R. D. Richtmeyer, K. W. Morton, *Difference Methods for Initial Value Problems*, Wiley-Interscience, New York, 1967.

[13] L. F. Shampine, *Numerical Solutions of Ordinary Differential Equations*, Chapman & Hall, 1994.

[14] A. H. Siddiqi and M. Kocvara (Eds), *Trends in Industrial and Applied Mathematics*, Kluwer, 2002.

[15] A. H. Siddiqi, I. Duff and O. Christensen (Eds.), *Modern Mathematical Methods, Models and Algorithms*, Anamaya-Anshan, 2006.

[16] A. H. Siddiqi, A. K. Gupta and M. Brokate, *Modelling of engineering and Technological Problems*, American Institute of Physics, Vol. 146, New York, 2009.

[17] A. H. Siddiqi, R. C. Singh, P. Manchanda (Eds.), *Mathematics in Science and Technology, Mathematical Methods, Models and Algorithms in Science and Technology*, World Scientific, Singapore, 2011.

[18] A. H. Siddiqi, *Emerging Applications of Wavelet Methods*, American Institute of Physics (AIP), Vol. 1463, New York, 2012.

[19] A. H. Siddiqi, P. Manchanda, R. Bhardwaj : *Mathematical Models, Methods and Applications*, Springer, Singapore, 2015.

Chapter 8

Complex Analysis

8.1 Motivation and Historical Development for Complex Analysis

As we know the real analysis or calculus is the study of functions of real variables ($f : R \to R$ or $f : R^2 \to R$). Similarly complex analysis is the study of functions of complex variables $f : C \to C$, where C denotes the set of complex numbers $z = x + iy$, $x \in R, y \in R$) and the imaginary number $i = \sqrt{-1}$ is the root of the algebraic equation $x^2 + 1 = 0$. It is well known that complex analysis was developed as a result of mathematical curiosity but subsequently it was found very useful in signal and image processing, fluid flow, quantum mechanics and many other areas of engineering.

It is common belief that real analysis is more useful than complex analysis but in practice the reverse is true. One strange application of complex analysis is in quantum mechanics, where complex numbers are used to represent approximately the probabilities of different states of a system. Complex analysis can be thought of as a subject that deals with the study of calculus of functions of imaginary numbers.

The story of imaginary numbers began in 1545 with the work of Italian mathematician Girolamo Cardano (1501-1576). The historical development of imaginary numbers is covered in John Stilwell's *Mathematics and Its History* (Springer, 2010). Cardano commented that arithmetic dealing with quantities involved mental tortures and noted that computations sometimes seemed useless. Many mathematicians of that time agreed with him.

In 1572, Rafael Bombelli, another Italian mathematician, showed the utility of roots of negative numbers. Interested readers are referred to Paul Nahin's book titled *An Imaginary Tale: The Story of* $\sqrt{-1}$ (Princeton University Press, 1998). See also the November 1999 issue of *Review Notices of AMS*, pages 1233 through 1236. History indicates that the seeds of imaginary numbers were planted in the 12th century when Arabian algebra was translated and introduced to Italy.

Rene Descartes (1596-1650), the French mathematician and philosopher, devised imaginary terms in complex numbers. One of the most distinguished mathematicians in history, Carl Friedrich Gauss (1777-1855), suggested the use of complex numbers instead of imaginary numbers. Euler was the first mathematician to propose the square root of 1 symbolized by i.

Complex numbers are ordered pairs of real numbers. They may be represented by points in a plane. The xy plane is known as a complex plane or Argand's plane (for Jean Robert Argand, 1768-1822) when displaying complex numbers. Gauss published a work on geometric representation of complex numbers as points in a plane. He also published the first proof of the fundamental theorem of algebra: every polynomial equation of degree n with complex coefficients different from zero has n roots in complex numbers.

Great controversy surrounded the acceptance of complex numbers in algebra and analysis in 1770, by 1830, the geometry of complex numbers was accepted by Gauss and then by the rest of the mathematics world. Gauss and Simon Denis Poisson (1781-1840) initiated studies of complex functions and their integrals. August Louis Cauchy (1789-1857) contributed to the field in a series of papers written between 1814 and 1845. He formulated his integral theorem and related concepts such as independence of path and integral representations of functions and their derivatives.

Pierre Alphonse Laurent (1813-1854) developed his Laurent series around

1843. Karl Weierstrass of Germany devised complex analysis on a rigorous foundation. Geoge Bernhard Riemann (1826-1866) introduced derivatives of functions of complex variables. His work and that of Cauchy led to naming of the Cauchy-Riemann equation that specifies conditions for a complex function to be differentiable at a point.

8.2 Functions of Complex Variables

8.2.1 Complex Numbers

A symbol of the form $x + iy$ or $x + yi$ or where x and y are real numbers and $i^2 = -1$, is called a **complex number.**
Let $z = x + iy$ and $w = u + iv$ and be two complex numbers, their sum $z + w$, their multiplication zw and division z/w, $w \neq 0$ are defined as follows:
Addition or sum:

$$z + w = (x + u) + i(y + v).$$

Multiplication:

$$zw = (x + iy)(u + iv).$$

Division:

$$\begin{aligned} \frac{z}{w} &= \frac{x + iy}{u + iv} = \frac{x + iy}{u + iv}\frac{u - iv}{u - iv} \\ &= \frac{(xu + yv) + (yu - xv)i}{u^2 + v^2} \\ &= \frac{xu + yv}{u^2 + v^2} + \frac{yu - xv}{u^2 + v^2}i. \end{aligned}$$

x is called the real part of the complex numbers z and often denoted Rez, while y is called the imaginary part of z and it is often denoted Imz. Two complex numbers z and w are called equal, that is, $z = w$ if real part of z is equal to real part of w, that is $x = u$ and imaginary part of z is equal to imaginary part of w, that is, $y = v$.
Let α be a scalar and $z = x + iy$ be a complex number then $\alpha z = \alpha x + i\alpha y$. In particular, if $\alpha = 1$ then

$$\begin{aligned} z - w &= (x + iy) + (-1)(u + iv) \\ &= x + iy - u - iv \\ &= (x - u) + i(y - v). \end{aligned}$$

Example 263. Let $z = 6 + 5i$ and $u = 4 + 3i$. Find

(a) $z+u$ (b) $z-u$

(c) zu (d) $\frac{z}{u}$

Solution: (a) $z+u=(6+4)+(5+3)i=10+8i$.
(b) $z-u=(6-4)+(5-3)i=2+2i$.
(c) $z.u=(6+5i)(4+3i)=(24-15)+(18+20)i=9+38i$.
(d) $\frac{z}{u}=\frac{6+5i}{4+3i}\frac{4-3i}{4-3i}=\frac{24+20i-18i+15}{16+9}=\frac{39+2i}{25}=\frac{39}{25}+\frac{2}{25}i$. The set of all complex numbers is often denoted by C.

Magnitude: The magnitude of $z=x+iy$ is denoted by $|z|$ or $|x+iy|$ and is defined by

$$|z|=|x+iy|=\sqrt{x^2+y^2}.$$

Conjugate: The complex conjugate of $z=x+iy$ is denoted by $\overline{x+iy}$ or $\overline{z}$ and is defined by

$$\overline{z}=\overline{x+iy}=x-iy.$$

Example 264. (a) Find the magnitude of $3+i8$ and $2-i4i$.
(b) Find the complex conjugate of $16+9i$ and $4-8i$.
(c) Show that

$$Re(z)=\frac{z+\overline{z}}{2}$$

and

$$Im(z)=\frac{z-\overline{z}}{2}$$

where $\overline{z}$ is the complex conjugate $z=x+iy$.
(d) Let $z=5-7i$. Find $|z|^2$.

8.2.2 Geometrical Representation of Complex Numbers

Complex numbers can be given two natural geometric interpretations. One can identify the complex number $x+iy$ with point (x,y) in the plane as shown in Figure 8.1. Each real number x or $x+i0$, is identified with the point $(x,0)$ on the x-axis (horizontal axis) often called the real axis. A number $0+iy$ or just iy is called a pure imaginary number and is associated with the point $(0,y)$ on the vertical axis, also called the imaginary axis. This correspondence between complex numbers and points in the plane is referred as the complex $(x-y)$ and also known as Argand's plane to pay tribute to Jean Robert Argand (1768-1822), a Swiss bank employee.

The second geometrical interpretation of complex numbers is through vectors. The complex number $z=x+iy$, or the point (x,y), may be considered a vector $x(1,0)+y(0,1)$ in the plane which may in turn be represented as an arrow from the origin to (x,y), see Figure 8.2.

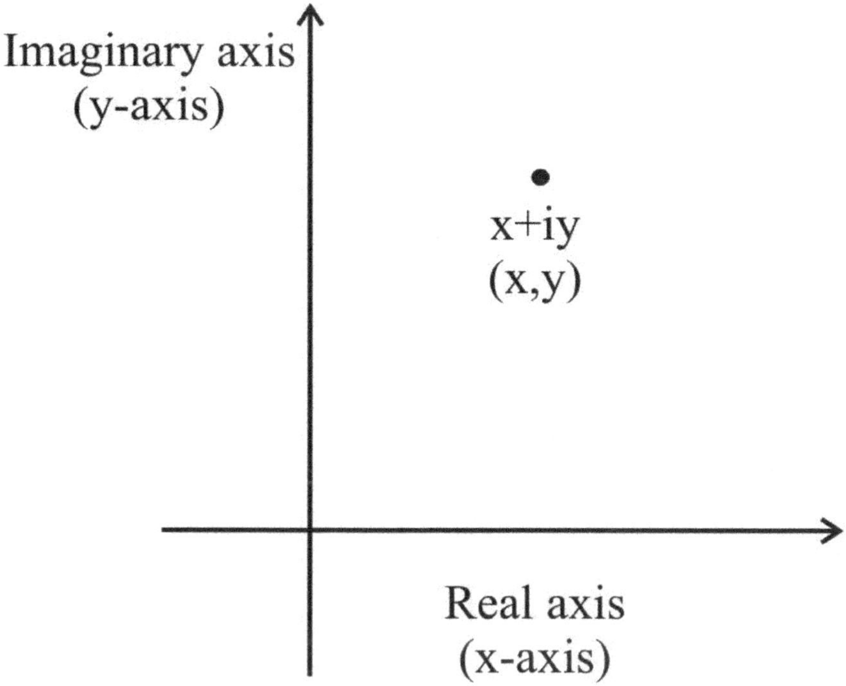

FIGURE 8.1: The Complex Plane

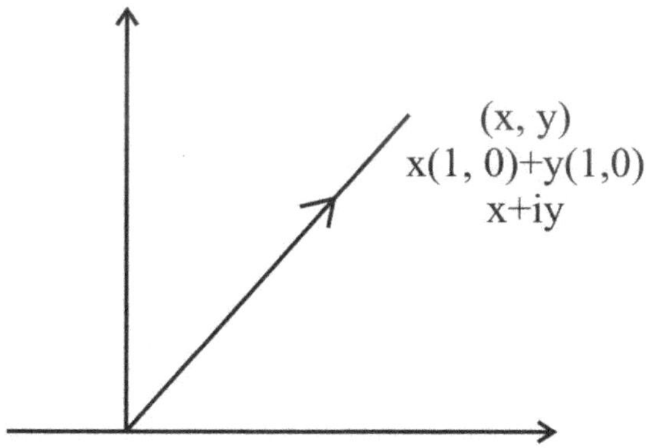

FIGURE 8.2: Complex Number as Vector in Plane

The first component of this vector is $Re(z)$ and the second component is $Im(z)$. In this interpretation, the definition of addition of complex numbers is equivalent to the parallelogram law for vector addition because we add two vectors by adding their respective components, see Figure 8.3.

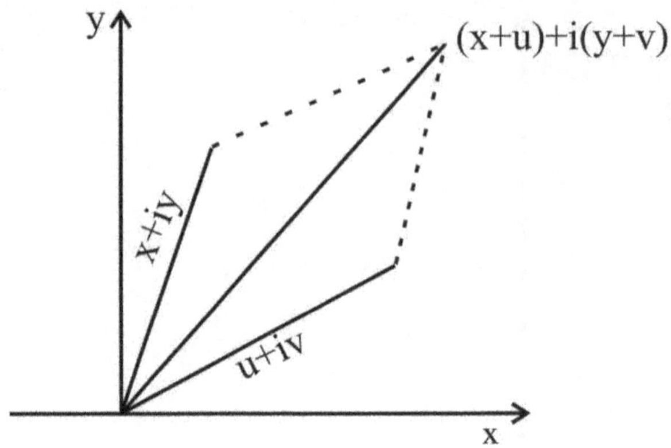

FIGURE 8.3: Parallelogram Law for Addition of Complex Numbers

Example 265. Draw Argand's diagram (pictorial representation) of $z = x + iy$ and $\overline{z} = x - iy$. See Figure 8.4.

Polar form of complex number

As we know rectangular coordinates (x, y) and polar coordinates (r, θ) are related as

$$x = r\cos\theta,\ y = r\sin\theta.$$

Thus

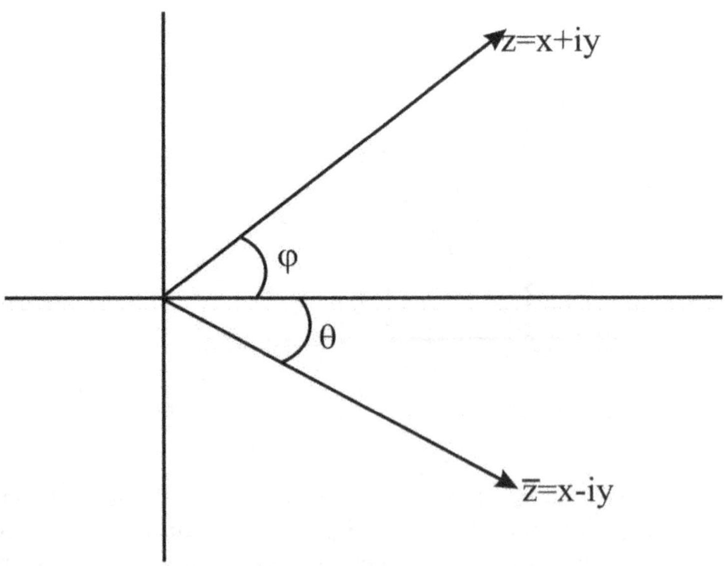

FIGURE 8.4: Argand's Diagram of z and $\overline{z}$

$$z = x + iy = r\cos\theta + ir\sin\theta = r(\cos\theta + i\sin\theta) = re^{i\theta},$$

applying Euler's formula. This representation of z is called the **polar form** of a complex number.

As we see in Figure 8.5, the polar coordinate r is the distance from the origin

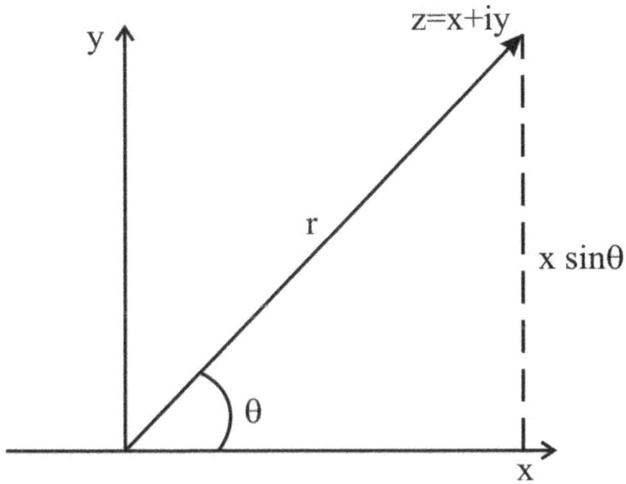

FIGURE 8.5: Polar Coordinates

to the point (x, y) in xy plane. It is clear that $r = \sqrt{x^2 + y^2} = |z|$. The angle θ of inclination of the vector z measured in radians from the positive real axis is positive when measured counterclockwise and negative when measured clockwise. Then angle θ is called an argument of complex number that must satisfy $\tan\theta = \dfrac{y}{x}$. The solution of this equation is not unique.

The angle θ argument of z is often denoted by $argz$. The argument of a complex number in the interval $-\pi < \theta \le \pi$ is called the principal argument of z and is denoted by $argz$. It may be noted that

$$Arg(i) = \frac{\pi}{2}.$$

Example 266. Express the following complex numbers in the polar form and find their arguments:

(a) $2 + 8i$.

(b) $6 + 4i$.

(c) $3 - \sqrt{3}i$.

Solution: (a) $-2 + 8i = x + iy \Rightarrow x = -2, y = 8$

$$z| = r = \sqrt{x^2 + y^2} = \sqrt{4 + 64} = \sqrt{68}$$

$$v = \tan^{-1}\frac{y}{x} = \tan^{-1}\frac{8}{-2} = \tan^{-1}(-4)$$

$$-2+8i = \sqrt{68} \quad (\text{as } \tan^{-1}(-4) + i\sin\tan^{-1}(-4))$$
$$= \sqrt{68}e^{i\tan^{-1}(-4)}.$$

(b) $r = \sqrt{36+16} = \sqrt{52}$

$$\theta = \tan^{-1}\frac{y}{x} = \tan^{-1}\frac{\sqrt{-3}}{3} = \tan^{-1}\frac{-1}{\sqrt{3}}$$

$$z = 2\sqrt{3}e^{i\tan^{-1}(-\frac{1}{3})}.$$

(c) Let

$$3-\sqrt{3}i = z \Rightarrow x = 3, y = -\sqrt{3},$$
$$r = \sqrt{x^2+y^2} = \sqrt{9+3} = \sqrt{12} = 2\sqrt{3}$$
$$\theta = \tan^{-1}\frac{y}{x} = \tan^{-1}\frac{-\sqrt{3}}{3} = \tan^{-1}(-\frac{1}{\sqrt{3}})$$
$$z = 2\sqrt{3}e^{i\tan^{-1}(-\frac{1}{\sqrt{3}})}.$$

Multiplication and Division in Polar Form

Let

$$z = x+iy = r(\cos\theta + i\sin\theta),$$
$$w = u+iv = \lambda(\cos\varphi + i\sin\varphi),$$

then

$$zw = r\lambda[(\cos\theta\cos\varphi - \sin\theta\sin\varphi) + i(\sin\theta\cos\varphi + \cos\theta\sin\varphi)]$$

and for $w \neq 0$

$$\frac{z}{w} = \frac{r}{\lambda}[(\cos\theta\cos\varphi + \sin\theta\sin\varphi) + i(\sin\theta\cos\varphi - \cos\theta\sin\varphi).$$

zw can also be written as

$$zw = r\lambda[\cos(\theta+\varphi) + i\sin(\theta+\varphi)] \tag{8.1}$$

and $\frac{z}{w}$, $w \neq 0$ is given by

$$\frac{z}{w} = \frac{r}{\lambda}[\cos(\theta-\varphi) + i\sin(\theta-\varphi). \tag{8.2}$$

It follows from (8.1) and (8.2) that

$$|zw| = |z||w|, \quad |\frac{z}{w}| = \frac{|z|}{|w|} \tag{8.3}$$

and

$$\arg(zw) = \arg z + \arg w, \quad \arg(\frac{z}{w}) = \arg z - \arg w. \tag{8.4}$$

Powers and roots of complex numbers

(8.1) and (8.2) can be used to find integral powers of z, z^n, where n is an integer. For

$$z = r(\cos\theta + i\sin\theta)$$

choose $z_1 = z$ and $z_2 = z$; then (8.1) implies

$$z^2 = r^2[\cos(\theta+\varphi) + i\sin(\theta+\varphi)] = r^2[\cos 2\theta + i\sin 2\theta]$$

$$\begin{aligned} z^3 &= z^2 z = r^2[\cos 2\theta + i\sin 2\theta]\ r\ [\cos\theta + i\sin\theta] \\ &= r^3[\cos 3\theta + i\sin 3\theta]. \end{aligned}$$

Since

$$\arg(1) = \tan^{-1}\frac{0}{1} = 0,$$

it follows from (8.2)

$$\frac{1}{z^2} = z^{-2} = r^{-2}[\cos(-2\theta) + i\sin(-2\theta)].$$

Continuing in this manner, we obtain a formula for z^n,

$$z^n = r^n(\cos n\theta + i\sin\theta).$$

Let $z = \cos\theta + i\sin\theta$, then $|z| = 1 = r$ and by (8.3) we get

$$(\cos\theta + i\sin\theta)^n = \cos n\theta + i\sin\theta. \tag{8.5}$$

Equation (8.5) is known as **DeMoivre's formula**.

Example 267. (a) Write the given complex numbers in the polar forms:

(i) $6i$ (ii)$-2-2\sqrt{3}$ (iii) $-1+5i$ (iv) $\frac{12}{\sqrt{3}+i}$

(b) Write $z = 10(\cos\frac{\pi}{5} + i\sin\frac{\pi}{5})$ in the form $x+iy$.

(c) Find $(1+\sqrt{3}i)^9$.

(d) Find (i) $(8)^{\frac{1}{3}}$ (ii) $i^{\frac{1}{3}}$ (iii) $(-1+i)^{\frac{1}{3}}$.

Solution: (a) (i) $6i = x+iy \Rightarrow x = 0, y = 6$

$$\begin{aligned} r &= \sqrt{0^2+6^2} = \sqrt{36} \\ \theta &= \tan^{-1}\frac{y}{x} = \tan^{-1}\frac{6}{0} = \tan^{-1}\infty = \frac{\pi}{2} \\ 6e^{\frac{i\pi}{2}} &= 6i = 6(\cos\frac{\pi}{2} + i\sin\frac{\pi}{2}) \end{aligned}$$

is the polar form of $6i$.

(ii)

$$\begin{aligned} -2-2\sqrt{3}i &= x+iy \\ \Rightarrow x=-2 \quad &, \quad y=-2\sqrt{3} \\ \Rightarrow r &= \sqrt{4+12}=4 \\ \theta &= \tan^{-1}\frac{-2\sqrt{3}}{-2}=\tan^{-1}\sqrt{3}=\frac{4\pi}{3} \end{aligned}$$

Polar form of $-2-2\sqrt{3}$ is

$$4(\cos\frac{4\pi}{3}+i\sin\frac{4\pi}{3})=4e^{\frac{i4\pi}{3}}.$$

(iii)

$$\begin{aligned} -1+5i &= x+iy \Rightarrow x=-1, y=5 \\ \Rightarrow r=\sqrt{6} \quad &, \quad \theta=\tan^{-1}\frac{5}{1} \\ -1+5i &= \sqrt{6}e^{i\tan^{-1}(-5)}. \end{aligned}$$

(iv)

$$\frac{12}{\sqrt{3}+i}=\frac{12(\sqrt{3}-i)}{(\sqrt{3}+i)(\sqrt{3}-i)}=\frac{12\sqrt{3}-i12}{3+1}=3\sqrt{3}-3i$$

$$\Rightarrow x=3\sqrt{3}, y=-3$$

$$r=\sqrt{27+9}=6, \theta=\tan^{-1}\frac{-3}{3\sqrt{3}}=\tan^{-1}(-\frac{1}{\sqrt{3}})$$

$$\frac{12}{\sqrt{3}+i}=6(\cos\tan^{-1}(-\frac{1}{\sqrt{3}})+i\sin\tan^{-1}(-\frac{1}{\sqrt{3}})$$

$$=6(\cos(-\frac{\pi}{6})+i\sin(-\frac{\pi}{6}))=6e^{\frac{-i\pi}{6}}.$$

(b)

$$\begin{aligned} r^2 &= \sqrt{x^2+y^2} \\ 100 &= x^2+y^2 \\ 100 &= x^2+x^2\tan\frac{\pi}{4} \\ 100 &= x^2\sec^2\frac{\pi}{4} \\ \tan\theta &= \frac{y}{x} \\ \tan\frac{\pi}{4} &= \frac{y}{x} \\ \text{or} \quad x &= \frac{\sqrt{100}}{\sec\frac{\pi}{4}} \end{aligned}$$

$\theta = 8.0902$ and $z = 8.0902 + 5.8779i$.

(c)

$$\begin{aligned}
(1+\sqrt{3}i)^9 &= r^9(\cos 9\theta + i\sin 9\theta)\\
\text{where } r &= \sqrt{1^2+(\sqrt{3})^2}\\
\tan\theta &= \frac{\sqrt{3}}{1}\\
\theta &= \tan^{-1}\sqrt{3} = \frac{\pi}{3}\\
(1+\sqrt{3}i)^9 &= 2^9(\cos 3\pi + i\sin 3\pi) = 2^9\cos\pi.
\end{aligned}$$

Remark 61. There is no ordering in the set of complex numbers.

Remark 62. The binomial expansion of the real numbers is also valid for complex numbers in the following form:

$$(z+w)^n = \Sigma_{j=0}^{n}\begin{pmatrix} n \\ j \end{pmatrix} z^{n-j}w^j.$$

8.2.3 Sets in Complex Plane

If $z = x + iy$ is any complex number, $|z| = \sqrt{x^2+y^2}$ is the distance from the origin to z (point (x, y) in the plane) in the complex plane.
If $w = u + iv$ is also a complex number, then

$$|z-w| = |(x-u)+i(y-v)| = \sqrt{(x-u)^2+(y-v)^2}$$

is the distance between z and w in the complex plane (Figure 8.6). If a is a complex number and ρ is a positive real number such that $|z - a| = \rho$, the locus of points satisfying this condition is the circle of radius ρ about a and often referred as "the circle $|z-a| = \rho$". The points z satisfying the inequality $|z - a| < \rho, \rho > 0$ lie within, but not on, a circle of radius ρ centred at a (Figure 8.7). This set is called a **neighborhood** of a or an open disk.
If $a = 0$, then any point on the circle $|z| = r$ has the polar form.

$$z = re^{i\theta}$$

where θ is the angle from the positive real axis to the line from the origin through z (Figure 8.8).

Definition 88. (a) A point a is said to be an **interior point** of a set S of the complex plane y entirely within S. A subset S of a complex plane is called an **open set** if every point z of S is an interior point.
(b) A point a is said to be a **boundary point** of a set of complex numbers S if every open disk about a contains at least one point in S and at least one point not in S.
S is called **closed** if it contains all of its boundary points.
(c) A complex number a is a limit point of a set S of complex numbers if every open disk about a contains atleast one point of S different from a.

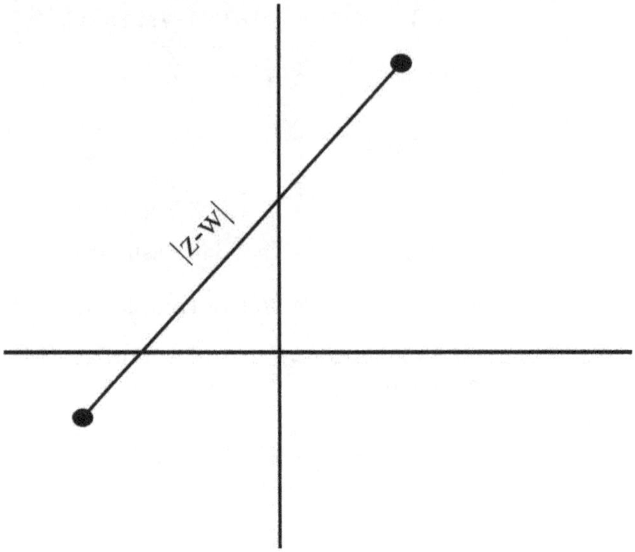

FIGURE 8.6: $|z - w|$ is Distance between z and w

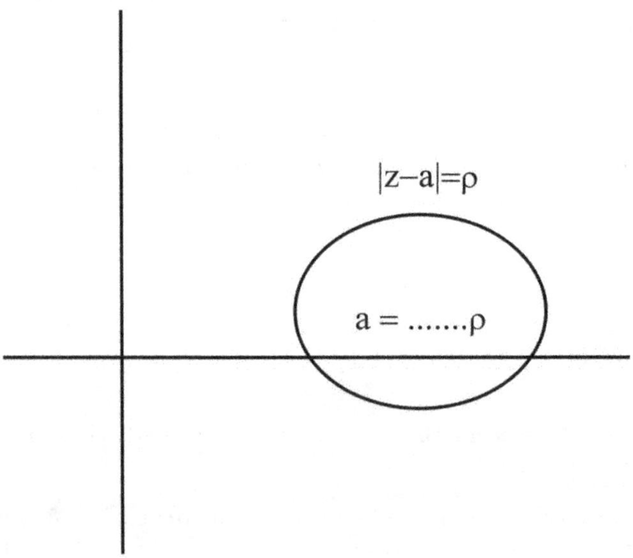

FIGURE 8.7: The Circle of Radius ρ about a

Remark 63. (a) A boundary point may or may not be in S. No point can

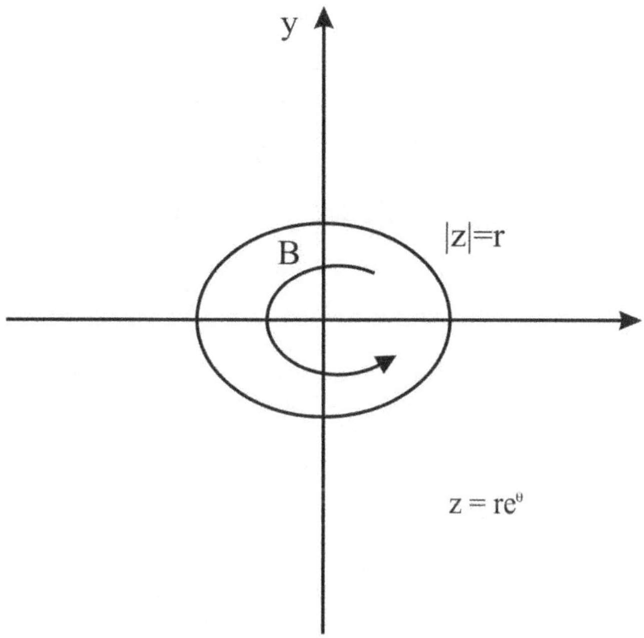

FIGURE 8.8: Circle of Radius r about Origin

simultaneously be interior and a boundary point as the definitions of interior point and boundary point are exclusive.

(b) The set of all boundary points of S is called the **boundary** of S, and is denoted by ∂S.

(c) Limit point differs from boundary point in that every open disk about the point contains something from S other than the point itself (see Example 268).

(d) Every point of a set is either an interior point or a boundary point.

Example 268. Let A consist of all complex numbers $z = x + iy$ with $y > o$, with the point $-23i$; see Figure 8.9. Then $-23i$ is a boundary point of A, because every disk about $-23i$ certainly contains points not in A, but also contains a point in A, namely $-23i$ itself. Every real number (point of x-axis) is also a boundary point of A.

Example 269. Let B consist of all points satisfying $|z - a| < \rho$. Every point of B in an interior point because about any point is B we can draw a disk of small enough radius to contain only points in B. Thus B is an open set which does not include any points on its bounding circle $|z - a| = \rho$.

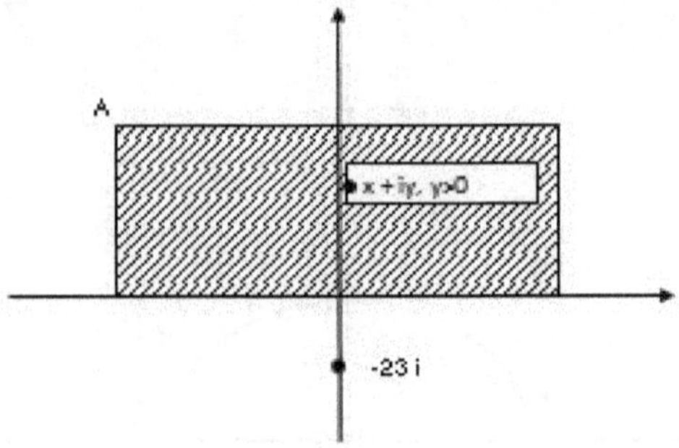

FIGURE 8.9: $-23i$ is Boundary Point of A

8.2.3.1 Complex Sequences and Series

Major parts of the theory of complex sequences and series are analogues to real calculus.

Definition 89. (a) A sequence z_n is a function whose domain is the set of positive integers, $n = 1, 2, 3, \ldots$; a complex number say, z_n, is assigned.
(b) A sequence of complex numbers z_n is said to be **convergent** to a complex number z if Rez_n and Imz_n respectively **converge** to Rez and Imz.
This can also be written as
The complex sequence z_n is called **convergent** to z if given any positive number ϵ, there is a positive number N such that

$$|z_n - z| < \epsilon,$$

if $n \geq N$. If z_n converges to z then we write $z_n \to z$ or $\lim_{n\to\infty} z_n = z$.
(c) z_n is called a **bounded sequence** if for some M, $|z_n| \leq M$ for $n = 1, 2, 3\ldots$ or equivalently a sequence is called **bounded** if there is a disk containing all of its terms.

Definition 90. (a) Let $\{z_n\}$ be a complex sequence; then $\sum_{n=1}^{\infty} z_n$ is called a **complex series.** $S_N = \sum_{n=1}^{N} z_n$ is called the Nth **partial sum** of this series. $\sum_{n=1}^{\infty} z_n$ is said to converge to z if S_N converges to z.
(b) For any a,

$$\sum_{n=1}^{\infty} az^n = a + az + az^2 + \cdots + az^n + \ldots$$

is called a **geometric series.**

Example 270. (a) Write the 100 th term of the sequences $2 + t^n$ and $\dfrac{i^{n+1}}{n+2}$.

(b) Examine whether $\dfrac{ni}{n+3i}$ is convergent or not.

(c) Give an example of a divergent complex sequence.

(d) Determine whether the given geometric series are convergent or not.

(i) $\sum_{n=1}^{\infty}\left(\frac{i}{2}\right)^n$.

(ii) $\sum_{n=0}^{\infty} 3\left(\frac{2}{1+2i}\right)^n$.

Solution: (a) $2+t^{100}$ and $\dfrac{i^{101}}{102}$.
(b) Convergent.
(c) $\{ni\}$.
(d) (i) Convergent as geometric series with common ratio < 1.

8.2.3.2 Functions of Complex Variable

Recall the definition of a general function: **function** from a set A to set B is a rule of correspondence that assigns to each element in A a unique element of B. If y is an element of B associated with an element of x of A, we write $y = f(x)$ and A is called a **domain** of f and the set of all $f(x)$ corresponding to each x is called the **range** of f. If A is a set of real numbers f is called a function of real variable. If A is a set of complex numbers, then $f(z)$ is called a function of **complex variables** z or simply a **complex function**. The value or image of complex number z will be some complex number $w = u + iv$, that is,

$$w = f(z) = u(x, y) + iv(x, y) \tag{8.6}$$

where u and v are the real and imaginary parts of w and are real valued functions.

Example 271. Examples of functions of a complex variable z.

(a) $f(z) = z + Im(z)$.

(b) $f(z) = z - Re(z) = x + iy - x = iy$.

(c) $f(z) = z^2 + 4z$.

(d) $f(z) = \dfrac{z+i}{z^2+4}$.

(e) $f(z) = \dfrac{z}{z^2+1}, z \neq i, z \neq -i$.

(f) Find the value of the line $Re(z) = 1$ under the mapping $f(z) = z^2$.

Remark 64. We cannot draw a graph of a complex function $w = f(z)$ since a graph would require four axes in a four-dimensional coordinate system. A complex function $w = f(z)$ can be interpreted as a mapping or transformation from the z-plane to the w-plane, see Figure 8.10.

Definition 91. (a) Suppose that function f is defined in some neighbourhood of z_0, except possibly at z_0 itself. Then f is said to posses a limit at z_0 written $\lim_{z \to z_0} f(z) = l$. if, for each $\varepsilon > 0$, there exists a $\delta > 0$ such that $|f(z) - l| < \epsilon$ whenever $0 < |z - z_0| < \delta$.

(b) A function f is **continuous** at a point z_0 if $\lim_{z \to z_0} f(z) = f(z_0)$.

(c) A function of the type

$$f(z) = a_n z^n + a_{n-1} z^{n-1} + \cdots + a_2 z^2 + a_1 z + a_0, a_n \neq 0 \tag{8.7}$$

is called a **polynomial of degree** n.

(d)

$$f(z) = \frac{g(z)}{h(z)}, \tag{8.8}$$

where g and h are polynomials is called a rational function. $f(z)$ is continuous except at those points for which $h(z)$ is zero.

Definition 92. Let a complex function f be defined in a neighborhood of a point z_0. The derivative of f at z_0, denoted by $f'(z_0)$ is defined as

$$f'(z_0) = \lim_{\Delta z \to 0} \frac{f(z_0 + \Delta z) - f(z_0)}{\Delta z} \tag{8.9}$$

provided this limit exists.

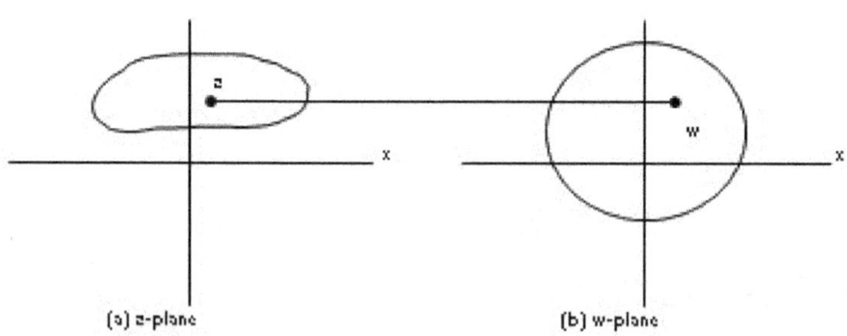

FIGURE 8.10: Complex Function as Mapping

If the limit (8.9) exists, the function f is called differentiable at z_0. The derivative of the function $w = f(z)$ is also written

$$\frac{dw}{dz}.$$

Remark 65. (i) As in real variables, differentiability implies continuity, that is, if f is differentiable at z_0 then f is continuous at z_0.

(ii) The rules of differentiation are the same as in the calculus of real variables. If f and g are differentiable at point z and c is a complex constant, then

(a)
$$\frac{dc}{dz} = 0, \ \frac{d}{dz}cf(z) = cf'(c). \tag{8.10}$$

(b)
$$\frac{d}{dz}[f(z) + g(z)] = f'(z) + g'(z). \tag{8.11}$$

(c)
$$\frac{d}{dz}[f(z)g(z)] = f(z)g'(z) + f'(z)g(z). \tag{8.12}$$

(d)
$$\frac{d}{dz}[\frac{f(z)}{g(z)}] = \frac{g(z)f'(z) - f(z)g'(z)}{[g(z)]^2}. \tag{8.13}$$

(e)
$$\frac{d}{dz}\left(f(g(z))\right) = f(g(z))g'(z). \tag{8.14}$$

(f)
$$\frac{d}{dz}z^n = nz^{n-1}, \quad \text{and } n \text{ is an integer.} \tag{8.15}$$

Definition 93. A complex function $w = f(z)$ is said to be **analytic** (holomorphic) at a point z_0 if f is differentiable at z_0 and at every point in some neighbourhood of z_0. A function f is analytic in a domain if it is analytic at every point in D. A function that is analytic at every point z is said to be an **entire** function.

Example 272. (a) Show that polynomial functions are entire functions.

(b) Find the image or value of the given line under mapping $f(z) = z^2$:

(i) $x = -3$.

(ii) $y = x$.

(iii) $y = -x$.

(c) Express the following functions in the form $f(z) = u + iv$:

(i) $f(z) = 6z + 9i$.

(ii) $f(z) = z^3 - 4z$.

(iii) $f(z) = \dfrac{z}{z+1}$.

(d) Evaluate the given function at the indicated points:

(i) $f(z) = 2x - y^2 + i(xy^3 - 2x^2 + 1)$.

(i) $2i$.

(ii) $2 + i$.

(ii) $f(z) = e^x \cos y + ie^x \sin y$.

(i) $\frac{\pi i}{4}$.

(ii) $3 + \frac{\pi i}{3}$.

(e) Find $\lim_{z \to i}(4z^3 - 5z^2 + 4z + 5i)$.

(f) $\lim_{z \to 1+i} \frac{z^2 - 2z + 2}{z^2 - 2i}$.

(g) Show that $\lim_{z \to z_0} \frac{z}{\overline{z}}$ does not exist.

(h) Use the definition of the derivative to show that

$$f'(z) = 2z \text{ if } f(z) = z^2.$$

(i) Use the basic rules of differentiation to find $f'(z)$ if

(i) $f(z) = 5z^4 - iz^3 + (8 - i)z^2 - 6i$.

(ii) $f(z) = (z^5 + 3iz^3)(z^4 + iz^3 + 2z^2 - 6iz)$.

(iii) $f(z) = \frac{5z^2 - z}{z^3 + 1}$.

(j) Show that $f(z) = \overline{z}$ is not differentiable at any point.

Solution: (a) A polynomial is differentiable at every point of the domain.

(b) (i) Substituting $x = -3$ into $u = x^2 - y^2, v = 2xy$, $f(z) = z^2 = u(x, y) + iv(x, y)$, or $(z + iy)^2 = u(x, y) - x^2 - y^2, v = 2xy$, given parametric equations $u = 9 - y^2, v = -6y$.
Using $y = -\frac{v}{6}$ the first equation gives $v = 9 - \frac{v^2}{36}$.
The graph is a parabola.

(ii) $y = x$ gives $u = 0, v = 2x^2$. Since $x^2 \geq 0$ for all real values of x, the image is the origin and the positive v-axis.

(iii) $y = -x$ gives $u = 0, v = -2x^2$. Since $x^2 \leq 0$ for all real values of x, the image is the origin and the negative v-axis.

(c) (i) $f(z) = 6z + 9i = 6(x+iy) + 9i = 6x + i(9+6y)$ where $u(x,y) = 6x$ and $v(x,y) = 9 + 6y$.
(ii) $z^3 - 4z = f(z)f(z) = (x^3 - 3xy^2 - 4x) + i(3x^2y - y^3 - 4y)$

$$f(z) = \frac{x^2+y^2+x}{(x+1)^2+y^2} + \frac{y}{(x+1)^2+y^2}.$$

$$\begin{aligned}
\text{(iii)}\quad f(z) &= \frac{z}{z+1} = \frac{x+iy}{x+iy+1} = \frac{x+iy}{(x+1)+iy} \\
&= \frac{x+iy}{(x+1)+iy}\frac{(x+1)-iy}{(x+1)-iy} \\
&= \frac{x^2+y^2+x}{(x+1)^2+y^2} + i\frac{(y+y^2)}{(x+1)^2+y^2}
\end{aligned}$$

$$\Rightarrow u(x,y) = \frac{x^2+y^2+x}{(x+1)^2+y^2},\ v(x,y) = \frac{y+y^2}{(x+1)^2+y^2}.$$

8.2.3.3 Cauchy-Riemann Equations

Theorem 58. *Let $f(z) = u(x,y) + iv(x,y)$ be differentiable at a point $z = x+iy$. Then first order derivatives of u and v exist and satisfy the Cauchy-Riemann equations*

$$\frac{\partial u}{\partial x} = \frac{\partial v}{\partial y} \text{ and } \frac{\partial v}{\partial x} = -\frac{\partial u}{\partial y}. \tag{8.16}$$

Proof. *We have*

$$f'(z) = \lim_{\Delta z} \frac{f(z+\Delta z) - f(z)}{\Delta z} \tag{8.17}$$

as $f(z)$ is differentiable at z. Putting

$$f(z) = u(x,y) + iv(x,y) \text{ and } \Delta z = \Delta x + i\Delta y$$

in (8.17), we get

$$f'(z) = \lim_{\Delta z\to 0} \frac{u(x+\Delta x, y+\Delta y) + iv(x+\Delta x, y+\Delta y) - u(x,y) - iv(x,y)}{\Delta x + i\Delta y}. \tag{8.18}$$

Since this limit exists, Δz can approach zero from any convenient direction. In particular, if $\Delta z \to 0$ horizontally then $\Delta z = \Delta x$ and so (8.18) becomes

$$f'(z) = \lim \Delta x \to 0 \frac{u(x+\Delta x, y) - u(x,y)}{\Delta x} + i \lim_{\Delta x\to 0} \frac{v(x+\Delta x, y) - v(x,y)}{\Delta x}. \tag{8.19}$$

Two limits in (8.19) exist as it is differentiable at the point z. By definition, limits in (8.19) are partial derivatives, that is ,

$$f'(z) = \frac{\partial u}{\partial z} + i\frac{\partial v}{\partial x}. \tag{8.20}$$

Now, if we let $\Delta z \to 0$ vertically, then $\Delta z = i\Delta y$ and (8.20) takes the form

$$f'(z) = \lim_{\Delta y \to 0} \frac{u(x, y + \Delta y) - u(x, y)}{i\Delta y} + i \lim_{\Delta y \to 0} \frac{v(x, y + \Delta y) - v(x, y)}{i\Delta y}$$

or equivalently

$$f'(z) = -i\frac{\partial u}{\partial y} + \frac{\partial v}{\partial y}. \tag{8.21}$$

Equating real and imaginary points of (8.20) and (8.21), we get the Cauchy-Riemann equations

$$\frac{\partial u}{\partial x} = \frac{\partial v}{\partial y}, \frac{\partial v}{\partial x} = -\frac{\partial u}{\partial y}.$$

□

Remark 66. The Cauchy-Reimann equations are not sufficient for ensuring analyticity of $f(z)$ unless derivatives

$$\frac{\partial u}{\partial x}, \frac{\partial v}{\partial x}, \frac{\partial u}{\partial y} \text{ and } \frac{\partial v}{\partial y}$$

are continuous. In fact the following result holds.

Suppose real valued functions $u(x, y)$ and $v(x, y)$ are continuous and have continuous derivatives $\frac{\partial u}{\partial x}, \frac{\partial v}{\partial x}, \frac{\partial u}{\partial y} and \frac{\partial v}{\partial y}$ in a domain D. If u and v satisfy the Cauchy-Riemann equation then the complex function $f(z) = u(x, y) + iv(x, y)$ is analytic in D.

Analyticity implies differentiability but not vice versa. We have an analogue of the above result that gives sufficient conditions for differentiability, namely if $u(x, y)$ and $v(x, y)$ are continuous and have continuous first order derivatives in a neighborhood of z, and if u and v satisfy the Cauchy-Reimann equations at the point z, then $f(z) = u(z, y) + iv(x, y)$ is differentiable at z and $f'(z)$ is given by

$$f'(z) = \frac{\partial u}{\partial x} + i\frac{\partial v}{\partial x} = \frac{\partial v}{\partial y} - i\frac{\partial u}{\partial y}.$$

Remark 67. It may be noted that if a complex function $f(z) = u(x, y) + iv(x, y)$ is analytic through a domain D, then the real functions $u(x, y)$ and $v(x, y)$ must satisfy Cauchy-Riemann equations at every point of the domain.

Example 273. (a) Find $u(x, y)$ and $v(x, y)$ if

(i) $f(z) = z^2$.

(ii) $f(z) = \frac{1}{z}$, $z \neq 0$.

(b) Show that $f(z) = (2x + y) + i(y^2 - x)$ is not analytic at any point.

(c) Show that $f(z) = Re(z)$ is not analytic for $z \neq 0$.

(d) Show that the following functions are analytic in an appropriate domain.

(i) $f(z) = e^x \cos y + ie^x \sin y$.

(ii) $f(z) = \frac{x-1}{(x-1)^2 + y^2} - i\frac{y}{(x-1)^2 + y^2}$.

Solution: (a) (i)
$$f(z) = z^2 = (x+iy)^2 = x^2 - y^2 + 2ixy$$
$$\Rightarrow u(x,y) = x^2 - y^2,\ v(x,y) = 2xy.$$

(ii)
$$\begin{aligned} f(z) = \frac{1}{z} &= \frac{1}{x+iy} = \frac{1}{x+iy}\frac{x-iy}{x-iy} = \frac{x-iy}{x^2+y^2} \\ &= \frac{x}{x^2+y^2} - i\frac{y}{x^2+y^2} \\ \Rightarrow u(x,y) &= \frac{x}{x^2+y^2},\ v(x,y) = -\frac{y}{x^2+y^2}. \end{aligned}$$

(b)
$$f(z) = (zx^2 + y) + i(y^2 - x)$$
$$u(x,y) = 2x^2 + y,\quad v(x,y) = y^2 - x$$
$$\frac{\partial u}{\partial x} = 4x,\ \frac{\partial v}{\partial x} = -1$$
$$\frac{\partial u}{\partial y} = 1,\ \frac{\partial v}{\partial y} = 2y$$
$$\Rightarrow \frac{\partial u}{\partial x} = 4x \neq \frac{\partial v}{\partial y},$$

$\Rightarrow f(z)$ is not analytic as the Cauchy-Riemann equation is not satisfied.

(c)
$$f(z) = Re(z) = x + i0 \Rightarrow u(x,y) = x,\ v(x,y) = 0$$
$$\Rightarrow \frac{\partial u}{\partial x} = 1,\ \frac{\partial v}{\partial y} = 0,$$

$\Rightarrow$ Cauchy-Riemann equation is not satisfied, for $z \neq 0$. Hence $f(z)$ is not analytic.

(d) (i)
$$u(x,y) = e^x \cos y,\ v(x,y) = e^x \sin y$$

$$\frac{\partial u}{\partial x} = e^x \cos y, \quad \frac{\partial v}{\partial y} = -e^x \sin y$$

$$\frac{\partial v}{\partial x} = e^x \sin y, \quad \frac{\partial v}{\partial y} = e^x \cos y$$

$$\Rightarrow \frac{\partial u}{\partial x} = \frac{\partial v}{\partial y} \text{ and } \frac{\partial v}{\partial x} = -\frac{\partial u}{\partial y},$$

$\Rightarrow$ Cauchy-Riemann equations are satisfied and partial derivatives of the first order are continuous. Thus the given function is analytic for all z.

(ii) $$u = \frac{x-1}{(x-1)^2 + y^2}, \quad v = \frac{y}{(x-1)^2 + y^2}$$

$$\frac{\partial u}{\partial x} \frac{\partial y^2 - (x-1)^2}{\partial[(x-1)^2 + y^2]} = \frac{\partial v}{\partial y},$$

$$\frac{\partial u}{\partial y} = -\frac{2y(x-1)}{[(x-1)^2 + y^2]^2} = -\frac{\partial v}{\partial x}$$

and f is analytic in any domain not containing $z = -1$.

Harmonic functions: A real valued $\varphi(x, y)$ is called **harmonic** in the domain D if $\frac{\partial^2 \varphi}{\partial x^2}, \frac{\partial^2 \varphi}{\partial y^2}$ and $\frac{\partial^2 \varphi}{\partial x \partial y}$ exist and are continuous and

$$\frac{\partial^2 \varphi}{\partial x^2} + \frac{\partial^2 \varphi}{\partial y^2} = 0.$$

Theorem 59. *Let $f(z) = u(x, y) + v(x, y)$ be analytic in a domain D. Then the functions $u(x, y)$ and $v(x, y)$ are **harmonic**.*

Proof. *Let $u(x, y)$ and $v(x, y)$ have continuous second partial derivatives. Since f is analytic, the Cauchy-Riemann equations are satisfied. Differentiating both sides of*

$$\frac{\partial u}{\partial x} = \frac{\partial v}{\partial y}$$

with respect to x and both sides of

$$\frac{\partial u}{\partial y} = -\frac{\partial v}{\partial x}$$

with respect to y we get

$$\frac{\partial^2 u}{\partial x^2} = \frac{\partial^2 v}{\partial x \partial y}, \frac{\partial^2 u}{\partial y^2} = -\frac{\partial^2 v}{\partial x \partial y}.$$

Under the assumption of continuity, the mixed partials are equal. By adding these equations we get

$$\frac{\partial^2 u}{\partial x^2} + \frac{\partial^2 u}{\partial y^2} = \frac{\partial^2 v}{\partial x \partial y} - \frac{\partial^2 v}{\partial x \partial y} = 0.$$

This implies that $u(x,y)$is harmonic. Similarly we can show that

$$\frac{\partial^2 v}{\partial x^2} + \frac{\partial^2 v}{\partial y^2} = 0.$$

That is, $v(x,y)$ is harmonic. □

Conjugate harmonic functions:
Let $u(x,y)$ be a harmonic function in D then $v(x,y)$ is called a **conjugate harmonic function** if $u(x,y) + iv(x,y)$ is analytic function in D.

Example 274. Show that the following functions are harmonic:

(i) $u(x,y) = x$

(ii) $u(x,y) = \log_e(x^2 + y^2)$

(iii) $u(x,y) = e^x(x\cos y - y\sin y)$

Solution: (i)
$$u(x,y) = x \Rightarrow \frac{\partial u}{\partial x} = 1$$

and
$$\frac{\partial^2 u}{\partial x^2} = 0 \quad \text{and} \quad \frac{\partial^2 u}{\partial y^2} = 0$$
$$\frac{\partial^2 u}{\partial x^2} + \frac{\partial^2 u}{\partial y^2} = 0,$$

$\Rightarrow$ $u(x,y)$ is harmonic.

(ii)
$$\frac{\partial^2 u}{\partial x^2} = 2e^x\cos y + e^x(x\cos y - y\sin y)$$
$$\frac{\partial^2 u}{\partial y^2} = e^x(-x\cos y + y\sin y - 2\cos y)$$
$$\Rightarrow \frac{\partial^2 u}{\partial x^2} + \frac{\partial^2 u}{\partial y^2} = 0.$$

Thus u is harmonic.

Example 275. Let $u(x,y) = x^2 - y^2$, Find a conjugate harmonic function of u.

Solution: $u(x,y) = x^2 - y^2$ is harmonic.
Now
$$\frac{\partial u}{\partial x} = 2x = \frac{\partial v}{\partial y}$$
$$\Rightarrow v = 2xy + h(x)$$

$$\frac{\partial u}{\partial y} = -2y = -\frac{\partial v}{\partial x} = -2y - h'(x)$$

implies $h'(x) = 0$ or $h(x) = c$. Therefore

$$f(z) = x^2 - y^2 + i(2xy + c)$$

$$\Rightarrow v(x) = (2xy + c)$$

is a conjugate harmonic function of u.

Exponential and logarithmic functions

$$e^z = e^{x+iy} = e^x(\cos y + i \sin y).$$

For $z \neq 0$,and $\theta = \arg z$

$$\ln z = \log_e |z| + i(\theta + 2n\pi), n = 0, \pm 1, \pm 2, \ldots.$$

Trignometric and hyperbolic functions

For any complex number $z = x + iy$

$$\sin z = \frac{e^{iz} - e^{-iz}}{2i} \text{ and } \cos z = \frac{e^{iz} + e^{-iz}}{2}$$

$$\sinh z = \frac{e^z - e^{-z}}{2} \text{ and } \cosh z = \frac{e^z + e^{-z}}{2}.$$

The inverse sine function, denoted by $\sin^{-1} z$ or $\arcsin z$, is defined by

$$\sinh^{-1} z = \ln[z + (z^2 + 1)^{\frac{1}{2}}]$$

$$\cosh^{-1} z = \ln[z + (z^2 - 1)^{\frac{1}{2}}].$$

8.3 Complex Integration

In complex variables, a piecewise smooth curve C is also called a **contour** or **path**. An integral of $f(z)$ on C is denoted by $\int_C f(z)dz$ or $\oint_C f(z)dz$. If the contour C is closed, it is referred to as a contour integral or a complex line integral. If f is continuous on a smooth curve C given by $z(t) = x(t) + iy(t),\ a \leq t \leq b$, then

$$\int_C f(z)dz = \int_a^b f(z(t))z'(t)dt. \tag{8.22}$$

Remark 68. If f is expressed in terms of the symbol z then to evaluate $f(z(t))$ we replace z by $z(t)$. If f is not expressed in terms of z, then to evaluate $f(z(t))$ we replace x and y wherever they appear by $x(t)$ and $y(t)$ respectively.

Example 276. (i) Evaluate $\int_C \bar{z}dz$, where C is given $x = 3t, y = t^2, -1 \leq t \leq 4$.

(ii) Evaluate $\int_C (z+3)dz$, where C is $x = 2t, y = 4t - 1, 1 \leq t \leq 3$.

(iii) Evaluate $\int_C z^2 dz$, where C is $z(t) = 3t + 2it, -2 \leq t \leq 2$.

Review of line integrals and independence of paths

The notion of the definite integral $\int_a^b f(x)dx$, that is, integration defined over an interval, can be generalized to integration of function defined along a curve. We require the following concepts. Suppose C is a curve parameterized by

$$x = f(t),\ y = g(t),\ a \leq t \leq b$$

and A and B are points $(f(a), g(a))$ and $(f(b), g(b))$ respectively.

A curve C is **smooth** if f' and g' are continuous on the closed interval $[a, b]$ and not simultaneously on the open interval (a, b), see Figure 8.11a.

C is **piecewise smooth** if it consists of a finite number of smooth curves $C_1, C_2, \ldots, C_n$ joined end to end, namely $C = C_1 \cup C_2 \cup C_3 \cup \ldots, C_n$ (see Figure 8.11b). C is closed if $A = B$, see Figure 8.12a.

C is a **simple closed curve** if $A = B$ and the curve does not cross itself, see Figure 8.12b.

If C is not a closed curve, then the positive direction on C is the direction corresponding to increasing values of t.

FIGURE 8.11a: Smooth Curve

Definition 94. Let $f(z)$ be a complex function defined along a curve C in the complex plane and let

$$F(z) = u(x, y) + iv(x, y)$$

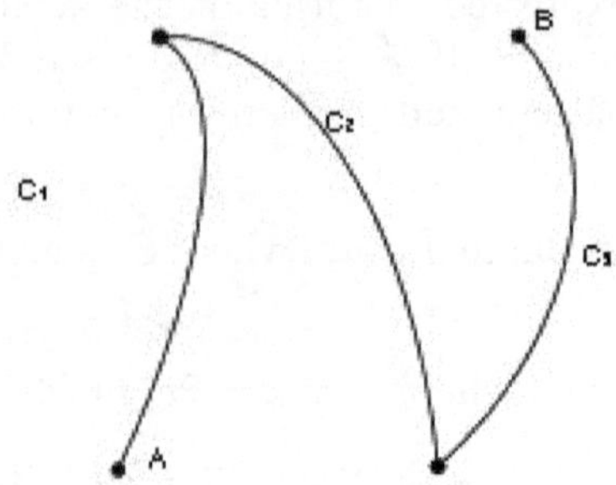

FIGURE 8.11b: Piecewise Smooth Curve

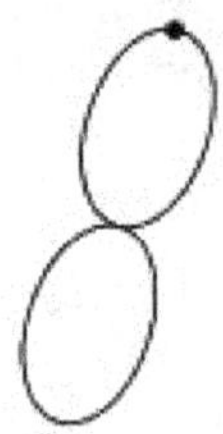

FIGURE 8.12a: Closed Piecewise Smooth Curve

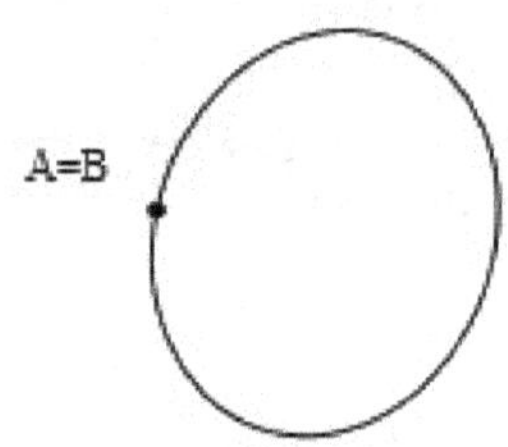

FIGURE 8.12b: Simple Closed Curve

and C be a smooth curve defined by

$$x = x(t),\ y = y(t),\ a \leq t \leq b.$$

Furthermore divide C into n sub arcs according to the partition

$$a = t_0 < t_1 < t_2 < \ < t_n = b$$

of $[a, b]$. The corresponding points on the curve C are

$$z_0 = x_0 + iy_0 = x_0(t) + iy_0(t),$$

$$z_1 = x_1 + iy_1 = x_1(t) + iy_1(t), \ldots, z_n(t) = x_n(t) + iy_n(t).$$

Let

$$\triangle z_k = z_k - z_{k-1}, k = 1, 2, \ldots, n.$$

Assume that $\|P\|$ denotes the norm of the partition, that is, $\|P\|$ = maximum of $|\triangle z_k|$. The contour integral or integral of $f(z)$ on C, denoted by $\int_C f(z)dz$ is defined by

$$\int_C f(z)dz = \lim_{\|P\|\to 0} \sum_{k=1}^{n} f(z_k^*)\triangle z_k.$$

Properties of contour integrals

Theorem 60. *Let f and g be continuous in a domain D and C be a smooth curve inside D. Then*

(a) $\int_C kf(z)dz = k\int_C f(z)dz$.

(b) $\int_C [f(z) + g(z)]dz = \int_C f(z)dz + \int_C g(z)dz$.

(c) *For $C = C_1 \cup C_2$, where C_1 and C_2 are smooth curve,* $\int_C f(z)dz = \int_{C_1} f(z)dz + \int_{C_2} f(z)dz$.

(d) $\int_{-C} f(z)dz = -\int_C f(z)dz$, *where $-C$ denotes the curve having opposite orientation of C.*

(e) *If f is continuous on C and $|f(z)| \leq M$ for all z on C, then* $|\int_C f(z)dz| \leq ML$, *where L is the length of C.*

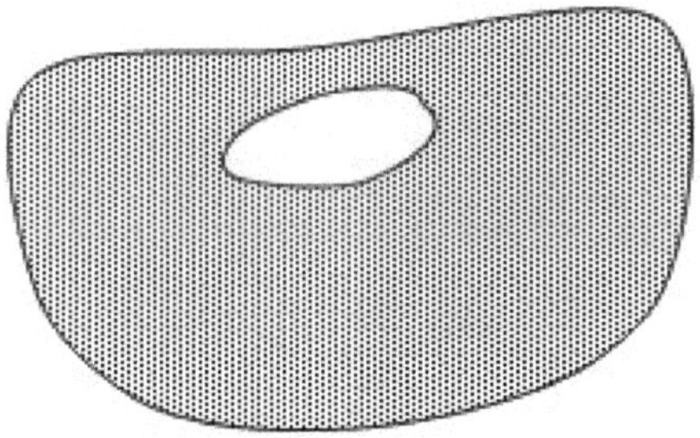

FIGURE 8.13a: Non Simply Connected Domain

We concentrate here on contour integrals, where the contour C is a simple closed curve. In this case the contour integral of $f(z)$ on C is denoted by $\oint_C f(z)dz$. A domain D, on which f is defined, is called **simply connected** (Figure 8.13a) if every simple closed curve C inside D can be shrunk to a point without leaving D. In other words a simply connected domain has no holes. A domain that is not simply connected is called a **multiply connected domain** (Figure 8.13b). A domain having one hole is called doubly connected while a domain with two holes is triply connected. It may be noted that the entire complex plane is simply connected.

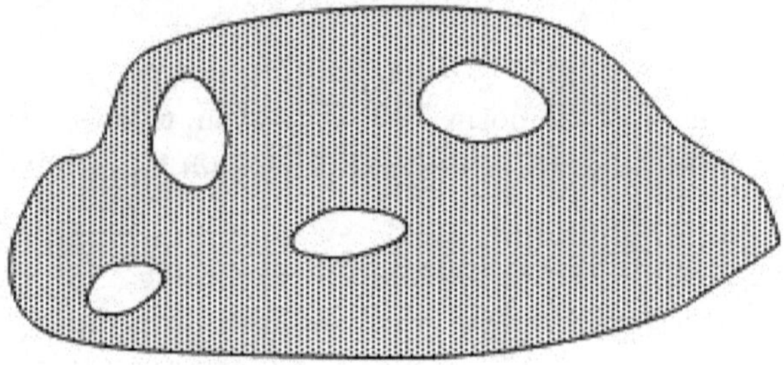

FIGURE 8.13b: Multiply Connected Domain

Theorem 61. *(Cauchy Theorem) Let f be as analytic function in a simply connected domain D and f' continuous in D. Then $\oint_C f(z)dz = 0$ for every simple closed contour C in D. See Figure 8.14.*
In 1883, another French mathematician named Èdovard Goursat generalized the Cauchy theorem by dropping continuity of f'.

Theorem 62. *(Goursat Theorem) Let f be analytic in a simply connected domain D. Then $\oint_C f(d)dz = 0$ for every simply closed contour C.*

It may be observed that neither the Goursat nor the Cauchy theorem can be extended for multiply connected domains in that form. However, the following result holds.

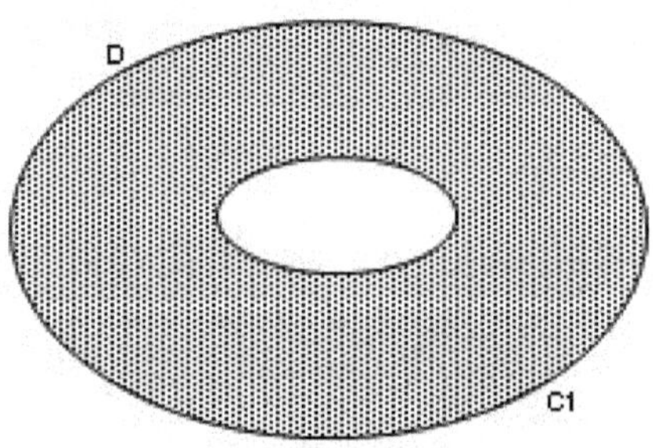

FIGURE 8.14: Proof of Cauchy Integral Formula

Theorem 63. *Let $f(z)$ be analytic in any domain $D1$ that contains D and its boundary curves, where D is a doubly connected domain with outer boundary C_1 and inner boundary C_2. Then*

$$\oint_{C_1} f(z)dz = \oint_{C_2} f(z)dz$$

where both integrals taken counterclockwise (or both taken clockwise).

Proof. *The proof is based on Green's theorem and the Cauchy-Reimann equation. Since f' is continuous throughout D, the real and imaginary parts of $f(z) = u(x,y) + iv(x,y)$ and their first derivatives are continuous on D. We have*

$$\begin{aligned}\oint_C f(z)dz &= \lim \sum (u+iv)(\triangle x + i\triangle y) \\ &= \lim \sum (u\triangle x - v\triangle y) + i\sum (v\triangle x + u\triangle y) \\ &= \oint_C u(x,y)dx - v(x,y)dy + i\oint_C v(x,y)dx + u(x,y)dy \\ &= \int\int_D (-\frac{\partial v}{\partial x} - \frac{\partial u}{\partial y})dA + i\int\int_D (\frac{\partial u}{\partial x} - \frac{\partial v}{\partial y})dA \qquad (8.23)\end{aligned}$$

by applying Green's theorem on each line integral. Since f is analytic, we have

$$\frac{\partial u}{\partial x} = \frac{\partial v}{\partial y}$$

and

$$\frac{\partial u}{\partial y} = -\frac{\partial v}{\partial x}$$

that clearly imply that both integrals on the right hand side of (8.23) are zero. Hence $\oint_C f(z)dz = 0$. □

Remark 69. It has been proved that analyticity implies path independence, namely the value of $\int_C f(z)dz$ is the same for every C if f is analytic in a simply connected domain.

Example 277. Evaluate the following integrals:

(a) $\oint_C e^z dz$; where C is a simple closed curve.

(b) $\oint_C e^{z^2} dz$; where C is a simple closed curve.

(c) $\oint_C (z^3 - 1 + 3i)dz$, where C is the unit circle $\mid z \mid = 1$.

Solution: (a) Since e^z is analytic in a domain containing C, where C is a simple closed curve, by Goursat's theorem

$$\oint_C e^z dz = 0.$$

(b) Since e^{z^2} is analytic in the domain containing a simple closed curve

$$\oint_C e^{z^2} dz = 0.$$

(c) $z^3 - 1 + 3i$ is an entire function being a polynomial.$\mid z \mid = 1$ is a simple closed curve (unit circle). Hence

$$\oint_{|z|=1} (z^3 - 1 + 3i)dz = 0.$$

Example 278. (a) Evaluate $\int_C 2zdz$, where C is the curve with initial point $z = -1$ and terminal point $z = -1 + i$.

(b) C is $z(t) = 2t^3 + i(t^4 - 4t^3 + 2), -1 \leq t \leq 1$.

Solution: (a) Since $f(z) = 2z$ is analytic in the entire complex plane, we can replace the curve C by any convenient curve C_1 joining $z = -1$ and $z = -1+i$. Let us choose C_1 to be the straight line segment $x = -1, 0 \leq y \leq 1$, shown in Figure 8.14 and we have $z = -1 + iy, dz = idy$. Therefore,

$$\int_c 2zdz = \int_{c_1} 2zdz = -2\int_0^1 ydy - 2i\int_0^1 dy = -1 - 2i\int_{-1}^{-1+i} 2zdz = -1 - 2i.$$

(b) The given integral is independent of the path and, therefore

$$\int_c 2zdz = \int_{-2+7i}^{2-i} 2zdz = [z^2]_{-2+7i}^{2-i} = 48 + 24i.$$

Example 279. Show that

$$\oint_C (z - z_0)^n dz = \begin{cases} 2\pi & \text{if } n = -1 \\ 0 & \text{if } n \neq -1, \end{cases}$$

where C is the circle with center z_0 and radius r (Figure 8.15).

Solution: C can be written in the form

$$z(t) = z_0 + r(\cos\theta + i\sin\theta) = z_0 + re^{i\theta}, \quad 0 \leq \theta \leq 2\pi.$$

Then

$$(z - z_0)^n = r^n e^{in\theta}$$

and

$$dz = ire^{i\theta}d\theta.$$

We get

$$\oint_C (z - z_0)^n dz = \int_0^{2\pi} r^n e^{in\theta} re^{i\theta} dt$$
$$= ir^{n+1}\int_0^{2\pi} e^{i(n+1)\theta}d\theta.$$

If we apply the Euler formula, the right hand side takes the form

$$ir^{n+1}\left(\int_0^{2\pi} \cos(n+1)\theta d\theta + i\int_0^{2\pi} \sin(n+1)\theta d\theta\right).$$

For $n = -1$,we have $r = 1, \cos 0 = 1, \sin 0 = 0$. Thus

$$\oint_c (z - z_0)^{-1} dz = 1\int_0^{2\pi} d\theta = 2\pi i.$$

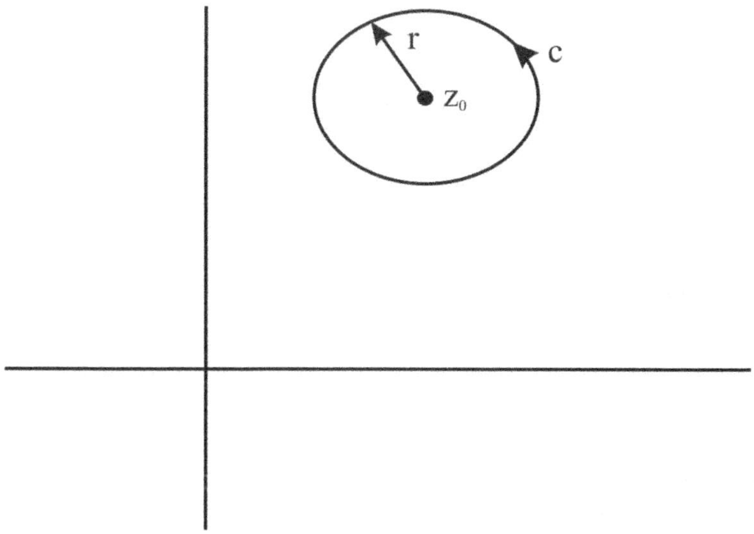

FIGURE 8.15: Path (Curve) of Example 279

For $n \neq -1$ each of the two integrals is zero as we integrate over an interval of length 2π, equal to a period of *sine* and *cosine*. Thus, we obtain the desired result.

Theorem 64. *(Cauchy's integral formula)*

(a) Let f be analytic in a simply connected domain D, and let C be a simple closed contour in D. If z_0 is any point of C, then the value of f at z_0 is given by

$$f(z_0) = \frac{1}{2\pi i} \oint_C \frac{f(z)}{z - z_0} dz. \tag{8.24}$$

(b) The value of nth derivative of f at z_0 is given as

$$f^{(n)}(z_0) = \frac{n!}{2\pi i} \oint_C \frac{f(z)}{(z - z_0)^{n+1}} dz. \tag{8.25}$$

(c) (Cauchy's inequality)

$$|f^{(n)}(z_0)| \leq \frac{n!M}{r^n}. \tag{8.26}$$

where $|z - z_0| = r$ and M is a real number such that $|f(z)| \leq M$ for all points z on C.

(d) (Liouville's theorem) The only bounded entire functions are constants.

(e) (Fundamental theorem of algebra) Let $P(z)$ be a non-constant polynomial, then equation $P(z) = 0$ has at least one root.

(f) *(Morera's theorem–Converse of Cauchy's Theorem) Let $f(z)$ be continuous in a simply connected domain D and if*

$$\oint_C f(z)dz = 0$$

for every closed contour C in D, then $f(z)$ is analytic in D.

Proof. (a) ***(Cauchy's integral formula)***
We can write

$$f(z) = f(z_0) + [f(z) - f(z_0)]$$

$$\oint_C \frac{f(z)}{z - z_0} dz = \oint_C \frac{f(z_0)}{z - z_0} dz + \oint_C \frac{[f(z) - f(z_0)]}{z - z_0} dz. \quad (8.27)$$

The first term of (8.27) is $f(z_0)2\pi i$ by Example 279. The theorem is proved if we show that the second term is zero. The integrand in the second term is analytic except at z_0. Replace C by a circle of radius r and centre z_0, see Figure 8.16. For $\varepsilon > 0$, we can find a $\delta > 0$ such that $|f(z) - f(z_0)| < \varepsilon$ for all

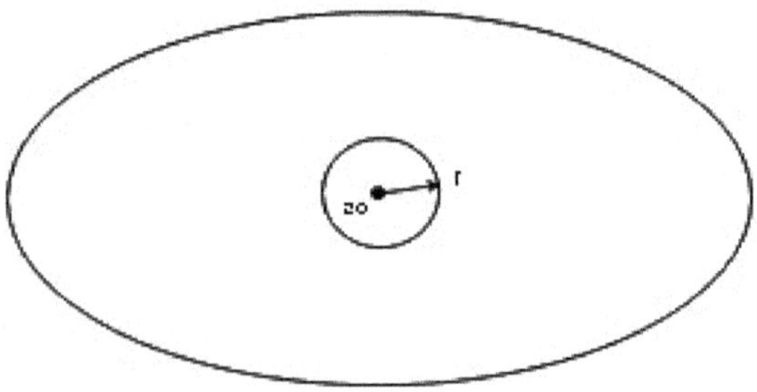

FIGURE 8.16: Proof of Cauchy's integral formula

z in the disk $|z - z_0| < \delta$ (since $f(z)$ is continuous as it is analytic). Choosing radius r smaller than ρ, we have

$$\left| \frac{f(z) - f(z_0)}{z - z_0} \right| < \frac{\varepsilon}{\rho}$$

at each point of the circle. The length of the circle is $2\pi\rho$. Hence by Theorem 60(e)

$$\left| \oint_{|z-z_0|=r} \frac{f(z) - f(z_0)}{z - z_0} dz \right| < \frac{\varepsilon}{\rho} 2\pi\rho = 2\pi\varepsilon.$$

This implies the second term of (8.27) is zero and the result is proved.

(b) The theorem is proved by the principle of mathematical induction. The statement is true for $n = 1$, that is,

$$f'(z_0) = \frac{1}{2\pi i}\oint_C \frac{f(z)}{(z-z_0)^2}dz \tag{8.28}$$

$$f'(z_0) = \lim_{\triangle z\to 0}\frac{f(z_0+\triangle z)-f(z_0)}{\triangle z} \tag{8.29}$$

$$= \lim_{\triangle z\to 0}\frac{1}{2\pi i\triangle z}\left[\oint_C \frac{f(z)}{z-(z_0+\triangle z)}dz - \oint_C \frac{f(z)}{z-z_0}dz\right]$$

$$\text{or} \quad f'(z_0) = \frac{1}{\lim_{\triangle z\to 0}}\frac{1}{2\pi i}\oint_c \frac{f(z)}{(z-z_0-\triangle z)(z-z_0)}dz. \tag{8.30}$$

Since f is continuous on C, it is bounded, that is, there exists a positive constant M such that $|f(z)| \leq M$ for all z on C. Further, let L be the length of C and let δ denote the shortest distance between points in C and the point z_0 implying

$$|z-z_0| \geq \delta$$

or

$$\frac{1}{|z-z_0|^2} \leq \frac{1}{\delta^2}$$

for all points z on C.
If we choose $|\triangle z| \leq \frac{\delta}{2}$, then

$$|z-z_0-\triangle z| \geq |\ |z-z_0| - |\triangle z| \geq \delta - |\triangle z|\ | \geq \frac{\delta}{2}$$

and so

$$\frac{1}{|\ z-z_0-\triangle z\ |} \leq \frac{2}{\delta}.$$

Now,

$$\left|\oint_C \frac{f(z)}{(z-z_0)^2}dz - \oint_C \frac{f(z)}{(z-z_0-\triangle z)(z-z_0)}dz\right|$$

$$= \left|\oint_C \frac{-\triangle z f(z)}{(z-z_0)^2(z-z_0-\triangle z)}dz\right| \leq \frac{2ML\ |\ \triangle z\ |}{\delta^3}. \tag{8.31}$$

Because the last expression approaches zero as $\triangle z \to 0$, we get

$$f'(z_0) = \lim_{\triangle z\to 0}\frac{f(z_0+\triangle z)-f(z_0)}{\triangle z} = \frac{1}{2\pi i}\oint_C \frac{f(z)}{(z-z_0)^2}dz \tag{8.32}$$

Assume that

$$f^{(n)}(z_0) = \frac{n}{2\pi i}\oint_C \frac{f(z)}{(z-z_0)^{n+1}}dz.$$

Then

$$\begin{aligned} f^{(n+1)}(z_0) &= \frac{(n+1)}{2\pi i}\oint_C \frac{f(z)}{(z-z_0)^{n+2}}dz \\ &= (n+1)\frac{n}{2\pi i}\oint_C \frac{f(z)}{(z-z_0)^{n+1}}\frac{1}{(z-z_0)}dz. \end{aligned}$$

(c) From (8.25), we get

$$| f^{(n)}(z) |= \frac{n!}{2\pi}\left|\oint_C \frac{f(z)}{(z-z_0)^{n+1}}dz\right| \le \frac{n}{2\pi}M\frac{1}{r^{n+1}}2\pi r$$

as $| f(z) |\le M$ *on* C. *Thus we have desired result*

$$| f^{(n)}(z_0) |\le \frac{nM}{r^n}.$$

(d) Suppose an entire function $f(z)$ *is bounded, say* $| f(z) |< k$ *for all* z. *Using the inequality of the part (c), we get*

$$| f'(z_0 |\le \frac{k}{r}.$$

Since $f(z)$ *is entire, this time for all* r*, normally we can choose* r *as large as we please so we conclude that* $f'(z) = 0$ *for all* z. *This means that* $f(z)$ *is constant.*
(e) Let $P(z) \neq 0$ *for all* z. *This implies that* $Q(z) = \frac{1}{P(z)}$ *is an entire function.* $| Q(z) |\to 0$ *as* $| z |\to \infty$ *implies* $Q(z)$ *is bounded for all finite* z. *Thus by part(d),* $Q(z)$ *is constant and so* $P(z)$ *is constant, thus our assumption leads to contradiction that* $P(z)$ *is non-constant. Hence* $P(z) = 0$ *for atleast one* z.
(f) Let

$$F(z) = \int_D f(z')dz'.$$

It can be checked that $F(z)$ *is analytic and by the part (b)* $F'(z)$ *is analytic and so* $f(z) = F'(z)$ *is analytic in* D. *This proves Morera's theorem.* □

Example 280. Evaluate the following integrals applying Cauchy's integral formula for derivatives and functions:

(a) $\oint_C \frac{z+1}{z^4+4z^3}dz$, where C is the circle $| z |= 1$.

(b) $\oint_C \frac{1+2e^z}{z}dz; \quad | z |= 1$.

(c) $\oint_C \frac{e^{z^3}}{(z-i)^3}dz$, where C is closed curve not passing through i.

Solution: (a)

$$f(z) = \frac{z+1}{z^4 + 4z^3}$$

is not analytic at $z = 0$ and $z = -4$ and only $z = 0$ lies in the contour $\mid z \mid = 1$. $f(z)$ can be written as

$$f(z) = \frac{\frac{z+1}{z+4}}{z^3}.$$

We can identify $z_0 = 0, n = 2$ and

$$f(z) = \frac{z+1}{z+4}.$$

By the quotient rule,

$$f''(z) = \frac{-6}{(z+4)^3}$$

and so by Cauchy's integral formula for derivatives we have

$$\oint_c \frac{z+1}{z^4 + 4z^3} dz = \frac{2\pi i}{2} f''(0) = -\frac{3\pi}{32} i.$$

(b) By Cauchy's integral formula for

$$f(z) = 1 + 2e^z,$$

$$\oint_c \frac{1 + 2e^z}{z} dz = 2\pi i(1 + 2e^0) = 6\pi i.$$

(c) If C does not enclose i then this integral is zero by Cauchy's theorem since the only point at which

$$f(z) = \frac{e^{z^3}}{(z-i)^3}$$

fails to be differentiable is i. Assume that C encloses i because the factor $z - i$ occurs to the third power in the denominator and use $n = 2$ in Cauchy's integral formula for derivatives with $f(z) = e^{z^3}$ to obtain

$$\oint \frac{e^{z^3}}{(z-1)^3} dz = \frac{2\pi i(n)}{2} f(i) = \pi i f''(i).$$

We have

$$f'(z) = 3z^2 e^{z^3}$$

and

$$f''(z) = 6ze^{z^3} + 9z^4 e^{z^3}$$

so

$$\oint_c \frac{e^{z^3}}{(z-i)^3} dz = \pi i[6ie^{-i} + 9e^{-i}] = (-6 + 9i)\pi e^{-i}.$$

8.4 Residues and Residue Theorem

8.4.1 Introduction

A function of complex variable $f(z)$ has two kinds of representation in series of powers of $(z - z_0)$. If $f(z)$ is differentiable at z_0 then expansion of $f(z)$ will contain only non-negative integer powers of $(z - z_0)$ and hence is a power series. If $f(z)$ is not differentiable at z_0 then the expansion of $f(z)$ will contain negative and positive powers of $(z - z_0)$. See Figure 8.17. The expansion is called the **Laurent expansion** or **Laurent series**. The open set of points between two concentric circles is called an **annulus**:
bounding the open annulus $r <| z - z_0 |< R$. Annulus is described by inequal-

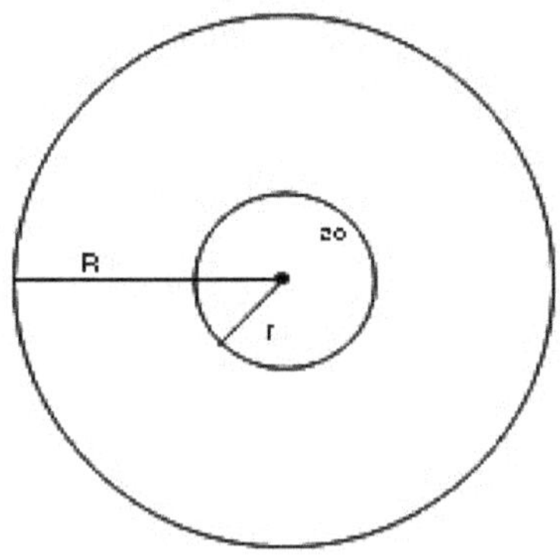

FIGURE 8.17: Circles $| z - z_0 |= r$ and $| z - z_0 |= R$

ities.

$$r <| z - z_0 |< R,$$

where r is the radius of the inner circle and R is the radius of the outer circle. If $r = 0$ the annulus is a punctured disk (open disk with the center removed). $0 <| z - z_0 |< \infty$, represents the entire complex plane except z_0 and $r <| z - z_0 |< R$, contains all points outside the inner circle of radius r. The notion of Laurent series leads to concept of a residue that in turn provides another way to evaluate complex and real integrals.
Taylor's theorem states that

$$f(z) = \sum_{k=-\infty}^{\infty} a_k \frac{f^k(z_0)}{k}(z - z_0)^k,$$

where $f(z)$ is analytic within a domain D and z_0 is a point in D for all z belonging to the largest circle C with the center at z_0 and radius R that lies entirely within D.

Laurent expansion

Let f be analytic within an annulus D defined by $r <| z - z_0 |< R$. Then f is represented by the series

$$f(z) = \sum_{k=-\infty}^{\infty} a_k(z - z_0)^k \tag{8.33}$$

valid for $r <| z - z_0 |< R$. The coefficients a_k are

$$a_k = \frac{1}{2\pi i} \oint_C \frac{f(t)}{(t - z_0)^{k+1}} dt, k = 0, \pm 1, \pm 2, \ldots \tag{8.34}$$

where C is a simple closed curve that lies entirely within D and has z_0 in its interior, see Figure 8.18

It may be noted that Laurent expansion is a generalization of Taylor's series as $a_{-k} - 0$ for $k = 1, 2, 3,.$ The Laurent expansion yields Taylor's series.

Remark 70. Finding the coefficient a_k of a Laurent expansion using (8.34) is not an easy task. Very often geometric series

$$\frac{1}{1 - z} = 1 + z + z^2 + z^3 + \ldots$$

and

$$\frac{1}{1 + z} = 1 - z + z^2 - z^3 + \ldots$$

for $| z |< 1$ are helpful.

8.4.2 Singularities and Residues

If a complex function $f(z)$ is not analytic at $z = z_0$ then z_0 is called a **singularity** or a **singular point** of $f(z)$. A singularity z_0 of $f(z)$ is called an **isolated singularity** if there exists a punctured disk (open disk), $0 <| z - z_0 |< R$ of z_0 throughout which $f(z)$ is analytic. The Laurent expansion of $f(z)$ can be written as

$$f(z) = \sum_{k=1}^{\infty} a_{-k}(z - z_0)^{-k} + \sum_{k=0}^{\infty} a_k(z - z_0)^k. \tag{8.35}$$

The first term in (8.35),

$$\sum_{k=1}^{\infty} \frac{a_{-k}}{(z - z_0)^k}$$

is a principal part of the Laurent expansion and will converge for

$$\left| \frac{1}{(z - z_0)} \right| < r^n$$

or for

$$| z - z_0 | > \frac{1}{r^n} = r.$$

The second term, $\sum_{k=0}^{\infty} a_k(z - z_0)^k$ in (8.35) is called the **analytic part** of the Laurent expansion and it converges for $| z - z_0 | < R$. Thus, the sum of these two parts converges, if $r < | z - z_0 |$ and $| z - z_0 | < R$, that is, the Laurent expansion converges in the annulus: $r < | z - z_0 | < R$.

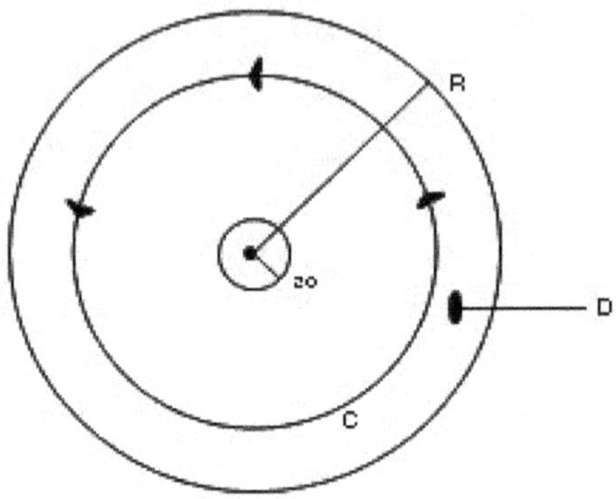

FIGURE 8.18: Closed Simple Curve Enclosing z_0 and Lying in Annulus $r < | z - z_0 | < R$

Example 281. Examine the nature of singularities of

$$f(z) = \frac{z}{z^2 + 9}.$$

Solutions: $z = 3i$ and $z = -3i$ are singularities of the function

$$f(z) = \frac{z}{z^2 + 9}.$$

Both are isolated singularities since f is analytic at every point in the neighborhood (open disk) $0 < | z - 3i | < 1$ and $0 < | z + 3i | < 1$.

Classification of isolated singular points

An isolated singularity $z = z_0$ of $f(z)$ is called a **removable singularity** if the principal part in the Laurent expansion is zero, that is, all coefficients a_{-k} in (8.35) are zero. An isolated singularity $z = z_0$ is called a **pole** if the principal part in the Laurent expansion contains a finite number of non-zero terms. It is called a **pole of order** n if it has n non-zero terms.
If $n = 1$,that is, a_{-1} is the only non-zero term then $z = z_0$ is called a **simple pole**. $z = z_0$ is called an **essential singularity** if the principal part has infinite non-zero terms. The forms of Laurent series (expansion) about $z = z_0$ are

1. If $z = z_0$ is a removable singularity then the Laurent expansion takes the form $a_0 + a_1(z - z_0) + a_2(z - z_0)^2 + \ldots$

2. If $z = z_0$ is a pole of order n then the Laurent series takes the form

$$\sum_{k=1}^{n} a_{-k}\frac{1}{(z-z_0)^k} + \sum_{k=0}^{\infty} a_k(z-z_0)^k$$

$$\frac{a_{-n}}{(z-z_0)^n} + \frac{a_{-(n-1)}}{(z-z_0)^{n-1}} + \cdots + \frac{a-1}{z-z_0} + a_0 + a_1(z-z_0) + \ldots$$

3. If $z = z_0$ is an essential singularity then the Laurent series takes the form

$$\cdots + \frac{a_{-2}}{(z-z_0)^2} + \frac{a_{-1}}{z-z_0} + a_0 + a_1(z-z_0) + a_2(z-z_0)^2 + \ldots$$

Let us recall that z_0 is a zero of a function f if $f(z_0) = 0$. An analytic function f has a zero of order n at $z = z_0$ if $f(z_0) = 0, f'(z_0) = 0, f''(z_0) = 0, \ldots f^{n-1}(z_0) = 0$ but $f^n(z_0) \neq 0$.

Example 282. (a) (i) Show that $\frac{\sin z}{z}$ has a removal singularity at $z = 0$.

(ii) Show that $f(z) = \frac{\sin z}{z^2}$ has a simple pole at $z = 0$.

(b) Show that $z = 0$ is a removal singularity of

$$f(z) = \frac{e^{2z} - 1}{z}.$$

(c) Determine the zeros and their order for the functions

(i) $f(z) = (z + 2 - i)^2$

(ii) $f(z) = z + \frac{9}{z}$

(iii) $f(z) = e^{2z} - e^z$

Solutions: (a) (i) We know that

$$\sin z = z - \frac{z^3}{3!} + \frac{z^5}{5!} + \ldots$$

or

$$\frac{\sin z}{z} = 1 - \frac{z^2}{3!} + \frac{z^4}{5!} \cdots$$

The right hand side is of the form

$$a_0 + a_1(z - z_0) + a_2(z - z_0) + \cdots\cdots \text{ for } z_0 = 0.$$

Hence $z = 0$ is a removal singularity of $\frac{\sin z}{z}$.
(ii) We have

$$\frac{\sin z}{z^2} = \frac{1}{z} - \frac{z}{3!} + \frac{z^3}{5!} \cdots$$

Since $0 <\mid z \mid$, we see that $a_{-1} = \frac{1}{z} \neq 0$, and so $z = 0$ is a simple pole of the function $f(z) = \frac{\sin z}{z^2}$.
(b) Since

$$e^{2z} = 1 + \frac{2z}{1!} + \frac{2^2 z^2}{2!} + \frac{2^3 z^3}{3!} + \dots,$$

we get

$$\frac{e^{2z} - 1}{z} = \frac{(1 + \frac{2}{1}z + \frac{2^2}{2}z^2 + \frac{2^3}{3}z^3 + ..) - 1}{z}$$
$$= \frac{2}{1} + \frac{2^2}{2}z + \frac{2^3}{3}z^2 + \dots$$

$z = 0$ is a removal singularity of the function $f(z) = \frac{e^{2z} - 1}{z}$ as it is of the form

$$a_0 + a_1(z - z_0) + a_2(z - z_0) + \dots$$

for $z = 0$.
(c) (i) $f(z) = (z + 2 - i)^2$ has a zero at $z = -2 + i$. It is of order 2 as

$$\begin{aligned} f'(z) &= 2(z + 2 - i) \\ =) &= 0 \text{ at } z = -2 + i \end{aligned}$$

but $f''(z) \neq 0$ for $z = -2 + i$. Hence $z = -2 + i$ is a zero of order 2.
(ii) We have

$$f(z) = \frac{z^2 + 9}{z} = \frac{(z + 3i)(z - 3i)}{z}$$
$$f'(z) = \frac{2z^2 - (z^2 + 9)}{z^2} = \frac{z^2 - 9}{z^2} = 1 - \frac{9}{z^2}.$$

Since $f(3i) = 0$ and $f(-3i) = 0$, $z = 3i$ and $z = -3i$ are zeros of the given function. The order of each of these zeros is 1 as $f'(3i) = 2 \neq 0$ and $f'(-3i) \neq 2$.
(iii)

$$f(z) = e^z(e^z - 1)$$

and

$$z = 2\pi n i, n = 0, \pm 1, \pm 2, \pm 3 \dots$$

are zeroes of $f(z)$. Since

$$f'(z) = 2e^{2z} - e^{z}, f(2\pi ni) = 2e^{4\pi ni} - e^{2\pi ni} = 1 \neq 0,$$

by the definition each of

$$z = 2\pi ni, n = 0, \pm 1, \pm 2, \pm 3 \ldots$$

is a zero of order 1.

Definition 95. (residue) The coefficient a_{-1} of $\dfrac{1}{(z-z_0)}$ in the Laurent series expansion of $f(z)$

$$\begin{aligned} f(z) &= \sum_{k=-\infty}^{\infty} a_k (z-z_0)^k \\ &= \cdots + \frac{a_{-2}}{(z-z_0)^2} + \frac{a_{-1}}{(z-z_0)} + a_0 + a_1(z-z_0) + a_2(z-z_0)^2 + \ldots. \end{aligned}$$

where f has removable singularities at the point z_0 is called the **residue** of f at z_0 and is often denoted by $Res(f(z), z_0)$.

The following theorems provide techniques for computing residues for a simple pole and pole of order n.

Theorem 65. *(a) If f has a simple pole at $z = z_0$ then*

$$Res(f(z), z_0) = \lim_{z \to z_0} (z - z_0) f(z). \tag{8.36}$$

(b) If $f(z) = \dfrac{g(z)}{h(z)}$ where g and h are analytic at $z = z_0$ if $g(z_0) \neq 0$ and if the function h has a zero of order 1, at z_0, then f has a simple pole at $z = z_0$ and

$$Res(f(z), z_0) = \frac{g(z_0)}{h'(z_0)}.$$

Theorem 66. *If f has a pole of order n at $z = z_0$, then*

$$Res(f(z), z_0) = \frac{1}{(n-1)} \lim_{z \to z_0} \frac{d^{n-1}}{dz^{n-1}} (z - z_0)^n f(z). \tag{8.37}$$

Proof. (a) For a simple pole $z = z_0$ the Laurent expansion of $f(z)$ about z_0 is given by

$$f(z) = \frac{a_{-1}}{(z-z_0)} + a_0 + a_1(z-z_0) + a_2(z-z_0)^2 + \cdots$$

Multiplying both sides by $(z - z_0)$ and taking the limit

$$\begin{aligned} \lim_{z \to z_0} (z - z_0) f(z) &= \lim_{z \to z_0} [a_{-1} + a_0(z-z_0) + a_1(z-z_0)^2 + a_2(z-z_0)^3 + \ldots] \\ &= Res(f(z), z_0). \end{aligned}$$

(b) By part (a) we have

$$\lim_{z\to z_0}(z-z_0)f(z) = Res(f(z), z_0)$$

and by the assumption on $h(z)$, *we see that* $h(z) = 0$ *and*

$$\lim_{z\to z_0}\frac{h(z)-h(z_0)}{z-z_0} = h'(z_0)$$

$$Res(f(z), z_0) = \lim_{z\to z_0}(z-z_0)\frac{g(z)}{h(z)} = \lim_{z\to z_0}\frac{g(z)}{\dfrac{h(z)-h(z_0)}{z-z_0}} = \frac{g(z_0)}{h'(z_0)}.$$

Since f *has a pole of order* n, *its Laurent expansion for* $0 <| z - z_0 |< R$ *is of the form*

$$f(z) = \frac{a_{-n}}{(z-z_0)^n} + \cdots + \frac{a_{-2}}{(z-z_0)^2} + \frac{a_{-1}}{z-z_0} + a_0 + a_1(z-z_0) + \ldots.$$

Multiplying the last expression by $(z-z_0)^n$ *and differentiating the result* $(n-1)$ *times we get*

$$\frac{d^{n-1}}{dz^{n-1}}(z-z_0)^n f(z) = (n-1)a_{-1} + na_0(z-z_0) + \ldots.$$

By taking the limit as $z \to z_0$ *of both sides we get*

$$\lim_{z\to z_0}\frac{d^{n-1}}{dz^{n-1}}(z-z_0)^n f(z) = (n-1)a_{-1}$$

as all terms on the right hand side will be zero except the first term. This gives us

$$a_{-1} = Res(f(z), z_0) = \frac{1}{(n-1)}\lim_{z\to z_0}\frac{d^{n-1}}{dz^{n-1}}(z-z_0)^n f(z).$$

This proves the theorem. □

Theorem 67. *(Residue theorem) Let* $f(z)$ *be analytic inside a simple closed curve* C *and on* C, *except for finitely many singularities* $z_1, z_2, z_3 \ldots, z_n$ *inside* C. *Then the integrals* $f(z)$ *taken counterclockwise around* C *equal* $2\pi i$ *times the sum of the residues of f(z) at* $z_1, z_2, z_3 \ldots, z_n$, *namely*

$$\oint_C f(z)dz = 2\pi i \sum_{k=1}^{n} Res(f(z), z_k).$$

Proof. *We enclose each of the singularities* z_k *in a circle* C_k *with radius small enough that those* n *circles and* C *are all separated (see Figure 8.19).*
Then $f(z)$ *is analytic in the multiply connected domain* D *banded by* C *and*

$C_1, C_2, \ldots, C_n$ and the entire boundary of D. By Cauchy's integral theorem we get

$$\oint_C f(z)dz + \oint_{C_1} f(z)dz + \oint_{C_2} f(z)dz + \cdots + \oint_{C_n} f(z)dz = 0$$

where the integral on C is taken counterclockwise and other integrals clockwise. By taking integrals on $C_1, C_2, \ldots, C_n$ also counterclockwise we get

$$\oint_C f(z)dz = \oint_{C_1} f(z)dz + \oint_{C_2} f(z)dz + \cdots + \oint_{C_n} f(z)dz. \tag{8.38}$$

By the Laurent expansion (8.35) and Cauchy's integral formula (8.34), residue $f(z)$

$$z = z_1 = a_{-1} = \frac{1}{2\pi i}\oint_{C_1} f(z)dz = Res(f(z), z_1)$$

or

$$\oint_C f(z)dz = 2\pi i Res(f(z), z_k), \quad k = 1, 2, \ldots, n. \tag{8.39}$$

By (8.38) and (8.39), we get the desired result

$$\oint_{C_k} f(z)dz = 2\pi i \sum_{k=1}^{n} Res(f(z), z_k), \quad k = 1, 2, \ldots, n. \tag{8.40}$$

□

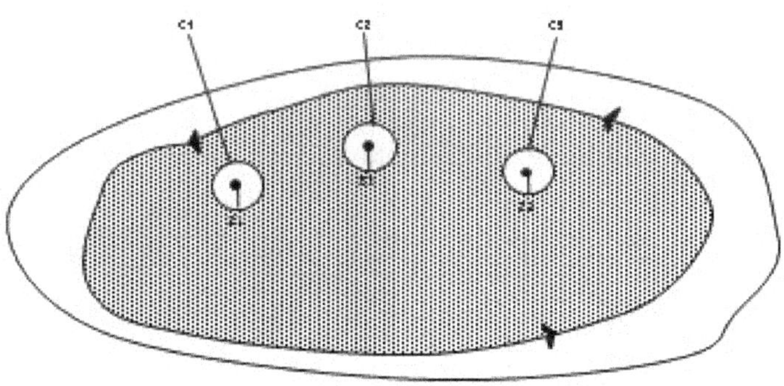

FIGURE 8.19: Residue Theorem

Remark 71. The evaluation of integrals using the residue theorem depends on the determination of residues at singular points. The residue theorem should be used to evaluate integrals of those functions and finding those residues at singular points is not tedious.

Example 283. (a) Find the residue of $f(z) = \dfrac{1}{2(z-1)^2}$ at $z = 1$.

(b) Find the residue of $f(z) = e^{3/z}$. at $z = 0$.

(c) Find the residue of $f(z) = \dfrac{z-6i}{(z-2)^2(z+41)}$ at $z = -4i$.

(d) Find the residue of $f(z) = \dfrac{4iz-1}{\sin z}$ at $z = \pi$.

Solution: (a) The Laurent expansion of $f(z)$ is

$$f(z) = \frac{1}{2(z-1)^2} - \frac{1}{4(z-1)} - \frac{1}{8} - \frac{z-1}{16}\cdots$$

The principal part is

$$\frac{1}{2(z-1)^2} - \frac{1}{4(z-1)},$$

implying a_{-1} is $\frac{-1}{4}$. Hence

$$Res(f(z),1) = \frac{-1}{4}.$$

(b) The Laurent expansion of

$$f(z) = e^{\frac{3}{z}}$$

is

$$e^{\frac{3}{z}} = 1 + \frac{3}{z} + \frac{3^2}{2z^2} + \frac{3^3}{3z^3} + \ldots$$

$a_{-1} = \text{Residue}(f(z),0) = 3$.
(c) The given function has a simple pole at $z = -4i$. By Theorem 65(a)

$$Res(f(z),-4i) = \lim_{z\to-4i}(z+4i) = \lim_{z\to-4i}\frac{3-6i}{(z-2)^2} = \frac{-4i-6i}{(-4i-2)^2} = -\frac{2}{5} + \frac{3}{10}i.$$

(d) By Theorem 65(b)

$$Res(f,z_0) = \frac{h(z_0)}{g'(z_0)}$$

where $h(z) = 4iz - 1, g(z) = \sin z, z_0 = \pi$. Hence

$$Res(f,\pi) = \frac{4i\pi - 1}{\cos\pi} = 1 - 4\pi i.$$

Example 284. (a) Find $Res(f(z),1)$, where

$$f(z) = \frac{2}{(z-1)(z+4)}.$$

(b) Find $Res(f(z), 0)$, where

$$f(z) = e^{-2/z^2}.$$

Solution: (a) The Laurent expansion of the given function at $z = 1$ is

$$f(z) = \frac{2}{5(z-1)} \frac{1}{1 + \dfrac{z-1}{5}} = \frac{2}{5(z-1)}(1 - \frac{z-1}{5} + \frac{(z-1)^2}{5^2} - \frac{(z-1)^3}{5^3} + \ldots$$

$$= \frac{2}{5} \frac{1}{(z-1)} - \frac{2}{25} + \frac{2(z-1)}{5^3} - \frac{2(z-1)^2}{54}.$$

$Res(f(z), 1) = a_{-1} = \dfrac{2}{5}$.
(b) Laurent expansion of $f(z)$ at $z = 0$ is

$$f(z) = \cdots - \frac{2^3}{3z^6} + \frac{2^2}{2z^4} - \frac{2}{1z^2} + 1 \ldots.$$

implying $Res(f(z), 0) = 0$.

Example 285. Evaluate by the residue theorem

$$\oint_C \frac{e^z}{z^4 + 5z^3} dz,$$

where C is the circle $\mid z \mid = 2$.

Solution: Since $z^4 + 5z^3 = z^3(z+5)$, the given function

$$f(z) = \frac{e^z}{z^4 + 5z^3}$$

has singularities at $z = 0$ and $z = -5$. Only $z = 0$ lies inside the given circle so

$$\oint_C f(z) dz = \oint_C \frac{e^z}{z^4 + 5z^3} dz = 2\pi i Res(f(z), 0)$$

by the residue theorem (Theorem 67). By Theorem 66,

$$Res(f(z), 0) = \frac{1}{2} \lim_{z \to 0} \frac{d^2}{dz^2} z^3 \frac{e^3}{z^4 + 5z^3}$$

or

$$Res(f(z), 0) = \frac{1}{2} \lim_{z \to 0} \frac{(z^2 + 8z + 17)e^z}{(z+5)^3} = \frac{1}{2} \frac{17\pi}{125}.$$

Therefore,

$$\oint_C f(z) dz = 2\pi \frac{1}{2} \frac{1}{2} \frac{17\pi}{125} = \frac{17}{125} i.$$

Example 286. Find the contour integral of terms over the circle $\mid z \mid = 2$.

(i) $\tan z = \dfrac{\sin z}{\cos z}$.

(ii) $\dfrac{1}{(z-1)(z+2)^2}$.

Solution: (i) The integrand

$$\tan z = \frac{\sin z}{\cos z}$$

has simple poles at the points where $\cos z = 0$. The only zeroes for $\cos z$ are real numbers

$$z = (2n+1)\frac{\pi}{2}, n = 0, \pm 1, \pm 2, \pm 3, \ldots.$$

and $-\frac{\pi}{2}$ and $\frac{\pi}{2}$ are in the given circle $|\, z \,| = 2$. Thus

$$\oint_{|z|=2} \tan z dz = 2\pi i (Res(f(z), -\frac{\pi}{2}) + Res(f(z), \frac{\pi}{2})).$$

By Theorem 65(b), for $h(z) = \cos z, g(z) = \sin z$,and $h'(z) = -\sin z$,

$$Res(f(z), \frac{\pi}{2}) = \frac{\sin\frac{\pi}{2}}{-\sin\frac{\pi}{2}} = -1$$

and

$$Res(f(z), \frac{-\pi}{2}) = \frac{\sin\frac{-\pi}{2}}{-\sin\frac{-\pi}{2}} = -1.$$

Hence

$$\oint_{|z|=2} \tan z dz = 2\pi i(-1-1) = -4\pi i.$$

(ii)

$$\oint_C \frac{1}{(z-1)(z+2)^2} dz = 2\pi i [Res(f(z), 1) + Res(f(z), -2)]$$

by the residue theorem

$$Res(f(z), 1) = \frac{1}{9}, Res(f(z), -2) = -\frac{1}{9}.$$

Thus

$$\oint \frac{1}{(z-1)(z+2)^2} dz = 2\pi i (\frac{1}{9} - \frac{1}{9}) = 0.$$

Example 287. Evaluate

$$\oint_C \frac{z+1}{z^2(z+2)^2} dz$$

using the residue theorem; C in the circle $|\, z \,| = 3$.

Solution:
$$\oint \frac{z+1}{z^2(z+2)^2}dz = 2\pi i Res((f(z),2i) + 2\pi i Res(f(z),0)$$
$$= 2\pi i[(-\frac{1}{4} - \frac{1}{2}i) + (\frac{1}{4} + \frac{1}{2}i)] = 0.$$

Example 288. Evaluate
$$\oint_C \frac{z+1}{z^2(z-2i)}dz$$
by applying the residue theorem over the circle C: $\mid z - 2i \mid= 1$.

Solution: $z = 0$ and $z = 2i$ are singular points but $z = 0$ does not lie inside $\mid z - 2i \mid$. $z = 2i$ is a simple pole.

$$Res(f(z), 2i) = L \lim_{z \to 2i} (z - 2i)\frac{z+1}{z^2(z-2i)}$$
$$= \frac{2i+1}{-4} = -\frac{1}{4} - \frac{1}{2}i$$

$$\oint f(z)dz = 2\pi i(-\frac{1}{4} - \frac{1}{2}i)$$
$$= \pi(1 - \frac{1}{2}i).$$

8.5 Application of Residue Theory for Evaluation of Real Integrals

We use here the residue theory discussed in Section 8.5 to evaluate integrals of the type

(a) $\int_0^{2\pi} F(\cos\theta, \sin\theta)d\theta$

(b) $\int_{-\infty}^{\infty} f(x)dx$

(c) $\int_{-\infty}^{\infty} f(x) \cos dx dx$ and $\int_{-\infty}^{\infty} f(x) \sin dx dx$

8.5.1 Evaluation of Integrals Involving Trigonometric Functions

We want to evaluate integrals of the type

$$\int_0^{2\pi} F(\cos\theta, \sin\theta)d\theta. \tag{8.41}$$

First we convert the formula into a complex integral over the unit circle center at the origin. The unit circle can be expressed in the form

$$z - \cos\theta + i\sin\theta = e^{i\theta}, \quad 0 \le \theta \le 2\pi.$$

Applying $dz = ie^{i\theta}d\theta$,

$$\cos\theta = \frac{e^{i\theta} + e^{-i\theta}}{2}, \quad \sin\theta = \frac{e^{i\theta} - e^{-i\theta}}{2i},$$

we replace $d\theta, \cos\theta$ and $\sin\theta$ by

$$d\theta = \frac{dz}{iz}, \cos\theta = \frac{1}{2}(z + z^{-1}), \sin\theta = \frac{1}{2i}(z - z^{-1}).$$

The integral (8.41) takes the form

$$\int_0^{2\pi} F(\cos\theta, \sin\theta)d\theta = \oint_{|z|=1} F(\frac{1}{2}(z + z^{-1})), \frac{1}{2i}(z - z^{-1})\frac{dz}{iz}.$$

Finding the real integral is equivalent to evaluating the integral

$$\oint_{|z|=1} f(z)dz, \text{ where } f(z) = F(\frac{1}{2}(z + z^{-1}), \frac{1}{2i}(z - z^{-1}))\frac{1}{iz}dz.$$

By the residue theorem,

$$\oint_{|z|=1} f(z)dz = 2\pi i \sum_p Res(f(z), p). \tag{8.42}$$

The sum on the right hand side is taken over all of the poles p of $f(z)$ enclosed by the unit circle.

Example 289. Evaluate

$$\int_0^{2\pi} \frac{1}{10 - 6\cos\theta} d\theta$$

using the residue theorem.

Solution:
$$\begin{aligned}\int_0^{2\pi}\frac{1}{10-6\cos\theta}d\theta &= \oint_{|z|}\frac{1}{10-\frac{6}{2}(z+z^{-1})}\frac{dz}{iz}\\ &= \oint_{|z|=1}\frac{dz}{10-3z+3z^{-1}}\frac{1}{iz}\\ &= \frac{1}{i}\oint_{|z|=1}\frac{dz}{10z-3z^2+3}\\ &= \frac{1}{2}\left(-\frac{2}{i}\right)\oint_{|z|=1}\frac{dz}{3z^2-10z-1}\\ &= \frac{1}{2}\left(-\frac{2}{i}\right)\oint_{|z|=1}\frac{dz}{(3z-1)(z-3)}.\end{aligned}$$

Note that $z=\frac{1}{3}$ and $z=3$ are singular points but $z=3$ does not lie inside the unit circle. Therefore by the residue theorem

$$\begin{aligned}\oint_{|z|=1}\frac{dz}{(3z-1)(z-3)} &= 2\pi i Res(f(z),\frac{1}{3})\\ &= 2\pi i\lim_{z\to 1/3}(z-\frac{1}{3})\frac{1}{(3z-1)(z-3)}\\ &= 2\pi i\frac{1}{3}\lim_{z\to 1/3}\frac{1}{(z-3)}\\ &= 2\pi i\frac{1}{3}(-\frac{3}{8})=-\frac{\pi}{4}i.\end{aligned}$$

Therefore,

$$\int_0^{2\pi}\frac{1}{10-6\cos\theta}d\theta=\frac{\pi}{4}.$$

Example 290. Evaluate

$$\int_0^{\pi}\frac{1}{\lambda+\mu\cos\theta}d\theta$$

where $0<\mu<\lambda$.

Solution: $\displaystyle\int_0^{2\pi}\frac{1}{\lambda+\mu\cos\theta}d\theta=\int_0^{\pi}\frac{1}{\lambda+\mu\cos\theta}d\theta+\int_{\pi}^{2\pi}\frac{1}{\lambda+\mu\cos\theta}d\theta.$

Let $w=2\pi-\theta$ in the second integral on the right hand side to obtain

$$\begin{aligned}\int_{\pi}^{2\pi}\frac{1}{\lambda+\mu\cos\theta}d\theta &= \int_{\pi}^{0}\frac{1}{\lambda+\mu\cos(2\pi-w)}(-1)dw\\ &= \int_0^{\pi}\frac{1}{\lambda+\mu\cos w}dw.\end{aligned}$$

Therefore,

$$\int_0^{2\pi}\frac{1}{\lambda+\mu\cos\theta}d\theta=2\int_0^{\pi}\frac{1}{\lambda+\mu\cos\theta}d\theta$$

$$\int_0^{\pi} \frac{1}{\lambda + \mu \cos\theta} d\theta = \frac{1}{2} \int_0^{2\pi} \frac{1}{\lambda + \mu \cos\theta} d\theta$$

or

$$\int_0^{2\pi} \frac{1}{\lambda + \mu \cos\theta} d\theta = \int_{|z|=1} f(z) dz,$$

where

$$f(z) = \frac{1}{\lambda + \frac{\beta}{2}(z + \frac{1}{2})} \frac{1}{iz} = -\frac{2i}{\mu z^2 + 2\lambda z + \mu}.$$

$f(z)$ has simple poles at

$$z = \frac{-\lambda \pm \sqrt{\lambda^2 - \mu^2}}{\mu}.$$

Since $\lambda > \mu$, these numbers are real. Only one of them,

$$z_1 = \frac{-\lambda + \sqrt{\lambda^2 - \mu^2}}{\mu},$$

is enclosed by the unit circle $| z |= 1$. The other is outside the unit circle and is irrelevant for our purpose. Then

$$\begin{aligned} \int_0^{\pi} \frac{1}{\lambda + \mu \cos\theta} d\theta &= \frac{1}{2} \int_0^{2\pi} \frac{1}{\lambda + \mu \cos\theta} d\theta \\ &= \frac{1}{2} 2\pi i (Res(f, z_1)) \\ &= \pi i \frac{-2i}{2\mu z_1 + 2\lambda} = \frac{\pi}{\sqrt{\lambda^2 - \mu^2}}. \end{aligned}$$

Example 291. Evaluate

$$\int_0^{2\pi} \frac{\sin^2\theta}{5 + 4\cos\theta} d\theta.$$

Solution: As discussed above,

$$\begin{aligned} \int_0^{2\pi} \frac{\sin^2\theta}{5 + 4\cos\theta} d\theta &= -\frac{1}{4i} \oint_{|z=1} \frac{(z^2-1)^2}{z^2(2z^2 + 5z + 2)} dz \\ &= -\frac{1}{4i} 2\pi i [Res(f(z), 0) + Res(f(z), -\frac{1}{2})] \\ &= \frac{\pi}{4}. \end{aligned}$$

8.5.2 Evaluation of Several Integrals

$$\int_{-\infty}^{\infty} f(x)dx$$

$$\int_{-\infty}^{\infty} f(x)dx = \lim_{R\to\infty} \int_{-R}^{R} f(x)dx.$$

The limit on the right hand side is called the Cauchy principal value and is denoted by P.V$\int_{-\infty}^{\infty} f(x)dx$.

$$\begin{aligned} P.V. \int_{-\infty}^{\infty} f(z)dz &= \lim_{R\to\infty} \int_{-R}^{R} f(z)dz \\ &= 2\pi i \sum_{k=1}^{n} Res(f(z), z_k). \end{aligned}$$

Example 292. Evaluate

$$\int_{0}^{\infty} \frac{dx}{1+x^4}.$$

Solution: Since integral $\dfrac{1}{1+x^4}$ is an even function

$$\int_{0}^{\infty} \frac{dx}{1+x^4} = \frac{1}{2}\int_{-\infty}^{\infty} \frac{dx}{1+x^4}.$$

Furthermore,

$$f(z) = \frac{1}{1+z^4}$$

has four singular points $z_1 = e^{\frac{\pi i}{4}}$, $z_2 = e^{\frac{3\pi i}{4}}$, $z_3 = e^{\frac{-3\pi i}{4}}$, $z_4 = e^{\frac{-\pi i}{4}}$ which are simple poles. See Figure 8.20.
The first two singular points lie in the upper half plane. From Theorem 65(b) we find the residues:

$$\begin{aligned} Res(f(z), z_1) &= \left[\frac{1}{(1+z^4)'}\right]_{z=z_1} = \left[\frac{1}{4z^3}\right]_{z=z_1} \\ &= \frac{1}{4}e^{\frac{-3\pi i}{4}} = -\frac{1}{4}e^{\frac{\pi i}{4}}. \end{aligned}$$

$$\begin{aligned} Res(f(z), z_2) &= \left[\frac{1}{(1+z^4)'}\right]_{z=z_2} = \left[\frac{1}{4z^3}\right]_{z=z_2} \\ &= \frac{1}{4}e^{-\frac{9\pi i}{4}} = \frac{1}{4}e^{-\frac{\pi i}{4}}. \end{aligned}$$

Here we used $e^{\pi i} = -1$ and $e^{-2\pi i} = 1$. By Theorem 67,

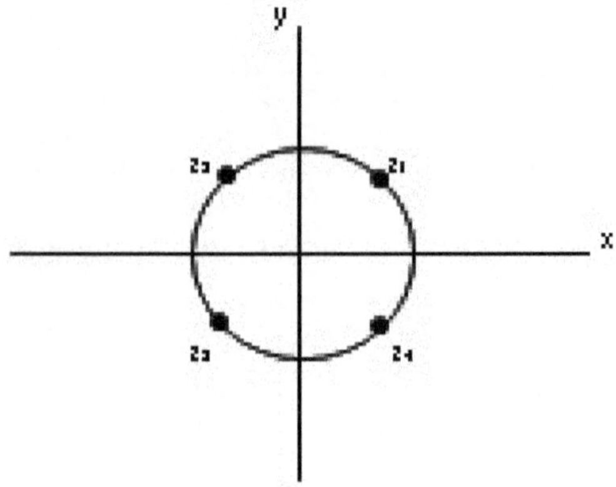

FIGURE 8.20: Poles

$$\begin{aligned}\int_{-\infty}^{\infty} \frac{dx}{1+x^4} &= 2\pi i(-\frac{1}{4}e^{\frac{\pi i}{4}} + \frac{1}{4}e^{-\frac{\pi i}{4}}) \\ &= -\frac{2\pi i}{4}(e^{\frac{\pi i}{4}} - e^{-\frac{\pi i}{4}}) \\ &= \frac{-2\pi i}{4} 2i \sin\frac{\pi}{4} = \pi \sin\frac{\pi}{4} = \frac{\pi}{\sqrt{2}}\end{aligned}$$

as

$$\frac{e^{i\theta} - e^{-i\theta}}{2i} = \sin\theta \quad (\text{Here } \theta = \frac{\pi}{4}).$$

Example 293. Find the principal value of

$$\int_{-\infty}^{\infty} \frac{dx}{(x^2 - 3x + 2)(x^2 + 1)}.$$

Solution: We have $z^2 - 3x + 2 = (x-2)(x-1)$ and

$$x^2 + 1 = (x - i)(x + i).$$

The integrand $f(x)$, considered for complex z, has simple poles at

$$\begin{aligned} z &= 1, \quad Res(f(z), 1) = \lim_{z\to 1}(z-1)f(z) = -\frac{1}{2} \\ z &= 2, \quad Res(f(z), 2) = \lim_{z\to 2}(z-2)f(z) = -\frac{1}{5} \\ z &= i, \quad Res(f(z), i) = lim_{z\to i}(z-i)f(z) = -\frac{3i}{20}.\end{aligned}$$

By Theorem 65, $z = -i$ is in the lower half plane, which is not relevant here. The final result is obtained by applying Theorem 67.

Example 294. Evaluate the Cauchy principal value of the improper integral

$$\int_{-\infty}^{\infty} \frac{x^2}{(x^2+1)^2} dx.$$

Solution: $\displaystyle\int_{-\infty}^{\infty} \frac{x^2}{(x^2+1)^2} dx = 2\pi i Res(f(z), i) = \frac{\pi}{2}.$

Fourier integrals: We use the following formulas to evaluate Fourier integrals:

$$\int_{-\infty}^{\infty} f(x) \cos \alpha x dx = -2\pi \sum_p Im Res(f(z)e^{i\alpha z}, p). \tag{8.43}$$

$$\int_{-\infty}^{\infty} f(x) \sin \alpha x dx = -2\pi \sum_p Re Res(f(z)e^{i\alpha z}, p). \tag{8.44}$$

where p represents poles in the upper half plane.

Example 295. Show that

$$\int_{-\infty}^{\infty} \frac{\cos \alpha x}{16+x^2} dx = \frac{\pi}{4} e^{-4\alpha}.$$

$$\int_{-\infty}^{\infty} \frac{\sin \alpha x}{16+x^2} dx = 0, \quad \alpha > 0.$$

Solution: $\dfrac{e^{i\alpha z}}{16+z^2}$ has only one pole in the upper half plane, namely a simple pole

at $z = 4i$. By Theorem 65 we obtain

$$Res\left(\frac{e^{i\alpha z}}{16+z^2}, 4i\right) = \left[\frac{e^{i\alpha z}}{2z}\right]_{z=4i} = \frac{e^{-4\alpha}}{8i}.$$

Thus

$$\int_{-\infty}^{\infty} \frac{e^{i\alpha x}}{16+x^2} dx = 2\pi i \frac{e^{-4\alpha}}{8i} = \frac{\pi}{4} e^{-4\alpha}$$

or

$$\int_{-\infty}^{\infty} \frac{\cos \alpha x + i \sin \alpha x}{16+x^2} dx = \frac{\pi}{4} e^{-4\alpha}.$$

This implies that

$$\int_{-\infty}^{\infty} \frac{\cos \alpha x}{16+x^2} dx = \frac{\pi}{4} e^{-4\alpha}$$

and

$$\int_{-\infty}^{\infty} \frac{\sin \alpha x}{16+x^2} dx = 0.$$

8.6 Conformal Mappings

8.6.1 Complex Functions as Mappings

A complex function $w = f(z)$ can be interpreted as a mapping of transformation from the z-plane to the w-plane, see Figure 8.21.

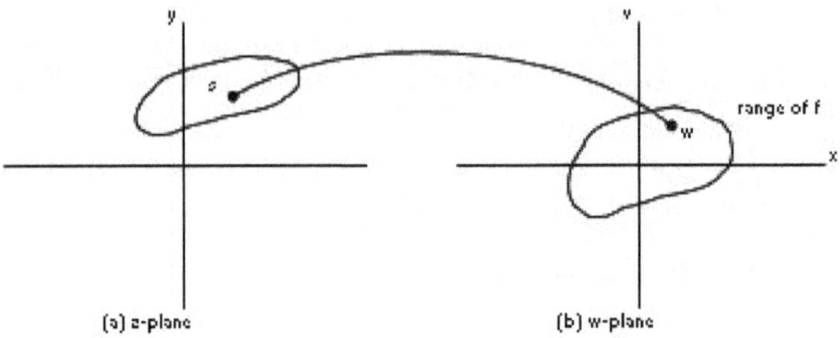

FIGURE 8.21: Mapping from z-Plane to w-Plane

Example 296. Find the images of the lines:
(a) $Rez = 2$ under the mapping $f(z) = z^2$.
(b) $Rez = 0$.
(c) $Imz = 0$ under the mapping $f(z) = z^2$.

Solution: (a) For the function

$$f(z) = z^2 = (x + iy)^2 = x^2 - y^2 + 2ixy,$$

$$w = u(x, y) + iv(x, y) = x^2 - y^2 + 2ixy$$

and

$$u(x, y) = x^2 - y^2, \quad v(x, y) = 2xy$$

by comparing real and imaginary parts.

$$Rez = 2, \quad u(x, y) = 4 - y^2$$

and

$$v(x, y) = 4y.$$

There are parametric equations of a curve in the w-plane. Substituting $y = \frac{1}{4}v$ into the first equation eliminates the parameter y to give

$$u = 4 - \frac{1}{16}v^2.$$

In other words, the image of the vertical line $x = 2$ in Figure 8.22(a) is the parabola shown in Figure 8.22(b)

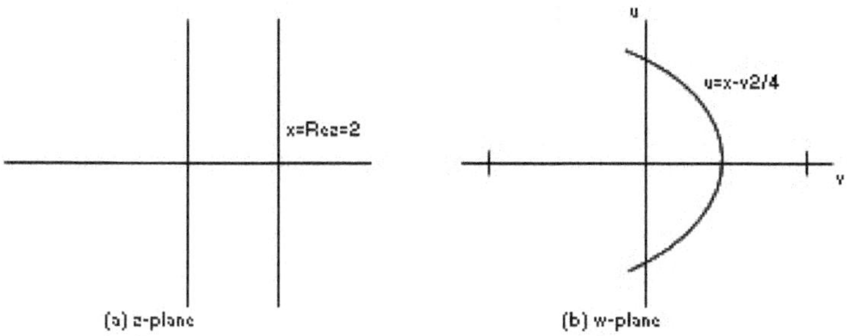

FIGURE 8.22: Image of $x = 2$ is Parabola

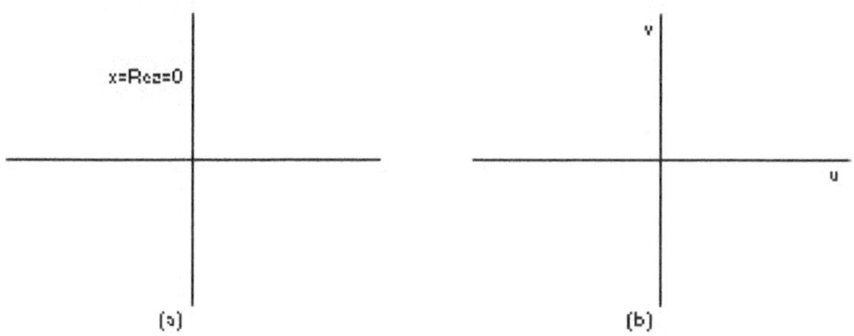

FIGURE 8.23: (a) $z-$plane (b) $w-$ plane

$$f(z) = z^2 = x^2 - y^2 + 2ixy = -y^2$$

as $Re z = x = 0$.

$$u(x, y) = -y^2, \quad v(x, y) = 0.$$

In short the complex function

$$w = f(z) = u(x, y) + iv(x, y)$$

may be considered the planar transformation $u = u(x, y)$, $v = v(x, y)$ and $w = f(z)$ is called the image of z under f.

Remark 72. (a) If $z(t) = x(t) + iy(t)$, describes a curve C in the region, then

$$w = f(z(t)), \quad a \leq t \leq b$$

is a parametric representation of the corresponding curve C in the w-plane.

(b) A point z on the level curve

$$u(x, y) = a$$

will be mapped to a point that lies on the vertical line $u = a$, and a point z on the level curve

$$v(x, y) = b$$

will be mapped to a point w that lies on the horizontal line

$$v = b.$$

Examples of complex functions

Example 297. (a) **Translation function:**

$$T(z) = w = z + b,$$

where b is a fixed complex number.
(b) **Magnification function:** It is a complex function defined as $f(z) = \alpha z$, where α is a fixed positive real number.
(c) **Exponential function:** It is a complex function defined as

$$f(z) = e^z.$$

(d) $g(z) = w = re^{i\theta_0}z$ elementary function is simply a rotation function through θ_0 degrees, for if

$$z = re^{i\theta_0},$$

then

$$w = g(z) = re^{i(\theta+\theta_0)}.$$

(e) $w = az$ with $\mid a \mid = 1$ is also a rotation function.
(f) $f(z) = w = az + b$ where a and b are real or complex numbers is called a linear function.
(g) $w = f(z) = \dfrac{1}{z}$ is called an inversion in the unit function. This function has domain $z \neq 0$ and real and imaginary parts:

$$\begin{aligned} w &= u(x, y) + iv((x, y) \\ &= \frac{1}{x + iy} = \frac{1}{x + iy}\frac{x - iy}{x - iy} \\ &= \frac{x - iy}{x^2 - y^2} = \frac{x}{x^2 - y^2} + i\frac{-y}{x^2 - y^2} \end{aligned}$$

$$u(x, y) = \frac{x}{x^2 - y^2} \quad \text{and} \quad v(x, y) = \frac{-y}{x^2 + y^2}.$$

For $a \neq 0$ the level curve $u(x, y) = a$ can be written as

$$x^2 - \frac{1}{a}x + y^2 = 0$$

or

$$(x - \frac{1}{2a})^2 + y^2 = (\frac{1}{2a})^2.$$

The level curve is therefore a circle with its center on the x-axis and passing through the origin. A point z on this circle other than zero is mapped to a point w on the line $u = a$. Similarly, the level curve $v(x, y) = b, b \neq 0$ can be written as

$$x^2 + (y + \frac{1}{2b})^2 = (\frac{1}{2b})^2,$$

and a point z on this circle is mapped to a point w on the line $v = b$; it is clear that

$$f^{-1}(w) = \frac{1}{w}$$

and

$$w = f(z) = \frac{1}{z}$$

or

$$z = \frac{1}{w}$$

and so $f^{-1} = f$. Therefore we conclude that it maps the horizontal line $y = b$ to the circle

$$u^2 + (v + \frac{1}{2}b^2)^2 = (\frac{1}{2}b)^2$$

and it also maps the vertical line to $x = a$ to the circle

$$(u - \frac{1}{2}a)^2 + v^2 = (\frac{1}{2}a)^2.$$

Example 298. (a) Find the image curve in the w-plane of the circle$|\ z\ |= 1$ under the complex function

$$w = f(z) = z + \frac{1}{z}.$$

(b) Find the image region in the w-plane of the rectangle $0 \leq x \leq 1, 0 \leq y \leq 1$ under the complex function

$$w = f(z) = e^z.$$

8.6.2 Conformal Mappings

Conformal mappings preserve the angles between two curves. Mappings of Example 297 are conformal. Let C_1 and C_2 be two curves in z-plane intersecting at $z = z_0$. Angle between C_1 and C_2 is θ (angle between tangent between those two curves at z_0). Let the images of C_1 and C_2 under $w = f(z)$ be respectively Γ^1 and Γ^1 and let the angle between Γ^1 and Γ^1 be ϕ. f is conformal mapping if $\theta = \phi$. See Figure 8.24.

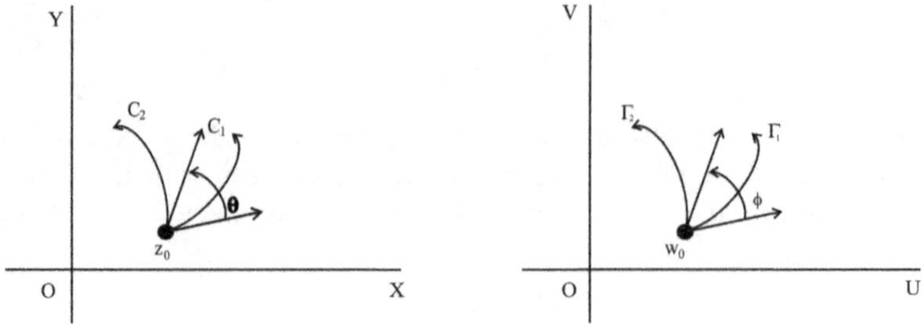

FIGURE 8.24: $w = f(z)$ is Conformal if $\theta = \phi$

Remark 73. Let z_1^1 and z_2^1 denote tangent vectors to curves C_1 and C_2, respectively, then

$$| z_1' - z_2' |=| z_1' |^2 + | z_2' |^2 -2 | z_1' || z_2' | \cos\theta$$

by applying the law of cosines to the triangle determined by z_1' and z_2'. Thus

$$\cos\theta = \frac{| z_1' |^2 + | z_2' |^2 - | z_1' - z_2' |}{2 | z_1' || z_2' |}$$

or

$$\theta = \cos^{-1}\frac{| z_1' |^2 + | z_2' |^2 - | z_1' - z_2' |}{2 | z_1' || z_2' |}. \tag{8.45}$$

Let w_1' and w_2' denote tangent vectors C_1' and C_1' respectively, then

$$\Phi = \cos^{-1}\frac{| w_1' |^2 + | w_2' |^2 - | w_1' - w_2' |}{2 | w_1' || w_2' |} \tag{8.46}$$

and $w = f(z)$ is a conformal mapping if angles given in (8.45) and (8.46) are equal to $\theta = \phi$.

The following theorem provides a simple criterion for a complex function to be conformal mapping.

Theorem 68. *A function of complex variable $f(z)$ defined on D is a conformal mapping at $z = z_0$ if $f(z)$ is analytic in the domain D and*

$$f'(z_0) \neq 0.$$

Proof. *Let a curve C in the domain D be parameterized by $z = z(t)$; then $w = f(z(t))$ describes the image curve in the w-plane. By the chain rule applied to w, we get*

$$w' = f'(z(t))z'(t).$$

Since curves C_1 and C_2 intersect at z_0 in D, we have

$$w_1' = f'(z_0)z_1'$$

and

$$w_2' = f'(z_0)z_2'.$$

Since $f'(z_0) \neq 0$, *we obtain*

$$\begin{aligned}
\Phi &= \cos^{-1}\left(\frac{| f'(z_0)z_1' |^2 + | f'(z_0)z_2' |^2 - | f'(z_0)z_1' - f'(z_0)z_2' |^2}{2 | f'(z_0)z_1' || f'(z_0)z_2' |}\right) \\
&= \cos^{-1}\left(\frac{| z_1' |^2| f'(z_0) |^2 + | z_2' || f'(z_0) |^2 - | f'(z_0) |^2| z_1' - z_2' |^2}{2 | f'(z_0) || z_1' | f'(z_0) || z_2' |}\right) \\
&= \cos^{-1}\left(\frac{| z_1' |^2 + | z_2' |^2 - | z_1' - z_2' |^2}{2 | z_1 | z_2 |}\right) = \theta.
\end{aligned}$$

Therefore $\Phi = \theta$ *and* f *is conformal at* $z = z_0$. □

Example 299. Examine whether the following complex mappings are conformal

(a) $w = f(z) = \sin z$

(b) $w = f(z) = \cos z$

(c) $w = f(z) = 1 + e^z + z$

(d) $w = f(z) = z^3 - 3z + 1$

(e) $w = f(z) = z + \dfrac{1}{z}$

8.6.3 Möbius Transforms

In this subsection we present a class of conformal transform, called the **Möbius transform** or **linear fractional transform**, which provides very efficient methods of finding one-to-one mappings of domains into one another.

Definition 96. For a, b, c, d real or complex numbers such that $| a+ | c |> 0$, $ad - bc \neq 0$, the transformation defined as

$$w = T(z) = \frac{az + b}{cz + d} \tag{8.47}$$

is called a Möbius or linear fractional transform.

Remark 74. (a) For $c = 0, w = T(z) = \dfrac{a}{d}z + \dfrac{b}{d}$; Möbius transform is linear.
(b) If $c \neq 0$ and $a = 0$ then the Möbius transform

$$T(z) = \frac{b}{cz + d}$$

is an inversion.
(c) In case $ac \neq 0$, we can write

$$w = T(z) = \frac{a}{c}\left(1 + \frac{bc - ad}{a(cs + d)}\right)$$

and so the Möbius transform is a decomposition of a linear transformation and inversion of mappping of (8.47) is obtained by solving for z, namely:

$$cwz + wd = az + b$$

$$z(cw - a) = b - wd$$

or

$$z = \frac{b - wd}{wc - a} = \frac{dw - b}{-cw + a}$$

when $c \neq 0$, then $-cw + a = 0$ for $w = \dfrac{a}{c}$, and let $\dfrac{a}{c}$ be the image of $z = \infty$. The complex together with the point ∞ is called the extended complex plane.
(d) Since

$$\begin{aligned} T(z) &= \frac{a(cz+d)}{-(az+b)c}(cz+d)^2 \\ &= \frac{ad - ac}{(cz+d)^2} \neq 0, \ \text{for z} \neq \frac{-\text{d}}{\text{c}}. \end{aligned}$$

Every Möbius transform is a conformal mapping. We can summarize the above properties of the Möbius transform in the form of theorems.

Fixed points of complex mappings

Fixed points of a complex mapping $w = f(z)$ are mapped, that is, $w = f(z) = z$.

(a) The identity mapping$I(z) = z = w$ has every point as a fixed point.

(b) The mapping $w = f(z) = z$ has infinitely many fixed points.

(c) The mapping $w = f(z) = \dfrac{1}{z}$ has two fixed points.

(d) A rotation mapping has one fixed point.

(e) A translation mapping has no fixed point in the complex plane.

Existence of fixed points for complex mappings can be described by the following theorems.

Theorem 69. *A Möbius transform different from the identity has at most two fixed points. Any Möbius transform having three or more fixed points must be the identity mapping.*

Theorem 70. *Let T be a Möbius transform. Then*

(a) T can be written as a composition of magnifications, rotations, translations and inversions.

(b) T maps the extended complex plane onto itself. It is a one-to-one mapping.

(c) T maps the class of circles and lines into itself.

(d) T is a conformal map at every point except its pole.

The following theorem (see proof in Appendix D if desired) describes how to find Möbius transformations by three different points.

Theorem 71. *Three given distinct points z_1, z_2, z_3 can always be mapped onto three prescribed distinct points w_1, w_2, w_3 by only one Möbius transform $w = f(z)$ given implicitly by the equation*

$$\frac{w - w_1}{w - w_3}\frac{w_2 - w_3}{w_2 - w_1} = \frac{z - z_1}{z - z_3}\frac{z_2 - z_3}{z_2 - z_1}.$$

Example 300. Construct a Möbius transform that maps points 1, 0 and -1 on the circle $|z| = 1$ x-axis to the points. -1, 0, and 1 on the real axis.

Example 301. (a) Find all Möbius transforms whose fixed points are -1 and 1.
(b) Find all Möbius transformations with no fixed points in the complex plane.

8.7 Applications

In this section we briefly indicate how the results of complex analysis can be used in areas such as Electrostatics potential, Heat Flow, Fluid Flow and Tomography.

8.7.1 Electrostatics Potential

We have discussed earlier solutions of Laplace equation under different initial and boundary conditions. The equation models gravitational fields, electrostatic fields, heat conduction, compressible fluid flows and problems of many other fields. The theory of solving the Laplace equation

$$\nabla^2\phi = \frac{\partial^2\phi}{\partial x^2} + \frac{\partial^2\phi}{\partial y^2} = 0 \tag{8.48}$$

is called **potential theory**, and solutions having continuous second partial derivatives are called **harmonic functions**.

Equation (8.48) can be solved by complex analysis because its solutions are closely related to analytic complex functions as seen in Section 8.2. Transitions from real to complex have the advantage that by the complex potential

$$F = \phi + i\psi$$

one can simultaneously handle equipotential lines $\phi =$ constant and their orthogonal trajectories, namely the lines or flow $\psi =$ constant.

Furthermore, in solving the Dirichlet problem of finding a potential with given boundary values the conformal mapping can be used. The electrical force of attraction or repulsion between charged particles is governed by Coulomb's law. This force is the gradient of a function ϕ called the **electrostatic potential**. At any points free of charges ϕ in a solution of Laplace equation

$$\nabla^2\phi = 0.$$

The surfaces $\phi(x, y) =$ constant are called equipotential surfaces. At each point P the gradient of ϕ is perpendicular to the surface $\phi =$ constant through P; that is, the electrical force has the direction perpendicular to the **equipotential surface**. We consider here two-dimensional problems, that is, we consider physical systems that lie in three-dimensional space, but are such that the potential ϕ is independent of one of the space coordinates, so that ϕ depends only on two coordinates, which are x and y and equipotential surfaces appear as equipotential lines (curves) in the xy plane.

Complex potential: Let $\phi(x, y)$ be harmonic in some domain D and $\psi(x, y)$ a conjugate harmonic of ϕ in D. Then

$$F(z) = \phi(x, y) + i\psi(x, y)$$

is an analytic function of $z = x + iy$. This function F is called the **complex potential** corresponding to the real potential ϕ. The use of F has two advantages, one technical and the other physical. Technically, F is easier to handle than real or imaginary parts. Physically, ψ has a meaning. By conformality, the curves $\phi =$ constant intersect the equipotential lines $\phi =$ constant in the xy-plane at right angles except where $F'(z) = 0$. Hence they have the direction of the electrical force and, therefore are called lines of force. They are the paths of moving charged particles (electrons in an electron microscope). As we see complex potentials relate potential theory closely to complex analysis.

Conformal mapping is used to map a complicated domain into a simpler one where solution of the Laplace equation is known or can be found more easily. This solution is then mapped back to the given domain. We obtain these results due to the fact that harmonic functions remain harmonic under conformal mapping.

Theorem 72. *Let ϕ^* be harmonic in a domain D^* in the w-plane. Suppose that*

$$w = u + iv = f(z)$$

is analytic in a domain D in the z-plane and maps conformally onto D^*. Then the function

$$\phi(x,y) = \phi^*(u(x,y), v(x,y)) \tag{8.49}$$

is harmonic in D.

Proof. *We know that the composition of analytic function is analytic (the chain rule). Thus by taking a harmonic conjugate $\psi^*(x,y)$ of ϕ^* and forming the analytic function*

$$F^* = \phi^*(v,u) + i\psi^*(u,v),$$

we find that

$$F(z) = F^*(f(z))$$

is analytic in D, Hence its real part, $\phi(x,y) = ReF(z)$ is harmonic in D. □

8.7.2 Heat Flow

Laplace's equation also models heat flow problems that are steady, that is, time independent. The heat equation

$$\frac{\partial T}{\partial t} = \alpha^2(\frac{\partial^2 T}{\partial x^2} + \frac{\partial^2 T}{\partial y^2}) \tag{8.50}$$

represents heat conduction in a body of homogeneous material, where T denotes temperature, t stands for time and α^2 is a positive constant depending on the material of the body. Therefore, if a problem is steady, so that,

$$\frac{\partial T}{\partial t} = 0,$$

the heat equation reduces to Laplace's equation.

$$\nabla^2 T = \frac{\partial^2 T}{\partial x^2} + \frac{\partial^2 T}{\partial y^2} = 0. \tag{8.51}$$

$T(x,y)$ is called the **heat potential**. It is the real part of the complex heat potential.

$$F(z) = T(x,y) + i\psi(x,y) \tag{8.52}$$

The curves $T(x,y)$ = constant are called **isotherms** (lines of constant temperature) and the curves $\psi(x,y)$ = constant heat flow lines, because heat flows from higher to lower temperatures along them. Discussions in Section 8.7.1 can be reinterpreted as problems on heat flow.

It may be observed that to have a steady flow, the boundary of the domain of the heat flow must be kept at constant temperature by heating or cooling. Usefulness of conformal mappings and complex potentials may be demonstrated in study of temperature distribution in a region comprising a cross section of a solid cylinder and heat conduction in the upper half plane; for details see for example Section 16.3 of [7].

8.7.3 Fluid Flow

Laplace's equation plays an important role in hydrodynamics and in steady nonviscous fluid flow under certain physical conditions. We discuss here the relevance of complex functions and their integrals by modeling and analyzing fluid flow. Let the velocity vector V, representing the motion of the fluid, be written as

$$V = V_1 + iV_2. \tag{8.53}$$

The magnitude V is denoted by $\mid V \mid$ and direction by

$$ArgV = \frac{V_2}{V_1}$$

at each point $z = x + iy$. Here V_1 and V_2 are components of V in the direction of the x and y axes respectively.

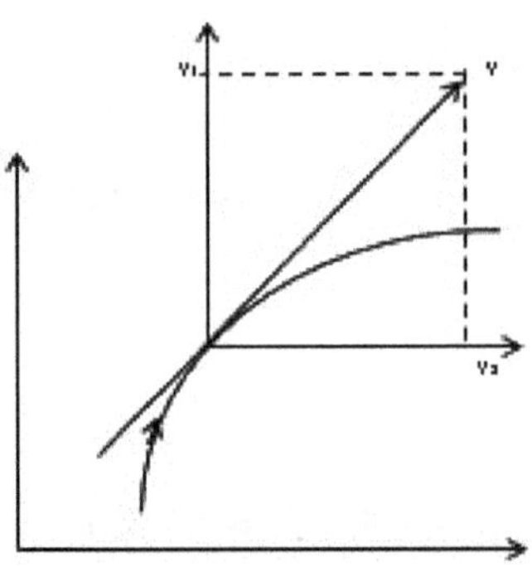

FIGURE 8.25: Streamline

V is tangential to the path of the moving particles, called a **streamline** (Figure 8.25) of the motion,

$$F(z) = \phi(x,y) + i\psi(x,y)$$

is called the **complex potential** of the flow, and $\psi(x,y)$ = constant are the streamlines. The velocity V is given by

$$V = V_1 + iV_2 = \overline{F'(z)}.$$

ψ is called the **stream function** and ϕ is called the **velocity potential**. The curves $\phi(x,y)$ = constant are called **equipotential lines**. V is the gradient ϕ,

$$V_1 = \frac{\partial \phi}{\partial x}, V_2 = \frac{\partial \phi}{\partial y}.$$

In fact, for $F = \phi + i\psi$,

$$F' = \frac{\partial \phi}{\partial x} + i\frac{\partial \psi}{\partial x}$$

with

$$\frac{\partial \psi}{\partial x} = -\frac{\partial \phi}{\partial y}$$

by the Cauchy-Riemann equation. These together yield

$$\overline{F'(z)} = \frac{\partial \phi}{\partial x} - i\frac{\partial \psi}{\partial x} = \frac{\partial \psi}{\partial x} + i\frac{\partial \phi}{\partial y} = V_1 + iV_2 = V.$$

Let us consider any smooth curve C in the z-plane, given by $z(z) = x(s)+iy(s)$, where s is the arc length of C. Let

$$\frac{\partial V}{\partial t}$$

denote the component of V tangent to C, see Figure 8.26.
$\int_C V_t ds$ is called the **circulation of the fluid** along C.

$$\frac{\partial V_2}{\partial x} - \frac{\partial V_1}{\partial y}$$

is called **velocity of the flow** and

$$\frac{1}{2}(\frac{\partial V_2}{\partial x} - \frac{\partial V_1}{\partial y}) = w(x,y)$$

is called the **rotation**. A fluid is called **irrotational** if $w(x,y) = 0$ throughout the flow. In other words a fluid is irrotational if

$$\frac{\partial V_2}{\partial x} - \frac{\partial V_2}{\partial y} = 0.$$

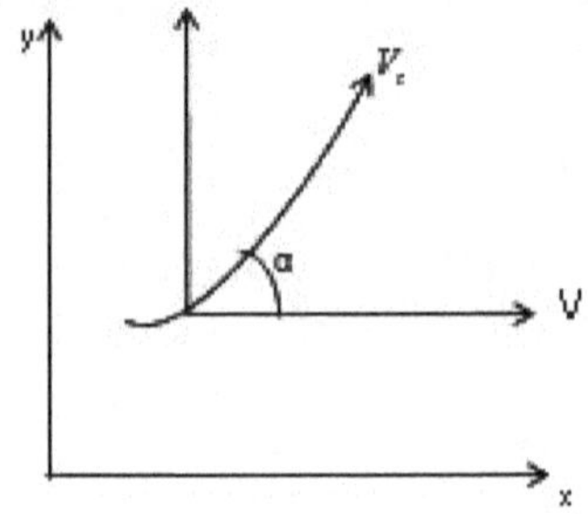

FIGURE 8.26: Tangential Component of Velocity with Respect to Curve C

If C is taken as a circle of radius r then the mean velocity of the fluid along C is equal to

$$\frac{1}{2}\frac{(\frac{\partial V}{\partial x}-\frac{\partial V_1}{\partial y})}{2\pi r}.$$

A fluid is called **incompressible** if

$$\frac{\partial V_1}{\partial x}+\frac{\partial V_2}{\partial y}=0$$

in region that is free of sources or sinks V that is, points at which fluid is produced or disappears.

$$v(x,y)=u(x,y)+iv(x,y)$$

$$\oint_c = -vdx+udy$$

is called the **flux** if it is non-zero. It is called solenoidal if this integral is zero for all closed curves.
The following theorem describes the connection between the velocity field of a fluid and complex functions.

Theorem 73. *Let u and v be continuous with continuous first and second partial derivatives in a simply connected domain D.*
Let $ui+vj$ be irrotational and solenoidal in D. Then u and $-v$ satisfy the Cauchy-Riemann equation in D, and

$$f(z)=u(x,y)-iv(x,y)$$

is a differentiable complex function on D.
Conversely, if u and $-v$ satisfy Cauchy-Riemann equations on D, then $ui+vj$ defines an irrotational, solenoidal flow on D.

Remark 75. Theorem 73 provides insight into irrotational and solenoidal flows. For irrotational flow

$$curl(ui + vj) = 0$$

and

$$div(ui + vj) = 0$$

for solenoid a flow.

The following theorem implies that any differentiable function

$$f(z) = \phi(x, y) + i\psi(x, y)$$

defined on a simply connected domain determines an irrotational solenoidal flow.

Theorem 74. *Let f be a differentiable function defined on a domain D. Then $f'(z)$ is an irrotational solenoidal flow on D. Conversely, if*

$$v(x, y) = u(x, y)i + v(x, y)j$$

is an irrotational solenoidal vector field on a simply connected domain D, then there is a differentiable complex function and defined on D such that $\overline{f'(z)} = V$.
Further, if

$$f(z) = \phi(x, y) + i\psi(x, y)$$

then

$$\frac{\partial \phi}{\partial x} = u, \frac{\partial \psi}{\partial y} = v, \frac{\partial \psi}{\partial x} = -v \text{ and } \frac{\partial \psi}{\partial y} = u.$$

8.7.4 Tomography

This section discusses the role of complex analysis in tomography. The word tomography comes from the Greek word for slice and the discipline deals with finding internal structures of non-transparent objects by transmitting signals such as electromagnetic waves at various frequencies. Examples of such frequencies are radio transmissions, microwaves, light waves in the visual spectrum, x-rays, and acoustic waves. Tomography deals with inverse problems in which the unknown parameters of a system are estimated from its known reactions to specific signals.

Austrian mathematician Johann Radon studied the integral transform and inversion of an integral of a real valued function over straight lines in 1917. The technique became known as the Radon transform and it plays an important role in tomography, specifically for computed axial tomography (CAT) scans, barcode scanners, electron microscopy of structures like viruses and proteins, reflection seismology, and other disciplines that involve applications of partial differential equations.

Roger Penrose introduced a more complex version of the Radon transform in 1967. The Penrose transform relates massless fields in space time to cosmology of sheaves in complex projective spaces and is a significant aspect of the Twistor theory. The inspiration for the Penrose transform was a 1904 paper by Harry Bateman in which a harmonic function of four variables was represented as a contour integral of a holomorphic function of three complex variables, namely:

$$\phi(w,x,y,z) = \oint f[w+ix+(iy+z)\xi, iy-z+(w-ix)\xi, \xi]d\xi,$$

Details may be found in a book titled *Introduction to the Penrose Transform* by Baston and Eastwood [12]. Gindikin [13] studied the construction of an explicit inversion formula for the Penrose transform focusing on connections to the Radon transform and multi-dimensional residues.

8.8 Exercises

8. 1 Write the following numbers in the form $x+iy$:

(a) $(5-9i)+(2-4i)$

(b) $(1-i)^3$

(c) $\dfrac{10-5i}{6+2i}$

(d) $\dfrac{(4+5i)+2i^3}{(2+i)^2}$

8.2. Let $z=x+iy$. Find the indicated expressions:

(a) $Rez(z^2)$

(b) $Im(\overline{z}^2+z^2)$

(c) $\mid z+5\overline{z}\mid$

8.3. Write the following complex numbers in the polar form:

(a) $\dfrac{12}{\sqrt{3}+i}$

(b) $(5-5i)$

(c) $-2-2\sqrt{3i}$

8.4. Write the polar form in the form $x+iy$:

(a) $z = 8\sqrt{2}(\cos\frac{11}{4}\pi + i\sin\frac{11}{4}\pi)$

(b) $z = 6(\cos\frac{\pi}{8} + i\sin\frac{\pi}{8})$

(c) $z = 10(\cos\frac{\pi}{5} + i\sin\frac{\pi}{5})$

8.5. (a) Find z_1z_2 and $\frac{z_1}{z_2}$ if

$$\begin{aligned} z_1 &= 2(\cos\frac{\pi}{8} + i\sin\frac{\pi}{8}), \\ z_2Z &= 4(\cos\frac{3\pi}{8} + i\sin\frac{\pi}{8}). \end{aligned}$$

(b) Find $(-1+\sqrt{3}i)^{\frac{1}{2}}$and $(-1-\sqrt{3}i)^{\frac{1}{4}}$.

8.6. If $z_1 = -1$ and $z_2 = 5i$, verify

$$\begin{aligned} \arg(z_1z_2) &= \arg(z_1) + \arg(z_2), \\ \arg(\frac{z_1}{z_2}) &= \arg(z_1) - \arg(z_2). \end{aligned}$$

8.7. Sketch the graphs of the following functions

(a) $Re(z) = 4$

(b) $Im(\overline{z} + 3i) = 6$

(c) $\mid z + 2 + 2i \mid = 2$

8.8. Find the images of the given lines under the mapping $f(z) = z^2$:

(a) $y = 2$

(b) $y = x$

8.9. Express the following functions in the form $u(x, y) + iv(x, y)$:

(a) $f(z) = 7z - 9i\overline{z} - 3 + 2i$

(b) $f(z) = z^3 - 4z$

(c) $f(z) = z^4$

(d) $f(z) = \frac{z}{z+1}$

8.10. Evaluate the functions at the indicated points:

(a) $f(z) = 2x - y^2 + i(xy^3 - 2x^2 + 1)$
(i) $2i$ (ii) $2 - i$ (iii) $5 + 3i$

(b) $f(z) = 4z + i\overline{z} + Re(z)$
(i) $4 - 6i$ (ii) $-5 + 12i$

(c) $f(z) = e^x \cos y + ie^x \sin y$

(i) $\pi\frac{i}{4}$ (ii) $3 + \pi\frac{i}{3}$

8.11. Find

(a) $\lim_{z\to 1+i} \frac{5z^2 - 2z + 2}{z+1}$

(b) $\lim_{z\to 1+i} \frac{z^2 - 2z + 2}{z^2 - 2i}$

8.12. Show that the following limits do not exist

(a) $\lim_{z\to 0} \frac{z}{\bar{z}}$

(b) $\lim_{z\to 1} \frac{x + y - 1}{z - 1}$

8.13. Using the definition of differentiability to show that:

(a) $f'(z) = 2z$ if $f(z) = z^2$

(b) $f'(z) = -\frac{1}{z^2}$ if $f(z) = \frac{1}{z}$

8.14. Using rules of differentiation for complex functions find $f'(z)$ if:

(a) $f(z) = (2z + 1)(z^2 - 4z + 8i)$

(b) $f(z) = (z^2 - 4i)^3$

(c) $f(z) = (z^2 - 4i)^3$

(d) $f(z) = \frac{3z - 4 + 8i}{2z + i}$

(e) $f(z) = \frac{5z^2 - z}{z^3 + 1}$

8.15. Find points at which the following functions are not differentiable:

(a) $f(z) = \frac{z}{z - 3i}$

(b) $\frac{z^3 + 3}{z^2 + 4}$

8.16. State and prove Cauchy-Riemann equations.

8.17. State a criterion for analyticity of a function of complex variable.

8.18. Prove that $u(x, y)$ and $v(x, y)$ are harmonic if $f(z) = u(x, y) + iv(x, y)$ is analytic in a domain D.

8.19. Show that the following functions are not analytic at any point:

(a) $f(z) = y + ix$

(b) $f(x) = \dfrac{x}{x^2 + y^2} + i\dfrac{y}{x^2 + y^2}$

(c) $f(z) = z$

8.20. Show that the following functions are analytic in an appropriate domain:

(a) $f(z) = e^x \cos y + ie^x \sin y$

(b) $f(z) = 4x^2 + 5x - 4y^2 + 9 + i(8xy + 5y - 1)$

8.21. Verify that the following functions are harmonic.

(a) $u(x, y) = 2x - 2xy$

(b) $u(x, y) = 4xy^3 - 4x^3y + x$

8.22. Evaluate the given integrals along the indicated contours.

(a) $\int_C \mid z \mid^2 dz$, where C is $x = t^2, y = \dfrac{1}{t}, 1 \leq t \leq 2$

(b) $\int_C e^z dz$, where C is the polygonal path consisting of the line segments from $z = 0$to$z = 2$ and from $z = 2$to $z = 1 + \pi i$.

8.23. State and prove the Cauchy-Goursat theorem.

8.24. Show that

$$\oint_c f(z)dz = 0,$$

where f is given by $f(z) = z^3 - 1 + 3i$ and C is the unit circle$\mid z \mid= 1$.

8.25. State and prove Cauchy's integral formula.

8.26. Evaluate

(a) $\oint_C \dfrac{4}{z - 3i} dz, \mid z \mid= 5$

(b) $\oint_C \dfrac{1 + 2e^z}{z} dz, \mid z \mid= 1$

8.27. State Laurent's theorem.

8.28. Expand the given function in a Laurent series valid for the indicated annular domains:

(a) $f(z) = \dfrac{z - \sin z}{z^5}, 0 <\mid z \mid$

(b) $f(z) = \dfrac{1}{z(z - 3)}, 0 <\mid z - 3 \mid< 3$

8.29. Determine the orders of the poles for the given functions:

(a) $f(z) = \dfrac{3z - 1}{z^2 + 2z + 5}$

(b) $f(z) = \dfrac{z-1}{(z+1)(z^3+1)}$

8.30. Find

(a) $Res(f(z), 0)$ if $f(z) = \dfrac{4z-6}{z(2-z)}$.

(b) $Res(f(z), 2)$ if $f(z) = \dfrac{e^{-z}}{(z-2)^2}$.

8.31. Find the residue at each pole of the given function

(a) $f(z) = \dfrac{z}{z^2+16}$

(b) $f(z) = \dfrac{2z-1}{(z-1)^4(z+3)}$

8.32. Use Cauchy's residue theorem to evaluate the following integrals along the indicated contours:

(a) $\oint_C \dfrac{1}{z^2+4z+13} dz, \quad C : \mid z - 3i \mid = 3$

(b) $\oint_C \dfrac{1}{z^3(z-1)^4} dz, \quad C : \mid z - 2 \mid = \dfrac{3}{2}$

8.33. Evaluate the following integrals:

(a) $\int_0^{2\pi} \dfrac{1}{1+3\cos^2\theta} d\theta$

(b) $\int_0^{2\pi} \dfrac{\sin^2\theta}{5+4\cos\theta} d\theta$

(c) $\int_{-\infty}^{\infty} \dfrac{1}{1+x^2} dx$

(d) $\int_{-\infty}^{\infty} \dfrac{\cos x}{1+x^2} dx$

(e) $\int_{-\infty}^{\infty} \dfrac{\sin x}{x} dx$

8.34. Show that $f(z) = e^z$ is a conformal mapping for all z.

8.35. Determine whether the given mappings are conformal.

(a) $f(z) = z^3 - 3z + 1$

(b) $f(z) = (z^2-1)^{\frac{1}{2}}$

8.36. Show that the mapping $w = \dfrac{1}{2}(z + \dfrac{1}{z})$ maps the circle $\mid z \mid = r$ onto an ellipse with foci 1 and -1 in the w-plane.

8.37. Find the image of the given circle $\Im(z) = 2$ under the linear fractional transformation

$$w = \frac{2z + i}{2 - i}.$$

8.38. Prove Theorem 65.

8.39. Prove Theorem 66.

8.40. Analyze the flow given by the complex potential $f(z) = \alpha z$ where α is a non-zero complex constant.

8.41. Solve Example 298.

8.42. Solve Example 299.

8.43. Solve Example 300.

8.44. Solve Example 301.

8.45. Find the temperature T in the sector $0 \leq \arg z \leq \frac{2\pi}{3}, \mid z \mid \leq 1$ if $T = 10°C$ on the x-axis, $80°C$ on $y = -\sqrt{3}x$, and the curved portion is insulated.

8.46. Find the complex potential of a parallel flow in the upward direction of $y = x$.

8.9 Suggestion for Further Reading

Applications of complex analyses to image processing, Radon transformations, inverse problems, wavelet and shape analyses, free boundary problems and vertex dynamics potential theories are popular areas for further research. The Penrose transform, a major component of Twistor theory, is a complex analogue of Radon transformation. The importance of complex analysis can be seen in studies of electrical independence tomography [2], [9].

Applications of complex wavelets to image processing are discussed in References [4] and [5]. Real wavelets have been applied to Radon transformations but complex wavelets have not yet been applied to complex analogues of Radon transforms.

The Penrose transform is a fascinating area for study by interested readers. Gindikin's 2014 article cited in Section 8.7.4 provides updated information about the transform. Kuchment's monograph [8] provides interesting data and several references on tomography. Krantz [6] describes applications of harmonic analysis to complex function analysis. Huggett's text [3] introduces

Twistor theory and describes its relationship to complex theory. Interesting discussions on the themes throughout this chapter appear in References [1], [8], [10], and [11].

Bibliography

[1] R. V. Churchill, *Complex Variables and Applications*, McGraw Hill, 2009.

[2] M. Hanke, N. Hyvonen and S. Revesing, *An Inverse Back Scatter Problem for Electrical Impedance Tomography*, 2008.

[3] S. A. Huggett, *An Introduction to Twistor Theory*, Cambridge University Press, 1994, on line publication, 2010.

[4] N. G. Kingsburg, Image processing with complex wavelets, *Phil. Trans. Royal Society*, London,1999.

[5] N. G. Kingsburg, Complex wavelets for shift invariant analysis and filtering signals, *J. Appl. and Harm. Anal.*, 10(3): 234-253, 2001.

[6] S. G. Krantz, *Explorations in Harmonic Analysis with Application to Complex Function Theory and the Heisenberg Groups*, Birkhauser, Berlin, 2009.

[7] E. Kreysszig, *Advanced Engineering Mathematics*, 10th Edition, Wiley, 2011.

[8] P. Kuchment, *The Radon Transform and Medical Imaging*, Vol. 85, Regional Conference Series in Applied Mathematics (CBMS-NSF), SIAM, 2014.

[9] K. Paridis and W. R. B. Lionheart, *Shape Corrections for 3DEIT, MIMS ePrint.* Manchester Institute of Mathematical Science, University of Manchester UK, 2010.

[10] A. D. Wunsch, *Complex Variable with Applications*, Pearson Education, 2005.

[11] D. G. Zill and P. D. Shanahan, *A First Course in Complex Analysis with Applications*, Jones and Bartlett, 2003.

[12] R. Baston and M. Eastwood, *Introduction to the Penrose transform*, Dover Publications, New York, 2016 (reprint).

[13] S. Gindikin, Inversion of Penrose transform and Cauchy-Fantappie formula, *J. Geom. Phys.*, 78:127-131, 2014.

Chapter 9

Inverse Problems

9.1 Introduction

In this chapter we explain the concept of inverse problems through concrete examples. A mathematical model enables us to understand a mathematical representation of a person, a building, a vehicle, a tree or a process such as weather pattern, traffic flow or air flow over a wing of an aeroplane. Models are created from a mass of data, equations and computations that mimic the actions of objects or processes presented.

Models usually include graphic displays that translate number crunching into animations that one can see on a computer screen or other visual device. Models can be simple images they can be complex, carrying all the characteristics of the object or process they represent. A complex model will simulate the action and reaction of the real world phenomenon. To make these models behave as they would be in real life, accurate, real time simulations require fast computers with a lot of number crunching power. Generally modeling is the process of representing a real world object or phenomenon by a set of mathematical equations. More specifically, the term is often used to describe the process of representing three-dimensional objects on a computer.

For proper understanding of any real world system, a mathematical formulation (mathematical modeling) of that system is essential. Usually mathematical formulation is an ordinary differential equation or a partial differential equation or in the form of an optimization problem (finding maxima or minima of a function) or matrix equation. Finding solutions of these equations (models) for given parameters and boundary or/and initial conditions is the response to a **direct problem**. Finding parameter or initial conditions or boundary conditions of a model with given solution is known as an **inverse problem**.

There exists vast literature on direct problems. But in spite of the importance of inverse problems, less attention has been paid to the topic in many parts of the world. The main objective of this chapter is to illustrate the concepts of inverse problems with the help of concrete examples from pre-calculus, calculus, matrix analysis, differential equations, partial differential equations, image processing and medical imaging.

9.2 Inverse Problems in Pre-calculus

9.2.1 Inverse Problem in Torricelli's Law

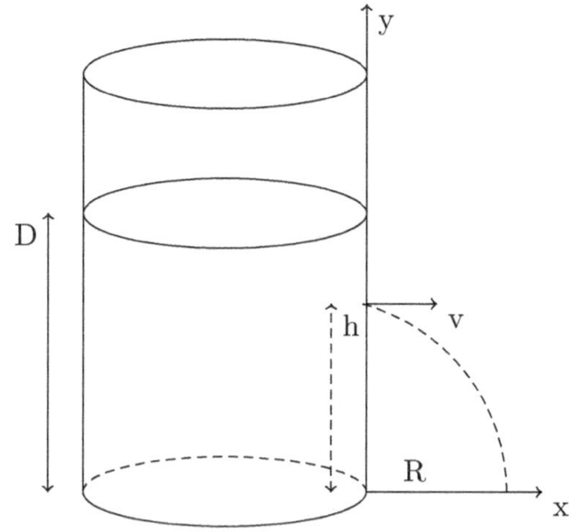

FIGURE 9.1: Tank Depicting Torricelli's Law

To illustrate an inverse problem, imagine filling a tank with water and then drilling a hole in its side. Water squirts from the hole, follows an arc and splashes down some distance from the tank we will designate R. The relationship between horizontal velocity v, height of the hole h, the height of column of water above the hole (hydraulic head) $D - h$ discovered by Evangelista Torricelli (1608-1647) a contemporary of Galileo, is known as Torricelli's law given by

$$v = \sqrt{2g(D - h)}$$

where D is the depth of the water in the tank, h is height of the hole above the bottom of the tank and g is the gravitational force.

The direct problem for Torricelli's law is to determine the range of the spurt R (see Figure 9.1), given D and h. A simple inverse problem is to find the height h of the hole based on R.

Example 302. Find height h of the hole if the horizontal velocity v with which water leaves the hole and D, the depth of water in the tank, are known.

9.2.2 Inverse Problem in Projectile Motion

Suppose a point projectile of unit mass is launched from the origin O at an angle θ to the positive x-axis with initial speed v (see Figure 9.2). OP is

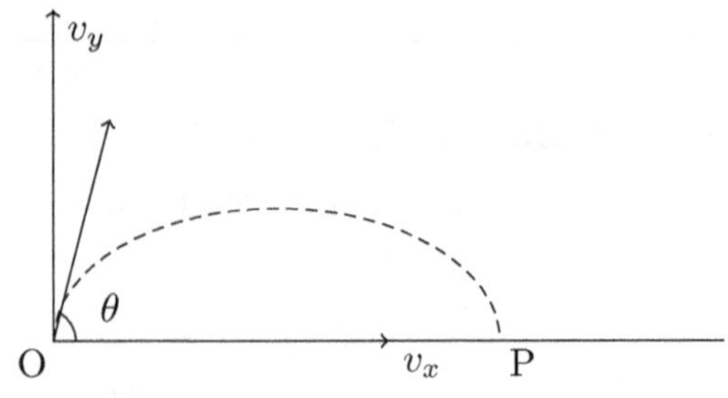

FIGURE 9.2: Projectile Motion

the horizontal range. The direct problem is to find OP if θ is given while the inverse problem is to find θ given OP (horizontal range).

Example 303. State the physical law that governs the relationship between the depth of the wall and the time it takes a dropped stone to reach the bottom of a well, neglecting air resistance.

9.2.3 Inverse Problem in Scattering

A certain type of signal is transmitted to strike an object called the scatterer and is bounced off the object or scattered. The scattered signal, which has been affected by the scatterer is then collected and the characteristics of the scatterer are inferred from the information contained in the scattered signal. Well known applications of this idea include radar, sonar and ultrasonic medical imaging. Consider the following simple scattering problem posed by Sir Isaac Newton, "Find the depth of a well from the sound of a stone falling to the well and striking the bottom." The direct problem is determination of the time at which the echo is heard and the inverse problem is to find a physical characteristic of the well, its depth, from one aspect of the reflected signal, the echo time.

Example 304. State the physical law that governs the relationship between the depth of well and the time it takes the dropped stone to reach the bottom, neglecting air resistance.

9.2.4 Inverse Problem of Location and Identification

We consider here a simplified model of reflection seismology. Suppose an explosive charge is detonated at some point on the flat surface of the earth.

The direct problem is to find the propagation time T for given velocity v, the depth d of the stratum, and location X of the receiver. The determination of v and d from X and T is an interesting inverse problem. For example, a reflected seismic signal is heard at a geophone 300 meters away from the source 3 seconds after detonation. What is the slowest possible velocity of propagation, irrespective of the depth of the stratum?

9.2.5 Inverse Problem Related to eigenvalues

Given a real $m \times m$ symmetric matrix P and m real numbers $\lambda_1, \lambda_2, \ldots\ldots \lambda_m$, finding a diagonal matrix D such that $P+D$ has eigenvalues $\lambda_1, \lambda_2, \ldots\ldots \lambda_m$. This problem is inverse to the direct problem of computing the eigenvalues of the given matrix $P+D$.

9.3 Inverse Problems in Calculus

9.3.1 Inverse Problem in Draining

Let a vessel be formed by revolving a curve $x = f(y)$ about y-axis. We can fill this vessel to various depths with water and allow water to flow out under the influence of gravity through an orifice of cross sectional area a at the base of the vessel (Figure 9.3). The depth y of water in the vessel can be observed, say, by means of a pilot tube of negligible volume. The time $T(y)$ for a given depth y that it takes for the vessel to drain completely can be measured with a stop watch. The drain time function depends on the shape of the vessel. Applying Torricelli's law, we can derive that

$$T(y) = -\int_0^T \frac{\pi(f(y))^2}{a\sqrt{2gy}}\frac{dy}{dt}dt = \frac{\pi}{a\sqrt{2g}}\int_0^T \frac{(f(u))^2}{\sqrt{u}}du. \tag{9.1}$$

This equation provides relationship between the drain time function T and the shape function f provided f is continuous and $\frac{(f(u))^2}{\sqrt{u}}$ is integrable on $[0, T]$. Computing the drain time function T for given f is known as the direct problem. The inverse problem consists of determining the shape f if the drain time T is given.

Example 305. A classical inverse problem in an ancient technology was design of a clepsydra or water clock. One design has depth y change at a constant rate with time. What is the shape function of such a clepsydra?

Example 306. Show that weight distribution is constant if shape is a parabola.

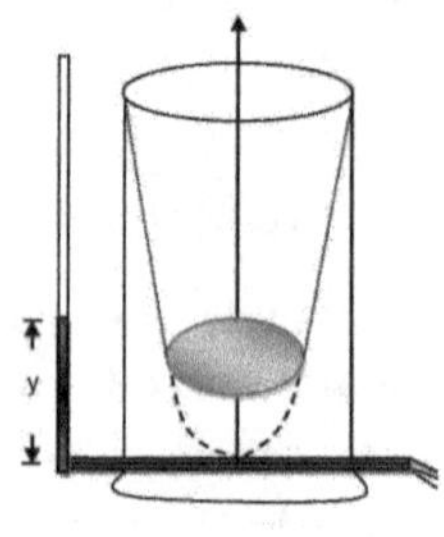

FIGURE 9.3: Influence of Gravity on Liquid Level

9.3.2 An Inverse Problem in Hanging Cable Models

Let us consider a cable that supports a horizontally distributed load, for example a suspension bridge (see Figure 9.4).

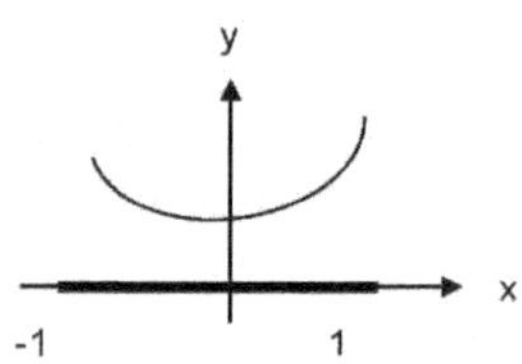

FIGURE 9.4: Loaded Cable

We neglect the weight of the cable. Assume that weight of the load is distributed nonuniformly along the interval $[-1, 1]$ of the x-axis. Let us assume that cable has a unique lowest point say, $x = 0$ in a simple case. $\int_{-1}^{x} w(u)du$ is the total weight of the segment $[-1, x]$ for each $x \in [-1, 1]$ for a weight distribution $w(x)$.

A given weight distribution w gives a shape $y = y(x)$ for the hanging cable. The direct problem is to find shape y caused by a given weight distribution w while its inverse problem is to determine a distribution w from observing the shape y.

9.3.3 Inverse Problems in Study of Trajectories

Consider again motion of a projectile with respect to a flat earth, see Figure 9.5.

We consider a projectile of unit mass that is fired horizontally with a given velocity v from a height h as shown in Figure 9.5. It can be derived that the

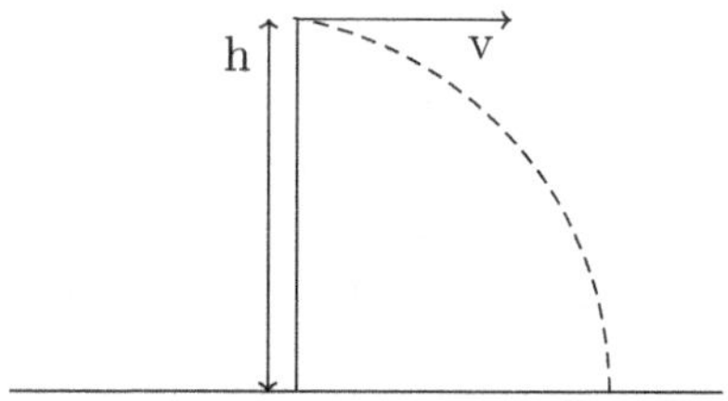

FIGURE 9.5: Horizontal shot from height

path of projectile is parabolic if there is no resistance and gravitational force is treated as a constant acting vertically, that is, the path of a projectile is a parabola under no resistance and constant gravity. Two related inverse problems are:

1. If the trajectory of a projectile is parabolic and there is no resistance, is gravitational force necessarily constant?
2. If the trajectory of a projectile is parabolic under constant gravitational force, can there be any resistance?

Example 307. Let a particle of unit mass be acted upon by a velocity gravitational force $g(y)$ and follow the arc of a circular trajectory $x^2 + y^2 = a^2$. Find the form of $g(y)$.

9.4 Inverse Problems in Matrix Equations

Consider a matrix equation of type $Ax = y$, where A is an $n \times n$ matrix and x and y are vectors. A **direct problem** is to find y if A and x are given. Finding x if A and y are given is known as an **inverse problem** in our terminology. The problem of solving this linear system is the type of inverse problem, often called a **causation problem**. If x and y are given, finding A is another inverse problem, called the **identification problem**. In courses of linear algebra, existence and uniqueness of a solution are critical, but no attention is paid to a third issue: inverse problem instability. In this section we illustrate inverse problems of matrix equations with special reference to elementary tomography, gravimetry, dynamical parameters of binary dynamical systems and stereography.

9.4.1 Inverse Causation Problem

Let A be an $m \times n$ matrix. Finding vector $y = Ax$ for given A and x is direct problem. The **inverse causation problem** involves finding all vectors

x such that $Ax = y$ for a given A and vector y. This problem gets a lot of attention in linear algebra but rarely seen as example of an inverse problem. Another interesting inverse problem, called the **identification problem** is the one in which we find matrix A such that $Ax = y$ for given vectors x and y. In this subsection we confine discussion to the inverse causation problem. A solution $x \in R^n$ of the inverse causation problem, where A is $m \times n$ matrix and $y \in R^m$ is a given vector, exists if and only if y belongs to the range of A, that is, in the subspace

$$R(A) = \{Ax | x \in R^n\}.$$

Determining whether $y \in R(A)$, that is, whether a solution exists, and finding all solutions can be accomplished by the method of Gaussian elimination. The uniqueness issue is addressed by the null space of A, that is

$$N(A) = \{x \in R^n | Ax = 0\}.$$

The Gaussian elimination method is an elegant algorithm for characterizing the null space and settling the uniqueness question. Let us assume that $Ax = y$ has a unique solution for each y, that is, A^{-1} exists. We are interested in knowing the error in the solution x for small errors in y. Let x be the unique solution of the system of equation $Ax = y$ for known y, and let $\tilde{x}$ be the unique solution of the system $Ax = \tilde{y}$, that is, $A\tilde{x} = \tilde{y}$ where $\tilde{y}$ is perturbation of y, then

$$\|x - \tilde{x}\| = \|A^{-1}y - A^{-1}\tilde{y}\| \leq \|A^{-1}\|\|y - \tilde{y}\|$$

$$\begin{aligned} \|x - \tilde{x}\| &\leq \|A^{-1}\|\|y\|\frac{\|y - \tilde{y}\|}{\|y\|} \\ &= \|A^{-1}\|\|A\|\|x\|\frac{\|y - \tilde{y}\|}{\|y\|}. \end{aligned}$$

Hence,

$$\frac{\|x - \tilde{x}\|}{\|x\|} \leq cond(A)\frac{\|y - \tilde{y}\|}{\|y\|}$$

$$\text{where } cond(A) = \|A\|^{-1}\|A\|.$$

This is called the condition number of the matrix A with respect to the norm $\|.\|$. This condition number gives an upper bound for the relative error in the solution caused by a relative error in y. For matrices with large condition numbers, that is, ill-conditioned matrices, relatively small changes in y yield relatively large changes in the solution x.

9.4.2 Identification Problems

The identification problem is finding $m \times n$ matrices A for given pairs of vectors $(\mathbf{x}, \mathbf{y})$ connected by

$$A\mathbf{x} = \mathbf{y}.$$

For each such pair we call $\mathbf{x}$ the input and $\mathbf{y}$ the output. An $m \times n$ matrix A is called identifiable from the matrix pair $(\mathbf{x}, \mathbf{y})$ if $A\mathbf{x} = \mathbf{y}$.

9.4.3 Inverse Problem for Eigenvalues and Eigenvectors

For a given matrix A finding a scaler λ such that $Ax = \lambda x$ for vector x is called an eigenvalue problem. It is a direct problem associated with an eigenvalue λ and eigenvector x.

The inverse problem for eigenvalues and eigenvectors is using eigenvalues and corresponding eigenvectors of a matrix find the matrix. For example, suppose that eigenvalues of

$$A = \begin{bmatrix} \alpha & -\beta \\ -\beta & \gamma + \eta \end{bmatrix}$$

are 2 and 7 and eigenvalues of the reduced matrix

$$A^{-} = \begin{bmatrix} \alpha & -\beta \\ -\beta & \gamma \end{bmatrix}$$

are 1 and 5. Find A and A^{-}.

9.4.4 Least Squares Solutions to Inverse Problems

In the case when $\mathbf{b} \in R^m$ is not in range of the $m \times n$ matrix A, there is no solution to the problem $A\mathbf{x} = \mathbf{b}$. A remarkable relationship between the null space, range, and transpose allows the development of a type of generalized solution, and such generalized solutions always exist. One type is called the **least squares** solution, that is a vector $\mathbf{u} \in R^n$ that minimizes the quantity $\|A\mathbf{x} - b\|$ over all $\mathbf{x} \in R^n$, where the norm is the usual Euclidean norm. Note that this minimum is zero if and only if the system has a solution. If $\mathbf{u}$ is a least square solution, then for any vector $\mathbf{v} \in R^n$, the function

$$g(t) = \|A(\mathbf{u} + t\mathbf{v}) - \mathbf{b}\|^2 = \|A\mathbf{u} - \mathbf{b}\|^2 + 2(A\mathbf{v}, A\mathbf{u} - \mathbf{b})t + \|A\mathbf{v}\|^2 t^2,$$

where $(\cdot, \cdot)$ is the familiar Euclidean inner product, has a minimum at $t = 0$. The ncessary condition $g'(0) = 0$ for a minimum then gives

$$(A\mathbf{v}, A\mathbf{u} - \mathbf{b}) = 0$$

and hence $(\mathbf{v}, A^T A\mathbf{u} - A^T\mathbf{b}) = 0$ for all $\mathbf{v} \in R^n$. That is, if $\mathbf{u}$ is a least square solution, then

$$A^T A\mathbf{u} = A^T\mathbf{b},$$

where A^T is the transpose of A. Conversely, if $A^T A\mathbf{u} = A^T\mathbf{b}$, then for any $\mathbf{x} \in R^n$,

$$\begin{aligned}
\|A\mathbf{x} - b\|^2 &= \|A(\mathbf{x}-\mathbf{u}) + A\mathbf{u} - \mathbf{b}\|^2 \\
&= \|A(\mathbf{x}-\mathbf{u})\|^2 + 2(A(\mathbf{x}-\mathbf{u}), A\mathbf{u}-\mathbf{b}) + \|A\mathbf{u}-\mathbf{b}\|^2 \\
&= \|A(\mathbf{x}-\mathbf{u})\|^2 + 2(\mathbf{x}-\mathbf{u}, A^T A\mathbf{u} - A^T\mathbf{b}) + \|A\mathbf{u}-\mathbf{b}\|^2 \\
&\geq \|A\mathbf{u}-\mathbf{b}\|^2,
\end{aligned}$$

that is, $\mathbf{u}$ is a least squares solution of $A\mathbf{x} = \mathbf{b}$.

So, least squares solutions of $A\mathbf{x} = \mathbf{b}$ coincide with ordinary solutions of the symmetric problem $A^T A\mathbf{x} = A^T\mathbf{b}$. Now this symmetric problem always has a solution since $R(A^T) = R(A^T A)$ and hence $A^T\mathbf{b} \in R(A^T A)$ for any $\mathbf{b} \in R^m$. Therefore, any linear system $A\mathbf{x} = \mathbf{b}$ has a least squares solution. However, least squares solutions need not be unique. Indeed, if $\mathbf{u}$ is a least squares solution, then so is $\mathbf{u} + \mathbf{v}$ for any $\mathbf{v} \in N(A)$, that is, the set of least squares solutions forms a hyperplane parallel to null space. Therefore, if A has a nontrivial null space, then $A\mathbf{x} = \mathbf{b}$ has infinitely many least squares solutions.

However, one least squares solution can be distinguished from the others, namely, the one that is orthogonal to the null space. There can be at most one least squares solution if $\mathbf{u}$ and $\mathbf{w}$ are both least squares solutions that are orthogonal to $N(A)$. Also, $A^T A(\mathbf{u} - \mathbf{w}) = A^T\mathbf{b} - A^T\mathbf{b} = 0$, and hence $\mathbf{u}-\mathbf{w} \in N(A^T A) = N(A)$. Therefore, $\mathbf{u}-\mathbf{w} \in N(A) \bigcap N(A)^{\perp}$, that is, $\mathbf{u} = \mathbf{w}$. On the other hand, there is always a least squares solution that is orthogonal to the null space, and hence any linear system has a unique least squares solution that is orthogonal to the null space of the coefficient matrix. If we agree to accept this notion of generalized solutions, then every linear system has a unique (generalized) solution.

9.5 Inverse Problems in Differential equations

9.5.1 Inverse Problems for Mixing Problems

Direct mixing problems for well-stirred solutions are well known and taught in an introductory course of ordinary differential equations. Given an initial concentration of a solute and certain inflow and outflow rates, we can determine the concentrations at future times. The derivation of the model relies on rate balancing. Let $x(t)$ denote the quantity of solute in the vessel at time t, then the rate of change of $x(t)$ with time, namely $\dfrac{dx}{dt}$, is the difference between the rate at which the solute enters the vessel (inflow rate) and outflow rate (the rate at which it leaves the vessel). Thus

$$\frac{dx}{dt} = ar - \frac{x}{V}r \tag{9.2}$$

where V represents volume of the vessel, a denotes concentration of the solute and r denotes the rate at which the liquid enters the vessel and goes out. The concentration of solute in the vessel is given by $c(t) = \frac{x(t)}{V}$ and so the differential equation (9.2) takes the following form in terms of the concentration:

$$\frac{dc(t)}{dt} = \frac{r}{V}(a - c). \tag{9.3}$$

Equation (9.3) has the unique solution

$$c(t) = a + (c_0 - a)e^{\frac{-rt}{V}},$$

where c_0 is the initial concentration of solute in the vessel. The direct problem is to find $c(t)$, concentration at time t for given parameters a, c_0, V and r. There are several related inverse problems like finding the initial concentration or finding parameters for given concentration at time t.

9.5.2 Inverse Problem in Newton's Law of Falling

Let $s(t)$ denote the position of a particle, that is, its distance down a ramp from its origin, see Figure 9.6. Since the gravitational force in the direction of

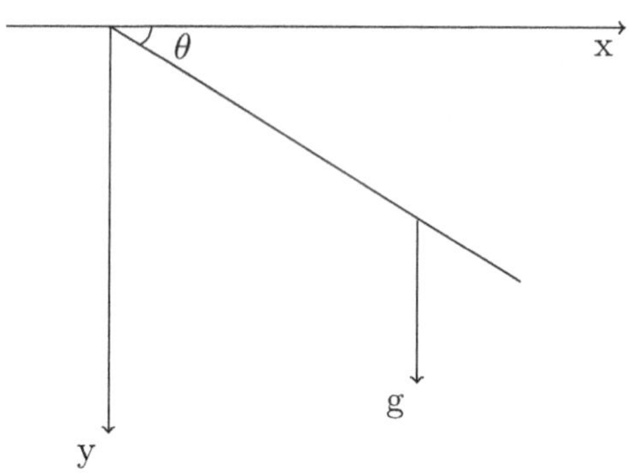

FIGURE 9.6: Descent of Ramp

the ramp is $g \sin\theta$, we have by Newton's law

$$\frac{d^2s}{dt^2} = g\sin\theta, s(0) = 0, \frac{ds}{dt}(0) = 0.$$

The solution of the initial value problem is

$$s = \frac{g}{2}(\sin\theta)t^2.$$

We review Galileo's law of falling bodies if $\theta = \frac{\pi}{2}$. For a given fixed time T, the curve of points reached in time T for various declination angles θ is a circle of radius $\frac{g}{4}T^2$:

$$\text{Let } x = s\cos\theta,\ y = s\sin\theta.$$

Using the equation for s, we get

$$x^2 = \frac{g}{2}T^2\sin^2\theta\frac{g}{2}T^2\cos^2\theta = y(\frac{g}{2}T^2 - y)$$

and hence

$$x^2 + (y - \frac{g}{4}T^2)^2 = (\frac{g}{4}T^2)^2.$$

The direct problem is the determination of the forms of the equitemporal curves, for certain given laws, while the inverse problem is to find the form of the resistance law from the knowledge of shapes of the equitemporal curves.

9.5.3 Inverse Problem in Newton's Cooling Law

According to the simplest version of Newton's law of cooling the rate at which the surface temperature changes in time is proportional to the difference between the ambient and surface temperatures. Let surface temperature at time t be $T(t)$ and ambient temperature be constant, say S. Newton's law of cooling implies that

$$\frac{dT}{dt} = \lambda(S - T)$$

where λ, the heat transfer coefficient, is a positive constant. The direct problem of determining the surface temperature has the unique solution:

$$T(t) = S + (T(0) - S)e^{-\lambda t}.$$

This solution depends on three parameters: the ambient temperature S, the initial temperature $T(0)$, and the heat transfer coefficient λ. Given surface temperature at appropriate times determination of parameters S, $T(0)$ and λ is an inverse problem.

Example 308. Surface temperatures of a body that cools according to Newton's law are given in the following table.

t (min)	$u(^oF)$
5	72
16	62
15	54

Find the ambient temperature, the initial surface temperature, and heat transfer coefficient $(S, T(0), \lambda)$.

9.5.4 Inverse Problem in Finance

In a continuously compounded interest model, the percentage rate of change of an investment is a given constant interest rate r. That is, the rate of change of the value, relative to the current value is a constant r. A mathematical formulation is

$$\frac{1}{f}\frac{df}{dt} = r$$

where f is the value history, that is $f(t)$ denotes the value of the investment at time t. This, of course is equivalent to

$$\frac{d}{dt}\ln f = r.$$

Solution of this equation is

$$f(t) = f(0)e^{rt}.$$

In the variable interest rate, the interest rate is a function of time, say $r(t)$. This implies

$$\frac{d}{dt}\ln f = r(t)$$

which gives

$$f(t) = f(0)e^{\int_0^t r(s)ds}.$$

Given $f(0)$ and $r(t)$, finding $f(t)$ is regarded as a direct problem while finding $r(t)$ given $f(t)$ for all t is an inverse problem. $f(t)$ is called value history and $r(t)$ is known as variable interest rate.

Example 309. Find the interest rate $r(t)$ so that the number of *Rs* 50,000 invested yields *Rs* 50,000 in 10 years.

9.5.5 Inverse Problem in Carbon dating

Solution of Equation (2.10) is the basis of an important technique developed by the chemist Willard Libby in 1950 to determine the ages of certain artifacts. Libby was awarded the 1960 Nobel Prize for chemistry for this work. The process of estimating the age of an artifact or fossil is called carbon dating. The theory of carbon dating is essentially based on the fact that the half life of radioactive C–14 is approximately 5600 years. Libby's method has been used to date furniture in Egyptian tombs, examine Van Meegeren Art forgeries and date different civilizations through archaeological excavation. In Example 128 direct problem is solved. The inverse problem is to find λ while T and $A(t)$ are given.

9.5.6 Inverse Problem in Population Growth

A direct problem is studied in Section 2.6.1 (Example 121). Finding λ when $P(t)$ and $P(0)$ are given is an inverse problem. In Example 122 finding the annual growth rate is an inverse problem.

Example 310. Consider the following one-dimensional initial value problem. Find $u(x)$ such that

$$\begin{aligned}-\frac{d}{dx}\left(a\frac{du}{dx}\right) &= f \text{ on } (0,1)\\ a(0)\frac{du}{dx}(0) &= g_0\\ a(1)\frac{du}{dx}(1) &= g_1.\end{aligned}$$

We are interested in the following inverse problem. Given z, a measurement of u, find a which together with z satisfies the boundary value problem.

Let $z = u + \mu(x)$, where $\int_0^1 |\mu(x)|^2 dx < \infty$. We choose $f = 0$, $g = g_1 = 1$. Then for $u(x) = x + 1$. The unique solution is $a(x) = 1$. If we have

$$z_n(x) = u(x) + \epsilon\frac{\cos(2n\pi x)}{2n\pi}, \quad 0 < \epsilon < 1,$$

The corresponding solution is

$$a_n = \frac{1}{1 - \epsilon\sin(2n\pi x)}.$$

It can be checked that $z_n \to n$ but $a_n \nrightarrow a$.

9.6 Inverse Problem in Partial Differential Equations

9.6.1 Inverse Problem for Heat Equation

Consider a rod of unit length and unit thermal conductivity which ends at a fixed temperature. The temperature distribution $u(x,t)$ satisfies the heat equation

$$a^2\frac{\partial^2 u}{\partial x^2} - \frac{\partial u}{\partial t} = 0, \; 0 < x < 1, \; t > 0$$

with the initial and boundary conditions

$$u(x,0) = u_0(x) \text{ and } u(0,t) = u(1,t) = 0.$$

The inverse problem is to find the initial temperature distribution $u_0(x)$ when the temperature distribution at time $T > 0$ is given. Finding A while u is known is another inverse problem as is determining $u_0(x)$ if $a$and u are given. This problem can be solved following Amita and Siddiqi [41].

9.6.2 Inverse Scattering Problem

In inverse scattering problems one is looking for an object D or a medium parameter from results of observations of a so-called field generated by (plane) incident waves $u'(x) = \exp(ik\hat{\theta}x)$, where $k > 0$ denotes **wave number** and $\hat{\theta}$ is a unit vector that describes the direction of the incident wave. The field (acoustic, electromagnetic or elastic) in the simplest situation of scattering by an obstacle D is a solution u to the Helmholtz equation

$$\Delta u + k^2 u = 0 \text{ in } R^N \backslash \overline{D} \quad (N = 2, 3)$$

satisfying the homogeneous Dirichlet boundary condition

$$u = 0 \text{ on } \partial D \text{ (soft obstacle)}$$

or another boundary condition like the Neumann condition

$$\partial_v u + bu = 0 \text{ on } \partial D \text{ (hard obstacle)}$$

along with

$$\frac{\partial u^s}{\partial r} - iku^s = O(r^{-(N+1)/2}) \text{ for } r = |x| \to \infty \text{ uniformly in } \frac{x}{|x|}.$$

This solution is assumed to be the sum of the plane incident wave u_i and a scattered wave u_s due to the presence of an obstacle, that is,

$$u = u' + u^s.$$

In most cases in scattering theory one assumes that $R^3 \backslash D$ is connected, and D is a bounded open set with Lipshitz boundary. In some situations spherical incident waves (depending only on $|x - a|$) are more useful and natural.

For, acoustic scattering problems, $v(x) = u(x)e^{i\omega t}$ describes the pressure and $k = \omega/c$ is the wave number with speed of sound c. For suitably polarized time harmonic electromagnetic scattering problems. Maxwell's equation reduces to the two-dimensional Helmholtz equation for the components of the electric (or magnetic) field u. The wave number k, is given in terms of the dielectric constant ϵ and permeability μ by $k = \sqrt{\epsilon\mu\omega}$.

In both cases, the radiation condition yields the following asymptotic behavior:

$$u^s(x) = \frac{\exp(ik|x|)}{|x|^{(n-1)/2}} u_\infty(\tilde{x}) + O(|x|^{-(N+1)/2}) \; as|x| \to \infty,$$

where $\tilde{x} = x/|x|$. The inverse problem is to determine the shape of D when the far field pattern $u_\infty(\tilde{x})$ is measured for all $(\tilde{x})$ is measured for all $\tilde{x}$ on the unit sphere in R^N.

9.6.3 Examples

These and related inverse scattering problems have various applications in computer tomography, seismic and electromagnetic exploration in geophysics, and nondestructive testing of materials.

9.6.4 Inverse Problem for Wave Equation

In the following equation,

$$\frac{\partial^2 u}{\partial t^2} = c^2 \frac{\partial^2 u}{\partial x^2},$$

$u(x,t)$ represents the displacement, for example, of a vibrating string from its equilibrium position, and c the wave speed.

These types of equations have been applied to model vibrating membranes, acoustic problems for the velocity potential of the fluid flow through which sound can be transmitted, longitudinal vibrations of an elastic rod or beam, and both electric and magnetic fields in the absence of charge and dielectric. Solving this equation under given initial or/and boundary conditions for a given parameter is called a direct problem. Find the parameter c for a given solution under the approximate boundary and initial conditions is an inverse problem. Similarly for a given solution and parameter, determining an initial or boundary condition is also known as an inverse problem.

9.6.5 Examples

The inverse problem for the following well known equations can be posed on similar lines.

(a) Laplace equation in R^2 (two-dimensional)

$$\nabla^2 u = \Delta u = \frac{\partial^2 u}{\partial x^2} + \frac{\partial^2 u}{\partial y^2} = 0,$$

where $\nabla^2 = \nabla.\nabla$ denotes the Laplacian

$$\nabla = \left(\frac{\partial}{\partial x}, \frac{\partial}{\partial y} \right).$$

The equation is satisfied by the electrostatic potential in absence of charges, by the gravitational potential in the absence of mass, by the equilibrium displacement of a membrane with a given displacement of its boundary and by the velocity potential for an inviscid, incompressible, irrotational homogeneous fluid in the absence of sources and sinks and many other real world systems.

(b) Poisson equation (non-homogeneous Laplace equation)

$$\nabla^2 u = -f(x,y).$$

One encounters this equation while studying the electrostatic potential in the presence of charge, the gravitational potential in the presence of distributed matter, the equilibrium displacement of a membrane under distributed forces, the velocity potential for an inviscid, incompressible, irrotational homogeneous fluid in the presence of distributed sources and sinks, and the steady state temperature in the presence of thermal sources or sinks.

(c) Transport equation in R (one-dimensional)

$$\frac{\partial u}{\partial t} + c\frac{\partial u}{\partial x} = 0,$$

where c is a constant and $u(x,t)$ denotes the location of a car at time t and position x.

(d) Traffic flow

$$\frac{\partial u}{\partial t} + a(u)\frac{\partial u}{\partial x} = 0,$$

where $u(x,t)$ denotes the density of cars per unit kilometer of expressway at location x and time t and $a(u)$ is a function of u, say the local velocity of traffic at location x at time t.

(e) Burger's equation in one dimension

$$\frac{\partial u}{\partial t} + u\frac{\partial u}{\partial x} = 0.$$

This equation arises in the study of streams of particles or fluid flow with zero viscosity.

(f) Eikonal equation in R^2 (two dimensional)

$$\left(\frac{\partial u}{\partial x}\right)^2 + \left(\frac{\partial u}{\partial y}\right)^2 = 0$$

models problems of geometric optics.

(g) Helmholtz equation

$$(\nabla^2 + k^2)u = 0$$

has been found useful in diffraction theory.

(h) Klein-Gordon equation

$$\frac{\partial^2 u}{\partial t^2} - c^2\nabla^2 u + m^2 u = 0.$$

This equation arises in quantum field theory, where m denotes the mass.

(i) Telegraph equation

$$\frac{\partial^2 u}{\partial t^2} + A\frac{\partial u}{\partial t} + Bu = \frac{\partial^2 u}{\partial x^2}.$$

A and B are constants. This equation arises in the study of propagation of electrical signals in a cable transmission line. Both the current I and voltage V satisfy equations of this type. This equation also arises in the propagation of pressure waves in the study of pulsatile blood flow in arteries.

(j) Schrodinger equation (time independent)

$$\frac{h^2}{2m}\left(\frac{\partial^2 u}{\partial x^2}+\frac{\partial^2 u}{\partial y^2}\right)+(E-V)u=0.$$

Note that m is the mass of the particle whose wave function is $u(x, y)$, h is the universal Planck's constant, V is the potential energy and E is a constant. This equation arises in quantum mechanics. If $V = 0$ then it reduces to the Helmholtz equation.

(k) Korteweg de Vries (KDV) equation in one dimension

$$\frac{\partial u}{\partial t}+cu\frac{\partial u}{\partial x}+\frac{\partial^3 u}{\partial x^3}=0.$$

This equation arises in shallow water waves.

(l) Euler equation in R^3

$$\frac{\partial u}{\partial t}+(u.\nabla)u+\frac{1}{\rho}\nabla p=0.$$

The u denotes the velocity field, and p the pressure.

(m) Navier-Stokes equation in R^3

$$u_t+(u.\nabla)u+\frac{1}{\rho}\nabla p=v\nabla^2 u.$$

The v denotes the kinematic viscosity, p stands for pressure and ρ for density of the fluid, and $u = (u_1, u_2, u_3)$ represents velocity of the fluid.

(n) Maxwell equations in R^3

$$\frac{\partial E}{\partial t}-\nabla\times H=0.$$

$$\frac{\partial H}{\partial t}-\nabla\times E=0.$$

E and H denote the electric and the magnetic fields, respectively.

9.6.6 Financial Mathematics

Option pricing also called options, are contracts that give an owner the right to buy or to sell a fixed number of underlying assets in a specified common market at a fixed price on or before a prescribed date. Options are widely traded on all major exchanges. Strike price or exercise price is the price at which the underlying asset is bought in options. Underlying assets are commodities, shares, foreign currencies, government or private enterprise bonds, stocks and stock indices. Equity is a share in the ownership of a company which usually guarantees the right to vote at meeting and share in dividends (payments to shareholders as return for investment). A derivative (financial derivative) is a contract or security whose payoffs are final (asset). In many cases a derivative is the price of the underlying equity. The termination time of a derivative (contract), usually the time when payoff value is calculated and paid, is called expiry (time). Volatility is a measure of standard deviations of returns. It is a function of underlying asset and time.

Let u denote the value or premium of the options which depends on a number of factors such as stock price say x, current time t, the maturity date T, the exercise price K, the risk free interest r and local volatility (infinitesimal standard deviation) σ (may be constant only time dependent or may satisfy relation of type $\sigma(x,t) = \sigma(x)\sigma(t)$. It has been proved that $u(x,t)$ is a solution of the partial differential

$$\frac{\partial u}{\partial t} + \frac{1}{2}x^2\sigma^2(x,t)\frac{\partial^2 u}{\partial x^2} + rx\frac{\partial u}{\partial x} - ru = 0. \tag{9.4}$$

This is the Black-Scholes equation.

Direct problem of option pricing: Given the local volatility function $\sigma(x,t)$ find the solution of (9.4) with boundary conditions:

$$u(x,t) = max(x - K, 0) \text{ and } \mathrm{u(0,t) = 0}. \tag{9.5}$$

Inverse problem: Find the local volatility function $\sigma(x,t)$ such that the solution to (9.4) with different strikes K and maturities T satisfy

$$u(x^*, t^*, K, T) = u^*(K, T) \tag{9.6}$$

where the right hand side denotes the current market price of the option with the corresponding strike and maturity at time t^* when the underlying price (asset) is x^*. Practitioners refer to this problem as model calibration.

9.7 Inverse Problems of Image Processing

9.7.1 Fundamental Steps in Digital Image Processing

Image acquisition: The types of images in which we are interested are generated by the combination of an illumination source and reflection or absorption of energy from that source. The illumination may originate from a source of electromagnetic energy such as radar, infrared or X-ray energy. Sensors are used to transform illumination energy into digital images. In a simple way incoming energy is transformed into voltage by a combination of input electrical power and sensor material that is responsive to the particular type of energy detected. The output voltage waveform is the response of the sensor, and a digital quantity is obtained from each sensor by digitizing its response.

Image enhancement: Brings out obscured details or highlights certain features of interest in an image.

Compression: Deals with techniques for reducing the storage required to save an image. Image compression is familiar to most users of computers in the form of image file extensions, such as jpg file extensions.

Morphological processing: Deals with the tools for extracting image components that are useful in the representation and description of shape.

Segmentation: Partitions images into their constituent parts.

Representation and description: They follow the output of a segmentation stage which is usually raw pixel data, constituting the boundary of a region or all points in a region.

9.7.2 Introduction to Medical Imaging

Gamma-ray imaging: Major uses of imaging based on gamma rays include nuclear medicine and astronomical observations. In nuclear medicine, the approach is to inject a patient with a radioactive isotope that emits gamma rays as it decays. Images are produced from the emissions collected by gamma ray detectors. The principle is the same as for X-ray tomography. However, instead of using an external source of X-ray energy, the patient is given a radioactive isotope that emits protons as it decays. When a proton meets an electron both are annhilated; image is created using the basic principles of tomography.

X-ray imaging: X-rays are among the oldest sources of electromagnetic radiation used for imaging. The best known use of X-rays is for medical diagnostics, but they are also used extensively in industry and other areas, like astronomy. X-rays for medical and industrial imaging are generated using a vacuum tube with a cathode and anode. The cathode is heated, causing free electrons to be released. These electrons flow at high speed to the positively charged anode. When the electrons strike a nucleus, energy is released in the form of X-ray radiation. The energy (penetrating power) of the X-rays is controlled by a voltage applied across the anodes, and the number of X-rays is controlled by a current applied to the filament in the cathode.

Imaging in ultraviolet band: Applications of ultraviolet light are varied. They include lithography, industrial inspection, microscopy, lasers, biological imaging and astronomical observation. We illustrate imaging in this band with examples from microscopy and astronomy.

Ultraviolet light is used in fluorescence microscopy, one of the fastest growing areas of microscopy. Fluorescence is a phenomenon discovered in the middle of the 19th century, when it was first observed that the mineral fluorspar fluoresces when ultraviolet light is directed upon it. The ultraviolet light is not visible, but when a photon of ultraviolet radiation collides with an electron in an atom of a fluorescent material, it elevates the electron to a higher energy level. Subsequently, the excited electron relaxes to a lower level and emits light in the form of a lower energy photon in the visible (red) light region. The basic task of a fluorescence microscope is to use an excitation light to irradiate a prepared specimen and then separate the much weaker radiation fluorescent light from brighter excitation light.

Imaging in visible and infrared bands: Considering that the visual band of the electromagnetic spectrum is the most familiar, it is not surprising that imaging in this band outweighs by far all the others in terms of scope of application. The infrared band is often used in conjunction with visual imaging in areas such as light microscopy, astronomy, remote sensing, industry and law enforcement.

Even in microscopy, the application areas are too numerous to detail here. It is not difficult to conceptualize the types of processes one might apply to these images, ranging from enhancements to measurements.

EEG: An electroencephalogram (EEG) is a test that measures and records the electrical activity of the brain. Special sensors (electrodes) are attached to a patient's head and hooked by wires to a computer. The computer records the brain's electrical activity on the screen or on paper as wavy lines. Certain conditions, such as seizures, can be revealed by the changes in the normal pattern of the brain's electrical activity.

ECG: The electrocardiogram (ECG) records the electrical activity of the heart, where each heart beat is displayed as a series of electrical waves characterized by peaks and valleys. Any ECG gives two kinds of information; the duration of the electrical wave crossing the heart which in turn decides whether the electrical activity is normal or slow or irregular and also the amount of electrical activity passing through the heart muscle which reveals whether the parts of the heart are too large or overworked. Normally, the frequency range of an ECG signal is 0.05 to 100 Hz and its dynamic range to 1–10 mV.

MRI: Magnetic resonance imaging (MRI) is a test that uses a magnetic fields and pulses of radio wave energy to depict organs and structures inside the body. In many cases MRI gives more detailed information about structures in the body that can be seen with an X-ray, ultrasound, or computed tomography (CT) scan. MRI also may show problems that cannot be seen with other imaging methods.

FMRI: Functional magnetic resonance imaging (FMRI) is a technique for measuring brain activity. It works by detecting the changes in blood oxygenation and flow that occur in response to neural activity. When a brain area is more active, it consumes more oxygen and to meet this increased demand blood flow increases flow to the active area. FMRI can be used to produce activation maps showing which parts of the brain are involved in a particular mental process.

9.7.3 Tomography

Scientists who developed the field of tomography have received several Nobel prizes. The word *tomography* comes from the Greek word for slice and the discipline deals with exposing internal structures of non-transparent objects by transmitting signals such as electromagnetic waves (such as radio waves, microwaves and X-rays) and acoustic waves. In computed tomography, scanner images do not provide ready images as X-ray systems do. Rather, the images result from intricate data measurements and mathematical processes. The scientists who have studied the relevance of wavelets and their variants to tomography include Mark Bottema, Bill Moran and Sofia Suvorova [5], Glenn Easley, Flavia Colonna, Demetrio Labate, and Kanghui Guo. Several publications focus on variants of wavelets such as curvelets, ridgelets, complex wavelets and vector valued wavelets.

9.8 Seismic Tomography

Seismic tomography is a methodology for analyzing and computing earth's activities. Seismology is the scientific study of earthquakes and the propagation of elastic waves through the earth or other planet-like bodies. It also includes studies of earthquake effects, such as tsunamis diverse seismic sources such as volcanic, tectonic, oceanic, atmospheric and artificial process (such as explosions). A record of earth motion as a function of time is called a seismogram. Experts treat seismic tomography as a part of seismic imaging, where they are mainly concerned with estimating properties such as propagating velocities of compressional waves (P-waves) and shear waves (S-waves).

Seismic migration is a branch of seismic imaging in which the properties to be estimated include the reflection coefficient or reflectivity.

A tomography is defined as a methodology where three-dimensional images are derived from the integrated properties of a medium that rays encounter along their paths through it. Thus seismic tomography may be thought of as the derivation of a three-dimensional velocity structure from seismic waves. Estimation of P-wave velocity is the simplest example of seismic tomography. Several methods such as refraction, travel time tomography, finite frequency travel time tomography reflection travel time tomography and wave form tomography have been developed to estimate seismic tomography.

Seismic tomography is normally formulated as an inverse problem. A beautiful account of this special theme and inverse problems in general can be found in Iske and Randen [17] and Vogel [42]. An application of wavelets in tomography has been discussed in Bottema et al. [5].

9.9 Exercises

9.1. Solve the inverse problem of determining the depth D of a well from the elapsed time t between the dropping of a stone and hearing of a splash. Does the equation determining D have a unique solution? Is there a unique feasible solution?

9.2. A stone is dropped into a well and 4.2 seconds later a splash is heard. How deep is well?

9.3. Find the weight distribution with total weight 1 that gives shape $y(x) = 1 - \cos\left(\frac{\pi}{x}\right)$.

9.4. Show that if a shape is a parabola its weight distribution is constant.

9.5. Find the solution of Exercises 304 through 307.

9.6. Find matrix A that satisfies $AX = B$, where

$$X = \begin{pmatrix} 1 & 1 & 1 \\ -1 & -1 & 0 \\ 0 & 1 & 1 \end{pmatrix} \quad \text{and} \quad B = \begin{pmatrix} -2 & 6 & 3 \\ -1 & 0 & 0 \\ 3 & 0 & \end{pmatrix}.$$

9.7. Find the solution of Exercise 308.

9.8. Show that a differential coefficient satisfying

$$-\frac{d}{dx}\left(k(x)\frac{du}{dx}\right) = f(x)$$
$$-k(0)u'(0) = A, \quad -k(1)u'(1) = B$$

exists only if $\int_0^1 f(s)ds = A.B.$

9.9. Find the variable interest rate corresponding to the value history

$$u(t) = \sqrt{t + 5}e^{10t}.$$

9.10. Show that the two value histories have the same interest rate if and only if their ratio is a positive constant.

9.10 Suggestion for Further Reading

Direct problems for equations in Section 9.6.5 are discussed in Chapter 5 and 7. Study of inverse problems related to these equations may be challenging. References [1], [4], [5], [7], [9], [11], [13], [17], [18], [24], [32] and [39] provide interesting results in this field. As we have seen there is a close connection of tomography, radon transform and medical imaging. Kutchment [31] presents updated developments in this area. Readers interested in applications of inverse problems in medical and biological science may pursue references [31, 37]. For advanced knowledge and application in these areas, we refer [2], [3], [6], [8], [10], [12], [14] through [16], [19] through [23], [25] through [30], [33] through [36], [38] through [41] and [42].

Bibliography

[1] G. Bal, Inverse Transport theory and applications, *Inverse Problems*, 25, 053001, 2009.

[2] G. Bal, Cauchy problem ultrasound modulated EIT, *Analysis of Partial Differential Equations*, 6: 751-775, 2013.

[3] G. Bal, D. Finch, P. Kuchment, P. Stefanov, G. Uhlmann (Eds.), *Tomography and Inverse Transport Theory*, AMS, Providence, RI, 2011.

[4] C. Börgers and F. Natterer (Eds), *Computational Radiology and Imaging. Therapy and Diagnostics*, Springer-Verlag, New York, 1999.

[5] M. Bottema, B. Moran and S. Suvorova, An application of wavelets in tomography, *Digital Signal Processing*, 8: 244-254, 1998.

[6] R. Ewing (Ed.), *The Mathematics of Reservoir Simulation*, SIAM, Philadelphia, 1983.

[7] T. M. Buzug, *Computed Tomography*, Springer, Berlin, 2008.

[8] D. Colton, R. Ewing, W. Rundell (Eds.), *Inverse Problems in Partial Differential equations*, SIAM, Philadelphia, 1990.

[9] D. Colton and R. Kress, *Inverse Acoustic and Electromagnetic Scattering Theory*, Second Edition, Springer, Berlin, 1998.

[10] P. Elbau, O. Scherzer, R. Schulze, Reconstruction formulas for photoacoustic sectional imaging, *Inverse Problems*, 28, 045004, 2012.

[11] H. W. Engl, M. Hanke, A. Neubauer, *Regularization of Inverse problems*, Kluwer, Dordrecht, 1996.

[12] K. M. Furati, P. Manchanda, A. H. Siddiqi, Wavelet methods in oil industry, *in Proceedings of IWW*, Istanbul Aydin University, pp.26-36, 2008.

[13] K. M. Furati, M. Z. Nashed and A. H. Siddiqi(Eds.), *Mathematical Models and Methods for Real World Systems*. Chapman & Hall, New York, 2006.

[14] K. M. Furati, H. Tawfiq and A. H. Siddiqi, Simulation and visualization of safing sensor by fast flow, *AM. J. Appl. Sci.*, 2: 1261-1265, 2005.

[15] L. Gao, K. J. Parker, S. K. Alam, Sonoelasticity imaging: theory and experimental verification, *J. Acoust. Soc. Amer.*, 97: 3875-3880, 1995.

[16] R. C. Gonzales and R. E. Wood, *Digital Image Processing*, Reading MA, 1993.

[17] A. Iske and T. Randen (Eds), *Mathematical Methods and Modelling in Hydrocarbon Exploration and Production.* Springer, New York, 2006.

[18] C. W. Groetsch, *Inverse Problems in Mathematical Sciences*, Viewag, Braunschweig, Wiesbaden, 1994.

[19] C. W. Groetsch, *Inverse Problems: Activities for Undergraduates*, Mathematical Association of America, 1999.

[20] I. Hazou and D. Solmon, Inverse of exponential Radon transform II, *Analysis, Math. Methods Appl. Sci.*, 13: 205-218, 1990.

[21] I. Hazou and D. Solmon, Inversion of the exponential Radon Transform II, *Numerics, Math. Methods Appl. Sci.*, 13: 109-119, 1990.

[22] S. Helgason, *The Radon Transform*, Second Edition, Cambridge University, 1999.

[23] V. Isakov, *Inverse Source Problems*, AMS, New York, 1990.

[24] V. Isakov, *Inverse Problems for Partial Differential Equations*, Second Edition, Springer, Berlin, 2005.

[25] A. C. Kak and M. Slaney, *Principles of Computerized Tomographic Imaging*, SIAM, Philadelphia, 2001.

[26] A. I. Katsevich, Local tomography for the generalized Radon Transform, *SIAM J. Appl. Math.*, 57: 1128-1162, 1997.

[27] A. Katsevich, Local tomography with nonsmooth attenuation, *Trans. Amer. Math. Soc.*, 351: 1947-1974, 1999.

[28] A. Kirsch, *An Introduction to the Mathematical Theory of Inverse Problem*, Springer, New York, 1996.

[29] A. Kirsch and N. Grinberg, *The Factorization Model for Inverse Problems*, Oxford University Press, London, 2008.

[30] A. Kirsch and O. Scherzer, Simultaneous reconstructions of absorption density and wave speed with photoacoustics measurements, *SIAM, J. Appl. Math.*, 72: 1508-1523, 2012.

[31] P. Kuchment, *The Radon Transform and Medical Imaging*, Vol.85, Regional Conference Series in Applied Mathematics, SIAM, 2014.

[32] P. Monk, *Finite Element Methods for Maxwell's Equations*, Oxford Science Publications, Oxford, 2003.

[33] F. Natterer, *The Mathematics of Computerized Tomography*, Wiley, New York, 1986, reprinted in 2001 by SIAM.

[34] F. Natterer and F. Wübbeling, *Mathematical Methods in Image Reconstruction*, SIAM, Philadelphia, 2001.

[35] H. Neunzert and A. H. Siddiqi, *Topics in Industrial Mathematics: Case studies and Related Mathematical Methods*. Kluwer, Dordrecht, 2000.

[36] A. H. Siddiqi, *Applied Functional Analysis*, Marcel Dekker, New York, 2004. Indian Edition by Anamaya, New Delhi, 2010.

[37] A. H. Siddiqi, R. C. Singh and P. Manchanda (Eds.), *Mathematics in Science and Technology*, World Scientific, Singapore, 2011.

[38] A. H. Siddiqi, A. K. Gupta and M. Brokate (Eds.), *Modelling of Engineering and Technological Problems*, American Institute of Physics, Melville, NY, 2009.

[39] A. H. Siddiqi, I. S. Duff and O. Christensen, *Modern Mathematical Models: Methods and Algorithms for Real World Systems*, Anshan and Anamaya, 2007.

[40] A. H. Siddiqi and K. M. Furati, *Fast Wavelet Algorithms for Simulation Safing Sensor Final Report*, KFUPM, Dhahran, Saudi Arabia, 2005.

[41] J. Tittman, *Geological Well Logging*, Academic Press, London, 1986.

[42] C. R. Vogel, *Computational Methods for Inverse Problems*, Frontiers in Applied Mathematics, SIAM, Philadelphia, 2002.

Chapter 10

Wavelets

10.1 Introduction

Wavelet theory is the outcome of a multidisciplinary endeavor that brought together mathematicians, physicists and engineers. This interaction created a flow of ideas that goes well beyond the construction of new bases and transforms.

In fact, wavelet theory is a refinement and extension of Fourier analysis. Short-comings of Fourier analysis were realized as early as 1946. To remove these deficiencies, Nobel Laureate of Physics, Dennis Gabor, introduced the windowed Fourier transform (short-time transform). This transform suffered from some algorithmic difficulties and to eliminate them wavelet theory was introduced.

During the 1980's, French geophysicist Morlet developed a new approach the wavelet transform, while studying problems in oil and gas exploration. The term "wavelet" arose because the function from which the wavelet transform is constructed can be thought of as a localized wave.

In the last 35 years wavelet methods have been applied to diverse fields of science, engineering and technology [1] through [5], [9] through [11], [14], [17] through [40]. Typical applications are presented in these references to: physics, geophysics, function approximation, signal processing, harmonic analysis, differential equations, electrical engineering, biomedical engineering, EEG, ECG,

FIGURE 10.1: Prof. Yves Meyer

MRI, finance and financial data analysis, mechanical engineering, civil engineering, remote sensing, forestry and historical document retrieval.

A leading worker in this field, Prof. Yves Meyer, has received the Abel Prize of 2017 for his significant contributions in wavelet methods.

In this chapter we give an overview of wavelet methods and their applications to several areas of engineering and technology, namely biometrics, computed axial tomography, seismic tomography, medical imaging and power systems.

10.2 Overview of Wavelet Methods

10.2.1 Definition and Example of Wavelets

The definition of a wavelet is based on the notion of an orthonormal basis of functions. (Without explicitly mentioning wavelets, we have used them in Fourier analysis already.) A system $\{f_1, f_2, \dots\}$ of real valued functions defined on $\mathbb{R}$ is said to be orthonormal, if

$$\langle f_m, f_n \rangle = \int_{-\infty}^{\infty} f_m(x) f_n(x)\, dx = 0\,, \quad \text{whenever } m \neq n, \tag{10.1}$$

$$\text{and } \langle f_n, f_n \rangle = 1 \text{ for all } n.$$

This is the same as Definition 77, except that we now consider $\mathbb{R}$ instead of an interval $[a, b]$ as the domain of the functions f_n. The system $\{f_n\}$ is called an **orthonormal basis** (of the space of all square integrable functions) if it is complete in the sense that every square integrable function f can be expressed either as a linear combination of functions from the system $\{f_n\}$ or as a limit of such linear combinations, namely

$$\lim_{N\to\infty} \int_{-\infty}^{\infty} \left(f(x) - \sum_{n=1}^{N} c_n f_n(x)\right)^2 dx = 0 \tag{10.2}$$

for suitable coefficients c_n.

Definition 97.

$$L_2(R) = \{f : R \to R / \int_{-\infty}^{\infty} |f(t)|^2 dt < \infty\}.$$

Definition 98. (Vanishing moments) Let N be a positive integer. A function $\psi : R \to R$ has N vanishing moments if

$$\int_{-\infty}^{\infty} x^l \psi(x) dx = 0 \text{ for } l = 0, 1, 2, \cdots\cdots, N-1.$$

Definition 99. (Wavelet, wavelet family) Let ψ be a real valued function on $\mathbb{R}$ which is square integrable. For any integers j, k we define the functions

$$\psi_{j,k}(t) = 2^{j/2}\psi(2^j t - k)\,, \quad x \in \mathbb{R}\,. \tag{10.3}$$

If $\{\psi_{j,k}\}_{j,k\in\mathbb{Z}}$ is an orthonormal basis of the space of all square integrable functions, then ψ is called a **wavelet** and the family $\{\psi_{j,k}\}$ is called a **wavelet family**.

Example 311. (Haar wavelet) Let ψ be defined as follows

$$\psi(t) = \begin{cases} 1\,, & 0 \leq t < \frac{1}{2} \\ -1\,, & \frac{1}{2} \leq t < 1\,, \\ 0\,, & \text{otherwise.} \end{cases}$$

It can be checked that $\{\psi_{j,k}\}$ is an orthonormal basis for the square summable functions.

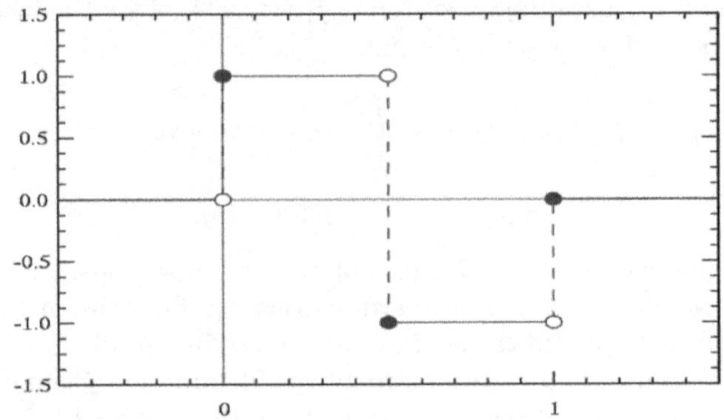

FIGURE 10.2: Haar Wavelet

The graph of ψ is given on the right side of Figure 10.2.

Example 312. (Mexican Hat Wavelet) Let the function ψ be defined as

$$\psi(t) = c(1 - t^2)e^{-t^2/2},$$

where c is a suitable constant to ensure orthonormality of the family $\{\psi_{j,k}\}$. Indeed, ψ is a wavelet, it is called the **Mexican hat wavelet** or the **Ricker wavelet**. Its graph (with $c = 1$) is given in Figure 10.3.

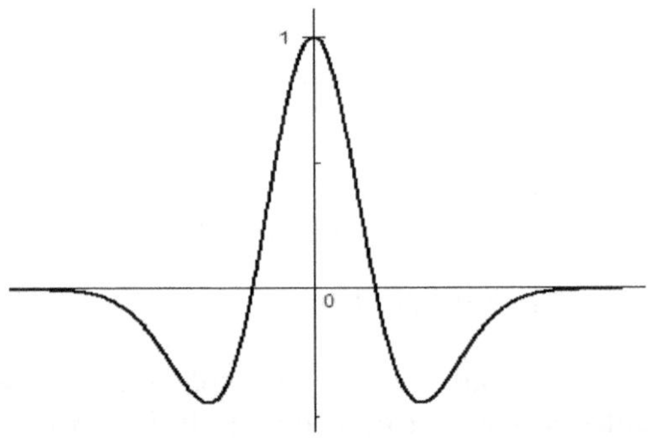

FIGURE 10.3: Mexican Hat Wavelet

Example 313. (Daubechies Wavelets) The Daubechies wavelets (Figure 10.4), based on the work of Ingrid Daubechies, are orthogonal wavelets defining a discrete wavelet transform and characterized by a maximal number of vanishing moments for some given support. With each wavelet type of this class, there is a scaling function (called the father wavelet) which generates an orthogonal multiresolution analysis. The Daubechies wavelets are not defined in terms of the resulting scaling and wavelet functions.

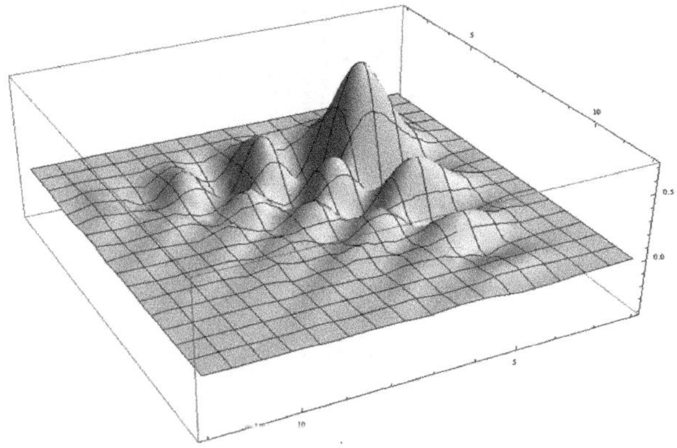

FIGURE 10.4: Daubechies 20 2-D Wavelet (Wavelet $Fn\times$ Scaling Fn)

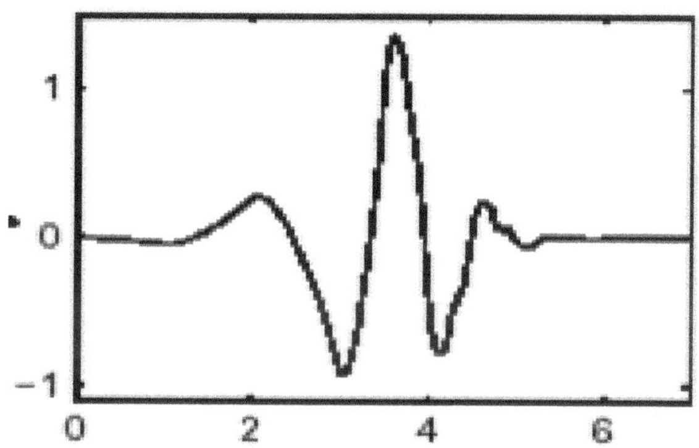

FIGURE 10.5: Db4

$Db4$ and $Db10$ (Figures 10.5 and 10.6) are from the Daubechies family of wavelets. Db wavelets have no analytical expression, they are constructed numerically.

Definition 100. Let ψ be a wavelet. The **wavelet coefficients** $d_{j,k}$ of a square integrable function f are defined as

$$d_{j,k} = \langle f, \psi_{j,k} \rangle = \int_{-\infty}^{\infty} f(t)\psi_{j,k}(t)\, dt\,. \tag{10.4}$$

Since the wavelet family is an orthonormal basis, one can show that for

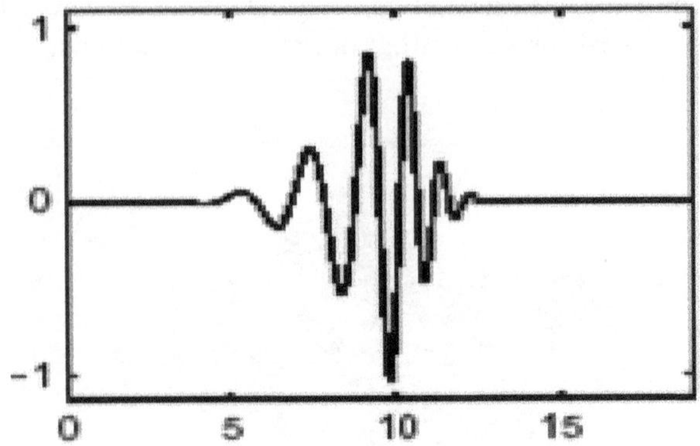

FIGURE 10.6: Db10

square integrable functions f

$$f(t) = \sum_{j=-\infty}^{\infty} \sum_{k=-\infty}^{\infty} d_{j,k}\psi_{j,k}(t)\,. \tag{10.5}$$

The right-hand side is called the **wavelet series** or the **wavelet representation** of f. The limit in the double sum is understood as convergence in the quadratic mean, as in (10.2),

$$\lim_{N\to\infty} \int_{-\infty}^{\infty} \left(f(t) - \sum_{j=-N}^{N} \sum_{k=-N}^{N} d_{j,k}\psi_{j,k}(t) \right)^2 dt = 0\,.$$

Remark 76. The wavelet coefficients $d_{j,k}$ measure the frequency content of f near the point $t = 2^{-j}k$. Small values of j correspond to low frequency components and high values of j correspond to high frequency components.

According to the above,

$$f(t) \approx \sum_{j=-N}^{N} \sum_{k=-N}^{N} d_{j,k}\psi_{j,k}(t) \tag{10.6}$$

if N is sufficiently large. One of the advantages of wavelets compared to other tools of signal processing is that it is often possible to obtain a good approximation of f using only relatively few coefficients $d_{j,k}$, so the computational effort is not large. This does not necessarily mean that N is small. But the localizing effect of wavelets often leads to situations where most of the wavelet coefficients $\{d_{j,k}\}_{|j|,|k|\le N}$ vanish or are very small, for example, when f varies only slowly on most parts of its domain and has high fluctuations in small regions only.

In Fourier analysis we studied the Fourier series the Fourier transform. For wavelets, have wavelet coefficients and wavelet transforms.

Definition 101. (Wavelet transform) Let ψ be a square integrable real valued function on $\mathbb{R}$. For $a, b \in \mathbb{R}$ with $a \neq 0$ we set

$$\psi_{a,b}(t) = \frac{1}{\sqrt{|a|}}\psi\Big(\frac{t-b}{a}\Big).$$

The **wavelet transform** of any square integrable function f with respect to ψ is denoted by $\mathcal{W}_\psi[f]$ and defined by

$$\mathcal{W}_\psi[f](a,b) = \langle f, \psi_{a,b}\rangle = |a|^{-1/2}\int_{-\infty}^{\infty} f(t)\psi\Big(\frac{t-b}{a}\Big)\,dt. \tag{10.7}$$

Remark 77. (a) The wavelet transform of f is a function of two variables (a and b), in contrast to the Fourier transform which is a function of a single variable. While both transforms yield information concerning the frequency domain, the wavelet transform exhibits information related to the time domain much more readily than the Fourier transform.
(b) The wavelet coefficients $d_{j,k}$ of f are related to the wavelet transform by

$$d_{j,k} = \langle f, \psi_{j,k}\rangle = \mathcal{W}_\psi[f](2^{-j}, k2^{-j}).$$

While the formulas defining the wavelet transform make sense for any square integrable function ψ, as in the definition above, the wavelet transform is of good use only if it is invertible, like the Fourier transform, because only in this case we can retrieve the original function from its wavelet transform. As an extreme case, setting $\psi = 0$ we have $\mathcal{W}_\psi[f] = 0$ for all f, so all information is lost when we apply $\mathcal{W}_\psi$. A sufficient condition for invertibility is given in the following definition.

Definition 102. Let ψ be integrable as well as square integrable on $\mathbb{R}$. We say that ψ is admissible if $\int_{-\infty}^{\infty}\psi(t)^2\,dt = 1$ and

$$c_\psi = \int_{-\infty}^{\infty}\frac{|\hat{\psi}(\xi)|^2}{|\xi|}\,d\xi < \infty. \tag{10.8}$$

Remark 78. A function ψ can be admissible according to this definition only if

$$\int_{-\infty}^{\infty}\psi(t)\,dt = 0.$$

Indeed, since ψ is assumed to be integrable on $\mathbb{R}$, its Fourier transform $\hat{\psi}$ is a continuous function according to Remark 50(ii); therefore the integral in (10.8) would be divergent if $\hat{\psi}(0) \neq 0$. Hence admissibility implies that

$$\int_{-\infty}^{\infty}\psi(t)\,dt = \hat{\psi}(0) = 0.$$

Remark 79. The function displays the time information and hides the information about frequencies while the Fourier transform displays information about frequencies and hides the time information. The energy of two signals represented by ψ $\hat{\psi}$ is the same, that is,

$$\|\hat{\psi}\|^2 = \int_{-\infty}^{\infty} |\hat{\psi}(\xi)|^2 d\xi = \int_{-\infty}^{\infty} |\hat{\psi}(t)|^2 dt = \|\psi\|^2.$$

Theorem 75. *(Inverse wavelet transform) Let ψ be admissible according to Definition 102. Then for any square integrable f we have*

$$f(t) = \frac{1}{c_\psi} \int_{-\infty}^{\infty} \int_{-\infty}^{\infty} \frac{1}{a^2} \mathcal{W}_\psi[f](a,b)\psi_{a,b}(t)\, da\, db\,. \tag{10.9}$$

The right hand side in (10.9) is to be understood as the limit for $\varepsilon \to 0$, in the mean quadratic sense, of the improper integrals

$$f_\varepsilon(t) = \frac{1}{c_\psi} \int_{-\infty}^{\infty} \int_{|a|\geq\varepsilon} \frac{1}{a^2} \mathcal{W}_\psi[f](a,b)\psi_{a,b}(t)\, da\, db\,.$$

During the above exposition, as well as previously for the Fourier transform, three concepts have appeared repeatedly, namely, dilation, translation (time shift) and modulation (frequency shift). These may be viewed as operators whose domains and ranges are sets of functions.

Definition 103. (a) **(Dilation)** For any real $a \neq 0$, the **dilation operator** D_a acts on functions $f : \mathbb{R} \to \mathbb{R}$ and yields the function $D_a f$ defined by

$$(D_a f)(t) = \sqrt{|a|} f(at)\,. \tag{10.10}$$

(b) **(Translation)** For any $b \in \mathbb{R}$, the **translation operator** T_b acting on f as in part (a) is defined as

$$(T_b f)(t) = f(t-b)\,. \tag{10.11}$$

(c) **(Modulation)** For any $c \in \mathbb{R}$, the **modulation operator** M_c acting on f as in part (a) is given by

$$(M_c f)(t) = e^{2\pi ict} f(t)\,. \tag{10.12}$$

Remark 80. (a) We observe that if $a > 1$, then $D_a f$ is a compressed version of f, and if $0 < a < 1$, then $D_a f$ is a spread-out version of f. If $a < 0$, $D_a f$ is, in addition, a reflected version of f.

(b) It can be checked from (4.29) that for every $a > 0$, $\widehat{D_a f}(\xi) = (D_{1/a}\hat{f})(\xi)$. This means that when a function is compressed by the factor $a > 1$, its Fourier transform is spread out by the factor $0 < 1/a < 1$.

(c) We can check that for every $c \in \mathbb{R}$, $\widehat{M_c f}(\xi) = (T_c \hat{f})(\xi)$. This means that modulation in the time variable corresponds to translation in the frequency variable, often referred to as **frequency shift**.

Dilation, translation and modulation have the following properties.

$$\begin{aligned} D_aT_bf(t) &= a^{1/2}f(at-b)\,, & D_aT_bf(t) &= T_{a^{-1}b}D_af(t)\,,\\ \langle f, D_ag\rangle &= \langle D_{a^{-1}}f, g\rangle\,, & \langle f, T_bg\rangle &= \langle T_{-b}f, g\rangle\,,\\ \langle f, D_aT_bg\rangle &= \langle T_{-b}D_{a^{-1}}f, g\rangle\,, & \langle D_af, D_ag\rangle &= \langle f, g\rangle\,,\\ \langle T_bf, T_bg\rangle &= \langle f, g\rangle\,, & T_bM_cf(t) &= e^{-2\pi ibc}M_cT_bf(t)\,,\\ \langle f, M_cg\rangle &= \langle M_{-c}f, g\rangle\,, & \langle f, T_bM_cg\rangle &= \langle T_{-b}M_{-c}f, g\rangle e^{2\pi ibc}\,. \end{aligned}$$

It may be seen that the Haar wavelet has only one vanishing moment. It can be proved that if ψ has large vanishing moments then few wavelet coefficients $\langle f, \psi_{j,k}\rangle$ will be large.

10.2.2 Multiresolution Analysis

We introduce here a concept called multiresolution analysis (MRA) that provides a method to construct wavelets. Multiresolution means that we want to use different degrees of resolution to approximate a given function f, ranging from coarse to fine structures, like a camera zooming in on an object. These different degrees of resolution are described mathematically by a nested sequence of vector spaces of functions,

$$\cdots \subset V_{-2} \subset V_{-1} \subset V_0 \subset V_1 \subset V_2 \subset \ldots. \tag{10.13}$$

The concept of a vector space is treated in linear algebra; a set of functions forms a vector space V if addition and scalar multiplication of functions from V never leads to functions which do not belong to V. The inclusions in (10.13) mean that the information on some level of resolution is also included in all finer resolutions.

Definition 104. (Multiresolution analysis) A sequence $\{V_j\}$ of vector spaces of square integrable functions is called a **multiresolution analysis** (MRA) with the **scaling function** φ if the following conditions are satisfied.

(a) The sequence $\{V_j\}$ is nested,

$$\cdots \subset V_{-2} \subset V_{-1} \subset V_0 \subset V_1 \subset V_2 \subset \ldots$$

(b) We have $\cap_j V_j = \{0\}$, that is, the zero function is the only function which belongs to the intersection of all spaces V_j. In other words, no nontrivial function belongs to all V_j.

(c) Every square integrable function can be expanded in a (finite or infinite) series $\sum_j a_j g_j$ of functions $g_j \in V_j$, where convergence is understood in the mean quadratic sense. Equivalently, every square integrable function can be approximated to arbitrary precision by elements of V_j.

(d) f is an element V_j if and only if $D_2 f$ (that is, f compressed by a factor 2) is an element of V_{j+1}.

(e) The set of translated functions $T_k\varphi$, for all integer k, forms an orthonormal basis for V_0. In particular, the scaling function φ is an element of V_0 and is orthogonal to all of its translates $(T_k\varphi)(t) = \varphi(t-k)$ for $k \neq 0$.

From the scaling function φ of an MRA one can construct a wavelet ψ. By (d), the function $\sqrt{2}D_{1/2}\varphi$ belongs to V_{-1}. Since $V_{-1} \subset V_0$ due to (a), by (e) we can expand $\sqrt{2}D_{1/2}\varphi$ in a series,

$$\varphi\left(\frac{t}{2}\right) = \sqrt{2}(D_{1/2}\varphi)(t) = \sum_{k\in\mathbb{Z}} a_k \varphi(t-k)\,,$$

such that $\sum_{k\in\mathbb{Z}} a_k^2$ converges. One then can prove (we will not do it here) that

$$\psi(t) = \sum_{k\in\mathbb{Z}} (-1)^k a_{1-k}\varphi(2t-k) \tag{10.14}$$

is a wavelet in the sense of Definition 99.

Example 314. (Haar MRA) Let V_0 be the space of all functions which are square integrable on $\mathbb{R}$ and constant on all intervals $(k, k+1)$ for integer k. According to (d) of Definition 104, V_j is taken as the space of square integrable functions which are constant on all intervals of the form $[k2^{-j}, (k+1)2^{-j})$. For φ we choose

$$\varphi(t) = \chi_{[0,1)}(t) = \begin{cases} 1\,, & 0 \le t < 1\,, \\ 0\,, & \text{otherwise.} \end{cases}$$

One can show that $\{V_j\}$ is a multiresolution analysis with the scaling function φ. The Haar wavelet

$$\psi(t) = \begin{cases} 1\,, & 0 \le t < \frac{1}{2}\,, \\ -1\,, & \frac{1}{2} \le t < 1\,, \\ 0\,, & \text{otherwise} \end{cases}$$

is related to φ by $\psi(t) = \varphi(2t) - \varphi(2t-1)$.

Example 315. (Shannon MRA) Let V_0 be the space of all square integrable functions having bandwidth not exceeding π, that is $\hat{f}(\xi) = 0$ whenever $|\xi| > \pi$. According to (c) of Definition 104, V_j is taken as the space of square integrable functions of bandwidth not exceeding $2^j\pi$. The Shannon scaling function is given by

$$\varphi(t) = \frac{\sin(\pi t)}{\pi t}\,.$$

It can be verified that the set of translates $T_k\varphi$ is orthonormal, and due to the Whittaker-Shannon formula (with $l = \pi$) for functions in V_0 we have

$$f(t) = \sum_{n=-\infty}^{\infty} f(n)\frac{\sin(\pi(t-n))}{\pi(t-n)}\,,$$

so the translates form an orthonormal basis of V_0. It can be verified that

$$\psi(t) = \varphi(t - \frac{1}{2}) - 2\varphi(2t - 1) = \frac{\sin(2\pi t) - \cos(\pi t)}{\pi(t - \frac{1}{2})}$$

is the **Shannon wavelet**.

The Shannon wavelet family is given by

$$\begin{aligned}\psi_{j,k}(t) &= 2^{j/2}\psi(2^j t - k) \\ &= \frac{2^{j/2}}{\pi(t - \frac{1}{2})}(\sin(2\pi(2^j t - k)) - \cos(\pi(2^j t - k)) .\end{aligned}$$

Moreover, it can be checked that ψ satisfies

$$\hat{\psi}(\xi) = e^{-i\xi/2}\chi_A(\xi) ,$$

where A consists of all $\xi \in [-2\pi, -\pi]$ together with all $\xi \in [\pi, 2\pi]$. That is, on each of these intervals $\hat{\psi}(\xi) = -e^{-i\xi/2}$, and for ξ outside of these intervals, $\hat{\psi}(\xi) = 0$.

Example 316. Show that $\psi(x) = (1 - x^2)e^{-x^2/2}$ is a wavelet.

Solution: It is Mexican hat basic wavelet, which is one of the oldest wavelets. Because of its fast decaying $e^{\frac{-x^2}{2}}$ factor, it vanishes to zero very fast. However, it does not vanish to identically equal to zero beyond a finite interval. Thus it does not have a (compact) support, but we may say that it is essentially with compact support as shown in Figure 10.7. It is easy to see that this

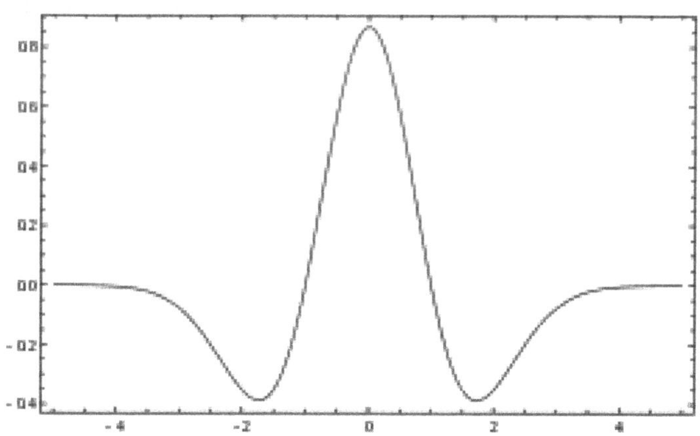

FIGURE 10.7: The Mexican Hat Wavelet $\psi(t)$

wavelet satisfies the basic condition for any wavelet, namely, that its integral over $(-\infty, \infty)$ vanishes. This is possible from the following expression:

$$(1 - x^2)e^{-\frac{x^2}{2}} = -\frac{d^2}{dx^2}e^{-\frac{x^2}{2}} ,$$

which can be shown easily,

$$
\begin{aligned}
-\frac{d^2}{dx^2}e^{-\frac{x^2}{2}} &= -\frac{d}{dx}\left[\frac{-2x}{2}e^{\frac{x^2}{2}}\right] \\
&= \left[x(-\frac{2x}{2})e^{-\frac{x^2}{2}} + e^{-\frac{x^2}{2}}\right] \\
&= (1-x^2)e^{-\frac{x^2}{2}}.
\end{aligned}
$$

With this result we have

$$
\begin{aligned}
\int_{-\infty}^{\infty} \psi(x)dx &= \int_{-\infty}^{\infty} (1-x^2)e^{-\frac{x^2}{2}}\, dx \\
&= \int_{-\infty}^{\infty} -\frac{d^2}{dx^2}e^{-\frac{x^2}{2}}\, dx \\
&= -\frac{d}{dx}e^{-\frac{x^2}{2}}\Big|_{x=-\infty}^{\infty} \\
&= 0-0=0
\end{aligned}
$$

after using L'Hopital's rule for the indeterminate form

$$
\lim_{x\to\pm\infty} \frac{x}{e^{x^2/2}} = \lim \frac{1}{xe^{x^2/2}} = 0.
$$

The Mexican hat wavelet did not prove useful for the discrete wavelet series analysis. However, its simple expression makes it a good example for illustrating the continuous wavelet transform, its inverse, and other properties.

Example 317. (Shannon wavelet) We present Shannon scaling function

$$
\phi(t) = \frac{\sin \pi t}{\pi t} = \text{sinc}\pi t, \quad -\infty < t < \infty,
$$

and its associated basic wavelets

$$
\psi(t) = \frac{\sin \pi\left(t-\frac{1}{2}\right) - \sin 2\pi\left(t-\frac{1}{2}\right)}{\pi\left(t-\frac{1}{2}\right)} \tag{10.15}
$$

as shown in Figures 10.8 and 10.9, respectively.

These wavelets are continuous. However, they are infinitely many times differentiable. Also, in a certain sense they are better localized than Haar ones. However, even though they tend to zero as $t \to \pm\infty$, they nevertheless do not die out to identically zero beyond a finite interval. Hence, they are denied a compact support.

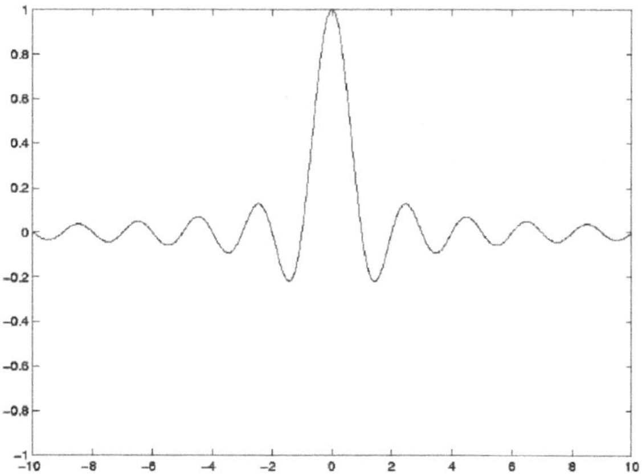

FIGURE 10.8: Shannon Scaling Function

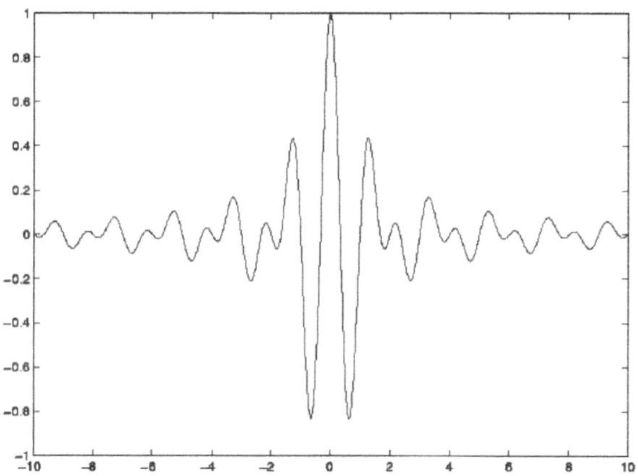

FIGURE 10.9: Shannon Wavelet Function

10.3 Applications of Wavelets

10.3.1 Applications of Wavelets to Biometrics

National Security - Automated methods capable of rapidly determining an individual's true identity, previously used identities and past activities.

Homeland Security and Law Enforcement - Technologies to secure the U.S. and while facilitating legitimate trade and movement of people and identifying criminals in the civilian law enforcement environment.

Enterprise and E-government Services - Administration of people, processes and technologies.

Personal Information and Business Transactions - Business plans that meet customer demands for service at any time, from any location and through multiple communication devices.The term "biometric" is derived from the Greek words "bios" (life) and "metron" (measurement); biometric identifiers are measurements from living human bodies.

Biometrics are automated methods of recognizing a person based on physiological or behavioral characteristic called biometric identifiers, traits or characteristics for automatically recognizing individuals.

Biometric technologies are becoming the foundations of an extensive array of highly secure identification and personal verification solutions.

Finger Print Compression

Most of the fingerprint recognition systems, especially those in commercial applications, do not store fingerprint images and only store numerical features extracted from the images.

However, in certain applications (e.g., law enforcement), it may be necessary to store the fingerprint images acquired during enrollment in a database so that a trained expert can verify the matching results output by an Automated Fingerprint Identification System (AFIS).

Storing millions of fingerprint images (as in a large AFIS), or transmitting these images through low-bandwidth networks is particularly demanding in terms of space and time.

Hence, several ad hoc compression techniques have been proposed and one of them, known as wavelet scalar quantization (WSQ), has been adopted as a standard by the FBI in the United States.

FIGURE 10.10

Figures 10.12 through 10.14 indicate how wavelets are used in image compression.

10.3.2 CAT Scan

A computerized axial tomography (CAT or CT) scan is generated from a set of thousands of X-ray beams, consisting of 160 or more beams at each of 180 directions. We will explain here just one beam in order to comprehend this large collection of X-rays. When a single X-ray beam of known intensity

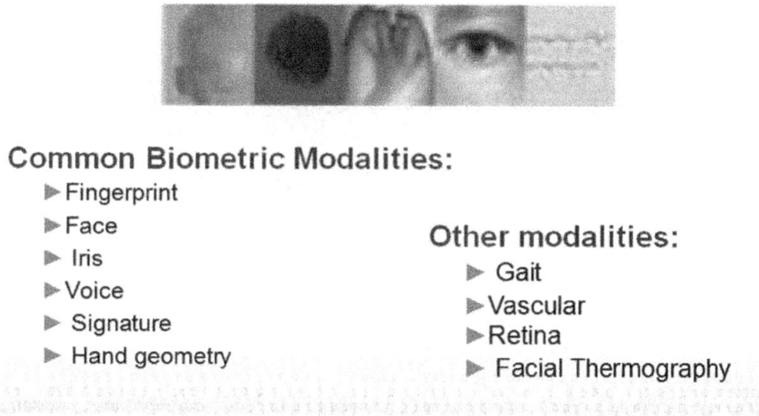

FIGURE 10.11

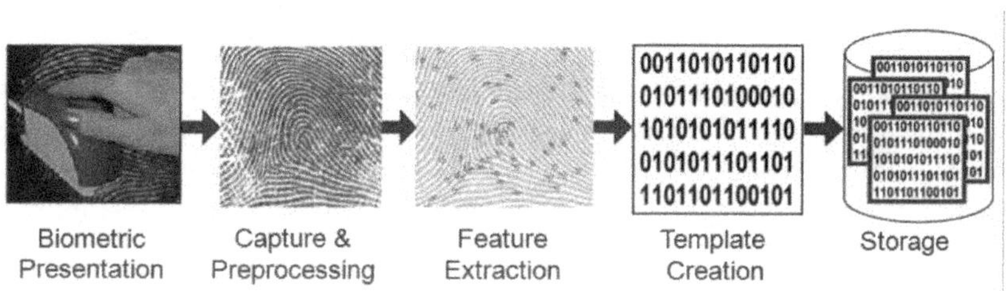

FIGURE 10.12: Biometric Compression System

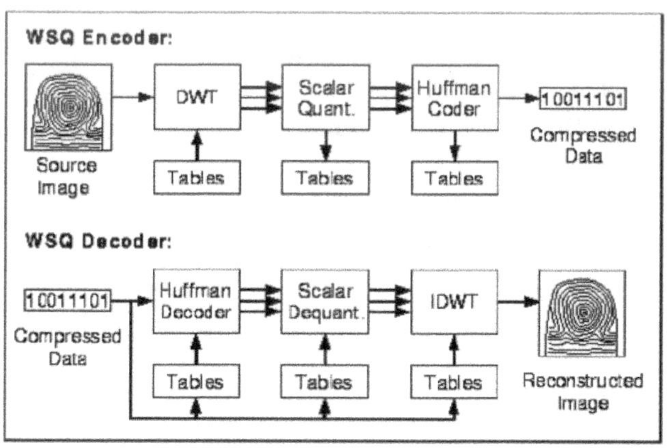

FIGURE 10.13: WSQ- Endcoder

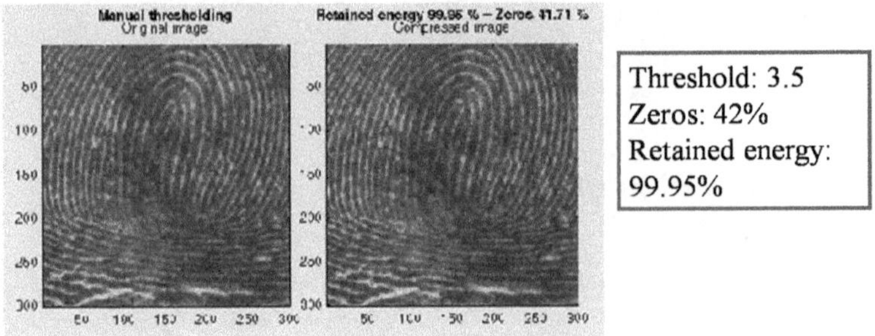

FIGURE 10.14: Results

passes through a medium, such as muscle or brain tissue, some of the energy in the beam is absorbed by the medium and some passes through. The intensity of the beam as it emerges from the medium can be measured by a detector. The initial and final intensities tells us about the ability of the medium to absorb energy.

For the sake of clarity we make some assumptions that present an idealized view of what an X-ray is and how it behaves. Assume X-ray beam composed of photons and is monochromatic. Each photon has the same energy level of E and the beam propagates at a constant frequency, with the same number of photons per second passing through every centimeter of the path of the beam. Let us denote $N(x)$ the number of photons per second passing through a point x. Then the intensity of the beam at the point x is

$$I(x) = E.N(x).$$

Every substance through which an X-ray passes has the property that each millimeter of the substance absorbs a certain proportion of the photons that pass through it. This proportion which is specific to the substance, is called the **attenuation coefficient** of the material. A Hounsfield unit of the medium is denoted by H medium and defined by

A denotes the true attenuation coefficient Beer's law:

$$\frac{dI}{dx} = -A(x)I(x)$$

$$\frac{dI}{I} = -A(x)dx.$$

If the beam starts at a location with initial intensity I and is detected, after passing through the medium, at the location x, with final intensity, then we get

$$\int_{x_0}^{x_1} \frac{dI}{I} = -\int_{x_0}^{x_1} A(x)dx.$$

Thus

$$\ln[I(x_1)] - \ln[I(x_0)] = -\int_{x_0}^{x_1} A(x)dx.$$

Here we know the initial and final values of intensity $I(x)$, and $A(x)$ providing special property of medium being sampled by the X-ray is unknown.

Thus from the measured intensity of X-ray we are able to determine not the values of A itself, but the values of the integral of A along the line of X-ray.

What we can measure: we can design an X-ray emission/detection machine that can measure the values of $I(x)$. Hence we can compute from the equation that is integral of the (unknown) attenuation coefficient function along the path of X-ray.

What we want to know: The value of $A(x)$ at each location depends on the nature of matter located at the point x. We wish to know $A(x)$.

Tomography is from a Greek word meaning slice. In this field we want to find internal structure of a non-transparent object by sending signals like electromagnetic waves of different frequencies and acoustic waves.

The current excitement in tomographic imaging originated with Hounsfield and Cormak's invention of X-ray computed tomography for which they jointly received 1979 Noble prize for physiology and medicine.

Radon transform invented by the Austrian mathematician Johann Radon is the integral transform consisting of a scalar valued function over the straight line. By computerized (computed) tomography (CT) we mean the reconstruction of a function from its line or plane integrals. Essentially it amounts to inverting the Radon transform. Wavelet methods have been used in the studies of Radon Transform [42].

10.3.3 Seismic Tomography

Seismic tomography uses mathematical modeling of P and S wave travel times to map velocity perturbations in the interior of the earth. The primary energy source used in global seismic tomography is seismic waves generated by earthquakes which pass through the Earth in all directions, and are recorded on seismograms around the world. Inversion of arrival time data is used to determine the speed of the waves at any given points. Use of seismic tomography to interpret the internal structure of the earth is similar in technique to a CAT scan. Computer assisted tomography (CAT) uses X-rays transmitted through the body in many different directions. A mathematical method is then applied to explain the loss in intensity of the X-rays due to the varying absorptive properties of different parts of the body. CAT scans and seismic tomography differ because X-rays travel in straight paths, whereas the ray paths of sound waves bend with changes in the velocity structure of the medium.

Seismic tomography has several applications in exploration and global geophysics. Seismic tomography can also be used to characterize fractured bedrock, map groundwater reservoirs, and locate ore bodies. Global seismic tomography is used to interpret ancient subducted slabs, locate the sources of hotspots, and model convection patterns in the mantle.

Global seismic tomography is limited by the irregularities in time and space of the source, and by the incomplete coverage of recording stations. The primary sources earthquakes, which are impossible to predict and only occur at certain locations around the world. In addition, the global coverage of recording stations is limited due to economic and political reasons. Because of these limitations, seismologists must work with data that contains crucial gaps. Experimental data cannot accurately replicate conditions deep in the earth's interior, making comparisons with real world data difficult. Another limitation in imaging deep structures is attenuation and absorption of energy due to the long distances waves travel, which reduces the resolution which can be attained. Due to the problem of attenuation, the minimum sizes of features in the mantle which can be resolved are blocks 100 to 200 km on each side.

In short, seismic tomography is a methodology for analyzing and computing earth properties. Seismologists treat seismic tomography as a part of seismic imaging, where they are mainly concerned with estimating properties such as propagating velocities of P-waves and S-waves. The seismic tomography may be thought as the derivation of three dimensional velocity structures of earth from seismic waves [43], [44] and [45]. Estimation of P-wave velocity is the simplest example of seismic tomography and seismic tomography is formulated as an inverse problem. See references

Main Ideas

Several types of faults occur in the earth's crust.

The faults break due to accumulated stress along the routes. The sudden release of energy is called an earthquake. The energy is released as seismic waves that travel away from the earthquake location. Two major types of waves produced are: body waves and surface waves.

The waves can be measured by a seismometer instrument. The timing and amplitude of the seismic waves can be used to determine the location and magnitude of the earthquake. Earthquakes commonly occur along plate boundaries.

These waves also provide information on the structure of the earth. A clear layering is visible.

Two very recent papers give current trends application of wavelet to seismology.

There is a general discussion how to use wavelet and fractal method in prediction of earthquakes. This is a challenging problem but scientists are optimistic.

10.3.4 Variants of Wavelets in Medical Imaging

PROPERTIES OF SHEARLETS

This section summarizes an important paper about shearlets published in the *Journal of Medical Imaging and Health Informatics* in 2016 [48]. Wavelets have several properties that make them useful in various disciplines such as medicine, electricity generation, and petroleum chemistry. For example, shearlets:

- Are well localized.
- Exhibit high levels of directional sensitivity.
- Satisfy the principles of parabolic scaling.
- Are spatially localized.
- Are optimally sparse.

Shearlets and Curvelets in Image Processing

Digital signal and image processing is an important technique for analyzing, manipulating and processing real world data and images. The technology can generate time series, collect data and calculate values. It can produce audio signals, video images, seismic activity records, rainfall data and biomedical test results.

Edges are prominent features of images. Analyzing and detecting edges are important aspects of image processing even though they represent low-level tasks in applications such as three-dimensional reconstruction, shape recognition, and image compression, enhancement and restoration.

Shearlets and curvelets represent a novel directional multiscale mathematical framework that is well adapted for identification and analysis of distributed discontinuities such as edges in natural images. Despite their value, wavelets have limited ability to deal with directional information.

The shearlet and curvelet [38] approach is designed to deal with directional and anisotropic features typically present in natural images, and have the ability to accurately and efficiently capture the geometric information of edges.

As a result, the shearlet framework provides effective algorithms for detecting both the locations and orientations of edges, and for extracting and classifying basic edge features such as corners and junctions.

The shearlet framework provides a unique combination of mathematical rigidity and computational efficiency when addressing edges.

Continuous Shearlet Transform (Figures 10.15 and 10.16)

For $\psi \in L^2(R^2)$, the continuous shearlet system $SH(\psi)$ is defined by

$$SH(\psi) = \{\psi_{a,s,t} = T_t D_{A_a} D_{S_s} \psi : a > 0, s \in R, t \in R^2\}.$$

$$A_a = \begin{pmatrix} a & 0 \\ 0 & a^{1/2} \end{pmatrix},$$

and shearing operator D_{S_s}, $s \in R$, where the shearing matrix S_s is given by

$$S_s = \begin{pmatrix} 1 & s \\ 0 & 1 \end{pmatrix}$$

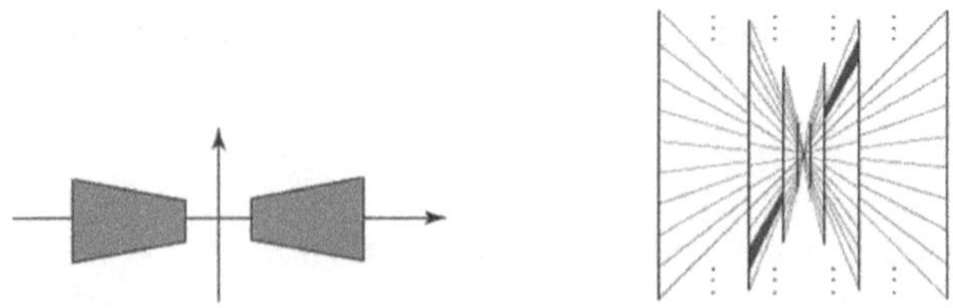

(a) Support of the Fourier transform of a classical shearlet.

(b) Fourier domain support of several elements of the shearlet system, for different values of a and s.

FIGURE 10.15

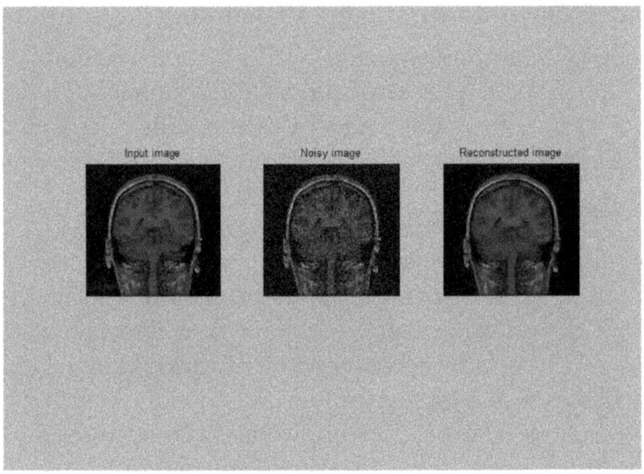

FIGURE 10.16: MRI of Brain

and T_t is the translation operator on $L^2(R^d)$, defined by

$$T_t\psi(x) = \psi(x - t), \quad \text{for } t \in R^d.$$

Shearlet in Magnetic Resonance Image (MRI) of Brain

Image denoising is the process of recovering an original image from an image corrupted with noises such as Gaussian, speckle and other types. Shearlets can be used effectively for image denoising by using various shrinkage rules. The main steps of image denoising are:

1. Compute shearlet transform of the noisy image.

2. Apply hard and soft threshold to the obtained shearlet coefficients.

3. The thresholded shearlet coefficients are subjected to reconstruction to recover the original image.

Reference [48] presents the results obtained from the experimentation. The proposed approach of image compression was explored with MRI image

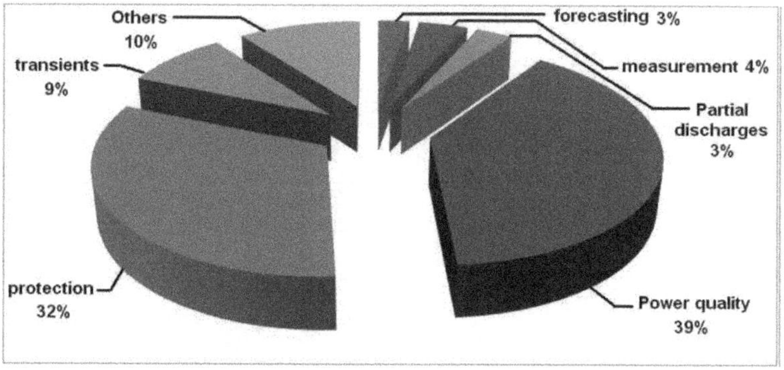

FIGURE 10.17: Wavelet Analysis Data Applied to Power System

datasets and the results evaluated with compression ratio, PSNR, average difference, cross correlation and normalized absolute error calculations.

10.3.5 Applications in Power Systems (Figure 10.17)

Wavelets were first applied to power system in 1994 by Robertson et al [49] and Ribeiro [50]. Since then number and scope of the applications have increased. The most popular wavelet analysis applications include power quality and measurement, partial discharges, load and revenue forecasting, protection, and transient interruptions. (www.intechopen.com).

Power Quality (Figure 10.17)

Several studies [50] focused on detecting power system disturbances using the wavelet transform (WT) to analyze sags, swells, and other slowly changing anomalies of non-stationary signals. The technique covers spectral contents at low frequencies. Examining WT coefficients at high decomposition levels can help predict onsets of disturbance events. Application of Wavelet Theory in Engineering Physics and Technology, Published in Tech, 2012, http://www.intechopen.com/books/advances-in-wavelet-theory-and-their-applications-in-engineering-physics-and-technology/ application-of-wavelet-analysis-in-power-systems.

Wavelets in Oil Industry (Figure 10.18)

Siddiqi [38] studied the use of wavelet methods to identify and characterize oil reservoirs. The petroleum product market is completely dependent on exploration, drilling, and production costs. Researchers in the oil industry focus on developing techniques to minimize exploration and production costs and frequently apply wavelet methods to achieve their goals. Siddiqis article discusses use of the wavelet-based solution of the Buckley-Leverett equation for reservoir modeling. The technique allows researchers to perform virtual anal-

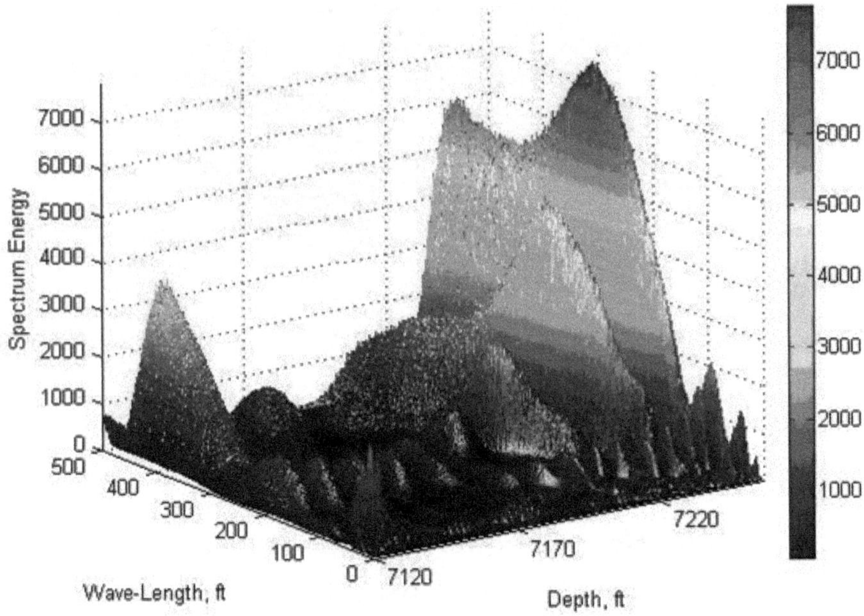

FIGURE 10.18: $3-D$ Morlet Scalogram for GR, SDGM 256

yses to determine locations, quantities, and other parameters of oil and gas reserves worldwide.

10.4 Exercises

10.1 Let $\psi(t) = \begin{cases} 1, & 0 <\leq t < 1/2 \\ -1, & 1/2 \leq t < 1 \\ 0, & \text{otherwise} \end{cases}$

and $\phi(t) = \begin{cases} 1, & 0 <\leq t < 1 \\ 0, & \text{otherwise.} \end{cases}$

$\psi(t)$ is known as the Haar mother wavelet and $\phi(t)$ is the called Haar scaling function or Haar father wavelet.

(a) Write the support of ψ and ϕ. Draw the graphs of

(b) $\psi(4t-3)$, $\psi(4t+3)$, $\psi(2^3t)$

(c) $\phi(2t)$, $\phi(2t-2)$, $\phi(2^3t)$ on intervals of your choice.

10.2 Show that for ψ and ϕ of Exercise 10.1 the following relation holds:

$$\psi(t) = \phi(2t) - \phi(2(t - 1/2)).$$

Is this relation true for any mother and father wavelets?

10.3 Show that the Haar mother wavelet ψ and the Haar father wavelet ϕ are orthogonal over $(-\infty, \infty)$, that is,

(a) $\int_{\infty}^{-\infty} \phi(t)\psi(t)dt = 0$

(b) Verify that $\int_{\infty}^{-\infty} \psi^2(t)dt = 1$

(c) Verify that $\int_{\infty}^{-\infty} \phi^2(t)dt = 1$.

10.4 Show that for above ϕ and ψ the following results hold:

(a) $\int_{\infty}^{-\infty} \phi(t)\phi(t-k)dt = \begin{cases} 0, & k \neq 0 \\ 1, & k = 0 \end{cases}$.

(b) $\int_{\infty}^{-\infty} \psi(t)\psi(t-k)dt = \begin{cases} 0, & k \neq 0 \\ \int_{-\infty}^{\infty} \psi^2(t)dt = 1, & k = 0 \end{cases}$.

10.5 For the Haar scaling function (Haar father wavelet) ϕ show whether or not the following translated sets re orthogonal on $(-\infty, \infty)$:
(a) $\{\phi(2t - k)\}$, $k \in Z$ and (b) $\{\psi(2t - k)\}$, $k \in Z$.

10.6 Examine the question of Exercise 10.5 for the Haar mother wavelet.

10.7 Introduce the concept of energy of the signal and discuss its relationship with the wavelet transform of the signal.

10.5 Suggestion for Further Reading

Wavelet analysis is an emerging field. We recommend that reader choose among the several references in the bibliography, particularly those covering applications of wavelet analysis.

Bibliography

[1] P. S. Addison, *The Illustrated Wavelet Transform Handbook*, Institute of Physics Publishing, 2002.

[2] A. Aldroubi and M.Unser, *Wavelets in Medical Biology*, CRC Press, 1996.

[3] H. Anton, *Calculus: A New Horizon*, Sixth Edition, Wiley, New York, 1999.

[4] A. Boggess and F. T. Narcowich, *A First Course: Wavelets with Fourier Analysis*, Prentice Hall, 2001.

[5] S. A. Broughton and K. Bryan, *Discrete Fourier Analysis and Wavelets: Application to Signal and Image Processing*, Wiley, 2009.

[6] M. Cartwright, *Fourier Methods for Mathematicians, Scientists and Engineers*, Ellis Horwood, New York, 1990.

[7] T. F. Chan and J. Shen, *Image Processing and Analysis: Variational, PDE, Wavelets and Stochastics Methods*, SIAM, 2005.

[8] S. J. Chapman, *MATLAB Programming for Engineers*, Second Edition, Brooks/ Cole, 2000.

[9] O. Christensen, *An Introduction to Frames and Riesz Bases*, Birkhäuser, Boston, 2003.

[10] O. Christensen, *Function Spaces and Expansions: Mathematical Tools in Physics and Engineering*, Birkhäuser, Boston, 2003.

[11] O. Christensen, K. L. Christensen, *Approximation Theory from Taylor Polynomials to Wavelets*, Birkhauser, Boston, 2004.

[12] K. M. O. Connor, *Calculus: Labs for MATLAB*, Jones and Bartlett, 2005.

[13] J. M. Cooper, *A MATLAB Companion for Multivariable Calculus*, Academic Press, 2001.

[14] I. Daubechies, *Ten Lectures on Wavelets*, SIAM, Philadelphia, 1992.

[15] J. H. Davis, *Methods of Applied Mathematics with MATLAB Overview*, Birkhäuser, 2004.

[16] K. Delvin, *An Electronic Companion to Calculus*, Cogito Learning Media, New York, 1997.

[17] W. Drongelen, *Signal Processing for Neuroscientists:An Introduction to the Analysis of Physiology Signals*, Elsevier, 2007.

[18] K. M. Furati , M. Z. Nashed, A. H. Siddiqi (eds.), *Mathematical Models and Methods for real world systems*, Chapman & Hall/CRC, Taylor and Francis, 2006.

[19] R. Gencay, F. Seluk and B. Whitcher, *Introduction to Wavelet and Filtering Methods in Finance and Economics*, Academic Press, 2002.

[20] R. C. Gonzales and R. E. Wood, *Digital Image Processing*, Addison-Wesley Reading, MA, 1993.

[21] B. B. Hubbard, *The World According to Wavelets*, A. K. Peters Natick, Second Edition, MA, 1998.

[22] A. Jerri, *Introduction to Wavelets*, Sampling Publishing, 2011.

[23] G. Kaiser, *A Friendly Guide to Wavelets*, Birkhäuser, 1994.

[24] M. Kobyashi, *Wavelets and their Applications*, SIAM, 1998.

[25] A. K. Louis, P. Maass and A. Reider, *Wavelet Theory and Applications*, Wiley, 1997.

[26] S. Mallat, *A Wavelet Tour of Signal Processing*, Academic Press, New York, 1998.

[27] Y. Meyer, *Wavelets: Algorithms and applications*, SIAM, Philadelphia, 1993.

[28] Y. Meyer, *Lecture Notes on Mathematical Problems in Image Processing*, 4-22 September, 2000 Abdus Salam Centre of Theoretical Physics, Trieste, Italy, 2000.

[29] P. V. O'Neil, *Advanced Engineering Mathematics*, Fifth Edition, Brooks/Cole, 2003.

[30] D. B. Percival and A. T Walden, *Wavelet Methods for Time Series Analysis*, Cambridge University Press, 2000.

[31] H. L. Resnikoff and R. O Wells Jr, *Wavelet Analysis*, Springer, 1998.

[32] R. H. Shumway and D. S Stoffer, *Time Series Analysis and Its Applications*, Springer, 2003.

[33] A. H. Siddiqi, *Applied Functional Analysis*, Marcel Dekker, 2004.

[34] A. H. Siddiqi, G. Korvin, et al. (Eds.), Theme Issue on Wavelet & Fractals in Science and Engineering, *Arab. J. Sci. Eng.*, 28–29, 2003–2004.

[35] A.H. Siddiqi, I. S. Duff and O. Christensen (Eds.), *Modern Mathematical Methods and Algorithms for Real Word Systems*, Anamaya and Anshan, 2006.

[36] A. H Siddiqi, A. K. Gupta and M. Brokate, *Modelling of Engineering and Technological Problems*, Americans Institute of Physics, 2009.

[37] A. H. Siddiqi, R. C. Singh, P. Manchanda (Eds.) *Mathematics in Science and Technology*, World Scientific, Singapore, 2011.

[38] A. H. Siddiqi, *Emerging Applications of Wavelet Methods*, American Institute of Physics, Melville, NY, 2012.

[39] A. H. Siddiqi, *Functional Analysis with Applications* (Chapter 12, 13), Springer Nature, 2017.

[40] A. H. Siddiqi, *Keynote Address, International Workshop on Wavelet and Applications*, September 21-24, 2016, Istanbul, Turkey.

[41] A. Garg and A.H. Siddiqi, Inverse estimation of 1-D heat problem with Neumann boundary condition by Morozov discrepancy principle, *Ind. J. Industr. Appl. Math.*, 7:58-64, 2016.

[42] M. Bottema, B. Moran and S. Suvorova, An application of wavelets in tomography, *Digital Sig. Proc.*, 8: 244-254, 1998.

[43] A. Iske and T. Randen, *Mathematical Methods and Modeling in Hydrocarbon Exploration and Production*, 16, pp.267-297, Springer, New York, 2006.

[44] G. Nolet, *A Breviary of Seismic Tomography*, Cambridge university Press, Cambridge, 2008.

[45] C. R. Vogel, *Computational Methods for Inverse Problem Frontiers in Applied Mathematics*, SIAM, 2002.

[46] S. Mostafa Mausavi and Charles A. Langston, Hybrid Seismic Denoising Using Higher-Order Statistics and Improved Wavelet Block Thresholding, *Bulletin of the Seismological Society of America*, Vol. 106, No. 4, 2016.

[47] S. Mostafa Mausavi and Charles A. Langston and Stefan P. Horton, Automatic mircoseismic denoising and onset detection using the synchosqueezed continuous wavelet transform, *Geophysics*, Vo. (81), No. 4, 2016.

[48] R. Aneja and A. H. Siddiqi, A Hybrid Shearlet based compression coefficients and ROI Detection, *Journal of Medical Imaging and Health Informatics*, ISSN: 2156-7018, pp. 506-517(12), 2016, American Scientific Publishers

[49] D. C. Roberston, O. Camps and I. Meyer, Wavelets and Power System Transients Feature Detection and Classification, *Proceddings of SPIE*, 2242: 474-487, 1994.

[50] P. F. Ribeiro, Wavelet Transform an advanced Tool for Analyzing nonstationary Harmonic Distortion in Power System, *Proceeding IEEE, Transcation of Power System*, Bologna, Italy, September 21-24, 1994.

Webpages for MATLAB and Wavelets

http://www.math.poly.edu/courses/matlab.html

http://faculty.olin.edu/bstorey/notes/Fourier.pdf

http://www.mathworks.com/products/signal

http://www.bearcave.com/misl/misl_tech/

http://oceanz.tamu.edu/~baum/wavelets.html

http://www-stat.stanford.edu/~wavelab/Wavelab_850/AboutWaveLab.pdf

http://www-stat.stanford.edu/~donoho/Reports/1995/wavelab.pdf

Chapter 11

Miscellaneous Topics Used for Engineering Problems

11.1 Fractals in Engineering Science

A fractal is a never-ending pattern. Fractals are infinitely complex patterns that are similar across different scales. They are created by repeating a simple process over and over in an ongoing feedback loop.

Benoit Mandelbrot (1924-2010) first used the word *fractals* to describe a set of models that accurately depict natural geometry. Mandelbrot then challenged the Euclidian geometry that relies on primitives such as spheres, cuboids and cones that is not sufficient to accurately represent complex naturally occurring objects. He put forward the geometry of fractals that contained infinite details that can accurately model objects such as trees, mountains, and clouds.

By 1980 Mandelbrot accomplished his goal while working at Harvard University, and started working on fractal images and exploring the self-similarity characteristic of all fractals. He also worked closely with IBM in its fractal project and added to an incredible visual field of mathematics. Over nearly seven decades, working with dozens of scientists, he contributed to the fields of geology, medicine, cosmology and engineering. He used the geometry of fractals to explain how galaxies cluster, how wheat prices change over time and how mammalian brains fold as they grow, among other phenomena.

11.1.1 Fractals and Interaction with Wavelets

Sharp signal transitions create large amplitude wavelets coefficients. Singularities are detected by following across scales the local maxima of the wavelet transform.

Fractals describe objects that are too irregular to fit into traditional geometrical settings. Fractals occur as graphs of functions. Indeed various phenomena display fractal features when plotted as functions of time. Examples include atmospheric pressure, oil and natural gas reservoir characteristics and stock market prices, usually when recorded over fairly long time spans. The zooming capability of the wavelet transform not only locates isolated singular events, but can also characterize more complex multi-fractal signals having non-isolated singularities. Multifractals are fractal objects which cannot be completely described using a single fractal (monofractals) dimension. They have in fact an infinite number of dimension measures.

The wavelet transform takes advantage of multifractal self-similarities to compute the distribution of their singularities. This singularity spectrum is used to analyze multifractal properties. Signals that are singular at almost every point are multi-fractals and they appear in the maintenance of economic records, physiological data including heart records, electromagnetic fluctuations in galactic radiation noise, textures in images of natural terrain and variations of traffic flow.

Hausdorff Measure

Let U be any non-empty subset of n-dimensional Euclidean space, R^n; the diameter of U is denoted by $|U|$ and defined as

$$|U| = \sup\{|x-y| : x, y \in U\},$$

that is, the greatest distance apart of any pair of points in U. If $\{U_i\}$ is a countable (or finite) collection of sets of diameter almost δ that cover F, that is $F \subset \bigcup_{i=1}^{\infty} U_i$ with $0 < |U_i| \leq \delta$ for each I, we say that $\{U_i\}$ as a δ-cover of F.

Suppose that F is a subset of R^n and s is a non-negative number. For any $\delta > 0$ we define $H_\delta^s(F) = \inf\{\sum_{i=1}^{\infty} |U_i|^s : \{U_i\} H_\delta^s(F) = \inf\{\sum_{i=1}^{\infty} |U_i|^s : \{U_i\}$ is a δ-cover of F.

Thus, one looks at all covers of F by sets of diameter at most δ and seek to minimize the sum of the sth powers of the diameter. As δ decreases, the class of permissible covers of F is reduced. Therefore, the infinimum $H_\delta^s(F)$ increases, and so approaches a limit as $\delta \to 0$. We write,

$$H_\delta^s(F) = \lim_{\delta \to 0} H_\delta^s(F).$$

This limit exists for any subset F of R^n, though the limit value can be 0 or ∞. $H_\delta^s(F)$ is called the s-dimensional **Hausdorff measure** of F. The Hausdorff dimension of F, denoted by $\dim_H F$, is defined as

$$\begin{aligned} \dim{}_H F &= \inf\{s : H^s(F) = 0\} \\ &= \sup\{s : H^s(F) = \infty\}. \end{aligned}$$

If $s = \dim_H F$, then $H^S(F)$ may be zero at infinity or may satisfy

$$0 < H^s(F) < \infty.$$

Dimension of the middle third Cantor set is s, if $s = \frac{\log 2}{\log 3} = 0.6309$. In general, the Hausdorff dimension of the middle Cantor set lies between 0.5 and 1.

singularity spectrum

Definition 105. (Self similar set) A set $S \subset R^n$ is said to be a self-similar if it is the union of disjoint subsets $S_1, S_2, \ldots, S_R$ that can be obtained from S with scaling, translation and rotation. It may be observed that the self-similarity often implies an infinite multiplication of details, which creates irregular structures.

The triadic Cantor set and the Von Koch curve are well known examples.

Definition 106. (Spectrum) Let S$_\alpha$ be the set of all points $t \in R$, where the pointwise Lipschitz regularity of f is equal to α. The spectrum of singularity, $D(\alpha)$, of f is the fractal dimension of S_α. The support of $D(\alpha)$ is the set of α

such that S_α is not empty. The singularity spectrum gives the proportion of Lipschitz singularities that appear at any scale a. Fractal dimension is a first order parameter of complexity which can degenerate. Very different structures share the same fractal dimension. Therefore, Nekka and Li [34], [35] studied the concept of the Hausdorff measure spectrum functions (HMSFs).

They introduce HMSF as a new way to distinguish sets having the same fractal dimension. HMSF is based on a Hausdorff measure of the translation of the set through itself in a continuous manner. Translation is made continuously on each point (local) and the Hausdorff measure (global) is estimated implying HMSF extracts the whole information set. The indicator function of the intersection of a set with its translate can be viewed as a two-point joint moment (autocovariance). This explains in a way why HMSF completes the information obtained from pointwise descriptors like fractal dimension. From the definition of capacity dimension as a special form of Hausdorff dimension, it follows that if we make a disjoint cover of the support of f with intervals of size s then the number of intervals that intersect is:

$$N_a(s) \approx s^{-D(\alpha)}.$$

A multifractal f is called **homogeneous** if all singularities have the same Lipschitz exponent which means the support of $D(\alpha)$ is restricted to α_0. Fractional Brownian motions [26] are examples of homogeneous multifractals.

Definition 107. (Fractional Brownian motion)
A fractional Brownian motion of Hurst exponent $0 < H < 1$ is a zero-mean Gaussian process B_H such that

$$\begin{aligned} B_H(0) &= 0 \\ E\{|B_H(t) - B_H(t-\Delta)|^2\} &= \sigma^2|\Delta|^{2H}. \end{aligned}$$

Partition Function

One cannot compute the Lipschitz regularity of a multifractal because its singularities are not isolated and finite numerical resolution is not sufficient to discriminate them. To overcome this difficulty Areodo, Bacry and Muzy [56] introduced the concept of wavelet transform modulus maximum using a global partition function. Let ψ be a wavelet with n vanishing movements. Mallat [57] states that if f has pointwise Lipschitz regularity $\alpha_0 \leq n$at v then the wavelet transform has a sequence of modulus maxima that converges toward v at fine scales. The set of maxima at the scale a can thus be interpreted as a cover of the support of f with wavelets of scale a. At these maxima locations

$$|T_\psi f(a,b)| \approx a^{\alpha_0+\frac{1}{2}}.$$

Let $\{U_p(a)\}_{p\in Z}$ be the position of all local maxima of $T_\psi g(a,b)$ at a fixed scale a. The partition function Z measures the sum at a power q of all these wavelet modulus maxima:

$$Z(q,a) = \sum_p |T_\psi f(a,u_p)|^q .$$

For each $q \in R$, the scaling $\tau(q)$ measures the asymptotic decay of $Z(q,a)$ at fine scales:

$$\tau(q) = \lim_{a\to 0} \frac{\log Z(q,a)}{\log a}.$$

This typically means that

$$Z(q,a) \approx a^{\tau(q)}.$$

Theorem 76. *(Areodo, Bacry, Jaffard, Muzy [56]) Let $\wedge = [\alpha_{\min}, \alpha_{\max}]$ be the support of $D(\alpha)$. Let ψ be a wavelet with vanishing moments. If f is a self-similar signal then*

$$\tau(q) = \min_{\alpha\ \in\wedge} (q(\alpha + 1/2) - D(\alpha)).$$

Theorem 77. *The scaling exponent $\tau(q)$ is a convex and increasing function of q. The Legendre transform is invertible if and only if $D(\alpha)$ is convex, in which case*

$$D(\alpha) = \min_{q\in R}(q(\alpha + 1/2) - \tau(q)).$$

The spectrum $D(\alpha)$ of self-similar signals is convex. For detail properties of singularities spectrum we refer to Meyer [28] and [29].

We first calculate $z(q,a) = \sum_p |T_\psi f(a,u_p)|^q$, then derive the decay scaling exponent $\tau(q)$ and finally compute $D(\alpha)$ with a Legendre transform. If $q<0$ then the value of $Z(q,a)$ depends mostly on the small amplitude

$$\max |T_\psi f(u_p,a)| .$$

Procedure for Numerical Calculation

1. Maxima: compute $T_\psi f(u_p,a)$ and the modulus maxima at each scale a. Chain the wavelet maxima across scales.

2. Partition: compute:

$$Z(q,a) = \sum_p |T_\psi f(a,u_p)|^q .$$

3. Scaling: compute $\tau(q)$ with a linear regression of $Z(a,q)$ as a function of $\log_2(a)$

$$\log_2(a) Z(q,a) \approx \tau(q) \log_2 a + C(q).$$

4. Spectrum: compute:

$$D(\alpha) = \min_{q\in R}(q(\alpha + 1/2) - \tau(q)).$$

Smooth Perturbations

Let f be a multi-fractal whose spectrum of singularity $D(\alpha)$ is calculated from $\tau(q)$. If a regular signal is added to f then the singularities are not modified and the singularity spectrum of $\tilde{f} = f + g$ remains unchanged. We study the effect of this smooth perturbation on the spectrum calculation.

The wavelet transform of $\tilde{f}$is $T\tilde{f}(u,s) = Tf(u,s) + Tg(u,s)$.

Let $\tau(q)$ and $\tilde{\tau}(q)$ be the scaling exponent of the partition functions $Z(q,s)$ and calculated from the modulus maxima respectively .

Theorem 78 (Arneodo, Bacry, Muzy). *Let ψ be a wavelet with exactly n vanishing moments. Suppose that f is a self-similar function.*

1. *If g is a polynomial of degree $p < n$ then $\tau(q) = \tilde{\tau}(q)$ for all $q \in \Re$.*
2. *if $g^{(n)}$ is almost every where non-zero then*

$$\tilde{\tau}(q) = \begin{cases} \tau(q) & q \geq q_c \\ (n + \frac{1}{2}) & q \leq q_c \end{cases}$$

where q_c is defined by

$$\tau(q_c) = (n + \frac{1}{2})q_c.$$

Hurst Exponent and Application

There are many processes that have random (stochastic) components and also exhibit some predictability between one element and the next. In statistics, this is sometimes described by the autocorrelation function (the correlation of data set with shifted version of the data set). The autocorrelation is one measure of whether a past value can be used to predict a future value.

A random process that has some degree of autocorrelation is referred to as long memory process (or long range dependence). River flow exhibits this kind of long-term dependence. A hydrologist, named Hurst, studied Nile River flows and reservoir modeling [27 and website]. The Hurst exponent, to be estimated in the sequel using wavelet method having various kinds of applications, is named for him. In the recent past, applications of the Hurst exponent have attracted attention of researchers working in different fields.

The Hurst exponent is also directly related to the fractal dimension, which gives a measure of the roughness of a surface. The fractal dimensions and their refinements have been used in diverse fields. The relationship between the Hausdorff fractal dimension D and the Hurst exponent H is $D = 2 - H$.

Correlations

Let us recall concept of cross correlation and auto-correlation which are closely related to Hurst exponents and analysis of real world systems.

Cross Correlation

The cross-correlation coefficient r, which is a measure of linear association between two variables, is defined as

$$r = \frac{\sum_{i=1}^{n}(X_i - \bar{X})(Y_i - \bar{Y})}{\sqrt{\sum_{i=1}^{n}(X_i - \bar{X})^2 \sum_{i=1}^{n}(Y_i - \bar{Y})}}.$$

A positive value of the coefficient r indicates that as one value increases, the other tends to increase whereas a negative value indicates that as one variable increases the other tends to decrease.

Auto-Correlation

We say that a data set exhibits auto-correlation if value x_i at time t_i is correlated with a value X_{i+d} at time t_{i+d} where some time increment is in the future. In a long memory process the auto-correlation decays over time following a power law namely,

$$p(k) = Ck^{-\alpha}.$$

where C is a constant and $p(k)$ is the auto-correlation function with lag k. For a given $X_1, X_2, \ldots\ldots, X_n$ at time $t_1, t_2, \ldots\ldots, t_n$ the k-lag auto-correlation function is defined as

$$r_k = \sum_{i=1}^{n-k}(X_i - \bar{X})(X_{i+k} - \bar{X}) / \sum_{i=1}^{n}(X_i - \bar{X})^2,$$

where $\bar{X} = \dfrac{X_1 + X_2 + \cdots + X_n}{n}$.

It may be remarked that in the above definition, the observations are uniformly sampled. Unlike cross correlation, the auto-correlation results in a correlation coefficient indicating the degree of similitude between two values of same variable at times t_i and t_{i+k}.

The auto-correlation function is used to detect non-randomness in data and to identify an appropriate time series model if the data are not random. When auto-correlation is used to detect non-randomness, it is usually only the 1-lag auto-correlation that is of interest. When the auto-correlation is used to identify appropriate time series model, the k-lag auto-correlation is plotted.

Relationship of Autocorrelation and Hurst Exponent

Exponent of α is related to the Hurst exponent by the equation

$$H = 1 - \alpha/2..$$

The value of Hurst exponent lies between 0 and 1. A value of 0.5 indicates a true random walk (a Brownian motion time series). In a random walk, there is no correlation between any element and future elements. A Hurst exponent value $0.5 < H < 1$ indicates persistent behavior (for example a positive auto-correlation). If there is increase of time step there will probably be an increase from t_i to t_{i+1}. The same is true of decreases, where a decrease will tend to follow a decrease. A Hurst exponent value $0 < H < 0.5$ will exist for a time series with anti-persistent behavior (or negative auto-correlation). Here an increase will tend to be followed by a decrease or a decrease will be followed by an increase. The larger the Hurst exponent, the smaller is the fractal dimension and the smoother is the surface.

Wavelet Method to Estimate Hurst Exponent

We describe here a method for estimating the Hurst exponent of a time series (signal and function) using the wavelet method.

The total energy contained in a signal, $f(t)$, is defined as

$$E = \int_{-\infty}^{\infty} |f(t)|^2 \, dt = \|f\|^2 .$$

Two-dimensional wavelet energy density functions is defined as $E = (a, b) = |T_\psi f(a, b)|^2$. It signifies the relative contribution of the energy contained at a specific scale a and location b. A plot of $E(a, b)$ is known as a scalogram (analogous to spectrogram). The scalogram can be integrated across a and b to recover the total energy in the signal using admissibility constant C_g as follows

$$E(a) = \frac{1}{C_g} \int_{-\infty}^{\infty} |T(a, b)|^2 db.$$

Peaks in $E(a)$ highlight the dominant energy scales within the signal. The wavelet power spectrum $P_w(f)$ is defined as

$$P_w(\xi) = \frac{1}{\tau} E_W(\xi) = \frac{1}{\tau \xi_c C_g} \int_0^{\tau} |T_\psi f(a, b)|^2 \, db$$

where

$$\xi_c = \sqrt{\frac{\int_0^\infty \xi^2 \left|\hat{\psi}(\xi)\right|^2 d\xi}{\int_0^\infty \left|\hat{\psi}(\xi)\right|^2 d\xi}}$$

τ is length of a signal, and $\xi = \dfrac{\xi_c}{a}$. The wavelet variance defined as

$$\sigma^2(a) = \frac{1}{\tau} |T(a, b)|^2 \, db$$

is often used in practice to determine dominant scale in the signal. We assume that τ is of sufficient length to gain a reasonable estimate of $\sigma^2(a)$.

Brownian walks can be generated from a defined Hurst exponent. If the Hurst exponent is $0.5 < H < 1$, the random walk will be a long memory process. Data sets like this are referred to as fractional Brownian motion (fbm). The motion can be generated by a variety of methods including wavelet transform. It may be recalled that in 1827 botanist R. Brown noticed that minute particles suspended in liquid moved on highly irregular paths. There are several phenomena of this nature such as smoke particles in air, fluctuations in the stock market recorded at intervals of 5 minutes during market hours. Einstein published a mathematical study of this motion, which eventually led Perrin's Nobel Prize winning calculation of Avogadro's number [43].

It may be noted that spectral density is proportional to $\frac{1}{\xi^{\beta}}$ where $\beta = 2H+1$,

$$P_W(\xi_f) \propto \frac{1}{\xi^{\beta}}$$

where $\xi = \frac{\xi_a}{a}$, and a is the scale parameter. For this reason, the *fbm* is sometimes referred to as noise. There are several methods [43] to estimate the Hurst exponent for real world systems. We present here one of those methods which is based on wavelet methodology. The wavelet coefficient variance scale index j denoted by σ_j^2 is defined as follows.

$$\sigma_j^2 = \sum_{k=0}^{2^{J-j}-1} (d_{j,k})^2/2^{J-j}.$$

It is known that the wavelet power spectrum $P_W(\xi)$ is related to the wavelet coefficient variance

$$P_W(\xi_j) = \frac{2^j \Delta t}{\tau \ln 2} \sum_{k=0}^{2^{J-j}-1} d_{j,k}^2 \Delta t = \frac{\Delta t}{\ln(2)} \sigma_{j,k}^2$$

as $\tau = 2^j \Delta t$, where Δt is sampling time (one may choose). That is,

$$P_W(\xi_m) \propto \sigma_{j,k}^2.$$

Combining these relations and the fact that frequency ξ is inversely proportional to the wavelet scale $a = 2^j$, we obtain the scaling relationship

$$\sigma_{j,k}^2 \propto a_j^{2H+1}$$

or simply,

$$\sigma_j^2 \propto a_j^{2H+1}$$

or

$$\sigma_m \propto a_m^{H+\frac{1}{2}}.$$

For an orthonormal multiresolution analysis using a dyadic grid, the scale a is proportional to 2^m. We can therefore take base 2 logarithms of both sides of $\sigma_{j,k}^2 \propto a_j^{2H+1}$ to get equation

$$\log_2(\sigma_m^2) = (2H+1)m + \mathrm{C}$$

where the constant C depends both on the wavelet used and the Hurst exponent. By plotting this, the slope of regression line λ will give the Hurst exponent.

$$H = \left| \frac{\lambda - 1}{2} \right| .$$

This method has been applied to estimate the Hurst exponents for financial and rainfall data [43] for different time intervals. A graphic user interface (GUI) has been developed.

Wavelet Correlation Coefficient

As mentioned above wavelets can detect both the location and the scale of a structure. Wavelets are parameterized both by $a > 0$ (dilation parameter) and translation parameter $b(-\infty < b < \infty)$ such that

$$\psi_{a,b} = \psi \left(\frac{x-b}{a} \right) .$$

The wavelet domain of one-dimensional function ψ is rather two-dimensional in nature; one dimension corresponds to scale and other to translation. The continuous wavelet transform for one dimension is defined as

$$\omega(a,b) = \int_{-\infty}^{\infty} f(x) \psi^* \left(\frac{|x-b|}{b} \right)$$

where a is the scale, $f(x)$ is a one-dimensional function and ψ^* (* is a complex conjecture) is an analyzing wavelet, also known as mother wavelet. We can choose a Mexican hat wavelet as a possible analyzing wavelet:

$$\psi \frac{(|x-b|)}{a} = \frac{1}{(2\pi)^{0.5} a} \left(2 - \frac{|x-b|^2}{a^2} \right) \exp \left(-\frac{|x-b|^2}{a^2} \right) .$$

Wavelet spectrum denoted by $M(a)$ is defined as

$$M(a) = \frac{1}{a} \int_{-\infty}^{\infty} w(a,b)^2 db.$$

The wavelet spectrum has power law behavior

$$M(a) \approx a^{\lambda} .$$

Wavelet spectrum $M(a)$ defines the energy of wavelet coefficient for scale a. Here λ is exponent, the value of which is decided by the power of a.
Wavelet Cross Correlation Coefficient is defined as

$$r_w = \frac{\int w_1(a,b) w_2^*(a,b)}{(M_1(a) M_2(a))^{1/2}} db.$$

The relation between correlation coefficient and wavelet correlation coefficient can be written as

$$r = \frac{\int \gamma_\omega(a) \left(M_1(a) M_2(a) \right)^{1/2} a^{-1/2} da}{\int (M_1(a) a^{-1} da \int M_2(a)^{-1} da)^{1/2}} db.$$

If r is negative then the variations at two stations are not correlated and for positive value the variations are correlated. In order to check the features of temperature variations at two metrological stations one can plot wavelet spectrum of temperatures at these stations and compare the features of the correlation coefficients. In order to check the correlation between any two parameters at two different stations we compute r with the aid of equation or r given by equations.

11.1.2 Fractal Image Processing

Image compression techniques based on fractals have been developed in the last decade and promise better compression performance. These techniques are based on the recognition that fractals can describe natural scenes better than shapes of traditional geometry. There are three main fractal image compression techniques:
(i) fractal image compression based on the iterated function systems (IFS), (ii) segment-based coding, and (iii) yardstick coding. Here, we confine ourselves to the first technique where images are compressed into compact IFS codes at the encoding stage, and fractal images are generated to approximate the original image at the decoding stage.

For this technique, we refer to Neunzert and Siddiqi [35] and references therein. The word *fractal* was coined by Benoit Mandelbrot from the Latin word *fractus*, meaning broken, for describing objects that were too irregular to fit into the traditional geometrical setting. There are several definitions of fractals but we shall treat fractals as fixed points of mappings on the metric spaces of all compact subsets of a complete metric space into itself. Cantor set, Sierpinski gasket, Sierpinski carpet and Von Koch curves are examples of fractals. An image can be approximated by these objects and some numbers or quantities, characteristic of these objects will be transmitted or stored and the original image can be retrieved from these characteristic numbers or quantities.

For example, the fern image can be represented by 24 integers which can be represented by 192 bits (if each integer is represented by 8 bits). On the other hand, if we store fern pixels, we need 307,200 bits (supposing that the image size is 480 lines by 640 pixels). Therefore, we achieved a compression ratio of 1600 to 1 by the IFS technique of fractal image compression. Here, we briefly introduce the fractal objects, IFS theorem and collage theorem. Methods for finding IFS for images can be found in [35]. Commercial software to implement the process is available.

Cantor Set

The Cantor set C is a subset of the metric space $X = [0,1]$, which is obtained by successive deletions of middle third open subintervals, as follows:

$$I_0 = [0,1]$$

$I_1 = [0, \frac{1}{3}] \bigcup [\frac{2}{3}, \frac{3}{3}]$

$I_2 = [0, \frac{1}{9}] \bigcup [\frac{2}{9}, \frac{3}{9}] \bigcup [\frac{6}{9}, \frac{7}{9}] \bigcup [\frac{8}{9}, \frac{9}{9}]$

$I_3 = [0, \frac{1}{27}] \bigcup [\frac{2}{27}, \frac{3}{27}] \bigcup [\frac{6}{27}, \frac{7}{27}] \bigcup [\frac{8}{27}, \frac{9}{27}] \bigcup [\frac{18}{27}, \frac{19}{27}] \bigcup [\frac{20}{27}, \frac{21}{27}] \bigcup [\frac{24}{27}, \frac{25}{27}] \bigcup [\frac{26}{27}, \frac{27}{27}]$.

I_4= I_3 minus the middle open third of each interval in I_3.

...

...

...

$I_N = I_{N-1}$ minus the middle open third of each interval in I_{N-1}.
The Cantor set C is defined as

$$C = \bigcap_{n=0}^{\infty} I_n,$$

where $I_0 \supseteq I_1 \supseteq I_2 \supseteq I_3 \supseteq I_{N-1} \supseteq I_N \supseteq I_{N+1}$.

$C \neq \emptyset$ as $0 \in C$. C is a perfect set.

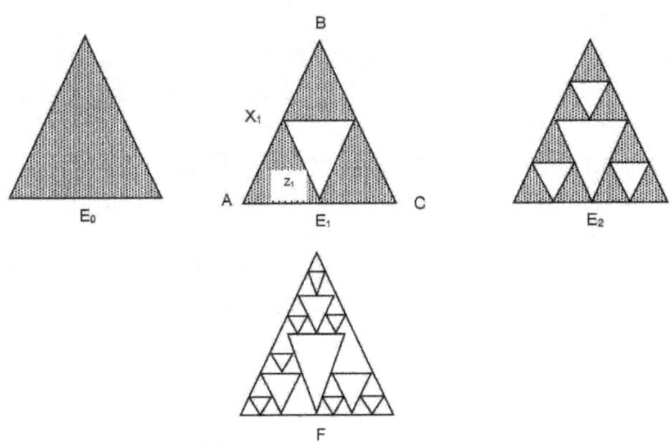

FIGURE 11.1: Sierpinski Gasket

In Figure 11.1, X_1, Y_1, Z_1 are middle points of the sides AB, BC and AC, respectively. Remove the triangle with vertices X_1, Y_1 and Z_1.

X_1'', X_1''', Y' are middle points of the triangle AX_1Z_1 and so on. Next we remove the central triangles of three left-over triangles of E_1.

Continue this process. Whatever is left is called the Sierpinski gasket or Sierpinski triangle (Figure 11.1). For an interesting account of the Sierpinski gasket, we refer to Stewart [45].

Iterated Function System

Theorem 79. *(Banach contraction fixed point theorem) Let X be a complete metric space and let $T : X \to X$ be a contraction mapping, that is, $d(T(x), T(y)) \leq \alpha d(x, y)$, $0 \leq \alpha < 1$. Then T has a unique fixed point, that is, $\exists$ unique $u \in X$ such that $Tu = u$. Furthermore, $T^{0n}(y) \to u$ as $n \to \infty$ where $T^{0n}(.)$ is defined as follows:*

$$T^{00}(x) = x$$

$$T^{01}(x) = T(x), T^{02}(x) = T(T(x)) = T(T^{01}(x))$$

$$T^{03}(x) = T(T^{02}(x), \ldots, T^{0n}(x) = T(T^{0(n-1)}(x))).$$

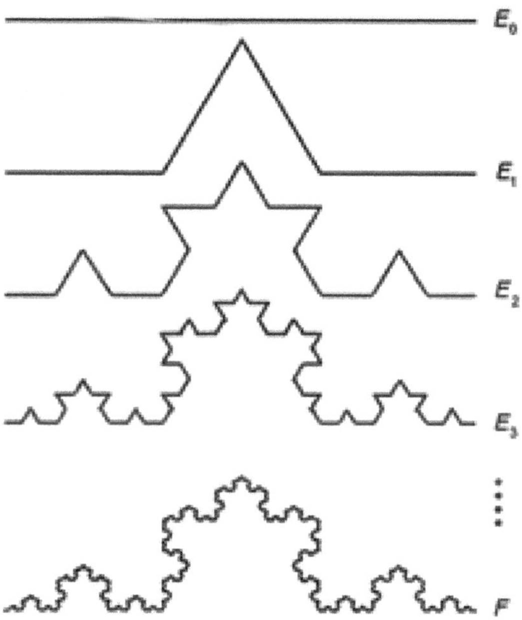

FIGURE 11.2: Von Koch Curve

This theorem is a key to the fractal image compression.

Definition 108. Let (X, d) be a complete metric space. X together with a finite set of contraction mappings W_n,n = 1,2. . . , N with contractivity factors $s_n, n = 1, 2, \ldots, N$ is called an "iterated function system" abbreviated as IFS, where

$$s = \max_n \, s_n.$$

An IFS will be denoted by $\{X, W_n, n = 1, 2, \ldots, N\}, s = \max_n S_n/$.

Example 318. $\{R, W(x) = 0,\ W(x) = \frac{2}{3}x + \frac{1}{3}\}$ is an IFS.

Let (X, d) be a complete metric space and $H(X) = \{K \subset X/K \text{ compact}\}$. For $B \in H(X)$ and $x \in X, d(x, B) = \min\{d(x, y)/y \in B\}$,

$$\begin{aligned}
&d(x, B) = \min\{d(x, y)/y \in B\},\\
&\mathrm{d}\,(\mathrm{A}, \mathrm{B}) = \max\{\mathrm{d}(\mathrm{x}, \mathrm{B})/\mathrm{x} \in \mathrm{A}\}\\
&\mathrm{d}\,(\mathrm{A}, \mathrm{B}) \neq \mathrm{d}\,(\mathrm{B}, \mathrm{A})\\
&\mathrm{h}\,(\mathrm{A}, \mathrm{B}) = \max\{\mathrm{d}\,(\mathrm{A}, \mathrm{B}), \mathrm{d}\,(\mathrm{B}, \mathrm{A})\},
\end{aligned}$$

has a unique fixed point, $A \in H(X)$ satisfying

$$A = W(A) = \bigcup_{n=1}^{N} W_n(A),$$

and is given by

$$A = \lim_{n \to \infty} W^{0n}(B), \text{ for any } B \in H(X).$$

Definition 109. The fixed point $A \in H(X)$ described in the IFS theorem is called the attractor of the IFS. The attractor is also called deterministic fractal or fractal.

Let $W : R^2 \to R^2$ be a contraction map, then A and $W(A)$ are shown in Figure 11.3

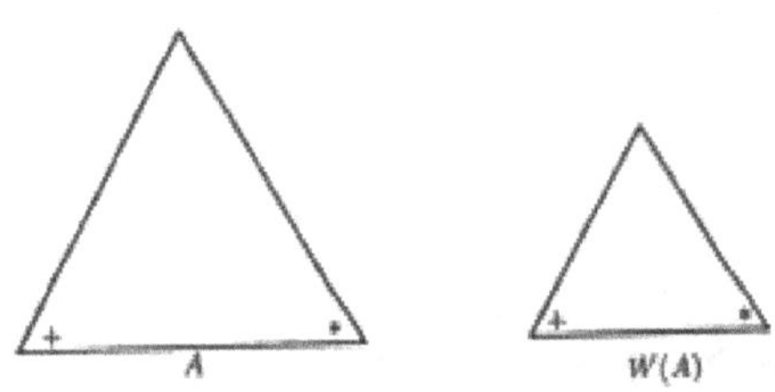

FIGURE 11.3: Contraction map

Relation between mappings on X and $H(\ X)$

Lemma 2. Let W_1 be a continuous mapping on the metric space (X, d) into itself. Then W_1 maps $H(K)$ into itself.

Lemma 3. Let $V : X \to X$ be a contraction mapping on the metric space (X, d) with contractivity factors. Then $\{W : H(X) \to H(X)$ defined by $W(B) = \{V(x)/x\} \in B, \forall B \in H(X)\}$ is a contraction mapping on $(H(X), h(.))$ with the same contractivity factor.

Lemma 4. Let (X, d) be a metric space and $\{W_n, n = 1, 2, 3, \ldots, N\}$ be an IFS. Let the contractivity factor for W_n be s_n Define $W : H(X) \to H(X)$ by

$$W(B) = W_1(B) \bigcup W_2(B) \bigcup \ldots \bigcup W_N(B) = \bigcup_{n=1}^{N} W_n(B)$$

for each $B \in H(X)$. Then W is a contraction mapping with contractivity factor $s = \max\{s_n : n = 1, 2, \ldots, N\}$.

Example 319. (a) X=[0,1], $W_1(x) = \frac{1}{3}$, $W_2(x) = \frac{1}{3}x + \frac{2}{3}$. W_1 and W_2 are contraction maps on X into itself. The deterministic fractal or fractal of the IFS system $\{ X, W_n, n = 1,2\}$ is the Cantor set.
(b) $X = [O, l]X[0, 1]$

$$W_i(x) - T_i \begin{pmatrix} x \\ y \end{pmatrix} + b_i = 1, 2, 3,$$

$$W_1 \begin{pmatrix} 1/2 & 0 \\ 0 & 1/2 \end{pmatrix}, \quad b_1 = \begin{pmatrix} 0 \\ 0 \end{pmatrix}$$

$$W_2 \begin{pmatrix} 1/2 & 0 \\ 0 & 1/2 \end{pmatrix}, \quad b_2 = \begin{pmatrix} 1/4 \\ 1/4 \end{pmatrix}$$

$$W_3 \begin{pmatrix} 1/2 & 0 \\ 0 & 1/2 \end{pmatrix}, \quad b_3 = \begin{pmatrix} 1/2 \\ 0 \end{pmatrix}.$$

The fixed point A of the map $W : H(X) \to H(X)$ defined $W(B) = \bigcup_{n=1}^{N} W_n(B)$ is the Sierpinski gasket.
(c) Sierpinski gasket and Von Koch curve are examples of fractals (deterministic fractals).

Theorem 80. *(Collage theorem) Let B be an arbitrary target image (an element of $H(X), X = R^2$) and $\{X, W_n = 1, 2, \ldots, N\}$ be an iterated function system with contractivity factor s, $0 < s < 1$. Further, let $W(B)$ be as in Lemma 4 such that the Hausdorff distance between B and $W(B)$ is less than ξ. Then the Hausdorff distance between B and the attractor A of the given IFS system is less than $(1 - s)^{-1}\xi$.*

Remark 81. (a) The iterated function theorem tells us that each IFS defines a unique fractal image (attractor) and that the attractor of the IFS is exactly the same as the union of the transformations of the attractor.

(b) Small (affine) deformed copies of the target are arranged so that they cover the target as exactly as possible. This collage of deformed copies determines an IFS. Theorem tells us that the better the collage, as measured by the Hausdorff distance, the closer will be the attractor of the IFS to the target.

(c) An interesting consequence of the collage theorem is that if the matrix entries in two codes are close, attractors of the codes are also close. In other words, small errors in codes lead to small errors in images.

11.1.3 Differential Equations on Fractals

A fine account of partial differential equations on fractals is given in a dissertation from Uppsala University [38] surveying fundamental research work to date. The author refers exciting work by Strichartz [46] through [49], Mosco [32] and Kigami [17] and [18]. In 1989, Kigami gave analytic construction of a Laplacian on the Sierpinski gasket. Around 2003, Strichartz showed first order linear differential equations based on the Laplacian that are not solvable on the Sierpinski gasket. Pelander gave a characterization on the polynomial p so that the differential equation $p(\Delta)u = f$ is solvable on any open subset of the Sierpinski gasket for any continuous function f on the subset. Mosco [32] studied energy functionals of certain fractal structures. These authors present an interesting exposition on the connections to the work of physicists. Pelander [38] indicated in his thesis how these studies are relevant to physics and engineering.

Understanding the complexities in a non-linear dynamical system is of great interest in contemporary sciences. Both chaos and fractals to offer important ingredients toward this effort. Even a simple quadratic nonlinearity as represented by the logistic map is known to generate, for a sufficiently large control parameter, a fractal attractor, indicating the onset of the deterministic chaos in the model.

Biologists studying the variability in populations of various species found an equation that predicted animal populations reasonably well. This was a simple quadratic equation called the logistic difference equation.

Example 320. The logistic difference equation is given by

$$x_{n+1} = rx_n(1 - x_n)$$

where r is the so-called driving parameter. The equation is used by starting with a fixed value of the driving parameter, r, and an initial value of x_0. One then runs the equation recursively, obtaining $x_1, x_2, \ldots, x_n$. For low values of r, x_n (as n goes to infinity) eventually converges to a single number. In biology, this number (x_n as n approaches infinity) represents the population of a species. When the driving parameter r is slowly turned up, interesting things happen. When $r = 3.0, x_n$ no longer converges, it oscillates between two values. This characteristic change in behavior is called a **bifurcation**. Turn up the driving parameter even further and x_n oscillates between not two, but four values. As one continues to increase the driving parameter, x_n goes through bifurcations of period eight, then sixteen, then chaos! When the value of the driving parameter r equals 3.57, x_n neither converges nor oscillates; its value becomes completely random. For values of r larger than 3.57, the behavior is largely chaotic. However, there is a particular value of r where the sequence again oscillates with a period of three.
The bifurcation diagram of the logistic difference equation is shown in Figure 11.4.

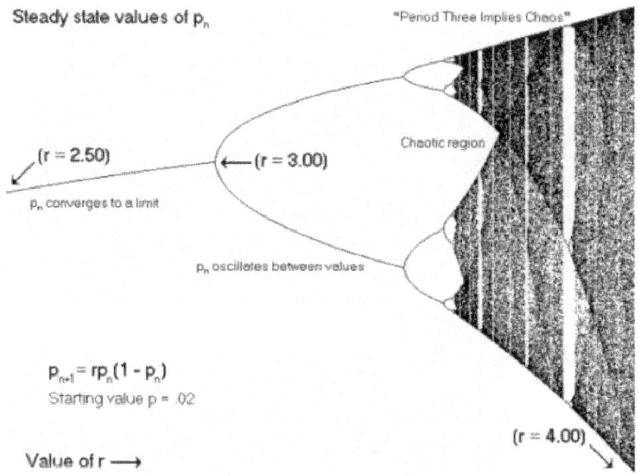

FIGURE 11.4: Bifurcation

11.1.4 Chaos and Fractals

Relation between Chaos and Fractals

Chaotic systems and fractals are both iterated functions. Each iteration takes the state of the previous iteration as an input to produce the next state.

Some chaotic systems like Figure 11.4 show fractal repetition when you zoom in. Asymmetric fractals on the other hand, can get chaotic and their fractal nature can be hard to spot.

Although fractals can show beautiful patterns, you can't actually predict their value in a specific iteration other than by calculating all the preceding iterations. So in a sense, fractals are just beautiful chaos.

11.2 Introduction to Time Series

Many statistical methods relate to data which are independent, or at least uncorrelated. There are many practical situations where data might be correlated. This is particularly so where repeated observations on a given system are made sequentially in time.

Definition 110. A time series is a sequence of numerical data points in successive order.

In investing, a time series tracks the movements of the chosen data points, such as a security's price, over a specified period of time with data points recorded at regular intervals. There is no minimum or maximum amount of time that must be included, allowing the data to be gathered in a way that provides the information being sought by the investor or analyst examining the activity. Data gathered sequentially in time are time series.

Definition 111. A time series is said to be stationary if it exhibits no systematic trend, no systematic changes in variance, and if it shows no periodic variations or seasonality.

Most processes in nature are non-stationary. Time series calculations have many applications, for example:

- Descriptions of data in the forms of summary statistics and graphs
- Analysis and interpretation of data via models that demonstrate time dependence of data
- Forecasting of future activity based on samples from aseries
- Controls made possible by adjusting and analyzing parameters to make a series fit closer to a target
- Adjustment of estimated variances or errors in a linear model to form a time series of correlated observations

11.2.1 Examples of Time Series

Time series are analyzed to demonstrate the underlying structures and functions that produce the observations. Understanding the mechanisms of a time series allows a mathematical model to be developed that explains the data in such a way that prediction, monitoring, or control can be achieved. Some examples in which time series arise are economics and finance, environmental modelling, meteorology and hydrology, demographics, medicine, engineering and quality control. The simplest form of data is a long series of continuous measurements at equally spaced time points. Observations are made at distinct points in time, and the observations may take values from a continuous distribution.

Example 321. Analyzing time series airline passenger data using MATLAB. First we create an array of monthly counts of airline passengers (measured in thousands) from January 1949 through December 1960. See Figure 11.5.

```
% 1949 1950 1951 1952 1953 1954 1955 1956 1957 1958 1959 1960
y = [112 115 145 171 196 204 242 284 315 340 360 417 % Jan
118 126 150 180 196 188 233 277 301 318 342 391 % Feb
132 141 178 193 236 235 267 317 356 362 406 419 % Mar
129 135 163 181 235 227 269 313 348 348 396 461 % Apr
121 125 172 183 229 234 270 318 355 363 420 472 % May
135 149 178 218 243 264 315 374 422 435 472 535 % Jun
148 170 199 230 264 302 364 413 465 491 548 622 % Jul
```

148 170 199 242 272 293 347 405 467 505 559 606 % Aug
136 158 184 209 237 259 312 355 404 404 463 508 % Sep
119 133 162 191 211 229 274 306 347 359 407 461 % Oct
104 114 146 172 180 203 237 271 305 310 362 390 % Nov
118 140 166 194 201 229 278 306 336 337 405 432]; % Dec

When we create a time series object, we can keep the time information along with the data values. We have monthly data, so we create an array of dates and use it along with the Y data to create the time series object. The Matlab Code is:

```
 yr = repmat((1949:1960),12,1);
mo = repmat((1:12)',1,12);
time = datestr(datenum(yr(:),mo(:),1));
ts = timeseries(y(:),time,'name','AirlinePassengers');
ts.TimeInfo.Format = 'dd-mmm-yyyy';
tscol = tscollection(ts);
plot(ts)
```

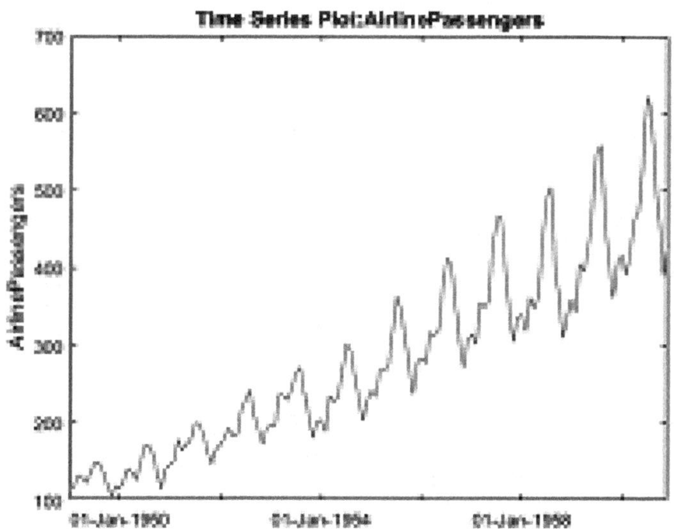

FIGURE 11.5: Time Series Plot: Airline Passenger data 1949 to 1960

Example 322. We can use above example to examine trend and seasonality (Figure 11.6). This series seems to have a strong seasonal component, with a trend that may be linear or quadratic. Furthermore, the magnitude of the seasonal variation increases as the general level increases. Perhaps a log transformation would make the seasonal variation be more constant. First we'll change the axis scale.

```
h\_gca = gca;
h\_gca.YScale = 'log';
```

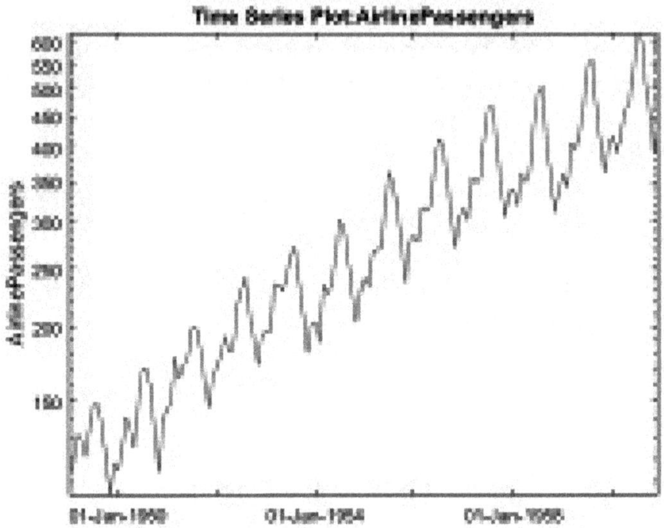

FIGURE 11.6: Time Series Plot: Airline Passenger Seasonality Trends

Time series can be also defined as follows.

Definition 112. A time series is a collection of data recorded over time (weekly, monthly, quarterly), to generate a history that can be used by management to make current decisions and plans based on long-term forecasting. It usually uses past patterns to project the future.

Components of Time Series

1. Secular trend: smooth long term direction of a time series (Figure 11.7).

2. Cyclical Variation: the rise and fall of a time series over periods longer than one year (Figure 11.8)

3. Seasonal Variation: Patterns of change in a time series within a year which tends to repeat each year (Figure 11.9).

4. Irregular variation: classified as episodic (unpredictable but identifiable)or residual (chance fluctuation and unidentifiable).

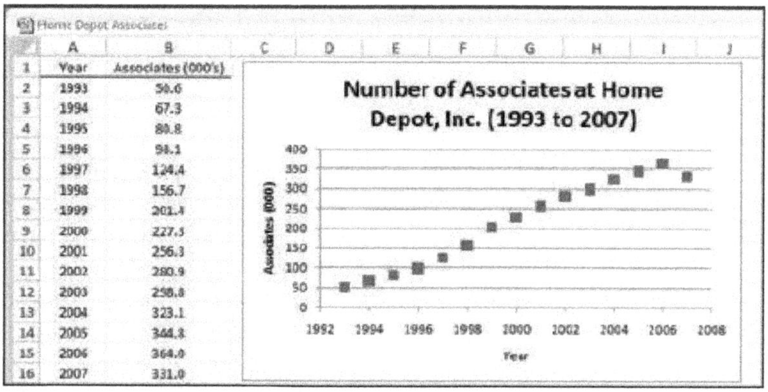

FIGURE 11.7: Secular Trend

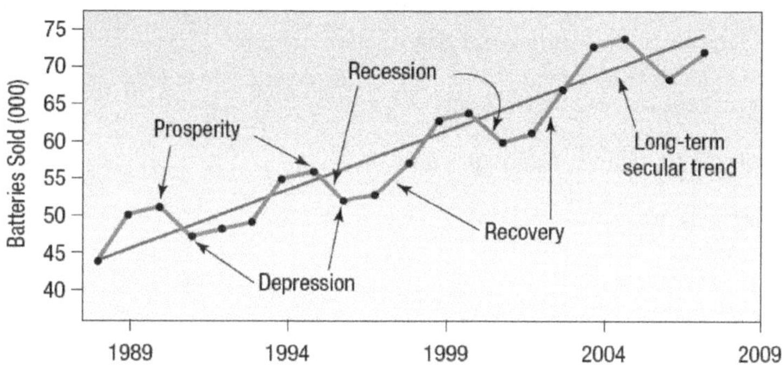

FIGURE 11.8: Cyclical Variation

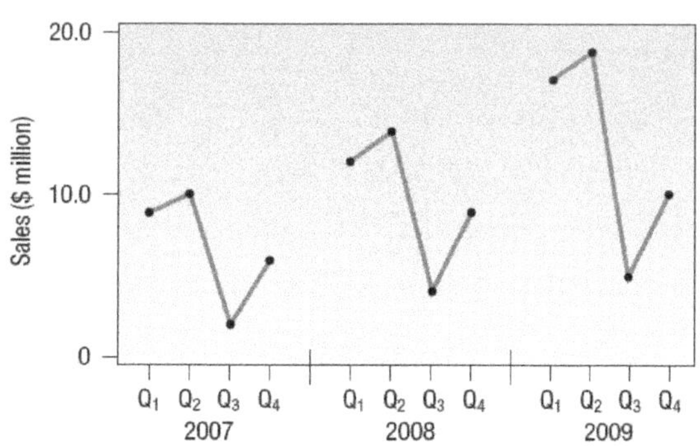

CHART 16–2 Sales of Baseball and Softball Equipment, Hercher Sporting Goods, 2007–2009 by Quarter

FIGURE 11.9: Seasonal Variation

11.2.2 Wavelets and Fractals in Time Series Analysis

Analysis of observed experimental data studied at different time points is known as time series analysis. The technique allows us to study trends, seasonal variations, business cycles, impacts of abrupt changes, impacts of unwanted components, drift factors, self-similarities, and Compression. Time series analysis is invaluable for forecasting future business parameters.

Economics and bankers use time series analysis to monitor stock market activities, unemployment figures and foreign exchange rates. Social scientists study population changes, birth rates, and educational statistics. In medicine, the method is used to study epidemics and pandemics and clinical trial of drugs. Electrocardiogram, magnetic resonance imagery, and other techniques are used to diagnose patients. Time series analysis is also useful for studying respiratory patterns, blood pressure fluctuations, and gastrointestinal problems. Geneticists use time series analysis in DNA. In the industrial arena, time series is used to monitor machine performance and prevent wear and breakage. Other uses include analysis of :

- Nuclear reactor performance
- Industrial equipment vibrations
- Global warming indicators
- Speech patterns
- El Niño impacts
- Wild life populations
- Earthquake prediction
- Explosion effects
- Physiological parameters that measure performance of cardiovascular, respiratory and other body system:
  - Electrocardiograms (ECG) to measure heart rate functioning
  - Electroencephalograms (EEGs) to measure brain waves
  - Electrooculograms (EOGs) to measure ocular performance
  - Electromyelograms (EMGs) to study muscle functioning
  - Ultrasound to measure various anatomic systems

Electroencephalogram (EEG)

Electroencephalogram is used to study brain activity. A seizure causes a sudden surge of electrical activity in the brain and may be an indicator of various types of brain disorders. Time series analysis measurements help diagnose and treat diseases of the nervous system such as epilepsy. More than 40 types of epilepsy characterized by different energy distributions in the brain may be distinguished by the time series analysis.

In recent years, wavelet methods have been used to diagnose and treat epilepsy. This common disease affects about 2% of the world's population, mostly children. In some epilepsy syndromes, interictal paroxysmal discharges of cerebral neurons indicate the severity of the disorder and may contribute to other disturbances cerebral functions such as speech impairment and behavioral changes. Studies of energy spikes within the brain serve as important diagnostic tools.

Electrooculography (EOG)

This technique measures the resting potential of the retina of the eye. The eyeball functions as a dipole with an anterior positive pole at the cornea and a posterior negative pole at the retina. Periorbital surface electrodes allow recording of eye movements during sleep. The two eye movement patterns identified in EOG signals are rapid eye movements (REMs) characterized by sharp deflections and slow eye movements (SEMs), indicated by somewhat sinusoidal pattern of low frequency ($\approx$ 0.2 to 0.6 Hz).

Recognition of the occurrence of these eye movements is important for clinicians for identifying and characterizing sleep disturbances. EOG provides reliable methods for identifying REMs and SEMs and separating them from other activities and artifacts, thus overcoming the problems related with manual scoring. Several algorithms based on filtering techniques have been developed for automatic detection of REMs showing better performance than conventional methods based on Fourier transform.

A new approach has been introduced for localizing SEM events based on wavelet decomposition and multiresolution framework that involves computing energies at the different scales of decomposition.

Wavelet methods

Wavelet methods have emerged as a key time frequency analysis and coding tool for EEG and other types of signals. Wavelet transforms and scalograms act like microscopes by zooming into small structures to reveal time events and into large structures to determine wave form trends. Energetic parameters of signals are analyzed to extract details of specific events. The techniques are based on the different frequenices of transient events that exhibit different energy contents.

A paper by Magosso et al. [58] summarizes recent developments and cites Daubechies technique and other types of wavelets such as Coeflets and Symlets. Performance of various wavelets have been studied and indicate that wavelet transforms out-perform short-time Fourier (Gabor) transform.

A wavelet is a quickly vanishing oscillation function localized as a continuous or discrete signal decomposing into a scaled and translated version $\psi_{a,b}(\mathrm{t})$ of a single function $\psi(t)$ called a mother wavelet:

$$\psi_{a,b}(t) = \frac{1}{\sqrt{|a|}}\psi\left(\frac{t-b}{a}\right)$$

where a and b are the scale and translation parameters, respectively, with $a, b \in R$ and $a \neq 0$.

The continuous wavelet transform (CWT) of a signal $f \in L_2(R)$ or space of finite energy signals or the space of the square integrable functions is defined as:

$$W(a,b) = C_{a,b} = \int_{-\infty}^{\infty} f(t)\sqrt{\frac{1}{|a|}}\psi^{\star}\left(\frac{t-b}{a}\right) dt = \langle f, \psi_{a,b}\rangle$$

where $\langle .,. \rangle$ denotes the inner product (dot product) and the symbol * means complex conjugate. CWT provides a redundant representation of the signal and requires a heavy burden of computation.

The discrete wavelet transform (DWT) is obtained by discretizing the parameters a and b. Choose $a = 2^{-j}, b = k2^{-j}$ with $j, k \in Z$. By substituting this we get

$$\psi_{j,k}(t) = 2^{j/2}\psi(2^j t - k).$$

The DWT can be written as

$$d_{j,k} = \int_{-\infty}^{\infty} f(t)2^{j/2}\Psi(2^j t - k)dt = \langle f, \psi_{j,k}\rangle$$

where $d_{j,k}$ are known as wavelet (or detailed) coefficients at scale j and location k.

By appropriate selection of mother wavelet ψ, the family $\{\psi_{j,k}(t)\}_{j,k\in Z}$ forms an orthonormal basis for $L_2(R)$. The original signal can be constructed from DWT. Further more, the DWT may be interpreted in terms of a multiresolution analysis, where a hierarchy of approximation and details of the signal is constructed in nested subspaces of $L_2(R)$.

Given a signal $f(t)$, its multiresolution decomposition at level h defined as $\psi(t)$ as mother wavelet, while it is a companion function, called a scaling function:

$$\varphi_{j,k}(t) = 2^{j/2}\varphi(2^j t - k)$$

representing approximation coefficients (or scaling coefficient) at level h defined as

$$a_{h,k} = \langle f, \phi_{h,k}\rangle.$$

Extending decomposition over all resolution levels, the complete wavelet expansion can be obtained as

$$f(t) = \sum_{j=-\infty}^{\infty}\sum_{k=-\infty}^{\infty} d_{j,k}2^{j/2}\psi(2jt - k).$$

This equation expresses the synthesis of the original signal from wavelet coefficients.

Let a signal be of finite length, say N and $N = 2^M$. The approximated and detailed signals at scale j will have only 2^{M-j} samples each, because of the downsampling operation. Coefficients at each scale j are placed at instant $t_{j,k} = k2^j(k = 0, 1, 2, ...2^{M-j})$.

Hence, the range of scales that can be investigated is $1 \leq j \leq M$, since the decomposition can proceed only until the individual details contains a single coefficient.

If the decomposition is carried out over all resolution levels M, the wavelet expansion will be

$$f(t) = \sum_{j-1}^{M} \sum_{k=0}^{2^M-1} d_{j,k} 2^{j/2} \psi(2jt - k).$$

Note that k starts from 0 since we assume, without loss of generality, that the signal starts from $t = 0$.

Scalograms

Scalograms are graphical representations of the squares of the wavelet coefficients for the different scales. They are isometric views of sequences of the wavelet coefficients versus wavelength. A scalogram clearly shows more details, identifies the exact location and time and detects low frequency cyclicity of the signal. The scalogram surface highlights the location (depth) and scale (wavelength) of dominant energetic features within the signal. The combination of the various vectors of coefficients at different scales (wavelengths) forms the scalogram. The depths (location and time) with the largest (strongest) coefficients indicate the position where the particular wavelength change is taking place. The scalogram provides a good space-frequency representation of the signal.

Wavelet energy

The energy contained in a one-dimensional signal $f(t)$ is

$$E(t) = \int_c^d |f(t)|^2 dt = \|f\|^2.$$

The total energy contained in a two-dimensional signal $f(x, y)$ is

$$\int_{c1}^{d1} \int_{c2}^{d2} |f(x, y)|^2 dx dy = \|f\|^2.$$

If f is a discrete signal then

$$E = \sum_n \sum_m |f(m, n)|^2 = \|f\|.$$

Wavelet transform decomposes a given signal into coefficients from which the original function can be reconstructed. Total energy of the given signal and its wavelet transform are identical.

A scalogram is a graphical representation of the square of the wavelet coefficient versus wavelength. A scalogram shares more details of a signal and detects low frequency cyclicities (Figure 11.10).

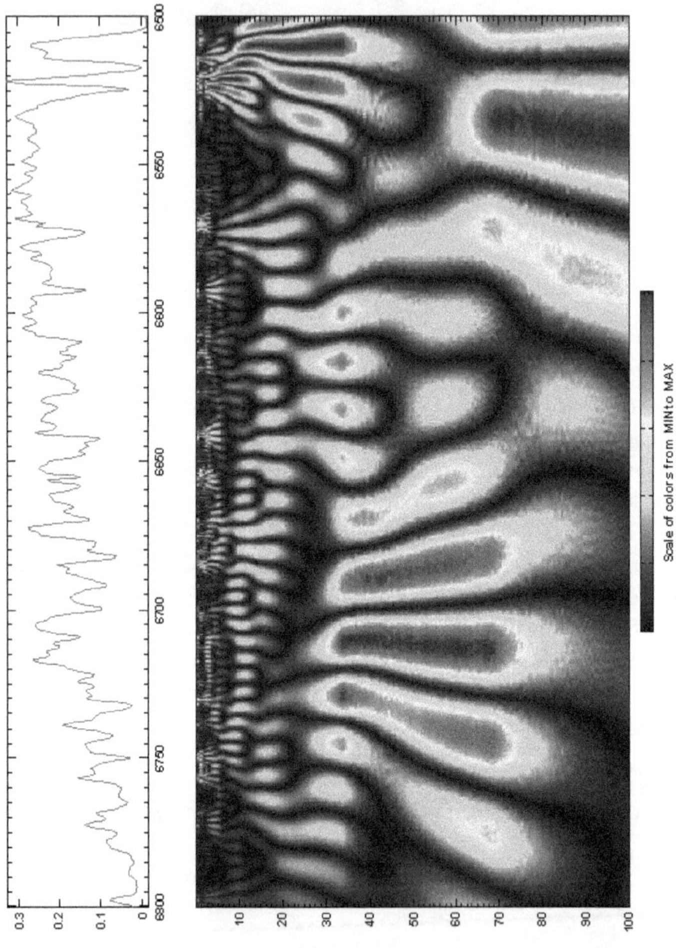

FIGURE 11.10: Scalogram

11.2.3 Prediction of Time Series Behavior Using Wavelets and Fractals

Fractal dimension and Hurst exponent analysis of a geophysical time series is popular, see [1], [7], [8], [21], [22] and [39] through [44]. However, these analyses are mainly restricted to obtaining the fractal dimensions of various time series. Only in the last couple of years has the linkage between these dimensions to the dynamics of the time series of meteorological parameters such as temperature, pressure and precipitation been studied.

Climate involves four major components, namely geographical parameters (latitude, longitude, distance from sea and height above mean sea level), temperature, pressure and precipitation. Since geographical parameters do not change significantly they are treated as constants. The climatic dynamics of any country or continent depend on the behavior of time series of temperature, pressure and precipitation. Mandelbrot noted that time series for a climatic variable correspond to a Brownian motion. Consider a discrete time series given by $x(t_i), i = 1, 2, 3, \ldots, N$, where t denotes the time and x the amplitude of the variable under consideration. For a fractional Brownian motion, the amplitude increments $x(t_j) - x(t_i)$ have a Gaussian distribution with variance:

$$\langle [x(t_j) - x(t_i)]^2 \rangle \ (t_j - t_i)^{2H}$$

where the symbol $\langle \ \rangle$ denotes the average over many samples of $x(t)$ and H denotes the Hurst exponent taking values between 0 and 1. If $H = 0.5$, we obtain the usual Brownian motion. The Hurst exponent is related to the fractal dimension D of the time series curve by:

$$D = 2 - H.$$

If the fractal dimension D for the time series is 1.5, we again get the usual Brownian motion. In this case there is no correlation between amplitude changes corresponding two successive time intervals. Therefore, no trend in amplitude can be discerned from the time series and the process is unpredictable. However, as the fractal dimension decreases to 1.0, the process becomes more predictable as it exhibits persistence. The future or the past trend is more likely to follow an established trend [52]. As the fractal dimension increases from 1.5 to 2.0, the process exhibits anti-persistence; a decrease in amplitude of the process is more likely to lead an increase in the future or past trend. Hence the predictability again increases. The scale independent unit thus gives us predictability of the acting process. References [39] through [40], [41] through [44] and [53] deal with the prediction of meteorological behavior using fractal methods for India, Saudi Arabia and Mexico data.

Fractal dimension (D)

Fractal dimension is a numerical measure of the roughness of an object. We are familiar with one dimension of a straight line, or two dimensions of a plane. What about a fractal dimension between the two. A mathematical definition of this concept is given in Section 11.1.1. A relationship between Hurst exponent (H) and fractal dimension (D) is given earlier. Computation of H is described in Section 11.1.1. There are several software programs available for the computation of fractal dimensions and Hurst exponents; see for example MATLAB and Benoit software.

Wavelet analysis is a tool for analyzing localized variations in power by decomposing a trace into time frequency space to determine both the dominant modes of variability and how those modes vary in time. This method is appropriate for analysis of non-stationary traces, that is, where the variance does not remain constant with increasing length of the data set. Fractal properties are present where the wavelet power spectrum is a power law function of frequency. The wavelet method is based on the property that wavelet transforms of self-affine traces have self-affine properties. Consider n wavelet transforms each with a different scaling coefficient a_i, where $S_1, S_2, \ldots, S_n$ are the standard deviations from zero of the respective scaling coefficients a_i. Define the ratio of the standard deviations $G_1, G_2, \ldots, G_{n-1}$ as:

$$G_1 = S_1/S_2, G_2 = S_2/S_3, \ldots, G_{n-1} = S_{n-1}/S_n.$$

Estimate the average value of G_i as:

$$G_{avg} = \sum_{i=1}^{n-1} G_i/n - 1.$$

The Hurst exponent (H) is $H = f(G_{avg})$, where f is a heuristic function which approximates the Hurst exponent by G_{avg} for stochastic self-affine traces.

11.2.4 Fractal Dimension and Predictability

$$D = 2 - H.$$

If the fractal dimension D for the time series is 1.5, there is no correlation between amplitude changes corresponding to two successive time intervals. Therefore, no trend in amplitude can be discerned from the time series and the process is unpredictable. However, as fractal dimension decreases, to 1, the process becomes more predictable as it exhibits persistence. Predictably indices (denoted by PI_T, PI_P and PI_R) for temperature, pressure, and precipitation are defined as follows:

$$PI_T = 2|D_T - 1.5|; PI_P = 2|D_P - 1.5|; PI_R = 2|D_R - 1.5|.$$

Concepts of fractal and multifractal and their relevance to real world systems were introduced by Benoit Mandelbrot, for updated references and interesting introduction of the theme we refer to Mandelbrot and Hudson [27]. In many real world systems represented by time series, the pattern of singularities shown by a graph of points changes abruptly and analysis is a challenging task. The time series of rainfall data are usually fractal or multifractal.

Wavelet-based Hurst exponent and fractal dimensional analysis

Saudi Arabia climatic dynamics
The climate predictability indices were obtained for nine meteorological stations using wavelet-based Hurst exponent and fractal dimensions [42]. The meteorological data which included the ambient temperature, barometric pressure, precipitation, relative humidity and wind speed covered the 16 years between 1990 and 2005. The major observations of the study are summarized as follows.

The temperature was found to be strongly predictable for all the stations except Abha when evaluated using the entire data set. The process became completely unpredictable during winter and summer time with PIT values much less than 0.5. In summer season the situation was almost the same except for Jeddah, Yanbu and Guryat where PIT was greater than 0.5. The pressure data showed a strong persistence behavior at all the stations when evaluated using entire data set and became anti-persistence at most of the stations during winter season with the exception of Hail, Guryat, Turaif and Riyadh. In summer season, the pressure predictability was found strong at most of the locations except for Gizan and Yanbu where PIT was less than 0.5. The relative humidity was strongly correlated with the previous values for all the stations when evaluated with entire data set and varied between 0.58 and 0.91. During winter the process was less predictive. The precipitation predictability indices PIR were found to be independent of length of the data set, i.e., complete, winter or summer sets with a few exceptions. Furthermore, the precipitation predictability indices were also found to be independent of the temperature and pressure predictability indices. The wind speed was always found unpredictable in Hail, Guryat and Turaif and predictable in Abha based on the entire data set.

Moreover, the wind speed predictability was found to be independent of the temperature and pressure predictability indices irrespective of seasonality. It is recommended that predictability indices for other meteorological stations be established and detailed monthly studies be conducted.

Software list

- Benoit^{TM} software developed by B. Mandelbrot
- Last Wave, http://www.cmap.polytechnique.fr/ bacry/Lastwave developed by E-Bacry et al.
- WaveLab, http://www.stat.standford.edu/ wavelab/ developed by D. Donoh et al.
- FracLab, http:// fractals.inria.fr, developed by J. Levy, Vehel et al.

11.3 Introduction to Neural Networks

Neural network simulations are recent developments. Currently, the neural network field enjoys a resurgence of interest and a corresponding increase in funding. Neural networks, with their remarkable ability to derive meaning from complicated or imprecise data, can be used to extract patterns and detect trends that are too complex to be noticed by humans or basic computer techniques. A trained artificial neural network (ANN) can be thought of as an expert in the category of information it has been given to analyze. This expert can then be used to provide projections given new situations of interest and answer "what if" questions. Other advantages are listed below:

Adaptive learning: An ability to learn how to do tasks based on the data acquired by training or initial experience.

Self-organisation: An ANN can create its own organization or representation of the information it receives during learning time.

Real time operation: ANN computations may be carried out in parallel, and special hardware devices take advantage of this capability.

Fault tolerance via redundant information coding: Partial destruction of a network leads to the corresponding degradation of performance. However, some network capabilities may be retained even with major damage.

Definition 113. A technical neural network consists of simple processing units called neurons and directed, weighted connections between those neurons. Here, the strength of a connection (or the connecting weight) between two neurons i and j is referred to as wi, j.

Artificial Neural networks (ANNs) are used for effective and efficiently modelling of large and complex problems and also in classification of problems (where the output is a categorical variable) or regressions (where the output variable is continuous). An ANN is a powerful non-linear modelling approach based on the function of the human brain. It identifies and learns the correlated patterns between input data sets and target values. An ANN is as a network of simple processing nodes or neurons interconnected in specific order to perform simple numerical manipulations

Definition 114. A neural network is a sorted triple (N, V, w) with two sets N, V and a function w, where N is the set of neurons and V a set $\{(i, j)ji, j{\in}N\}$ whose elements are called connections between neuron i and neuron j.

The function $w : V{\to}R$ defines n. A network = neurons + weighted connection, where $w((i, j))$, the weight of the connection between neuron i and neuron j, is shortened to wi, j. Depending on wi, j is either undefined or 0 for connections that do not exist in the network. The weights can be implemented

in a square weight matrix W or in a weight vector W with the row number of the matrix indicating where the connection begins, and the column number of the matrix indicating which neuron is the target. Indeed, in this case the 0 marks a non-existing connection. This matrix representation is also called a Hinton diagram.
These networks are consist of an input layer consisting of nodes representing different input variables, a hidden layer consisting of many hidden nodes and an output layer consisting of output variables. An ANN has four common assumptions:

1. Information processing occurs at many simple elements called neurons.

2. Signals are passed between neurons over connection links.

3. Each connection link has an associated weight that in a typical neural net multiplies the signal transmitted.

4. Each neuron applies an activation function (usually non-linear) to its net input (sum of weighed input signals) to determine its output signal.

Definition 115. An artificial neural network can be defined as an information processing system consisting of many processing elements joined in a structure inspired by the cerebral cortex of the brain.

The processing elements are usually organized in a sequence of layers with full connections between layers. Typically, there are three (or more) layers: an input layer where data are presented to the network through an input buffer, an output layer with a buffer that holds the output response to a given input, and one or more intermediate or hidden layers. The operation of an artificial neural network involves processes of learning and recall. Learning is the process of updating the connection weights in response to external stimuli presented at the input buffer. The network "learns" in accordance with a rule governing the adjustment of connection weights in response to learning examples applied at the input and output buffers. Recall is the process of accepting an input and producing a response determined by the geometry and synaptic weights of the network.

Processing units

Each processing element (or neuron) receives inputs (signals) from neighbors or external sources and uses them to compute an output signal which is propagated to other units. Along with this processing, the weights are adjusted. The system is inherently parallel in the sense that many units can carry out their computations at the same time. During operation, units can be updated synchronously or asynchronously. For synchronous updating, all units update their activation simultaneously and for asynchronous updating, each unit has a (usually fixed) probability of updating its activation at a time t and usually only one unit will be able to do this at a time.

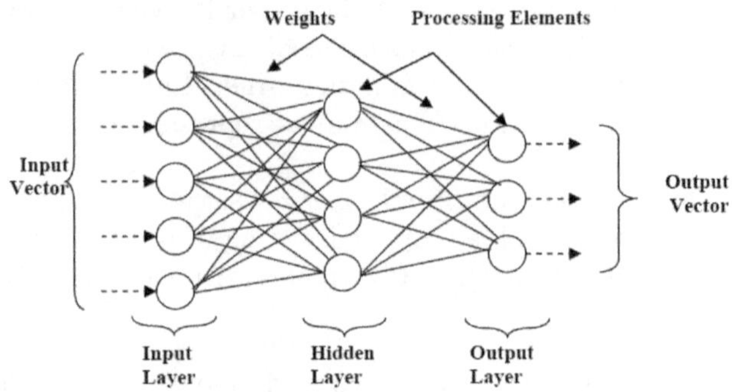

FIGURE 11.11: Neural Network Layer Diagram

Artificial neural networks

Activation and output functions

Signals are processed by neurons using two functions. The activation function determines the received total signal neurons. In most cases, a linear combination of the incoming signals is used. For neuron i connected to neurons j (for $j = 1, \ldots, N$) sending signals x_j with the strength of the connections W_{ij}, the total activation signal I_i is

$$I_i(x) = \sum_{n=1}^{N} w_{ij}(t)x_j.$$

The output function determines the neuron's signal processing and is denoted by $o(I)$. Together these two functions determine the values of the outgoing neuron signals. The total function acts in the input space of N-dimension called the parameter space. The composition of these two functions is called the transfer function $o(I(x))$. The activation and the output functions of the input and the output layers (Figure 11.11) may be of different types than those of the hidden layer. In general, linear functions are used for inputs, outputs and non-linear functions for hidden layers.

Single neuron

The basic unit of computation in a neural network is the neuron (Figure 11.12) often called a **node** or **unit**. It receives input from other nodes or from an external source and computes an output. Each input has an associated weight (w), which is assigned on the basis of its relative importance to other inputs. The node applies a function f (defined below) to the weighted sum of its inputs as shown in the figure:

The network takes numerical inputs X_1 and X_2 and has weights w_1 and w_2 associated with those inputs. Additionally, another input 1 with weight b (called the bias) is associated with it. The output Y from the neuron is computed as

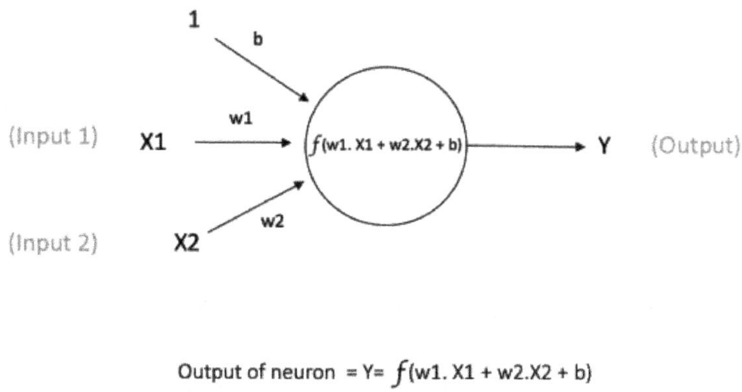

FIGURE 11.12: Single Neuron

shown in the figure. The function f is non-linear and is called the activation function (Figure 11.13). The purpose of the activation function is to introduce non-linearity into the output of a neuron. This is important because most real world data is non-linear and we want neurons to learn non-linear representations. Every activation function (or non-linearity) takes a single number and performs a certain fixed mathematical operation on it. There are several activation functions you may encounter in practice:

1. The sigmoid function takes a real valued input to a range between 0 and 1:
$$\sigma(x) = 1/(1 + exp(-x))$$
.

2. The tanh function squashes a real-valued input and squashes it to the range $[-1, 1]$:
$$\tanh(x) = 2\sigma(2x) - 1$$
.

3. The rectified linear unit (ReLU) threshold a real valued input by replacing negative values with 0:
$$f(x) = \max(0, x).$$

Figure 11.13 shows the three activation functions above.

Network topologies

Feed forward network: The data flow from input to output units is strictly feed forward. The data processing can extend over multiple (layers of) units,

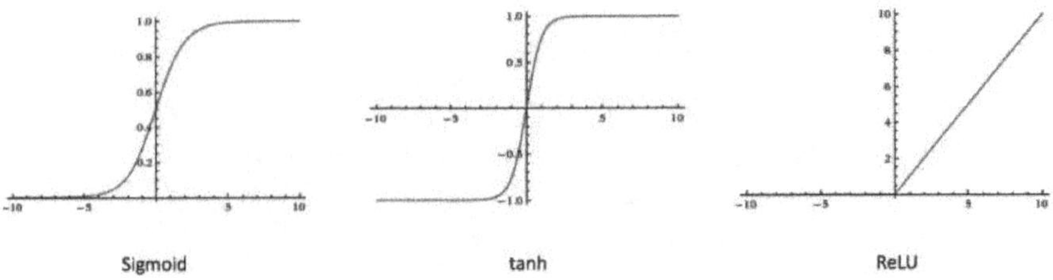

FIGURE 11.13: Different Activation Functions

but no feedback connections extending from outputs of units to inputs of units in the same layer or previous layers are present.

Recurrent networks can contain feedback connections. The dynamical properties of the network are important. In some applications, the activation values of the units undergo a relaxation process such that the network will evolve to a stable state in which these activations do not change. In other applications the changes of the activation values of the output neurons are significant with the dynamical behavior that constitutes the output of the network.

An example of a feedforward neural network is shown in Figure 11.14. A

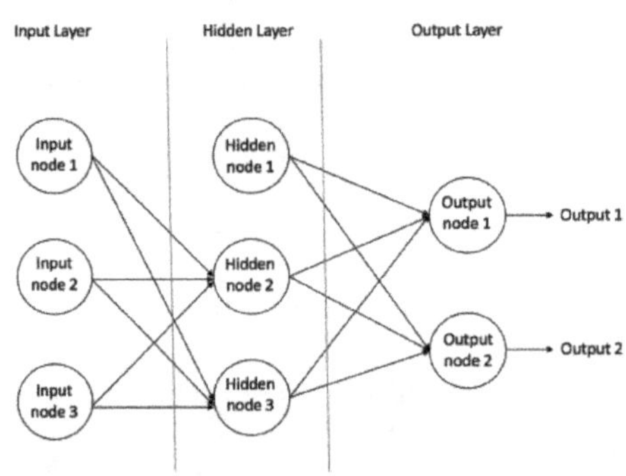

FIGURE 11.14: Example of Feedforward Neural Network

network can consist of three types of nodes:

Input nodes provide information from the outside world to the network and are referred to as the input layer. No computation is performed in input nodes; they just pass on the information to the hidden nodes.

Hidden nodes have no direct connection with the outside world (hence the name "hidden"). They perform computations and transfer information from the input nodes to the output nodes. A collection forms a hidden layer. While a feedforward network will only have a single input layer and a single output layer, it can have zero or multiple Hidden Layers.

Output nodes are collectively referred to as the output layer and are responsible for computations and transferring information from the network to the outside world.
In a feedforward network, the information moves in only one direction: forward from the input nodes, through the hidden nodes (if any) to the output nodes. There are no cycles or loops in the network. This property of feedforward networks is different from recurrent neural networks in which the connections between the nodes form a cycle.

Feedforward Neural Networks
The **Single Layer Perceptron** is the simplest type and has no hidden layer and can learn only linear functions. A multilayer perceptron (MLP) contains one or more hidden layers along with an input layer and an output layer. It can handle linear and non-linear functions. Figure 11.15 shows a multilayer perceptron with a single hidden layer. All connections of an MLP have associated weights. The figure shows only three weights (w_0, w_1, w_2).

Layer Description
The **input layer** has three nodes. The bias node has a value of 1. The other two nodes have external inputs values of X_1 and X_2 and depend on the input data set. No computations are performed in the input layer. The 1, X_1 and X_2 outputs are fed into the hidden layer.

The **hidden layer** also has three nodes and the bias node has an output of 1. The outputs of the other nodes depend on the 1, X_1, X_2 outputs and weights associated with the connections (edges). The output connection from one of the hidden nodes is highlighted in Figure 11.15. The output from other hidden node can be calculated. Remember that f denotes activation function. Outputs from the other hidden layer are fed to the output layer.
The **output layer** has two nodes that receive inputs from the hidden layer and perform computations similar to those shown for the highlighted hidden node in the figure. The Y_1 and Y_2 values calculated act as outputs of a multilayer perceptron.

Given a set of features $X = (x_1, x_2, \ldots)$ and a target y, a Multi Layer Perceptron can learn the relationship between the features and the target,for either classification or regression.

Example 323. (XOR Problem) Some problems cannot be solved by any perceptron. In fact there are more such problems than problems which can be solved using perceptrons. The most often quoted example is the XOR problem of building a perceptron which takes two Boolean inputs and outputs the XOR of them. What we want is a perceptron which will output one if the two inputs are different and 0 otherwise.

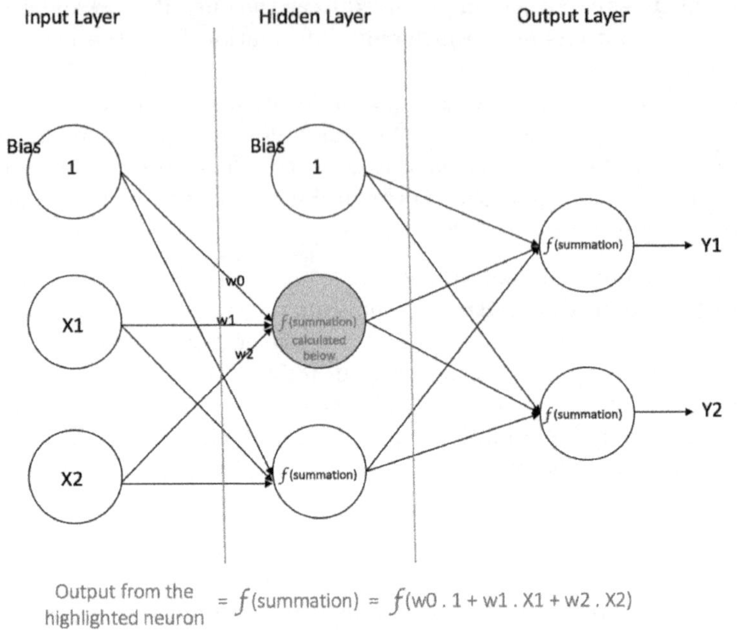

FIGURE 11.15: Multilayer Perceptron with One Hidden Layer

Input	Desired Output
0 0	0
0 1	1
1 0	1
1 1	0

Consider the following perceptron as an attempt to solve the problem.
If the inputs are both 0, the net input is 0 which is less than the threshold.

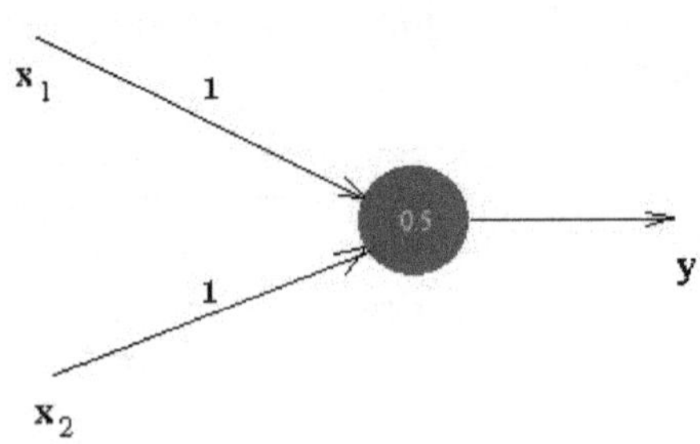

FIGURE 11.16

So the output is the desired 0 output. If one of the inputs is 0 and the other is

1, then the net input is 1. This is above the threshold, and so the output 1 is obtained. But the given perceptron fails in that case. To see that no perceptron can be built to solve the problem, try to build one. Now we need to cover an important point regarding the input data and the desired output. We use the binary OR operator as an example to explain the function of the weights and threshold. With OR we want a binary output so a single perceptron with two inputs is created. Now the search space for the neural network can be drawn as shown in Figure 11.17.

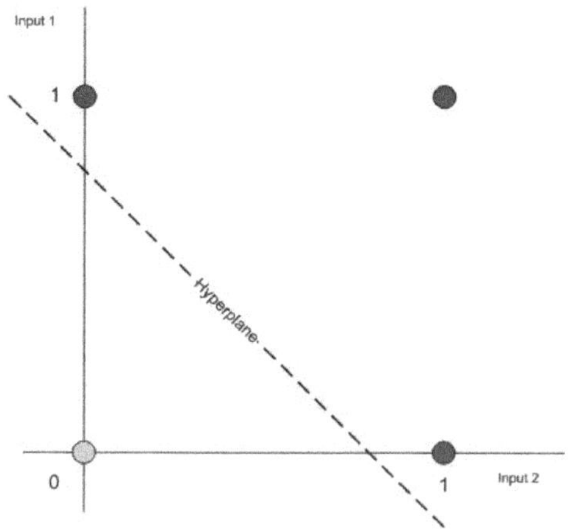

FIGURE 11.17: OR Operator Modelled by Single Perceptron

The dark dots represent values of true and the light dot represents a value of false; you can clearly see how the two classes are separable. We can draw a line separating them as in the above example. This separating line is called a hyperplane. A single neuron can create a single hyperplane and the above function can be solved by a single neuron. Another important point is that the hyperplane above is a straight line and that means we used a linear activation function (i.e. a step function) for our neuron. If we used a sigmoid or similar function the hyperplane would resemble a sigmoid shape as seen in Figure 11.18. The hyperplane generated by the image depends on the activation function used.

Remember the threshold (bias) value cited earlier? What does that do? Simply put, it shifts the hyperplane left and right while the weights orientate the hyperplane. In graphical terms the threshold translates the hyperplane while the weights rotate it. This threshold also needs to be updated during the learning process. The basic procedure is as follows:

1. Run input pattern through function.

2. Calculate error (desired versus actual value).

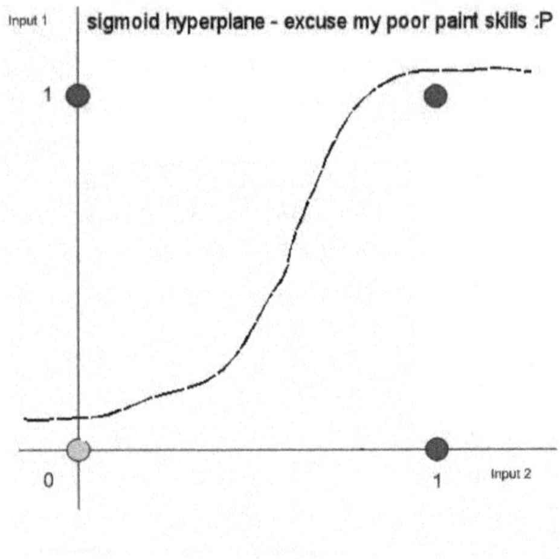

FIGURE 11.18: Hyperplane OR Operator Modelled by Single Perceptron

3. Update weights according to learning rate and error.
4. Move onto next pattern.

The learning rate term is very important because it greatly affects the performance and accuracy of a network. For more details we refer to [20], [23] ad [63].

11.4 Introduction to Fuzzy and Neuro-fuzzy

Definition 116. (Fuzzy sets) A classical crisp set is a collection of distinct objects. Crisp set theory was founded by the German mathematician George Cantor (1845-1918). It is defined as dividing the elements of a given universe into two groups: members and non-members. It is defined by a characteristic function. Let U be a universe of discourse. The characteristic function $\mu(x)$ of a crisp set A in U is defined as:

$$\mu_A(x) = \begin{cases} 1 \text{ iff } x \in A \\ 0 \text{ iff } x \notin A \end{cases}.$$

Zadeh introduced fuzzy sets with flexible membership. In fuzzy sets, many degrees of membership are allowed. The degree of membership to a set is indicated by a number between 0 and 1. Fuzzy sets can be considered extensions and generalizations of the crisp sets.

A fuzzy set A in the universe of discourse U can be defined as a set of ordered pairs,

$$A = \{(x, \mu_A(x))|x \in U\}$$

where μ_A is called the membership function of A and $\mu_A(x)$ is the degree of membership of x in A which indicates the degree that x belongs to A. The membership function maps U to the membership space M that is $\mu_A : U \rightarrow M$.

When $M = \{0, 1\}$, set A is non-fuzzy and μ_A is the characteristic function of the crisp set A. For a fuzzy set the range of the membership function is a subset of the non-negative real numbers. In general, M is the set of unit interval [0, 1].

Definition 117. Membership functions (MFs) can be represented by following functions:

Triangular: A triangular MF is a function with three parameters defined by

$$triangle(x, a, b, c) = \max\left(\min\left(\frac{x-a}{b-a}\right), \left(\frac{c-x}{c-b}\right), 0\right).$$

Trapezoidal: A trapezoidal MF is a function with four parameters defined by

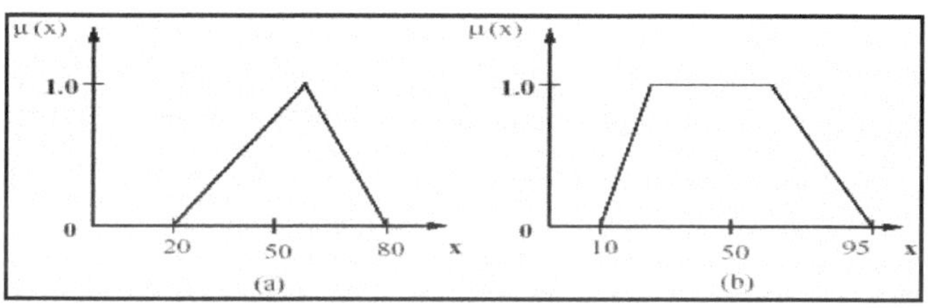

FIGURE 11.19: Membership Functions: (a) Triangle (b) Trapezoid

Gaussian: A Gaussian MF is a function with two parameters defined by

$$gaussian(x, \sigma, c) = e^{-\left(\frac{x-c}{\sigma}\right)^2},$$

where c is the center and the width of membership function.

Bell: A bell MF (Figure 11.20) is a function with two parameters defined by:

$$bell(x, a, b, c) = \frac{1}{1 + \left|\frac{x-c}{a}\right|^{2b}}.$$

Sigmoidal: A sigmoidal MF is a function with two parameters defined by

$$sigmmoid(x, k, c) = \frac{1}{1 + e^{-k(s-c)}},$$

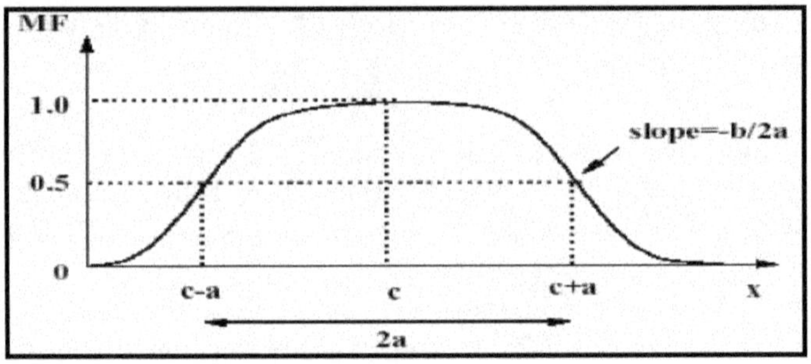

FIGURE 11.20: Bell Membership Function

where parameter k influences sharpness of function in the point where $a = c$. If $k > 0$ the function is open on the right site, if $k < 0$ the function is open on the left site and thus the function can be used for describing the concept like "very big" or "very small". Sigmoidal function is used in neural networks as on activation function.

Fuzzy rules and fuzzy reasoning

Fuzzy rules and fuzzy reasoning are the main frames of fuzzy inference systems used as modelling tools based on fuzzy set theory. They are used in real-world problems such as expert systems, pattern recognition and data classification. Fuzzy reasoning, also known as approximate reasoning, is an inference procedure that derives conclusions from a set of fuzzy if-then rules and known facts.

Fuzzy if-then rules

Fuzzy if-then rules (also known as fuzzy conditional statements) are expressions of the form

$$\text{If } x \text{ is } A, \text{ then } y \text{ is } B$$

where A and B are linguistic labels defined by fuzzy sets on universes of discourse X and Y, respectively. Often "x is A" is called the antecedent or premise while "y is B" is called the consequence or conclusion. Fuzzy if-then rules are used to capture the imprecise modes of reasoning and play an essential role in the human ability to make decisions in an environment of uncertainty and imprecision. Fuzzy if-then rules have been used extensively in both modelling and control. Also, due to the qualifiers on the premise parts each fuzzy if-then rule can be viewed as a local description of the system under consideration.

Fuzzy inference systems

The fuzzy inference system (Figure 11.21) is a popular computing framework based on the concepts of fuzzy set theory, fuzzy if-then rules and fuzzy reasoning. It is used in automatic control, data classification, decision analysis, expert systems, robotics and pattern recognition. The fuzzy inference system is also known as the fuzzy expert system, fuzzy model, fuzzy-rule-based system, fuzzy logic controller and fuzzy system. The basic structure of a fuzzy inference system consists of five functional components:

1. Rule base which contains a selection of fuzzy rules.
2. Database which defines the membership functions used in the fuzzy rules.
3. Reasoning mechanism which performs the inference procedure upon the rules and given facts to derive a reasonable conclusion.
4. Fuzzification interface which transforms the crisp inputs into degrees of match with linguistic values.
5. Defuzzification interface which transforms the fuzzy results of the inference into a crisp output.

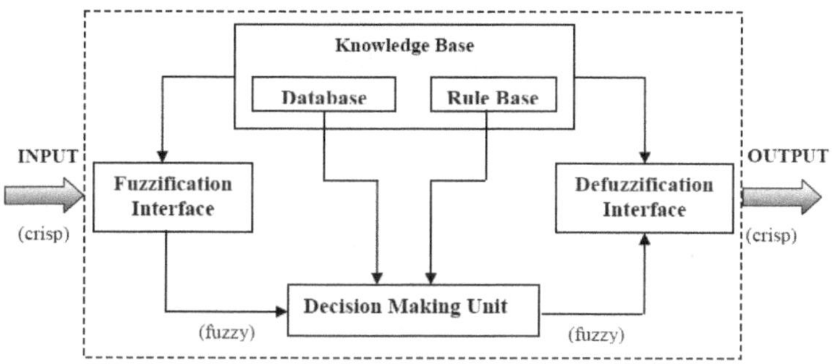

FIGURE 11.21: Fuzzy Inference System

The steps of fuzzy reasoning (inference operations upon fuzzy if-then rules) performed by fuzzy inference systems are:

1. Compare the input variables with the membership functions on the antecedent part to obtain the membership values of each linguistic label (fuzzification step).
2. Combine (through a specific T-norm operator, usually multiplication or min) the membership values on the premise part to get firing strength (weight) of each rule.
3. Generate the qualified consequents (either fuzzy or crisp) of each rule depending on the firing strength.
4. Aggregate the qualified consequents to produce a crisp output (defuzzification step).

Mamdani fuzzy model

The Mamdani fuzzy inference system was proposed as the first attempt to control a steam engine and boiler combination by a set of linguistic control rules obtained from experienced human operators. An example of if-then rules is *if the pressure is high, the volume is low*, where pressure and volume are linguistic variables and high and small are linguistic values or labels characterized by membership functions.

Tsukamoto fuzzy model

In the Tsukamoto fuzzy models the consequent part of each fuzzy if-then rule is specified by a membership function of a step function centered at a constant. As a result the inferred output of each rule is defined as a crisp value induced by the rule's firing strength. The overall output is taken as the weighted average of each rule's output. Fuzzy models avoid the time consumed by the defuzzification process since it aggregates each rule's output by the method of weighted average. This fuzzy model has the limitation that it is not transparent.

Sugeno fuzzy model

The Sugeno fuzzy model (also known as the TSK fuzzy model) was proposed by Takagi, Sugeno and Kang in an effort to develop a systematic approach to generating fuzzy rules from an input-output data set. The Sugeno fuzzy model is implemented into the neural fuzzy system ANFIS. A typical fuzzy rule has the format *if x is A, then y is B, then $Z = f(x,\ y)$* where A and B are fuzzy sets in the antecedent, $z = f(x,y)$ is a crisp function in the consequent part, $f(x,y)$ is a polynomial in the input variables x and y that can appropriately describe the output of the system within the fuzzy region specified by the antecedent part of the rule.

If $f(x,y)$ is a first-order polynomial, the first-order Sugeno fuzzy model is obtained. If f is a constant the zero-order Sugeno fuzzy model can be viewed either as a special case of the Mamdani fuzzy inference system where each rule's consequent part is specified by a fuzzy set or a special case of Tsukamoto's fuzzy model. A zero-order Sugeno fuzzy model is functionally equivalent to a radial basis function network under certain minor constraints. Using the Takagi and Sugeno's fuzzy if-then rule the resistant force on a moving object is defined:

If velocity is high, then force $= k * (\textit{velocity})^2$ where high in the premise part is a linguistic label characterized by an appropriate membership function. The consequent part is described by a non-fuzzy equation of the input variable velocity.

Neuro-fuzzy inference system

The basic concepts and rationale involve integrating fuzzy logic and neural networks into a working functional system. The combination of the techniques of fuzzy logic and neural networks suggests the novel idea of transforming. ANFIS is a universal estimator which may be able to approximate any real continuous function on a compact set. The basic structure of this type of fuzzy inference system is to map input characteristics to input membership functions. The system then relates input membership function to rules and the rules to a set of output characteristics. It then maps output characteristics to output membership functions and output membership function to a single output. Each fuzzy system contains a fuzzifier, fuzzy database and defuzzifier. A fuzzy database includes fuzzy rule base and an inference engine.

ANFIS architecture

According to the Takagi and Sugeno types, the fuzzy inference system has two inputs x and y and one output f as shown in Figure 11.22:

Rule 1: If x is A_1 and y is B_1 then $f_1 = p_1x + q_1y + r_1$. Rule 2: If x is A_2 and y is B_2 then $f_2 = p_2x + q_2y + r_2$.

Layer 1: Every node i in this layer is a square node with a node function

$$O_i^1 = \mu A_i(x)$$

where x is the input to the node i and A_i is the linguistic label associated with this node function. O_i^1 is the membership function of A_i and it specifies the degree to which the given x satisfies the quantifier A_i. Assume $\mu A_i(x)$ is bell-shaped with a maximum equal to 1 and minimum equal to 0 such as the generalization bell function

$$\mu A_i(x) = \frac{1}{1 + \left[\left(\frac{x-c_i}{a_i}\right)^2\right]^{b_i}}$$

or Gaussian function

$$\mu A_i(x) = e^{-\left(\frac{x-c}{\sigma}\right)^2}$$

where $\{a_i, b_i, c_i\}$ is the parameter set. As these parameters change, the bell shaped functions vary accordingly. Any continuous and piecewise differentiable functions such as commonly used trapezoidal or triangular-shaped membership functions are also qualified candidates for node functions in this layer. Parameters in this layer are referred as premise parameters.

Layer 2: Every node in this layer is a circle node which multiplies the incoming signals and sends the product out. For instance

$$O_i^2 = w_i = \mu A_i \times \mu B_i, \ i = 1, 2, 3 \ldots.$$

Each node output represents the firing strength of a rule.

Layer 3: Every node in this layer is a circle node labeled N. The ith node calculates the ratio of the ith rule's firing strength to the sum of all rules firing strengths:

$$O_i^3 = w_i = \frac{w_i}{w_1 + w_2}.$$

The outputs of this layer will be called normalized firing strength.

Layer 4: Every node i in this layer is a square node with a node function

$$O_i^4 = w_i f_i = w_i = w_i(p_i x + q_i y + r_i)$$

where is the output of layer 3 and (p_i, q_i, n_i) is the parameter set. Parameters in this layer will be referred as consequent parameters.

Layer 5: The single node in this layer is a circle node that computes the overall output as the summation of all incoming signals, i.e.,

$$O_i^5 = \sum w_i f_i = \frac{\sum_i w_i f_i}{\sum_i w_i}.$$

Thus, it construct an adaptive network which is functionally equivalent to a type 3 fuzzy inference system.

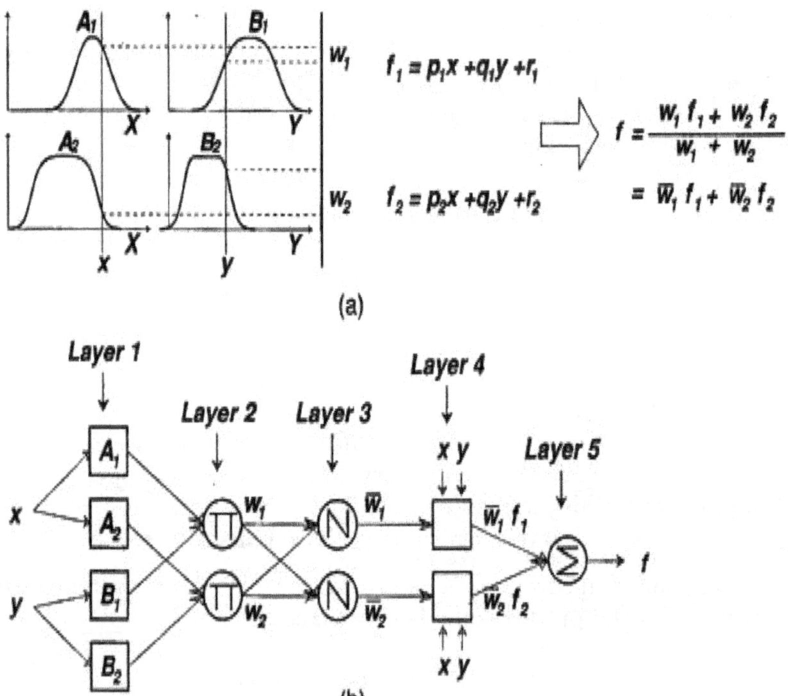

FIGURE 11.22: Neuro-Fuzzy

Example 324. In some applications, such as expert systems, it is necessary to introduce formal methods capable of dealing with such expressions so that a computer using rigid Boolean logic can still process them. This is what the theory of fuzzy sets and fuzzy logic tries to accomplish.

Example 325. Consider the statement: if the food is delicious, then the tip is high. The tip variable belongs to the *high* fuzzy set to a degree that depends on the level of the validity of the premise, i.e., the membership degree of the *food* variable to the *delicious* fuzzy set. The concept is that the more propositions in a premise are checked, the more suggested output actions must be applied. To determine the degree of truth of the *tip is high fuzzy* proposition, we must define the fuzzy implication. The result of application of a fuzzy rule thus depends on three factors:

1. The definition of fuzzy implication chosen
2. The definition of the membership function of the fuzzy set of the proposition located at the conclusion of the fuzzy rule
3. The degree of validity of propositions located premise

Since we understand the fuzzy operators AND, OR and NOT, the premise of a fuzzy rule may well be formed from a combination of fuzzy propositions. All the rules of a fuzzy system constitute the decision matrix. Here is the decision matrix for our tip example:

If the service is bad or the food is awful then the tip is low.
If the service is good then the tip is average.
If the service is excellent or the food is delicious then the tip is high.

11.5 Software for Time Series, Neural Network, Neuro-fuzzy and Fuzzy

Software programs relevant to the three categories above are listed in the following table:

Time Series	Neural Networks	Neuro-Fuzzy and Fuzzy Systems
CRAN Task View	Brian	Fuzzy Clustering
Eviews	Emergent	Learning Vector Quantization Demonstration
MATHEMATICA	Genesis	MATHEMATICA
MATLAB Simulink	MATHEMATICA	MATLAB Simulink
R	MATLAB Simulink	Multilayer Perceptron Training Demonstration
SAS	Neural Lab	Neuro-Fuzzy Control
SPSS	Neuron	Neuro-Fuzzy Function Approximation
Time Series Analysis	Next	R
	R	
	Stuttgart Neural Network Simulator (SNNS)	

11.6 Introduction to Graph Theory with Applications

Graphs

A linear graph (or simply a **graph**) $\mathbf{G} = (\boldsymbol{V}, \boldsymbol{E})$ consists of a set of objects $\boldsymbol{V} = \{v_1, v_2, \ldots\}$ called **vertices,** and another set $E = \{e_1, e_2, \ldots\}$, whose elements are called **edges,** such that each edge $\boldsymbol{e_k}$ is identified with an unordered pair (v_i, v_j) of vertices. The vertices (v_i, v_j) associated with edge e_k are called the **end vertices** of $\boldsymbol{e_k}$.

Remark 82. The most common representation of a graph is by a diagram, in which the vertices are represented as points and each edge as a line segment joining its end vertices. The object shown in Figure 11.23 is a graph.

Observe that this definition permits an edge to be associated with a vertex pair (v_p, v_t).

Definition 118. An edge having the same vertex as both its end vertices is called a **self-loop** (or simply a **loop**).

Example 326. Edge $\boldsymbol{e_1}$ in Figure 11.23 is a self-loop.

Note that the definition allows more than one edge associated with a pair of vertices, for example, edges $\boldsymbol{e_4}$ and $\boldsymbol{e_5}$ in Figure 11.23. Such edges are referred to as **parallel edges.**

Incidence and Degree

Definition 119. When a vertex v_i is an end vertex of some edge e_j, v_i and e_j are said to be incident with (on or to) each other.

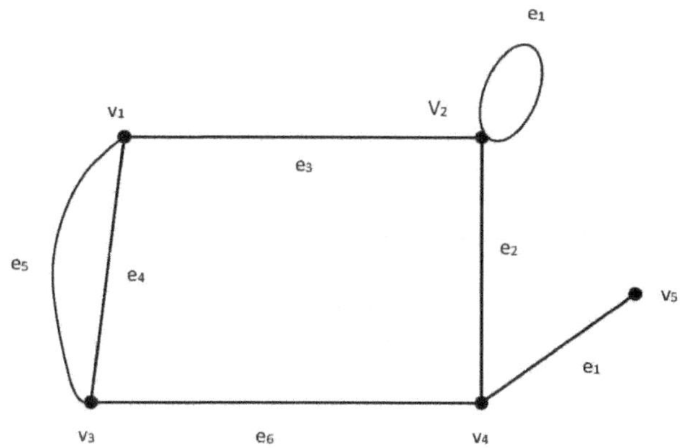

FIGURE 11.23: Graph with Five Vertices and Seven Edges

Example 327. In Figure 11.23 edges e_2, e_6 and e_1 are incident with vertex v_4.

Definition 120. Two nonparallel edges are said to be adjacent if they are incident on a common vertex.

Example 328. e_2 and e_7 in Figure 11.23 are adjacent. Similarly, two vertices are said to be adjacent if they are the end vertices of the same edge. In the figure, v_4 and v_5 are adjacent, but v_1, and v_4 are not. The number of edges incident on a vertex v_i, with self-loops counted twice, is called the degree, $d(v_i)$, of vertex v_i.

Example 329. In Figure 11.23, $d(v_1) = d(v_3) = d(v_4) = 3, d(v_2) = 4$, and $d(v_5) = 1$. The degree of a vertex is sometimes also referred to as its valency.

Let us now consider a graph G with e edges and n vertices $v_1, v_2, \ldots, v_n$. Since each edge contributes two degrees, the sum of the degrees of all vertices in G is twice the number of edges in G. That is

$$\sum_{i=1}^{n} d(v_i) = 2e.$$

Taking Figure 11.23 as an example once more, $d(v_1) + d(v_2) + d(v_3) + d(v_4) + d(v_5) = 3 + 4 + 3 + 3 + 1 = 14 =$ twice the number of edges. We shall derive the following interesting result.

Theorem 81. *The number of vertices of odd degree in a graph is always even.*

Definition 121. A graph that has neither self-loops nor parallel edges is called a **simple graph.** In some graph theory literature, a graph is defined to be simple, but in most engineering applications it is necessary that parallel edges and self-loops be allowed; this is why our definition includes graphs with self loops and/or parallel edges.

It should also be noted that in drawing a graph, whether the lines are drawn straight or curved, long or short is immaterial. What is important is the incidence between the edges and vertices. For example, the two graphs drawn in figure below are the same, because incidence between edges and vertices is the same in both cases.

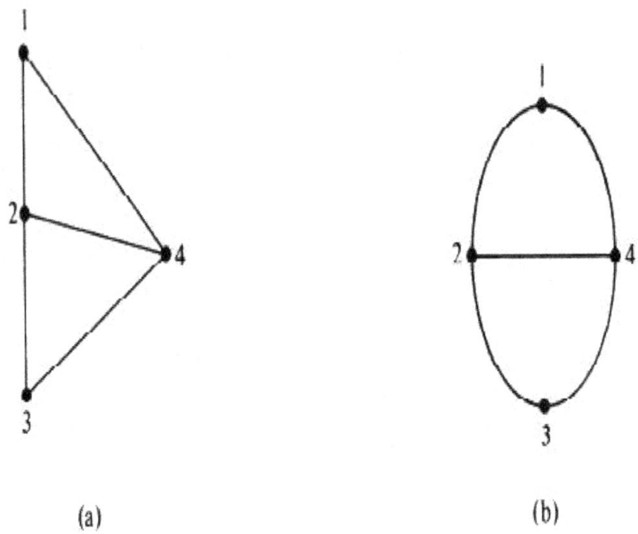

FIGURE 11.24: Same graph drawn differently

In a diagram of a graph, two edges may seem to intersect at a point that does not represent a vertex, for example, edges e and f in Figure 11.24. Such edges should be thought of as in different planes, thus having no common point. (Some authors break one of the two edges at such a crossing to emphasize this fact.)

Definition 122. A graph is also called a **linear complex,** a **1-complex,** or a **one-dimensional complex.** A vertex is also referred to as a node, a **junction,** a point, **0-cell,** or an **0-simplex**. Other terms used for an edge are a **branch,** a **line,** an **element,** a **1-cell**, an **arc,** and a **1-simplex.** In this book we generally use the terms **graph**, **vertex**, and **edge**.

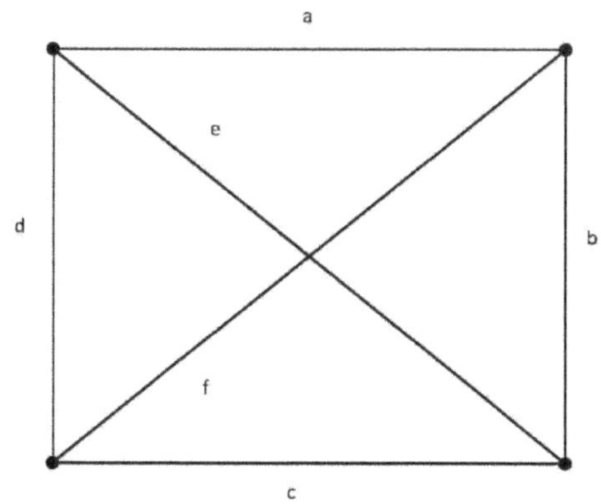

FIGURE 11.25: Edges e and f Lack Common Points

Applications of Graphs

Graph theory has a very wide range of applications in engineering and other areas. A graph can be used to represent almost any physical situation involving discrete objects and a relationship among them. The following are four examples from hundreds of such applications.

Konigsberg Bridge problem

The Konigsberg bridge problem is perhaps the best-known example in graph theory. It was a long-standing problem until solved by Leonhard Euler (1707-1783) in 1736, by means of a graph. Euler wrote the first paper in graph theory and thus became the originator of the theory the rest of topology. The problem is depicted in Figure

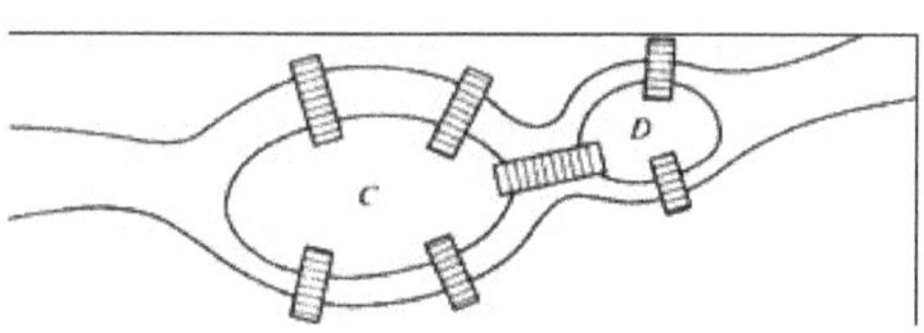

FIGURE 11.26: Konigsberg Bridge Problem

Two islands, C and D, formed by the Pregel River in Konigsberg were connected to each other and to the banks A and B with seven bridges, as shown in Figure The problem was to start at any of the four land areas of the city, A, B, C, or D, walk over each of the seven bridges exactly once, and return to the starting point (without swimming across the river, of course). Euler represented this situation by means of a graph, as shown in Figure 11.26. The vertices represent the land areas and the edges represent the bridges.

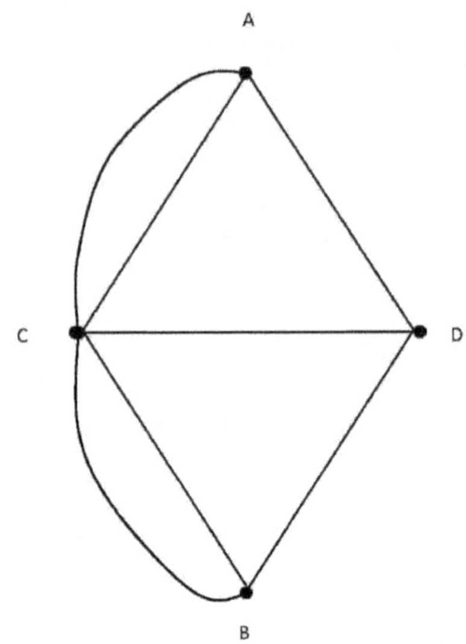

FIGURE 11.27: Graph of Königsberg bridge problem

Example 330. (Utilities problem) There are three houses H_1, H_2 and H_3 to be connected to three utilities: water (W), gas (G), and electricity (E) by means of conduits. Is it possible to make such connections without any crossovers of the conduits?

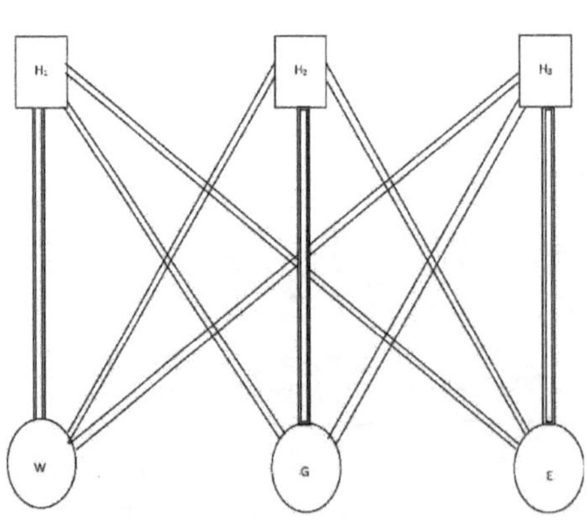

FIGURE 11.28: Three-Utilities Problem

In Figure 11.28 it shows how this problem can be represented by a graph; the conduits are shown as edges while the houses and utility supply centers are vertices. Thus the answer to the problem is no.

Electrical network problems: Properties (such as transfer function and input impedance) of an electrical network are functions of only two factors:

1. The nature and value of the elements forming the network, such as resistors, inductors, transistors, and so forth.
2. The way these elements are connected, that is, the topology of the network.

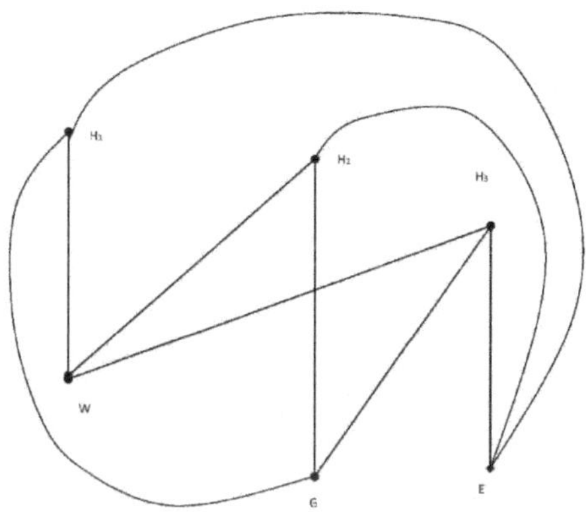

FIGURE 11.29: Graph of Three-Utilities Problem

Since there are only a few different types of electrical elements, the variations in networks are chiefly due to the variations in topology. Thus electrical network analysis and synthesis are mainly studies of network topology. In the topological study of electrical networks, factor 2 is separated from 1 and is studied independently. The topology of a network is studied by means of its graph. In drawing a graph of an electrical network the junctions are represented by vertices, and branches (which consist of electrical elements) are represented by edges, regardless of the nature and size of the electrical elements. An electrical network and its graph are shown in Figures 11.29 (a) and 11.29 (b).

Definition 123. A graph in which all vertices are of equal degree is called a **regular graph** (or simply a regular).

Example 331. The graph of three utilities shown in Figure 11.29b is a regular of degree 3.

Example 332. (Seating problem) Nine members of a new club meet each day for lunch at a round table. They decide to sit such that every member has different neighbors at each lunch. How many days can this arrangement last?

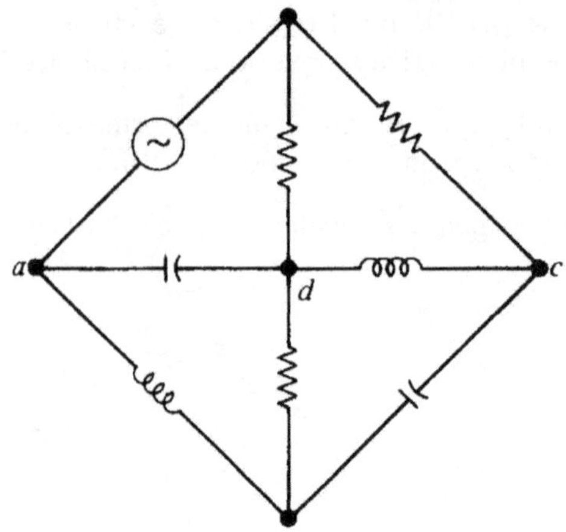

FIGURE 11.30a: Electrical Network and Graph

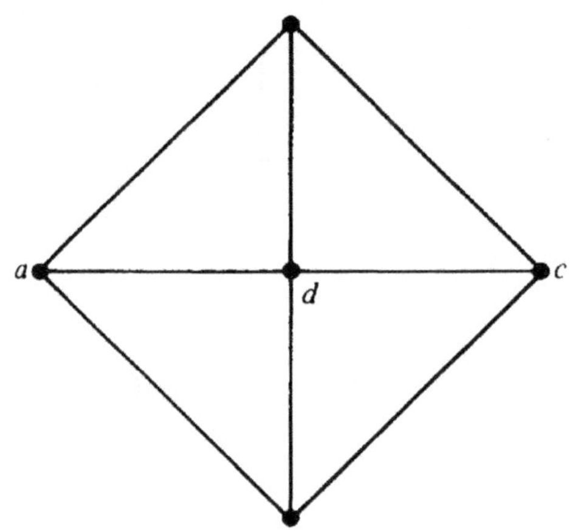

FIGURE 11.30b: Electrical Network and graph

This situation can be represented by a graph with nine vertices such that each vertex represents a member, and an edge joining two vertices represents the relationship of sitting next to each other. Figure 11.31 shows two possible seating arrangements: 1 2 3 4 5 6 7 8 9 1 (solid lines) and 1 3 5 2 7 4 9 6 8 I (dashed lines). It can be shown by graph-theoretic considerations that there are only two more arrangements possible. They are 1 5 7 3 9 2 8 4 6 1 and I 7 9 5 8 3 6 2 4 1 . In general it can be shown that for n people the number of such possible arrangements is $(n-1)/2$ if n is odd and $(n-2)/2$ if n is even.

Finite and Infinite Graphs

Although in our definition of a graph neither the vertex set V nor the edge

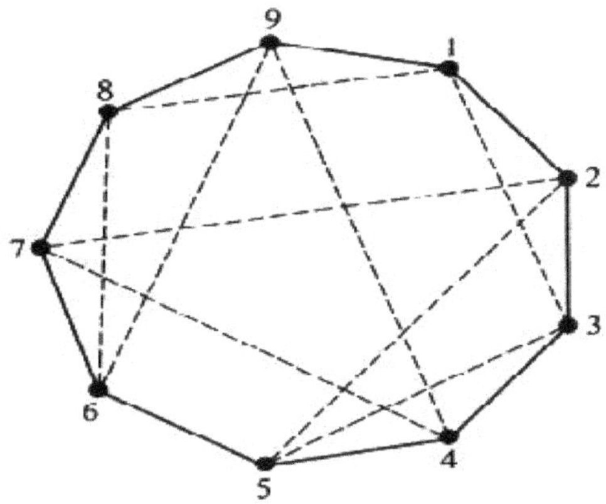

FIGURE 11.31: Dinner Table Arrangements

set E need be finite, in most theories and most application sets are finite. A graph with a finite number of vertices and a finite number of edges is called a **finite graph**; otherwise, a graph is infinite. The graphs in Figures 11.23, 11.26, 11.27 and 11.29(b) are all examples of finite graphs. Portions of two infinite graphs are shown in Figure 11.31

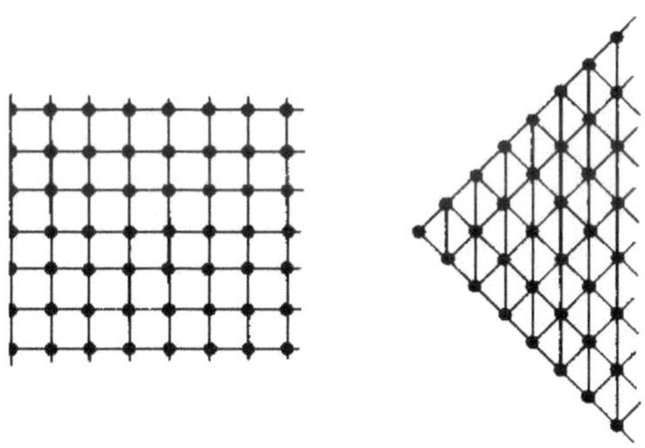

FIGURE 11.32: Portions of Two Infinite Graphs

Isolated Vertex, Pendant Vertex, And Null Graph

Definition 124. A vertex having no incident edge is called an **isolated vertex**. In other words, isolated vertices are vertices with zero degree.

Example 333. Vertices v_4 and v_7 in Figure 11.32 are isolated vertices.

Definition 125. A vertex of degree 1 is called a **pendant vertex** or an **end vertex**. The vertex v_3 in Figure 11.32 is a pendant vertex.

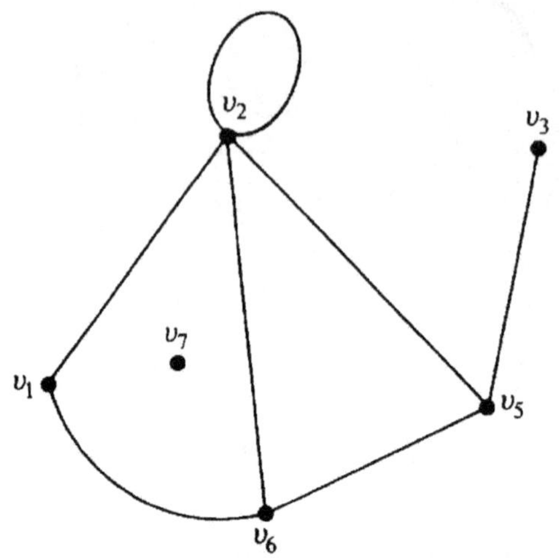

FIGURE 11.33: Graph Containing Isolated Vertices, Series Edges, and Pendant Vertex

Definition 126. Two adjacent edges are said to be in series if their common vertex is of degree 2.

Example 334. In Figure 11.32 the two edges incident on v_1 are in series.

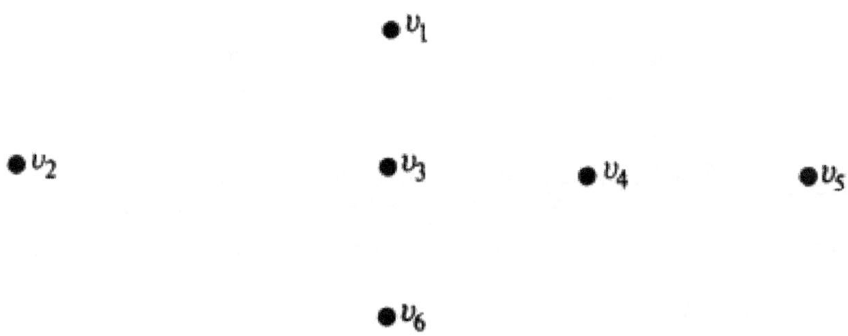

FIGURE 11.34: Null graph of Six Vertices

In the definition of a graph $G = (V, E)$, it is possible for the edge set E to be empty. Such a graph, without any edges, is called a **null graph**. In other words, every vertex in a null graph is an isolated vertex. A null graph of six vertices is shown in Figure 11.33. Although the edge set E may be empty, the vertex set V must not empty; otherwise, there is no graph. In other words, by definition, a graph must have at least one vertex.

Application of graph theory to linguistic has attracted attention of fairly good number of scholars in non-engineering fields, for example, Morgan Sanderegger, an assistant linguistics professor at McGill University in Montreal [mcgill.ca/~morgen/rlhijmeGraphs.pdf].

11.7 Applications of Spline Polynomials

In mathematics a spline is a special function defined piecewise by polynomials. In computer science the spline term refers to a piecewise polynomial curve. A piecewise polynomial function $f(x)$ is obtained by dividing of X into contiguous intervals, and representing $f(x)$ by a separate polynomial in each interval. The polynomials are joined at the interval endpoints (knots) in such a way that a certain degree of smoothness of the resulting function is guaranteed.

11.7.1 Polynomial Interpolation

In numerical analysis, polynomial interpolation is the interpolation of a given data set by a polynomial: given some points, find a polynomial which goes exactly through these points. Given a set of $n+1$ data points (x_i, y_i) where no two x_i are the same, one is looking for a for a polynomial p of degree at most n with the property $p(x_i) = y_i, i = 1, 2, \ldots, n$.

Applications

Polynomials can be used to approximate complicated curves, for example, the shapes of letters in typography, given a few points. A relevant application is the evaluation of the natural logarithm and trigonometric functions: pick a few known data points, create a lookup table, and interpolate between those data points. This results in significantly faster computations. Polynomial interpolation also forms the basis for algorithms in numerical quadrature and ordinary differential equations and secure multiparty computation and secret sharing schemes.

Polynomial interpolation is also essential to perform sub-quadratic multiplication and squaring such as Karatsuba multiplication and Toom-Cook multiplication, where an interpolation through points on a polynomial which defines the product yields the product itself. For example, given $a = f(x) = a_0x_0 + a_1x_1 + \ldots$ and $b = g(x) = b_0x_0 + b_1x_1 + \ldots$, the product ab is equivalent to $W(x) = f(x)g(x)$. Finding points along $W(x)$ by substituting x for small values in $f(x)$ and $g(x)$ yields points on the curve. Interpolation based on those points will yield the terms of $W(x)$ and subsequently the product ab. In the case of Karatsuba multiplication this technique is substantially faster than quadratic multiplication, even for modest-sized inputs. This is especially true when implemented in parallel hardware.

Definition 127. Polynomial interpolation is a method of estimating values between known data points. When graphical data contains a gap, but data is available on either side of the gap or at a few specific points within the gap, an estimate of values within the gap can be made by interpolation.

The simplest method of interpolation is to draw straight lines between the known data points and consider the function a combination of those straight lines. This method, called linear interpolation, usually introduces considerable error. A more precise approach uses a polynomial function to connect the points. A polynomial is a mathematical expression comprising a sum of terms, each term including a variable or variables raised to a power and multiplied by a coefficient. The simplest polynomials have one variable. Polynomials can exist in factored form or written out in full. For example:

$$(x-4)(x+2)(x+10)$$

$$x^2 + 2x + 1$$

$$3y^3 - 8y^2 + 4y - 2.$$

The value of the largest exponent is called the degree of the polynomial.

If a set of data contains n known points, then there exists exactly one polynomial of degree $n-1$ or smaller that passes through all of those points. The polynomial's graph can be thought of as "filling in the curve" to account for data between the known points. This methodology, known as polynomial interpolation, often (but not always) provides more accurate results than linear interpolation. The main problem with polynomial interpolation arises from the fact that even when a certain polynomial function passes through all known data points, the resulting graph might not reflect the actual state of affairs.

It is possible that a polynomial function, although accurate at specific points, will differ wildly from the true values at some regions between those points. This problem most often arises when "spikes" or "dips" occur in a graph, reflecting unusual or unexpected events in a real-world situation. Such anomalies are not reflected in the simple polynomial function which, even though it might make perfect mathematical sense, cannot take into account the chaotic nature of events in the physical universe.

Example 335. Given data points at $x_0 = 2, y_0 = 3$ and at $x_1 = 5, y_1 = 8$, find the following: At $x = 4, y = ?$

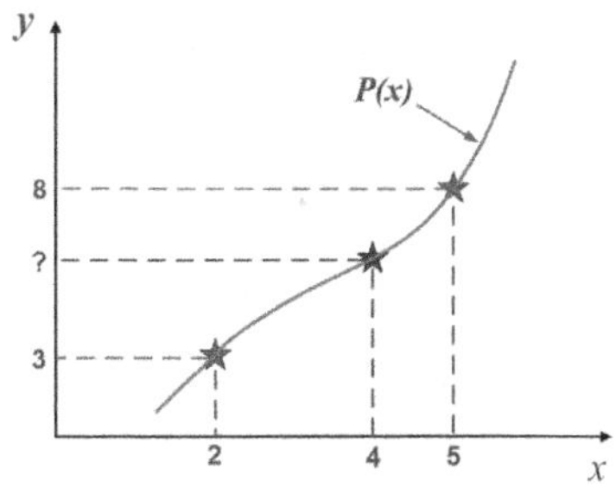

FIGURE 11.35

TABLE 11.1

x	f(x)
2.0	0.85467
2.3	0.75682
2.6	0.43126
2.9	0.22364
3.2	0.08567

$P(x)$ should satisfy the following conditions: $P(x = 2) = 3$ and $P(x = 5) = 8$. $P(x)$ can satisfy the above conditions if at $x = x_0 = 2, L_0(x) = 1$ and $L_1(x) = 0$ and at $x = x_1 = 5, L_0(x) = 0$ and $L_1(x) = 1$.

The Lagrange interpolating polynomial passing through three given points; $(x_0, y_0), (x_1, y_1)$ and (x_2, y_2) is:

At $x_0, L_0(x)$ becomes 1. At all other given data points $L_0(x)$ is 0.

At $x_1, L_1(x)$ becomes 1. At all other given data points $L_1(x)$ is 0.

At $x_2, L_2(x)$ becomes 1. At all other given data points $L_2(x)$ is 0.

General form of the Lagrange interpolating polynomial

Newton's Interpolating Polynomials

Newton's equation of a function that passes through two points (x_0, y_0) and (x_1, y_1) is

$$P(x) = a_0 + a_1(x - x_0).$$

The coefficients of Newton's interpolating polynomial are:

Example 336. Find Newtons interpolating polynomial to approximate a function whose 5 data points are given below $P(2.8) = 0.275$.

$f(x)$	First divided differences	Second divided differences	Third divided differences
$f[x_0]$			
	$f[x_0,x_1]=\dfrac{f[x_1]-f[x_0]}{x_1-x_0}$		
$f[x_1]$		$f[x_0,x_1,x_2]=\dfrac{f[x_1,x_2]-f[x_0,x_1]}{x_2-x_0}$	
	$f[x_1,x_2]=\dfrac{f[x_2]-f[x_1]}{x_2-x_1}$		$f[x_0,x_1,x_2,x_3]=\dfrac{f[x_1,x_2,x_3]-f[x_0,x_1,x_2]}{x_3-x_0}$
$f[x_2]$		$f[x_1,x_2,x_3]=\dfrac{f[x_2,x_3]-f[x_1,x_2]}{x_3-x_1}$	
	$f[x_2,x_3]=\dfrac{f[x_3]-f[x_2]}{x_3-x_2}$		$f[x_1,x_2,x_3,x_4]=\dfrac{f[x_2,x_3,x_4]-f[x_1,x_2,x_3]}{x_4-x_1}$
$f[x_3]$		$f[x_2,x_3,x_4]=\dfrac{f[x_3,x_4]-f[x_2,x_3]}{x_4-x_2}$	
	$f[x_3,x_4]=\dfrac{f[x_4]-f[x_3]}{x_4-x_3}$		$f[x_2,x_3,x_4,x_5]=\dfrac{f[x_3,x_4,x_5]-f[x_2,x_3,x_4]}{x_5-x_2}$
$f[x_4]$		$f[x_3,x_4,x_5]=\dfrac{f[x_4,x_5]-f[x_3,x_4]}{x_5-x_3}$	
	$f[x_4,x_5]=\dfrac{f[x_5]-f[x_4]}{x_5-x_4}$		
$f[x_5]$			

I	x_i	$f[x_i]$	$f[x_{i-1},x_i]$	$f[x_{i-2},x_{i-1},x_i]$	$f[x_{i-3},x_{i-2},x_{i-1},x_i]$	$f[x_{i-4},x_{i-3},x_{i-2},x_{i-1},x_i]$
0	2.0	0.85467				
			-0.32617			
1	2.3	0.75682		-1.26505		
			-1.08520		2.13363	
2	2.6	0.43126		0.65522		-2.02642
			-0.69207		-0.29808	
3	2.9	0.22364		0.38695		
			-0.45990			
4	3.2	0.08567				

11.7.2 Spline Interpolation

This method is used to construct a function that crosses through a discrete set of known data points. In the mathematical field of numerical analysis, **spline interpolation** is a form of interpolation where the interpolant is a special type of piecewise polynomial called a **spline**.

Given $n+1$ distinct knots x_i such that: $x_0 < x_1 < \cdots < x_n$, with $n+1$ knot values y_i find a spline function with each $S_i(x)$ a polynomial of degree at most

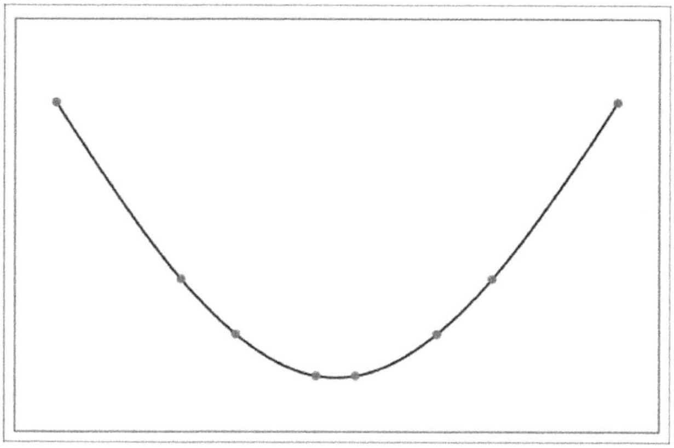

FIGURE 11.36

n.

$$S(x) := \begin{cases} S_0(x) & x \in [x_0, x_1] \\ S_1(x) & x \in [x_1, x_2] \\ \cdot & \cdot \\ \cdot & \cdot \\ S_{n-1}(x) & x \in [x_{n-1}, x_n]. \end{cases}$$

Linear spline interpolation: Points connected by lines: each S_i is a linear function constructed as must be continuous at each data point:

$$S_i(x) = y_i + \frac{y_{i+1} - y_i}{x_{i+1} - x_i}(x - x_i).$$

Quadratic spline interpolation: The quadratic spline can be constructed as:

$$S_i(x) = y_i + z_i(x - x_i) + \frac{z_{i+1} - z_i}{2\left(x_{i+1} - x_i\right)}(x - x_i)^2.$$

The coefficients can be found by choosing a z_0 and then using the recurrence relation

$$z_{i+1} = -z_i + 2\frac{y_{i+1} - y_i}{x_{i+1} - x_i}.$$

Cubic spline

All splines considered on this page are cubic splines; they are all piecewise cubic functions. However, anyone who says "cubic spline" usually means a special cubic spline with continuous first and second derivatives. The cubic spline is given by the function values in the nodes and derivative values on the edges of the interpolation interval (either of the first or second derivatives).

- If the exact values of the first derivative in both boundaries are known,

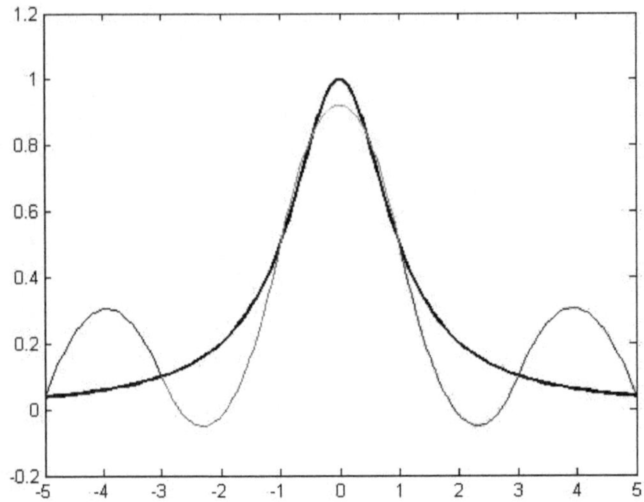

FIGURE 11.37: Quadratic Spline for $f(x) = 1/(1 + x^2)$

the spline is called a clamped spline, or spline with exact boundary conditions. This spline has interpolation error $O(h^4)$.

- If the value of the first (or second) derivative is unknown, we can set the so-called natural boundary conditions $S''(A) = 0,\ S''(B) = 0$. Thus, we get a natural spline with interpolation error $O(h^2)$. The closer to the boundary nodes, the greater the error becomes. In the inner nodes the interpolation accuracy is much better.
- One more boundary condition we can use when boundary derivatives are unknown is the parabolically terminated spline. The boundary interval is represented as the second (instead of the third) degree polynomial (for inner intervals, third degree polynomials are still used). In a number of cases this provides better accuracy than natural boundary conditions.
- We can also set periodic boundary conditions (this kind of conditions is used to model periodic functions).

Finally, we can combine different types of boundary conditions for different boundaries. It does make sense if we have only partial information about the function behavior at the boundaries (e.g., we know the left boundary derivative, and have no information about the right boundary derivative).

Akima spline

The Akima spline is a special type which is stable to the outliers. The disadvantage of cubic splines is that they may oscillate in the neighborhood of an outlier. On the graph you can see a set of points having one outlier. On the intervals which are next to the outlier, the spline noticeably deviates from the given function because of the outlier. We can see that in contrast to the cubic spline, the Akima spline is less affected by the outliers. An important property of the Akima spline is its locality. Function values in $[x_i, x_{i+1}]$ depend on $f_{i-2}, f_{i-1}, f_i, f_{i+1}, f_{i+2}, f_{i+3}$ only. The second property which should be taken into account is the non-linearity of the Akima spline interpolation: the result of interpolation of the sum of two functions doesn't equal the sum of the interpolation schemes constructed on the basis of the given functions. No fewer than five points are required to construct the Akima spline. In the inner area (i.e. between x_2 and x_{N-3} when the index goes from 0 to $N-1$) the interpolation error has order $O(h^2)$.

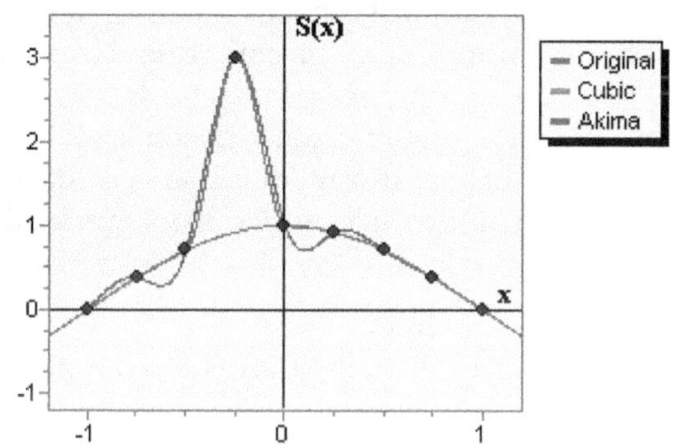

FIGURE 11.38

11.8 Compression Sensing

Compressed sensing (also known as compressive sensing, compressive sampling, or sparse sampling) is a signal processing technique for efficiently acquiring and reconstructing a signal, by finding solutions to underdetermined linear systems. This is based on the principle that, through optimization, the sparsity of a signal can be exploited to recover it from far fewer samples than required by the Shannon-Nyquist sampling theorem. There are two conditions under which recovery is possible. The first one requires the signal to be sparse in some domain. The second one is incoherence which is applied through the isometric property which is sufficient for sparse signals.

A common goal of the engineering field of signal processing is to reconstruct a signal from a series of sampling measurements. In general, this task is impossible because there is no way to reconstruct a signal during the times that the signal is not measured. Nevertheless, with prior knowledge or assumptions about the signal, it is possible to perfectly reconstruct a signal from a series of measurements. Over time, engineers have improved their understanding of which assumptions are practical and how they can be generalized.

An early breakthrough in signal processing was the Nyquist–Shannon sampling theorem. It states that if the signal's highest frequency is less than half of the sampling rate, then the signal can be reconstructed perfectly. The main idea is that with prior knowledge about constraints on the signal's frequencies, fewer samples are needed to reconstruct the signal.

Underdetermined linear system

An underdetermined system of linear equations has more unknowns than equations and generally has an infinite number of solutions. In order to choose a solution to such a system, one must impose extra constraints or conditions (such as smoothness) as appropriate. In compressed sensing, one adds the constraint of sparsity, allowing only solutions which have a small number of non-zero coefficients. Not all underdetermined systems of linear equations have sparse solutions. However, if there is a unique sparse solution to the underdetermined system, then the compressed sensing framework allows the recovery of that solution.

Solution reconstruction method

Compressed sensing (CS) takes advantage of the redundancy in many interesting signals—they are not pure noise. In particular, many signals are sparse, that is, they contain many coefficients close to or equal to zero, when represented in some domain. This is the same insight used in many forms of lossy compression.

Compressed sensing typically starts with taking a weighted linear combination of samples also called compressive measurements on a basis different from the basis in which the signal is known to be sparse. Therefore, the task of converting the image back into the intended domain involves solving an underdetermined matrix equation since the number of compressive measurements taken is smaller than the number of pixels in the full image. However, adding the constraint that the initial signal is sparse enables one to solve this underdetermined system of linear equations.

The least-squares solution to such problems is to minimize the L^2 norm, that is, minimize the amount of energy in the system. This is usually simple mathematically (involving only a matrix multiplication by the pseudo-inverse of the basis sampled). However, this leads to poor results for many practical applications, for which the unknown coefficients have non-zero energy. To enforce the sparsity constraint when solving for the underdetermined system of linear equations, one can minimize the number of non-zero components of the solution. The function counting the number of non-zero components of a vector was called the L^0 norm.

Applications: The field of compressive sensing is related to several topics in signal processing and computational mathematics, such as underdetermined linear-systems, group testing, heavy hitters, sparse coding, multiplexing, sparse sampling, and finite rate of innovation. Its broad scope and generality enabled several innovative CS-enhanced approaches in signal processing and compression, solution of inverse problems, design of radiating systems, radar and through-the-wall imaging, and antenna characterization. Imaging techniques having a strong affinity with compressive sensing include coded aperture and computational photography. Implementations of compressive sensing in hardware at different technology readiness levels is available.

Photography: Compressed sensing is used in mobile phone camera sensors. The approach allows a reduction in image acquisition energy per image by as much as a factor of 15 at the cost of complex decompression algorithms; the computation may require an off-device implementation. Image quality improves with the number of snapshots, and generally requires a small fraction of the data of conventional imaging, while eliminating lens and focus-related aberrations.

Holography: Compressed sensing can be used to improve image reconstruction in holography by increasing the number of voxels; it is also used for image retrieval from under sampled measurements in optical and millimeter-wave holography.

Facial recognition: Compressed sensing is used in facial recognition applications.

Magnetic resonance imaging: Compressed sensing has been used to shorten magnetic resonance imaging scanning sessions using conventional hardware.

Compressed sensing addresses the issue of high scan time by enabling faster acquisition by measuring fewer Fourier coefficients. This produces a high-quality image with relatively lower scan time. Another application (also discussed ahead) is for Computerized Tomography (CT) reconstruction with fewer X-ray projections. Compressed sensing, in this case, removes the high spatial gradient parts, mainly, image noise and artifacts. This holds tremendous potential as one can obtain high-resolution CT images at low radiation doses (through lower current-mA settings).

Network tomography: Compressed sensing has showed outstanding results in the application of network tomography to network management. Network delay estimation and network congestion detection can both be modeled as underdetermined systems of linear equations where the coefficient matrix is the network routing matrix.

Shortwave-infrared cameras: Commercial shortwave-infrared cameras based upon compressed sensing are available. These cameras have light sensitivity from 0.9μm to $1.7\ \mu$m, which are wavelengths invisible to the human eye.

11.9 Applications of Lozi Mappings

In 1978, Lozi introduced a two-dimensional map of equations and attractors of which resemble those of the celebrated Hénon map. Simply, a quadratic term in the latter is replaced with a piecewise linear contribution in the former. This allows one to rigorously prove the chaotic character of some attractors and generate a detailed analysis of their basins of attraction. The equation for the iterated Lozi map $L(x, y)$ is

$$L(x,y) = \begin{pmatrix} 1 - a\,|x| + y \\ bx \end{pmatrix}$$

or

$$L_1(x,y) = \begin{pmatrix} 1 - a\,|x| + by \\ x \end{pmatrix}.$$

These two formulas are equivalent dynamically, which means that their dynamical behaviors are the same. The Lozi attractor with its basin of attraction obtained for $a = 1.7052$ and $b = 0.5896$. For details, see [9].

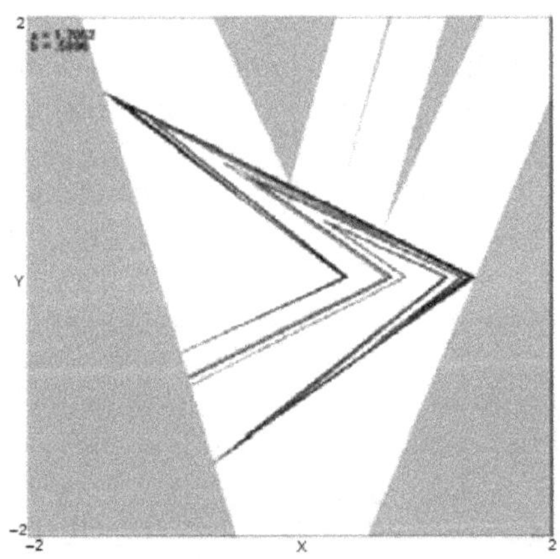

FIGURE 11.39

11.9.1 Lozi Mappings and Secure Communications

First, basic application of chaos theory concerns control of irregular behaviors in devices and systems. The list of applications includes but not limited to: engineering, computers, communications, medicine and biology, management and finance, consumer electronics, ... etc. The potential application types of chaos are: control, synthesis, synchronization and information processing.

Particle swarm optimization (PSO) is a population-based swarm algorithm developed by Eberhart and Kennedy [59], [60].PSO is an evolutionary optimiation approach based on a population in which the position of each member or particle is a potential solution to a problem. Each member of particle is associated with a randomized velocity that moves throughout the problem space [61].

Araujo and Coelho [62] proposed a chaotic PSO (CPSO) approach intertwined with a Lozi map chaotic sequences to develop the Takagi–Sugeno (TS) fuzzy model to represent dynamical behaviors. Their method was used to optimize the premise part of the *if-then* rules of the TS model that utilizes the **least mean squares technique**

As a practical example, the CPSO approach was utilized for a thermal vacuum system used in space environment emulation and satellite qualification. The results succeeded in eliciting a TS fuzzy model for non-linear and time-delay applications. Future research in this area should investigate the hybridization of the PSO and SPSO and other search methods such as simulated annealing and quasi-Newton methods.

11.10 Introduction to Maxwell Equations with Applications

Maxwell's equations are a set of fundamental relationships which govern how electric and magnetic fields interact. The equations explain how these fields are generated and interact and their relationships to charge and current. They form the backbone of modern electrical and telecommunication technology and are often quoted as being the most important equations of all time. The equations consist of a set containing Gauss' electric field law, Gauss' magnetic field law, Faraday's law and the Ampere Maxwell law.

1. **Gauss' electric field law** relates the distribution of electric charge to the field the charge creates. If you know the shape of an object and how the charge is distributed, you can use Gauss' law to find an expression for the electric field. This law is generally used when there's a degree of symmetry, making the equation simpler.

2. **Gauss' magnetic law** states that magnetic monopoles do not exist. It's really more of a statement than a formula we might use to derive expressions. Charges exist as positive or negative. But in magnetism, whenever you have a south pole, you also have a north pole. No single poles, or monopoles have been yet discovered.

3. **Faraday's law** states that any change to the magnetic environment of a coil of wire will cause a voltage to be induced in the coil. If the magnetic field strength changes, the magnet moves, the coil moves, or the coil is rotated, the change will create a voltage in the coil.

4. **Ampere Maxwell law** says that the magnetic field created by an electric current is proportional to the size of that electric current, with a constant of proportionality equal to the permeability of free space. Stationary charges produce electric fields, proportional to the magnitude of that charge. But moving charges produce magnetic fields, proportional to the current (the charge and movement).

   (a) $\nabla \cdot D = \rho_V$

   (b) $\nabla \cdot B = 0$

   (c) $\nabla \times E = -\frac{\partial B}{\partial t}$

   (d) $\nabla \times H = \frac{\partial D}{\partial t} + J$

where D, B, E, H and J are explained in Section 3.8.

Applications of Maxwell Equations

FIGURE 11.40

The uses and applications of Maxwell's equations are too many to count. By understanding electromagnetism we're able to create images of the body using MRI scanners; we've created magnetic tape, generated electricity, and built computers. Any device that uses electricity or magnets is on a fundamental level built upon the original Maxwell equations. To describe the weak force, physicists drew analogies to electromagnetism, and eventually found themselves a step higher up the unification ladder. Their ideas suggested that the two forces were, in fact, just two sides of the same coin: the unified **electroweak force**. The Universe would be completely different if the weak force were not weak. The idea of unification suggests that the similarity of the two forces, electromagnetism and the weak force, was only apparent right after the Big Bang, when the Universe was incredibly hot. As temperatures cooled, the forces crystallized and became different. Weird as it might seem, the concept isn't entirely unfamiliar: think of the dramatic change that happens to water when it freezes to ice.

The laws of nature responsible for the behavior of water are the same everywhere and do not favor any particular direction in space. This is why a patch of ocean looks much like any other, and appears the same no matter from what direction you look at it. The icebergs that form when the water freezes, however, display none of that symmetry: no two look the same. Those with rotational symmetry are extremely rare. As the Universe cooled, the messenger particles of the weak force (and other particles) acquired mass, while the messenger particles of electromagnetism remained massless.

Lightning is electrostatic discharge between electrically charged regions of a cloud during an electrical storm. Electroweak unification was a real triumph of theoretical physics and led to Nobel Prizes in Physics for Sheldon Glashow, Abdus Salam, and Steven Weinberg for the unified electroweak framework, and for François Englert and Peter Higgs for the description of the mass-related symmetry breaking mechanism. For detailed study we refer to Neunzert and Siddiqi [35] and Monk [31].

11.11 Stochastic Calculus for Engineering Problems

Stochastic calculus is a branch of mathematics that operates on stochastic processes. It allows a consistent theory of integration to be defined for integrals of stochastic processes and is used to model systems that behave randomly. The main types of stochastic calculus are the Itô calculus and its variational relative known as the Malliavin calculus. For technical reasons the integral is the most useful for general classes of processes. We present here the Itô lemma. Interested readers may pursue references [19], [30], [50] and [54] for detailed study.

Brownian Motion

Brownian motion also called **Brownian movement** involves various physical phenomena in which some quantity is constantly undergoing small, random fluctuations. It was named for the Scottish botanist Robert Brown, the first scientist to study such fluctuations (1827).

Example 337. 1. The random motion of particles suspended in a fluid (a liquid or a gas) resulting from their collision with the fast-moving atoms or molecules in the gas or liquid.

2. Random Walk

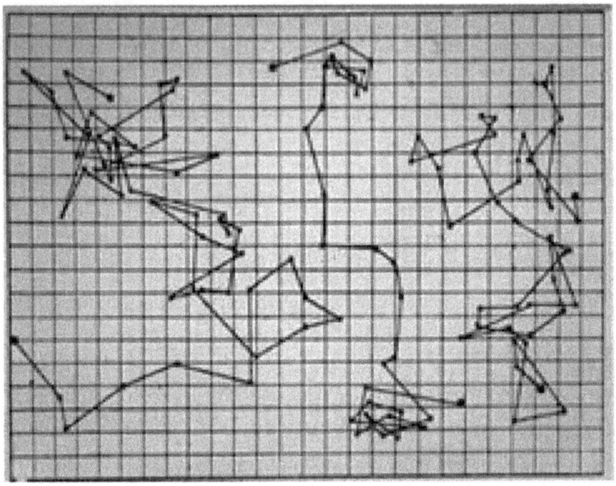

FIGURE 11.41: Random Walk

Brownian motion is among the simplest of the continuous-time stochastic (or probabilistic) processes, and serves as a limit of simpler and more complicated stochastic processes. This universality is closely related to the universality of the normal distribution. In both cases, mathematical convenience rather than the accuracy of the models motivates their use.

Brownian motion is a random continuous time process denoted $X(t)$ or X_t, defined for $t \geq 0$ such that $W(0)$ takes some predetermined value, usually 0, and for each $0 \leq s < t, W(t) - W(s)$ has a normal distribution with mean $\mu(t-s)$ and variance $\sigma^2(t-s)$. The parameters μ and σ are the drift and the diffusion parameters of the Brownian motion and in the special case $\mu = 0$ and $\sigma = 1$, $W(t)$ are often referred to as a standard Brownian motion or a Wiener process. Another property of the Brownian motion process that the sample paths are continuous functions of t with probability 1.

11.11.1 Stochastic Integration

The stochastic integral arose from attempts to use the techniques of Riemann-Stieltjes integration for stochastic processes. However, Riemann integration requires the integrating function to have locally bounded variation in order that the Riemann-Stieltjes sums converge.

We define a stochastic integral as:

$$W(t) = \int_0^t f(\tau)\, dX(\tau) = \lim_{n\to\infty} \sum_{j=1}^{n} f(t_{j-1})(X(t_j) - X(t_{j-1}))$$

with

$$t_j = \frac{jt}{n}.$$

The integration is evaluated in the summation at the left-hand point $t_{j-1.}$ It is important that each function evaluation does not know about the random increment that multiplies it, i.e., anticipate integration is **non-anticipatory**.

11.11.2 Stochastic Differential Equations

We say that (X_t) satisfies the stochastic integral equation:

$$\mathrm{X_t} = \mathrm{X_0} + \int_0^t \mu(s, X_s)\, ds + \int_0^t \sigma(s, X_s)\, dB_s$$

or in differential form:

$$d\mathrm{X_t} = \mu(t, \mathrm{X_t})dt \ + \sigma(t, X_t)dB_t$$

where μ and σ so that the integrals make sense.

Itô's lemma

$$dF = \frac{dF}{dX}dX + \frac{1}{2}\frac{d^2F}{dX^2}dt,$$

where $F(X)$ is a function of Brownian motion $X(t)$.

Application: An important application of stochastic calculus is in quantitative finance, in which asset prices are often assumed to follow stochastic differential equations. In the Black-Scholes model, prices are assumed to follow geometric Brownian motion.

A fine account of stochastic calculus useful for financial engineering is presented in Wilmot [54]. It may be observed that ordinary rules of calculus do not generally hold in a stochastic environment. For example: If $F(X) = X^2$ then $dF = 2XdX$ in calculus, but $dF = 2XdX + dt$ in stochastic calculus. This follows from Itô's lemma:

$$F(X) = X^2$$
$$\frac{dF}{dX} = 2X, \quad \frac{d^2F}{dX^2} = 2.$$

We obtain the result, $dF = 2XdX + dt$. For a comprehensive account, see [19], [30], [50] and [54].

11.12 Exercises

11.1. (a) Write the fractal dimensions of well-known fractals like
(i) Cantor set
(ii) Sierpinski triangle
(iii) Von Koch curve

(b) Can you give examples of two physical situations which are not necessarily similar (same) but have the same fractal dimension?

11.2. Write a note on irregular geometry.

11.3. Explain the concept of IFS with a concrete example.

11.4. Define the Hausdorff metric. Is Hausdorff metric space is a metric space.

11.5. Discuss connections between fractal, Lorenz attractor, Strange attractor, Rossler attractor and Henón attractor.

11.6. Write an essay on relations between fractals and chaos.

11.7. Is it possible to use fractals for weather forecasting or predictions of earthquakes and tsunamis?

11.8. Give examples of time series representing physical phenomena.

11.9. What is the Hurst exponent? What is the relation between Hurst parameter and fractal dimension? Who initiated the study of fractals? What do you know about multifractals?

11.10. Give example of feedforward networks.

11.11. Prove that the number of vertices of odd degrees in a graph is always even.

11.12. Indicate areas of science, social science, linguistics, engineering and technology where graph theory has been used. Is there any relevance of graph theory to big data?

11.13. Draw simple graphs of one, two and three vertices.

11.14. Show that the maximum number of edges in a simple graph with n vertices is $n(n-1)/2$.

11.15. Let $X(t)$ denote Brownian motion then show that:

(i) $\int_0^t X(\tau)dX(\tau) = \frac{1}{2}X^2(t) - \frac{1}{2}t$

(ii) $\int_0^t f(\tau)dX(t) = f(t)X(t) - \int_0^t X(\tau)df(\tau)$, where $f(t)$ is bounded and continuous function on [0,1].

Relation (ii) is known as the integration by parts of stochastic calculus.

11.16. Let B-splines be defined as

$$B_0(x) = \chi_{[-1/2,1/2]}(x)$$

and for any natural number n define

$$B_n(x) = B_{n-1} * B_0(x) = \int_{x-1/2}^{x+1/2} B_{n-1}(t)dt$$

where $*$ denotes convolution. Show that $B_1(x) = \chi_{[}0, 1]$ and $B_2(x)$ is a piecewise polynomial of degree 2.

11.17. A spline is a function $f : R \to R$ for which one can divide R into intervals in such a way that f is a polynomial on each interval. The points at which function changes from one polynomial to another are called knots. For example

$$f(x) = \begin{cases} 0, & \text{if } x^2 \in (-\infty, 0) \\ 2x^2, & \text{if } x^2 \in (0, 1) \\ 2 - x, & \text{if } x^2 \in (1, 4) \\ \frac{1}{16}x^3, & \text{if } x^2 \in (4, \infty) \end{cases} .$$

Find knots. Show that $B-$ splines are splines.

11.18. Find the Fourier transform of $B_1(x)$ ($B-$ spline of order 1).

11.19. Explain how splines can be used in finite element methods.

11.20. Write an essay on B-splines and wavelet expansions.

11.13 Suggestion for Further Reading

Partial differential equations provide models for important problems in natural and medical sciences and all engineering fields. A natural question concerns phenomena modelled by partial differential equations on fractals such as Cantor set, Sierpinski gasket and Von Koch curve. In this connection we recommend Mosco [32], Pelander [38] and Strichartz [46] through [49]. For an interesting introduction of the Sierpinski gasket we suggest Stewart [45]. We recommend Falconer [10], Nekka and Li [33,34], Siddiqi [44] and Triebel [51] for mathematical foundations and effective applications.

Applications of fractal dimension analysis and predictions of meteorological data of India, Saudi Arabia and Mexico are presented respectively in Rangarajan [39], Siddiqi [43], Rehman and Siddiqi [42] and Velasquenz et al. [53].

A study of these references may be helpful to carry out further investigation in this field.

For application of fractals to earth sciences, specially to earthquake and tsunami prediction we refer to Dimri [6,7] and Turcotte [52]. For a thorough study of stochastic calculus we refer to Klebaner [19], Shreve [50] and Wilmot [54]. For deeper study of splines we recommend Holling [15] and Christensen [53, Chapter 10]. For detail study we refer to Neunzert and Siddiqi[35] and Monk [31]. A wide range of emerging topics useful for engineers have been introduced in this chapter. For extensive discussion we refer to [2] through [5], [9], [11] through [14], [16], [20], [23] through [25], [36], [37] and [55].

Bibliography

[1] P. S. Addison, *The Illustrated Wavelet Transform Handbook*, Institute of Physics Publishing, Philadelphia, 2002.

[2] A. Z. Averbuch, *Spline and Spline Wavelet Methods with Applications to Signal and Image Processing*, Volume I, Springer, 2014.

[3] M. Brokate and A. H. Siddiqi, *Functional Analysis with Current Applications in Science, Technology and Industry*, Second edition, Longman/CRC Press, 2014.

[4] J. Chakravarty, *Introduction to MATLAB Programming, Tool Box and Simulink,* Universities Press, 2014.

[5] O. Christensen, *Functions, Spaces and Expansions* , Birkhauser, 2010.

[6] N. Deo, *Graph Theory with Applications to Engineering and Computer Science*, Dover Publications, 2016.

[7] V. P. Dimri, *Application of Fractals in Earth Sciences*, Oxford and IBH Publishing, Calcutta, 2000.

[8] V. P. Dimri (Ed.) *Fractal Behaviour of the Earth System*, Springer, 2005.

[9] Z. Elhadj, *Lozi Mappings: Theory and Applications*, CRC Press, 2013.

[10] K. Falconer, *Fractal Geometry: Mathematical Foundations and Applications*, Second Edition, Wiley-Blackwell, 2003.

[11] D. P. Feldman, *Chaos and Fractals: An Elementary Introduction*, Oxford University Press, 2012.

[12] E. Foufoula-Georgiou and P. Kumar (Eds), *Wavelets in Geophysics*, Academic Press, San Diego, 1994.

[13] K. M. Furati, Z. Nashed and A. H. Siddiqi, *Mathematical Models & Methods for Real World Systems*, Chapman & Hall, New York, 2006.

[14] Igor, Gorban, *Randomness and Hyper-Randomness*, Springer, 2017.

[15] K. Holling, *Finite Element Methods with B.Splines*, Applied Mathematics, Philadelphia, 2003.

[16] M. Javadov and A. Kh. Janahmadov, *Synergetics and Fractals in Tribology*, Springer, 2016.

[17] J. Kigami, A harmonic calculus on the Sierpinski spaces. *Japan J Appl. Math.* 6, 259-290. 1989.

[18] J. Kigami, *Analysis on Fractals*, Cambridge University Press, 2001.

[19] F. C. Klebaner, *Introduction to Stochastic Calculus with Applications*, Imperial College Press, 2005.

[20] B. Kosko, *Neural Networks and Fuzzy Systems: A Dynamical Systems Approach To Machine Intelligence*, Prentice-Hall, 2009.

[21] J. Kumar and E. Foufoula-Georgiou, Wavelet analysis for geophysical applications, *Rev. GeoPhysics*, 35, 385-412, 1997.

[22] J. Kumar, P. Manchanda and N. A. Sontake, Wavelet methods for Indian rainfall data, In *Mathematical Models and Methods for Real World Systems*, Chapman & Hall, New York, 2006, pp. 179-210.

[23] R. S. T. Lee, *Fuzzy-Neuro Approach to Agent Applications*, Springer, 2008.

[24] P. Manchanda, J. Kumar, F. Khene and A. H. Siddiqi, Wavelet based multifractal analysis of Indian rainfall data, In *Modern Mathematical Models, Methods and Algorithms for Real World Systems*, Anamaya and Anshan, www.anshan.co.uk, 2006.

[25] B. B. Mandelbrot, *Fractals: Form, Chance and Dimension.* Freeman, 1977.

[26] B. B. Mandelbrot and J. Van Ness, Fractional Brownian motion, fractional noises and applications, *SIAM Rev.* 10: 422-427, 1968.

[27] B. B. Mandelbrot and R.L. Hudson, *The (Mis)Behaviour of Markets*, Basic Books, New York, 2004.

[28] Y. Meyer, *Wavelet Vibrations and Scalings*, Vol. 9, CRM Monograph Series AMS, 1998.

[29] Y. Meyer (with S. Jaffard and R. D. Ryan), *Wavelets: Tools for Science and Technology*, SIAM, Philadelphia, 2001.

[30] G. Mircea, *Stochastic Calculus*, Applications in Science and Engineering, Springer, 2002.

[31] P. Monk, *Finite Element Methods for Maxwell's Equations*, Oxford Science Publications, 2003.

[32] U. Mosco, Energy functional on certain fractal structures, *J. Convex. Anal.* 9: 581-600, 2002.

[33] F. Nekka and J. Li, Introduction of triadic Cantor sets with their translates. Fundamental properties, *Chaos, Solitons Fractals*, 13(9): 1807-1817, 2002.

[34] F. Nekka and J. Li, A continuous translation based method to reveal the fine structure of fractal sets, *Arabian Journal for Science and Engineering*, 28, Part I, 169-188, 2003.

[35] H. Neunzert and A. H. Siddiqi, *Topics is Industrial Mathematics Case Studies and Related Mathematical Methods*, Kluwer, 2000.

[36] A. M. Ovrutsky and A. S Prokhoda, *Computational Materials Science: Surfaces, Interfaces, Crystallization*, Kindle Edition, Elsevier, 2013.

[37] B. Oksendal, *Stochastic Differential Equations: An Introduction with Applications*, Springer, reprint, 2003.

[38] A. Pelander, *A Study of Smooth Functions and Differential Equations on Fractals*, Dissertation, Uppsala university, Sweden, 2007.

[39] G. Rangarajan, and D.A. Sant, Fractal dimension analysis of Indian climatic dynamics, *Chaos, Solitons Fractals* 19: 285-291, 2004.

[40] G. Rangarajan, M. Ding (Eds.) *Processes with Long-Range Correlations Theory and Applications*, Springer-Berlin, Heidelberg and New York, 2003.

[41] S. Rehman, A. H. Siddiqi, Wavelet-based correlation coefficient of time series of Saudi meteorological data, *Chaos, Solitions Fractals*, 39: 1764-1789, 2009.

[42] S. Rehman, A. H. Siddiqi, Wavelet-based Hurst exponent and fractal dimensional analysis of Saudi climatic dynamics, *Chaos, Solitons, Fractals* , 40: 1081-1090, 2009.

[43] A. H. Siddiqi, *Research Project on Wavelet and Fractal Methods for the Analysis of Meteorological Data*, King Fahd University of Petroleum and Minerals, Dhahran Saudi Arabia, 2006.

[44] A. H. Siddiqi, Theme issues: wavelet and fractal methods is science and engineering, *Arab. J. Sci. Eng.*, 28-29, 2003-2004.

[45] I. Stewart, Four encounters with Sierpinski Gasket, *Math. Intel.*, 17: 52-64, 1995.

[46] R. S. Strichartz, Fratafolds based on the Sierpinski gasket and their spectra, *Trans. Amer. Math. Soc.* 355: 4019-4043, 2003.

[47] R. S. Strichartz, Analysis on products of fractals, *Trans. Amer. Math. Soc.* 357: 571-615, 2005.

[48] R. S. Strichartz, Laplacians on fractals with spectral gaps have nicer Fourier series, *Math. Res. Lett.* 12: 269-274, 2005.

[49] R.S. Strichartz, *Differential Equations on Fractals: A Tutorial*, Princeton University Press, 2006.

[50] S. Shreve, *Stochastic Calculus for Finance II: Continuous-Time Models*, Springer, reprint, 2010.

[51] H. Triebel, *Fractals and Spectra. Monographs in Mathematics 91*, Birkhäuser, 1997.

[52] D. L. Turcotte, *Fractals and Chaos in Geology and Geophysics*, Second Edition, Cambridge University Press, 1997.

[53] M. A. Valle Velasquenz, G. M. Garcia, I. S. Cohon, L. K. Oleschko, J. A. R. Correl and G. Koruin, Spatial variability of Hurst exponent for the daily scale rainfall series in the state of Zacatecas Mexico, *J. Am. Meterol. Soc.*, 2771-2780, 2013.

[54] P. Wilmot, *Derivatives, The Theory and Practice*, John Wiley & Sons, Chichester, 1998.

[55] Y. W. Biao Biao et al., Fuzzy logic and neuro-fuzzy Systems: a systematic introduction, *Int. J. of Artificial Intel. Expert Sys.* (IJAE), 2(2), 2011.

[56] A. Areodo, E. Bacry and J. Muzy, Solving the inverse fractal problem from wavelet analysis, *European Physics Letter*, 25: 7, 479-556, 1994.

[57] S. Mallat, *Wavelet Tour of Dignal Processing*, Academic Press, 1998.

[58] E. Magosso, M. Ursino, A. Zaniboni et al., A wavelet based energetic approach for the analysis of biomedical signals, *J. Appl. Math. Comput.*, 207:42-62, 2009.

[59] R. Eberhart and J. Kennedy, A new optimizer using particle swarm theory, MicroMachine and Human Science, IEEE, DOI:101109/MHS 1995.

[60] J. Kennedy and R. Eberhart, Particle swam optimization, Neural Networks Procceding IEEE Conference, DOI:101109/ICNN1995488968, IEE, 1995.

[61] D. E. Goldberg, *Genetic Algorithm in Search, Optimization and Machine Learning*, Reading, M. A. Addison Wesley, 1989.

[62] E. Araujo, L. S. Coelho, Particle swam approaches using Lozi map Chaotic sequences to fuzzy modelling of an experimental thermal-vacuum system, *Applied Soft Computing* 8: 1354-1364, 2008.

[63] M. Hassoun, *Fundamentals of Artificial Neural Networks*, MIT Press, 2003.

Appendix A

Basic Concept of Calculus

A.1 Number System

The simplest numbers are the **natural numbers** $1, 2, 3, 4, 5, \ldots$. The numbers $\ldots, -5, -4, -3, -2, -1, 0, 1, 2, 3, 4, 5, \ldots$ are called **integer numbers** or simply **integers**. We denote the set of all natural numbers by $\mathbb{N}$ and the set of all integers by $\mathbb{Z}$. It is clear that $\mathbb{N} \subset \mathbb{Z}$, that is, the set of natural numbers is a subset of the set of integers.

The numbers of the form $\frac{p}{q}$, where p and q are integers with $q \neq 0$, are called **rational numbers**. For example $\frac{3}{4}, \frac{-2}{5}, \frac{11}{2}$ are rational numbers. We denote the set of all rational numbers by $\mathbb{Q}$. We have $\mathbb{Z} \subset \mathbb{Q}$, since any integer p can be written as the ratio $\frac{p}{1}$.

Numbers which cannot be written as the ratio of two integers are called **irrational**. For example $\sqrt{2}$, $1 + \sqrt{2}$, $\sqrt{5}$ and π are irrational numbers. (In Section 1.6 we present Euclid's proof that $\sqrt{2}$ is irrational.) Together, rational and irrational numbers form what is called the **real number system**. Thus, a real number is either rational or irrational. The set of all real numbers is denoted by $\mathbb{R}$, and $\mathbb{Q} \subset \mathbb{R}$.

We assume that the reader is familiar with the algebraic operations (addition, subtraction, multiplication and division) for numbers. Recall that division by zero is not possible.

Since the square x^2 of a real number x is non-negative, there cannot be any real number which satisfies the equation $x^2 = -1$, or $x^2 + 1 = 0$. Nevertheless, it has been found very useful in mathematics to introduce the number $i = \sqrt{-1}$, so that $i^2 = -1$. i is called the imaginary unit. Numbers of the form $a + bi$, where a and b are real numbers, are called **complex numbers**. For example, $2 + 5i$ and $\sqrt{3} - 6i$ are complex numbers. A complex number $a + bi$ is called **imaginary** (or **pure imaginary**) if $a = 0$, so $9i$ is an imaginary number. Since $a = a + 0i$, every real number a can be interpreted as a complex number, as for example $\frac{2}{3}$, thus $\mathbb{R} \subset \mathbb{C}$. The algebraic operations are defined for complex numbers in the natural way, for example

$$\begin{aligned}
(3 + 5i) + (4 - \sqrt{2}i) &= 3 + 4 + 5i - \sqrt{2}i = 7 + (5 - \sqrt{2})i\,, \\
(3 + 5i) - (4 - \sqrt{2}i) &= 3 - 4 + 5i - (-\sqrt{2}i) = -1 + (5 + \sqrt{2})i\,, \\
(3 + 5i) \cdot (4 - \sqrt{2}i) &= 3 \cdot 4 + 5i \cdot 4 - 3\sqrt{2}i - 5i \cdot \sqrt{2}i \\
&= (12 + 5\sqrt{2}) + (20 - 3\sqrt{2})i\,.
\end{aligned}$$

The division of two complex numbers is a bit more complicated, in general one has

$$\frac{a + bi}{c + di} = \frac{(a + bi)(c - di)}{(c + di)(c - di)} = \frac{ac + bd}{c^2 + d^2} + \frac{bc - ad}{c^2 + d^2}i\,.$$

Rational and irrational numbers can be distinguished by their decimal representations. Decimals like $12.345 = 12.3450000\ldots$, where only zeroes appear from some point onward, are called **terminating decimals**. They correspond to those rational numbers p/q whose denominator q has only 2's and 5's as prime factors. Every other rational number has a periodic decimal representation, that is, a representation where a certain string of numbers is repeated from some point onward, for example

$$\begin{aligned}
\frac{31}{60} &= 0.51666666\ldots\,, \\
\frac{13}{17} &= 0.7647058823529411\,76470\ldots\,.
\end{aligned}$$

Irrational numbers have non-periodic decimal representations. For example, the decimal representations

$$\begin{aligned}
\pi &= 3.141592653\ldots\,, \\
\sqrt{2} &= 1.414213562\ldots\,.
\end{aligned}$$

do not exhibit any periodic repetition. Moreover, if we truncate a non-terminating decimal representation (periodic or non-periodic) of a real number x at some point, the resulting terminating decimal will be only an approximation to x.

A.2 Intervals, Absolute Value and Inequalities

We say that a real number x is less than another real number y, and write $x < y$, if $x - y$ is less than zero, that is, $x - y$ is negative or $y - x$ is positive. Writing $x \leq y$ (read "x is less than or equal to y") means that $x - y$ is less than or equal to 0, that is, $x - y$ is negative or equal to 0.

The set of all real numbers x such that $a \leq x \leq b$ is called the **closed interval** from a to b and denoted as $[a, b]$. The set of all real numbers x such that $a < x < b$ is called the **open interval** from a to b and denoted as (a, b).

Definition A.1. The **absolute value** or **magnitude** of a real number x is denoted by $|x|$ and is defined as

$$|x| = \begin{cases} x\,, & x > 0\,, \\ 0\,, & x = 0\,, \\ -x\,, & x < 0\,. \end{cases}$$

Properties of the absolute value. For any real numbers x and y, $|x - y|$ represents the distance of x and y on the real line. (Thus $|x|$ equals the distance of x from 0.) Moreover, the following properties hold.

(a) $\sqrt{x^2} = |x|$.

(b) $|-x| = |x|$. (A number and its negative have the same absolute value.)

(c) $|xy| = |x||y|$. (The absolute value of a product is the product of the absolute values.)

(d) $|x + y| \leq |x| + |y|$. (This is called the triangle inequality.)

(e) $||x| - |y|| \leq |x - y|$. (This is called the inverse triangle inequality.)

A.3 Binomial Formula and Quadratic Formula

Binomial Formulas

(a) $(x + y)^n = x^n + nx^{n-1}y + \frac{n(n-1)}{1.2}x^{n-2}y^2 + \frac{n(n-1)(n-2)}{1.2.3}x^{n-3}y^3 + \ldots + nxy^{n-1} + y^n$.

(b) $(x - y)^n = x^n - nx^{n-1}y + \frac{n(n-1)}{1.2}x^{n-2}y^2 - \frac{n(n-1)(n-2)}{1.2.3}x^{n-3}y^3 + \ldots \pm nxy^{n-1} + \mp y^n$.

Quadratic Formula The solution of the quadrat equation $ax^2 + bx + c = 0$ is

$$x = \frac{-b \pm \sqrt{b^2 - 4ac}}{2a}.$$

A.4 Analytic Geometry and Trigonometry

Rectangular coordinates. The real line consists of all real numbers; each point on the real line is associated with a real number. Let us now consider the plane. A **rectangular coordinate system** (also called **Cartesian coordinate system**) consists of two perpendicular coordinate lines, called coordinate axes. See Figure A.1 The intersection of the two axes is called the **origin** of the coordinate system and denoted by O. Usually, the coordinate axes are chosen along the horizontal and vertical direction; the horizontal axis is called the x-axis, the vertical axis is called the y-axis. In this case the plane and the axis together are referred to as the xy-plane. Any point P in the plane

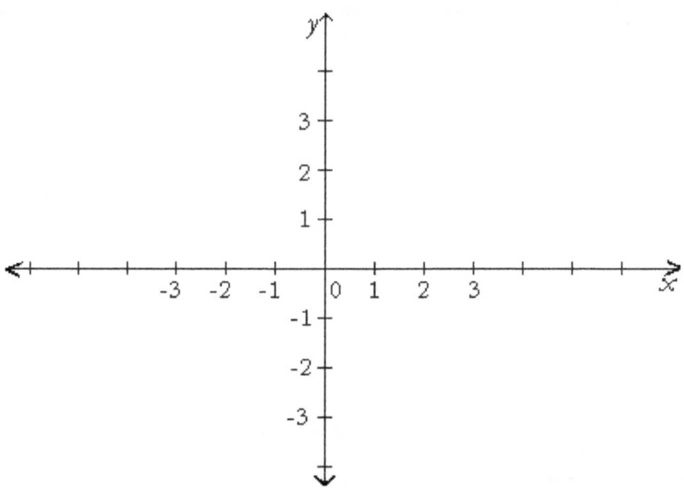

FIGURE A.1: Rectangular Coordinate System

is represented by a pair (x_1, y_1) of real numbers, x_1 is called the x-coordinate of P and y_1 is called the y-coordinate of P. Points on the x-axis have the form $(x_1, 0)$, while points on the y-axis have the form $(0, y_1)$. The origin O has the coordinates $(0, 0)$. The vertical line through a point $(x_1, 0)$ and the horizontal line through $(0, y_1)$ are represented, respectively, by the equations

$$x = x_1 \qquad \text{and} \qquad y = y_1.$$

Lines, distances, circles. Let P_1 and P_2 be two points in the xy-plane, having coordinates (x_1, y_1) and (x_2, y_2) respectively. Consider the straight line passing through P_1 and P_2. Whenever $x_1 \neq x_2$,

$$m = \frac{y_2 - y_1}{x_2 - x_1}$$

is called the slope of this line; the line is described by the equation

$$y = m(x - x_1) + y_1 \,.$$

The distance between P_1 and P_2 is defined by

$$d(P_1, P_2) = d((x_1, y_1), (x_2, y_2)) = \sqrt{(x_2 - x_1)^2 + (y_2 - y_1)^2} \,.$$

Indeed, this number yields the length of the line segment which connects P_1 and P_2, by the theorem of Pythagoras. As a particular case, the distance of $P_1 = (x_1, 0)$ and $P_2 = (x_2, 0)$ equals

$$d(P_1, P_2) = \sqrt{(x_2 - x_1)^2} = |x_2 - x_1| \,.$$

The equation of a circle with center (x_0, y_0) and radius r is given by

$$(x - x_0)^2 + (y - y_0)^2 = r^2 \,.$$

If $(x_0, y_0) = (0, 0)$ the equation becomes $x^2 + y^2 = r^2$ and describes the circle centered at the origin with radius r.

Polar coordinates. Instead of Cartesian coordinates, it is often more convenient to use polar coordinates (Figure A.2) which we usually denote by (r, θ). The Cartesian coordinates (x, y) and the polar coordinates (r, θ) of a given point P in the plane are related by

$$x = r \cos\theta \,, \quad y = r \sin\theta \,. \tag{A.1}$$

The number r equals the length of the line segment joining the origin O and the point P, and θ equals the angle between the line OP and the x-axis. Note that θ is usually measured in radians, that is, an angle of 90° has $\theta = \pi/2$, the full angle of 360° has $\theta = 2\pi$ and so on. Formula (A.1) expresses the Cartesian coordinates of P in terms of the polar coordinates of P Conversely, we obtain the polar coordinates of P from its Cartesian coordinates by the formula

$$r = \sqrt{x^2 + y^2} \,, \quad \theta = \tan^{-1} \frac{y}{x} \,. \tag{A.2}$$

Trigonometric functions

The trigonometric functions are directly related to the geometry of the circle.

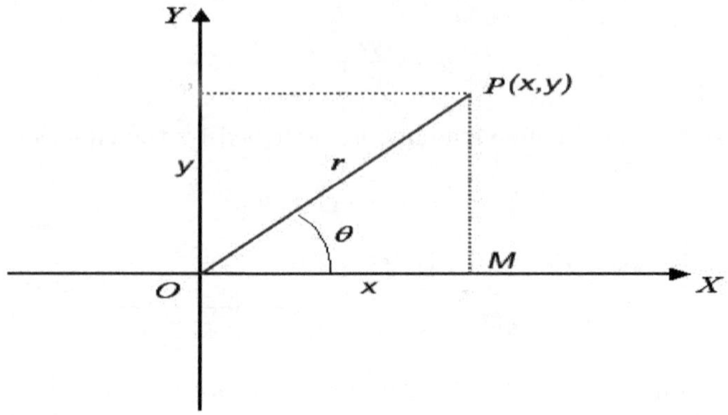

FIGURE A.2: Polar Coordinates

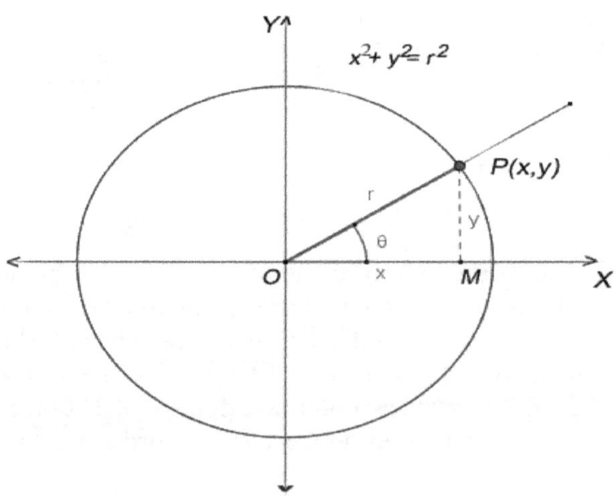

FIGURE A.3: Trigonometric Functions and Circle

Consider the point P with Cartesian coordinates (x, y) and polar coordinates (r, θ) as in Figure A.3. The standard trigonometric functions are defined as follows.

$$\sin\theta = \frac{y}{r} \qquad \text{(read as "sine of } \theta\text{")},$$
$$\cos\theta = \frac{x}{r} \qquad \text{(read as "cosine of } \theta\text{")},$$

and analogously

$$\tan\theta = \frac{y}{x}, \quad \cot\theta = \frac{x}{y}, \quad \sec\theta = \frac{r}{x}, \quad \csc\theta = \frac{r}{y},$$

called the tangent, the cotangent, the secant and the cosecant, respectively. From elementary geometry we see that the definitions above do not depend on the chosen value of r, as long as $r > 0$, so it suffices to consider the case $r = 1$ (the unit circle) See Figure A.4.

Recall that θ is usually measured in radians ($\theta = \pi/2$ corresponds to an angle of 90° and so on).

Trigonometric Identities

A trigonometric identity is an equation involving trigonometric functions that is true for all angles for which both sides of the equation are defined. We state a few useful trigonometric identities.

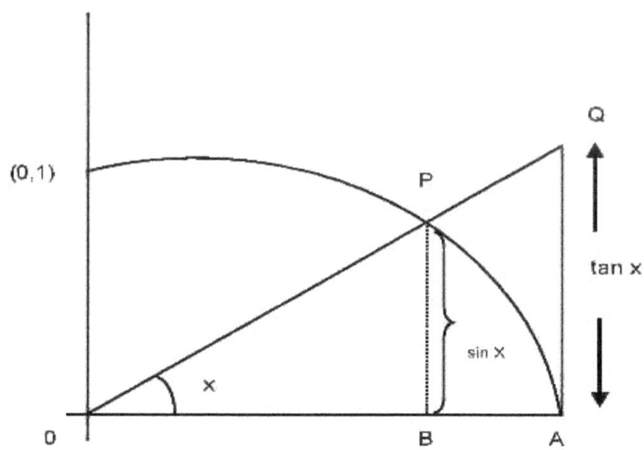

FIGURE A.4: Trigonometric Functions and the Circle

$$\cos^2\theta + \sin^2\theta = 1\,. \tag{A.3}$$

$$1 + \tan^2\theta = \sec^2\theta\,. \tag{A.4}$$

$$1 + \cot^2\theta = \csc^2\theta\,. \tag{A.5}$$

$$\sin(\theta + 2\pi) = \sin(\theta - 2\pi) = \sin\theta\,. \tag{A.6}$$

$$\cos(\theta + 2\pi) = \cos(\theta - 2\pi) = \cos\theta\,. \tag{A.7}$$

In general,

$$\begin{aligned} \sin(\theta \pm 2n\pi) &= \sin\theta\,, \quad \text{for } n = 0, 1, 2, 3, \ldots, \\ \cos(\theta \pm 2n\pi) &= \cos\theta\,, \quad \text{for } n = 0, 1, 2, 3, \ldots. \end{aligned} \tag{A.8}$$

$$\tan(\theta + \pi) = \tan\theta\,, \quad \tan(\theta - \pi) = \tan\theta\,. \tag{A.9}$$

Next we give formulas which involve sums or multiples of angles:

$$\sin(\alpha + \beta) = \sin\alpha\cos\beta + \cos\alpha\sin\beta\,. \tag{A.10}$$

$$\cos(\alpha + \beta) = \cos\alpha\cos\beta - \sin\alpha\sin\beta\,. \tag{A.11}$$

$$\sin 2\theta = 2\sin\theta\cos\theta\,. \tag{A.12}$$

$$\cos 2\theta = \cos^2\theta - \sin^2\theta\,. \tag{A.13}$$

$$\tan 2\theta = \frac{2\tan\theta}{1 - \tan^2\theta}\,. \tag{A.14}$$

$$\cos 2\theta = 2\cos^2\theta - 1\,. \tag{A.15}$$

$$\cos 2\theta = 1 - 2\sin^2\theta\,. \tag{A.16}$$

$$\tan(\alpha + \beta) = \frac{\tan\alpha + \tan\beta}{1 - \tan\alpha\tan\beta}\,. \tag{A.17}$$

$$\tan(\alpha - \beta) = \frac{\tan\alpha - \tan\beta}{1 + \tan\alpha\tan\beta}\,. \tag{A.18}$$

$$\sin\alpha + \sin\beta = 2\sin\frac{\alpha+\beta}{2}\cos\frac{\alpha-\beta}{2}\,. \tag{A.19}$$

$$\sin\alpha - \sin\beta = 2\cos\frac{\alpha+\beta}{2}\sin\frac{\alpha-\beta}{2}\,. \tag{A.20}$$

$$\cos\alpha + \cos\beta = 2\cos\frac{\alpha+\beta}{2}\cos\frac{\alpha-\beta}{2}\,. \tag{A.21}$$

$$\cos\alpha - \cos\beta = -2\sin\frac{\alpha+\beta}{2}\sin\frac{\alpha-\beta}{2}\,. \tag{A.22}$$

Theorem A.1. *We have*

$$\cos x < \frac{\sin x}{x} < 1\,, \quad \textit{for any } x \textit{ satisfying } 0 < |x| \le \frac{\pi}{2}\,. \tag{A.23}$$

Geometric proof. Assume first that $x > 0$. The proof is based on the relations between certain areas in Figure A.4. For a circle of radius 1, the figure shows a sector of angle x with corners O, A and P, and additional auxiliary points B and Q, connected to P and A by vertical lines of length $\sin x$ and $\tan x$ respectively. The distance between O and B equals $\cos x$. We then have

$$\begin{aligned} \text{area of triangle OAP} &= \frac{1}{2} \cdot 1 \cdot \sin x\,, \\ \text{area of sector OAP} &= \frac{1}{2} x\,, \\ \text{area of triangle OAQ} &= \frac{1}{2} \cdot 1 \cdot \tan x\,. \end{aligned}$$

We see that

$$\frac{1}{2} \sin x < \frac{1}{2} x < \frac{1}{2} \tan x = \frac{1}{2} \frac{\sin x}{\cos x}\,,$$

since the triangle OAP is contained in the sector OAP, which in turn is contained in the triangle OAQ. Multiplying with 2 and dividing by $\sin x$ results in

$$\frac{1}{\sin x} < \frac{x}{\sin x} < \frac{1}{\cos x}\,.$$

Using reciprocals yields the assertion (A.23) for $x > 0$. For $x < 0$ it is enough to observe that

$$\cos(-x) = \cos x\,, \qquad \frac{\sin(-x)}{-x} = \frac{-\sin x}{-x} = \frac{\sin x}{x}\,.$$

An alternative proof, for small values of *x* uses the power series representation of the sine and the cosine. We have

$$\sin x = x - \frac{x^3}{3!} + \frac{x^5}{5!} - \ldots\,.$$

Therefore

$$\frac{\sin x}{x} = 1 - \frac{x^2}{3!} + \frac{x^4}{5!} - \ldots\,. \tag{A.24}$$

On the other hand,

$$\cos x = 1 - \frac{x^2}{2!} + \frac{x^4}{4!} - \ldots\,. \tag{A.25}$$

If we consider just the first two terms of the series, we get for $x > 0$

$$1 - \frac{x^2}{2!} < 1 - \frac{x^2}{3!} < 1\,. \tag{A.26}$$

If x is small enough, inequality (A.26) continues to hold even after we add the remaining terms of the series (A.24) and (A.25), because those remaining terms have exponents of x greater than 2, and hence the remaining sums have the form $x^2 r(x)$ for some function r with $\lim_{x \to 0} r(x) = 0$.

Inverse Trigonometric Functions

The **inverse trigonometric functions** or **cyclometric functions** are obtained as inverse functions of the trigonometric functions. However, to ensure that this works out correctly in the sense of the general definition of an inverse function, one has to restrict the domain of the original function to an interval where the latter is increasing resp. decreasing. For example, the sine function is increasing on the interval $[-\pi/2, \pi/2]$; thus by a theorem it has an inverse defined on its range $[-1, 1]$ with values in $[-\pi/2, \pi/2]$. This inverse function is called an **arcsine**; more precisely, it is called the **principal branch** of the arcsine. Indeed, on the interval $[\pi/2, 3\pi/2]$ the sine is decreasing (again with range $[-1, 1]$), so we could also define an inverse on $[-1, 1]$ with range $[\pi/2, 3\pi/2]$, and similarly on other intervals.

In Table A.1 we list the principal branches of the six standard trigonometric functions. Their names are obtained by putting the syllable "arc" in front of the original function.

TABLE A.1

Name	**Standard notation**	**Domain**	**Range in radians**	**Range in degrees**
arcsine	$y = \arcsin(x)$	$x \in [-1, 1]$	$-\pi/2 \leq y \leq \pi/2$	$-90^o \leq y \leq 90^o$
arccosine	$y = \arccos(x)$	$x \in [-1, 1]$	$0 \leq y \leq \pi$	$0^o \leq y \leq 180^o$
arctangent	$y = \arctan(x)$	$x \in \mathbb{R}$	$-\pi/2 < y < \pi/2$	$-90^o < y < 90^o$
arccotangent	$y = \text{arccot}(x)$	$x \in \mathbb{R}$	$0 < y < \pi$	$0^o < y < 180^o$
arcsecant	$y = \text{arcsec}(x)$	$x \geq 1$ or $x \leq -1$	$0 \leq y < \pi/2$ or $\pi/2 < y \leq \pi$	$0^o \leq y < 90^o$ or $90^o < y \leq 180^o$
arccosecant	$y = \text{arccsc}(x)$	$x \leq -1$ or $x \geq 1$	$-\pi/2 \leq y < 0$ or $0 < y \leq \pi/2$	$-90^o \leq y < 0^o$ or $0^o < y \leq 90^o$

The inverse function notations $\sin^{-1}$, $\cos^{-1}$ etc. are natural, but one must be aware of the following notational conflict. Since we commonly write $\sin^2 x$ instead of $(\sin x)^2$, an unspecified use of $\sin^{-1}(x)$ might mean $\arcsin(x)$ as well as $1/(\sin x)$.

Inverse Hyperbolic Functions

The **inverse hyperbolic functions** are the inverses of the hyberbolic functions sinh, cosh, tanh, coth, sech and csch. Similarly as in the case of trigonometric functions, one has to consider suitable domains and ranges in order to define inverses in the sense of definition of the inverse problem (In contrast to the trigonometric functions, this is not always necessary; sinh is invertible on its entire domain $\mathbb{R}$.) They are called **area hyperbolic functions**, and for the mathematical notation the prefix "ar" precedes the name of the original function, for example "arsinh" denotes the inverse of sinh. Table A.2 lists the principal branches of the inverse hyperbolic functions.

TABLE A.2

Function	Domain	Range
$y = \text{arsinh}(x)$	$x \in (-\infty, \infty)$	$y \in (-\infty, \infty)$
$y = \text{arcosh}(x)$	$x \in [1, \infty)$	$y \in [0, \infty)$
$y = \text{artanh}(x)$	$x \in (-1, 1)$	$y \in (-\infty, \infty)$
$y = \text{arcoth}(x)$	$x \in (-\infty, -1) \cup (1, \infty)$	$y \in (-\infty, 0) \cup (0, \infty)$
$y = \text{arsech}(x)$	$x \in (0, 1]$	$y \in [0, \infty)$
$y = \text{arcsch}(x)$	$x \in (-\infty, 0) \cup (0, \infty)$	$y \in (-\infty, 0) \cup (0, \infty)$

The inverse hyperbolic functions are expressible in terms of natural logarithms. The formulas in Table A.3 hold for all x in the domains of the inverse hyperbolic functions.

A.5 Logarithmic and Exponential Functions

If $b > 0$ and $b \neq 1$, then for positive values of x the **logarithm** to the base b of x is denoted by $\log_b x$. If $b = 10$, logarithms are called **common logarithms**. If $b = e \approx 2.718282$ ($\approx$ means approximately) then logarithms are called **natural logarithms**.

$f(x) = \log_{10} x$ or $f(x) = \log_e x$ is called the **logarithm function**. $\log_e x$

TABLE A.3

$\text{arsinh}(x) = \ln\left(x + \sqrt{x^2+1}\right)$	$\text{arcosh}(x) = \ln\left(x + \sqrt{x^2-1}\right)$
$\text{artanh}(x) = \frac{1}{2}\ln\left(\frac{1+x}{1-x}\right)$	$\text{arcoth}(x) = \frac{1}{2}\ln\left(\frac{x+1}{x-1}\right)$
$\text{arsech}(x) = \ln\left(\frac{1+\sqrt{1-x^2}}{x}\right)$	$\text{arcsch}(x) = \ln\left(\frac{1}{x} + \frac{\sqrt{1+x^2}}{\lvert x\rvert}\right)$

is often denoted by $\ln x$ or $\ln|x|$. $y = e^x$ if and only if (equivalent to) $x = \ln y$ if $y > 0$ and x is any real number. $f(x) = e^x$ is called the **exponential function**.

(a) $\ln(x.y) = \ln x + \ln y$

(b) $\ln \frac{x}{y} = \ln x - \ln y,\ y \neq 0$

(c) $\ln(x^r) = r \ln x$

(d) $\ln\left(\frac{1}{x}\right) = -\ln x$

(e) $e^{x+y} = e^x e^y$

(f) $\ln(e^x) = x$ for all real values of x

(g) $e^{ln x} = x$ for $x > 0$

(h) $\ln 1 = 0$, $\ln e = 1$, $\ln\left(\frac{1}{e}\right) = -1$ and $\ln(e^2) = 2$

A.6 Integration by Parts

Let $u = f(x),\ du = f'(x)dx,\ v = G(x),\ dv = G'(x)dx = g(x)dx$. Then

$$\int u dv = uv - \int v du.$$

If $G(x)$ is an anti-derivative of a function $g(x)$ and $f(x)$ is a differentiable function, then the formula for integration by parts can be written as

$$\int f(x)g(x)dx = f(x)G(x) - \int f'(x)G(x)dx.$$

A.7 Integration Formulas

1. $\int u^n du = \frac{u^{n+1}}{n+1} + c, \quad n \neq 1$
2. $\int u^{-1} du = \int \frac{du}{u} = \ln u + C$
3. $\int e^u du = e^u + C$
4. $\int u e^u du = e^u(u-1) + C$
5. $\int \sin u du = -\cos u + C$
6. $\int \cos u du = \sin u + C$
7. $\int \tan u du = \ln \sec u + C$
8. $\int \cot u du = \ln \sin u du + C$
9. $\int e^{au} \sin mu du = \frac{e^{au}(a \sin mu - m \cos mu)}{m^2+a^2} + C$
10. $\int e^{au} \cos mu du = \frac{e^{au}(m \sin mu + a \cos mu)}{m^2+a^2} + C$
11. $\int \sec u du = \ln(\sec u + \tan u) + C$
12. $\int \csc u du = \ln(\csc u - \cot u) + C$
13. $\int \sec^2 u du = \tan u + C$
14. $\int \csc^2 u du = -\cot u + C$
15. $\int \sec u \tan u du = \sec u + C$
16. $\int \csc u \cot u du = -\csc u + C$
17. $\int \frac{du}{a^2-u^2} = \arcsin \frac{u}{a} + C$
18. $\int \frac{du}{a^2+u^2} = \frac{1}{a} \arctan \frac{u}{a} + C$
19. $\int \frac{du}{\sqrt{a^2-u^2}} = \ln(u + \sqrt{u^2 \pm a^2}) + C$
20. $\int \frac{du}{a^2-u^2} = \frac{1}{2a} \ln \frac{a+u}{a-u} + C$
21. $\int u \sin u du = \sin u - u \cos u + C$
22. $\int u \cos u du = \cos u + u \sin u + C$
23. $\int u du = u^{n+1} \left[\frac{\ln u}{n+1} - \frac{1}{(n+1)^2} \right] + C$
24. $\int \sin^2 u du = \frac{1}{2}u - \frac{1}{4} \sin 2u + C$

25. $\int \cos^2 u du = \frac{1}{2}u + \frac{1}{4}\sin 2u + C$

26. $\int \tan^2 u du = \tan u - u + C$

27. $\int u dv = uv - \int v du + C$

A.8 Gamma Function

The gamma function is defined as

$$\Gamma(x) = \int_0^\infty t^{x-1} e^{-t} dt. \tag{A.27}$$

Convergence of the integral requires $x - 1 > 1$ or $x > 0$. Integrating (A.27) by parts, we get the recurrence relation

$$\Gamma(x+1) = x\Gamma(x).$$

Now, $\Gamma(1) = \int_0^\infty e^{-t} dt = 1$. Therefore, $\Gamma(2) = 1\Gamma(1) = 1$ and $\Gamma(3) = 2\Gamma(2) = 2.1$ and so on.
We see that n is a positive integer, $\Gamma(n+1) = n!$ $\Gamma\left(\frac{1}{2}\right) = \sqrt{\pi}$ can be derived from (A.27) by $x = 1/2$.

Remark A.1. Although the integral form (A.27) does not converge for $x < 0$, it can be shown by means of alternative definitions that the gamma function is defined for all real and complex numbers except $x = -n,\ n = 0, 1, 2, \ldots$.

A.9 Definition of Multiple Integrals

In this section we present the definitions of double and triple integrals,

$$\iint_Q f(x, y)\, dA\,, \quad \text{respectively} \quad \iiint_Q f(x, y, z)\, dV\,,$$

over rectangular regions Q. The exposition closely parallels that given in [6] for the case of a single variable. We therefore shorten it somewhat and refer the reader to [7] for more details.

Double integral

Let the function f be defined in a rectangular region $Q = [a, b] \times [c, d]$ of the xy-plane. We partition Q into rectangular subregions as follows. Choose a partition of $[a, b]$ of the form $a = x_0 < x_1 < \cdots < x_n = b$, and a partition of $[c, d]$ of the form $c = y_0 < y_1 < \cdots < y_m = d$. This gives us a partition Δ of Q into nm rectangles:

$$Q_{ij} = \{(x, y) : x_{i-1} \leq x \leq x_i \,, \, y_{j-1} \leq y \leq x_j\} \,, \quad 1 \leq i \leq n \,, 1 \leq j \leq m \,.$$

A Riemannian sum for the partition Δ is defined as

$$s_\Delta = \sum_{i=1}^{n} \sum_{j=1}^{m} f(\xi_i, \eta_j)(x_i - x_{i-1})(y_j - y_{j-1}) \,. \tag{A.28}$$

Here, each point (ξ_i, η_j) lies somewhere in the rectangle Q_{ij}. We define the oscillation of f on the rectangles Q_{ij} as

$$O_{ij}(f) = \operatorname{osc}_{Q_{ij}}(f) = \max_{x,z \in Q_{ij}} |f(x) - f(z)| \,. \tag{A.29}$$

Again, we have to replace the maximum by the supremum if the former does not exist. Next, we consider two different Riemannian sums for the same partition Δ,

$$\begin{aligned} s_\Delta &= \sum_{i=1}^{n} \sum_{j=1}^{m} f(\xi_i, \eta_j)(x_i - x_{i-1})(y_j - y_{j-1}) \,, \\ \tilde{s}_\Delta &= \sum_{i=1}^{n} \sum_{j=1}^{m} f(\tilde{\xi}_i, \tilde{\eta}_j)(x_i - x_{i-1})(y_j - y_{j-1}) \,. \end{aligned} \tag{A.30}$$

Their difference can be estimated as

$$|s_\Delta - \tilde{s}_\Delta| = \sum_{i=1}^{n} \sum_{j=1}^{m} O_{ij}(f)(x_i - x_{i-1})(y_j - y_{j-1}) =: V_\Delta(f) \,. \tag{A.31}$$

The number $V_\Delta(f)$ defined in (A.31) is called the **oscillation sum** of f for the partition Δ.

Definition A.2. [Integrable Function] A bounded function $f : Q \to \mathbb{R}$ is said to be **integrable on Q**, if for every $\varepsilon > 0$ there exists a partition Δ of Q such that $V_\Delta(f) \leq \varepsilon$.

The latter condition means that, in view of (A.31), we can enforce the difference between different Riemannian sums for the same partition Δ to become as small as we want, if we choose Δ fine enough.

Let now $f : Q \to \mathbb{R}$ be integrable. In order to define its integral, one goes through the same three steps as we did in can case of single variable.

1. One proves that
$$|s_\Delta - s_{\tilde{\Delta}}| < V_\Delta(f) \,,$$
whenever the partition $\tilde{\Delta}$ is a refinement of the partition Δ (that is, $\tilde{\Delta}$ is obtained from Δ by adding partition points), for all Riemannian sums s_Δ and $s_{\tilde{\Delta}}$ for those partitions.

2. One proves that
$$|s_\Delta - s_{\tilde{\Delta}}| < V_\Delta(f) + V_{\tilde{\Delta}}(f)\,,$$
for arbitrary partitions Δ and $\tilde{\Delta}$ and all Riemannian sums for those partitions.

3. One proves that there exists a unique number I such that
$$|I - s_\Delta| \le V_\Delta(f) \tag{A.32}$$
holds for all partitions and all corresponding Riemannian sums.

We then define the **integral of f on Q** as
$$\iint_Q f(x,y)\,dA = I\,,$$
where I is the number obtained in step 3 above.

Theorem A.2. *Let $f : Q \to \mathbb{R}$ be continuous. Then f is integrable on Q.*

Triple integral. Let the function f be defined in a rectangular region $Q = [a_1, b_1] \times [a_2, b_2] \times [a_3, b_3]$ of three-dimensional space. We partition each of those intervals,
$$a_1 = x_0 < \ldots x_n = b_1\,, \quad a_2 = y_0 < \ldots y_m = b_2\,, \quad a_3 = z_0 < \ldots z_l = b_3\,.$$
The partition Δ now consists of rectangular regions
$$Q_{ijk} = [x_i, x_{i-1}] \times [y_j, y_{j-1}] \times [z_k, z_{k-1}]\,.$$
The Riemannian sums have the form
$$s_\Delta = \sum_{i=1}^{n} \sum_{j=1}^{m} \sum_{k=1}^{l} f(\xi_i, \eta_j, \zeta_k)(x_i - x_{i-1})(y_j - y_{j-1})(z_k - z_{k-1})\,,$$
where (ξ, η_j, ζ_k) lies somewhere in Q_{ijk}. From this point onward, the definition of
$$\iiint_Q f(x,y,z)\,dV$$
proceeds in a completely analogous manner as in the case of the double integral.

Appendix B

Summary of Properties of Matrices

B.1 Properties of Matrix

1. A matrix **A** is any rectangular array of numbers or functions:

$$\mathbf{A} = \begin{pmatrix} a_{11} & a_{12} & \dots & a_{1n} \\ a_{21} & a_{22} & \dots & a_{2n} \\ \vdots & \vdots & \vdots & \vdots \\ a_{m1} & a_{m2} & \dots & a_{mn} \end{pmatrix}.$$

   If a matrix **A** has m rows and n columns, we say A is an $m \times n$ matrix. If $m = n$, A is called a **square matrix of order** n.

   The element in the ith row and jth column is denoted by a_{ij}. We write $\mathbf{A} = (a_{ij})_{m\times n}$. Each of the mn numbers is called **an element of the matrix**.

2. Two $m \times n$ matrices $\mathbf{A} = (a_{ij})$ and $B = (b_{ij})$ are equal if $a_{ij} = b_{ij}$ for each i and j.

3. A matrix having m rows and one column is called a column matrix. It is also called a **column vector or a vector**. For example $A = \begin{pmatrix} a_{11} \\ a_{21} \\ \vdots \\ a_{m1} \end{pmatrix}$ is a column vector. A matrix having a single row is called a **row matrix**.

4. A multiple of a matrix **A** by a number k is defined as $kA = (ka_{ij})$.

5. The sum of two $m \times n$ matrices **A** and **B** is defined to be the matrix $\mathbf{A} - \mathbf{B} = (a_{ij} + b_{ij})_{m\times n}$, that is, we add the corresponding elements. Differential of **A** and **B** can be defined as $\mathbf{A} - \mathbf{B} = \mathbf{A} + -\mathbf{B}$.

6. The determinant having the same elements as the square matrix **A** is called the **determinant** of the matrix and is denoted by $|A|$ or det **A**.

7. A square matrix is said to be **singular** if its **determinant** is zero, otherwise it is **non-singular**.

8. A square matrix, all of whose elements except those in the leading diagonal (a_{ii}), are zero is called the **diagonal matrix**. A diagonal matrix where all the leading diagonal elements are equal is called a **scalar matrix**. For example $\begin{pmatrix} 3 & 0 & 0 \\ 0 & 1 & 0 \\ 0 & 0 & -1 \end{pmatrix}$ and $\begin{pmatrix} 2 & 0 & 0 \\ 0 & 2 & 0 \\ 0 & 0 & 2 \end{pmatrix}$ are diagonal and scalar matrices respectively.

9. A diagonal matrix of order n, where all diagonal element are unity is called a **unit matrix** of order n and is denoted by **I**. For example $\begin{pmatrix} 1 & 0 & 0 \\ 0 & 1 & 0 \\ 0 & 0 & 1 \end{pmatrix}$ is a unit matrix of order 3.

10. A square matrix $\mathbf{A} = [a_{ij}]$ is said to be **symmetric** (skew symmetric) if $a_{ij} = a_{ji}$ $(a_{ij} = -a_{ji})$ for all i and j. For a skew symmetric matrix, leading diagonal elements are zero.

11. Let $\mathbf{A} = \begin{pmatrix} a_{11} & a_{12} & \dots & a_{1n} \\ a_{21} & a_{22} & \dots & a_{2n} \\ \vdots & \vdots & \vdots & \vdots \\ a_{m1} & a_{m2} & \dots & a_{mn} \end{pmatrix}$ be an $m \times n$ matrix, and $\mathbf{B} = \begin{pmatrix} b_{11} & b_{12} & \dots & b_{1n} \\ b_{21} & b_{22} & \dots & b_{2n} \\ \vdots & \vdots & \vdots & \vdots \\ b_{m1} & b_{m2} & \dots & b_{mn} \end{pmatrix}$ be a $n \times p$ matrix. Then the product **AB** is the $m \times p$ matrix $\mathbf{AB} = \begin{pmatrix} c_{11} & c_{12} & \dots & c_{1n} \\ c_{21} & c_{22} & \dots & c_{2n} \\ \vdots & \vdots & \vdots & \vdots \\ c_{m1} & c_{m2} & \dots & c_{mn} \end{pmatrix}$ where $c_{ij} = a_{i1}b_{1j} + a_{i2}b_{2j} + \ldots + c_{in}b_{nj}$.

   The expression for c_ij is known as the inner product of the ith row with the jth column. In general matrix multiplication is not commutative that is $\mathbf{AB} \neq \mathbf{BA}$.

   Note : For the product **AB** to be defined the number of columns in **A** should be same as number of rows in **B**.

12. In the product **AB**, the matrix **A** is said to be post-multiplied by the

matrix **B**. In the product **BA** the matrix **A** is said to be pre-multiplied by **B**.

13. Multiplication of matrices is associative, that is $(\mathbf{AB})\mathbf{C} = \mathbf{A}(\mathbf{BC})$, where $\mathbf{A}$, $\mathbf{B}$, $\mathbf{C}$ are $m \times n$; $n \times p$ and $p \times r$ matrices respectively.

14. If all products are defined, multiplication of matrices is distributive over addition that is

$$\mathbf{A}(\mathbf{B}+\mathbf{C}) = \mathbf{AB}+\mathbf{AC}\text{and}(\mathbf{B}+\mathbf{C})\mathbf{A} = \mathbf{BA} + \mathbf{CA}.$$

15. If $\mathbf{A}$ is a square matrix, then the product $\mathbf{AA}$ is defined as $\mathbf{A}^2$. Similarly we define higher powers of A, i.e., $\mathbf{AA}^2 = \mathbf{A}^3$.

16. A matrix consisting of all entries zero is called a **zero matrix** and is denoted by $\mathbf{O}$.

17. If $\mathbf{A}$ and $\mathbf{O}$ are $m \times n$ matrices then

$$\mathbf{A} + \mathbf{O} = \mathbf{O} + \mathbf{A} = \mathbf{A}.$$

If $\mathbf{A}$ is a square matrix of the same order as $\mathbf{I}$ then $\mathbf{IA} = \mathbf{AI} = \mathbf{A}$. $\mathbf{I}$ is called the multiplicative identity of matrix $\mathbf{A}$.

18. The transpose of the $m \times n$ matrix $\mathbf{A}$ is the $n \times m$ matrix $\mathbf{A}'$ or $\mathbf{A}^T$ obtained from $\mathbf{A}$ by interchanging rows and columns.

    In other words the rows of a matrix $\mathbf{A}$ become the columns of its transpose $\mathbf{A}'$ or $\mathbf{A}^T$. **Observation:** The transpose of the product of two matrices is the product of their transposes taken in reverse order, that is $(\mathbf{AB})' = \mathbf{B}'\mathbf{A}'$.

19. If $\mathbf{A}$ is an $n \times n$ matrix, then an $n \times n$ matrix $\mathbf{B}$ if it exists such that $\mathbf{AB} = \mathbf{BA} = \mathbf{I}$ where $\mathbf{I}$ is the multiplicative identity, is called the inverse of $\mathbf{A}$ and is denoted by $\mathbf{B} = \mathbf{A}^{-1}$.

20. An $n \times n$ matrix $\mathbf{A}$ has inverse $\mathbf{A}^{-1}$ if and only if A is **non-singular**.

21. Formula for $\mathbf{A}^{-1} = \frac{Adj A}{det(A)}$, where $Adj\mathbf{A}$ (adjoint of $\mathbf{A}$) is the transposed matrix of cofactors of $\mathbf{A}$.

22. If $\mathbf{A}(t) = (a_{ij}(t))_{m\times n}$ is a matrix whose entries are functions differentiable on a common interval then the derivative of $\mathbf{A}$ is defined as

$$\frac{dA}{dt} = \left(\frac{d}{dt}a_{ij}\right)_{m\times n}.$$

The derivative of a matrix **A(t)** is also denoted by **A'(t)**.

23. If $\mathbf{A(t)} = (a_{ij}(t))_{m\times n}$ is a matrix whose entries are continuous on a common interval containing t and t_0 then

$$\int_{t_0}^{t} A(s)ds = \left(\int_{t_0}^{t} a_{ij}(s)ds\right)_{m\times n}.$$

To differentiate (integrate) a matrix of functions we simply differentiate (integrate) each entry.

24. Matrices are used in solving algebraic system of n linear equations in n unknowns.

$$\begin{aligned} a_{11}x_1 + a_{12}x_2 + \ldots + a_{1n}x_n &= b_1. \\ a_{21}x_1 + a_{22}x_2 + \ldots + a_{2n}x_n &= b_2. \\ &\vdots \\ a_{n1}x_1 + a_{n2}x_2 + \ldots + a_{nn}x_n &= b_n. \end{aligned}$$

If $\mathbf{A}$ denotes the matrix of coefficients of the above system, Cramer's rule can be used to solve the system whenever $det\mathbf{A} \neq 0$. The above system of equations can be written in the form of matrix equation

$$\mathbf{AX} = \mathbf{B}$$

where $\mathbf{A} = \begin{pmatrix} a_{11} & a_{12} & \ldots & a_{1n} \\ a_{21} & a_{22} & \ldots & a_{2n} \\ \vdots & \vdots & \vdots & \vdots \\ a_{m1} & a_{m2} & \ldots & a_{mn} \end{pmatrix}$, $\mathbf{X} = \begin{pmatrix} x_1 \\ x_2 \\ \vdots \\ x_n \end{pmatrix}$, $\mathbf{B} = \begin{pmatrix} b_1 \\ b_2 \\ \vdots \\ b_n \end{pmatrix}$.

The matrix $\begin{pmatrix} a_{11} & a_{12} & \ldots & a_{1n} & b_1 \\ a_{21} & a_{22} & \ldots & a_{2n} & b_2 \\ \vdots & \vdots & \vdots & \vdots & \vdots \\ a_{m1} & a_{m2} & \ldots & a_{mn} & b_n \end{pmatrix}$ is called the **augmented matrix.** Gauss elimination or Gauss-Jordan elimination can be used to solve the system of equations.

25. **Rank** of a matrix is the largest order of any non-vanishing minor of the matrix.

26. If $\mathbf{A}$ is any square matrix of order n and $\mathbf{I}$ is the nth order unit matrix, the determinant of the matrix $\mathbf{A} - \lambda\mathbf{I}$ equated to zero is called the characteristic equation of $\mathbf{A}$, that is $|\mathbf{A} - \lambda\mathbf{I}| = 0$ gives a characteristic equation of A. The roots of the characteristic equation are called the **characteristic roots** or **eigenvalues** of the matrix A.

27. **Cayley-Hamilton Theorem** Every square matrix satisfies its characteristic equation.

28. Let $\mathbf{A}$ be an $n \times n$ matrix. A number λ is called an eigenvalue of $\mathbf{A}$ if there exists a non-zero solution vector $\mathbf{X}$ of the linear system

$$\mathbf{AX} = \lambda\mathbf{X}.$$

The column vector $\mathbf{X}$ is said to be an eigenvector corresponding to the eigenvalue λ.
Observation: Corresponding to n distinct eigenvalues, we get n linearly independent eigenvalues. But when two or more eigenvalues are equal, it may or may not be possible to get linearly independent eigenvectors corresponding to the repeated roots.

B.2 Metric Space

Definition 128. Let X be a non-empty set and d a function defined on $X \times X$ into the set of real numbers $R, d : X \times X \to R$, satisfying the following conditions:

(i) $d(x, y) \geq 0 \; \forall x, y \in X$.

(ii) $d(x, y) = 0$ if and only if $x = y$.

(iii) $d(x, y) = d(y, x) \; \forall x, y \in X$.

(iv) $d(x, y) \leq d(x, z) + d(z, y)$ for all $x, y, z \in X$.

The number $d(x, y)$ is called the distance between x and y, d is called the metric and the pair (X, d) is called a metric space.

Example B.1. The set of all real numbers with distance

$$d(x, y) = |x - y|$$

is the metric space R^1.

Example B.2. The set of all ordered $n-$tuples

$$x = (x_1, x_2, \ldots.., x_n)$$

of real numbers with distance

$$d(x, y) = \sqrt{\sum_{i=1}^{n} (x_i - y_i)^2}$$

is the metric space R^n.

B.3 Hilbert Space

Definition 129.

$$L_2(a,b) = \left\{ \text{Real functions on } [a,b] / \int_a^b |f(x)|^2 dx < \infty \right\}$$

are examples of Hilbert space (complete inner product space)

B.4 Sobolev Space

Definition 130.

$$H'(0,1) = \{f \in L_2(0,1)/f' \in L_2(0,1)\}$$

is called Sobolev space of order 1.

$$H_0'(0,1) = \{f \in H'(0,1)/f(0) = f(1) = 0\}.$$

Appendix C

Proof of Selected Theorems

C.1 Fundamental Theorem of Calculus

In this section we present the precise definition of the integrals and the proofs of the fundamental theorem of calculus and related results.

Definition of integral

The integral

$$\int_a^b f(x)\,dx$$

of the function f (which we assume to be bounded) over the interval $[a, b]$ is defined through an approximation procedure which involves Riemannian sums

$$s_\Delta = \sum_{k=1}^{n} f(\xi_k)(x_k - x_{k-1})\,. \tag{C.1}$$

Here, Δ is a partition of the interval $[a, b]$ of the form $a = x_0 < x_1 < \cdots < x_n$, and each point ξ_k lies somewhere in the subinterval $[x_{k-1}, x_k]$. Thus, the value of s_Δ depends on the choice of Δ as well as on the choice of the points ξ_k. One can estimate the influence of the latter choice through the notion of **oscillation** of a bounded function. If I is a subset of the domain of f (here, of the interval $[a, b]$), the oscillation of f on I is defined as the maximum possible difference of two function values on I, that is,

$$\operatorname{osc}_I(f) = \max_{x,z\in I} |f(x) - f(z)|\,. \tag{C.2}$$

We remark that it may happen that this maximum does not exist (that is, there are no points $x, z \in I$ where a maximum value is attained). One then replaces the maximum by the so-called **supremum**, which in this case is equal

to the smallest number η such that $|f(x) - f(z)| \leq \eta$ for all $x, z \in I$.
Let us now consider two different Riemannian sums for the same partition Δ,

$$s_\Delta = \sum_{k=1}^{n} f(\xi_k)(x_k - x_{k-1}), \quad \tilde{s}_\Delta = \sum_{k=1}^{n} f(\tilde{\xi}_k)(x_k - x_{k-1}). \tag{C.3}$$

We estimate their difference as

$$\begin{aligned} |s_\Delta - \tilde{s}_\Delta| &= \left| \sum_{k=1}^{n} f(\xi_k)(x_k - x_{k-1}) - \sum_{k=1}^{n} f(\tilde{\xi}_k)(x_k - x_{k-1}) \right| \\ &\leq \sum_{k=1}^{n} |f(\xi_k) - f(\tilde{\xi}_k)|(x_k - x_{k-1}) \\ &\leq \sum_{k=1}^{n} O_k(f)(x_k - x_{k-1}) =: V_\Delta(f), \end{aligned} \tag{C.4}$$

where $O_k(f)$ denotes the oscillation of f on the subinterval $I = [x_{k-1}, x_k]$. The number $V_\Delta(f)$ defined in (C.4) is called the **oscillation sum** of f for the partition Δ.

Definition C.1. [Integrable function] A bounded function $f : [a, b] \to \mathbb{R}$ is said to be **integrable on $[a, b]$**, if for every $\varepsilon > 0$ there exists a partition Δ of $[a, b]$ such that $V_\Delta(f) \leq \varepsilon$.

The latter condition means that, in view of (C.4), we can enforce the difference between different Riemannian sums for the same partition Δ to become as small as we want, if we choose Δ fine enough.

Let $f : [a, b] \to \mathbb{R}$ be integrable. In order to define its integral, one goes through the following three steps.

1. One proves that
$$|s_\Delta - s_{\tilde{\Delta}}| < V_\Delta(f),$$
whenever the partition $\tilde{\Delta}$ is a refinement of the partition Δ (that is, $\tilde{\Delta}$ is obtained from Δ by adding partition points), for all Riemannian sums s_Δ and $s_{\tilde{\Delta}}$ for those partitions.

2. One proves that
$$|s_\Delta - s_{\tilde{\Delta}}| < V_\Delta(f) + V_{\tilde{\Delta}}(f),$$
for arbitrary partitions Δ and $\tilde{\Delta}$ and all Riemannian sums for those partitions.

3. One proves that there exists a unique number I such that
$$|I - s_\Delta| \leq V_\Delta(f) \tag{C.5}$$
holds for all partitions and all corresponding Riemannian sums.

We then define the **integral of f on $[a, b]$** as

$$\int_a^b f(x)\,dx = I\,,$$

where I is the number obtained in step 3 above.

First we prove that every continuous function on a closed interval $[a, b]$ is integrable. Its proof uses the notion of uniform continuity. A function $f : [a, b] \to \mathbb{R}$ is **uniformly continuous on $[a, b]$**, if for every $\epsilon > 0$ there exists $\delta > 0$ such that $|f(x) - f(z)| < \varepsilon$ holds for all $x, z \in [a, b]$ with $|x - z| < \delta$. This is similar to, but not the same as the definition of continuity. It is a theorem (not treated in this book) that every continuous function defined on a closed and bounded interval $[a, b]$ is uniformly continuous on that interval.

Let $f : [a, b] \to \mathbb{R}$ be continuous. According to the definition of integrable function for an arbitrarily given $\varepsilon > 0$ we have to find a partition Δ such that $V_\Delta(f) < \varepsilon$. Choose $\delta > 0$ small enough such that $|f(x) - f(z)| < \varepsilon/(b - a)$ whenever $|x - z| < \delta$. (This is possible since, by what we said above, f is also uniformly continuous.) Next, choose a natural number n large enough such that $(b - a)/n < \delta$. Take as partition Δ the equidistant partition of $[a, b]$ with $x_k - x_{k-1} = (b - a)/n$. Then the oscillation $O_k(f)$ of f on $[x_{k-1}, x_k]$ satisfies $O_k(f) \le \varepsilon/(b - a)$ for every k and

$$V_\Delta(f) = \sum_{k=1}^{n} O_k(f)(x_k - x_{k-1}) \le n\frac{\varepsilon}{b - a} \cdot \frac{b - a}{n} = \varepsilon\,.$$

Therefore f is integrable.

Let us remark that there are different ways of defining the integral. We have chosen a method which can be generalized conveniently to the cases of double and triple integrals.

Fundamental theorem

Let f be a continuous function on $[a, b]$ then

$$F(x) = \int_a^x f(t)dt$$

has derivative at every point in $[a, b]$ and $\dfrac{d}{dx} \int_a^x f(t)dt = f(x)$.

Also $\int_a^b f(x)dx = F(b) - F(a)$.

Proof. Since f is continuous, it attains its maximum and minimum on $[a, b]$. Let

$$M = \max_{x \in [a,b]} f(x)\,, \quad m = \min_{x \in [a,b]} f(x)\,.$$

Since $m \le f(x) \le M$ for all $x \in [a, b]$, due to continuity, we have

$$m(b - a) = \int_a^b m\,dx \le \int_a^b f(x)\,dx \le \int_a^b M\,dx = M(b - a)\,.$$

We divide by $b-a$ and obtain

$$m \le \frac{1}{b-a}\int_a^b f(x)\,dx \le M\,.$$

By the intermediate value theorem, f attains every value between m and M. Therefore there exists $c \in [a,b]$ such that

$$f(c) = \frac{1}{b-a}\int_a^b f(x)\,dx\,.$$

To establish the first part of this theorem, we must show that if x is an arbitrary point in $[a,b]$, then $F'(x) = f(x)$, that is,

$$\lim_{h\to 0} \frac{F(x+h)-F(x)}{h} = f(x)\,.$$

In order to prove this, fix $x \in [a,b]$ and let h be any number such that $x+h \in [a,b]$. Using the definition of F together with property of indefinite integral yields

$$\begin{aligned}
F(x+h)-F(x) &= \int_a^{x+h} f(t)\,dt - \int_a^x f(t)\,dt \\
&= \int_a^x f(t)\,dt + \int_x^{x+h} f(t)\,dt - \int_a^x f(t)\,dt \\
&= \int_x^{x+h} f(t)\,dt\,.
\end{aligned}$$

Consequently, if $h \neq 0$, then

$$\frac{F(x+h)-F(x)}{h} = \frac{1}{h}\int_x^{x+h} f(t)\,dt\,.$$

In the case $h > 0$, by the mean value theorem, we can find a number $z = z(h)$ in the open interval $(x, x+h)$ such that

$$\int_x^{x+h} f(t)\,dt = f(z(h))\cdot(x+h-x) = f(z(h))\cdot h$$

and, therefore,

$$\frac{F(x+h)-F(x)}{h} = f(z(h))\,. \tag{C.6}$$

Since $x < z < x+h$, we have $\lim_{h\to 0^+} z(h) = x$. It then follows from the continuity of f that

$$\lim_{h\to 0^+} f(z(h)) = f(x)\,,$$

and we conclude from (C.6) that

$$\lim_{h\to 0^+} \frac{F(x+h)-F(x)}{h} = f(x)\,.$$

If $h < 0$, we may prove in a similar way that

$$\lim_{h\to 0^-} \frac{F(x+h)-F(x)}{h} = f(x)\,.$$

The two preceding one-sided limits imply that

$$F'(x) = \lim_{h\to 0} \frac{F(x+h)-F(x)}{h} = f(x)\,.$$

This completes the proof of first part.
Let F be any antiderivative of f and let

$$G(x) = \int_a^x f(t)\,dt\,. \tag{C.7}$$

By well known results of Calculus we can show that

$$G(x) = F(x) + C$$

for every x in $[a, b]$. Together with (C.7) this implies that

$$\int_a^x f(t)\,dt = F(x) + C$$

for every x in $[a, b]$. If we let $x = a$ and use the fact that $\int_a^a f(t)\,dt = 0$, we obtain $0 = F(a) + C$, or $C = -F(a)$. If we let $x = b$, we arrive at

$$\int_a^b f(t)\,dt = F(b) + C = F(b) - F(a)\,.$$

(Recall that whether the integration variable is denoted by x or by t is irrelevant.)

Integral test
For continuous positive non-increasing functions f the series $\sum_{n=1}^{\infty} f(n)$ converges if and only if the improper integral $\int_1^{\infty} f(x)\,dx$ converges.
Proof. First, assume that $\int_1^{\infty} f(x)\,dx$ converges. We define a function $\varphi : [1,\infty) \to \mathbb{R}$ by setting $\varphi(x) = f(n)$ if $n-1 \le x < n$. Since f is non-increasing, we have $\varphi(x) \le f(x)$ for all $x \ge 1$. For every natural number $N \ge 2$ we therefore obtain, using property of integral,

$$\begin{aligned} 0 \le \sum_{n=2}^{N} f(n) = \sum_{n=2}^{N} \int_{n-1}^{n} \varphi(x)\,dx = \int_1^N \varphi(x)\,dx \le \int_1^N f(x)\,dx \\ \le \int_1^{\infty} f(x)\,dx < \infty\,. \end{aligned}$$

Thus, the partial sums of the series $\sum_{n=1}^{\infty} f(n)$ are bounded by $\int_1^{\infty} f(x)\,dx$, and hence the series converges. To prove the converse, assume that $\sum_{n=1}^{\infty} f(n)$ converges. We define a function $\psi : [1,\infty) \to \mathbb{R}$ by setting $\psi(x) = f(n-1)$ if $n-1 \leq x < n$. Since f is non-increasing, we have $\psi(x) \geq f(x)$ for all $x \geq 1$. For every natural number $N \geq 2$ we therefore obtain, using property of integral,

$$
\begin{aligned}
0 \leq \int_1^N f(x)\,dx \leq \int_1^N \psi(x)\,dx = \sum_{n=2}^{N} \int_{n-1}^{n} \psi(x)\,dx = \sum_{n=2}^{N} f(n-1) \\
\leq \sum_{n=1}^{\infty} f(n) < \infty\,.
\end{aligned}
\tag{C.8}
$$

The function $F : [1,\infty) \to \mathbb{R}$ defined by

$$F(t) = \int_1^t f(x)\,dx$$

is non-decreasing and, due to (C.8), bounded by the finite number $\sum_{n=1}^{\infty} f(n)$. Therefore, the improper limit $\lim_{t\to\infty} F(t)$ exists which means that the improper integral converges.

C.2 Green-Ostrogradski Theorem

Theorem C.1. *Let D be a bounded domain in the plane whose boundary C is a closed, simple and positively oriented curve. Let $\mathbf{F} = (f,g)$ be a vector field whose components are continuously differentiable. Then*

$$\oint_C f(x,y)\,dx + g(x,y)\,dy = \iint_D \left(\frac{\partial g}{\partial x} - \frac{\partial f}{\partial y}\right) dA\,. \tag{C.9}$$

Proof. In this proof we restrict ourselves to the case where the region D has the form indicated in Figure C.1. First, we consider the case $g = 0$. The boundary C of D consists of a lower part C_1 which is the graph of some function $y = h(x)$, and of an upper part C_2 which is the graph of some function $y = k(x)$; again we refer to Figure C.1. The line integral becomes

$$\oint_C \mathbf{F}\cdot d\mathbf{r} = \int_{C_1} \mathbf{F}\cdot d\mathbf{r} + \int_{C_2} \mathbf{F}\cdot d\mathbf{r}\,,$$

where, due to the counterclockwise orientation, C_1 is traversed from left to right, and C_2 from right to left.

$$\int_{C_1} \mathbf{F}\cdot d\mathbf{r} = \int_a^b f(x,h(x))\,dx\,,$$

and analogously we obtain (note the reversal of the integration limits)

$$\int_{C_2} \mathbf{F} \cdot d\mathbf{r} = \int_b^a f(x, k(x))\, dx\,.$$

Taken together the formulas above yield

$$\oint_C \mathbf{F} \cdot d\mathbf{r} = \int_a^b f(x, h(x)) - f(x, k(x))\, dx\,. \tag{C.10}$$

Now we consider the double integral. The region D has the form

$$D = \{(x, y) : a \le x \le b\,,\ h(x) \le y \le k(x)\}\,.$$

We compute the double integral on the right hand side of (C.9) for $g = 0$, using the fundamental theorem of calculus,

$$\begin{aligned}\iint_D -\frac{\partial f}{\partial y}(x, y)\, dA &= -\int_a^b \int_{h(x)}^{k(x)} \frac{\partial f}{\partial y}(x, y)\, dy\, dx \\ &= -\int_a^b f(x, k(x)) - f(x, h(x))\, dx\,.\end{aligned}$$

From (C.10) we see that in the case $g = 0$

$$\oint_C \mathbf{F} \cdot d\mathbf{r} = \oint_C f(x, y)\, dx = \iint_D -\frac{\partial f}{\partial y}(x, y)\, dA\,. \tag{C.11}$$

An analogous proof for the case $f = 0$, taking $\tilde{F} = (0, g)$, shows that

$$\oint_C \tilde{\mathbf{F}} \cdot d\mathbf{r} = \oint_C g(x, y)\, dy = \iint_D \frac{\partial g}{\partial x}(x, y)\, dA\,. \tag{C.12}$$

(In that proof, we decompose the boundary C of D into a left part and a right part, described by some functions $x = \tilde{h}(y)$ and $x = \tilde{k}(y)$, respectively.) For the general case, we decompose $(f, g) = (f, 0) + (0, g)$, and add (C.11) and (C.12). The theorem is proved for domains D of the form in Figure C.1.

For more general domains D, the two-dimensional variant of the divergence theorem of Gauss can be used conveniently to prove Theorem C.1.

C.3 The Divergence Theorem of Gauss

The Gauss divergence theorem, Theorem 42 asserts that

$$\iint_\Sigma \mathbf{F} \cdot \mathbf{n}\, d\sigma = \iiint_D \operatorname{div} \mathbf{F}\, dV\,. \tag{C.13}$$

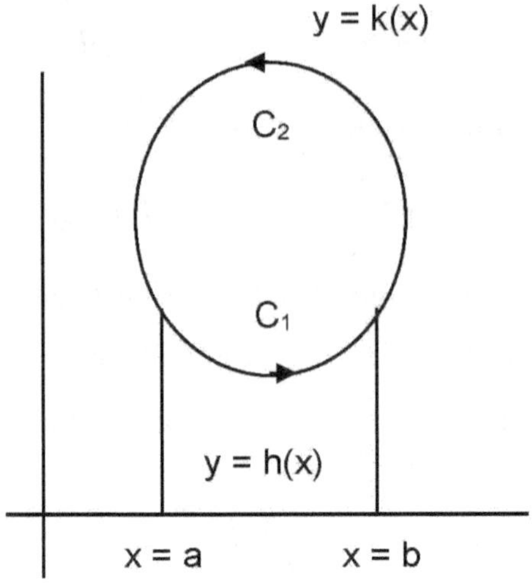

FIGURE C.1: Domain for Green-Ostrogradski Theorem

D is a domain in $\mathbb{R}^3$ with boundary Σ and outer unit normal field $\mathbf{n}$. We present a proof of Theorem 42 for the special case where D and Σ can be represented, with respect to the coordinate directions x, y and z, in the following way. With respect to the z-coordinate, we assume that D has the form

$$D = \{(x, y, z) : k(x, y) < z < h(x, y)\,,\ (x, y) \in D_z\}\,, \tag{C.14}$$

where $D_z = p_z(D)$ with $p_z(x, y, z) = (x, y, 0)$ being the projection of D onto the xy-plane, and k and h are certain functions. Moreover, we assume that the boundary surface Σ of D consists of an upper part Σ_+ described as $z = h(x, y)$, and a lower part Σ_- described as $z = k(x, y)$, and possibly a part Σ_0 which is vertical to the xy-plane (that is, the normals to Σ_0 are perpendicular to the z-direction).

Before proceeding further, let us illustrate this situation with two examples. If D is a ball bounded by a sphere Σ, the surfaces Σ_+ and Σ_- are the upper and lower half spheres, respectively, while Σ_0 is empty. If D is a rectangular box parallel to the coordinate axes, Σ_+ and Σ_- are the rectangles forming the top and the bottom, respectively, while Σ_0 consists of the four vertical sides.

We assume moreover that D admits analogous representations with respect to the x- and y-coordinates.

Proof. Recall that for $\mathbf{F} = f_1\mathbf{i} + f_2\mathbf{j} + f_3\mathbf{k}$, its divergence is given by

$$\operatorname{div}\mathbf{F} = \frac{\partial f_1}{\partial x} + \frac{\partial f_2}{\partial y} + \frac{\partial f_3}{\partial z}\,.$$

Due to the form of D described above, we can decompose a volume integral over D into an outer integral over the two-dimensional projection D_z and an integral over intervals in the z direction. We compute (using Fubini's theorem and the fundamental theorem of calculus):

$$\begin{aligned} \iiint_D \frac{\partial f_3}{\partial z}\, dV &= \iint_{D_z} \left[\int_{k(x,y)}^{h(x,y)} \frac{\partial f_3}{\partial z}(x,y,z)\, dz \right] dx\, dy \\ &= \iint_{D_z} \Big[f_3(x,y,h(x,y)) - f_3(x,y,k(x,y)) \Big]\, dx\, dy \\ &= \iint_{D_z} f_3(x,y,h(x,y))\, dx\, dy - \iint_{D_z} f_3(x,y,k(x,y))\, dx\, dy\,. \end{aligned} \tag{C.15}$$

The surface integral of $\mathbf{F} \cdot \mathbf{n} = f_1 n_1 + f_2 n_2 + f_3 n_3$ decomposes into

$$\iint_\Sigma \mathbf{F} \cdot \mathbf{n}\, d\sigma = \iint_\Sigma f_1 n_1\, d\sigma + \iint_\Sigma f_2 n_2\, d\sigma + \iint_\Sigma f_3 n_3\, d\sigma\,. \tag{C.16}$$

For the third term on the right hand side we consider the partition of Σ as described above,

$$\iint_\Sigma f_3 n_3\, d\sigma = \iint_{\Sigma_+} f_3 n_3\, d\sigma + \iint_{\Sigma_-} f_3 n_3\, d\sigma + \iint_{\Sigma_0} f_3 n_3\, d\sigma\,. \tag{C.17}$$

On Σ_0, the unit outer normal $\mathbf{n}$ is perpendicular to the z-direction, so $n_3 = 0$ on Σ_0 and the corresponding integral vanishes. We consider Σ_+. Since Σ_+ is given as $z = h(x,y)$ and D lies below Σ_+ the unit outer normal $\mathbf{n}$ points upward. In the point (x,y,z) with $z = h(x,y)$ it is given by

$$\mathbf{n}(x,y,z) = \frac{1}{\nu}\big(-\partial_x h \mathbf{i} - \partial_y h \mathbf{j} + \mathbf{k} \big)\,, \quad \nu = \sqrt{1 + (\partial_x h)^2 + (\partial_y h)^2}\,, \tag{C.18}$$

where the partial derivatives $\partial_x h$ and $\partial_y h$ are evaluated at (x,y). We thus obtain

$$n_3 = \frac{1}{\nu} \tag{C.19}$$

for its third component. We transform the surface integral over Σ_+ into a two-dimensional integral over the region D_z, according to Definition C.1.

$$\begin{aligned} \iint_{\Sigma_+} f_3 n_3\, d\sigma &= \iint_{\Sigma_+} \frac{f_3}{\nu}\, d\sigma = \iint_{D_z} \frac{f_3}{\nu} \cdot \sqrt{1 + (\partial_x h)^2 + (\partial_y h)^2}\, dA \\ &= \iint_{D_z} f_3(x,y,h(x,y))\, dx\, dy\,. \end{aligned} \tag{C.20}$$

The surface integral over Σ_- is treated analogously; there, however, the outer normal vector points downward, so we have $n_3 = -1/\nu$ instead of (C.19). Consequently,

$$\iint_{\Sigma_-} f_3 n_3\, d\sigma = -\iint_{D_z} f_3(x,y,k(x,y))\, dx\, dy\,. \tag{C.21}$$

Putting together (C.15), (C.17), (C.20) and (C.21) we arrive at

$$\iint_\Sigma f_3 n_3 \, d\sigma = \iiint_D \frac{\partial f_3}{\partial z} \, dV \,. \tag{C.22}$$

Working with the representations of D with respect to the x- and y-coordinates, one obtains in an analogous manner that

$$\iint_\Sigma f_1 n_1 \, d\sigma = \iiint_D \frac{\partial f_1}{\partial x} \, dV \,, \quad \iint_\Sigma f_2 n_2 \, d\sigma = \iiint_D \frac{\partial f_2}{\partial y} \, dV \,. \tag{C.23}$$

Adding the three equations in (C.22) and (C.23) yields (C.13). This completes the proof for the special form of D as considered.

To prove the divergence theorem for domains D of general form, one employs so-called **partitions of unity** of the vector field $\mathbf{F}$, which reduces the general situation to a situation where similar computations can be done as in the proof presented above. In addition, let us remark that while we have treated the situation in three-dimensional space, more or less the same proof works for the case of an n-dimensional region D bounded by an $n-1$-dimensional surface Σ, where n is an arbitrary number greater than or equal to 2. Both these developments are, however, outside the scope of this book.

C.4 Stokes Theorem

This section is devoted to the proof of the theorem of Stokes, It states that under suitable assumptions.

$$\oint_C \mathbf{F} \cdot d\mathbf{r} = \iint_\Sigma (\operatorname{curl} \mathbf{F}) \cdot \mathbf{n} \, d\sigma \,, \tag{C.24}$$

where $\mathbf{F}$ is a vector field, Σ is a surface with unit normal field $\mathbf{n}$ and boundary curve C, suitably oriented.

Proof of Theorem 45. The strategy of the proof is to transform the situation to the xy-plane, in order to apply the Green-Ostrogradski theorem. Let us first consider the surface integral on the right hand side of (C.24). The surface Σ is described as $z = S(x, y)$, that is, as the graph of a function S defined on D, the projection of Σ onto the xy-plane. The unit normal at a point $(x, y, S(x, y))$ of Σ is given by

$$\mathbf{n} = \frac{1}{\nu}\big(- \partial_x S \mathbf{i} - \partial_y S \mathbf{j} + \mathbf{k} \big) \,, \quad \nu = \sqrt{1 + (\partial_x S)^2 + (\partial_y S)^2} \,, \tag{C.25}$$

where the right hand side is evaluated at (x, y). Since

$$\operatorname{curl}\mathbf{F} = \nabla \times \mathbf{F} = \begin{vmatrix} \mathbf{i} & \mathbf{j} & \mathbf{k} \\ \partial_x & \partial_y & \partial_z \\ f_1 & f_2 & f_3 \end{vmatrix} \tag{C.26}$$

$$= (\partial_y f_3 - \partial_z f_2)\mathbf{i} + (\partial_z f_1 - \partial_x f_3)\mathbf{j} + (\partial_x f_2 - \partial_y f_1)\mathbf{k}\,, \tag{C.27}$$

the integrand of the surface integral becomes the scalar function

$$(\operatorname{curl}\mathbf{F}) \cdot \mathbf{n} = \frac{1}{\nu}\Big((\partial_y f_3 - \partial_z f_2) \cdot (-\partial_x S) + (\partial_z f_1 - \partial_x f_3) \cdot (-\partial_y S) \\ + (\partial_x f_2 - \partial_y f_1)\Big)\,, \tag{C.28}$$

to be evaluated at points of Σ. Using Definition 75 we can transform the surface integral into a double integral over the domain D,

$$\iint_\Sigma (\operatorname{curl}\mathbf{F}) \cdot \mathbf{n}\, d\sigma = \iint_D (\operatorname{curl}\mathbf{F}) \cdot \mathbf{n} \cdot \nu\, dA \\ = \iint_D \Big((\partial_y f_3 - \partial_z f_2) \cdot (-\partial_x S) + (\partial_z f_1 - \partial_x f_3) \cdot (-\partial_y S) \\ + (\partial_x f_2 - \partial_y f_1)\Big)\, dA. \tag{C.29}$$

Let us now consider the line integral on the left side of (C.24). If C is parameterized by $\mathbf{r} : [a, b] \to \mathbb{R}^3$, then by the definition of the line integral we have

$$\oint_C \mathbf{F} \cdot d\mathbf{r} = \int_a^b \mathbf{F}(\mathbf{r}(t)) \cdot \mathbf{r}'(t)\, dt\,. \tag{C.30}$$

Since Σ is the graph of S defined on D, C is the graph of S restricted to the boundary Γ of D. Let Γ be positively oriented by the parametrization $\mathbf{q} : [a, b] \to \mathbb{R}^2$, set

$$\mathbf{r}(t) = q_1(t)\mathbf{i} + q_2(t)\mathbf{j} + S(q_1(t), q_2(t))\mathbf{k}\,. \tag{C.31}$$

Using the chain rule we obtain

$$\mathbf{r}'(t) = q_1'(t)\mathbf{i} + q_2'(t)\mathbf{j} + \Big(\partial_x S \cdot q_1'(t) + \partial_y S \cdot q_2'(t)\Big)\mathbf{k}\,, \tag{C.32}$$

where $\partial_x S$ and $\partial_y S$ are evaluated at $\mathbf{q}(t) = (q_1(t), q_2(t))$. Let us now define the plane vector field $\tilde{\mathbf{F}}$ by

$$\tilde{\mathbf{F}}_1(x, y) = f_1(x, y, S(x, y)) + f_3(x, y, S(x, y))\partial_x S(x, y) \\ \tilde{\mathbf{F}}_2(x, y) = f_2(x, y, S(x, y)) + f_3(x, y, S(x, y))\partial_y S(x, y)\,. \tag{C.33}$$

Setting $x = q_1(t)$ and $y = q_2(t)$ we see in view of (C.31) and (C.32) that

$$\mathbf{F}(\mathbf{r}(t)) \cdot \mathbf{r}'(t) = \tilde{\mathbf{F}}(\mathbf{q}(t)) \cdot \mathbf{q}'(t) \tag{C.34}$$

for all $t \in [a, b]$. (This is the reason for defining $\tilde{\mathbf{F}}$ by (C.33.))

The theorem of Green and Ostrogradski, Theorem C.1,asserts that

$$\oint_\Gamma \tilde{\mathbf{F}} \cdot d\mathbf{r} = \iint_D \left(\partial_x \tilde{\mathbf{F}}_2 - \partial_y \tilde{\mathbf{F}}_1\right) dA\,. \tag{C.35}$$

We compute the partial derivatives from (C.33) with the aid of the chain rule,

$$\begin{aligned}
\partial_y \tilde{\mathbf{F}}_1 &= \partial_y f_1 + \partial_z f_1 \cdot \partial_y S + (\partial_y f_3 + \partial_z f_3 \cdot \partial_y S)\partial_x S \\
&\quad + f_3 \cdot \partial_y \partial_x S\,, \\
\partial_x \tilde{\mathbf{F}}_2 &= \partial_x f_2 + \partial_z f_2 \cdot \partial_x S + (\partial_x f_3 + \partial_z f_3 \cdot \partial_x S)\partial_y S \\
&\quad + f_3 \cdot \partial_x \partial_y S\,.
\end{aligned} \tag{C.36}$$

Since the function S was assumed to have continuous second partial derivatives, we can interchange their order, so we have $\partial_x \partial_y S = \partial_y \partial_x S$. Thus we obtain from (C.36) that

$$\partial_x \tilde{\mathbf{F}}_2 - \partial_y \tilde{\mathbf{F}}_1 = (\partial_z f_2 - \partial_y f_3) \cdot \partial_x S + (\partial_x f_3 - \partial_z f_1) \cdot \partial_y S + (\partial_x f_2 - \partial_y f_1)\,. \tag{C.37}$$

We compare this expression with the corresponding one in (C.29) and find that

$$\iint_D \left(\partial_x \tilde{\mathbf{F}}_2 - \partial_y \tilde{\mathbf{F}}_1\right) dA = \iint_\Sigma (\operatorname{curl} \mathbf{F}) \cdot \mathbf{n}\, d\sigma\,. \tag{C.38}$$

We now put together the previous calculations and finally conclude that

$$\begin{aligned}
\oint_C \mathbf{F} \cdot d\mathbf{r} &= \int_a^b \mathbf{F}(\mathbf{r}(t)) \cdot \mathbf{r}'(t)\, dt = \int_a^b \tilde{\mathbf{F}}(\mathbf{q}(t)) \cdot \mathbf{q}'(t)\, dt \\
&= \oint_\Gamma \tilde{\mathbf{F}} \cdot d\mathbf{r} = \iint_D \left(\partial_x \tilde{\mathbf{F}}_2 - \partial_y \tilde{\mathbf{F}}_1\right) dA \\
&= \iint_\Sigma (\operatorname{curl} \mathbf{F}) \cdot \mathbf{n}\, d\sigma\,.
\end{aligned}$$

The proof is complete.

C.5 Conservative Fields

In this section, we prove the two theorems 36 and 37 concerning conservative vector fields.

Proof of Theorem 36.

Let $\mathbf{r} : [a, b] \to \mathbb{R}^3$ be a parametrization of C, so $\mathbf{A} = \mathbf{r}(a)$ and $\mathbf{B} = \mathbf{r}(b)$. We set $g(t) = \psi(\mathbf{r}(t))$. From the chain rule we get

$$g'(t) = \nabla \psi(\mathbf{r}(t)) \cdot \mathbf{r}'(t)\,.$$

We compute

$$\int_C \mathbf{F}\cdot d\mathbf{r} = \int_C \mathbf{F}(\mathbf{r}(t))\cdot\mathbf{r}'(t)\,dt = \int_a^b \nabla\psi(\mathbf{r}(t))\cdot\mathbf{r}'(t)\,dt$$
$$= \int_a^b g'(t)\,dt = g(b) - g(a) = \psi(\mathbf{r}(b)) - \psi(\mathbf{r}(a))$$
$$= \psi(\mathbf{B}) - \psi(\mathbf{A})\,.$$

In the middle line of this computation we have used the fundamental theorem of calculus. Indeed, one may view Theorem 36 as a generalization of the fundamental theorem of calculus to line integrals.

Proof of Theorem 37.

Since we already know that every conservative vector field is circulation free, it remains to prove that every vector field, which is circulation free, is conservative. Let $\mathbf{F}$ be circulation free. We first show that $\mathbf{F}$ has the property of path independence. If C_1 and C_2 are two curves with initial point A and end point B, let C denote the curve which first connects A to B via C_1 and then B to A via C_2 in the opposite direction (the latter curve we denote by $-C_2$). Since $\mathbf{F}$ is circulation free,

$$0 = \oint_C \mathbf{F}\cdot d\mathbf{r} = \int_{C_1} \mathbf{F}\cdot d\mathbf{r} + \int_{-C_2} \mathbf{F}\cdot d\mathbf{r} = \int_{C_1} \mathbf{F}\cdot d\mathbf{r} - \int_{C_2} \mathbf{F}\cdot d\mathbf{r}\,,$$

thus

$$\int_{C_1} \mathbf{F}\cdot d\mathbf{r} = \int_{C_2} \mathbf{F}\cdot d\mathbf{r}\,.$$

Therefore, the line integral is path independent. We now fix a point P in D and define a function ψ by

$$\psi(\mathbf{x}) = \int_{C_x} \mathbf{F}\cdot d\mathbf{r}\,, \quad \mathbf{x}\in D\,, \tag{C.39}$$

where C_x is a curve which connects P to $\mathbf{x}$. (Since the line integral is path independent, it does not matter which curve we choose.) We claim that ψ is a potential for $\mathbf{F}$ in D. To this end, fix $\mathbf{x}\in D$ and choose $h>0$ so small that the line segment L from $\mathbf{x}$ to $\mathbf{x}+h\mathbf{e}_1$ lies in D, where $\mathbf{e}_1$ denotes the unit vector in the x-direction. Since we obtain a curve from P to $\mathbf{x}+h\mathbf{e}_1$ by first traversing C_x and then L, we see from the definition of ψ that

$$\psi(\mathbf{x}+h\mathbf{e}_1) = \psi(\mathbf{x}) + \int_L \mathbf{F}\cdot d\mathbf{r}\,. \tag{C.40}$$

We parametrize L by $\mathbf{r}(t) = \mathbf{x} + t\mathbf{e}_1$ with $0\le t\le h$, then $\mathbf{r}'(t) = \mathbf{e}_1$ and

$$\int_L \mathbf{F}\cdot d\mathbf{r} = \int_0^h \mathbf{F}(\mathbf{r}(t))\cdot\mathbf{e}_1\,dt = \int_0^h f_1(\mathbf{r}(t))\,dt\,.$$

Therefore

$$\frac{\psi(\mathbf{x}+h\mathbf{e}_1)-\psi(\mathbf{x})}{h}=\frac{1}{h}\int_0^h f_1(\mathbf{r}(t))\,dt\,.$$

Since the limit of the right hand side exists as $h \to 0$ and is equal to $f_1(\mathbf{r}(0)) = f_1(\mathbf{x})$, we obtain

$$\frac{\partial \psi}{\partial x}(\mathbf{x}) = f_1(\mathbf{x})\,, \quad \mathbf{x} \in D\,.$$

An analogous argument works for the other coordinate directions, so we finally conclude that $\nabla\psi = \mathbf{F}$. The proof is complete.

C.6 Proofs of Properties of Determinants

Proof of Theorem 6: It is a consequence of the result that $detA = -detB$, where B is obtained from A by interchanging two rows or columns. We prove it for 2×2 matrices.

Let $A = \begin{pmatrix} a_{11} & a_{12} \\ a_{21} & a_{22} \end{pmatrix}$ and $B = \begin{pmatrix} b_{11} & b_{12} \\ b_{21} & b_{22} \end{pmatrix}$. Then

$$detA = a_{11}a_{22} - a_{12}a_{21},$$

$$detB = a_{21}a_{12} - a_{11}a_{22},$$

that is, $detA = -detB$.
It two rows or columns are same then $detA = -detA$ implying $detA = 0$.

Proof of Theorem 7: It follows immediately by definition. In fact, $detA = 0$ if A has a zero row.

Proof of Theorem 9: Suppose the entries in the ith row of A are multiplied by the number k. Call the resulting matrix B. We obtain

$$detB = ka_{i1}C_{i1} + ka_{i2}C_{i2} + \cdots + ka_{in}C_{in}.$$

By expanding $detB$ by cofactors along ith row or

$$\begin{aligned} detB &= k(a_{i1}C_{i1} + a_{i2}C_{i2} + \cdots + a_{in}C_{in}) \\ &= kdetA. \end{aligned}$$

Proof of Theorem 11: We prove it for a 3×3 lower triangular matrix

$$A = \begin{pmatrix} a_{11} & 0 & 0 \\ a_{21} & a_{22} & 0 \\ a_{31} & a_{32} & a_{33} \end{pmatrix}.$$

$$\det A = a_{11}\begin{vmatrix} a_{22} & 0 \\ a_{32} & a_{33} \end{vmatrix} = a_{11}(a_{22}a_{33} - 0.a_{32}) = a_{11}a_{22}a_{33}.$$

Proof of Theorem 71 (three point theorem):
Let $w = T(z)$ be the solution for w in terms of z and the six given points in the equation

$$(w_1 - w)(w_3 - w_2)(z_1 - z_2)(z_3 - z) = (z_1 - z)(z_3 - z_2)(w_1 - w_2)(w_3 - w)$$

$$\text{or} \quad \frac{(w - w_1)}{(w - w_3)}\frac{(w_2 - w_3)}{(w_2 - w_1)} = \frac{(z - z_1)}{(z - z_3)}\frac{(z_2 - z_3)}{(z_2 - z_1)}.$$

Appendix D

Basic Concepts in Medical Imaging and Oil Exploration

D.1 Fundamental Steps in Digital Image Processing

Image acquisition: The types of images in which we are interested are generated by the combination of an illumination source and reflection or absorption of energy from that source. The illumination may originate from a source of electromagnetic energy such as radar, infrared or X-ray energy. Sensors are used to transform illumination energy into digital images. In a simple way incoming energy is transformed into a voltage by combination of input electrical power and sensor material that is responsive to the particular type of energy being detected. The output voltage waveform is the response of the sensor, and a digital quantity is obtained from each sensor by digitizing its response.

Image enhancement reveals the details that are obscured or highlights certain features of interest in an image.

Compression deals with techniques for reducing the storage required to save an image. Image compression is familiar to most users of computers in the form of image file extensions, such as jpg extensions.

Morphological processing deals with the tools for extracting image components that are useful in the representation and description of shapes.

Segmentation partitions an image into its constituent parts.

Representation and description follow the output of a segmentation stage which is usually raw pixel data constituting the boundary of a region or all points in the region itself.

D.2 Introduction to Medical Imaging

Gamma-ray imaging: Major uses of imaging based on gamma rays include nuclear medicine and astronomical observations. In nuclear medicine, the approach is to inject a patient with a radioactive isotope that emits gamma rays as it decays. Images are produced from the emissions collected by gamma ray detectors. The principle is the same as with X-ray tomography. However, instead of using an external source of X-ray energy, the patient is given a radioactive isotope that emits protons as it decays. When a proton meets an electron both are annhihiated; the image is created using the basic principles of tomography.

A star in the constellation of Cygnus exploded about 15,000 years ago, generating a superheated stationary gas cloud (known as the Cygnus loop) that glows in a spectacular array of colors.

X-ray imaging: X-rays are among the oldest sources of electromagnetic (EM) radiation used for imaging. The best known use of X-ray is medical diagnostics, but X-rays are also used extensively in industry and other areas, like astronomy. X-rays for medical and industrial imaging are generated using X-ray vacuum tubes with a cathode and anode. The cathode is heated, causing free electrons to be released. These electrons flow at high speed to the positively charged anode. When the electrons strike a nucleus, energy is released in the form of X-ray radiation. The energy (penetrating power) of the X-rays is controlled by a voltage applied across the anodes,and the number of X-rays is controlled by a current applied to the filament in the cathode.

Imaging in ultraviolet band: Applications of ultraviolet light are varied. They include lithography, industrial inspection, microscopy, lasers, biological imaging and astronomical observation. We illustrate imaging in this band with examples from microscopy and astronomy. Ultraviolet light is used in fluorescence microscopy, one of the fastest growing areas of microscopy. Fluorescence is a phenomenon discovered in the middle of the 19th century with the observation that the fluorspar minerals fluoresces when ultraviolet radiation is directed on it. The ultraviolet light is not visible until a photon of ultraviolet radiation collides with an electron in an atom of a fluorescent material. The photon elevates the electron to a higher energy level. The excited electron relaxes to a lower level and emits light as a lower energy photon in the visible (red) light region. A fluorescence microscope uses excitation light to irradiate a prepared specimen and then separates the much weaker radiation fluorescent light from brighter excitation light.

Imaging in visible and infrared bands: Imaging in these bands is useful in various disciplines because of its wide scope and number of applications. The infrared band is used in conjunction with visual imaging in areas such as light microscopy, astronomy, remote sensing, industrial operations, law enforcement, and other fields that utilizes images to generate enhancement and measurements.

EEG: An electroencephalogram (EEG) measures and records the electrical activity of the brain. Special sensors (electrodes) are attached to head and hooked by wires to a computer. The computer records brain electrical activity on a screen or on paper as wavy lines. Certain conditions, such as seizures, can be seen by the changes in the normal pattern of brain electrical activity.

ECG: The electrocardiogram (ECG) records the electrical activity of the heart. Each heart beat is displayed as a series of electrical waves characterized by peaks and valleys. An ECG reveals (1) the duration of the electrical waves crossing the heart which determines whether the electrical activity is normal or slow or irregular and (2) reveals is the amount of electrical activity passing through the heart muscle which enables to find whether the parts of the heart are too large or overworked. Normally, the frequency level of an ECG signal is 0.05100 Hz and its dynamic level is 110 mV.

MRI: Magnetic resonance imaging (MRI) uses a magnetic field and pulses of radio wave energy to depicts organs and structures inside the body. In many cases MRI gives different information about the body that can be seen with an X-ray, ultrasound, or computed tomography (CT) scan. For more details we refer to [8] and [9].

D.3 Core Data and Well Loggings

A petroleum reservoir is composed of hydrocarbons in porous rock formations. The crude oil found in the petroleum reservoir forms from the remains of organism. As we know a hydrocarbon is a chemical compound consisting of elements carbon (C) and hydrogen (H).

Hydrocarbons contain backbones of carbon atoms, known as carbon skeleton with hydrogen atoms attached to the backbones. Hydrocarbons, which are combustible, are the main components of fossil fuels, that include petroleum, coal and natural gas. Petroleum is nothing but crude oil.

An oil field is a region with an abundance of oil wells extracting crude Oil (petroleum) from below ground. There are more than 40,000 oil fields scattered around the globe. The largest are the Ghawar fields in Saudi Arabia and Burgan oil field in Kuwait. Famous companies engaged in the service of creating infrastructure and providing specialized services to operate a field profitably are Schlumberger, Esso, Bechtel, Baker-Hughes, Weatherford and Halliburton.

Core data is obtained using special hollow drill bits. Core data from rock provides information for indirect measurements of seismic and well logs. Samples can be taken from the cores and measurements are made in the laboratory.

One of the most important uses of core data is to help geologists identify depositional environment as a function of distance along the wells. Well logs record parameters such as spontaneous potential (SP), resistivity logs, dielectric logs, passive gamma-ray logs, active gamma-ray logs, neutron logs, NMR logs, sonic logs and dip meter logs. Thus, a well log is a set of time series data.

These logs are very well presented in Iske and Randen [1]. For detailed discussion one may refer to Luthi [2] and Selly [3]. An intensive review of reservoir modelling is also given in [2] while [3] deals with the updated petroleum geology.

The procedure of well logging involves sending a package of instruments known as a sonde-suspended from a cable or wire, into the formation. Well logging is used to determine the type and properties of any field in the rocks and to determine the geometric and physical properties of the rock. A major aim of well logging is to tie horizons observed in wells to horizons. Well logging helps to locate and quantify potential depths zones containing hydrocarbons. This appendix discusses applications of wavelet methods to well logs for determination and interpretation of cyclicity, zonation and abrupt changes in sedimentary successions. Stratigraphy is a branch of geology studying rock layers and layering. Sequence stratigraphy is the study of cyclic sedimentary deposits. Gamma rays, porosity and bulk density are the most important properties in oil exploration.

Appendix E

Solution of Odd Number Exercises

E.1 Exercises

Exercise Set - Chapter 1

1.1 (i) $(2, 4, 12)$

(ii) $(3, -19, -2)$

(vi) $22\sqrt{14}$

1.3 (i) 12 (ii) 72

1.5 (i) $\alpha = 9$

1.7 $||a|| = 3\sqrt{2}, \quad \cos\theta = \frac{1}{\sqrt{5}}$

1.9 $||\alpha a + \beta b|| =< \alpha a + \beta b, \alpha a + \beta b >$

$$= \alpha^2 < a, a > +\beta^2 < b, b > +2\alpha\beta ab = \alpha^2||a||^2 + \beta^2||b||^2 + 2\alpha\beta a.b$$

1.11 $a.(a \times b) = 0$ and $b.(a \times b) = 0$

1.13 Linearly independent.

1.17 None of these matrices are equal.

1.19 $c_{23} = 9$, $c_{12} = 0$.

1.21 (i) $(A^T)^T = \begin{pmatrix} 2 & 4 \\ -3 & 2 \end{pmatrix} = A$

(ii) $(A+B)^T = \begin{pmatrix} 6 & -1 \\ 14 & 7 \end{pmatrix}$ and $A^T + B^T = \begin{pmatrix} 6 & -1 \\ 14 & 7 \end{pmatrix}$

1.23 The given system of equations can be written in the matrix form $AX = B$ where $A = \begin{pmatrix} 2 & 6 & 1 \\ 1 & 2 & -1 \\ 5 & 7 & -4 \end{pmatrix}, B = \begin{pmatrix} 7 \\ -1 \\ 9 \end{pmatrix}$ and $X = \begin{pmatrix} x_1 \\ x_2 \\ x_3 \end{pmatrix}$.

1.25 (i) We have $x_1 = x_3,\ \ 2x_2 = x_3{+}2x_4,\ x_2 = x_3$. We find that $x_1 = x_2 = x_3$ so the second equation becomes $2x_1 = x_1 + 2x_4$ or $x_1 = 2x_4$. A solution of the system is $x_1 = x_2 = x_3 = 2t,\ x_4 = t$. Letting $t = 1$ we obtain the balanced equation

$$2Na + 2H_2O \rightarrow 2NaOH + 1H_2.$$

(ii) From $x_1C_5H_8 + x_2O_2 \rightarrow x_3CO_2 + x_4H_2O$ we obtain the system $5x_1 = x_3, 8x_1 = 2x_4, 2x_2 = 2x_3 + x_4$. Choosing $x_1 = t$ we see that $x_3 = 5t, x_4 = 4t$ and $x_2 = 7t$. Taking $t = 1$ we obtain the balanced equation

$$C_5H_5 + 7O_2 \rightarrow 5CO_2 + 4H_2O.$$

1.27 (i) Interchange row 1 and row in I_3. It is elementary.

(ii) Multiply row 3 by c in I_3. It is elementary.

(iii) Add row 4 to row 1 in I_4. It is elementary.

1.29 Either use definition of linearly independence or the following theorem [6]. It is elementary.

Let $S = v_1, v_2, v_3$ be a subset of R^3. Let M be the matrix whose columns (rows) are the elements of S. Then S or vectors of S are linearly independent of R^3 if $detA \neq 0$.

$$M = \begin{vmatrix} 1 & 2 & 3 \\ 1 & 0 & 1 \\ 1 & -1 & 5 \end{vmatrix} = 1(0+1) - 2(5-1) + 3(-1-0) = 1 - 8 - 3 = -10 \neq 0$$

hence vectors $(1,2,3)$, $(1,0,1)$ and $(1,-1,5)$ are linearly independent.

1.31 $detA = 48$, and $detB = 40$

1.33 $detA = a_{11}a_{22} - a_{21}a_{12}$

$$A^t = \begin{pmatrix} a_{11} & a_{21} \\ a_{12} & a_{22} \end{pmatrix}$$

$detA^t = a_{11}a_{22} - a_{21}a_{12}$

Hence $detA = detA^t$.

1.35 $detA^2 = detA.detA = (detA)^2 = detI = 1$ or $detA = \pm 1$.

1.37 $A = \begin{pmatrix} 4 & 3 \\ 3 & 2 \end{pmatrix}$

1.39 $A^{-1} = \begin{pmatrix} \sin\theta & -\cos\theta \\ \cos\theta & \sin\theta \end{pmatrix}$

1.41 (i) Eigenvalues are $\lambda_1 = 4, \lambda_2 = -2$. Eigenvectors corresponding to $\lambda_1 = 4$ are $\begin{pmatrix} 0 \\ 4 \end{pmatrix}$ and eigenvector corresponding to $\lambda_1 = -2$ is $\begin{pmatrix} 6 \\ -1 \end{pmatrix}$.

(ii) $\lambda_1 = \frac{3+\sqrt{14}i}{2}, \lambda_2 = \frac{3-\sqrt{14}i}{2}$. Eigenvector corresponding to λ_1 is

$\begin{pmatrix} -1+\sqrt{47}i \\ 4 \end{pmatrix}$ and Eigenvector corresponding to λ_2 is $\begin{pmatrix} -1-\sqrt{47}i \\ 4 \end{pmatrix}$.

(iii) $\lambda_1 = \lambda_2 = 0$. There is only one associated eigenvector $\begin{pmatrix} 1 \\ 0 \end{pmatrix}$.

(iv) $\lambda_1 = 0$, corresponding eigenvector is $\begin{pmatrix} 0 \\ 1 \\ 0 \end{pmatrix}$.

$\lambda_2 = 2$, corresponding eigenvector is $\begin{pmatrix} 2 \\ 1 \\ 2 \end{pmatrix}$.

$\lambda_3 = 3$, corresponding eigenvector is $\begin{pmatrix} 0 \\ 2 \\ 3 \end{pmatrix}$.

(v) $\lambda_1 = 0, \lambda_2 = 1, \lambda_3 = 7$

Eigenvector corresponding to λ_1 is $\begin{pmatrix} 14 \\ 7 \\ 10 \end{pmatrix}$.

Eigenvector corresponding to λ_2 is $\begin{pmatrix} 6 \\ 0 \\ 5 \end{pmatrix}$.

Eigenvector corresponding to λ_3 is $\begin{pmatrix} 0 \\ 0 \\ 1 \end{pmatrix}$.

1.43 (i) Matrix $P = \begin{pmatrix} -1 & 3 \\ 1 & 1 \end{pmatrix}$ diagonalizes $A = \begin{pmatrix} 5 & 3 \\ 1 & 3 \end{pmatrix}$.

(ii) $A = \begin{pmatrix} 1 & 0 \\ -4 & 1 \end{pmatrix}$ is not diagonalizable.

(iii) $M = \begin{pmatrix} 0 & 5 & 0 \\ 1 & 1 & -3 \\ 0 & 0 & 2 \end{pmatrix}$ diagonalizes $A = \begin{pmatrix} 5 & 0 & 0 \\ 1 & 0 & 3 \\ 0 & 0 & -2 \end{pmatrix}$.

1.45 See [6, pp. 483-486].

1.47 $AX = \lambda X, \Sigma_{j=1}^{n} a_{k_j} x_j = \lambda x_k$

$|\lambda x_k| \leq \Sigma_{j=1}^{n} a_{k_j} |x_j| \leq \Sigma_{j=1}^{n} a_{k_j} |x_k| = |x_k|$ or $\lambda \leq 1$.

1.49 See [10].

1.51 See [10].

Exercise Set - Chapter 2

2.1 (a) Order 1 and linear

(b) First order and nonlinear

(e) Order 2 and linear

(f) Order 2 and nonlinear

2.3 $y' = -\frac{1}{2}e^{-x/2}$, so $-\frac{1}{2}e^{-x/2} + e^{-x/2} = 0$.

2.5 $y' = -\frac{1}{x^3}(-2) = \frac{2}{x^3}$ so $x^2\left(\frac{2}{x^3}\right) + 2x\left(-\frac{1}{x^2}\right) = 0$.

2.7 $y' = e^x$, $e^x = (e^x + 1) - 1 = e^x$.

2.9 Let $f(x, y) = \sqrt{xy}$, then $\frac{\partial f}{\partial y} = \frac{1}{2}\sqrt{\frac{x}{y}}$. The differential equation $\frac{dy}{dx} = \sqrt{xy}$ will have unique solution in any region where $x > 0$ and $y > 0$, where $x < 0$ and $y < 0$.

2.11 We have $\frac{\partial f}{\partial y} = 1$ if we choose $f(x, y) = x + y$. Therefore, the differential equation will have unique solution in the entire plane.

2.13 For $f(x, y) = x^2 \cos y$, $\frac{\partial f}{\partial y} = -x^2 \sin y$.

The equation has unique solution in the entire plane.

2.15 (i) $\frac{dy}{dx} = -e^{-3x}$

$$\text{or} \quad dy = -e^{-3x}dx$$

$$y = \frac{1}{3}e^{-3x} + c$$

(v) $NdN = (te^{t+2} - 1)dt$ or $log|N| = te^{t+2} - e^{t+2} - t + c$

(vi)

$$\begin{aligned}
\frac{dy}{dx} &= \frac{(y-1)(y+1)}{(x-1)(x+1)} \\
\text{or} \quad \frac{dy}{(y-1)(y+1)} &= \frac{dx}{(x-1)(x+1)} \\
\text{or} \quad dy\left[\frac{1}{y-1} - \frac{1}{y+1}\right]\frac{1}{2} &= \frac{1}{2}\left[\frac{1}{yx-1} - \frac{1}{x+1}\right]dx \\
\text{or} \quad \ln|y-1| - \ln|y+1| &= \ln|x-1| - \ln|x+1| + \ln c \\
\text{or} \quad \frac{y-1}{y+1} &= \frac{c(x-1)}{x+1}
\end{aligned}$$

Using $y(2) = 2$ we find $c = 1$. the solution of the given initial value problem is

$$\frac{y-1}{y+1} = \frac{x-1}{x+1} \text{or} \quad y = x$$

2.17 (i) $\frac{dy}{dx} - \frac{x}{y} = 2y, \quad y(1) = 5.$
2.19 (i) $4m^2 + 8m + 16 = 0$. this gives $m_1 = -4$ and $m_2 = -4$, that is, $m_1 = m_2$ so the general solution is

$$y = c_1 e^{-4x} + c_2 x e^{-4x}.$$

(ii) $m^2 - 4m + 5 = 0, \quad m_1 = 2 + i, \ m - 2 = 2 - i$, The general solution is $y(x) = e^{2x}(c_1 \cos x + c_2 \sin x)$.

(iii) $2m^2 - 3m + 4 = 0, \ m_1 = 3/4 \pm \frac{\sqrt{23i}}{4}$, so the general solution

$$y(x) = e^{3\pi} 4(c_1 \cos \sqrt{23}x/4 + c_2 \sin \sqrt{23}x/4).$$

2.21 (i) $y = \sqrt{2} \sin 2x - \frac{1}{2}$. (ii) $y = 200e^{-x/5} - 200 - 3x^2 + 30x$.
2.23 (i) The auxiliary equation is $m^2 - 8m + 41 = 0$ so that

$$y = x^4[c_1 \cos(5 \ln x) + c_2 \sin(5 \ln x)].$$

2.25 Desired differential equation is

$$\frac{dx}{dt} = r - kx(t), \quad \text{where} \quad k > 0.$$

2.27 The mathematical model is

$$\frac{dm}{dt} = km,$$

where $m(t)$ is the mass at time t and k is the constant of proportionality depending on the substance.

2.28 Let $P(t)$ be the number of owls present at time t. Then

$$\frac{dP}{dt} = k(P - 200 + 10t).$$

2.31 Let $S(t)$ denote the sale in rupees, then desired model is

$$\frac{dS(t)}{dt} = k(P - Q),$$

where P is the original amount by which sale decline after removing the advertising budget. **2.33** Desired model is

$$\frac{dP(t)}{dt} = kP(t), \quad P(0) = P_0,$$

$$P(2) = 2P_0$$

2.35 $\dfrac{dP(t)}{dt} = \lambda(P(t))$ or $P(t) = P(0)e^{\lambda t}$

$$\begin{aligned}
\frac{3}{2}P_0 &= P_0 e^{\lambda} \\
\lambda &= \ln\frac{3}{2} \\
3P_0 &= P_0 e^{\lambda t} \\
3 &= e^{\lambda t} \text{ or } t = \frac{1}{\lambda}\ln 3 \\
\text{or } t &= \frac{1}{\ln\frac{3}{2}}\ln 3 = \ln 3 - \ln\frac{3}{2} \\
t &= \ln 3 - \ln 3 + \ln 2.
\end{aligned}$$

2.37 $100000 = 50,000e^{\lambda}$ or $\lambda = \ln 2.$

$$P(t) = P_0 e^{2\ln 2} = 50000e^{2\ln 2}.$$

2.39 Similar to Example 126

2.41 Similar to Example 128.

2.43 Volume V depends on r^3 and surface area S on r^2, where r is the radius of the given spherical drop. Therefore, surface area depends on

$$V^{2/3},\ V = k_1 r^3,\ S = k_2 r^2 = k_1\left(\frac{V}{k_1}\right) = kV^{2/3}.$$

Desired differential equation is

$$\frac{dV}{dt} = -cV^{2/3}, \text{where} V \text{is a constant.}$$

2.47 Similar to Example 140.

2.49 $\dfrac{dI}{dt} = k(T-10)$ so $T = 10 + ce^{kt}$. If $T(0) = 70^0$ and $T(\frac{1}{2}) = 50$ then $c = 60$ and $k = 2\ln(2/3)$ so $T(1) = 36.67^0$. If $T(t) = 15^0$ then $t = 3.06$.

2.51 $\dfrac{dI}{dt} + \dfrac{IR}{L} = \dfrac{V}{L}$. Solve as a linear differential equation with integrating factor $= e^{\frac{Rt}{L}}$, and using initial conditions. Check $I = \dfrac{V}{R}(1 - e^{\frac{-Rt}{L}})$ satisfies the differential equation.

2.53 Let S be the desired distance. We can determine S by solving the differential equation $\dfrac{dS}{dt} = At + B$, Where A and B are constants to be determined by the given equations.

2.55 $\lim_{t\to\infty} e^{-bt}t^n = \lim_{t\to\infty} \dfrac{t^n}{e^{bt}} = \lim_{t\to\infty} \dfrac{n!}{b^n e^{bt}} = 0.$

$t^n = O(e^{bt}), t \to \infty$ for any positive value of b. Hence $t^n = O(e^{bt}), t \to \infty$ and t^n is of exponential order.

2.57 $e^{-3t}\cos 2t - \frac{3}{2}e^{-3t}\sin 2t$

2.59 $y(t) = 2e^{3t} + 11e^{3t} + \frac{1}{12}t^4e^{3t}$

2.61 $f(t) = 4e^{-4} - 7te^{-t} + 4t^2e^{-t}$

2.63 $f(t) = \cos\sqrt{2}t\cosh\sqrt{2}t + \frac{1}{2\sqrt{2}}(\sin\sqrt{2}t\cosh\sqrt{2}t + \cos\sqrt{2}t\sinh\sqrt{2}t) + \sin\sqrt{2}t\cosh\sqrt{2}t$

2.65 $y_1(t) = \frac{-\sqrt{2}}{10}\sin\sqrt{2}t + \frac{\sqrt{3}}{5}\sin 2\sqrt{3}t$

$y_2(t) = \frac{-\sqrt{2}}{5}\sin\sqrt{2}t \quad \frac{\sqrt{3}}{10}\sin 2\sqrt{3}t$

2.67 $y(t) = \frac{16}{39}e^{-2t} + \frac{13}{60}e^{t} - \frac{13}{20}e^{-t} + \frac{3}{130}\cos 3t - \frac{2}{130}\sin 3t$

2.69 [Ref [1] of Chapter 5, p. 87]. Apply principle of finite mathematical induction.

2.71 $y = c_0y_1(x) + c_1y_2(x)$, where

$y_1(x) = 1 + \frac{1}{2}x^2 + \frac{1}{6}x^3 + \frac{1}{24}x^4 + \frac{1}{30}x^5 \ldots$

$y_2(x) = x + \frac{1}{6}x^3 + \frac{1}{12}x^4 + \frac{1}{120}x^5 + \ldots$

2.73 $y = c_0(1 - \frac{1}{2!}x^2 + \frac{1}{4!}x^4 + \ldots) + c_1(x - \frac{1}{3!}x^3 + \frac{1}{5!}x^5 + \ldots) + \frac{1}{2!}x^2 + \frac{1}{3!}x^3 + \ldots$

2.75 $y_1(x) = 1 + \frac{1}{2}x^2 + \frac{1}{6}x^3 + \frac{1}{6}x^4 \ldots$

$y_2(x) = x + \frac{1}{2}x^2 + \frac{1}{2}x^3 + \frac{1}{4}x^4 + \ldots$

2.77 $y = x - \frac{1}{3}x^3 + \frac{1}{5}x^5 - \frac{1}{7}x^7 + \ldots y_1(x) = 1 + \frac{1}{2}x^2 + \frac{1}{6}x^3 + \frac{1}{24}x^4 + \frac{1}{30}x^5 \ldots$

$y_2(x) = x + \frac{1}{6}x^3 + \frac{1}{12}x^4 + \frac{1}{120}x^5 + \ldots$

2.79 $y = c_1y_1 + c_2y_2$, where

$y_1 = c_0(1 + \frac{1}{4}x^2 + \frac{1}{48}x^4 + \ldots)$

$y_2 = -\frac{1}{2}y_1\ln x + y[-\frac{1}{2x^2} + \frac{7}{96}x^2 - \frac{19}{2304}x^4 + \ldots]$

2.83 Since $v^2 = 0$ in the Bessel equation the general solution is $c_1J_0(x) + c_2Y_0(x)$, See [13].

2.85 It is Legendre's equation of order 0. It can be seen from solution of Legendre's equation in Section 2.8.5 that it is $y_0(x) = c_0 = 1$.

Exercise Set - Chapter 3

3.1 (i) Let $x = (x_1, x_2, x_3), y = (y_1, y_2, y_3)$ and $z = (z_1, z_2, z_3)$. Then

$$\begin{aligned}(x+y)+z &= [(x_1,x_2,x_3)+(y_1,y_2,y_3)]+(z_1,z_2,z_3)\\ &= (x_1,x_2,x_3)+(y_1+z_1,y_2+z_2,y_3+z_3)\\ &= x+(y+z)\end{aligned}$$

(vi) Since

$$\begin{aligned}y.(x\times y) &= \begin{vmatrix} y_1 & y_2 & y_3\\ x_1 & x_2 & x_3\\ y_1 & y_2 & y_3\end{vmatrix}\\ &= y_1(x_2y_3-x_3y_2)-y_2(x_1y_3-x_3y_1)+y_3(x_1y_2-x_2y_1)\\ &= y_1x_2y_3-x_3y_2y_1-y_2x_1y_3-x_3y_1y_2+y_3x_1y_2-x_2y_1y_3\\ &= 0,\end{aligned}$$

y is perpendicular to $x\times y$ or $x\times y$ is perpendicular to y.

3.3 (a) (i) We have $x_0=4, y_0=2, a=-1$ and $b=5$; therefore the parametric equations are $x=4-t, y=2+5t$.

(ii) $x=1+4t, y=2+5t, z=-3-7t$.

3.5 (a) We know that $\frac{d}{dt}[F.G]=F\frac{dG}{dt}+\frac{dF}{dt}G$. Taking $F=G=\gamma$ yields

$$\frac{d}{dt}(\gamma(t).\gamma'(t))=\gamma(t).\gamma'(t)+\gamma'(t).\gamma(t)$$

so $\frac{d}{dt}[||\gamma(t)||^2]=2\gamma(t).\gamma'(t)$.

Since $||\gamma(t)||$ is constant, $\gamma(t).\gamma'(t)=0$. Therefore, $\gamma(t)$ is orthogonal to $\gamma'(t)$.

3.7 (b) We know that $\nabla.(F\times G)=G.(\nabla\times F)-F.(\nabla\times G)$.

Choosing $F=\nabla\phi,\ G=\nabla\psi$.

Using $\nabla\times\nabla\phi=0,\ \nabla\times\nabla\psi=0$, we get

$$\nabla.(\nabla\phi\psi)=\nabla\psi.0-\nabla\phi.0=0$$

or $div(\nabla\phi\times\nabla\psi)=0$.

3.9

$$\begin{aligned}
\int^{C} F.dr &= \int_{\pi}^{0} (i - xj + k).(-\sin ti - \cos tj)dt \\
&= \int_{0}^{\pi} (-\sin t - x(t)\cos t)dt, \quad \text{where} x(t) = \sin t \\
&= \int_{0}^{\pi} (-\sin t - \frac{1}{2}\sin 2t)dt \\
&= [[\cos t]_0^{\pi} + [\cos 2t]_0^{\pi} = 0
\end{aligned}$$

3.11 The work done is equal to $\oint_C F.dr$. Let D be the disk enclosed by the circle C and let $A(D)$ denote its area. Then

$$\begin{aligned}
\oint_C F.dr &= \oint (e^x - y + x\cosh x)dx + (y^{3/2} + x)dy \\
&= \int\int_D \left[\frac{\partial}{\partial x}(y^{3/2} + x) - \frac{\partial}{\partial y}(e^x - y + x\cosh x\right] dA \\
& \quad \text{by the theorem of Green Ostrogradski} \\
&= \int\int_D (1+1)dA = 2A(D) = 2\pi(12)^2 = 288\pi
\end{aligned}$$

3.13 (a) Since $divF = 0$ for $F = 2yzi - 4xzj + xyk$,
$\int\int\int_D divF dV = 0$ for the ball D bounded by the sphere Σ.

3.15 By Gauss Divergence theorem

$$\int\int_{\Sigma} C.nd\sigma = \int\int\int_D divCdV = \int\int\int_D 0dV = 0$$

as C is constant, $divC = 0$.

3.17 By Storke's theorem, $\oint_C F.dr = \int\int_{\Sigma}(curlF).nd\sigma$, where n is suitably orient normal to the disk $x^2 + y^2 \leq 1$. We have

$$curlF = \begin{vmatrix} i & j & k \\ \frac{\partial}{\partial x} & \frac{\partial}{\partial y} & \frac{\partial}{\partial z} \end{vmatrix} = -zaj + (2xy+1)k$$

Since the disk in horizontal, it is parameterized by $z = S(x,y) = 0$ and $n = k$. Therefore $curlF.n = 2xy + 1$ and $\sqrt{1 + (\partial_x S)^2 + (\partial_y S)^2} = \sqrt{1} = 1$. We compute

$$\oint_C F.dr = \int\int_{\Sigma}(curlF).nd\sigma = \int\int_{\Sigma} 1d\sigma = \int\int_D 1dA = \pi,$$

since $x = y = 0$ on Σ and area of disk D equals π.

3.19 It can be shown that the line integral of given F is $\dfrac{-5}{4}\pi$ and the surface integral of given vector field is also equal to $\dfrac{-5}{4}\pi$. This verifies the Stroke's theorem.

Exercise Set - Chapter 4

4.1 Yes. An orthogonal set can be converted into an orthonormal system by dividing each vector by its magnitude.

4.3 (a) $a_0 = 2\pi, a_n = 0, b_n = \frac{2}{n}(-1)^{n+1}$ and $f(x) = \pi + \Sigma_{n=1}^{\infty}\frac{2}{n}(-1)^{(n+1)}\sin nx$.

4.5 Graph of the third partial sum of the Fourier series can be seen as the screen of a computer using MATLAB and MATHEMATICA.

4.7 $f(x) = \Sigma_{n=1}^{\infty}(\frac{2(\pi+1)}{n\pi}(-1)^{n+1} + \frac{2}{n\pi})\sin n\pi$

4.9 Part (a) is Bessel's inequality and its proof can be found in [3] of Chapter 4.

(c) It is famous Riemann-Lebsegue lemma and its proof can be found in [1,3]

4.13 Follows from definition.

4.15 We find $\hat{f}$ for $f = \chi_{[-\frac{1}{2},\frac{1}{2}]}$

$$\hat{f} = \frac{2}{s}\sin\frac{5}{2}$$

Exercise Set - Chapter 5

Separation of variables method for heat equation

5.1 $u(x,t) = \frac{8}{\pi}\sum\limits_{n=1}^{\infty}\frac{1}{(2n-1)^2}e^{-(2n-1)^2}\sin(2n-1)x$

5.3 $u(x,t) = \frac{4}{\pi^2}\sum\limits_{n=1}^{\infty}\frac{(2n-1)\pi-8(-1)^n}{(2n-1)^2}e^{-\frac{(2n-1)^2\pi^2 t}{64}}\sin\frac{(2n-1)\pi}{8}x$

5.5 $u(x,t) = \frac{40}{\pi}\sum\limits_{n=1}^{\infty}\frac{1-\cos\frac{n\pi}{2}}{n}e^{-\frac{n^2\pi^2 t}{64}}\sin\frac{n\pi}{2}x$

Laplace transform for heat equation

5.7 $u(x,t) = u_1 + (u_0 - u_1)\,\mathrm{erf}\,c\left(\dfrac{x}{2\sqrt{t}}\right)$

Fourier transform for heat equation

5.9 $u(x,t) = \frac{1}{2}\,\mathrm{erf}\left(\frac{1-x}{2c\sqrt{t}}\right) + \frac{1}{2}\,\mathrm{erf}\left(\frac{1+x}{2c\sqrt{t}}\right)$

5.11 $u(x,t) = \frac{1}{t\sqrt{2\pi}}\,\mathrm{erf}\int\limits_{-\infty}^{\infty} f(\tau)\, e^{-\frac{(x-\tau)^2}{2t^2}}\, d\tau$

Separation of variables method for wave equation

5.17 $u(x,t) = \frac{2}{\pi}\sum\limits_{n=1}^{\infty} \frac{1}{n^2}\sin\frac{2n\pi}{3}\sin nx\cos nt$

5.19 $u(x,t) = \frac{4}{\pi}\sum\limits_{n=1}^{\infty} \frac{(-1)^{n+1}}{(2n-1)^2}\sin(2n-1)\,x\cos(2n-1)\,t$

Fourier transform for wave equation

5.21 $u(x,t) = \frac{1}{2}\int\limits_{-\infty}^{\infty} e^{-|\omega|}\cos\omega t e^{i\omega x}$

5.23 $u(x,t) = \frac{1}{2c}\int\limits_{0}^{t}\left[\int\limits_{x-c(t-\tau)}^{x+c(t-\tau)} f(s,\tau)\, ds\right] d\tau$

Laplace transform for wave equation

5.25 $u(x,t) = (t-x)\sinh(t-x)\,H(t-x) + xe^{-x}\cosh t - te^{-t}\sinh t$, $H(.)$ is the unit step Heaviside function.

5.27 $u(x,t) = \sin\pi x\cos\pi t - \frac{1}{\pi}\sin\pi x\sin\pi t.$

Separation of variables method for Laplace equation

5.29 $u(x,y) = \frac{2}{\pi}\sum\limits_{n=1}^{\infty} \frac{1-(-1)^n\sinh n\pi}{n}\sinh ny\sin nx + \frac{2}{\pi}\sum\limits_{n=1}^{\infty} \frac{1-(-1)^n\sinh n\pi}{n}$

$$[\sinh ny + \sinh n(\pi - x)]\sin ny$$

5.31 $u(x,y) = \frac{400}{\pi}\sum\limits_{n=1}^{\infty} \frac{\sin\frac{(2n-1)\pi}{2}x}{(2n-1)\sinh\frac{(2n-1)\pi}{2}}\sinh\left[\frac{(2n-1)\pi}{2}(1-y)\right]$

$+\frac{200}{\pi}\sum\limits_{n=1}^{\infty} \frac{1}{n\sinh 2n\pi}\sinh n\pi x\sin n\pi y$

5.33 $u(x,y) = \frac{\sin 7\pi x\sinh(7\pi(1-y))}{\sinh 7\pi} + \frac{\sin\pi x\sinh\pi y}{\sinh\pi}$

$+\frac{\sin 3\pi y\sinh(3\pi(1-x))}{\sinh 3\pi} + \frac{\sinh 6\pi x\sin 6\pi y}{\sinh 6\pi}$

5.35 $u(x,y) = \frac{1}{2}x + \frac{2}{\pi^2}\sum\limits_{n=1}^{\infty} \frac{1-(-1)^n}{n^2\sinh n\pi}\sinh n\pi x\cos n\pi y$

5.37 $u(x,y) = A_0 y + \sum_{n=1}^{\infty} A_n \sinh\left(\frac{n\pi}{a}y\right)\cos\frac{n\pi}{a}y$; $A_0 = \frac{1}{ab}\int_0^a f(x)\,dx$,

$A_n = \frac{1}{a\sin\frac{n\pi}{a}b}\int_0^a f(x)\cos\frac{n\pi}{a}x dx.$

Fourier transform for Laplace equation

5.41 $u(x,y) = \frac{2}{\pi}\int_0^{\pi} F(\omega)\frac{\sinh\omega(2-y)}{(1+\omega^2)\sinh 2\omega}\sin\omega x d\omega$

5.43 $u(x,y) = \frac{1}{\pi}\left[\arctan\left(\frac{1+x}{y}\right) + \arctan\left(\frac{1-x}{y}\right)\right]$

Exercise Set - Chapter 6

6.6 The two iterations of Jacobi's method give the following results.

a. $(0.1428571, -0.3571429, 0.4285714)^t$

b. $(0.97, 0.91, 0.94)^t$

c. $(-0.65, 1.65, -0.4, -2.475)^t$

6.7 **b.** $r_0 = (1,0,-2)^t$, $x_1 = (5,0,-1)^t$,

$r_1 = (-1,-2,-5)^t$, $x_2 = (.51814, -.72539, .-1.94301)^t$

$r_2 = (1.28497, -.80311, .64249)^t$, $x_2 = (1, -1.4, -2.2)$

c. $r_0 = (1,2,0,-1)^t$, $x_1 = (.2, .4, 0, -.2)^t$

$r_1 = (1.2, -.8, -.8, -.4)^t$, $x_2 = (.90654, .46729, -.33645, -.57009)^t$

$r_2 = (-1.45794, -.59813, -.26168, -2.65421)^t$,

$x_3 = (4.56612, .40985, -2.92409, -5.50820)^t$

$r_3 = (-1.36993, 1.11307, -3.59606, .85621)^t$,

$x_4 = (9.50, 1.25, -10.25, -13.00)^t$

Exercise Set - Chapter 7

7.1

x_n	y_n
1.00	5.0000
1.10	3.9724
1.20	3.2284
1.30	2.6945
1.40	2.3163

1.50	2.0533

7.3

x_n	y_n	Actual Value
0.00	2.0000	2.0000
0.10	2.1230	2.1230
0.20	2.3085	2.3085
0.30	2.5958	2.5958
0.40	3.0649	3.0650
0.50	3.9078	3.9082

7.5

x_n	y_n
1.00	1.0000
1.10	1.0101
1.20	1.0417
1.30	1.0989
1.40	1.1905
1.50	1.3333

7.7 (a) Write the equation in the form

$$\frac{dv}{dt} = 32 - 0.125v^2 = f(t, v).$$

t_n	v_n
0.0	0.0000
1.0	25.2570
2.0	32.9390
3.0	34.9770
4.0	35.5500
5.0	35.7130

(b) Separating variables and using partial fractions we have

$$\frac{1}{2\sqrt{32}}\left(\frac{1}{\sqrt{32}-\sqrt{0.125}v}+\frac{1}{\sqrt{32}+\sqrt{0.125}v}\right)dv = dt$$

$$\frac{1}{2\sqrt{32}\sqrt{0.125}}\left(\ln\left|\sqrt{32}\right| + \sqrt{0.125}v - \ln\left|\sqrt{32}\right| - \sqrt{0.125}\right) = t + c$$

Since $v(0) = 0$, we find $c = 0$. Solving for v we obtain

$$v(t) = \frac{16\sqrt{5}\left(e^{3.2t}-1\right)}{\left(e^{3.2t}+1\right)}$$

and $v(5) \approx 35.7678$. Alternatively, the solution can be expressed as

$$v(t) = \sqrt{\frac{mg}{k}}\tanh\sqrt{\frac{kg}{m}}t.$$

7.9 The figure shows the values of $u(x, y)$ along the boundary. We need to determine u_{11} and u_{21}. The system is

$$u_{21} + 2 + 0 + 0 - 4u_{11} = 0$$
$$1 + 2 + u_{11} + 0 - 4u_{21} = 0$$

or

$$-4u_{11} + u_{21} = -2$$
$$u_{11} - 4u_{21} = -3.$$

Solving we obtain $u_{11} = 11/15$ and $u_{21} = 14/15$.

7.11 The figure shows the values of u(x, y) along the boundary. We need to determine u_{11}, u_{21}, u_{12},and u_{22}. By symmetry $u_{11} = u_{21}$ and $u_{12} = u_{22}$. The system is

$$3/2 + 0 + u_{11} - 4u_{12} = 0$$
$$u_{21} + u_{12} + 0 + 0 - 4u_{11} = 0$$
$$0 + u_{22} + u_{11} + 0 - 4u_{21} = 0$$
$$u_{22} + \sqrt{3/2} + 0 + u_{11} - 4u_{12} = 0$$
$$0 + \sqrt{3/2} + u_{12} + u_{21} - 4u_{22} = 0$$

or

$$3u_{11} + u_{12} = 0$$
$$u_{11} - 3u_{12} = -\sqrt{\frac{3}{2}}.$$

Solving we obtain $u_{11} = u_{21} = \sqrt{3}/16$ and $u_{12} = u_{22} = 3\sqrt{3}/16$.

7.13. The results of the computation are given in the table

Time	$x = 0.25$	$x = 0.50$	$x = 0.75$	$x = 1.00$	$x = 1.25$	$x = 1.50$	$x = 1.75$
0.000	1.0000	1.0000	1.0000	1.0000	0.0000	0.0000	0.0000
0.100	0.3728	0.6288	0.6800	0.5904	0.3840	0.2176	0.0768
0.200	0.2248	0.3942	0.4708	0.4562	0.3699	0.2517	0.1239
0.300	0.1530	0.2752	0.3448	0.3545	0.3101	0.2262	0.1183
0.400	0.1115	0.2034	0.2607	0.2757	0.2488	0.1865	0.0996
0.500	0.0841	0.1545	0.2002	0.2144	0.1961	0.1487	0.0800
0.600	0.0645	0.1189	0.1548	0.1668	0.1534	0.1169	0.0631
0.700	0.0499	0.0921	0.1201	0.1297	0.1196	0.0914	0.0494
0.800	0.0387	0.0715	0.0933	0.1009	0.0931	0.0712	0.0385
0.900	0.0301	0.0555	0.0725	0.0785	0.0725	0.0554	0.0300
1.000	0.0234	0.0432	0.0564	0.0610	0.0564	0.0431	0.0233

7.15 The table in this section gives a selection of the total number of approximations

Time	$x = 0.25$	$x = 0.50$	$x = 0.75$	$x = 1.00$	$x = 1.25$	$x = 1.50$	$x = 1.75$
0.000	1.0000	1.0000	1.0000	1.0000	0.0000	0.0000	0.0000
0.100	0.4015	0.6577	0.7084	0.5837	0.3753	0.1871	0.0684
0.200	0.2430	0.4198	0.4921	0.4617	0.3622	0.2362	0.1132
0.300	0.1643	0.2924	0.3604	0.3626	0.3097	0.2208	0.1136
0.400	0.1187	0.2150	0.2725	0.2843	0.2528	0.1871	0.0989
0.500	0.0891	0.1630	0.2097	0.2228	0.2020	0.1521	0.0814
0.600	0.0683	0.1256	0.1628	0.1746	0.1598	0.1214	0.0653
0.700	0.0530	0.0976	0.1270	0.1369	0.1259	0.0959	0.0518
0.800	0.04130	0.0762	0.0993	0.1073	0.0989	0.0755	0.0408
0.900	0.0323	0.0596	0.0778	0.0841	0.0776	0.0593	0.0321
1.000	0.0253	0.0466	0.0609	0.0659	0.0608	0.0465	0.0252

7.17(a) Identifying $h = 1/4$ and $k = 1/10$ we see that $\delta = 2/5$.

Time	$x = 0.25$	$x = 0.5$	$x = 0.75$
0.00	0.1875	0.2500	0.1875
0.10	0.1775	0.2400	0.1775
0.20	0.1491	0.2100	0.1491
0.30	0.1066	0.1605	0.1066
0.40	0.0556	0.0938	0.0556
0.50	0.0019	0.0148	0.0019
0.60	-0.0501	-0.0682	-0.0501
0.70	-0.0970	-0.1455	-0.0970
0.80	-0.1361	-0.2072	-0.1361
0.90	-0.1648	-0.2462	-0.1648
1.00	-0.1802	-0.2591	-0.1802

(b) Identifying $h = 1/4$ and $\Delta t = 1/10$ we see that $\delta = 1/4$

Time	$x = 0.4$	$x = 0.8$	$x = 1.2$	$x = 1.6$
0.00	0.0032	0.5273	0.5273	0.0032
0.10	0.0194	0.5109	0.5109	0.0194
0.20	0.0652	0.4638	0.4638	0.0652
0.30	0.1318	0.3918	0.3918	0.1318
0.40	0.2065	0.3035	0.3035	0.2065
0.50	0.2743	0.2092	0.2092	0.2743
0.60	0.3208	0.1190	0.1190	0.3208
0.70	0.3348	0.0413	0.0413	0.3348
0.80	0.3094	-0.0180	-0.0180	0.3094
0.90	0.2443	-0.0568	-0.0568	0.2443
1.00	0.1450	-0.0768	-0.0768	0.1450

[**c:**(i)]The solution of the wave equation is shown to be

$$u(x,t) = \sum_{n=0}^{\infty} (a_n \cos n\pi t + b_n \sin n\pi t) \sin n\pi x$$

where

$$a_n = 2\int_0^1 \sin n\pi x \sin n\pi x dx = \begin{cases} 1, & n = 1 \\ 0, & n = 2, 3, 4 \end{cases}$$

and

$$b_n = \frac{2}{n\pi}\int_0^1 0 dx = 0.$$

Thus

$$u(x,t) = \cos \pi t \sin \pi x$$

[**c**:(i) and (ii)]:

We have $h = 1/4, \Delta t = 0.5/5 = 0.1$ and $\delta = 0.4$. Now $u_{0,n} = u_{4,n} = 0$ or $n = 0, 1, ..., 5$, and the initial values of u are $u_{1,0} = u(1/4, 0) = \sin \pi/4 \approx 0.7071$, $u_{2,0} = u(1/2, 0) = \sin \pi/2 = 1$, $u_{3,0} = u(3/4, 0) = \sin 3\pi/4 \approx 0.7071$. From equation (7.93) in the text we have $u_{j,1} = 0.8(u_{j+1,0} + u_{j-1,0}) + 0.84u_{j,0}$ $+0.1(0)$. Then $u_{1,1} \approx 0.6740, u_{2,1} = 0.9531, u_{3,1} = 0.6740$.From equation (7.86) in the text we have for $n = 1, 2, 3, ...$

$$u_{j,n+1} = 0.16u_{j+1,n} + 2(0.84)u_{j,n} + 0.16u_{j-1,n} - u_{j,n-1}$$

The results of the calculations are given in the table.

Time	$x = 0.25$	$x = 0.50$	$x = 0.75$
0.0	0.7071	1.0000	0.7071
0.1	0.6740	0.9531	0.6740
0.2	0.5777	0.8169	0.5777
0.3	0.4272	0.6042	0.4272
0.4	0.2367	0.3348	0.2367
0.5	0.0241	0.0340	0.0241

i, j	Approx	Exact	Error
1,1	0.6740	0.6725	0.0015
1,2	0.5777	0.5721	0.0056
1,3	0.4272	0.4156	0.0116
1,4	0.2367	0.2185	0.0182
1,5	0.0241	0.0000	0.0241
2,1	0.9531	0.9511	0.0021
2,2	0.8169	0.8090	0.0079
2,3	0.6042	0.5878	0.0164
2,4	0.3348	0.0000	0.0340
3,1	0.6740	0.6725	0.0015
3,2	0.5777	0.5721	0.0056
3,3	0.4272	0.4156	0.0116
3,4	0.2367	0.2185	0.0182
3,5	0.0241	0.0000	0.0241

Exercise Set - Chapter 8

8.1 (a)

$$\begin{aligned} (5-9)i + (2-4i) &= x + iy \\ 7 - 13i &= x + iy \\ x &= 7 \\ y &= -13 \end{aligned}$$

(b)

$$\begin{aligned} \frac{10-5i}{6+2i} &= \frac{(10-5i)(6-2i)}{(6+2i)(6-2i)} \\ &= \frac{60 - 30i - 20i - 10}{36+4} \\ &= \frac{1}{40}(50 - 50i) \\ &= \frac{5}{4} - \frac{5}{4}i \\ &= x + iy \end{aligned}$$

where $x = \frac{5}{4}, y = -\frac{5}{4}$

8.3 (b) $5\sqrt{2}(\cos\frac{7\pi}{4} + i\sin\frac{7\pi}{4})$

8.5 (a)

$$z_1z_2 = 8[\cos(\frac{\pi}{8} + 3\frac{\pi}{8}) + i\sin(\frac{\pi}{8} + 3\frac{\pi}{8})]$$
$$= 8i$$
$$\frac{z_1}{z_2} = \frac{1}{2}[\cos(\frac{\pi}{8} - 3\frac{\pi}{8}) + i\sin(\frac{\pi}{8} - 3\frac{\pi}{8})]$$
$$= \frac{\sqrt{2}}{4} - i\frac{\sqrt{2}}{4}$$

8.9 (a) $f(z) = (7x-9y-3)+i(7y-9x+2)$ where $u(x,y) = 7x-9y-3, v(x,y) = (7y-9x+2)$

8.11 $\lim_{z\to 1+i} \frac{5z^2-2z+2}{z+1} = \frac{5(1+i)^2-2(i+1)+2}{1+i+1}$

$$= \frac{5(1+i^2+2i)-2i}{2+i}$$

8.13 (a) $f'(z) = \lim_{\Delta z\to 0} \frac{(z+\Delta z)^2-z^2}{\Delta z} = \lim_{\Delta z\to 0} \frac{2z\Delta z+(\Delta z)^2}{\Delta z}$

$$= \lim_{\Delta z\to 0}(2z+\Delta z) = 2z$$

8.15 (a) 3i (b) $2i$ and $-2i$

8.17 Cauchy Riemann Equations

8.19 (a)$u = y, v = x, \frac{\partial u}{\partial x} = 0 = \frac{\partial v}{\partial y}, \frac{\partial u}{\partial y} = 1, -\frac{\partial v}{\partial x} = -1.$

Since $1 \neq -1$, f is not analytic at any point.

(b) $\frac{\partial u}{\partial x} = \frac{y^2-x^2}{(x^2+y^2)2}, \frac{\partial v}{\partial y} = \frac{x^2-y^2}{(x^2+y^2)2}, \frac{\partial u}{\partial y} = \frac{-2xy}{(x^2+y^2)2} = \frac{\partial v}{\partial x}$

The Cauchy - Riemann equations hold only at (0,0). Since there is no neighbourhood about $z = 0$ within which f is differentiable , we conclude f is nowhere analytic.

8.21 (a) $\frac{\partial^2 u}{\partial x^2} = 0, \frac{\partial^2 u}{\partial y^2} = 0$ or $\frac{\partial^2 u}{\partial x^2} + \frac{\partial^2 u}{\partial y^2} = 0$. Thus, $u(x,y) = 2x - 2y$ is harmonic.

8.23 See [1]

8.25 See [1]

8.27 See [11]

8.28 (a)

$$\begin{aligned} f(z) &= \frac{1}{z^5}[z-(z-\frac{z^3}{3!}+\frac{z^5}{5!}-\frac{z^7}{7!}+\ldots)] \\ &= \frac{1}{3!z^2}-\frac{1}{5!}+\frac{z^2}{7!}-\frac{z^4}{9!}+\ldots \end{aligned}$$

8.29 (a) From $f(z) = \dfrac{3z-1}{[z-(-1+2i)][z-(-1-2i)]}$

$-1+2i$ and $-1-2i$ are simple poles.

8.31 (a)

$$\begin{aligned} Res(f(z),4i) &= \lim_{z\to 4i}\frac{(z-4i)z}{(z-4i)(z+4i)} \\ &= \lim_{z\to 4i}\frac{z}{(z+4i)} \\ &= \frac{1}{2} \\ Res(f(z),4i) &= \lim_{z\to -4i}\frac{z}{(z-4i)} \\ &= \frac{1}{2} \end{aligned}$$

8.33(a) $\displaystyle\int_0^{2\pi}\frac{1}{1+3\cos^2\theta}d\theta = \frac{4}{i}\oint_c\frac{z}{3z^4+10z^2+3}dz$

$$= (\frac{4}{i})2\pi i[Res(f(z),(\frac{\sqrt{3}i}{z}))] + [Res(f(z),-(\frac{\sqrt{3}i}{z}))]$$

$$= \pi$$

8.35 (a) f is conformal at all points except $z=\pm 1$ as $f'(z) = 3(z^2-1) \neq 0$ for $z \neq \pm 1$

8.37 The image of the circle $Imz = 2$ under the linear fractional transformation $w = \dfrac{2z+i}{z-i}$ is the circle with center at $(\frac{7}{2},0)$ and radius $\frac{3}{2}$.

8.39 See [1] and [11]

8.41 See Example 3 of section 20.2, p. 901 0f [11]

Exercise Set - Chapter 9

9.1 The relevant equation

$$D^2 - 2(ct + c^2/g)D + (ct)^2 = 0$$

has two roots.

$$ct + c^2/g \pm \sqrt{2c^3t/g + c^4/g^2}$$

Clearly $D < ct$ (t is total time), so roots corresponding to the + is physically impossible.

9.3 $w(x) = \frac{\pi}{4}\cos\frac{\pi x}{2}$.

9.5 See, [19]

9.7 See, [19]

9.9 $\gamma(t) =$ the variable interest rate $= \frac{1}{2t+10} + .02$

Exercise Set - Chapter 10

10.1 (a) [0,1], [0,1]

(b) For Haar wavelet, see Figure 10.2 . The function $\varphi(x-k)$ has the same graph as φ but translated to the right by k units (assuming k is positive integer.) Let V_0 denote the space of all functions of the form

$$\sum_{k\in Z} a_k\varphi(x-k), a_k \in R.$$

$\psi(t) = \varphi(2t) - \varphi(2t-1)$ is Haar Wavelet.

It can be checked that $\psi(2t-k), k \in Z$ is orthonormal .

The graph of function $\phi(2x-k) = \phi(2(x-\frac{k}{2}))$ is the same as the graph of the function $\phi(2x)$ but shifted to right by $\frac{k}{2}$ units. The graph of $\phi(2^jt)$ is a spike of width $\frac{1}{2^j}$.

(c) $\psi(t) = \phi(2t) - \phi(2t-1)$

Graph of $\psi(t-3)$, note shift of 3 units to right.

graph of $\psi(2t-3) = \psi 2(t-\frac{3}{2})$; note shift of $\frac{3}{2}$ units to right and compression by factor 2.

graph of $\psi(4t-3) = \psi 4(t-\frac{3}{2})$; note shift of $\frac{3}{4}$ units to right and compression by factor 2^2.

graph of $\psi(4t-3) = \psi 2^2(t-\frac{3}{2})$; note shift of $\frac{3}{4}$ units to left and compression by factor 2^2.

Graph of $\psi(2^3t)$, not compression by a factor of 2^3.

$$\begin{aligned}\psi(2t-k) &= \varphi(2(2t-k)) - \varphi(2(2t-k)-1)\\ &= \varphi(4t-2k) - \varphi(4t-2m-1)\end{aligned}$$

For any integer j and k

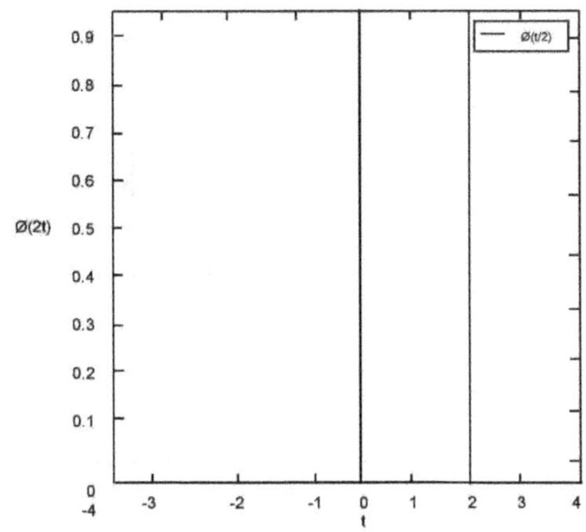

FIGURE E.1: A dilated Haar scaling function $\phi(t/2)$

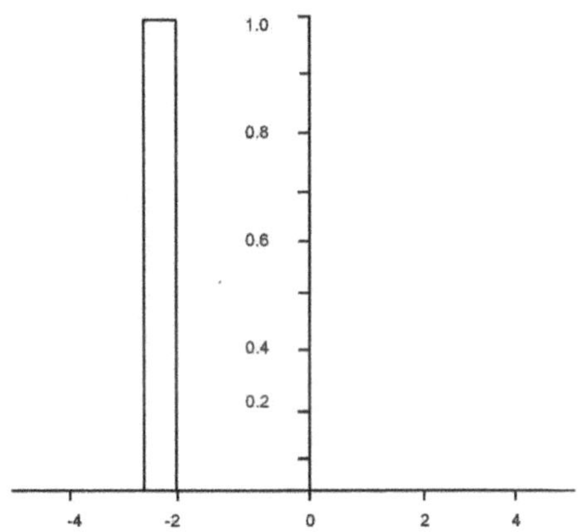

FIGURE E.2: A translated Haar scaling function $\phi(t+3)$ to left by $t_0 = -3$.

$\int_{-\infty}^{\infty} \psi(2t-k)\psi(2t-j)dt = 0$ if $j \neq k$. See Figures E.1 to E.4. For more details, see [31]

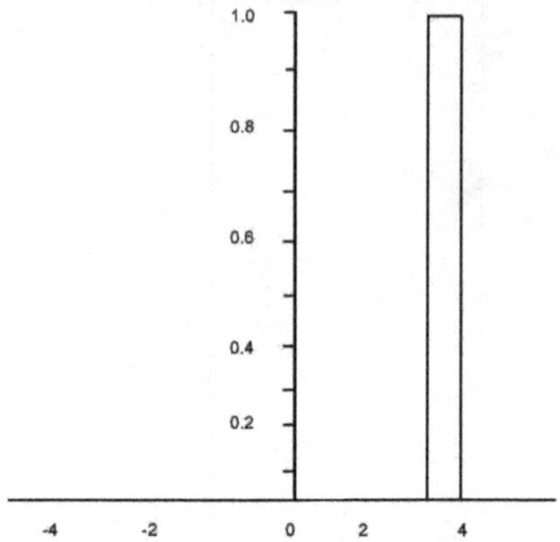

FIGURE E.3: A translated Haar scaling function $\phi(t-3)$ to t right by $t_0 = 3$.

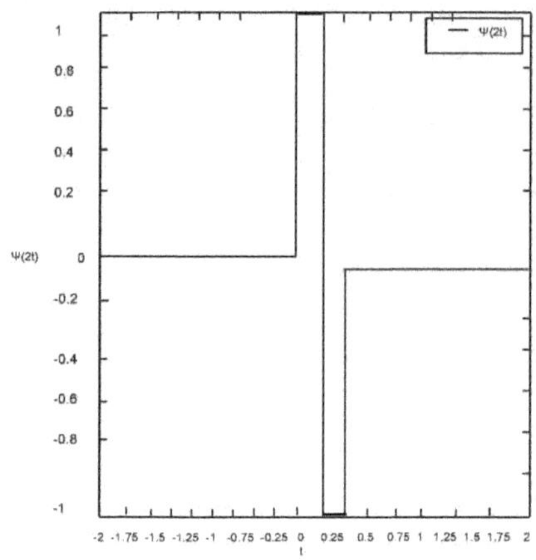

FIGURE E.4: The scaled (contracted) Haar wavelet: $\psi(2t)$

10.3 (i)
$$\begin{aligned}
<\varphi, \psi> &= \int_{-\infty}^{\infty} \varphi(t)\psi(t)dt \\
&= \int_{-\infty}^{0} \varphi(t)\psi(t)dt + \int_{0}^{\frac{1}{2}} \varphi(t)\psi(t)dt \\
&\quad + \int_{\frac{1}{2}}^{1} \varphi(t)\psi(t)dt + \int_{1}^{\infty} \varphi(t)\psi(t)dt \\
&= 0 + \int_{0}^{\frac{1}{2}} 1dt - \int_{\frac{1}{2}}^{0} 1dt \\
&= 1 - 1 = 0.
\end{aligned}$$

Hence φ and ψ are orthogonal.

(iii) $\int_{-\infty}^{\infty} \psi^2(t)dt = \int_0^1 \psi^2(t)dt = \int_0^{\frac{1}{2}} \psi(t)\psi(t)dt + \int_{\frac{1}{2}}^1 \psi(t)\psi(t)dt$

$= \int_0^{\frac{1}{2}} 1dt + \int_{\frac{1}{2}}^1 (-1)(-1)dt$

$= 1.$

10.5 $\varphi(x-k)$ is an orthonormal system, that is

$$\langle \phi(x-k)\phi(x-j)\rangle = 0 \text{ for } j \neq k$$

$$||\varphi(x-k)||_{L_2}^2 = 2^k.$$

Infact, $2^j\varphi(2^jt-k)$ is an orthonormal system. $\varphi(x-j)$ and $\varphi(x-k)$ have disjoint support.

10.7 Let $f(t)$ denote a signal over interval $[a,b]$ then its energy is defined as $E_f = (\int_a^b |f(t)|^2 dt)^{\frac{1}{2}}$. If $f(t)$ is discrete signal then $E_f = (\sum_{n=1}^{N} |f(t_n)|)^{\frac{1}{2}}$. Energy is the characteristic of a signal . If a function is represented by a wavelet series then $E_f = (\sum_{J=1}^{N} \sum_{k=1}^{M} |d_{j,k}|^2)^{\frac{1}{2}}$ where $d_{j,k}$ are wavelet coefficients of f and

$$E_f = (||W_f||^2)^{\frac{1}{2}}$$

where W_f is the continuous wavelet transform of f. For more details see [1], [14], [18], [22], [28], [34] and [33] through [40].

Exercise Set - Chapter 11

11.1 (a)(i) The fractional dimension of Cantor set is $\frac{\log 2}{\log 3} \approx .631$

(ii) The fractional dimension of Sierpinski triangle is $\frac{\log 3}{\log 2} \approx 1.585$

(iii) The fractional dimension of von koch curve is $4 \ln 3 = 1.26186$.

(b) See, [33] and [34]

11.3 See, [35]

11.5 See, [9] and [11]

11.7 Yes, See, [7], [8], [22], [24] through [29], [32], [39] through [44], [52] and [53]

11.9 Hurst is the name of hydrologist who studied Nile river flows and reservoir modeling [27].

Let H denote Hurst parameter and let D denote (Hausdoff) fractional dimension then $D = 2 - H.B.B.$ Mandelbrot

11.11 See, [6, p.8]

FIGURE E.5

FIGURE E.6

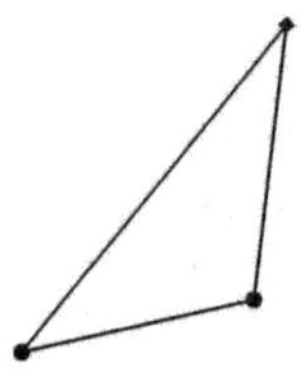

FIGURE E.7

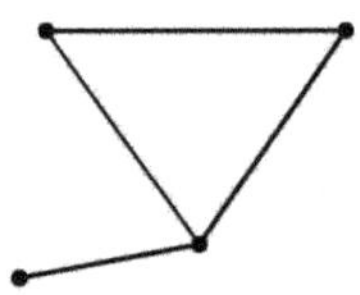

FIGURE E.8

11.13 A graph that has neither self-loops nor parallel edges is called simple graph.
11.15 Choosing $F = X^2$ in Itô Lemma we get

$$\begin{aligned} dF &= 2XdX + dt \\ X^2 &= F(X) = F(0) + \int_0^t 2XdX + \int_0^1 1dt \\ &= \int_0^t 2XdX + t \end{aligned}$$

Therefore $\int_0^t X(\Gamma)dX(\Gamma) = \frac{1}{2}X^2(t) - \frac{1}{2}t$

11.17 The knots are $x = 0, x = 1$ and $x = 4$

11.19 See, [15].

Bibliography

[1] A. Iske and T. Randen (Eds), *Mathematical Methods and Modelling in Hydrocarbon Exploration and Production.* Springer, Berlin, New York, 2006.

[2] S. M. Luthi, *Geological Well Logs: Their Use in Reservoir Modeling*, Springer, Berlin, 2001.

[3] R. C. Selly, *Elements of Petroleum Geology*, Academic Press, San Diego, 1998.

[4] J. Stewart, *Calculus*, Brooks/Cole Publishing Company, 1999.

[5] G. Strang, *Calculus*, PDF file at MIT http://onlinebooks.library.upenn.edu/book/lookpid. QA303. 57.

[6] E. W. Swokowski, M. Olinick, D. Pence, J. A. Ecole, *Calculus*, Sixth Edition, PWS Publishing, Boston, 1994.

[7] S. T. Tan, *Applied Calculus*, Second Edition, PWS, Kent Publishing, Boston, 2007.

[8] T.M. Buzug, Computed Tomography, Springer, 2008.

[9] T.G. Feeman, The Mathematics of Medical Imaging, Springer, 2010.

Computer Programs Used

Chapter 1

Section 1.8 (MATLAB and MATHEMATICA)

Chapter 2

(i) Visualization of scalar and vector valued function

Let us discuss an important feature of MATLAB that is handling powerful graphics. MATLAB has capability of presenting our output and helping us in interpreting the data graphically with help of various types of curves and plots.

Plotting scalar functions with MATLAB

$2D$ dimensional Plots: To plot a function, we have to create two arrays (vectors), one containing abscissa, and the other corresponding function values. Let us plot $f(x) = \sin(x)$. Type following commands on command prompt.

`>> x=-2*pi:pi/100:2*pi;` % range of x from $-2p$ to $+2p$ in steps of $p/100$.

`>> fx=sin(x);` % function sine() to compute sine of al x.

`>> plot(x,fx)` % plot function, which plot fx vs x.

`>> grid` % creates grid in plot.

After giving all these commands, and pressing Enter Key, we get following curve window as shown in Figure E.9.

Let us plot now $f(t) = e^{t/10}.\sin(t)$. Type the following commands on command prompt.

```
>> t=0:0.01:50;
>> ft=exp(-t/10).*sin(t);
>> plot(t,ft)
>> grid
```

In above two examples, we have used plot() function for plotting. Now, we will see that how we can combine two or more plots in one window. Just for simplicity, let us combine plot of function $f_1(t) = e^{t/10}.\sin(t)$ with another function $f_2(t) = e^{t/10}$ in same figure window.

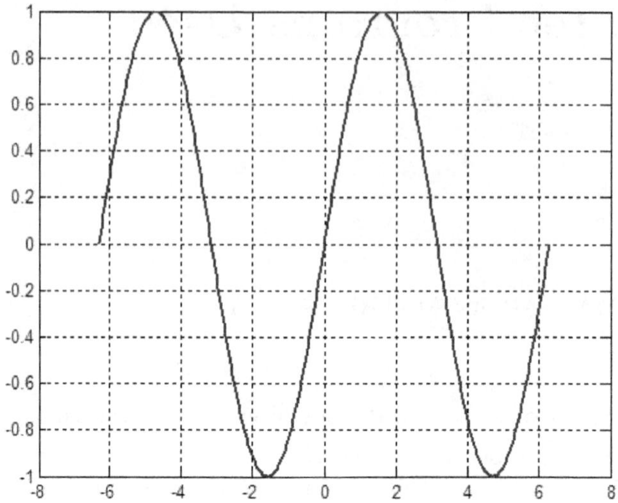

FIGURE E.9

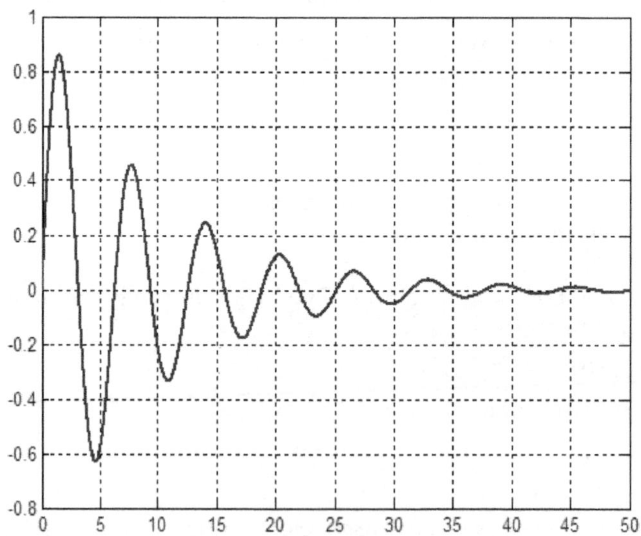

FIGURE E.10

Type following commands

```
>> t=0:0.01:50;
>> ft1=exp(-t/10);
>> ft2=exp(-t/10).*sin(t);
>> plot(t,ft1,t,ft2)       % This is how we can combine two plots.
>> grid
```

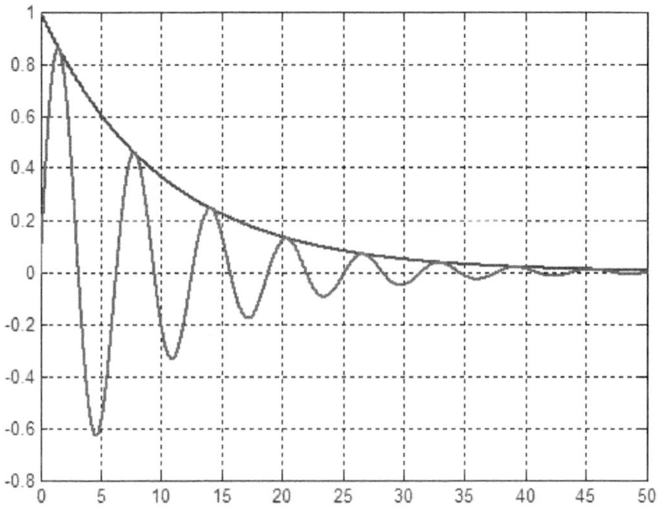

FIGURE E.11

Similarly, we can add several graphs in same window with the command plot($x, fx, y, fy, z, fz, ..$). Please see MATLAB help for more detail.

We can plot several other types of 2D plots in MATLAB with different attributes such as colors, styles etc. here we are taking some more examples -

Function	MATLAB Code

Stairs()	`>> x = 0:0.1:10;`
	`>> y = exp(-x).*sin(x)`
	`>> stairs(x,y)`

Area()	`>> x = 0:0.1:10;`
	`>> y = exp.*sin(x)`
	`>> area(x,y)`

Stem()	`>> x = -4*pi:pi/5:4*pi;`
	`>> y =sin(x)./x;`
	`>> y((length(y)-1)/2+1) =1;`
	`>> stem(x,y)`

Bar()	`>> x = -4*pi:pi/5:4*pi;`
	`>> y = sin(x)./x;`
	`>> y((length(y)-1)/2+1) =1;`
	`>> bar(x,y)`

Semilogx()	`>> x =0:0.1:10;`
	`>> y =x.*exp(-x);`
	`>> semilogx(x,y)`
	`>> grid`

Semilogy	`>> x =0:0.1:10;`
	`>> y =x.*exp(-x);`
	`>> semilogy(x,y)`
	`>> grid`

Loglog()	`>> x =0:0.1:10;`
	`>> y =x.*exp(-x);`
	`>> loglog(x,y)`
	`>> grid`

Polar()	`>> theta =0:pi/100:2*pi;`
	`>> r = sqrt(abs(sin(4*theta)));`
	`>> polar(theta, r)`

Output graphs of these examples are shown in Figure E.12. There are several other types of plots also, one can go through matlabs help.

Three dimensional ($3D$) plots: Let us have a look on 3D plots. Three dimensional graphs, we can plot for functions of two variables such as $z = f(x, y)$. Here, we will plot $x(t) = e^{t/10}.\sin(t)$ verses $y(t) = e^{t/10}.\cos(t)$ along with 't' axis (see Figure E.13). Type the following commands

```
>> t=0:0.01:30;
>> x=exp(-0.1*t).*sin(t);
>> y=exp(-0.1*t).*cos(t);
>> plot3(x,y,t)     % 3D plot command.
>> grid
```

We will get the following figure.

MATLAB has several specialized $3D$ plots. Just try this one-

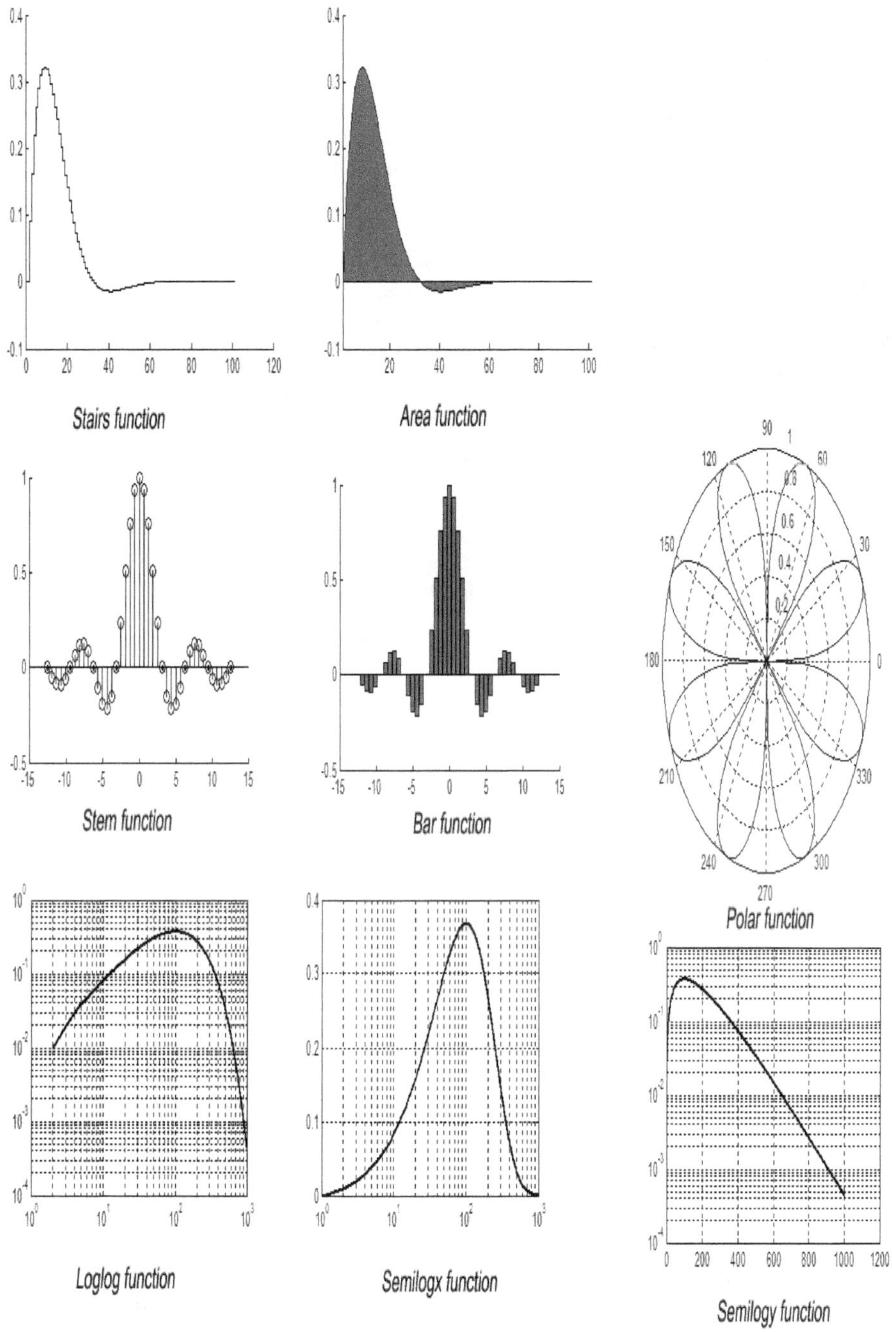
0.4
0.3
0.2
0.1
0
-0.1
0 20 40 60 80 100 120
Stairs function
0.4
0.3
0.2
0.1
0
-0.1
20 40 60 80 100
Area function
1
0.5
0
-0.5
-15 -10 -5 0 5 10 15
Stem function
1
0.5
0
-0.5
-15 -10 -5 0 5 10 15
Bar function
90
1
120
60
0.8
0.6
150
30
0.4
0.2
180
0
210
330
240
300
270
Polar function
Loglog function
0.4
0.3
0.2
0.1
0
Semilogx function
0 200 400 600 800 1000 1200
Semilogy function

FIGURE E.12

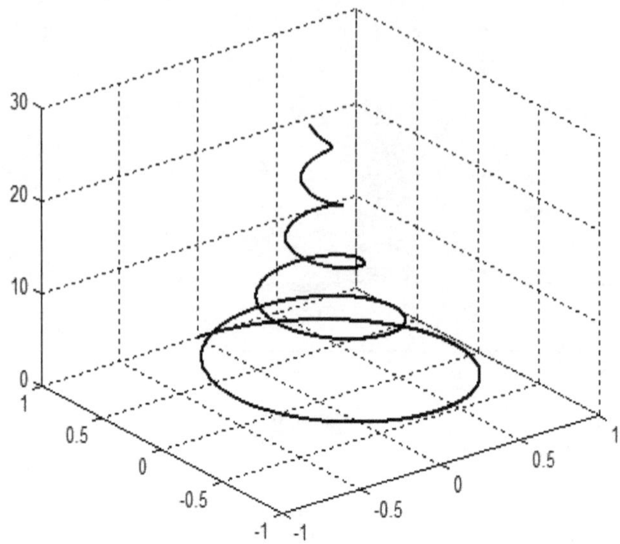

FIGURE E.13

```
>> [x,y]=meshgrid(-8:0.5:8);
>> r=sqrt(x.^2+y.^2) + eps;
>> z=sin(r)./r;
>> mesh(x,y,z)
```

This will give, following result-

One can try following plot commands on same function code and see the results.

```
>> surf(x,y,z)
>> contour(z)
>> surfc(x,y,z)
>> surf1(x,y,z)
>> meshz(x,y,z)
>> waterfall(z)
```

Please see help on each command in MATLAB help
(ii) MATLAB and MATHEMATICA for Differential equation Section 2.3.

Chapter 3

Plots for vector valued functions in $2D$ and $3D$

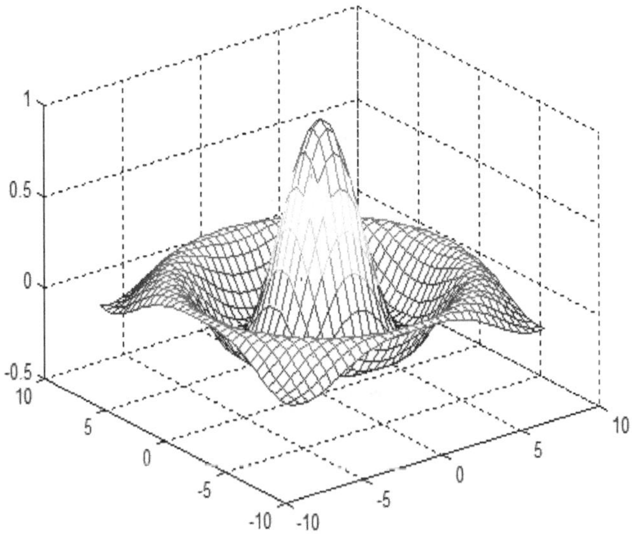

FIGURE E.14

In various applications, we need to visualize the vector valued functions (vector fields). MATLAB has several functions to visualize vector field in $2D$ and $3D$ such as quiver, quiver3, stream, stream3, streamline, streamline3, streamslice, streamtube, streamribbon, streamparticles, coneplot, divergence and curl etc. Here we plot for the function $z = xe^{-(x^2+y^2)}$ see.

First, we take quiver plot. Just write following code on command prompt

```
>> [x,y] = meshgrid(-2:0.1:2);
>> z = x.*exp(-x.^2 - y.^2);
>> [dx,dy] = gradient(z);
>> quiver (x,y,dx,dy)
```

We get following plot-

Now we use streamline function to plot a vector $ui + vj$ where $u = f(x, y)$ and $v = g(x, y)$ are given as-

$$u = x + y - x(x^2 + y^2) \text{ and } v = -x + y - y(x^2 + y^2)$$

This, we can achieve by writing this code on command prompt

```
>> [x,y] = meshgrid(-2:0.1:2);
>> u=x+y-x.(x.^2+y.^2);
>> v=-x+y-y.*(x.^2+y.^2);
```

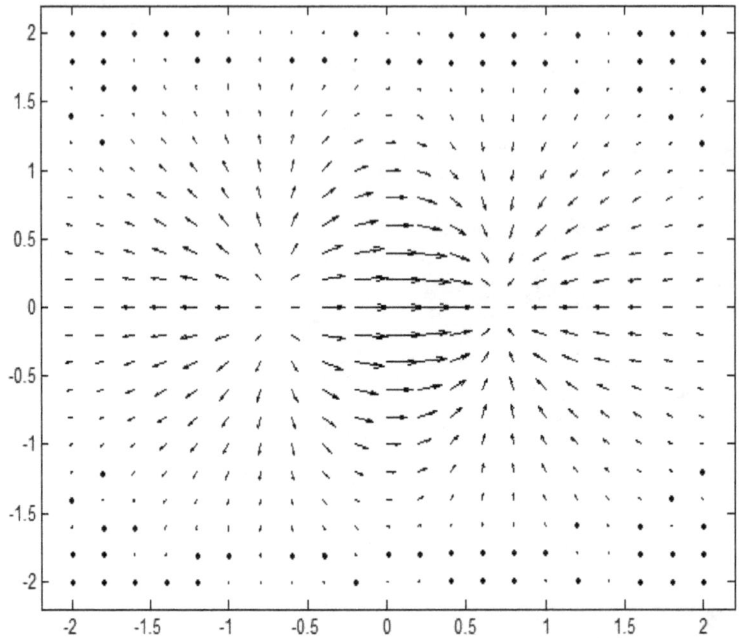


```
>> x0=[-2 -2 -2 -2 -.5 -.5 .5 .5 2 2 2 2 -.1 -.1 .01 .01];
>> y0=[-2 -.5 .5 2 -2 2 -2 2 -2 -.5 .5 2 -.01 .01 -.01 .01];
>> streamline(x,y,u,v,x0,y0)
>> axis square
```

We get this result-
For other types of plots, kindly go through the matlab help.

Chapter 4

Here we present one very versatile program written in MATLAB to compute Fourier series of the given function. This program can be used to compute Fourier series of any function, just by doing small modification in main program body.

The program code is given below to find Fourier series of the square function which is defined as:

$$f(x) = \begin{cases} -1 & -\pi \leq x < 0 \\ 1 & 0 < x \leq \pi \end{cases}$$

The program code is given below-% Fourier Analysis of square functions for first 15 harmonics.

```
t0=-pi;      % initial time
t0\_ T=pi;      % final time
```

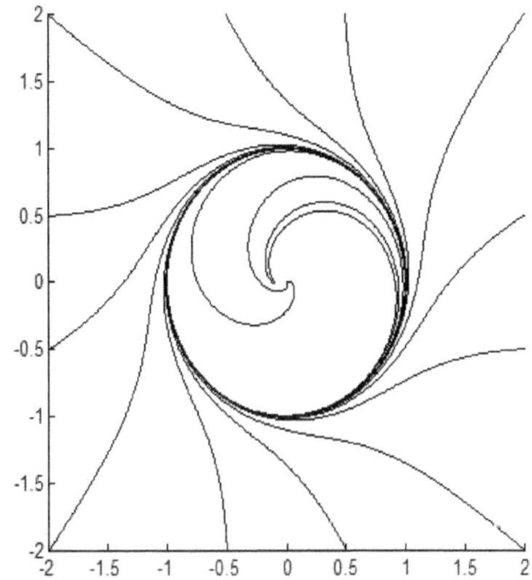


```
mp=0;      % mid point
T=t0\_ T-t0;      % time period
syms t;      % sym variable declaration
ft = -diff(t);      % −1 part of function
ftt= diff(t);      % 1 part of function
w0=2*pi/T;      % frequency
n=1:15;      % number of Harmonics
% computation of Trigonometric Fourier series Coefficients
a0=1/T*(int(ft,-pi,0)+int(ftt,0,pi));
an=2/T*(int(ft*cos(n*w0*t),-pi,0)+int(ftt*cos(n*w0*t),0,pi));
bn=2/T*(int(ft*sin(n*w0*t),-pi,0)+int(ftt*sin(n*w0*t),0,pi));
ann=an.*\cos(n*w0*t);
bnn=bn.*\sin(n*w0*t);
avg=double(a0);      % converting sym variable to value
t=-pi:pi/100:pi;
suma=0; sumb=0;
for j=1:15      % taking 15 harmonics
\hspace{.5cm}sumb=sumb+bnn(j);
\hspace{.5cm}suma=suma+ann(j);
bnsum=eval(sumb);
```

```
ansum=eval(suma);
plot(t, avg+bnsum+ansum)    % plot of truncated harmonics function
hold on
% plotting actual function
t1=-pi:pi/1000:mp;
plot(t1,-1,'r')
t2= mp:pi/1000:pi;
plot(t2,1,'r')
grid on
% formatting plot
x label('Time')
y label('Amplitude')
title('Fourier approximation plot for 15 harmonics for
      square function')
legend('Fourier Approximation','Actual Function')
```

The result is shown below

Here is another program to plot harmonics of the ramp function which is defined as:

$$f(x) = \begin{cases} 0 & -1 \leq x \leq 0 \\ x & 0 \leq x \leq 1 \end{cases}$$

One can see the required modifications in this program as compared with previous one.

```
% Fourier Analysis of ramp function for first 5 harmonics.
t0=-1;    % initial time
t0\_ T=1;    % final time
mp=0;    % mid point
T=t0\_ T-t0;    % time period
syms t;    % sym variable declaration
ft=diff(diff(t));    % zero part of function
ftt=t;    % t part of function
w0=2*pi/T;    % frequency
n=1:5;    % number of Harmonics
% computation of Trigonometric Fourier Series Coefficients
a0=1/T*(int(ft,-1,0)+int(ftt,0,1));
an=2/T*(int(ft*cos(n*w0*t),-1,0)+int(ftt*cos(n*w0*t),0,1));
bn=2/T*(int(ft*sin(n*w0*t),-1,0)+int(ftt*sin(n*w0*t),0,1));
```

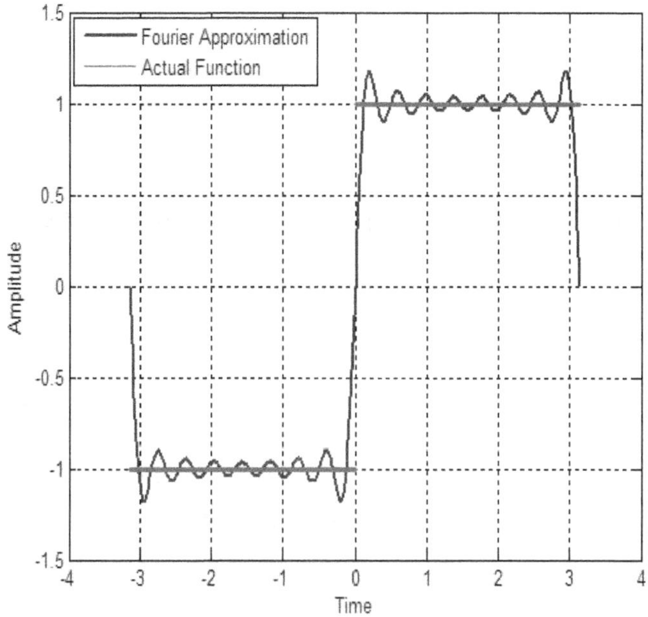

FIGURE E.15: Fourier Approximation Plot for 15 Harmonics for Square Function

```
ann=an.*cos(n*w0*t);
bnn=bn.*sin(n*w0*t);
avg=double(a0);  % converting sym variable to value
t=-1.5:0.010:1.5;
suma=0; sumb=0;
for j=1:5     % First 5 harmonics
      sumb=sumb+bnn(j);
      suma=suma+ann(j);
end
bnsum=eval(sumb);
ansum=eval(suma);
plot(t, avg+bnsum+ansum)      % plot of truncated harmonics function
hold on
% plotting actual function
t1=-1:0.010:0;
plot(t1,0,'r')
```

```
t2=0:0.010:1;
plot(t2,t2,'r')
grid on
% formatting plot
x label('Time')
y label('Amplitude')
title('Fourier approximation plot for 5 harmonics for ramp
      function')
legend('Fourier Approximation','Actual Function')
```

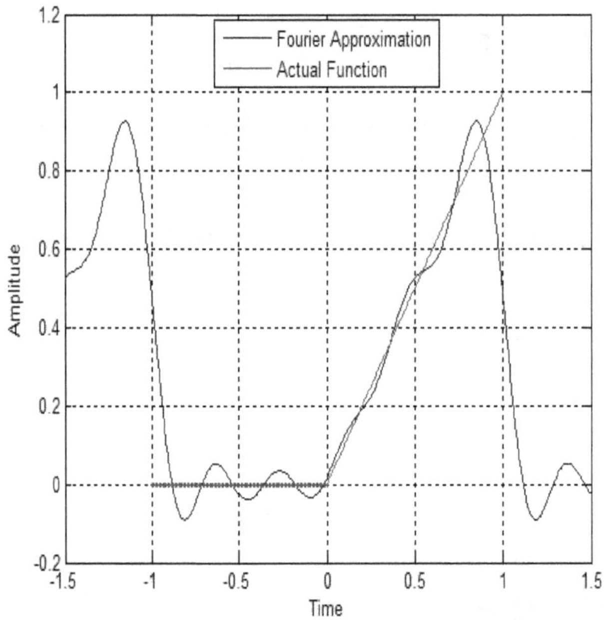

FIGURE E.16: Fourier Approximation Plot for 5 Harmonics for Ramp Function

Chapter 5

Simulation of Heat Equation, Wave Equation and Laplace Equation by MATLAB 404-423, Section 5-6.

Chapter 6

Algorithms in Section 6.9.1, 6.9.2, 6.9.3 and 6.9.5.

Chapter 9

For MATLAB programme of computation in this chapter we refer to [19].

Chapter 10 and Chapter 11

Following computer programmes have been used in these two chapters.

1. Program to Extract the Variables from the Raw Data file

```
clear all
close all
clc
STN = input('Enter Station Name: ','s');
R = input('Enter Station Data File name with Extension: ','s');
[ndata, headertext] = xlsread(R);
disp(headertext(1,:));
fid = fopen('test.txt','r+');
g1 = fgetl(fid); g2 = fgetl(fid);
disp(sprintf('\n \n %s \n \n %s \n', g1, g2));
pos=ftell(fid); fclose(fid);
a1 = input ('Please Input the Column Representing Year: ');
a2 = input ('Please Input the Column Representing Month: ');
a3 = input ('Please Input the Column Representing Day: ');
a=input('Enter the Date of first reading [DD-MM-YYYY]: ','s');
b=input('Enter the Date of last reading [DD-MM-YYYY]: ', 's');
c=datenum(b)-datenum(a);
d=datenum(a)+31+28;
datestr(d+365);
a4=datenum(d)-datenum(a);
N = input('Enter The Number of Variables of Interest: ');
for i=1:1:N
disp(sprintf('\n Pass the Column Number of Variable %d: \n', i))
B(1,i) = input('Column Number: ');
if (i==N)
fid = fopen('test.txt','r');
fseek(fid, pos, 'bof'); g1 = fgetl(fid);
disp(sprintf('%s \n ',g1));
pos=ftell(fid); fclose(fid);
disp('THE VARIABLES YOU ENTERED ARE AS FOLLOWS:-')
```

```
disp(headertext(1 , B(1,:)));
fid = fopen('test.txt','r');
fseek(fid, pos, 'bof'); g1 = fgetl(fid);
disp(sprintf('%s \n ',g1));
pos=ftell(fid); fclose(fid);
disp('If the Variables are correct, Enter 1, else Enter 2')
tt= input('ENTER YOUR CHOICE: ');
if (tt==1)
C(:,i)= ndata(:,B(1,i));
elseif (tt==2)
fid = fopen('test.txt','r');
fseek(fid, pos, 'bof'); fgetl(fid)
pos=ftell(fid); fclose(fid);
Disp ('Please Choose the Correct Column Numbers');
else
error('INPUT CAN BE 1 OR 2 ONLY');
end
end
end
x1=[]; y1=[];
for k = 1:length(B)
x1=[x1 seasonal(a4, ndata(:,B(1,k)), k)];
end
```

2. Program to Extract Seasonal Data

```
function [x , y] = seasonal(a4, lmd, t)
S=[];
for i = 1:1:length(lmd)/365
if (i==3||i==7||i==11||i==15)
S=[S lmd(a4+365*(i-1)+2:a4+365*(i-1)+215 )];
else
S=[S lmd(a4+365*(i-1)+1:a4+365*(i-1)+214 )];
end
end
x = [];
for i =1:2:length(lmd)/365
```

```
x =[x
     S(:,i)
      S(:,i+1)];
if (i==15)
break
end
end
WA = []; WB = [];
for i = 1:1:length(lmd)/365
WA = [WA lmd(365*(i-1)+1:a4+365*(i-1))];
WB = [WB lmd(365*(i-1)+a4+215:365*i )];
end
y = [];
for i =1:2:length(lmd)/365
y =[y
     WA(:,i)
     WB(:,i)
WA(:,i+1)
WB(:,i+1)];
if (i==15)
break
end
end
```

3. Program for Wavelet Spectrum Calculation

```
clear all
close all
clc
STN1 = input('Enter the 1st Location: ','s');
STN2 = input('Enter the 2nd Location: ','s');
load (STN1)
load (STN2)
fx1=_speed(:,1);
fx2=_speed(:,1);
a=100:100:20000;
b1=1:50:8000;
```

```
b2=1:50:8000;
x=1:1:length(fx1);
x=x';
for k = 1:1:length(a)
for i =1:1:length(b1)
c1=b1(i).*ones(size(x));
c2=b2(i).*ones(size(x));
y1=abs(x-c1);
y1=(y1.*y1)./(a(k).^2);
y2=abs(x-c2);
y2=(y2.*y2)./(a(k).^2);
z1=exp(-y1./2)/(a(k)*(2*pi)^0.5);
z1=(2-y1).*z1;
z2=exp(-y2./2)/(a(k)*(2*pi)^0.5);
z2=(2-y2).*z2;
phi1=conj(z1);
phi2=conj(z2);
w1(k,:)=fx1.*phi1;
w2(k,:)=fx2.*phi2;
m1(k,:)=abs(w1(k,:)).^2;
m2(k,:)=abs(w2(k,:)).^2;
end
n1(k)=sum(m1(k,:))/a(k);
n2(k)=sum(m2(k,:))/a(k);
a(k)
end
figure; plot(a,n1,'k:','LineWidth',2);hold on;
plot(a,n2,'r-','LineWidth',2);
xlabel('Scale'); ylabel('Wavelet Spectrum');
set(gca, 'XTick', 0:5000:20000);
set(gca, 'XTickLabel', 0:5000:20000);
legend(STN1,STN2);
```

4. Program for Wavelet Correlation Calculation

```
clear all
close all
clc
```

```
STN1 = input('Enter the 1st Location: ','s');
STN2 = input('Enter the 2nd Location: ','s');
STN3 = input('Enter the Variable: ','s');
load (STN1)
load (STN2)
fx1=riyadh_pr(1:5800,1);
fx2=guriat_pr(1:5800,1);
x=1:1:length(fx1);
x=x';
w1=[];tmp2=[];tmp1=[];
a=1:100:20000;
b1=-750:10:750;
for k = 1:1:length(a)
for i =1:1:length(b1)
c1=b1(i).*ones(size(x));
y1=abs(x-c1);
y1=(y1.*y1)./(a(k).^2);
z1=exp(-y1./2)/(a(k)*(2*pi)^0.5);
z1=(2-y1).*z1;
phi1=conj(z1);
w1(k,i)=phi1'*fx1;
w2(k,i)=phi1'*fx2;
tmp1=(w1.^2)';
tmp2=(w2.^2)';
m1=sum(tmp1)/a(k);
m2=sum(tmp2)/a(k);
end
a(k)
end
p1=(sum(w1*conj(w2)'));
p2=(m1.*m2).^0.5;
r=p1./p2;
figure; plot(a,r,':r','LineWidth',2); xlabel('Scale');
ylabel('Wavelet Correlation Coefficient');
set(gca, 'XTick', 0:5000:20000);
set(gca, 'XTickLabel', 0:5000:20000);
```

```
legend(STN3);
```

5. Program for Pointwise holder exponent using DWT.

```
clear all
close all
clc
STN = input('Enter the Station Name: ','s');
STNFILE = [STN,'_1dexp_nonparametric']
load (STNFILE)
a=0:1000:6000;
SUBPLOT(3,1,1), plot(__humid,'-b');
Title('Daily Humidity (1990-2005)','FontSize',8);
SUBPLOT(3,1,2), plot(__humid_HDt0,'-r');hold on;
plot(__humid_HDt1,'-b');
legend ('db2','db20');
Title('DWT-based pointwise Holder exponent','FontSize',8);
SUBPLOT(3,1,3), plot(__humid_HDt2,'-r'); hold on;
plot(__humid_HDt3,'-b');
legend ('coef6','coef24');
Title('DWT-based pointwise Holder exponent', 'FontSize',8);
SUBPLOT(3,1,1), plot(__pr,'-b');
Title(''Daily Pressure (1990-2005)','FontSize',8);
SUBPLOT(3,1,2), plot(__pr_HDt0,'-r');
hold on; plot(__pr_HDt1,'-b');
legend ('db2','db20');
Title('DWT-based pointwise Holder exponent','FontSize',8);
SUBPLOT(3,1,3), plot(__pr_HDt2,'-r');
hold on; plot(__pr_HDt3,'-b');
legend ('coef6','coef24');
Title('DWT-based pointwise Holder exponent', 'FontSize',8);
SUBPLOT(3,1,1), plot(__rain,'-b');
Title('Daily Rain (1990-2005)','FontSize',8);
SUBPLOT(3,1,2), plot(__rain_HDt0,'-r');
hold on; plot(__rain_HDt1,'-b');
legend ('db2','db20');
Title('DWT-based pointwise Holder exponent','FontSize',8);
```

```
SUBPLOT(3,1,3), plot(__rain_HDt2,'-r');
hold on; plot(__rain_HDt3,'-b');
legend ('coef6','coef24');
Title('DWT-based pointwise Holder exponent', 'FontSize',8);
SUBPLOT(3,1,1), plot(__speed,'-b');
Title('Daily Speed (1990-2005)','FontSize',8);
SUBPLOT(3,1,2), plot(__speed_HDt0,'-r');
hold on; plot(__speed_HDt1,'-b');
legend ('db2','db20');
Title('DWT-based pointwise Holder exponent','FontSize',8);
SUBPLOT(3,1,3), plot(__speed_HDt2,'-r');
hold on; plot(__speed_HDt3,'-b');
legend ('coef6','coef24');
Title('DWT-based pointwise Holder exponent', 'FontSize',8);
SUBPLOT(3,1,1), plot(__temp,'-b');
Title('Daily Temperature (1990-2005)','FontSize',8);"
SUBPLOT(3,1,2), plot(__temp_HDt0,'-r');
hold on; plot(__temp_HDt1,'-b');
legend ('db2','db20');
Title('DWT-based pointwise Holder exponent','FontSize',8);
SUBPLOT(3,1,3), plot(__temp_HDt2,'-r');
hold on; plot(__temp_HDt3,'-b');
legend ('coef6','coef24');
Title('DWT-based pointwise Holder exponent', 'FontSize',8);
```

Index

Supplier name: THE BS.

year: 2018

price: 9563.

Recomm name: B. K. Majhi

Signature Bichitra

Phone : 4647

DIVISION: CLSMC

Designation

BS/S NO. 36